PRACTICAL HANDBOOK OF GROUTING

PRACTICAL HANDBOOK OF GROUTING

Soil, Rock, and Structures

James Warner, P.E.

WILEY

John Wiley & Sons, Inc.

A NOTE TO THE READER
This book has been electronically reproduced from digital information stored at John Wiley & Sons, Inc. We are pleased that the use of this new technology will enable us to keep works of enduring scholarly value in print as long as there is a reasonable demand for them. The content of this book is identical to previous printings.

This book is printed on acid-free paper. ♾

Published by John Wiley & Sons, Inc., Hoboken, New Jersey
Published simultaneously in Canada.

Library of Congress Cataloging-in-Publication Data:

Warner, James, 1935–
Practical handbook of grouting : soil, rock, and structures / James Warner.
p. cm.
Includes bibliographical references and index.
ISBN 0-471-46303-5 (cloth)
1. Grouting—Handbooks, manuals, etc. I. Title.
TA755.W37 2004
624.1′83—dc22

2004000732

Printed in the United States of America

10 9 8 7 6 5 4

Contents

Preface xiii
Acknowledgments xv
Contributors Biographies xvii

PART I THE PRESSURE GROUTING PROCESS **1**

Chapter 1 Introduction **3**
1.1 History of Grouting 4
1.2 Why Grout? 8

Chapter 2 Types of Grouting **15**
2.1 Permeation Grouting 16
2.2 Compaction Grouting 19
2.3 Fracture/Claquage Grouting 31
2.4 Mixing/Jet Grouting 33
2.5 Fill Grouting 37
2.6 Vacuum Grouting 38

Chapter 3 The Wide World of Grout Materials **40**
3.1 Cementitious Materials 40
3.2 Admixtures for Cementitious Grouts 42
3.3 Noncementitious Grouts 42
3.4 Material Selection 43
3.5 Basic Material Properties 45

Chapter 4 Grout Rheology **50**
4.1 Grout Consistency 51
4.2 Material Behavior 51
4.3 Other Rheological Properties 59

Chapter 5 Cementitious Grout Materials **62**

5.1 Portland Cement 62
5.2 Special Cements 70
5.3 Supplementary Cementing Materials 72
5.4 Blended Hydraulic Cements 75
5.5 Ultrafine Cement 76
5.6 Admixtures for Cementitious Grouts 80
5.7 Fillers 87
5.8 Blocking Agents 89

Chapter 6 Cementitious Grout Mixtures **93**

6.1 Bleed Water 93
6.2 Water-to-Cement Ratio 95
6.3 Mix Design 96

Chapter 7 Noncementitious Grouts **115**

7.1 Chemical Solution Grouts 115
7.2 Chemical Solution Grouts for Strengthening 117
7.3 Chemical Solution Grouts for Water Control 128
7.4 Resinous Grouts 135
7.5 Miscellaneous Grout Materials 141

Chapter 8 Tests for Evaluation of Grouts and Grouted Masses **144**

8.1 Flow Cones 144
8.2 Specific Gravity 146
8.3 Evaluation of Bleed 146
8.4 Pressure Filtration 147
8.5 Slump 148
8.6 Strength Tests 149

Chapter 9 Injection Fundamentals **156**

9.1 Principles of Flow 156
9.2 Effects of Grout Rheology 163
9.3 Grout Penetration 165
9.4 Injection Pressure 166
9.5 Pumping Rate 170
9.6 Pressure Behavior 171
9.7 Vacuum 177

Chapter 10 Grout Holes **179**
10.1 Establishing the Holes 179
10.2 Hole Drilling 186
10.3 Hole Closure and Abandonment 205

Chapter 11 Grouting in Soil **207**
11.1 Compaction Grouting 209
11.2 Permeation Grouting in Soil 260
11.3 Fracture (Claquage) Grouting 286
11.4 Jet Grouting (Mixing) 290

Chapter 12 Grouting in Rock **296**
12.1 Know the Rock Properties 297
12.2 Grouting for Water Control 298
12.3 Grouting Rock for Strengthening 309
12.4 Grout Holes 310
12.5 The Grout Mix 317
12.6 Injection Considerations 319

Chapter 13 Grouting in Concrete and Masonry Structures **329**
13.1 Filling Structural Cavities and Voids 330
13.2 Filling Cracks and Joints 336
13.3 Contact Grouting 345
13.4 Posttension Tendon Grouting 345
13.5 Grouting Bedplates 349
13.6 Abandoning Conduits and Tanks 352
13.7 Filling Annular Space/Backfilling 354
13.8 Filling Fabric Forms 354
13.9 Preplaced Aggregate Concrete 355
13.10 An Unusual Structural Application 363

Chapter 14 Groutjacking **367**
14.1 Slabjacking 367
14.2 Groutjacking Structures 379

Chapter 15 Grouting in Pipes and Conduits **387**
15.1 Access Restrictions 387
15.2 Safety Issues 388

15.3 Small Inaccessable Conduits 389
15.4 Worker-Accessible Conduits 392

Chapter 16 Grouting of Underground Structures (Raymond w. Henn) **398**
16.1 Geotechnical and Structural Grouting 399
16.2 Excavation Methods 403
16.3 Lining Systems 406
16.4 Conclusions 406

Chapter 17 Leakage Control in Structures **410**
17.1 The Leakage Path 410
17.2 Grout Holes/Injection Ports 414
17.3 The Grout Material 417
17.4 Grout Injection and Monitoring 419
17.5 Examples of Use 421

Chapter 18 Grouting in Extreme Environments **427**
18.1 Temperature Effects on Grout 427
18.2 Extreme Chemical Exposures 428
18.3 Provisions for Temperature Extremes 429

Chapter 19 The Use of Explosives in Grouting **436**
19.1 Introduction 436
19.2 Principles of Explosives Use 438
19.3 Formation of Expanded Head Piles and Ground Anchors 440
19.4 Clearing Obstructions to Grout Flow 444
19.5 Accessing Subsurface Fissures and Voids 445
19.6 In-Place Perforation of Injection Casing 448
19.7 In Conclusion 450

Chapter 20 Emergency Response Grouting **451**
20.1 Before Injection Starts 451
20.2 Typical Emergency Situations 453

PART II GROUTING DESIGN AND CONTROL OF GROUTING **463**

Chapter 21 Understanding Geology **465**
21.1 Geologic Age 466

21.2	Types of Rock	467
21.3	Geologic Structure	469
21.4	The Rock Configuration/Geometry	474
21.5	Stress State	475
21.6	Derivation of Soil	476
21.7	Classification of Soil	479
21.8	Geotechnical Considerations	483
Chapter 22	Investigations for Grouting	**484**
22.1	Preliminary Considerations	484
22.2	Soil Grouting	491
22.3	Rock Grouting	495
22.4	Grouting in Structures	496
Chapter 23	Design of Grouting in Soil (Michael J. Byle)	**497**
23.1	Introduction	497
23.2	Grout Functions in Soil	498
23.3	Mass Improvement	499
23.4	Modulus/Strength Improvement	501
23.5	Permeability Reduction	505
23.6	Structural Elements	505
23.7	Karst Mitigation	508
23.8	Design Summary	510
Chapter 24	Design of Grouting in Rock	**511**
24.1	The Existing Geology	512
24.2	Strengthening Applications	514
24.3	Seepage Control	516
Chapter 25	Quality Control, Records, and Verification	**519**
25.1	Quality Control	519
25.2	Injection Records	524
25.3	Verification of Grouting Effectiveness	529
Chapter 26	Numerical Analysis of Grouting (Dawn Shuttle and Mike Jefferies)	**531**
26.1	About Numerical Analysis	531
26.2	Fractured Rock Grouting	532
26.3	Compaction Grouting	538

Chapter 27 Specifications and Contracts **543**

27.1 Specification Requirements 544

27.2 Contracting Arrangements 559

27.3 Full-Scale Trials 563

Chapter 28 The Games Contractors Play **564**

28.1 Marketing Efforts 565

28.2 Proposal Trickery 568

28.3 Once on the Job 568

28.4 Defending Bad Work 569

28.5 The Good Players 570

PART III DRILLING AND GROUTING EQUIPMENT **573**

Chapter 29 Understanding Power Transmission **575**

29.1 Power Transmission Fundamentals 575

29.2 Mechanical Systems 577

29.3 Hydraulic Systems 582

29.4 Pneumatic Systems 587

Chapter 30 Drilling Equipment and Accessories **589**

30.1 Drill Rigs 589

30.2 Drill Hole Casing 597

30.3 Drill Rod 605

30.4 Drill Bits 610

30.5 Sleeve Port Pipes (Tube-à-Manchette) 613

30.6 Special Drive Needles 614

30.7 Circulation (Mud) Pumps 614

Chapter 31 Pump Mechanics **615**

31.1 Pumping Fundamentals 615

31.2 Classification of Pumps 616

31.3 Types of Pumps 617

Chapter 32 Grout Pumps **627**

32.1 Power Source 627

32.2 Types of Grout Pumps 628

Chapter 33 Batchers, Mixers, and Agitators **635**

33.1 Small Mixers 635

33.2 Mixing Equipment for Pourable Grouts 637

33.3 Agitators 641
33.4 Material Storage and Transfer Equipment 641
33.5 Equipment for Low-Mobility Grouts 643

Chapter 34 Unitized Grout Plants **646**
34.1 Basic Considerations 646
34.2 Plants for Pourable Cementitious Grouts 647
34.3 Grout Plants for Low-Mobility Grouts 652
34.4 Custom-Built Grout Plants 652

Chapter 35 Delivery Lines and Fittings **654**
35.1 Delivery Lines 654
35.2 Valves 661

Chapter 36 Monitoring Equipment **663**
36.1 Pressure Gauges 663
36.2 Gauge Savers 665
36.3 Flow Meters 668
36.4 Automated Monitoring Equipment 669

Chapter 37 Packers **672**
37.1 Types of Packers 672
37.2 Packer Deployment 675
37.3 Double Packers 677

Chapter 38 Some Closing Thoughts **678**

Bibliography and References 685
Index 691

Preface

When I entered grouting in 1952, it was a generally undeveloped art with most of the players professing magical powers. Work was performed on a somewhat hit or miss basis with results that varied from excellent to complete failure. Little equipment specifically built for grouting was available, and most work was accomplished with simple water-cement suspensions, typically based on high water : cement ratios subject to large bleed, shrinkage, and poor durability.

Fortunately, that has all changed. Although we still have players that profess magical powers, grouting has become well-established science, and work today can be designed and performed according to sound engineering principles with a good understanding of the processes that occur.

To cover all pertinent subject matter and facilitate its review, the handbook is divided into three parts. Part I discusses the grouting process in a general way and should be of interest to all readers. Here, the many conditions for which grouting can be a solution are described, as are both grouting materials and injection fundamentals. The five general types of grouting—*permeation, compaction, fracture, replacement*, and *fill*—are discussed as is the pertinent methodology for the main applications of grouting, *in rock, soil, and into concrete and masonry structures.* Separate chapters discuss specialized applications such as Grout Jacking, Grouting in Pipes and Conduits, Leakage Control in Structures, Grouting of Underground Structures, and Grouting in Extreme Environments. Following a discussion of the very specialized but seldom used application of Use of Explosives in Grouting, the section ends with a discussion of Emergency Response Grouting.

Part II will be of particular interest to designers, engineers, and those responsible for quality control, although the material presented should be useful to all grouting participants. This part discusses the criteria upon which a grouting program should be based. Design issues including required surface and subsurface investigation are covered, as are monitoring and control of the actual work. Included are the numerous activities that ensure proper performance, including chapters on Understanding Geology, design and specification of grouting, quality control and verification, and Numeric Analysis for Grouting. This part ends with discussion relative to the preparation of contract documents, pay items, and The Games Contractors Play, which must be considered when preparing those documents.

Part III addresses the details of both drilling and grouting equipment and is certainly essential to contractors as well as field personnel. Because many grouters build their own equipment or significantly alter that which is commercially available, the section starts with a chapter on Understanding Power Transmission, which includes thorough coverage of the many facets of direct, hydraulic, and pneumatic power transmission. The specifics of drilling equipment, including details of the many different types of drill rod, casing, bits, and core barrels are provided. In Pump Mechanics, a thorough review of the different types of pumps used, and their relative advantages and limitations is given. Ample space is devoted to discussion of the minor though critical items of equipment, including Delivery Lines and Fittings, Gauges, Gauge Savers, and Flow Meters, and Packers.

Chapter 38 describes some personal experiences as well as some thoughts on the future of grouting technology.

James Warner, P.E.
Consulting Engineer
Mariposa, California

Acknowledgments

This book would not have been possible without the counsel of many great mentors. Although they are no longer with us, I want to recognize and thank them.

Ralph Stewart, a master mason who allowed me to "help" him as a child, and later as an apprentice, passing along his vast knowledge and enthusiasm for the material we call cement.

Max Kemper, an underground and tunneling contractor, and manufacturer of FloCrete Equipment, who allowed me free reign of his projects, yard, and shop and instilled in me the importance of understanding mechanics and ability to design and build custom equipment.

Dick and Marvin Bennett, fathers of the modern concrete pump and mechanical geniuses who passed along much valuable knowledge.

Ray Maurseth, an outstanding engineer who in the midst of a disastrous job, remained calm and explained to me beneficial soil compaction was occurring, the discovery of *compaction grouting,* and further sparked my interest to study geotechnical engineering.

Douglas Brown, a brilliant engineer and geologist, from whom I learned the importance of geologic origin and who challenged me to research every thought and theory and not form opinions without documentation for their support.

And special thanks to:

Professor Norbert Schmidt, who in 1979 organized the Annual Short Course on Grouting, greatly elevating knowledge and where, as a member of the faculty each year, I have had the opportunity to meet and learn with many great grouters.

Clive Houlsby, a special friend and colleague, for his leadership in improving rock grouting technology, with whom I have shared much time and lively dialog on the fine points of grouting.

Special thanks also to my contributing authors:

Mike Byle (Chapter 23), for his leadership in understanding the mechanics of working with limited mobility grouts.

Ray Henn (Chapter 16), who has labored tirelessly toward linking grouting technology within the special restraints of underground construction.

Mike Jefferies and Dawn Shuttle (Chapter 26), for improving grouting technology through their pioneering efforts to develop analytical models based on empirical grouting knowledge.

And completion of this book would not have occurred without the help of many other fine friends and colleagues. For sharing information and photos, my appreciation and thanks are extended to Sam Bandimere, Dick Berry, Don Bruce, Bruce Cossell, Steve Garner, Bob Gronowicz, Ron Hall, Joe Harris, Doug Heenan, Don Hegebarth, Dan Hourihan, Larry Jean, Dan Johnson, Hitaka Kanematsu, Bert Kriekemans, Paul Krumm, Fred Pico, Ron Lee, Rick Quiles, Monica Rourke, Joe Schatz, Fred Sherrill, Ken Tholin, Normand Tetreault, Peter Town, Dave Ward, and Peter White.

For review and suggested improvements to portions of the text, I thank Hank Baski, Steve Bellonomi, Paul Branagan, Bert Kriekemans, David Magill, Reinhard Muller, Paul Knight, Sten-Ake Pettersson, John Richards, Brian Skaggs, and Rob Taylor.

And thanks to the many grouting colleagues I have had the privilege to work with and learn from, among them Ernst Ahrens, George Bergemann, Dave Dorsch, Ed Graff, Gary Greenfield, Clay Griffin, Mick Honeck, Brian Iski, Mike Jaques, Larry Johnsen, Rube Karol, Ulrich Kuhner, Stuart Littlejohn, Mike Miluski, Alex Naudts, Larry Olson, Tony Plaisted, Steve Scherer, Lois Schwarz, Glen Smoak, Nate Snorteland, Joe Solomon, Al Stilley, John Tims, Kurt Von Fay, "Vipu" Vipulanandan, Ken Weaver, Peter Yen, and many others.

Contributors Biographies

Michael J. Byle (Chapter 23)
Michael J. Byle is a registered Professional Engineer and a Fellow of the American Society of Civil Engineers who has been practicing in the field of grouting for more than 20 years. He has served as the chair of the Geo-Institute Committee on Grouting, authored more than 20 technical papers on grouting and karst remediation, and was chair of the Third International Conference on Grouting and Ground Improvement in 2003. He is the Geotechnical Manager for Gannett Fleming, Inc., in Valley Forge, PA 19335. Phone: (610) 650-8101. Fax: (610) 650-8190. E-mail: mbyle@gfnet.com.

Raymond W. Henn (Chapter 16)
Raymond W. Henn is a principal of Lyman Henn, Inc., in Denver, Colorado. He has more than 32 years of heavy and underground construction experience with a concentration in tunneling, transportation, water and wastewater, and hydroelectric projects and holds degrees in geology, engineering, and construction management. During his career, Mr. Henn has held positions from field engineer through superintendent to construction manager on both direct-hire construction and construction management assignments. He has a strong background in general civil construction, mass and specialty concrete placement, grouting, deep foundations, and shaft and tunnel excavation and lining. Mr. Henn served as president of the American Underground Construction Association for the 2003–2005 term. He is the author of the ASCE-published *Practical Guide to Grouting of Underground Structures* and *AUA Guidelines for Backfilling and Contact Grouting of Tunnels and Shafts.* Mr. Henn is the recipient of the American Society of Civil Engineers 2002 Roebling Award.

Dawn Shuttle (Chapter 26)
Dr. Dawn Shuttle, C.Eng., MICE, MASCE was formerly an Associate of Golder Associates in Seattle and is now Associate Professor of Civil Engineering at the University of British Columbia. Ms. Shuttle has been grouting for 15 years with an emphasis on modeling and numeric analysis to better understand the processes. She is chair of the ASCE/G-I Standards Committee on Compaction Grouting.

Mike Jefferies (Chapter 26)
Mike Jefferies, C.Eng, MICE, is an Associate with Golder Associates in Vancouver, Canada and Visiting Professor at the University of Manchester Institute of Science and Technology (UMIST)in England. He has been involved in grouting for 25 years and is the author of several technical papers.

PART I

The Pressure Grouting Process

1 CHAPTER

Introduction

1.1 HISTORY OF GROUTING
- **1.1.1** Foreign Development
- **1.1.2** U.S. Development

1.2 WHY GROUT?
- **1.2.1** Grouting in Rock
  - **1.2.1.1** Reducing Water Flow
  - **1.2.1.2** Strengthening Rock
  - **1.2.1.3** Mitigation of Sinkhole Formation
- **1.2.2** Grouting in Soil
  - **1.2.2.1** Soil Densification
  - **1.2.2.2** Groutjacking
  - **1.2.2.3** Increasing Cohesion of Granular Soils
  - **1.2.2.4** Strengthening by Reinforcement
  - **1.2.2.5** Reducing Permeability/ Water Control
- **1.2.3** Structural Applications
  - **1.2.3.1** Preplaced Aggregate Concrete
  - **1.2.3.2** Filling of Large Annular Space
  - **1.2.3.3** Filling of Fabric Forms
  - **1.2.3.4** Structural Repair and Rehabilitation
  - **1.2.3.5** Control of Leakage into Structures
  - **1.2.3.6** Sealing of Leaking Sewers and Pipes

PRESSURE GROUTING consists of forcing a material (grout) under pressure, so as to fill joints and other defects in rock, soil, concrete, masonry, and similar materials. It can also modify soil through the filling of pore space or compaction into a denser state. In addition, grouting is used to fill and bond cracks and defects in structural concrete and masonry. In new construction, it is employed to ensure complete filling under precast members, base plates, and similar assemblies. It can be used to cast concrete in place, wherein a cementitious grout is pumped into previously placed large aggregate (preplaced aggregate concrete). Several specialty geotechnical techniques use pressure grouting as a principal component. These involve a wide spectrum of operations, ranging from the grouting of foundation piles to encapsulation or immobilization of contaminated soil through *jet grouting* or *deep soil mixing* with grout. The installation of soil anchors, rock bolts, and foundation piles can be facilitated, and their capacity significantly increased, with supplementary grouting.

The main ingredient of a grout may be cementitious material, a liquid or solid chemical including hot bitumen, or any one of a number of different resins. A combination of two or more components is often used. A wide variety of different filler materials may be included. Grout can have virtually any consistency, ranging from a true fluid to a very stiff mortarlike state. It will generally harden at some point after injection, so as to become immobile, and can be designed to have a wide variety of both bond and compressive strengths. It may harden to a very stiff mass or, by design, remain somewhat flexible. Grouts are available that will cure into a solid, a flexible gel, or a lightweight foam.

The application of grout and the use of grouting methods have proven to be advantageous in many endeavors. For the beneficial modification of geomaterials, they are now routinely used to achieve any of the following:

- Block the flow of water and reduce seepage
- Strengthen soil, rock, or combinations thereof
- Fill massive voids and sinkholes in soil or rock
- Correct settlement damage to structures
- Form bearing piles
- Support soil and create secant-pile walls
- Install and increase the capacity of anchors and tiebacks
- Immobilize hazardous materials and fluids

Applications in the construction and repair of structures include:

- Returning disintegrated concrete and masonry into a monolithic mass
- Repair and welding of cracks in structural concrete members
- Securing of bolts, rods, and anchors in drilled holes
- Filling cavities under and providing complete support for precast members, base plates, and the like
- Corrosion protection for prestress tendons and anchors
- Casting of preplaced aggregate concrete
- Filling cracks, splits, and other defects in the repair of timber members

As is apparent, the beneficial applications of pressure grouting are no longer limited to the control of water seepage, but now facilitate strengthening and improving virtually all geo- and other materials. They are also useful, and in some cases mandatory, for the repair and rehabilitation of concrete and masonry structures. Although the reasons for grouting, as well as the materials and procedures available, are many, the principles of injection are the same.

1.1 HISTORY OF GROUTING

Grouting is not new. In fact, records abound, documenting grouting projects throughout the 1800s and even before. Most of the early work involved injection of aqueous suspensions or slurries, often containing cement, lime, or clay, into joints and seams in bedrock underlying dams in order to reduce water leakage. As early practice involved the filling of seams or voids, the grout had to be very flowable, with a maximum particle size considerably smaller than the thickness of the particular discontinuity.

Because the pore spaces in soils are generally much smaller than the apertures of typical rock joints, injection of particulate grout was of limited success. Accordingly, low-viscosity *chemical solution grouts,* which could permeate granular soil formations and then chemically harden, were developed. In 1887 a patent was issued to Jeziorski for a sodium silicate–based formulation, which could be mixed on-site and injected. Unfortunately, the chemicals reacted soon after mixing, requiring very rapid injection, and, all too often, hardened in the pump and delivery system, which restricted their application.

To overcome the problem of early hardening, Hugo Joosten, a Dutch mining engineer, developed a two-shot sodium silicate–based system, patented in the United States in 1925. In this system the sodium silicate base chemical was first injected into the soil, followed by injection of a reactant, commonly calcium chloride, which would cause the silicate to harden. Although Joosten's system was used until the late 1960s, difficulty in achieving complete mixing of the components in the ground discouraged widespread application. Interestingly, however, sodium silicate remains a predominant component of chemical solution grouts used for strengthening, the earlier limitations of rapid setting having been overcome with the availability of many user-friendly reactant systems.

Another significant contribution to soil grouting was the development of the "mudjack" machine in 1933. The original objective was the filling of voids under, and the raising of, settled concrete pavement. Initially, a mixture of "loam" or clayey soil was used. As experience was gained, the addition of portland cement was found to result in a stronger and more durable hardened grout. It was also found that varying the grout consistency allowed a wider range of work to be accomplished. As a natural outgrowth of their work, some of the more inventive operators attempted stabilization of soil by various means, including pumping relatively stiff mixes into predrilled holes. Although early work was performed on a somewhat hit-or-miss basis with little engineering input, a great deal of knowledge was gained, and this procedure was a forerunner of the "compaction grouting" process.

Compaction grouting, which is the only major grouting technology to originate in the United States, is now practiced throughout the world, although its major use and most dramatic advancements have remained in this country. I discovered the compactive mechanism quite by accident in the mid 1950s. While I was attempting to pump a mortarlike grout into a saturated clayey soil underlying a settled swimming pool, water exuded from the ground. This indicated reduction of the soil pore space through compaction. Subsequent research, which involved injection and recovery of more than 100 injected masses, revealed the validity of ensuring orderly compaction of the soil, with a clear interface separating the grout.

The use or attempted use of common cementitious suspension grouts in soil did not end, however, most likely because of their costs being significantly less than those of their chemical cousins and the wide availability of equipment for their mixing and injection. Yet because they cannot permeate any but the coarsest of soils, their suitability is limited. In more recent times, the ability to accurately initiate fractures in the ground has facilitated use of these grouts to form a network of interconnected lenses, which are believed to increase formation strength. With the development of ultrafine cement in the last few decades, suspension grouts can now permeate and solidify even the finest sand.

As need developed to repair structural defects, injection of the then available cementitious suspension grouts was performed. Early structural grouting was used in large voids of massive masonry sections. These often did not require significant bond or compressive strength. The technology evolved, however, to treatment of defects in concrete, usually in the form of cracks, for which high strength is crucial.

This led to development of resinous injection grouts such as epoxies and urethanes in the 1950s. These more advanced materials could be formulated to be highly penetrable, allowing injection into the finest of cracks, as shown in Figure 1.1. Further, they could attain very high bond and compressive strengths. The more recently developed ultrafine cements are also highly penetrable in even the finest fissures. Grouts made of these materials can obtain relatively high compressive strength, usually in excess of 1000 psi (6.9 MPa). Grout strength, especially the bond strength of cementitious grouts, can further be increased by the use of modern admixtures.

In the 1940s it was found that mixing a source of calcium with clay soil was beneficial to strengthen and/or reduce the soil's expansiveness. High-calcium hydrated lime became the material of choice for such mixing because of its economy and wide availability. The practical depth for physical mixing was limited, however, and a method for deeper application was needed. This resulted in the use of pressure

FIGURE 1.1 Epoxy injection to repair cracks in concrete.

injected lime slurries, using grouting techniques. More recently, this beneficial mechanism, which consists of cation or base exchange, has become better understood, and a variety of other chemicals are now available. Because of familiarity with its use and its generally lower cost, lime slurry continues to be the most widely employed material, and its use for injection continues, primarily in geographic areas where swelling clay soils predominate.

1.1.1 Foreign Development

A significant amount of our present grouting knowledge is the result of foreign development, traditionally in European countries and more recently in Japan. In these countries there has been extensive use of both solution chemical and particulate suspension grouts. Unlike American practice, much European work is performed on a design-build basis, and very large projects are commonplace. This has resulted in large multidiscipline firms possessing the capability for research and development, design, and actual construction, as well as the ability to design, build, and maintain the very specialized equipment often used. The ability to integrate the various disciplines, combined with strong financial capability, has resulted in many technical advances. There has, however, been some reluctance to share knowledge, and it is common for the proposals of such firms to be void of any mention of the exact procedures or materials to be used, but to state only the end product to be achieved.

Unique to European work is the use of large central grout plants (Figure 1.2), often containing a number of different types of mixing and pumping units. Although a grout plant is gen-

FIGURE 1.2 Large central grout plant unique to European work.

erally sited as near to the center of injection as practicable, relatively long injection lines are required. For this reason, circulating grout delivery systems, combined with grouts with very long set times, are usually employed. Such plants are virtually always equipped with automated continuous monitoring and recording equipment, and the operation is normally under the direct supervision of a professional engineer.

With the exception of the development of techniques that rely on the use of massive equipment such as jet grouting, the European approach to grouting has remained largely unchanged in recent times, and although improved and refined, solution and fluid suspension grouts continue to be used almost exclusively.

1.1.2 U.S. Development

Contrary to the European experience, the development of grouting in the United States was somewhat erratic and primarily achieved by small, widely dispersed specialty contractors. Most of the more common grouting of rock dam foundations was under the control of one of three large federal government agencies: the Army Corps of Engineers, the Bureau of Reclamation, or the Tennessee Valley Authority. These agencies generally designed and supervised grouting operations in-house, to the point that contractors basically furnished only labor, equipment, and materials, performing work strictly as directed.

This practice left the contractors little incentive for innovation or advancement of the technology. The result was virtually no interaction between the contractors doing the more traditional dam foundations and those performing other work, which included most grouting in soils and structures. The latter group, although composed of small firms widely dispersed geographically, nonetheless contributed many innovations and greatly advanced grouting technology. Unfortunately, they often worked under a shroud of secrecy, which retarded the sharing of experiences, so that neither widespread understanding nor use of their developments occurred.

It is only within the last few decades that extensive knowledge of grouting technology has become readily available. A major contributor to this change has, no doubt, been the week-long Short Course on Fundamentals of Grouting, now sponsored by the Colorado School of Mines and held annually since 1979. In addition, the American Society of Civil Engineers Geo-Institute has sponsored specialty conferences, as well as a number of sessions on grouting, at its other gatherings. Both the American Society of Civil Engineers and the American Concrete Institute maintain active committees on grouting, the membership of which includes a broad spectrum of researchers, as well as designers and field personnel. This has greatly increased the interaction of the various players and has contributed enormously to a better understanding of the technology.

Yet despite the dissemination and sharing of knowledge, some firms continue within the shroud of secrecy in an effort to convey the idea that only they are capable of the "magic touch." Even worse is the sponsorship of university research, publications, and even educational seminars by some large firms, which provide information favorable only to the sponsoring firms. Although designed and promoted to appear authoritative, such marketing efforts often lack technical accuracy, are highly biased, and frequently disseminate technically incorrect information. Grouting technology is now well developed and available, but unfortunately, many players, especially design engineers, tend to rely on the often inaccurate and biased information more easily obtained from contractors with special interests. Thus, because of the extensive promotion of a few "geotechnical contractors," many remain unaware and/or misinformed of the well-established science of grouting.

Unlike European practice, much of the most advanced technology in the United States has originated from relatively small, often regional contractors. Likewise, most of the equipment commonly employed is portable and easily moved so it can be set up to operate quickly near the site of injection. This eliminates the need for long delivery lines and facilitates communication and flexibility of job site operations. It also makes practicable the execution of small jobs, and it is not unusual for a contractor to set up and complete a project in a single day.

Separate mixers, agitators, and pumps, which can be easily handled, are common, and these are set up as required, near the work area. Alternately, all of the units are sometimes assembled and combined into a compact grout plant. Unlike the Europeans and Japanese, U.S. contractors have been slow to adopt real-time computer monitoring and control of grouting. Such monitoring dramatically improves both the performance and final quality of virtually all grouting and, it is hoped, will soon become routine for all work. Computer monitoring and control is amply discussed in Chapters 25 and 36.

1.2 WHY GROUT?

Although pressure grouting is a fairly specialized procedure, its usage is now widespread and its range of application virtually unlimited. The primary purpose of grouting as originally conceived, and as practiced until the middle of the twentieth century, was control of water flow through defects in the rock under dams. Although this is still a valid use, it represents but a small proportion of the work now performed. Pressure grouting involves the filling of cracks, fissures, voids, and pore space—usually of unknown size, volume, or configuration—in rock, soil, concrete, masonry, and similar formations. The predominant reason for grouting is strengthening or inhibiting the flow of water through a mass.

A grouted formation may be required to last for only a short period of time, such as temporary improvement to aid ground support during excavation or the control of water seepage during construction. Many applications, however, require permanent improvement, wherein long-term durability becomes important. Grouting is often used in remedial work to control seepage, and/or strengthen soil or rock, concrete, or masonry. In new construction it is routinely employed to ensure complete filling of cracks, voids, and spaces under objects such as base plates and precast elements. The annulus around posttensioning tendons is commonly filled by injection of cement grout. It is also used to inject adhesive material such as epoxy, to integrate previously cast concrete elements, or to bond a variety of other materials, either to themselves or to other surfaces.

1.2.1 Grouting in Rock

Virtually all rock contains geologic defects that affect not only its propensity for water transmission, but its strength as well. The nature and extent of defects vary and are influenced by the origin, age, and stress history of the particular rock. Faults, joints, and bedding planes are typical to some degree in all rock. Whereas these elements can be quite tight and of minor significance, in many instances they are large enough to significantly affect both formation strength and permeability. Large subsurface channels commonly occur in rock of volcanic origin. Dissolution of minerals such as calcium carbonate, anhydrite, and gypsum can result in large, cavernous voids in soluble rock. All of these defects can be remediated with grouting.

Filling voids behind tunnel supports or adjacent to hard surfaces, which is known as *backpack* and *contact* grouting, is usually performed during underground construction. It is sometime referred to as *rock grouting,* but in this book such work is treated separately as structural grouting, as discussed in Chapter 16.

1.2.1.1 REDUCING WATER FLOW

Seepage reduction continues to be the most common reason for grouting in rock. Permanent reduction may be required, such as in prevention of flow under a dam, as illustrated in Figure 1.3, or into a tunnel or other subsurface structure. Conversely, water control may be required for only a short time period to enable construction of underground works. Typical examples are the prevention of excessive flow into a tunnel being mined below the water table and the control of subsurface inflows during shaft sinking or other similar excavations.

Many different materials are used in water control grouting. Both cementitious and chemical solution grouts see frequent use, either alone or in some combination. Where a large flow or fast moving water must be stopped, grouts that set very rapidly and provide strong resistance to washout are required. Typical examples are combinations of sodium silicate and common cement, hot bitumen, and ready-mixed concrete containing anti-washout admixtures. Fillers such as ribbons or punchings from plastic bags, which tend to gather and plug openings, may be added to enhance the grout's plugging ability.

1.2.1.2 STRENGTHENING ROCK

Badly jointed or fractured rock can be returned to a monolithic condition by grouting. This may be done for improvement of bearing capacity, but is more typically done to increase the capacity of embedded rock bolts, ground anchors, and such. It can also be used to strengthen weak rock under and around the tips of foundation piles. In many cases, the grouting intent is a combination of both water control and strengthening. This is particularly true where the improvement is made as an aid to construction of underground structures.

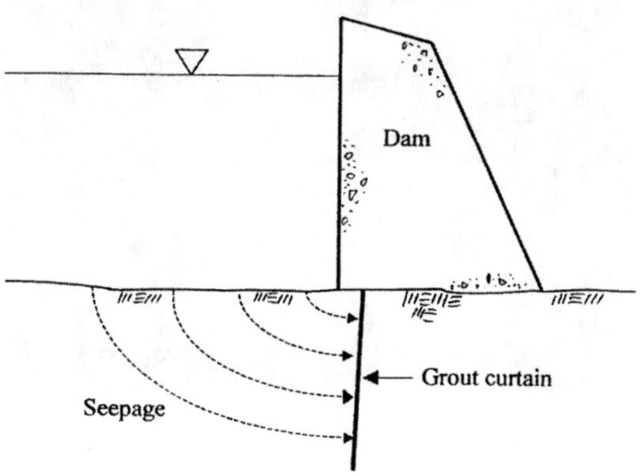

FIGURE 1.3 Forming an impermeable curtain under dams is one of the oldest uses of grouting.

1.2.1.3 MITIGATION OF SINKHOLE FORMATION

Subsurface voids are common in limestone and other soluble rock. When they are sufficiently large, overlying soil can run into them, resulting in surface sinkholes. Although such voids are sometimes discovered during geotechnical investigation for improvements, more often they occur suddenly under or adjacent to structures. In either event they must be capped to prevent further loss of soil. This is usually accomplished by grout injection through casings penetrating the disturbed mass. If the expected volume exceeds 3–5 yd^3, (2.3–3.8 m^3), it is most expedient and economical simply to use pumped concrete. Once a void is capped, densification of the overlying disturbed soil is typically accomplished by compaction grouting.

1.2.2 Grouting in Soil

Grouting is commonly used to strengthen soil formations, either temporarily during construction or permanently for increased strength and load-bearing capacity. There are four distinctly different mechanisms by which this is accomplished: densification, cohesion, reinforcement, and chemical exchange. As in rock, grouting is often performed in soil to lower the permeability and inhibit the movement of water.

1.2.2.1 SOIL DENSIFICATION

Virtually all soils other than clays can be strengthened through densification. Density can be increased through *compaction grouting,* which involves injection of a stiff grout that remains in a cohesive growing mass. The grout does not mix with the soil, but forms a distinct interface. Compaction grout must behave as a growing solid to provide controlled compaction. Regardless of consistency, should the grout behave as a fluid, hydraulic fracturing of the soil will result, and control of the densification will be lost. The principal mechanism is densification, and it has little influence on cohesion or other soil properties.

The primary use of compaction grouting is remedial improvement of low-density soil, and it is used extensively for improvement of soils in new construction, especially for mitigation of the liquefaction potential during earthquakes. Its greatest use is the correction of settlement damage to structures of virtually all types and configurations. A great advantage is its ability to be easily performed in the most restricted and isolated spaces, and it is often performed inside structures.

In practice, injection holes are located in a grid of 4–12 ft (1.2–3.6 m), extending to the desired depth of treatment. Grout injection is done in discrete vertical stages, which are typically of 1–6 ft (0.3–1.8 m) in height. Injection can be made from the bottom up or from the top down. Because the added material will increase the weight of the treated soil mass, it is imperative to grout all the way to a suitable supporting layer. There is virtually no technical limit to the depth for compaction grouting, and it has been performed to about 400 ft (120 m). The procedure is thoroughly discussed in Chapters 2 and 11.

1.2.2.2 GROUTJACKING

Groutjacking is an extension of compaction grouting that uses essentially the same equipment and grout mixtures to raise or level settled structures and other improvements. It has been used to successfully lift pavements as well as larger structures, including concrete buildings as high as eight stories. The procedure may be used alone to raise the settled improvements or, more typically, as an extension of a compaction grouting program to remedy structural settlement, where the deficient soil is first improved.

1.2.2.3 INCREASING COHESION OF GRANULAR SOILS

Interparticulate bonding of the soil grains can be provided through soil solidification by *permeation grouting,* which involves partial or complete filling of the intergranular pore space to increase the soil's *cohesion.* A variable increase in the soil unit weight will also occur, depending on the type of grout used. Chemical solution grouts can impart modest bonding with a minimal weight increase, whereas cementitious suspensions provide good adhesion but also add considerable weight.

The largest usage of soil solidification is in connection with underground construction where it is often used to facilitate earth retention, as illustrated in Figure 1.4. Most chemical solution grouted masses suffer strength regression as they age, but such grouts are entirely satisfactory and are often used for this work, which requires only short-term or temporary solidification. The strength of chemically grouted soil is very sensitive to the rate of loading, moisture

FIGURE 1.4 Solidification of granular soil facilitates shoring and provides soil retention during construction.

content, and many other factors. A reliable strength of up to about 100 psi (690 kPa) is obtainable with these grouts. Most formulations will retain very near their initial strengths for at least several months, and in some cases, many years.

Although the particle size of common cements is too great to enter all but the coarsest of sands, suspension grouts composed of appropriate ultrafine cement can penetrate the void matrix of virtually all sands and even some silty sands. Unlike their chemical solution grout cousins, the cementitious grouts are not subject to strength regression with time and are usually durable long term. Depending on the particular mix, the cementitious grouts can provide reliable unconfined strengths of several hundred psi, and strengths similar to those of concrete in clean, dense sands are possible. Although these high strengths are obtainable, they often are not needed and should not be inappropriately specified. This is especially important where the purpose of the grouting is to aid soil retention during construction and hand trimming of the grouted face will follow. Likewise, even where permanent improvement is being performed, one should consider the possibility of future excavation and apply appropriate limitations to the strength of the grouted formation.

1.2.2.4 STRENGTHENING BY REINFORCEMENT

Reinforcement of the soil can be provided with an interconnected system of thin but relatively strong grout lenses. These are formed by deliberate hydraulic fracturing of the soil, usually with a cementitious suspension grout. The intent of *fracture grouting*, also known as *claquage* in Europe where it originated, is to provide improvement using more traditional pourable grout mixes. Injection must be made so as to provide the maximum achievable control of the grout deposition, although accurate control of the fracture locations or orientation is not possible.

Another use of fracture grouting involves stabilization of clay soils, which are subject to volumetric instability with moisture variation. Clay particles are laminar, and moisture surrounding them becomes part of the soil structure and volume. Moisture changes thus result in significant volume changes. The effect can be moderated with exposure to a source of calcium such as lime slurry, which causes a *cation* or *base* exchange. This can stabilize the soil moisture, reducing its propensity for volume change. The beneficial mechanism of *chemical exchange* is not well understood, and the results not always predictable, but the technique is often performed as few alternatives are available.

1.2.2.5 REDUCING PERMEABILITY/ WATER CONTROL

The permeability of sands and sandy soils can be greatly reduced by injection of a grouted curtain, not unlike that which has been performed traditionally in rock. The pore size of most soils, however, is significantly less than the width of typical rock joints and, except in the case of coarse sands and gravels, is insufficient for intrusion of a common cementitious, particulate grout. Yet ultrafine cement grout will effectively permeate sands and sandy soils for water control just as they will for strengthening applications. Thus, seepage reduction in soils is almost always accomplished with either chemical solution or cementitious suspension grouts based on ultrafine cement. Whereas hard, rigid chemical gels are preferable for strengthening applications, some flexibility is often desirable in water control work, especially adjacent to concrete or other hard underground structures. Thus, chemical solution grouts especially formulated for water control applications are frequently preferred.

Ultrafine cementitious grouts are perhaps the most durable for water control, but are subject to dilution and/or washout unless they contain special admixtures. Further, they develop into brittle compositions that are subject to cracking when subjected to movement. They are thus usually avoided in applications where this may be a problem. Chemical solution grouts are available that provide a variety of special properties especially important in water control work, such as rapid setting and resistance to dilution. As in dealing with rock, in those rare instances where significant amounts of rapidly moving water are encountered, a flash setting combination of sodium silicate and cement grout, stream mixed at the grout header, may be used.

FIGURE 1.5 Structural grouts are injected to provide support to base plates.

1.2.3 Structural Applications

Pressure grouting is used in new construction as well as in the repair and rehabilitation of all types of structures. In geotechnical grouting, there is similarity of the various applications, whether in rock or soil, and most contractors will perform in either. In structural work, however, there is enormous variation in the many different types of grouting, and it is not common to find specialist contractors that perform in all of the established areas. In fact, it is not unusual to encounter specialists who limit their operations to only a single specialty area of application.

Grouting is part of virtually all posttensioned concrete work, wherein the spaces remaining in the tendon ducts after tensioning are filled with a cementitious grout. Construction joints of new massive concrete structures such as dams are frequently filled with a cementitious grout once the initial major shrinkage has occurred. Void spaces under and around precast elements are also frequently filled with cementitious grouts in order to attain monolithic construction. The bearing space below base plates of all types and sizes are supported by injected grout, as shown in Figure 1.5, as are the bases of a wide variety of tanks, machinery, and mechanical equipment. Filling of voids between soil or rock and a hard surface such as concrete or masonry, which is known as *contact* grouting, is a common requirement in tunneling and other underground construction.

Resinous grouts such as epoxy are injected into narrow joints to bond different elements of concrete, masonry, and sometimes other materials. Both cementitious and resinous grouts are widely used in the repair and rehabilitation of concrete and masonry structures. Many preblended packaged systems, which are formulated for a variety of special requirements, are commercially available for this work.

1.2.3.1 PREPLACED AGGREGATE CONCRETE

New concrete can be cast by pumping either a cementitious or a resinous grout into large aggregate that has been previously placed in appropriate forms. This so-called *preplaced aggregate concrete* has many benefits, in that it can be cast in spaces of virtually any size, volume, or configuration, in areas of difficult access or other restriction, and even under water. Because each piece of the large aggregate is in physical contact with its neighbors, shrinkage is greatly minimized and dimensional stability of the final mass is ensured. Both thermal and shrinkage cracks, which are common in conventionally cast concrete, are almost entirely eliminated.

Because the aggregate will not be in a mixer, there is virtually no limit as to the maximum size that can be used, which is advantageous in that larger aggregate leaves fewer voids to be filled with grout. Not only does the aggregate not shrink, but it costs much less than the grout, so maximizing the proportion of large aggregate adds economy as well. As the grout is injected under pressure, complete filling under penetrations and "roof" areas is ensured. The process allows massive placements to be cast free of construction joints by initiating injection at one extremity and continuing in an orderly manner to the opposite limit. Because of its low shrinkage and good dimensional stability, preplaced aggregate concrete finds extensive use in repair of structures.

1.2.3.2 FILLING OF LARGE ANNULAR SPACE

In construction of large pressure conduits in shafts and tunnels, it is common to place a heavy steel lining, which results in an annular space of a few inches to a foot or more. Such voids are commonly filled with grout, in a process referred to as *backfill grouting.* A constant problem that must be avoided is floating the new liner upward during the injection. Two distinctly different procedures are used. Where large or irregular spaces must be treated, they are often filled with pea gravel, which is pneumatically placed as the lining is installed. This helps to anchor and stabilize it in the correct position. Once the lining is placed, a cementitious grout is injected into the gravel so as to result in preplaced aggregate concrete filling the entire space.

Where the void is filled only with grout, the new conduit is usually blocked and wedged into proper position prior to injection. To minimize the buoyancy of the liner, low-density cellular grouts are often used. Similarly, when an existing pipeline is relined with a smaller tube, commonly referred to as *sliplining,* the resulting annulus must be filled. In this work, as well, low-density grouts are often employed, not only to minimize the risk of floatation, but to minimize forces against the liner, which is often of minimal strength.

1.2.3.3 FILLING OF FABRIC FORMS

Fabric forms can be fabricated in a wide range of different shapes, the only absolute requirement being that they must be completely closed to provide confinement for the injected grout. They are most commonly employed for shoreline protection and the encapsulation of pilings in water, but can be custom fabricated for any special configuration or application. Originally developed to enable repair with the use of fluid grouts, most of the work now being accomplished uses ready-mixed concrete or mortar as fill.

1.2.3.4 STRUCTURAL REPAIR AND REHABILITATION APPLICATIONS

Virtually all major repair and rehabilitation of concrete and masonry structures involves at least some pressure grouting, and in many cases grouting is the main task. The existence of sometimes large voids is common in most older masonry construction. These are often filled in order to strengthen the structure. New reinforcing bars are sometimes placed in holes drilled into the masonry prior to grout injection to provide additional tensile capacity to the section. Honeycomb and other porous areas in concrete are similarly filled by grout injection. As with masonry, new reinforcing steel is sometimes grouted into drilled holes in existing concrete, as are long anchor bolts.

Epoxy injection into cracks represents one of the largest areas of structural grouting. It is common in all types of concrete and masonry and is widely performed on structures that have experienced seismic damage. Extensive research and experience has shown that virtually all of the strength of an element, reduced by damage, can be reestablished. The procedure can even be used for the repair of damage to timber members.

1.2.3.5 CONTROL OF LEAKAGE INTO STRUCTURES

The control of water leakage into structures, and from within water-retaining structures, represents a huge use of grouting. In new construction waterstops can be formed by injection of special tubing, which is preplaced into a joint prior to concrete placement. Once the concrete has cured, the tubing is injected with a chemical solution grout. There are several systems on the market, and they all operate similarly, with one exception. The tubing of some systems is equipped with one-way valves so that it can be flushed out following injection, thus allowing for reinjection.

Many proprietary chemical solution grouts, formulated specially for this type of work, are typically used. The most commonly used are based on urethane resin, which can be formulated into a solid, flexible gel or either rigid or flexible foam. Several urethanes are water activated and will react as soon as they come in contact with water. Although the urethane-based grouts are the most commonly used, acrylic resins are also employed, especially for injection into low-permeability, porous areas and very fine crack networks.

Because this work is usually is done with the specialty grouts and equipment that is quite different and much smaller than that used for most other grouting (Figure 1.6), it is often performed by specialist contractors that do no other types of grouting. Many like to think of grouting as an art, in spite of its now being a well-developed technology. Control of water leaking into and through structures is, however, one area in which much art remains and prior experience is paramount. Often, a visible leak is a considerable distance from its source, and a good knowledge of the way things are built and intuition about water movement are prerequisite to successful stoppage.

FIGURE 1.6 Water control grouting requires special ability and equipment.

1.2.3.6 SEALING OF LEAKING SEWERS

Another type of grouting that is nearly always performed by specialized operators is the sealing of small-diameter sewers and similar pipes to correct water leakage or infiltration. Because these lines are generally too small in diameter to allow physical entry, they are repaired with the use of remotely controlled, closed circuit television cameras, together with similar remotely controlled packers for the injection. The technology is highly advanced, and it is now possible to seal leaks not only in main lines, but at lateral connections as well. As this work requires rather sophisticated equipment and specially trained operating crews, it is typically performed by operators who work exclusively in this area.

CHAPTER 2

Types of Grouting

2.1 PERMEATION GROUTING
- **2.1.1** Injection Systems
- **2.1.2** Water Control
- **2.1.3** Strengthening of Soils
- **2.1.4** Permeation Grouting in Structures

2.2 COMPACTION GROUTING
- **2.2.1** Development of the Technology
- **2.2.2** Grout Holes
- **2.2.3** Injection Staging
- **2.2.4** Injection Line Requirements
- **2.2.5** Grout Injection
- **2.2.6** Limitations

2.3 FRACTURE/CLAQUAGE GROUTING
- **2.3.1** Lime Injection
- **2.3.2** Limitations and Risks

2.4 MIXING/JET GROUTING
- **2.4.1** Jet Grouting Application
- **2.4.2** Risks and Limitations
- **2.4.3** Deep Soil Mixing

2.5 FILL GROUTING
- **2.5.1** Filling Fabric Forms

2.6 VACUUM GROUTING

HISTORICALLY, only one type of grouting was used. This was *permeation grouting,* whereby open voids or pores are filled. Such filling is requisite when the object of grouting is water control or reduction of seepage. However, much work now involves strengthening of the formation, and this can be accomplished with other methods. Regardless of the purpose, virtually all grouting in rock or structures involves permeation grouting, but for strengthening soil, four different methods are available: *permeation, compaction, fracture/claquage,* and *mixing.* As with other substrates, permeation grouting in soil fills voids in the form of the pore system, wherein the principal improvement mechanism is increasing *cohesion.* Compaction grouting involves injection of a growing mass of stiff mortarlike grout to compact or *densify* the soil. Fracture grouting, also known as claquage, involves hydraulically fracturing the soil with grout so as to provide *reinforcement* through the resulting network of strong, interconnected grout lenses. Mixing is accomplished by high-pressure jetting of the soil with grout so that the two are combined to form a *mixed* composition, *soilcrete.*

With the availability of modern concrete pumps, the filling of huge cavities, such as abandoned pipelines, tanks, old mine works, and so on, is easily performed with ready-mixed concrete or mortar. I prefer to consider this simply as *fill grouting.* Although grout is most often impelled by pressure, it can also be sucked into voids that have been placed under a vacuum. This is referred to simply as *vacuum grouting.*

Following is a description of the various methods, the mechanics by which they work, their areas of usage, and their relative advantages and limitations. Details of the materials and mixes are discussed in Chapters 3 through 7. Examples of usage for a variety of different needs are provided in Chapters 11 through 17.

2.1 PERMEATION GROUTING

Permeation grouting is the longest-established and most widely used grouting technique. Sometimes referred to as *penetration grouting,* it involves the filling of cracks, joints, or other generally small defects in rock, concrete, or masonry, or the pore spaces of soils, aggregates, or other porous media. The objective is to fill a void space without displacement of the formation or any change in the void configuration or volume. This can be done for the purpose of strengthening the host formation, halting the flow of water through it, or a combination of the two. Permeation is the only type of grouting that can be used in all of the different media into which grout may be pumped.

2.1.1 Injection Systems

There are two injection arrangements used in permeation grouting. The first, and certainly most frequently used historically, involves a *circulating* grout system, as illustrated in Figure 2.1. In this system the grout is constantly circulated from an agitator and back again. A "T" fitting is located in the line, with a connection made to the grout hole. This is typically by way of a *header,* as seen in Figure 2.2, that contains the "T" fitting, a pressure gauge, and valves to control the grout flow. One valve is located on the branch of the "T" that feeds the grout hole, and another on the grout return line. Through valve manipulation, the flow rate, and thus the pressure of the grout entering the hole, can be controlled. This allows the grout flow rate to be varied so as to maintain a given pressure level, regardless of the pumping rate.

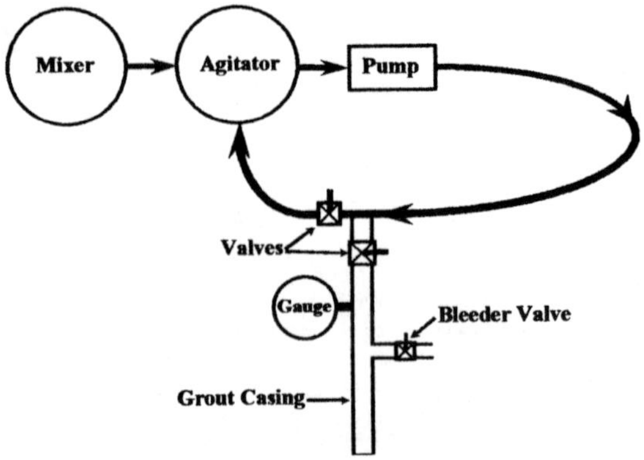

FIGURE 2.1 Typical circulating grout injection system.

A *direct* system involves only a single grout line directly from the pump to the grout hole, as illustrated in Figure 2.3. Because the grout pressure is directly related to the injection rate, it is controlled by adjusting the pump output. A direct injection system has the advantage of simplicity and the use of a minimal amount of hose. The greatest advantage, however, is the constant supply of fresh grout provided. Because circulating systems return old grout to the agitator, the

FIGURE 2.2 Grout header containing regulating valves for circulating system.

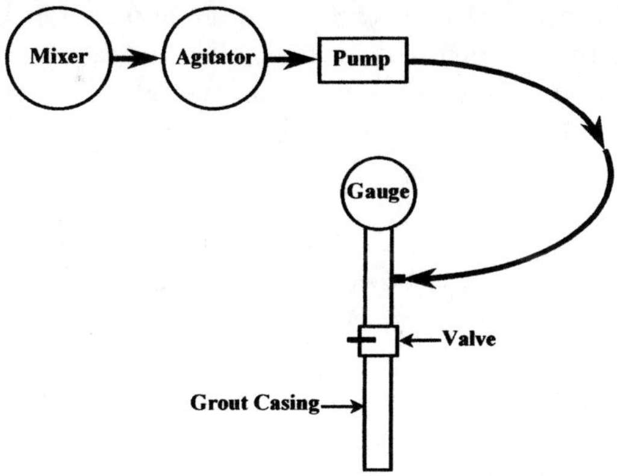

FIGURE 2.3 Direct injection system.

grout properties can change with time, especially in the event of very slow acceptance of the grout in the hole commonly referred to as take.

2.1.2 Water Control

When an objective of injection is control of water seepage through the formation, nearly all water passages must be filled. This requires that grout hole spacing be sufficiently close to provide some overlap of grout penetration. Although the injection of a single row of appropriately spaced holes can decrease subsurface flow, two or more rows are usually required where a high level of reduction or complete blockage is required.

Most permeation grouting for water control is performed in rock, and a large part is in connection with dam foundations. In such work a grout *curtain,* which typically consists of one or more rows of grout holes, is constructed. The curtain usually starts at the bottom elevation of the dam structure, where the holes are usually drilled. They will extend to a depth that is typically related to the reservoir head, but certainly into a low-permeability zone of rock. The movement of water in rock is through joints and other discontinuities, so it is imperative that the holes intersect major defects. This often means drilling inclined holes, and in some cases it may be necessary to drill them in multiple, or opposed, directions. The cost and effectiveness of a grouting program are closely related to the existing geology and the accuracy of the geological information on which the program is based.

Grouting effectiveness is usually determined by the level of reduced permeability achieved, which is determined by water pressure tests often in successive grout holes that are injected. Subsurface flow can also be measured by way of downstream drainage holes. Most grouting of this type is done with a *split spacing* hole layout. In such a layout the primary holes are widely spaced, usually at about 40 ft (12 m). After injection, the space is split and secondary holes are placed midway between the first holes. The space is further split for tertiary, and sometimes quaternary, holes, as required. The final spacing is based on the effectiveness of the earlier grout injection as indicated by water tests in the holes as they are grouted. Higher pressures are usually allowed at greater depths, which increases the effective distance of grout flow. Where more than a single row of holes is used, the later rows are often upstream of the primary row.

2.1.3 Strengthening of Soils

In soils, the procedure involves permeating and filling the soil pore spaces without any significant disturbance to, or movement of, the individual soil grains. This is important so as not to change the soil structure. Care must also be taken to control the injection rate and resulting pressure in order to avoid the occurrence of hydraulic fracturing of the soil. Should fracturing occur, the grout will flow so as to continuously propagate the fracture, which will be the course of least resistance. It then will not permeate the intended zone of the formation, and control will be lost.

The largest single use of permeation grouting is for temporary solidification of low-cohesion granular soils as an aid to construction or to reduce or eliminate soil disturbance or failure resulting from excavation. It is possible to solidify a self-supporting buttress of soil so as to allow a shear cut, as illustrated in Figure 1.4. Permeation grouting is also used for mitigation of the liquefaction potential of soil, as well as permanently increasing its load bearing capacity.

Grout holes used for injection into soil are typically much more closely spaced than those used for injection into rock, and they are usually on a grid pattern, covering the area to be solidified. The alternate primary holes are grouted first, followed by secondary holes, which are located in a split space manner. Further splitting to the tertiary level is not frequently performed, except where unusual conditions exist. Both direct and circulating injection systems are used. In the United States, a direct injection system is the most frequently employed, whereas European operators tend to prefer circulating systems.

Strengthening is generally limited to sands and sandy soils containing minor amounts of finer particles. Although suitable grouts can be injected into finer materials, significantly slower injection rates are required. This adds appreciably to both the time and cost of the work, which usually precludes the use of this method. Permeation grouting of soil is generally more expensive than other types of grouting. Its use is thus commonly limited to applications where a considerable increase in *cohesion* is required or in clean, coarse materials for which other methods are not as appropriate.

The structure and size of the soil pores dictate the type of grout that can be effectively used. Because of excessive particle size, ordinary portland cement grouts are usually limited to use in clean, coarse sands and gravels. Accordingly, permeation grouting of most soil employs either a highly penetrable chemical solution or ultrafine cement grout. Outside the United States, principally in Europe, clay or combinations of clay and cement or chemical solution grouts have also been used.

2.1.4 Permeation Grouting in Structures

As with other formations, permeation grouting into structures is performed for both seepage control and strengthening. Unlike geomaterials, which tend to be confined, concrete and masonry sections are typically unconfined and subject to constant movement. Expansion and contraction are continual, due to variations in ambient temperature, changes in loading and load patterns, and foundation movements. In contrast to water movement in geomaterials, structural leakage typically occurs through cracks and defects, which are usually visible. These factors often call for a grout material that will provide some flexibility and yet set rapidly. As structural work is nearly always limited to discrete defects, relatively small quantities of grout are required, so the use of more effective though costly compounds is justified.

Strengthening structural elements often requires the use of high-strength grout, as concrete, masonry, and similar materials are significantly stronger than typical geomaterials. A common objective in grouting concrete is to fill cracks so as to reintegrate broken sections. To be effective, the repaired crack must withstand tensile forces that are at least as great as the tensile strength of the concrete. This requires the grout to possess both high bonding properties and high tensile strength, which often dictates the use of resinous materials such as epoxy.

Unlike concrete, masonry often contains a large amount of open void space. This can include many small spaces, such as incompletely filled mortar joints, as well as substantial volumes of open cavities, such as between separate wythes of masonry. Between these extremes, there can also be unfilled cells of concrete block or decorative units, such as the terra-cotta cladding shown in Figure 2.4. It is not unusual to find incompletely filled joints, as well as large internal voids, in older masonry and stone structures.

Masonry presents further challenges to the grouter, because it is usually exposed to view and the surface must be kept clean and unblemished. This can be a particular problem, considering that masonry is usually porous, so that grout injected into interior defects can readily leak onto the surface. For this reason, many experienced structural grouters will not make injections unless complete sealing of the exposed surfaces has first been accomplished.

In some cases open voids exist vertically for a considerable distance, so that grout injected at higher elevations will leak all the way to the base. In one instance, during repair of earthquake damage, a cementitious grout was being injected into a spandrel beam on the third floor of a building. Abnormally large amounts of grout were being accepted at virtually no pressure. After injection of about 25 ft^3 (0.7 m^3) of grout, work was stopped and the structure carefully inspected. There was no sign of the grout and, most interestingly, little leakage or even moisture on the face of the wall, until the basement was entered. And there it was, virtually all of the grout that had been injected.

FIGURE 2.4 Large unfilled cells are typical of older terra-cotta masonry cladding.

Another potential problem is the additional weight of the added grout, which can be considerable. Many structures, particularly older ones, have already experienced foundation settlement; in fact, this is often the reason the work is being performed. Significant additional weight can cause renewed settlement, actually increasing the original problem.

2.2 COMPACTION GROUTING

Compaction grouting involves controlled injection of very stiff, mortarlike grout, at high pressure, into discrete soil zones. Properly placed, the grout remains in a homogeneous, expanding mass, which displaces, and thus compacts, the adjacent soil, as illustrated in Figure 2.5. The improvement mechanism is *densification,* which increases the bearing capacity. Because the volume of the soil's pore structure is reduced, a modest reduction in permeability will result. Compaction grouting in itself, however, cannot significantly reduce seepage. The soil's cohesion

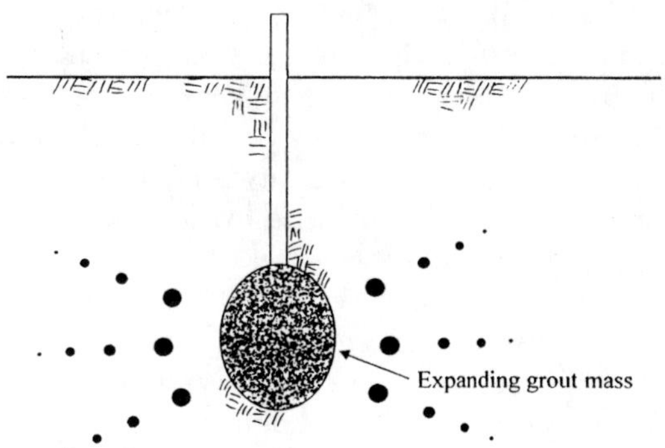

FIGURE 2.5 Compaction grouting densifies soil around the injected grout mass.

may be slightly increased by way of the *densification*, but such improvement will be minimal.

Because of the grout's required low mobility, much higher pressures are used than in traditional grouting. It is not unusual for several hundred psi (a few MPa) to be required when injection is made at depths of only 5 or 10 ft (1.5 or 3.0 m) from the surface. The primary direction of pressure influence is initially horizontal until a significant horizontal cross section has developed during injection. Because of the low mobility of the grout, a fairly high pressure loss is experienced in the injection system as well as in the ground. Fundamental to the success of the procedure is deposition of the grout in a globular mass with a distinct grout–soil interface. Indeed, one of the primary advantages of the technique is the ability to pinpoint the deposition location, allowing absolute control of the placement.

Regardless of the slump or stiffness of the grout, if it is unduly mobile or behaves as a fluid in the ground, hydraulic fracturing of the soil can occur, resulting in loss of control. Should this occur near a downslope, retaining wall, or other substructure or within a water-retaining embankment, severe displacement or complete failure can result. Compaction grouting has been successfully used in most types of soil, and to depths on the order of 400 ft (120 m). Injection into clay soils is possible, but because of slow drainage, extremely slow injection rates are required. This significantly lengthens the injection time, greatly increasing its cost and precluding its use in most high-plasticity soils. The procedure also has limited effectiveness in clean, coarse sands and gravels.

The work can be executed without a great deal of mess or interference with the normal operations of the facility being grouted. Large equipment is not required near the injection location, and drilling with handheld equipment can be done in most soils. This allows the work to be performed in the most confined areas, even those subject to restricted access. It is accomplished in many otherwise difficult situations, which no doubt accounts for the procedure's widespread use.

2.2.1 Development of the Technology

Compaction grouting originated in California, where the procedure has been used for about 50 years. Its application is now extensive throughout North America and many other countries, especially in Asia. The first publication on the procedure (Graf, 1969) presented a theoretical description of the process, hypothesizing that the injected masses would be generally spherical and would densify the injected soil radially, in all directions. The first report describing the actual mechanism and providing data derived from the excavation and removal of full-scale test injections was by Brown and Warner (1973). It reported columnar injected masses with essentially horizontal, radial densification.

Subsequent work has included a determination of the increase in soil density, utilizing standard penetration, cone penetrometer, and laboratory tests of split spoon specimens, both before and after grout injection. Large-scale load tests and long-term monitoring of structures overlying treated soils have also been performed on many applications. Additional studies have been directed at the effect of lateral forces exerted in the soil mass and on adjacent structures. The result of this substantial research and investigation is that compaction grouting has become one of the best-understood grouting technologies. It is unquestionably advantageous for the strengthening of soils, which is its only use.

The single most important requirement to ensure effective densification is to obtain a globular grout mass, which will typically be either columnar or tear shaped. Cracking or hydraulic fracturing of the soil mass, with resulting thin lenses of grout, must be avoided. One measure of the regularity of the obtained mass is the travel index (TI) of the grout. The TI is the maximum radial travel of the grout from the point of its injection, divided by the minimum radial distance to a grout–soil interface, and is thus representative of the grouts propensity to remain in a controlled mass and at the intended location. Grouts with a low travel index (less than about 3) remain in relatively symmetrical masses, with clear interfaces of the surrounding soil. Loss of placement control and hydrofracture of the soil, however, are virtually ensured with travel indexes exceeding about 5.

The importance of the shape of the injected grout mass was first reported by Brown and Warner (1973). They described a research program performed in the 1950s that involved the injection and subsequent excavation of more than 100 test grout masses. The effort employed a variety of mix designs, consistencies, aggregate materials, and injection rates. A photograph of two of the excavated grout columns depicting the desirable shape was provided, which appears here as Figure 2.6. It is interesting to note the

FIGURE 2.6 Extricated masses of grout from tests. (From Brown and Warner, 1973).

much smaller diameter of the lower portion of the mass shown on the right. Whereas the upper deposits of the test site consisted of mixed soils, the particular grout hole extended into an underlying clean sand layer, which was not subject to the degree of compaction experienced by the overlying soil. The report concluded with examples of projects successfully completed with grout that the research found to be optimal. It was reported to consist of "fine sand combined with about 12% cement and water to form a very stiff mortar like mixture." A number of criteria were emphasized: "The greatest amount of grout was injected, and thus greatest densification achieved, resulted from the use of very stiff mixtures" and "a slower pumping rate resulted in significantly higher grout takes."

These investigators expanded on their experience in a further report, *Planning and Performing Compaction Grouting* (Warner and Brown, 1974). In this report the significance of the grout composition and a strict limitation on any clay content were further emphasized. An entire section was devoted to describing the "sand material," which included a gradation envelope, "Preferred Limits of Gradation for Sand Used for Compaction Grout," provided here as Figure 2.7.

The zero allowance for clay size constituents is noteworthy. The significance of grout consistency was again emphasized: "It is preferable to use the least amount of water that will provide a very stiff plastic consistency grout. A rule of thumb is the stiffer the grout, the more effective its injection will be."

An accompanying photograph showed such grout extruding from a grout hose, included here as Figure 2.8. Also emphasized were the risks of excessive pumping rates, which could result in "rupture" of the soil. The first two of the report's concluding statements for proper performance stressed the importance of the aggregate gradation and resulting grout rheology, as well as the injection rate: "Proper gradation of the sand material which accounts for 80%–90%

of the total grout volume is imperative" and "absolute control of grout rate is imperative . . . within a range of 0.3 cu ft (0.009 m^3)/min to 2 cu ft (0.06 m^3)/min."

Quite surprisingly, the salient conclusions relative to the grout material in that early research have proven applicable to the present day. And these have been substantiated by extensive subsequent research and literally thousands of successfully completed projects, yet contractors frequently violate these well-established material requirements.

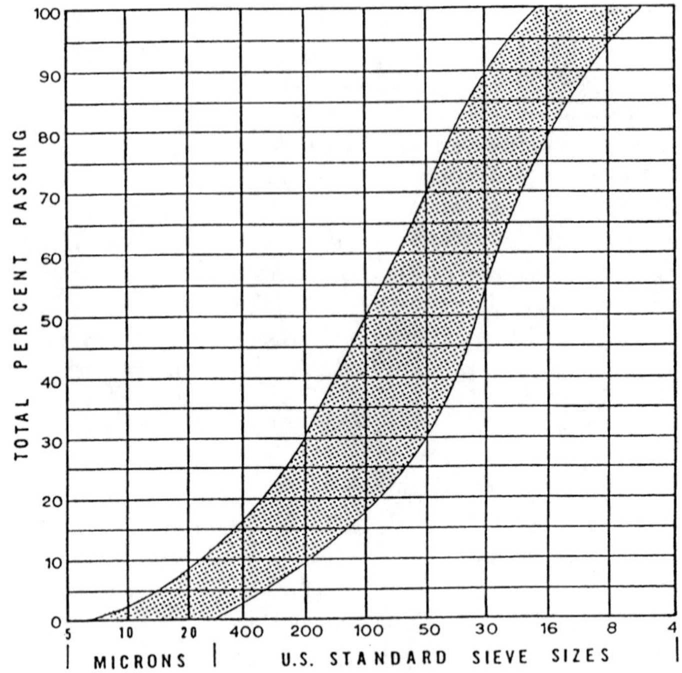

FIGURE 2.7 Acceptable compaction grouting aggregate. (From Warner and Brown, 1974).

As part of the grouting demonstration included in the 12th annual short course, Fundamentals of Grouting, now sponsored by the University of Florida and held in Denver, Colorado, in 1990, ten grout test injections were made and exposed, as illustrated in Figure 2.9. Two different grout mixtures, designated "A" and "B," were utilized. They were identical, except that 5 percent bentonite by weight of the sand material was included in the "B" mixes. Notwithstanding the recognized inappropriateness of the slump test for such grouts, which is thoroughly discussed in Chapter 8, such tests were made in as careful a manner as possible. The two grouts were injected at slumps of 1, 2, 3, and 4 in. (25, 51, 76, and 102 mm) and at a constant injection rate of 1.5 ft^3 (42 L) per minute.

The test site consisted of highly stratified stiff to dense clayey silts and sands, which exhibited negligible settlement potential. Untreated, they would provide adequate support for most normal foundations and thus provided an extreme condition for the evaluation of compaction grouting. Several of the injections resulted in hydraulic fracturing of the soil; however, the incidence and extent of the fracturing was directly related to the grout rheology. All of the "B" mixes resulted in hydraulic fracturing and revealed very high TIs. Of particular interest was the markedly better performance of the "A" grout at a 4 in. (100 mm) slump than the "B" grout at a slump of only 1 in. (25 mm).

FIGURE 2.8 Stiff, mortarlike, low-mobility grout pumped from hose.

FIGURE 2.9 Test site excavation to allow visual inspection of grout masses.

An extensive and very well documented research effort was conducted in San Diego, California, in 1991 (Warner et al., 1992). Eighteen grout masses were injected, with three different grout mix designs, using different grout consistencies and injection rates. The soils at the test site were predominantly fine- to medium-grained sands, with approximately 20 percent of the material passing a No. 200 (0.007 mm) sieve. The minus No. 200 (0.007 mm) portion consisted of approximately 70 percent silt and 30 percent low-plasticity clay. The soils were of waste fines from a quarry and were deposited as an uncontrolled fill. Their in-place densities varied from 79 to 96 percent of maximum dry density as determined by the American Society for Testing and Materials (ASTM) Standard D 1557.

The grout mixes, which were designated A, B, and C, were identical except for the aggregate fraction, which was of different gradations. The A and B aggregates as obtained from the pit were identical. They contained about 7 percent clay, and the minus No. 40 sieve (0.04 mm) fraction was plastic with a liquid limit of 35 and plasticity index of 10. The B grout mix contained the material as received from the pit, whereas the A mix had about 30 percent minus 3/4 in. (19 mm) gravel added. With the exception of a component of about 4 percent nonplastic clay, the C aggregate was close to the gradation envelope recommended by Warner and Brown (1974); see Figure 2.7. Each of the grout mixtures was injected using four different injection rates, 1, 2, 3, and 4ft^3 (28, 57, 95, and 113 L) per minute.

Of interest was the development of three different basic shaped masses of the injected grout. These were closely related to the aggregate material gradation, especially the clay content. They were radially symmetrical columnar shapes, as shown in Figure 2.10, with four vertical "wings," extending at about 180 degrees from a columnar mass, at the hole alignment (Figure 2.11) and two wings, with or without formation of an initial grout column, resulting from hydraulic fracturing, as illustrated in Figure 2.12. The detrimental effect of clay in the grout was again clearly illustrated by the thin wings of grout, shown in Figure 2.12, and the extremely high travel indices, shown in Figure 2.13.

An extension of the 1991 program involved an additional 11 grout holes. These were made on the same site and immediately adjacent to those of the 1991 effort. In order to establish the

FIGURE 2.10 Uniform globule of grout.

FIGURE 2.11 Four-winged grout mass. (From Warner et al., 1992).

FIGURE 2.12 Two-winged grout mass resulting from hydraulic fracturing of the soil. (From Warner et al., 1992).

influence, if any, of the beginning soil density, the entire test site was excavated and backfilled with minimal compaction. As in 1991, the grout mixes were identical except for the gradation of the aggregate. Three different aggregates, designated D, E, and F, were used, which contained 0, 1, and 4.5 percent bentonite clay, respectively. In spite of the obvious benefit of gravel in the aggregate, as demonstrated by the earlier work, because many contractors do not have the ability to pump larger aggregate, it was not included. Exposure revealed grout masses very similar to those of the previous work. Again, there was a close correlation between the clay content of the aggregate and the shape of the resulting grout mass. Figures 2.14 and 2.15 show typical examples of the D and F grout masses, respectively. Note the very thin, long wing length in Figure 2.15, which resulted from hydraulic fracturing. This mass had a very high travel index of 20.

A mill building at a diamond mine in South Africa had experienced serious differential set-

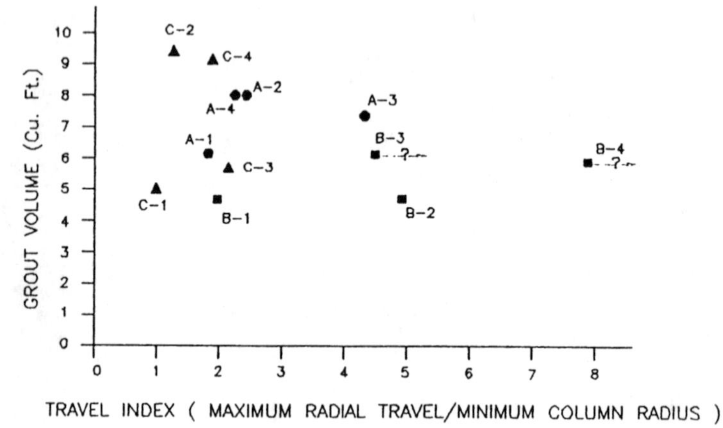

FIGURE 2.13 Grout travel index. (From Warner et al., 1992).

FIGURE 2.14 Globular D grout mass has low travel index.

tlement. The structure was founded on a marginally compacted mine waste fill, composed of sandy silt containing about 6 percent gravel. Compaction grouting was determined to be the best remedial method, but because the technique had not previously been used in the country, initial injections were exposed to ensure proper performance. The grout was a very stiff mixture of aggregate, falling within the envelope shown in Figure 2.7, and about 10 percent cement. It was injected at about 1.5 ft^3 (42 L) per minute. As can be observed in Figure 2.16, the resulting grout mass was a nearly perfect column.

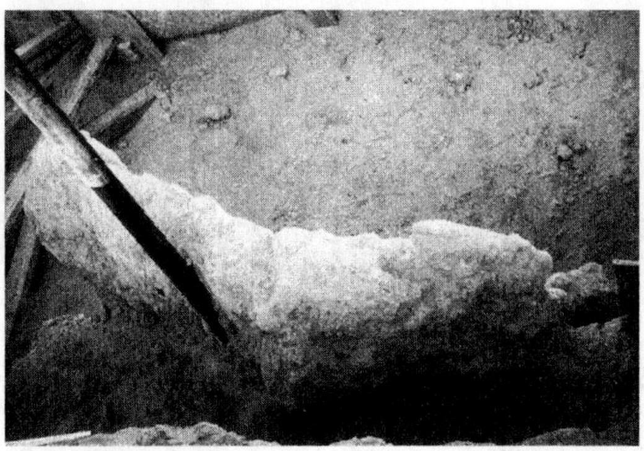

FIGURE 2.15 F grout mass exhibits long, thin wings typical of uncontrolled hydraulic fracturing.

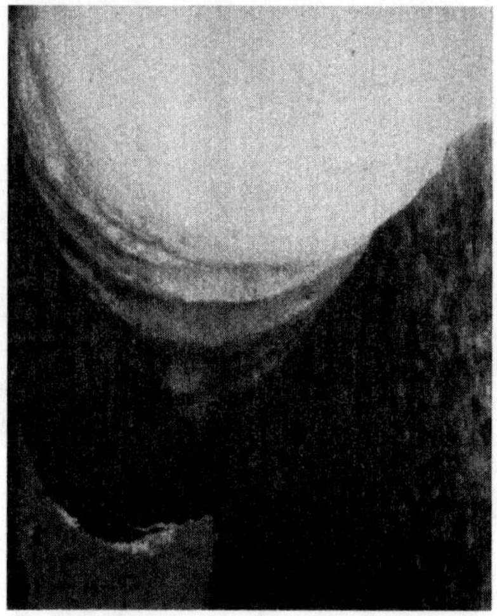

FIGURE 2.16 Near-perfect columnar mass.

Recent experimental work (1996) in Korea to develop a method to contain underwater bay mud involved many experimental injections, both in the dry material and under water. The grout typically used was a stiff mixture similar to that reported by Warner and Brown (1974) and was injected at rates of less than 57 L (2 ft^3) per minute. Columnar masses, of nearly 1 m (3.3 ft) in diameter were routinely obtained, as illustrated in Figures 2.17 and 2.18. Figure 2.17

FIGURE 2.17 Grout mass resulting from underwater injection.

FIGURE 2.18 Grout mass resulting from injection into loose sand.

FIGURE 2.19 Grout mass being extricated from soil box in testing machine.

shows the extricated mass from an underwater injection being raised by derrick barge; the mass shown in Figure 2.18 was injected into loose sand. Upon exposure, the sand fell off, revealing a near perfect column.

In a 2001 research program at the Bureau of Reclamation laboratory in Denver, Colorado, grout was injected into saturated gravelly sand at a simulated depth of 60 ft (18 m) by way of a load placed on a box filled with the soil. The watertight soil box measured 7 ft (2.1 m) square by 4 ft (1.2 m) deep and was placed in a 1 million lb (455 t) compression machine. It was carefully filled in 1 ft (28 L) layers, and a 2 in. (50 mm) injection pipe was embedded in the center so as to terminate 1 ft (0.3 m) from the bottom.

After the box was filled with the saturated soil, a nominal load was placed on the surface and the casing withdrawn 1 ft (0.3 m), followed by grout injection. The load was increased to the equivalent pressure of the soil at a depth of 60 ft (18 m). Following injection, the confining pressure was maintained for several hours to allow the grout to harden. The soil was then removed and the grout mass extricated, as illustrated in Figure 2.19. Many tests were run, including the beginning soil density as well as that resulting from the injection. These indicated a huge improvement, which was obvious from simple observation of the grout mass.

I have witnessed exposed grout masses on numerous projects. In the early use of the procedure, excavation and visual inspection were routine to increase knowledge and ensure quality. Exposure of full-scale test injections and/or early production work are often required as a qualification for contractors. The long history of diverse and successful application, combined with the very extensive research and confirmation evaluation, render properly performed compaction grouting an exceptionally reliable method for improvement of soil.

2.2.2 Grout Holes

Grout holes are usually spaced on a grid of 6 to 12 ft (1.8 to 3.6 m), although closer spacing is occasionally used. Alternate *primary* holes should be injected before the intermediate *secondary* holes. As a general rule, the injection should start at the outside of the area to be improved, working toward the interior portions. When grouting near a downslope or retaining wall, the points nearest these features should be injected first. Holes should generally be vertical, as inclined holes provide a greater horizontal ef-

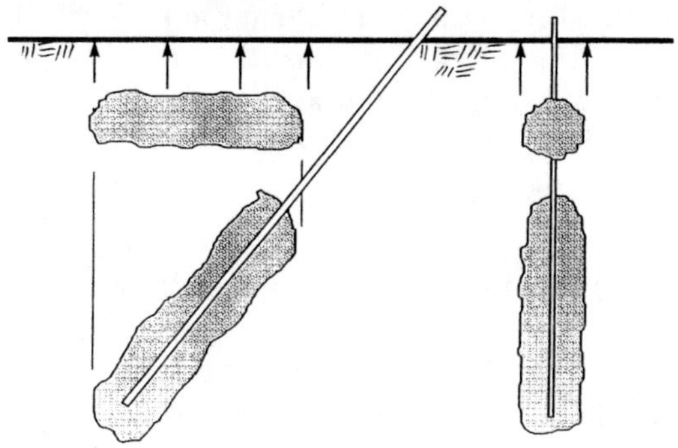

FIGURE 2.20 Inclined holes result in large horizontal area, causing jacking of the surface.

fective area, as illustrated in Figure 2.20. This limits the injected quantity of grout, as greater surface heave forces will develop at lower grout pressure, and thus provide less beneficial compaction. Further, a vertical column of grout and compacted soil provides better support than an inclined one. Where inclined holes are used, they should generally not be more than about 20 degrees off vertical.

2.2.3 Injection Staging

The injection work is virtually always done in stages; that is, only a few feet of the grout hole are injected at any one time. Staging can proceed from the top down (downstage) or from the bottom up (upstage). The upstage method is the fastest and most economical and thus the most frequently used, especially for deep injection. For injection at a depth less than about 15 ft (4.5 m), however, working downstage has the distinct advantage that each injected stage provides additional restraint and containment for those that follow. Consequently, higher pressure, enabling injection of more grout and thus greater densification, can be used in the succeeding deeper stages. Whereas upstage injection is nearly always accomplished in one continuous operation, each stage is allowed to harden before the next one is drilled when working downstage. A combination of up- and downstage injection can be used for holes that start at shallow depths but extend to a considerable depth.

Grouting upstage involves the following steps:

1. Drill a hole to the bottom of the zone to be grouted.
2. Place casing to within a few feet of the bottom of the hole. The casing should be a snug fit and may require pushing or driving into place. Sometimes it is driven entirely, the predrilling being eliminated.
3. Inject grout; continue until essential "refusal" is reached. Refusal is usually considered to be (1) a slight movement of the overlying ground surface or improvements, (2) injection of a predetermined amount of grout, or (3) reaching a given maximum pressure at a given pumping rate.
4. Raise the casing a regular increment, usually 1 or 2 ft (0.3 or 0.6 m).
5. Resume grout injection until refusal.
6. Repeat steps 4 and 5 until the top of the zone to be grouted has been reached.

Grouting downstage involves the following steps:

1. Drill an oversized (usually about 3 in. (75 mm) diameter) hole, to the top of the zone to be densified or a minimum of about 4 ft (1.2 m) deep.
2. Insert a casing (usually 2 in. (50 mm) internal diameter) into the hole and fill the annular space outside the casing with rapid-setting grout.
3. Drill through the casing and advance the hole several feet for the first stage. Typical stage lengths are on the order of 3–6 ft (0.9–1.8 m).
4. Inject grout until refusal is reached, as described earlier.

5. Repeat steps 3 and 4 after the previously placed grout has hardened until the bottom of the zone to be injected is reached.

Compaction grout casing is typically 2–4 in. (50–100 mm) internal diameter (ID) steel tubing. Some contractors use proprietary casing, whereas others employ either standard flush wall drill casing or iron pipe. When working upstage, the casing, and the couplings in particular, must have sufficient strength to resist the considerable extraction forces that are often required. Extraction is typically performed by pairs of hydraulic jacks, as illustrated in Figure 2.21. It is important to use a sensible casing length with such equipment, which means no longer than 5 ft (1.5 m), so that the workers can remain on the ground surface to make the necessary joint removals upon withdrawal. From the standpoint of grout effectiveness, larger casing can be used. However, with casing greater than 3 in. (75 mm), the weight of the grout will exceed the restraint provided by line friction and may cause excessive static grout pressure at the bottom of the hole. A few contractors attempt to perform the work with casing of less than 2 in. (50 mm) ID, but this is not advisable as it can be difficult to establish grout flow and the risk of blockage is greater.

FIGURE 2.21 Extraction jacks and proprietary casing with grout header and wide-sweep bends.

2.2.4 Injection Line Requirements

Because of the very low mobility of the grout used, all valves and fittings must provide full flow openings. Appropriate gauge savers must be supplied for all pressure gauges and should have a minimum dial size of 3 in. (76 mm) so they can be easily read. Standard pipe fittings must be avoided, and wide sweep bends must be used. A high-pressure 2 in. (50 mm) internal diameter hose is most often employed; however, 1-1/2 in. (38 mm) hose is sometimes used. Where especially long runs are required, rigid pipe will decrease the resulting pressure loss within the delivery system. Appropriate quick-connect couplings, such as those used for concrete pumping, should be used. Figure 2.21 shows a typical header connection during grouting. The flush joint proprietary casing shown is in 3 ft (0.9 m) lengths. As can be observed, it has been pulled almost to the extent that a joint can be removed.

2.2.5 Grout Injection

The effective radius of the injection improvement is strongly affected by both the nature of the soil and the injection rate. Obviously, to maximize economy, the fastest appropriate pumping rate should be used. As with all grouting, the injection pressure is directly related to the pumping rate. It increases with faster pumping and lowers as the pumping rate is slowed. The rate of pressure buildup behaves similarly. The optimal injection rate is nearly always within a range of 1–2 ft^3 (28–57 L) per minute. It will be the fastest rate that can be maintained at a grout pressure increase of no more than 5–8 psi (0.3–0.5 bars) per minute. As the rate of pressure increase exceeds about 8 psi (0.5 bars) per minute, the amount of beneficial compaction drops off markedly. Research by Brown and Warner (1973) found that in a given soil, although a spacing of 10–12 ft (3.0–3.6 m) was optimal for a pumping rate of 1-1/2 ft^3/min (42 L/min), it would have to be reduced to about 8 ft (2.4 m) for equivalent densification if the pumping rate increased to 2 ft^3 (57 L) per minute.

Grout is typically injected into an individual hole stage until one of three different refusal criteria occurs:

- There is jacking or heaving of the ground surface.
- A given pressure level has been reached at a given injection rate.
- A fixed grout quantity has been injected at a given injection rate.

The most frequent basis for refusal is surface jacking when the injection depth is less than about 30 ft (9 m). It is thus crucial to vigilantly monitor the ground surface and any structures within about 20 ft (6 m) of the injection point and to cease pumping upon the detection of even minute movement. When work progresses from the top down, there is a considerable time lag between the injection of consecutive stages. Minor heave that has occurred will often relax during this time interval so that there is no accumulation of the heave from other stages. When working from the bottom up, however, the entire hole is injected in a single operation. The heave, resulting from each of the stages, is therefore cumulative. Stage lengths of 1 ft (0.3 m) are common, but it can be advantageous to reduce the number of stages by using longer stage lengths where cumulative heave is a problem. The importance of continuous and meticulous surface monitoring, together with terminating injection at even minuscule displacements, cannot be overstated.

Injection to a pressure limit is used in situations where the existing soil conditions vary from place to place. By continuing densification until a uniform pressure is reached *at a given pumping rate,* the resulting soil density will be generally uniform. In this regard, it is important to understand that the injection pressure is directly influenced by the pumping rate. An increase of the pumping rate will always raise the pressure, whereas a reduction will lower it. Thus, injection to a pressure level cutoff must always be at a constant pumping rate, and this must not be changed from one hole to another. This fundamental is not well understood by many grouters, and although the practice is wrong, it is not uncommon to see pressure level cutoffs used without regard to the pumping rate.

Injection to a volume limit is applicable to those situations where the density of the existing soil is reasonably uniform but deficient. Perhaps the most common examples are some wind-deposited sediments and hydrocollapsible deposits. In these situations, the amount of improvement required is determined on the basis of the starting density of the soil. An amount of grout equivalent to the volume reduction required to provide an acceptable condition, including appropriate safety factors, is then injected. Volume cutoffs, which are easy to specify and control, are often used. They are, however, inappropriate for anisotropic or nonuniform soil conditions, where the low-density portions would receive insufficient improvement while those that are already more compact would be given unneeded treatment.

2.2.6 Limitations

Unfortunately, not all compaction grouting has performed satisfactorily. Many instances of less than acceptable performance have been investigated. In virtually every case, the deficiency was linked to one of five faults:

1. *Failure to treat the faulty material for its full depth.* Compaction grouting adds considerable weight to the treated zone, so it is imperative not to grout over a soil formation unable to support the weight of the improved soil. In apparent fill failures, experience has repeatedly revealed the culprit to be the soil in the bottom of the fill or the original soil immediately thereunder.
2. *An inappropriate injection sequencing.* In general, peripheral holes should be injected prior to those on the interior. Soil settlement often results in lateral spreading, causing open cracking on the ground surface or in structures. Where this has occurred, injection should *always* start at the farthest limits away from the surface cracking. For example, settlement in near proximity to a retaining wall or unrestrained downslope usually exhibits lateral movement in the direction of that feature. Initial injection should thus be in holes nearest to the wall or on the slope, progressing toward the cracks. With proper sequencing, the soil can usually be pushed so as to close such cracks.
3. *Excessively mobile grout.* As previously discussed, grout that is excessively mobile or that behaves as a fluid in the ground will cause hydraulic fracturing. Such fractures will be parallel to and near an area of weakened restraint, which is most often a retaining wall or downslope. The importance of the grout aggregate and rheology cannot be overemphasized and is amply covered in Chapter 6.
4. *A too rapid pumping rate.* Soil deformation is time dependent. Forcing it to occur too rapidly will cause disruption and possible hydraulic fracturing. Again, movement will be in the direction of least restraint.
5. *Excessive jacking of the ground surface or overlying structures.* It is imperative to rigorously monitor the surface and improvements during injection and to cease pumping upon detection of any movement. Excessive jacking will happen only when this simple action is neglected, and there is no excuse for its occurrence.

Use of grout aggregate containing clay, or the deliberate addition of clay or other deleterious materials to improve pumpability, has been a continuing problem within the industry. This is done to impart lubricity or water retention to grout, allowing otherwise inadequate pumps to function. The consequence of inappropriate aggregate and resulting poor grout rheology are clearly established, and there is no excuse for continuation of such inept practice.

Excessive injection rates are always a temptation for both contractors and their working crews, as less pumping time means earlier job completion and higher profits, but this practice has caused much defective work. In one case, a residential structure continued to settle following injection of a "huge amount of grout" about a year before. During an inspection, an excavation on an adjacent property was observed. There, exposed, was a vertical fracture, filled with grout, as illustrated in Figure 2.22. It ran more than 12 ft (3.6 m) onto the adjoining property. Legal considerations precluded further investigation, but the grout injected "under" the structure had clearly traveled far from that location. Other than a notation concerning the quantity of the grout, which was delivered in ready-mix trucks and injected with a standard concrete pump, no records had been kept. The fines in an acceptable compaction grout aggregate tend to make a sticky mix that builds up on a truck's mixer fins, so that it is rarely possible to obtain compaction grout of proper rheology with such mixers. A reasonable assumption is that the de-

FIGURE 2.22 Hydraulic fracture in the soil some 12 ft (3.6 m) away from injection hole.

fective work resulted from the use of inappropriate grout in combination with an excessive injection rate.

Displacement of downslopes and retaining walls occurs as a result of inappropriate injection sequencing and excessive pumping rates. The exterior wall of the building shown in Figure 2.23 was dislodged some 8 in. (203 mm) as a result of clay in the grout, an excessive pumping rate, and poor injection sequencing. The occupant, who was present when the damage occurred, explained: "The pumping was going real good in the middle of the room, when all of a sudden the floor split and everything opened up."

The relatively high grout pressures used in compaction grouting suggest the development of high lateral forces in the ground. This is often given as an excuse for damage, as discussed

FIGURE 2.23 Wall laterally displaced several inches during blowout of adjacent slope.

earlier, and the refusal of some contractors to work in such situations. Yet literally hundreds of successful applications in near proximity to retaining walls or unsupported downslopes indicate otherwise. Unfortunately, defective work is seldom discussed by those involved, and because the results of investigations are often sealed in legal proceedings, the lessons learned are not disseminated.

The occurrence of these poor practices is inexcusable, as proper performance of compaction grouting is relatively easy to achieve. It does, however, require proper planning, use of a correct mixture, an appropriate pumping rate and injection sequencing, and quality monitoring of the work.

2.3 FRACTURE/ CLAQUAGE GROUTING

Fracture grouting involves the intentional hydrofracture of soil, with a fluid suspension or slurry grout, to produce a network of interconnected grout lenses, as illustrated in Figure 2.24. Soil immediately adjacent to the fractures is obviously densified, but not nearly to the extent achieved

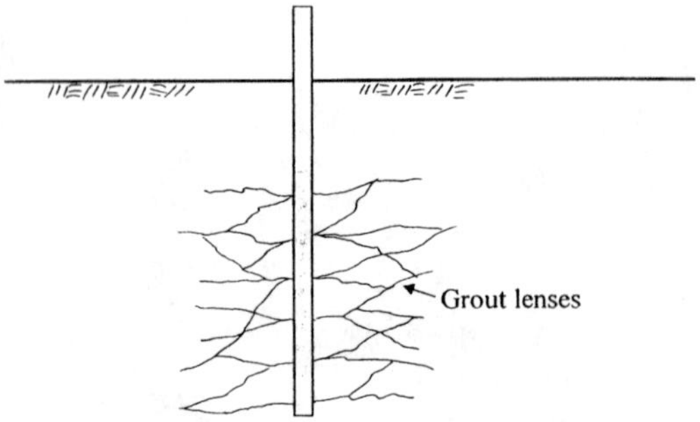

FIGURE 2.24 Fracture grouting should produce an interconnected network of grout lenses.

in compaction grouting. To prevent uncontrolled fractures from propagating excessively, a limited amount of grout is injected at any single location. In most soils the plane of minor principal stress is in a vertical orientation, and therefore the initial fractures should be vertical. It is felt the horizontal stress is decreased, however, by the early vertical fractures to the point that both horizontal and randomly oriented fractures occur. Recognizing that there is some, although limited, densification, it is thought that *reinforcement* of the soil with a network of stronger lenses is the primary improvement mechanism.

The procedure was developed in France (thus, the origin of the term *claquage*) in an effort to overcome the inability of common suspension grouts to penetrate soils. In theory, the procedure should be quite effective; in practice, however, control of the direction and extent of the fracture system is nearly impossible. The procedure has not been extensively used outside Europe, although some notable exceptions exist. Because uncontrolled hydrofracture can damage both the soil and the adjacent structures, it is crucial to maintain absolute control of the grout deposition. This is facilitated by relatively close hole spacing and the use of sleeve port pipes, which are described in Chapter 30, along with strict limitation of the amount of grout at any one pumping point. Obviously, meticulous observation and control of the injection parameters is required. Fracture grouting can be performed to any depth and in virtually any soil, although its most advantageous use is generally in clays that are not otherwise treatable by other grouting processes.

2.3.1 Lime Injection

Hydrated lime slurry is sometimes injected into clay soils to *chemically* stabilize the moisture and reduce expansive potential. The beneficial reaction relies on contact of the lime and soil, requiring closely spaced fractures. Such injection is typically done on a large scale by contractors who specialize in the procedure. Injection equipment tends to be massive, with several perforated probes pushed into the soil simultaneously, as illustrated in Figure 2.25. The process can be quite messy, and avoiding lime deposition on the equipment and surrounding surfaces is nearly impossible. It is a distinctively white and quite alkaline substance, with a pH of about 12, thus producing no serious health or environmental risks. With rainfall or washing, it will be readily

FIGURE 2.25 Multiprobe lime injection rig.

absorbed into the soil. The high alkalinity, however, can cause burns on human flesh and is especially hazardous to the eyes. Spills should therefore be promptly washed off with clean water.

The depth of treatment depends on the expansive potential of the subject soil. It is usually on the order of 5–20 ft (1.5–6 m), but applications to depths as great as 40 ft (12 m) have been reported. The slurry is composed of water mixed with about 25–30 percent, by weight, of dry hydrated lime. A surfactant wetting agent is often included to improve penetrability. Injection pressure is within a range of 50–200 psi (3.5–13.7 bars), and injection continues until refusal at the designated pressure, surface leakage, or when a predetermined amount of slurry, typically about 10 gal (38 L) per vertical ft (0.3 m) of hole, has been placed. Because the slurry consists primarily of water, which can be absorbed by clay, some accelerated expansion immediately after injection often occurs unless the initial soil moisture was near its maximum level. This expansion tends to stabilize shortly after injection, however, usually within a period of two or three weeks.

Lime injection has been subject to considerable promotion by those who perform the work and by the National Lime Association, a trade organization for manufacturers of commercial quicklime and hydrated lime. This organization has an available publication (Boynton and Blacklock, 1986), which includes a comprehensive bibliography of 51 references.

2.3.2 Limitations and Risks

Because permeation grouts are fluid, virtually the full pumping pressure is felt by the substrate. This can easily result in hydraulic fracture of geomaterials and/or leaks into utilities and other hidden features. Jacking can easily occur when such grouts are injected into the defects of rigid materials. Continuous monitoring of the injection parameters is thus required.

2.4 MIXING/JET GROUTING

Jet grouting involves erosion of the soil by a high-pressure jet of grout, water, or air-enshrouded grout or water, and the simultaneous injection of cement suspension grout into the disturbed soil by means of a jet *monitor.* The special drill stem and monitor are simultaneously raised and rotated so as to combine the grout with a portion of the original soil to form *soilcrete,* or, alternatively, to replace most of the soil with grout. The end product is cemented round columns, as illustrated in Figure 2.26, and the mechanism of improvement is physical *mixing* of the in situ soil with grout. It is effective in virtually any type of soil. Efficiency, however, is soil dependent; it is lowest in cohesive soils and increases as the soil becomes more granular. Large rocks or other obstructions complicate mixing and can cause a shadow effect, interrupting continuity.

FIGURE 2.26 Exposed columns of soilcrete resulting from jet grouting.

2.4.1 Jet Grouting Application

In a jet grouting application, a hole, typically about 4 in. (100 mm) in diameter is bored separately or by a jet grouting monitor or bit. This is followed by one of three major variations of operation. In the *single-fluid system,* which is the simplest, a special hollow drill rod, equipped with a monitor containing a horizontal jet nozzle at the tip, is lowered into the hole. Cement suspension grout is pumped down the drill rod at very high pressure, up to about 9000 psi (620 bars), while the drill rod and monitor are simultaneously rotated and withdrawn. The grout, which exits the jet nozzles at high velocity (Figure 2.27), disintegrates the soil and mixes with it to form *soilcrete.* The effective radius depends on the properties of the soil and the jet grouting parameters used. Column diameters on the order of 16–24 in. (0.4–0.6 m) are typically obtained in cohesive soils, and diameters up to about 4 ft (1.2 m) can be achieved in granular materials.

In the *two-fluid system* the grout is encased within a shroud of compressed air, which results in more effective cutting of the soil. This requires the use of a special coaxial drill string and jet monitor. The air acts as a buffer between the groundwater and the grout, greatly increasing the cutting efficiency. It also creates turbulence in the waste spoil, improving the efficiency of its removal. The diameter of the columns resulting from the two-fluid system is on the order of 2–3 ft (0.6–0.9 m) in cohesive soil and up to about 6 ft (1.8 m) in granular soils.

The *three fluid-system* is the most complicated, in that it requires a triaxial drill stem and monitor with appropriate nozzles. In this system an air-enshrouded jet of water erodes the soil while the grout is simultaneously injected through separate nozzles. The cutting jets are located above the grout supply, which allows a nearly complete replacement of the soil with grout as the monitor is withdrawn. The triple-fluid system enables formation of the largest-diameter columns. Effective diameters of 2–5 ft (0.6–1.5 m) in clay and up to 12 ft (3.6 m) or more in sands, as shown in Figure 2.28, have been formed. It is, however, the most complicated system to use and requires a very costly and complex drill string and monitor.

A more recently developed variation of the two-fluid system known as *Jumbo Jet* or *Super Jet* involves two or more streams of air-enshrouded grout blasted from multiple nozzles, which are often at different elevations and

FIGURE 2.27 High-pressure jet for one-fluid jet grouting system.

FIGURE 2.28 Very large columns of soilcrete are possible with a three-fluid system.

mounted in opposing directions on the monitor. With this system, soilcrete columns of similar diameter to those of the three-fluid system can be formed, but without complication of the triple tube setup. The advanced system displaces very large amounts of both air and grout, and imparts considerably greater energy into the soil. While it is unquestionably the most productive and economical method, careful consideration of its use is warranted as the very high energy produced can be damaging to nearby improvements. Typical parameters for the various jet grouting systems are provided in Table 2.1.

Jet grouting can be employed to depths of 200 ft (60 m) or more. It is fast and can be performed in virtually all soil types. In granular soils, the resulting soilcrete can reach strengths of 2000 psi (13.8 MPa) or more, and by overlapping probes a nearly continuous curtain can be formed. Generally, large and special equipment is required. While standard drill rigs can be used, very high pressure pumps and, of course, the special monitor and drill stem are required. The effective diameters of the resulting masses are dependent on the individual system used, the jetting and injection parameters, and the specifics of the site's soil. Thus, trial injections are usually made to establish the optimal hole spacing and operating parameters.

2.4.2 Risks and Limitations

The high pressures used and the fact that the soils within influence of the high velocity jet are always fluid, present risks to both the overlying soil, and any adjacent structures. Heaving of the surface and displacement of adjacent improvements have occurred with many installations. Another potential risk is plugging of the spoil return which will, of course, direct all pressure into the soil mass. This is particularly problematic in clayey soils and worsens with depth and underwater injection.

Most literature on jet grouting has been authored by various special interests, and they tend not to mention any problems or limitations of the procedure. A conspicuous exception is a paper by Fukuoka University professors Imanishi and Yamauchi (1996). The authors described research, where the ground response to the jet grouting of several elements was monitored, through extensive instrumentation of the surrounding soils. Development of excessive pore water pressure, ground settlement, horizontal deformation, and changes in the effective stress of the soil were observed. The zone of affected pore pressure extended 4 m (13 ft 4 in.) above, 2 m (6 ft 8 in.) below, and 6 m (20 ft) to the side of the jet monitor.

Risk obviously increases as size of the resulting element grows, as well as with increasing energy imparted to the soil. The combination of very high energy and large element size of the Jumbo/Super Jet system makes its use particularly questionable other than in wide open areas free of improvements. Although cost of the work typically lowers as the element diameter grows, one must carefully balance the risks of damage against these cost savings, especially when the work is adjacent to other improvements.

2.4.3 Deep Soil Mixing

Deep soil mixing involves physically mixing a cementing material with in situ soils to form continuous elements of soilcrete. The cementing material can be either cementitious grout or dry cement or other stabilizer, and the different processes are known as *wet mix* and *dry mix*, respectively. This work requires custom, and usually very large, mixing units, which are typically mounted on a standard excavator or suspended from a crane. They typically contain several mixing elements mounted on one or more vertical shafts. The diameter of the individual mixing elements varies from about 3 to 5 ft (0.9 to 1.5 m).

TABLE 2.1 Typical Injection Parameters for Different Jet Grouting Systems

Jet Grouting System	1-Fluid	2-Fluid	3-Fluid	Jumbo/Super Jet
Grout injection rate	10–30 GPM (38–113 L/min)	18–35 GPM (68–132 L/min)	18–35 GPM (68–132 L/min)	150–175 GPM (567–660 L/min)
Grout Pressure	2900–8700 psi (200–600 bar)	4200–8700 psi (290–600 bar)	500–2500 psi (34–170 bars)	3500–4500 psi (241–310 bars)
Air flow	—	130–210 CFM (3679–5943 L/min)	130–210 CFM (2679–5943 L/min)	400–500 CFM (11,320–14,150)
Air pressure	—	90–180 psi (6–12 bars)	90–180 psi (6–12 bars)	120–200 psi (8–14 bars)
Water flow	—	—	18–40 GPM (68–151 L/min)	—
Water pressure	—	—	2900–7250 psi (200–500 bars)	—
Rotation rate (RPM)	10–25	5–10	2–10	3–4
Withdrawal rate	4–20 in/min (10–50 cm/min)	3–12 in./min (7–30 cm/min)	2–12 in./min (5–30 cm/min)	3–4 in./min (7.6–10 cm/min)

Either individual pile-like elements or continuous walls can be formed to depths of 150 ft (45 m) or more.

The strength of the mixed mass will vary according to the particle size distribution of the soil, as with jet grouting, as well as by the proportion of cementing material added. Because of mechanical mixing, the resulting soilcrete is quite homogeneous, but the procedure is restricted to soils free of large rocks or other obstructions. Setup cost is very high, so use of the procedure is generally limited to large-scale applications.

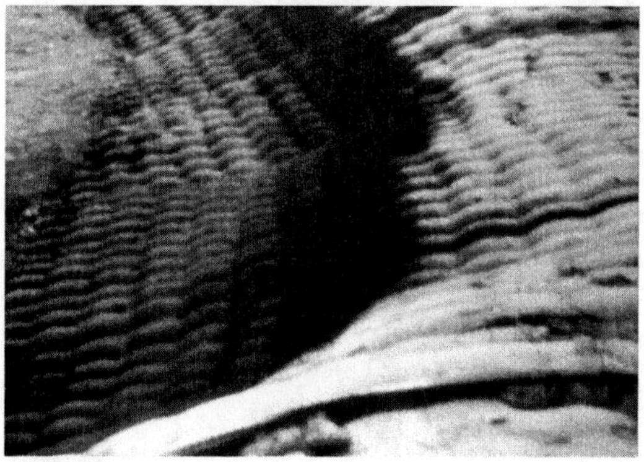

FIGURE 2.29 Fabric forms filled with grout to provide shore protection.

2.5 FILL GROUTING

Cavities of all descriptions, ranging from continuous channels produced in the abandonment of pipes, to massive voids resulting from abandoned mine works, solution cavities, and eroded areas, sometimes require filling. When such filling was required in earlier times, cementitious bituminous, or clay grouts were often placed with the same general type of equipment employed in other grouting. Since the development and wide availability of concrete pumping equipment, however, this earlier practice has become obsolete. In most instances the fastest and least costly method of filling is with commercially available concrete pumps using ready-mixed concrete or mortar. Although many grouters promote this type of work as being something special that only they can do, the fact is that, it can be easily performed with quite standard and widely available equipment and material. Thus, it is referred to here simply as *fill grouting.*

2.5.1 Filling Fabric Forms

Another large area of simple filling involves the use of fabric forms. Perhaps the most common usage is for erosion control, in which two sheets are sewed so as to form a quilted pattern when filled, with grout as illustrated in Figure 2.29. They can however, be fabricated in any configuration or for any special requirement. An example is the zipper-equipped sheets wrapped around the concrete piles shown in Figure 2.30,

FIGURE 2.30 Zippered sheets are easily wrapped around piles and closed to allow grout filling for high-quality encapsulation.

zipped up to closure and filled with a quality grout, encapsulating the pile.

These forms are especially useful in underwater operations. Originally developed to contain otherwise highly mobile suspension grouts, they are now most often filled with ready-mixed concrete or mortar. Ideally, they are fabricated from a material that will allow the water to penetrate but retain all of the solid ingredients. They thus compensate for the excess water in most grouts, often resulting in a hardened material of high strength and quality. Although there are less costly materials for their fabrication, nylon has proven by experience to provide superior strength and cost-effective performance.

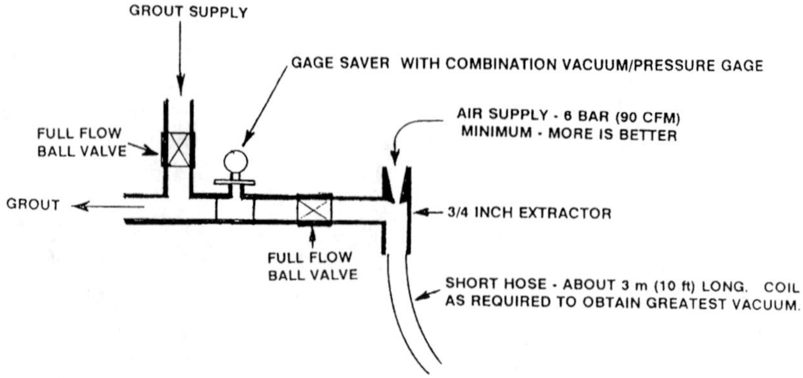

FIGURE 2.31 Vacuum ejector attached to grout header.

2.6 VACUUM GROUTING

It is usual to think of grout being impelled into its host formation under considerable pressure. There are situations, however, where pressure alone will not be satisfactory, and in those cases, a vacuum is used to suck the grout into the defect. In such work the defects must first be isolated from the barometric pressure of the environment, so a vacuum is drawn and the grout is then allowed to be sucked in. (The mechanics of vacuums are amply explained in Chapter 9.)

There are two general areas in which this procedure can be advantageous. First, it can be used for filling small, essentially airtight voids through a single hole where difficult access complicates the provision of vent holes. This application was first presented by Houlsby (1990) to facilitate the filling of contraction joints in a concrete dam. It has also been employed for filling small voids under steel liners, as well as defects in the original grouting of posttensioning ducts. In application, a vacuum pump or venturi ejector is attached to the grout header, and a vacuum drawn. The venturi ejector illustrated in Figure 2.31 is the simplest to use. Therein initially, the valve to the grout supply is closed and that to the ejector, open. Compressed air is circulated through the ejector, which creates a vacuum in the void. Once this has been accomplished, the two valves are switched simultaneously and the grout is sucked into the evacuated void.

Another quite different application is the

FIGURE 2.32 Vacuum grouting used to fill fine crack network in a parking deck slab.

filling of closely spaced fine defects in concrete or masonry. In this application, the area of defects to be filled is encased in plastic sheeting, a vacuum pump is connected, and the space is evacuated. Once a vacuum has been created, the grout, typically some type of low-viscosity resin, is introduced. It will saturate the entire area under the plastic sheeting and be sucked into any void spaces as well. In Figure 2.32 a resinous grout is being sucked into defects in the plastic-covered slab of a parking structure. The process can be advantageously applied in any configuration, subject only to the ability to draw a satisfactory vacuum.

CHAPTER 3

The Wide World of Grout Materials

3.1 CEMENTITIOUS MATERIALS
- **3.1.1** Grain Size
- **3.1.2** Cementitious Compositions
- **3.1.3** Special Cements
  - **3.1.3.1** Oil Well Cements

3.2 ADMIXTURES FOR CEMENTITIOUS GROUTS

3.3 NONCEMENTITIOUS GROUTS
- **3.3.1** Specialty Water Control Grouts
- **3.3.2** Resinous Grouts

3.4 MATERIAL SELECTION
- **3.4.1** Material Types

3.5 BASIC MATERIAL PROPERTIES
- **3.5.1** Dimensional Stability
- **3.5.2** Strength
- **3.5.3** Modulus of Elasticity
- **3.5.4** Coefficient of Thermal Expansion

ALTHOUGH, TRADITIONALLY, many grouters have thought of grout only as cement and water, the fact is that a huge assortment of grout materials are available. The most frequently used grouts are based on hydraulic cement—that is, a cement that will harden and cure under water—but there are many different types of cement, which can be combined with a nearly endless variety of different fillers and modifiers. In addition, many noncementitious grouts are available, which are based on an extensive range of different chemical solutions and resinous compositions. Moreover, in some special situations, especially those involving control of the movement of water, cementitious and noncementitious grouts may be combined.

Because grouting is a rather specialized application that represents but a tiny proportion of the total sales of cement and other common grout components, the manufacturers and distributors are not always eager to perform research and development or to provide technical support to the grouting industry. To make matters worse, many established testing methods and standards for materials are not applicable to grouts, so otherwise qualified and experienced technical staffs are not aware of grouting requirements and are thus often ill equipped to deal with grouters. It has been said that grouters are not unlike orphans, in that the normal attention given to other construction applications are seldom afforded them. Because the requirements for cementitious materials used in grouting can be fundamentally different from those for most other uses, much of the readily available technical data is not applicable.

3.1 CEMENTITIOUS MATERIALS

Traditionally, and to this day, portland cement is the most widely used compound in grouting. Although standards are well established and there are several different types of common cement, none have been designed with grouting in mind. Consequently, neither the composition nor the particulate grain size is well suited for much grouting work.

3.1.1 Grain Size

The grain size distribution of materials used in grout, and especially that of portland cement, is fundamental to the resulting grout's injectability. The prevailing standards for portland cement, however, contain no mention of grain size. Rather, they consider only the *specific surface*—that is, the surface area of all the grains in a given unit—which is known as the *Blaine fineness.* This is because the amount of surface area available for reaction, rather than the size of the individual grains, is what affects the performance of a cement when it is used in concrete and mortar. However, it is extremely difficult to obtain such information from the cement manufacturers. In fact, much of the grain size data that appears in this handbook has been obtained as a result of litigation and provided only as required by legal subpoena.

3.1.2 Cementitious Compositions

Even with portland cement, which is the primary constituent of the most frequently used grouts, a wide array of mixtures and different consistencies is common. The composition can be water thin or thick, like putty. The mixtures may be formulated to set rapidly or, perhaps, be retarded for several days. The details of mix design and the use of cementitious grouts are discussed further in Chapter 5.

3.1.3 Special Cements

Contrary to popular misconception, standard portland cements are not the only hydraulic cements available. A variety of different cements, which can provide mixtures with a wide assortment of special properties, are also to be had and are regularly used. Products that provide either very rapid or greatly delayed set and strength gain are available, as are ultrafine grinds of a variety of different cement types, which can be made into extremely penetrable grout. An example is illustrated in Figure 3.1, where a 2:1 water:cement ratio ultrafine cement grout is being poured out of the beaker on the right. For comparison, plain tap water is being poured on the left. One can readily comprehend the great penetrability of ultrafine cement grout. Other special cements are also marketed especially for use in very high or very low temperature environments.

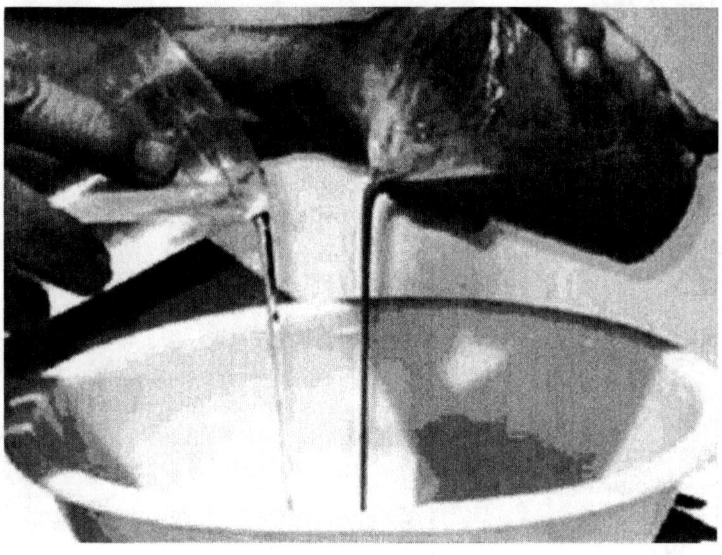

FIGURE 3.1 Ultrafine cement suspension (right) has flow properties similar to those of plain water (left).

3.1.3.1 OIL WELL CEMENTS

A whole suite of special cements has been developed and standardized in the oil well cementing industry. In these applications, very high injection pressures under greatly elevated temperature regimes are common. Although such attributes can be advantageous in a small number of highly specialized grouting applications, such as around blast furnaces, there are other beneficial properties as by-products. These, as well as other special cements, are discussed in greater detail in Chapter 5.

3.2 ADMIXTURES FOR CEMENTITIOUS GROUTS

Lack of support from material producers is not limited simply to a dearth of information about the use of cement. Many admixtures to modify the properties of cementitious compositions, such as ordinary concrete and mortar, have long been available and widely used. These materials have not seen much use in grouting, however, primarily because the manufacturers and suppliers have not made an effort to promote them for such use, and many grouters are simply unaware of their capability or availability. As with cement, the requirements for modification of grout can be quite different from those for more widely used conventional concrete and mortar, so the available technical data are not always applicable to grout.

For example, in the 1960s, I found the use of massive doses of lignin-based water-reducing admixtures to be effective in reducing shrinkage of cementitious grouts. Although these admixtures were widely used in concrete, the available technical data called for rather small dosages. When inquiry was made to the manufacturers, they replied that large doses of their products were not recommended, but were unable to give a reason in support of that advice, other than such doses would retard the set and strength gain. However, what is excessive for conventional concrete and mortar is not necessarily excessive for many grouting applications.

Because an extensive research program, detailed in Chapter 6, found no adverse effects from such use, many projects were performed very successfully with large doses of the material. Their use allowed a greatly reduced amount of water in the mix, resulting in a significant reduction of shrinkage and increasing both the strength and long-term durability. There is now available an extensive array of *midrange* and *high-range* water-reducing admixtures that can result in even greater water reduction. These are routinely employed in concrete but have seen only minimal use in grout, although they can provide an enormous improvement in the hardened properties. They will permit a substantial reduction of mix water without sacrifice of fluidity, thus resulting in a significant decrease in bleed (the settlement of solids out of suspension) and shrinkage, as well as improved strength and durability.

3.3 NONCEMENTITIOUS GROUTS

There is also a lack of supplier support for the use of many non-cementitious grouts. Perhaps the most widely used chemical solution grout is based on sodium silicate. As with portland cement, the amount used in grout represents but an infinitesimal amount of total sodium silicate sales, and thus only minimal support from the manufacturers has been received. In addition, the many chemical reagents that are used with sodium silicate are simply generic chemical products that have innumerable other uses. Because grouting represents such a small area of application, little research or technical development of these chemicals for use in grouts has been forthcoming.

3.3.1 Specialty Water Control Grouts

An exception to this trend (lack of research and development) involves specialty grouts for water control, especially in connection with leakage into underground conduits and structures. In this work, resistance to dilution and close control of setting times can be crucial, and nearly instantaneous set is a frequent requirement. An example of this behavior is illustrated in Figure 3.2, where the fluid grout was freely passed from one cup to the other until, in less than a minute, it gelled midway between. High-performance grouts of this type are usually subject to complex chemistry.

Because of their complexity, grout systems such as these are usually proprietary. Depending on the individual producer, they may be available on the open market or sold only to properly trained and equipped contractors. Unfortunately, even in this area, the exact chemical makeup is seldom provided, and effective support for their application by competent material technicians is not always forthcoming. This is somewhat understandable, however, in that very complex chemistry is often involved, which can be beyond the ability of most engineers and grouters to comprehend. The effectiveness of water control grout is usually quickly ascertained, in that the leakage is either stopped during an injection or continues.

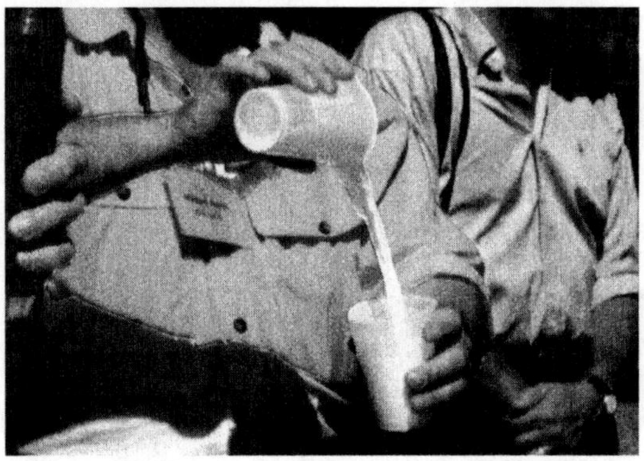

FIGURE 3.2 Tongue of grout gelled while being poured between two cups.

3.3.2 Resinous Grouts

Lack of support is not a problem in the case of resinous grouts such as epoxy. These materials are widely marketed for a variety of construction applications, of which injection is but one. Like their chemical solution grout cousins, they are subject to somewhat complex chemistry that is beyond the ability of most grouters and specifiers to readily understand. The producers, however, routinely provide product data sheets, which include simple-to-understand physical properties of the cured material.

Unfortunately, there are few standard tests developed for such materials, so the evaluation methods to establish reported values are often subject to modification of other test standards. Although product data provide values resulting from a standard test method, such information often includes a note stating that the test was modified, but provides no details as to the nature of the modification. Not only are test standards modified, but different manufacturers sometimes use completely different beginning standard tests. With such a variety of test protocols, meaningful comparison of the properties provided in otherwise similar reports is difficult at best and often impossible.

3.4 MATERIAL SELECTION

A huge variety of grout materials and mixes, representing an almost endless range of properties, is available. These extend from the waterlike fluids shown in Figure 3.1 to nearly immobile low-mobility mortarlike mixtures. In between are a variety of chemical solution and resinous compositions with properties varying from cohesive to adhesive, slippery to sticky, flexible to near rigid, and, of course, highly mobile to almost immobile. As is evident, many different generic origins, consistencies, degrees of user-friendliness, and cost are found in the broad spectrum of available materials. Although selection of the most appropriate grout material for a given requirement is obviously fundamental to optimal performance, it can be a daunting task, especially where unusual project requirements occur.

Penetrability is often considered the single most important characteristic of grout. This is understandable, in that the more penetrable a material, the faster it can be injected and the farther it will permeate. Although such properties can be beneficial, there are situations in which high penetrability can be a disadvantage, such as in applications where the grout travel is to be restricted. An example is the filling of portions of large cavernous voids or old mine workings. Often, there is a requirement to fill such defects under a structure, but in order to limit the cost, filling beyond the borders of the structure is not desired. Another example is solidification of an especially coarse sand, where there is a requirement to limit of the zone of permeation. In such a case, a highly penetrable grout will tend to permeate beyond the needed zone of improvement.

Because it is often difficult to assess the permeability and volume of the voids, it can be useful to postpone selection of the exact type or consistency of grout until full-scale test injections are made. These may be accomplished as part of the actual work when it first begins, or on large projects separate full-scale evaluations may be performed as part of the planning process.

3.4.1 Material Types

With such a broad array of materials, it should be no surprise that there is a considerable amount of confusion in their classification. For instance, virtually all grouts that are not cementitious are referred to as *resins* in many parts of Europe. Although this can be technically correct nomenclature, in the Americas and much of the East, formulations that appear to be more waterlike solutions are referred to differently than those of thick fluid or paste consistency. Simplification and standardization of the terminology used in grouting is badly needed. This is especially pertinent now, as many individual grouting materials and procedures are used internationally.

The need has not been unnoticed, and in fact the Grouting Committee of the Geotechnical Engineering Division (now known as the Geo Institute) of the American Society of Civil Engineers has been grappling with the problem for more than three decades. Following several years of debate and discussion, it published a document, *Preliminary Glossary of Terms Relating to Grouting* (1980). There were differences of opinion as to the meaning of various terms. In fact, it was for this reason that "Preliminary" was attached as the first word in the title. It was hoped that the document would encourage discussion and greater agreement and standardization of the terminology used, after which a final document would be produced. Unfortunately, this has not occurred and the committee is still actively struggling to produce an update. At present (2003) it appears that rather than simplification, greater confusion may result.

In this handbook, available grouts are classified in the following four very broad categories, which I hope will become universally used in order to clear some of the present confusion:

- Cementitious
- Chemical solution
- Resinous
- Miscellaneous

The term *cementitious* is used to describe any grout in which a hydraulic cement is used as a primary binding component. Such materials range from thin fluid suspensions to the low-slump, plastic consistency concrete and mortar grouts. *Chemical solution grouts* are defined as those compounds that have a basically waterlike appearance prior to injection. These are often referred to simply as *chemical grouts,* but with the addition of the term *solution,* they can be more easily differentiated from the resinous compounds, which are, strictly speaking, also chemical compositions, as are cements.

The *resinous* category is reserved for those polymer formulations that are usually solvent based and normally supplied in two or more components that must be mixed in correct ratio to properly harden and cure. These comprise epoxies, polyesters, and some urethanes, including compounds that will expand into a low-density foam upon injection. The foaming polymer grouts, although not widely used, do offer some distinct advantages, in that they are of low density, which can be of great benefit in instances when the addition of dead weight must be restricted. Moreover, because the original resin–to–final foam ratio is usually low, they can fill a void volume significantly greater than that of their original mass. This can be significant in cases where transportation of the material is especially difficult or expensive or where storage space for the raw material is scarce, such as in mining and other underground operations.

Obviously, there will be some crossover in terminology between the various solution chemical and resinous grouts. Finally, any grout that does not fit into one of the aforementioned categories is termed *miscellaneous.* This includes materials such as bitumen, clays, and the like.

Although it is not possible to provide an absolute recipe for the use of all of the available products, let alone all the possible project requirements, Table 3.1 has been prepared in an effort to provide a feel for the types of work that may be executed with the various available classes of material. The different substrates in which grouting is performed—rock, soil, and structures—are delineated at the top. Similarly, the four main categories of grout are grouped on the left. The data presented are not to be taken absolutely, but they do give a general idea of how different grouts are used.

Often the objective of a given grouting effort can be adequately fulfilled through the use of a single grout material. However, a grout mixture composed of two or more components from one or more categories is sometimes advantageous. For example, cementitious suspensions may be combined with sodium silicate solution grout, or perhaps hot bitumen in the miscellaneous category, in order to promote rapid setting for water control grouting where rapidly moving water is present. Likewise, two or more different grouts may be injected separately in order to optimally fulfill a given requirement. An example is the initial tightening of a soil with compaction grouting, followed by further increasing the stiffness with a permeation grout. This sort of requirement is common in changing the natural frequency of granular soils to correct problems of excessive vibration of overlying machine foundations.

Fundamental to the effort and equipment required to inject a given grout is its consistency. A great many terms have been used to describe this property, some technically accurate, others downright erroneous. This subject is more thoroughly discussed in Chapters 4, 6, and 7; suffice it to say at this point, however, *fluids* are materials that can be poured and that, given enough time, will seek their own level. At the opposite end of the scale are compositions said to be of *plastic* consistency. These are materials that will not flow unless a considerable external pressure is applied. Plastic consistency materials can generally be remolded and tend to hold the shape into which they have been worked.

3.5 BASIC MATERIAL PROPERTIES

Before delving into the finer details of the many grout materials, it is important to possess a basic understanding of the fundamental properties that relate to all materials, which are also significant to grouts. Of common importance in grouting are

TABLE 3.1 Types and Typical Uses of Grout

	ROCK			SOIL					STRUCTURES							
	Strengthen rock	Reduce water flow	Fill voids/sinkholes	Solidify soil	Densify soil	Reduce permeability	Alter soil chemistry	Groutjacking	Preplaced aggr. concr.	Fill annular spaces	Integrate masonry	Repair cracked concrete	Fill small voids	Fill large voids	Control leakage	Stop pipe leakage
CEMENTITIOUS																
Cementitious suspensions	C	C	O	S	N	S	N	S	C	S	C	S	O	S	S	S
Ultrafine suspensions	O	O	N	O	N	O	N	N	S	N	S	S	S	N	S	N
Cementitious slurries	S	S	O	N	N	S	N	S	O	C	C	N	C	C	S	N
Stiff, mortarlike	N	N	O	N	C	S	N	C	N	N	N	N	N	S	N	N
Low-density foamed	N	N	O	N	N	N	N	W	N	C	S	N	N	S	N	N
Ready-mixed concrete/mortar	N	N	C	N	S	S	N	S	N	O	N	N	N	N	N	N
CHEMICAL																
Sodium silicate base	S	S	N	C	N	O	N	N	N	N	N	N	N	N	S	N
Acrylates/acrylamides	N	O	N	O	N	O	N	N	N	N	N	N	N	N	C	C
Urethanes	O	O	N	S	N	S	N	N	N	N	O	O	S	N	C	O
RESINOUS																
Epoxies	S	S	N	N	N	S	N	N	S	N	O	C	O	N	S	N
Polyesters	S	S	N	N	N	N	N	N	N	N	O	S	S	N	N	N
Resinous foams	S	N	N	N	S	S	N	N	N	N	N	N	S	C	O	N
MISCELLANEOUS																
Hot asphalt/bitumen	N	S	S	N	N	N	N	N	N	N	N	N	N	N	N	N
Clay	N	S	N	N	N	N	N	N	N	N	N	N	N	N	N	N
High-calcium lime	N	N	N	N	N	N	C	N	N	N	N	N	N	N	N	N

Types of Usage: C = Common, O = Occasional, S = Seldom, N = Never

the dimensional stability, strength, elastic modulus, and the thermal properties of the grout once it has hardened and cured. In some work, none of these factors will be of great significance, but in many applications they must be considered to obtain acceptable final performance.

3.5.1 Dimensional Stability

The dimensional stability of a material refers to its propensity to change shape or volume. All materials experience volume changes with variation of temperature; many grouts and most cementitious mixtures shrink as they harden and dry out. This propensity is especially problematic with cementitious suspension grouts. It is possible, however, to offset or prevent shrinkage by the use of special cements or inclusion of chemical admixtures. In fact, it is actually possible to compound expansive cementitious grouts, but such expansion comes at the price of a reduction in strength, permeability, and durability.

Controlled low-strength material (CLSM) grouts, sometimes referred to as flowable fill, are nearly self-leveling and have densities as low as about 20 lbs/ft^3 (320 kg/m^3). These are especially advantageous for the filling of annular space between tunnel excavations and encased liners or similar voids resulting from slip lining of pipelines. Their light weight greatly minimizes potential problems of flotation of the new lining. And, because they can be nearly self-leveling, they can flow great distances at relatively low pressures, making them ideal candidates for the filling of abandoned pipes and such. Also available are many expansive resinous grouts, several of which provide structural properties. These have been extensively used in structural applications and are particularly suited to fill the sometimes large voids found in masonry.

Although most host materials into which a grout may be injected are essentially dimensionally stable, most grouts will experience at least some shrinkage once in place. With many grouts, especially higher water : cement ratio cementitious suspensions, the shrinkage can be substantial and should be considered. Obviously, reduction of the mix water will lessen shrinkage, and in many cases it is possible to largely eliminate it through thoughtful mix design, including the judicious use of admixtures.

3.5.2 Strength

The strength of a material refers to the magnitude of a stress, or load, the material can withstand without rupture. In some grouting, especially that which is in connection with water control, strength is not of much importance. However, there are applications where it is of fundamental significance, such as in the strengthening of rock or soil to enable it to withstand greater loads. In the case of rock, the strength of the grout should usually be at least as great as that of the formation. In soil, however, it is not the strength of the raw grout, but rather that of the resulting grouted soil, that is pertinent.

In structural applications, both compressive and tensile/bond strength are important. The required strength levels are typically on the order of the strength of the existing construction that is being upgraded. In most geotechnical applications, however, such high strengths are not often required, especially when the injection is into soil. A common mistake engineers and specifiers make when dealing with grouting is to require greater strengths than are actually needed for the particular work being accomplished. Calling for strength levels appreciably greater than those rationally required for the work can complicate the operation and virtually always increases the cost—and should thus be avoided.

3.5.3 Modulus of Elasticity

The modulus of elasticity, often called the *elastic modulus* or simply the *modulus*, refers to the stiffness of a material. For example, the elastic modulus of concrete is relatively high, whereas that of rubber is usually quite low. High-modulus materials do not deform under load as much as those of low modulus. When materials of widely differing moduli are in contact with each other, the lower-modulus material will tend to yield or bulge under load, as illustrated in Figure 3.3, whereas the high-modulus component will be unaffected. When the load is perpendicular to the bond line, as in Figure 3.3a, a difference in modulus usually does not cause great problems.

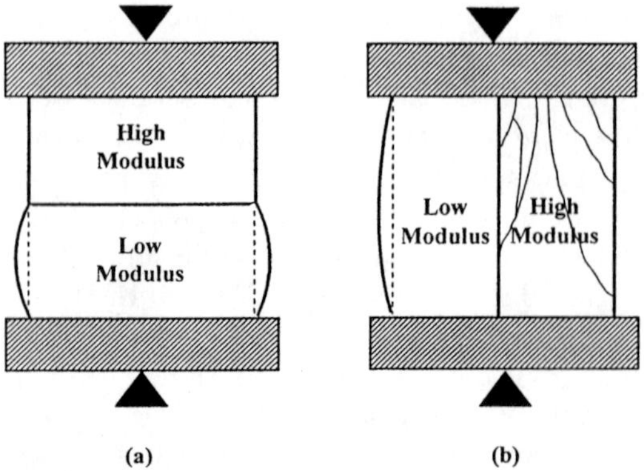

FIGURE 3.3 The importance of the modulus of elasticity depends on the direction of stress.

However, when the load is applied parallel to the bond line, as illustrated in Figure 3.3b, deformation of the lower-modulus material results in a transfer of some, or all, of the load it should support to the higher-modulus material. This can cause an overstress condition and, in extreme cases, failure. The stiffness of a grout material can thus be of considerable importance, as can the modulus of a grouted composite, especially if it involves structural components or unrestrained soils.

For example, many chemical solution grouts will cure to very flexible, low-modulus solids. Should these grouts be used to solidify a granular soil, the solidified mass can likewise have a low modulus. Where the purpose of the solidification is to provide temporary support to allow an adjacent excavation, the yielding of the grouted soil will transfer most of the vertical load into the adjacent retained soil in a manner similar to the example shown in Figure 3.3b. This has resulted in damage to structures supported by grouted soil and should be thoughtfully considered when selecting materials for such applications. In this regard, the use of ultrafine cement grout for solidification of granular soils is especially advantageous, as it results in a relatively high-modulus composition and can achieve concrete-like strengths in clean sands.

Similar problems have been experienced in structural repairs. Figure 3.4 shows a 10 ft (3 m) high test specimen that was removed from a strengthened column inside an industrial structure. Several such columns had been strengthened to increase their load-bearing capacity. The work involved augmenting the cross section of the column, which was accomplished with an epoxy-sand mortar pumped into place. Although the epoxy had a compressive strength of 12,000 psi (82.7 MPa), it had such a low modulus that the entire load was transferred back into the original concrete section. When distress of the "strengthened" columns in the structure was noted, the specimen shown was procured and tested. As expected, the failure was due to excessive yielding of the resinous augmentation, which directed the entire load into the original parent concrete, causing rupture. Although the

FIGURE 3.4 The lack of stiffness of the sand-filled epoxy grout used to strengthen this column resulted in a failure.

resinous material was of great strength, it yielded excessively because of its low elastic modulus, resulting in failure. Thus, where strength is important, it is imperative to consider the elastic modulus of the proposed material as well.

3.5.4 Coefficient of Thermal Expansion

Virtually all materials expand and contract with changes in temperature. For a given temperature change, the amount of expansion or contraction depends on the *coefficient of thermal expansion* of the particular material. The coefficient of thermal expansion represents the change in length over a unit length divided by the temperature change. It is usually given in millionths of an inch (cm) per linear inch (cm) per degree of temperature change. For example, concrete has a coefficient of thermal expansion of about 6 millionths per degree Fahrenheit. This sensitivity to temperature is often of little significance in grouting, but can be important in some structural applications. Resinous grouts usually have thermal coefficients much different from those of the host materials into which they may be injected. For this reason, their application is usually limited to thin cracks and joints, usually no greater than about 1/4 in. (6 mm) in width.

There are other influences of temperature that also warrant consideration in grouting. Virtually all grouts will set and/or harden more rapidly at high temperatures than at low. Most cementitious grouts will not set at temperatures much below freezing, so usage is commonly limited to a minimum temperature of about 40°F (4°C). Conversely, the set time of cementitious grouts is greatly accelerated at high temperatures, and limitation of the grout temperature to a level no greater than 90°–100° F (32°–38° C) is common. The effect of temperature on chemical solution and resinous grouts varies greatly, but must be considered. In situations where temperature extremes will be encountered, special formulations or amending admixtures should be used, or if these are unavailable, alternate grout materials should be employed. These factors are thoroughly discussed in Chapter 18.

Temperature extremes can also affect the performance and durability of many hardened and cured grouts. Of particular importance is the propensity of most polymer resin grouts to soften at elevated temperatures, as discussed in Chapter 7. This was a factor in the failure illustrated in Figure 3.4, wherein the plant process involved high temperatures. Conversely, these same materials can become brittle at very low temperatures. It is thus important to consider not only the temperature levels during grout injection, but also the greatest temperature gradient that may occur during the life of the grouted median.

CHAPTER 4

Grout Rheology

4.1 GROUT CONSISTENCY
- **4.1.1** Fluids
- **4.1.2** Plastic Consistency
- **4.1.3** Pastes and Slurries

4.2 MATERIAL BEHAVIOR
- **4.2.1** Thixotropy
- **4.2.2** Flow Properties
- **4.2.3** Viscosity
- **4.2.4** Mobility
  - **4.2.4.1** Slump
- **4.2.5** Penetrability
  - **4.2.5.1** Polar Intensity
  - **4.2.5.2** Grain Size of Solids
  - **4.2.5.3** Mixing Shear

4.3 OTHER RHEOLOGICAL PROPERTIES
- **4.3.1** Cohesion
- **4.3.2** Bleed
- **4.3.3** Temperature
- **4.3.4** Setting Time
- **4.3.5** Solubility

SIMPLY STATED, rheology is study of the deformation and flow of materials. Perhaps more germane to grouting, the American Concrete Institute, in ACI 116 "Cement and Concrete Terminology" (1990), defines rheology as "the science dealing with flow of materials, including studies of deformation of hardened concrete, the handling and placing of freshly mixed concrete, and the behavior of slurries, pastes, and the like." The term *rheology* is thus used to describe all the properties of a particular grout, both as initially mixed and in the hardened state. As a science, rheology is used to describe the properties of a wide variety of substances, including foods, polymers, and lubricants. Although rheologists come from many technical areas, most have a strong background in both chemistry and physics.

Akin to grouting, the oil well cementing industry has long understood the importance of rheological principles in the design of the cementitious formulations it uses. Although the compositions must often withstand high pressures and elevated temperatures unique to the work, the industry's rheological work with cementitious compositions is of value to grouting. Interestingly, two tests that are routinely used in grouting (and described in Chapter 8), the *mud balance* and the *marsh funnel,* originated in the oil well industry.

Within the grouting community, leadership in understanding the importance of rheology has come from Europe. Although pioneering work had been done by others, a paper by Lombardi (1985), presented at the Fifteenth Congress on Large Dams held in Lausanne, Switzerland, caught the interest of many. The importance of *cohesion,* which is but one of many important rheological properties, was presented and extensively discussed. Since that work, many have reported on various rheological subjects. Several references are available for those wishing to thoroughly understand the subject; among the more pertinent reports are those of Heenan and Naudts (2000), Wilson and Dreese (1998), De Paoli et al. (1992), Hakansson, Hassler, and Stille (1992), and Al-Manaseer and Keil (1990).

Satisfactory rheological properties of grout, in both the mixed and cured state, are fundamental to the successful completion of any project. The mixed grout must provide appropriate properties to enable proper injection and travel within the formation being improved. Obviously, durability and long-term performance are essential to achieve the intended goals. The flow properties of grout materials vary widely, and a basic knowledge of both fluid and plastic flow is requisite to understanding grout behavior.

4.1 GROUT CONSISTENCY

The consistency of a mix is one of the most important parameters in grout selection for a given application. There are, however, no clearly defined standards or categories for grout consistency and considerable confusion prevails, especially with the terms used in different countries. Some in the industry describe all grouts as fluids, further defining them as viscous, very viscous, and so forth. Such descriptions are confusing and, in some cases, technically incorrect. Thus, the following three categories are proposed:

1. Fluid
2. Plastic consistency
3. Paste or slurry

4.1.1 Fluids

Simply stated, a fluid is a substance that, given enough time, will always seek its own level. Yet one must recognize that even for fluids, there are almost limitless consistencies. Very thin water-like liquids are said to be *Newtonian.* If these are pressurized in a pipe or other space, the pressure exerted on the boundaries will be about the same, regardless of location. Such a fluid will find its level very quickly. Conversely, fluids we think of as thick or viscous will develop considerable resistance to flow and positive pressure will be required to impel them through a pipe or conduit. Although it may take a considerable amount of time, such liquids will nonetheless return to their given level. This behavior is known as *non-Newtonian* or *Bingham* flow.

4.1.2 Plastic Consistency

Plastic is defined by the American Concrete Institute (1990) as a "condition of freshly mixed cement paste, mortar, or concrete such that deformation will be sustained continuously in any direction without rupture." In simple terms, grout with a plastic consistency is capable of being physically remolded into any shape and tends to hold that shape absent application of a physical force. Most concrete, mortar, and very thick mortarlike grout possess this consistency.

4.1.3 Pastes and Slurries

Paste or slurry is a term that describes those compositions that fall between the two extremes of fluid and plastic. These include both cementitious and thick or nonsag resinous grouts. Many, if not all, grouts that are described as paste-like can be properly placed in either the fluid or plastic category, however, because *slurry* is a widely used term in grouting, it is included here.

4.2 MATERIAL BEHAVIOR

Fluids are defined by *viscosity;* these are materials that will not store energy, but always return to a level state, given enough time. Virtually all solid materials are elastic in that they will deform and yield when placed under a load. They will store the energy of that load until it is released, at which time they rebound to their near original state, as illustrated in Figure 4.1. The value of such yield is defined by the *elastic modulus,* which denotes the stiffness of a substance. Between viscous fluids and elastic solids are an intermediate group of materials, which are described as *viscoelastic.* These will store just some energy and will flow just a little, but will rebound partially when a load is released. The behavior of materials is a complex science. Here, only those properties that are fundamental to a basic understanding for grouting are discussed, and as simply as possible. For more technical explanations, see Tattersall and Banfill (1983) and Mehta and Monteiro (1993).

4.2.1 Thixotropy

Many grouts are subject to *thixotropy,* which is the behavior of a fluid as an immobile paste or gel when at rest, but as a fluid upon application of sufficient energy to start it moving. A common illustration is catsup being poured from a bottle. The catsup has a high resistance to flow (out of the bottle) at first attempt, but upon applied energy such as rapid shaking, flows readily. Many grouts are thixotropic in that they require a positive pressure of some magnitude to initiate movement, but flow freely at lower sustained pressure. This is important when considering allowable injection pressure for a given application, and an adequate initial pressure level to start flow must be provided.

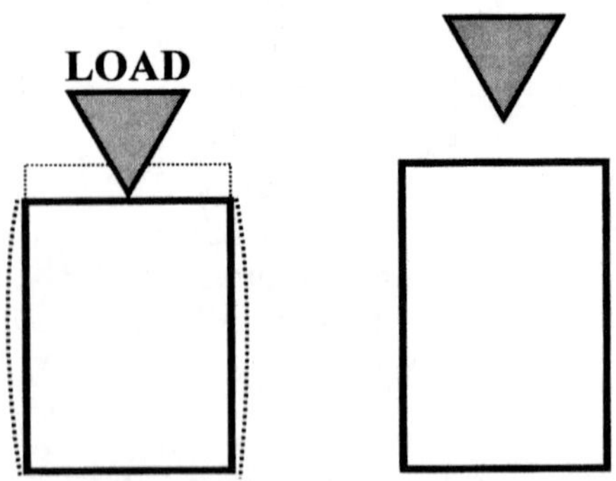

FIGURE 4.1 Solid materials will deform under load but return to their original state when the load is released.

4.2.2 Flow Properties

Fluid flow is described as either *Newtonian* or *non-Newtonian.* In Newtonian flow, the *shear stress,* which is the force required to move the fluid, is essentially constant, regardless of the rate of movement, which is technically referred to as the *shear rate* or *shear strain.* Water is an example of a Newtonian fluid as it exhibits the same shear stress, regardless of the shear rate. Non-Newtonian fluids, however, possess some thixotropy, requiring a measurable force (shear stress), which in grouting is the injection pressure, to become mobile. The magnitude of the force required to initiate movement is referred to as the *Bingham yield stress,* or simply the *yield,* and sometimes as *cohesion,* in reference to grout. Typical stress-strain curves for Newtonian and non-Newtonian, or *pseudoplastic,* flow are shown in Figure 4.2. Although a few chemical solution grouts perform as Newtonian fluids, most grouts exhibit non-Newtonian behavior.

Non-Newtonian behavior is not always as simple as indicated by the straight line relationship shown in Figure 4.2. Although the ratio of shear stress to shear rate can remain constant, it can also vary either up or down, depending on the tendency of the material to thicken or thin with an increase of the shear rate. *Shear thickening* occurs when resistance to flow (pressure) increases with an increase in the shear (flow) rate. This behavior, known as *rheopectic,* is not

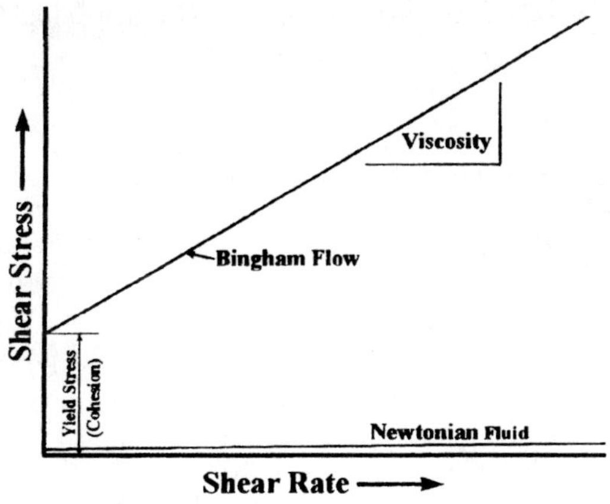

FIGURE 4.2 Newtonian and Binghamian behavior.

commonly experienced with cementitious grouts, but is experienced with many that are resinous. Conversely, in *shear thinning,* the resistance to flow of some grouts will decrease as the shear or flow rate increases.

4.2.3 Viscosity

Viscosity refers to a fluid's resistance to flow, which is the result of internal molecular friction. It is established by dividing the shear stress (pumping pressure) by the shear rate (pumping rate) at a specific temperature. The slope of the non-Newtonian curve in Figure 4.2 represents viscosity. At a constant shear rate, thixotropic fluids will exhibit a reduction in *apparent* viscosity, or resistance to flow. The term *apparent* is used because the viscosity of a non-Newtonian fluid will change at different shear rates, and thus the apparent viscosity is that experienced at any one particular rate.

The viscosity of a Newtonian fluid will always be the same, regardless of the shear rate. Conversely, the viscosity of a non-Newtonian (Binghamian) fluid will vary with time and level of shear rate (mixing energy and/or pumping rate). Fluids that become more viscous with increasing shear rate are described as *dilatant,* and those that become less, as *pseudoplastic.* Most, but not all, grouts are pseudoplastic, in that their viscosity lowers as the shear rate increases. An exception is compaction grout, which is dilatant and thus stiffens once it enters the formation being improved. At a given shear rate, the viscosity of both a Newtonian fluid and a pseudoplastic material can be the same. The value will vary, however, at different shear rates.

Temperature has a significant influence on viscosity as a result of its influence on molecular activity. For example, the viscosity of all fluids decreases as the temperature increases and, conversely, increases with lowering temperature. The sensitivity of a particular grout to temperature varies but can be significant, especially for the polymer resins. These grouts are sometimes heated prior to and/or during injection for this reason. Viscosity is also affected by pressure, but not as dramatically. It will normally go down as pressure increases. Although the sensitivity of a liquid to pressure is not as great as its sensitivity to temperature, it can become significant at very high pressures.

4.2.4 Mobility

Mobility denotes the propensity of a grout material to travel through the delivery system and into the desired voids or intended deposition zone of the formation being grouted. Of equal importance is its ability to achieve limited travel, so as *not* to flow beyond the desired zone. Thus, for proper performance, the grout should be of sufficient mobility to penetrate and fill the *desired* defects, but sufficiently limited so as *not* to flow beyond them. When an injection is made into soil, the mobility will also affect the manner in which the grout behaves during injection.

When grout is placed in soil under pressure, it behaves in one of three ways:

1. As a penetrant filling pore space, if the pumping rate is equal to or less than the rate at which the pore structure will accept the grout
2. As a fluid causing hydraulic fracture of the formation, if the pumping rate exceeds the permeation rate or if the grout consistency does not allow permeation
3. As an expanding mass pushed by fluidlike pressure at the source of injection, so as to compress or compact the soil.

A common misconception is that a thick grout will always be of low mobility, whereas a thin one will be highly mobile. Depending on the individual grout constituents and mixing parameters, thin grouts can be of relatively low mobility, whereas even very thick low-slump plastic consistency grouts can be highly mobile. Examples of the former are fluid suspension grouts, which contain blocking agents as described in the Bureau of Reclamation Technical Memorandum 646, (1957). Prior to development of the ability to pump low-mobility plastic consistency grouts, such fluid mixtures designed to restrict flow were quite common, and they are still sometimes encountered.

Even very thick, or essentially *no-slump*, grout mixtures can be of sufficiently high mobility to behave as fluid when injected. Such behavior would typically be expected of grouts containing clay components or admixtures such as some concrete pumping aids. When such grout behaves as a fluid during injection into soil, hydraulic fracturing occurs. This subject is discussed more extensively in Chapter 11, in relation to compaction grouting, in which the use of grout of the proper rheology is imperative.

4.2.4.1 SLUMP

Perhaps the first reference to slump in grouting was made in connection with the very stiff mortarlike grouts used in compaction grouting. In 1980, the American Society of Civil Engineers (ASCE) published the *Preliminary Glossary of Terms Relating to Grouting,* wherein they defined compaction grout as follows:

> COMPACTION GROUT—Grout injected with less than 1 in. (25 mm) slump. Normally a soil-cement with sufficient silt sizes to provide plasticity together with sufficient sand sizes to develop internal friction. The grout generally does not enter soil pores but remains in a homogeneous mass that gives controlled displacement to compact loose soils, gives controlled displacement for lifting of structures, or both.

Although a standard test to define "slump" was not provided, many grouters have assumed that, the well-established slump cone used in ASTM C 143 for concrete, which is the most common slump test, was intended. The C 143 test involves filling a truncated conical mold in a prescribed manner and measuring the shortening of the specimen when the mold is removed. Procedures for this as well as other slump tests are provided in Chapter 8. Many have ignored the inclusion of "sand sizes" and "internal friction" in the definition and have defined any grout with a low slump, using the C 143 cone, to be appropriate.

Notwithstanding the definition by ASCE, and the interpretation by grouters, slump alone is *not* a valid measurement of a grout's rheological properties or appropriateness for compaction grouting, or, for that matter, any grouting application. The test was originally developed for appraising the workability of concrete, which includes only clean sands and large aggregate. Even for use with that material, for which the test was developed, ASTM states in its document, *ASTM Special Technical Publication 169B* (1978):

> Slump Test—The slump test . . . is the most commonly used method of measuring consistency or wetness of concrete. It is not suitable for very wet or very dry concrete. It does not measure all factors contributing to work-

ability, nor is it always representative of the placeability of the concrete.

A mixture with a slump of only 1 in. (25 mm) would certainly be "very dry," a mixture that ASTM says is "not suitable" for its 143 test. Typical low-mobility grouts, such as those used for compaction grouting, contain silt-size particles, which tends to make them hold considerable moisture and appear sticky. They can behave as thixotropic fluids and require a considerable shear force to initiate slump. The gravity force under which slump is determined, however, can be too small to overcome the Bingham yield point, whereas the greater force provided by a grout pump can easily start fluid movement. A further limitation of the slump test is that it is nearly impossible to properly fill the cone with a sticky grout. This is discussed fully in Chapter 8.

Unless one is dealing with *concrete* that is not "very wet or very dry," the C 143 slump test should be neither specified nor used, as such would be outside the intent of the test method. It is certainly *not* appropriate to judge the rheology of thick mortarlike grout mixtures, such as those used for compaction grouting. The use of the C 143 mold to prepare specimens that can be used to judge grout consistency may be valid, but filling the mold or conducting the test in accordance with the test procedure usually cannot be done because of stickiness and other rheological properties.

4.2.5 PENETRABILITY

Penetrability defines the ability of a grout to permeate a porous mass, such as sand or soil, or to fill thin fractures or small voids. As previously discussed, viscosity alone is sometimes equated to the penetrability of a given grout. In actual practice, however, the penetrability of grout is dependent on a combination of viscosity and *wetability.* The wetability of a fluid is a function of its interaction with or affinity to a solid and is defined as *surface tension.* Within the viscosity ranges commonly found in solution grouts, wetability is a far more important and significant property than viscosity.

Although incorrect, frequently only *one,* or a few, of the rheological properties pertaining to a particular grout mixture, are used to describe the performance of that grout. Using viscosity alone is of little value unless other important properties such as surface tension and the resulting wetability are known. I suggest that the use of all-inclusive, simple terms to describe the properties of grout would be of practical value and thus propose the descriptive terms *mobility,* for relatively thick grouts not intended to permeate a formation, and *penetrability,* for fluid-like grouts that are to permeate soil, rock, or structural defects. Each of these terms obviously includes many rheological influences.

4.2.5.1 POLAR INTENSITY

Wetability is a function of the surface tension between the mineral surface of a formation and the grout solution at their interface. Surface tension is the result of molecular attraction within a liquid, as well as between a liquid and a solid. The degree of molecular attraction is dependent on the polar intensity of the material. All molecules contain a number of positive as well as negative sites. If equal in number, these sites will cancel each other out. This seldom happens in the real world, however, and the difference between the number of positive and negative sites determines the polar mode and intensity. As we know, magnetic likes repel whereas opposites attract. A common example is the tenacity with which ice sticks to the windshield of a car in freezing weather. Ice is strongly positive, and glass is strongly negative. The polar intensities of both are very strong, and *opposite.*

Because of its strong polar intensity, liquid with very high surface tension will form the most compact possible shape, which is a perfectly spherical drop, because a sphere has the least surface area per volume. Liquids with lesser surface tension will still form drops, but they will be somewhat flatter, whereas a fluid with no surface tension at all will spread out in a thin film. A good example is illustrated by sprinkling water on an automobile. It will form spherical drops on a newly waxed surface, but spread out as a film on one that is oxidized, for which it has good affinity. Wetability of a particular fluid to a solid is defined by the contact angle of a drop of that fluid with the receiving surface, as shown in Figure 4.3. A liquid is said to have good wetability to a particular surface if the contact angle is less than 90 degrees. In spite of the great importance of wetability, even experienced chemists and rheologists sometimes make the mistake of citing only a material's viscosity. In remarks to his fellow rheologists, Vossoughi (1999) stated:

> Anomalous behavior of non-Newtonian fluid flow through porous media could be due to the fluid, the nature of porous media, or the interaction of fluid and porous media. Therefore, flow study should be carefully designed to distinguish between these effects or at least to acquire knowledge of which aspect of the non-Newtonian fluid flow is being studied. . . .
>
> In addition to the complexity of the pore geometry, rock/fluid interaction becomes more pronounced in the case of naturally occurring rocks. Surface wetability could make a significant difference in flow behavior of wetting phase versus non-wetting phase. Rock/fluid interaction plays a more important role in injection of polymer solution into a naturally occurring rock.

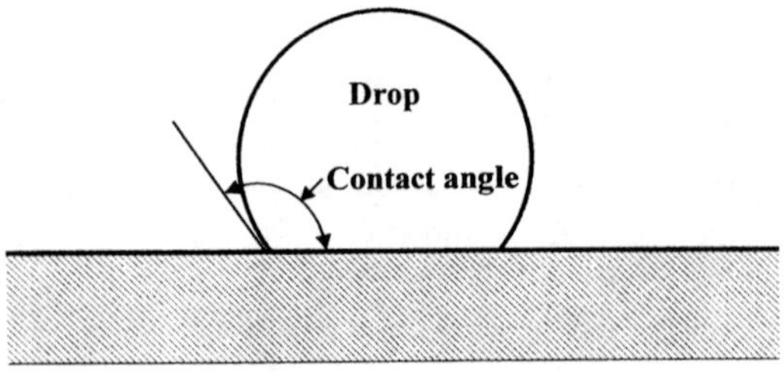

FIGURE 4.3 The contact angle of a fluid drop and a surface determines the wetability.

Wetability depends on the chemical composition of the grout and the chemistry and physical conditions of the individual formation into which it is being pumped. Affinity of the grout for the formation into, or thorough which, it is being driven thus determines penetrability and is of critical importance.

The polarities of the elements on a molecular level are primary contributors to the surface behavior of the formation, and although we usually cannot change the formation chemistry, we can alter the properties of the grout. The surface tension of the grout can be amended through the use of *surfactants*, which are chemicals whose molecules have polar "heads" and nonpolar "tails." They will literally encapsulate the individual components so as to reduce or eliminate the polar intensity. Although inclusion of a surfactant will not change a liquid's viscosity, it can increase its ability to penetrate more readily and behave as if it were thinner.

An excellent example of this occurred during an experimental test program to develop the optimal grout for injection into several thousand feet of 3/8 in. (9 mm) outside diameter (OD) tubing embedded within the embankment of a earth dam. The tubing had been installed in the embankment during original construction as part of an extensive instrumentation system. The test program was to include pumping of grout through several thousand feet (meters) of tubing to determine the injection parameters that would be required for the actual work. Although the original tubing was made of saran plastic, this material was no longer available and PVC plastic tubing of identical dimensions was used in the laboratory program.

As part of the evaluation, a length of tubing 8 ft (2.4 m) long was secured to a slanted board, as shown in Figure 4.4. Various trial grouts were then run through the tubing, with the flow times noted. Because of a concern that the different tubing materials might react differently to the grout, an 8 ft (2.4 m) long piece of the original saran tubing was salvaged from a gallery within the dam. It was secured to the slant board adjacent to the PVC tube and fed with grout from a common funnel. Interestingly, the grout flowed more easily in the PVC tubing, as confirmed in Figure 4.5, wherein the graduated cylinder on the right is filled much higher than that on the left, below the saran tubing.

The most promising grout mixes were then pumped through several thousand feet (meters) of the PVC tubing to evaluate the fastest pumping rate that could be used without exceeding the allowable pressure. As a result of the slant board tests, a correction factor was developed to account for the differential wetability of the saran tubing. As confirmed by these tests, the penetrability of a grout is not dependent solely on its viscosity. Low-viscosity grouts can have poor penetrability, whereas highly viscous grouts can be very penetrable in a given delivery system or formation.

FIGURE 4.4 Tubing attached to a slant board for grout evaluation.

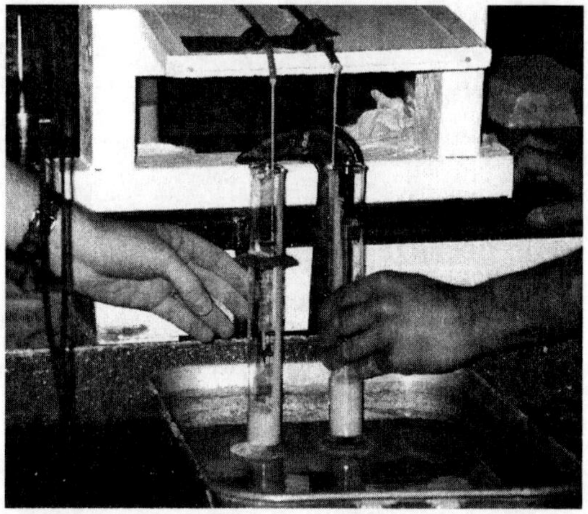

FIGURE 4.5 The cylinder on the right is filling faster than that on the left.

4.2.5.2 GRAIN SIZE OF SOLIDS

In the case of grouts that contain solids, such as fluid suspensions and slurries, the grain size of the solids also affects the grout's ability to penetrate. Theoretical research and discussion abound relative to the ratio of maximum grain size that can penetrate a given crack and/or joint width in the grouting of rock. Although the subject is somewhat contentious, in rock grouting there is general agreement that the minimum dimension of the defect must be at least three to four times that of the maximum grain size in the grout. In the injection of grout into soils, however, the maximum grain size becomes more important, as the largest grains will be the first to become lodged in the pore space near the point of injection. They will be followed by increasingly finer particles so as to form a filter, which will progressively reduce continued permeation.

It is important for grouters to have a good understanding of the particle size distribution of the materials with which they work. There is considerable variation in the size distribution of most cementing materials due to variation of the beginning components and the manufacturing process. This is especially so with common cements and fly ash. Grain size limits, within which several common material types fall, are presented in Figure 4.6. Materials from a few sources may exceed the boundaries shown, but the major portion of most available materials will fall within the range of grain sizes indicated.

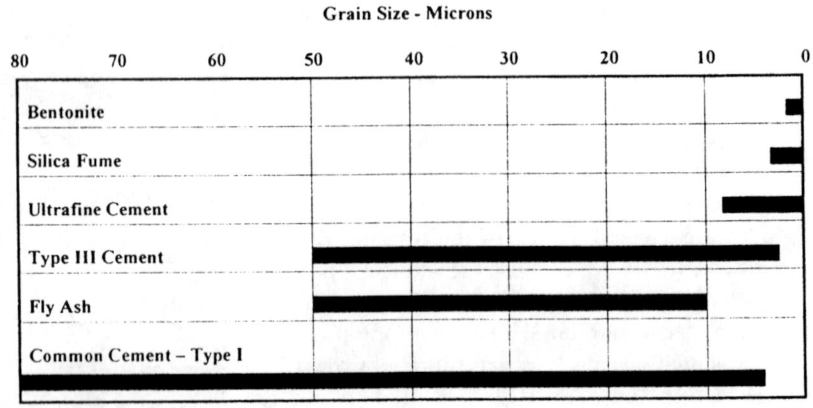

FIGURE 4.6 Relative grain size of various grout ingredients.

Although the maximum grain size unquestionably affects the penetrability of some defects, in many instances it is only one consideration and perhaps not the most important. For example, if a grout contains only a small percentage of the maximum-size particles, and those particles are well dispersed so they are not adjacent to or in near proximity of each other, as shown at the top of Figure 4.7, the maximum size will not be of great concern as long as it is somewhat less than that of the void being filled. Conversely, if the grout contains a larger proportion of the large-size particles and/or they clump together, as illustrated at the bottom of Figure 4.7, the risk of blockage will be high.

As previously mentioned, of critical importance regardless of grain size is thorough and complete dispersion of all grains. This is of particular importance when dealing with grouts that contain portland cement, as it is highly polar with a strong tendency to agglomerate. Obviously, a grout that contains clusters of many individual grains will not provide good penetrability.

The shape of the particles and the condition of their surfaces are also important. Rough, angular particles are far more likely to form blockages within a pore space or defect than smooth, round ones. Experience has repeatedly shown cementitious grouts containing large amounts of pozzolanic materials, such as fly ash and silica fume, to be more penetrable than those containing only cement. This is, no doubt, the result of the near spherical shape of most by-product pozzolans. Round grains tend to roll past and over each other, rather than bind together as do those of rough, angular shapes.

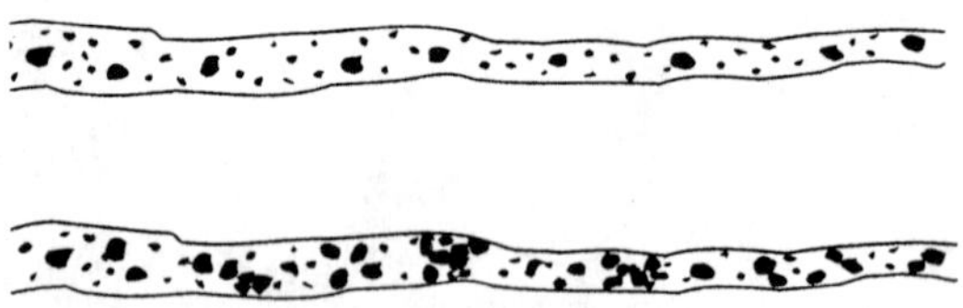

FIGURE 4.7 Occasional large-size particles are acceptable as long as they are well disbursed (top). Large particles will block openings if they can gather together (bottom).

4.2.5.3 MIXING SHEAR

At one time a cardinal rule of grouting was "The more mixing, the better." In the case of common cementitious suspension grouts, high shear mixing is, in fact, required to thoroughly disperse the individual grains. Now, however, it is not uncommon for grout mixtures to contain a variety of amendments, and that old rule is no longer valid, even for some cementitious mixtures. Rheologists have found that a given amount of mixing shear; not too much nor too little, is essential for grouts that contain many common ingredients. Variable shear laboratory mixers are thus widely used, which measure and display the actual value of the viscous shear occurring.

4.3 OTHER RHEOLOGICAL PROPERTIES

Although the terms *mobility* and *penetrability* are perhaps the most all-inclusive to describe the rheological properties of a grout, there are many factors that contribute to such a description, and these must also be individually considered.

4.3.1 Cohesion

As previously mentioned, Lombardi (1985) reported on a simple field test to determine what he called the "cohesion" of a cementitious suspension grout. What Lombardi was actually evaluating was the Bingham yield value for the grout, allbeit with a very simple test. It involves dipping a 100 mm (4 in.) square, slightly roughened metal plate, approximately 1.5 mm (1/16 in.) thick, into the grout and measuring the weight of that which adheres upon removal. The weight is divided by the total area of the plate (both sides), which provides a cohesion (C) value. The cohesion is usually divided by the unit weight of the grout to provide a relative cohesion (C_r), which is the normally reported value. This value, which is given in millimeters, will be on the order of 0.2–0.4 mm for high-cohesion grouts and as low as about 0.06 mm for very low cohesion mixes.

The measurement requires a very precise scale, and the needed accuracy requires the apparatus to be shielded from both wind and the sun. Cohesion is an important grout property, as it has a strong influence on the distance a grout can penetrate. In most cases it is desirable to have high grout penetrability, which calls for low cohesion. Alternatively, there are instances in which it is desired to limit the travel of the grout in the formation. In such circumstances, the travel can be limited through an increase in its cohesion.

4.3.2 Bleed

When at rest, the individual particles of a fluid suspension grout tend to settle out of the solution, leaving excess mix water on the top of the settled solids. As previously discussed, this is referred to as *bleed*, which can be greatly reduced by good dispersion of the solids through high shear mixing. To prevent bleed of the grout prior to injection, it is usually agitated continuously after mixing. Although excessive bleed will not occur under proper agitation, it will when the grout is essentially at rest, either during or after injection. As the solids settle to the bottom of a void or defect, the top portion will become filled initially with water, and eventually with air, leaving void space, as illustrated in Figure 4.8.

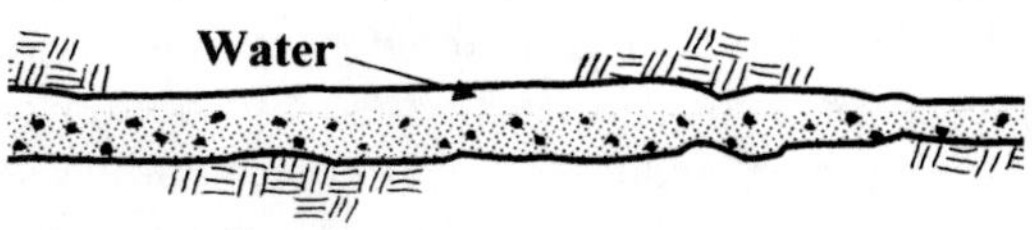

FIGURE 4.8 Void resulting from space originally filled with bleed water.

Bleed is of particular significance where large quantities of grout are injected into a single location or void, or where the injection rate is very slow. It can be a particular problem with grouts that are moving very slowly in an injection line, wherein the solids settle to the bottom of the line, where they become essentially immobile. The result is expulsion of an excessively high water content grout into the hole. The bleed potential of a grout can be greatly reduced by very thorough high shear mixing, as well as through thoughtful mix design, including the judicious use of admixtures that alter the fluid phase of the mix.

4.3.3 Temperature

The temperature of a grout can have an important bearing on viscosity. In addition, it can have a dramatic influence on the setting time, as well as the developed strength and long-term durability. Monitoring of grout temperature is thus important, especially when working in extremes of environmental temperature.

4.3.4 Setting Time

Control of the time required for grout to set or harden can be crucial to proper performance. Depending on the individual conditions, either rapid or delayed setting may be desired. Where many holes are to be grouted, grout injected into a given hole must set or become immobile before the adjacent holes are drilled. Occasionally, because of an excessive loss of drilling fluid, the drill is removed and the hole grouted so as to control the loss. In such cases, resumption of drilling is delayed until the grout sets. Rapid setting times are also often desirable when injection is into moving water, so that the grout will set before being excessively diluted or washed away. Even where the water is not moving, a rapid set will minimize the opportunity for dilution of the grout. In many cases of soil strengthening and groutjacking, rapid strength development is required to limit the downtime of the facility in which the work is being done. Likewise, minimizing downtime is a frequent requirement for grouting done for structural repairs.

Conversely, where injection is to be made through a very long delivery system, extension of the set time may be required so as to prevent hardening of the grout within the system or in the initially filled portions of the formation. Likewise, where large linear void spaces require filling, delay of the initial set is usually desirable until filling is complete. For cementitious grouts, admixtures are commercially available that can either accelerate or delay the setting time. In those rare cases where very long set times are required, an admixture is available that effectively prevents the hydration reaction from starting for a period of up to about 72 hours.

In the case of chemical solution and resinous grouts, the setting time can usually be controlled by adjustments in the proportions of the mix ingredients or by the inclusion of set-controlling components. Many of these grouts

FIGURE 4.9 A sampling valve facilitates the gathering of set time specimens.

start as two-component systems, with the set or hardening time being a function of the proportion of the reactant, sometimes referred to as the *hardener,* to the base compound. Variation of the components is readily accomplished through the use of adjustable proportioning pump systems. Provision of a sampling valve in the injection system, as shown in Figure 4.9, facilitates the easy acquisition of specimens to evaluate set time. Where very rapid set is desired, separate material lines can feed the grout header, where the components are mixed immediately ahead of injection.

4.3.5 Solubility

When preparing a grout, rapid dilution of all the components is beneficial and decreases the amount of time and energy required for mixing. If the grout is to be placed in a saturated formation or void, however, it is undesirable for the mixed grout to easily separate or dissolve. Under such placement conditions, and especially in placement in moving water, grout with high cohesion and low solubility is required. Also important is assurance the particular grout will not be adversely affected by the chemistry of the formation or groundwater into which it is injected. An example, of such an adverse effect is the attack of the grout curtain under a tailings dam by retained acidic water.

The durability of the hardened grout under the conditions that exist at the injection location should also be evaluated. The chemistry of both the formation and the groundwater to which the grout may be exposed should be appraised, as should any unusual chemical exposures that may occur during the lifetime of the installation. Most reacted water-control grouts will shrink upon drying. Some will swell upon being rewetted; however, there is usually a substantial time lag for complete expansion, and often the grout will not expand to the full volume that originally existed. Although this is of little concern where water will always be present, it is important in arid regions and other locations where drying may occur.

CHAPTER 5

Cementitious Grout Materials

5.1 PORTLAND CEMENT
- **5.1.1** Types of Portland Cement
- **5.1.2** The Hydration Reaction
- **5.1.3** Temperature Effects
  - **5.1.3.1** Hot Environments
  - **5.1.3.2** Cold Temperatures
- **5.1.4** Mix Water

5.2 SPECIAL CEMENTS
- **5.2.1** Expansive Cements
- **5.2.2** Calcium Aluminate Cement
- **5.2.3** Plastic Cement
- **5.2.4** Masonry Cement
- **5.2.5** Rapid Setting and Hardening Cement
- **5.2.6** Oil Well Cements

5.3 SUPPLEMENTARY CEMENTING MATERIALS
- **5.3.1** Natural Pozzolans
- **5.3.2** Fly Ash
- **5.3.3** Silica Fume
- **5.3.4** Ground Granulated Blast Furnace Slag
- **5.3.5** Rice Hull Ash

5.4 BLENDED HYDRAULIC CEMENTS

5.5 ULTRAFINE CEMENT

5.6 ADMIXTURES FOR CEMENTITIOUS GROUTS
- **5.6.1** Water-Reducing Admixtures
- **5.6.2** Set-Retarding Admixtures
- **5.6.3** Set Accelerators
- **5.6.4** Hydration Control Admixtures
- **5.6.5** Antiwashout/Viscosity Modifiers
- **5.6.6** Pumping Aids
- **5.6.7** Air- or Gas-Generating Admixtures
  - **5.6.7.1** Expansive Grout
  - **5.6.7.2** Low-Density Grouts
- **5.6.8** Bond Strength Enhancers
- **5.6.9** Air-Entraining Admixtures
- **5.6.10** Special Combinations of Admixtures

5.7 FILLERS

5.8 BLOCKING AGENTS
- **5.8.1** Lost Circulation Additives
- **5.8.2** Synthetic Fibers
- **5.8.3** Plastic Tabs and Punchings
- **5.8.4** Feed Grains

CEMENTITIOUS GROUTS, which were the first to be injected historically, continue to be the most frequently used. They can be formulated to provide a wide range of consistencies and other properties. These grouts always contain either portland cement or a specialty cement and water. Although not necessary, they can also contain supplementary cementing materials, fillers, and one or more property-modifying admixtures, resulting in the ability to compound grout with a wide variety of different properties.

5.1 PORTLAND CEMENT

Portland cement consists of a mixture of calcareous materials such as limestone, chalk, or shells and argillaceous materials such as clay or shale. Appropriate proportions of these raw materials are combined, crushed, pulverized, and burned in a rotary kiln at temperatures of 2600°F–3000°F (1430°C–1650°C). The resulting material, known as *clinker,* is pulverized upon cooling. It contains four principal components: tricalcium silicate, dicalcium silicate, tricalcium aluminate, and tetracalcium aluminoferrite. Because of the natural variations in the mineral materials of which clinker is composed, the exact proportions of these ingredients will vary. This is especially true with compounds from different geographical areas and those with raw materials of different origins, but can also occur within different batches of clinker from the same cement manufacturing plant. Typical proportions of these components are shown in Table 5.1.

The hardening rate of compositions made with a cement is largely dependent on the proportions of the aforementioned compounds, particularly the tricalcium aluminate. It is, however, virtually always too rapid. Therefore, to control the set time, an appropriate amount of gypsum is introduced into the product as it is further ground into a fine powder, most of which will pass through a No. 200 sieve. The required amount of gypsum will vary, depending on the exact proportions of the other constituents, so as to provide an acceptable hardening rate.

Ordinary portland cement is very active, in that a chemical reaction develops immediately upon its coming in contact with water. This reaction, which is known as *hydration,* is exothermic; that is, it produces heat and is responsible for the rate of hardening of the cementitious composition. Gypsum is the last ingredient introduced during manufacture, and its proportion, which is variable, is dictated by the exact chemistry of the other ingredients, which make up well over 90 percent of the total product.

5.1.1 Types of Portland Cement

The composition of portland cement is delineated in ASTM C 150, "Standard Specification for Portland Cement," wherein it is defined as "a hydraulic cement produced by pulverizing clinker consisting essentially of hydraulic calcium silicates, usually containing one or more of the forms of calcium sulfate as an interground addition."

The term *hydraulic* in the definition signifies that the product will set and harden under water as a result of a chemical reaction.

Five different types of cement are enumerated in C 150:

Type I—For use when the special properties specified for any other type are not required
Type II—For general use, more especially when moderate sulfate resistance or moderate heat of hydration is desired
Type III—For use when high early strength is desired

TABLE 5.1 Typical Values for the Principal Chemical Compounds in Common Portland Cement

		RANGE OF COMPOUND %			
Cement Type	**Common Description**	**C_3S**	**C_2S**	**C_3A**	**C_4AF**
I	General purpose	45–55	20–30	8–12	6–10
II	Moderate sulfate resistant	40–50	25–35	5–7	6–10
III	High early strength	50–65	15–25	8–14	6–10
V	Sulfate resistant	40–50	25–35	0–4	10–20

Type IV—For use when a low heat of hydration is desired

Type V—For use when high sulfate resistance is desired

These basic types of cement are often classified as *common* or *ordinary* portland cement, as opposed to specialty cements, of which there are many, as are described subsequently.

Most grouting involves the use of Type I, II, or III cement. For grouting purposes, Types I and II perform similarly, although Type II cement is somewhat more resistant to sulfate exposure. Sulfates can attack portland cement compositions and cause deterioration by way of uncontrolled internal expansion. Where high levels of sulfates exist in the formation to be grouted, Type V cement should be used in order to ensure good durability of the work. If sulfates are present, but in only low to moderate amounts, Type II cement should suffice. Type V cement is not available in all geographic locations, but is often the only type of cement stocked in those areas where high-sulfate geomaterials exist. Type III cement, commonly referred to as *high-early-strength cement,* is ground much finer than other common cements. It is thus often used for grouting in soil and rock, where the finer cement particles facilitate greater penetration of the grout into relatively fine fissures and defects.

Of great importance to grouters, but not necessarily to most other cement users, are the maximum grain sizes, as well as the grain size distribution of a given cement. Unfortunately, because grouting involves only a minute portion of the total cement market, such information is not commonly available. It should be noted that there are no standards that regulate the grain size distribution for common cements.

Nevertheless, research has documented significant variation in performance of cementitious compositions containing cement of various grain sizes. Bentz et al. (1999) concluded that grain size does indeed matter, and delineated variations in both setting time and heat release, as well as other properties of compositions made with cement of varying grain size. Relative to gradation, ASTM C 150 specifies only the required Blaine fineness, which is a measure of the specific surface area of all the grains in a given amount of cement. The Blaine surface area is usually expressed either as cm^2/g or m^2/kg. Typical Blaine finenesses of cements expressed in cm^2/g are as follows:

Ordinary portland cements—	
Type I	3000–5000
High-early-strength cement—	
Type III	4000–6000
Ultrafine cements	> 8000

Although Type III high-early-strength cement is required to be finer than type I or II to provide the required specific surface area, neither the maximum grain size nor the grain size distribution is mandated. Even though, on average, Type III cement is finer than Type I, some Type I cements have been encountered in which the largest particles were actually finer than those of some Type III cements.

5.1.2 The Hydration Reaction

When portland cement comes in contact with water, a chemical reaction, known a *hydration,* occurs. This creates a new substance, commonly referred to as cement *paste, hydrate,* or *gel.* It does not occur all at once; rather, the water contacts the individual cement grains in layers over an extended period of time. The process has been compared to the removal of layers from an onion. The hydration process will continue for a very long time, up to a year or more, as long as free moisture is in contact with the cement grains. Should they be allowed to dry, however, all progress stops.

The products of hydration occupy slightly

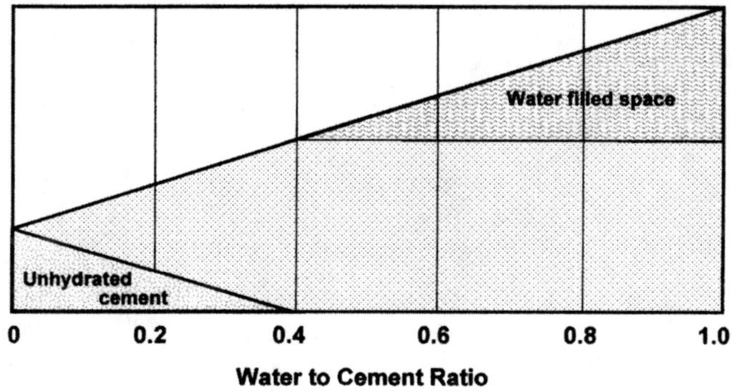

FIGURE 5.1 Varying effects of different water : cement ratios.

more space than the materials from which they are derived. When a unit of cement hydrates, the resulting hydration product or gel occupies a space 2.4 times its original volume. In order to achieve 100 percent hydration of all of the cement in a grout mix, an amount of water equal to at least 0.44 percent of its weight must be available (Powers, 1948). Thus, a 94 lb (42.7 kg) bag of portland cement requires about 41 lb or 5 gal (18.6 kg or 19 L) of water to completely react. Such complete hydration will consume an amount of water equal to only about 22 percent of the cement weight. Thus, whereas a water : cement ratio of 0.22 by weight will provide enough water for 100 percent hydration, it would be physically impossible for all of the cement to hydrate, because of insufficient space for the resulting expansive hydration product, as illustrated in Figure 5.1.

The existence of unhydrated cement is common in virtually all grout that is injected above the water table and, in fact, in most concrete as well. As long as the amount of unreacted product is not excessive, it does not detract from the quality of the grouting. Because all water in excess of 22 percent is surplus water that is not chemically combined, it will result in shrinkage and reduced strength, adversely affecting the quality and durability of the final grout. Because 100 percent hydration of the cement is not required, it follows that the amount of surplus water should be limited to the greatest extent practicable.

5.1.3 Temperature Effects

Setting time of grout is dependent on the temperature of the composition, as illustrated in Figure 5.2. As can be deduced, mixed grouts can be stored and/or circulated for longer time periods at low temperatures than at higher temperatures. Mixed grout should be fully injected prior to the development of significant hydration. As illustrated in Figure 5.2, this period can range from as little as about one hour at high temperatures to more than six hours when cooler. It is thus important to control the temperature of the grout to an acceptable level. Ideally, a rather moderate temperature within a range of about 45°–80°F (7°–27°C) is desirable. Because grout temperature is important in the determination of allowable dwell time, it is rational to set more than one dwell time requirement for projects that will be accomplished under a variety of different temperature conditions.

The rate at which hydration progresses is controlled not only by temperature, but also by the amount of surface area of the cement grains exposed to water at a given time. Grouts made with Type III cement will thus react faster than grouts made with the other types. Further, the hydration reaction itself is *exothermic,* in that heat is produced as a by-product. The maximum level of reaction temperature is thus dependent on several factors, including:

- Type of cement used
- Temperature of the beginning grout
- Temperature of the environment
- Temperature of the formation into which the grout is injected
- Volume of the grout mass

Where large amounts of grout are involved, the exothermic heat can be significant. It follows, then, that it is important to be keenly aware of temperature, and where unacceptable levels are anticipated, to initiate actions to minimize their effect.

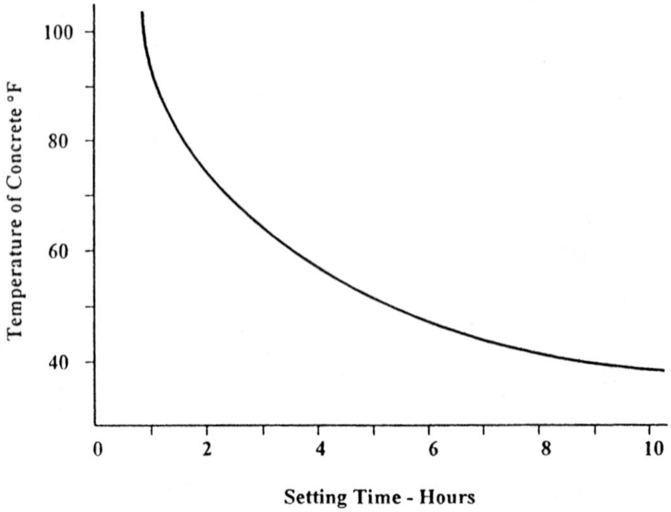

FIGURE 5.2 Setting time of portland cement at different temperatures.

5.1.3.1 HOT ENVIRONMENTS

High temperatures will result in an overrapid rate of hydration, an increase in drying shrinkage, and a marked reduction in both strength and grout durability. As shown in Figure 5.3, the type of cement used has a significant influence on the rate of heat development. Type III cement produces a high rate of temperature rise, due to the greater surface area available for hydration. Because of its finer particle size, it is often used in grouting to maximize permeation. It is best avoided in unusually warm environments, however, unless the finer gradation is truly needed to facilitate penetration into a tight formation or small-size defects.

Using chilled water or substituting crushed ice for part of the mix water can dramatically reduce the resulting mix temperature. Storing the grout materials in a cool or shady environment is also helpful. Where bulk materials are used, the storage silos should be painted either silver or white so as to reflect as much heat as possible. Obviously, the grout mixers and pumps should be in shaded locations. Long injection lines should be insulated, preferably with a light-colored reflective material.

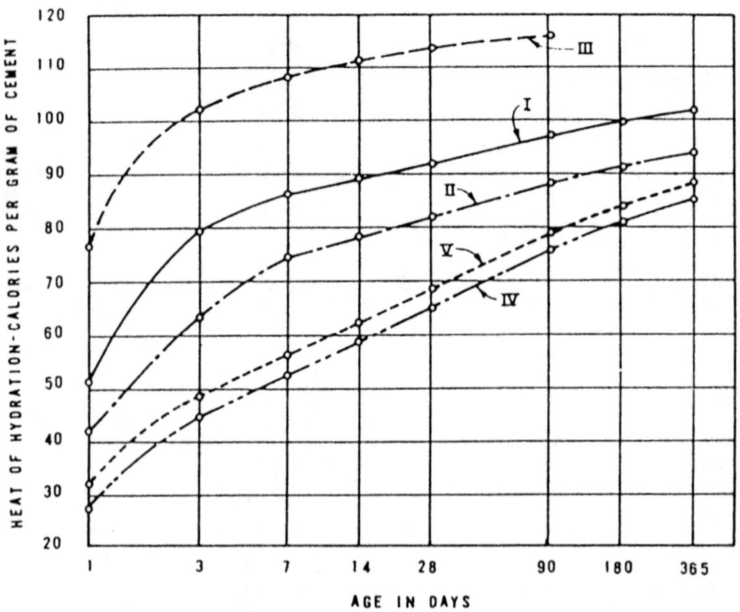

FIGURE 5.3 Cement type affects exothermic heat.

The mechanical energy of mixing and pumping, as well as the friction of the grout moving through the delivery system, generates heat. Mixing and pumping energy should therefore be minimized to the greatest extent possible. This is particularly important where circulating injection systems are used. The mixing and pumping equipment, delivery lines, and the resulting batch size should be matched to the anticipated rate of grout take, so that mixed grout is not held in the agitators or circulated through the delivery system for more time than absolutely necessary.

The effects of temperature and time were dramatically demonstrated during construction of the Eastside Reservoir Project by the Metropolitan Water District of Southern California in 1997. The specifications for the project appropriately placed limits on both the age and the temperature of the grout. The temperature was to "not fall below 35°F, not be injected at temperature in excess of 90°F," and it was not to be "held in the agitator longer than two hours."

The contractor had set up a high-production grout plant, the output of which far exceeded the low grout takes of most of the hole stages, resulting in circulation of the mixed grout for an extended time. This caused an excessive increase in the temperature of the grout. With the use of chilled water and other temperature-reducing measures, the grout had an acceptable temperature as mixed. The heat of hydration, as well as that generated during the long circulation period, however, often raised the temperature to excessive levels. Frequent interruption of injection was occurring, to clean unacceptable grout out of the system, and was severely hindering the progress of the work.

A proposal was made to raise the allowable grout temperature to 100°F (38°C). In order to evaluate the influence of such an increase, tests were run on grouts of different ages and temperatures. The grout was composed of Type III cement, with a water : cement ratio of 4:1 by volume, and contained 1 percent by weight of a plasticizer. Groups of nine grout specimens each were taken immediately upon mixing and at intervals of one, two, and three hours thereafter. During this time the grout temperature increased from an initial 66°F (19°C) all the way to 96°F (36°C), as shown in Table 5.2. The specimens were cast in standard 2 × 4 in. (51 × 102 mm) plastic mortar molds.

These specimens were allowed to cure for 13 days, after which they were evaluated for both strength and any change of volume or the length of 4 in. (102 mm) as cast. A reduction from the cast length was expected as a result of bleed of the high water : cement ratio of 4:1, which is at the upper limit of quality grout, and those specimens cast shortly after mixing did settle into a generally dense condition, with a final length of about 1.8 in. (46 mm). In addition, they obtained an average compressive strength of 1897 psi (13 MPa).

The remaining specimens performed progressively more poorly, and measurement of their lengths confirmed that significant hydration had occurred prior to casting. This resulted in hardened grout with a lower density, and thus both lower strength and lower durability. The three-hour specimens grew to an average length of 3.47 in. (88.1 mm), but their average strength

TABLE 5.2 Temperature Increase of Grout Over a Three-Hour Test Period

Time	Grout Temperature	Ambient Temperature
9:15	66	81
9:45	76	85
10:15	84	89
10:45	90	93
11:15	94	96
11:45	95	95
12:15	96	101

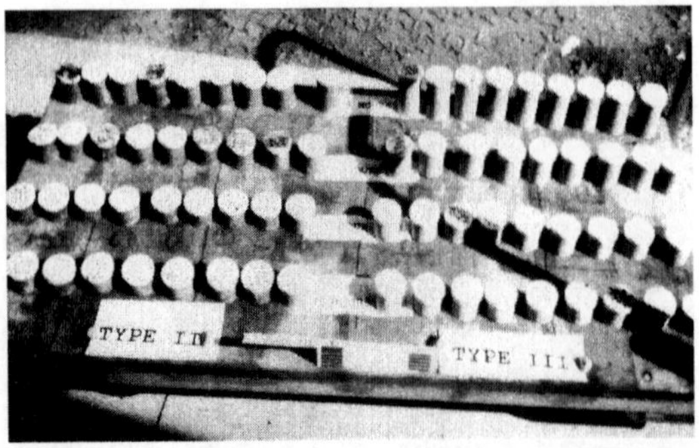

FIGURE 5.4 Seventy-two specimens of grout grouped according to age show length differences.

was only 1374 psi (9.5 MPa). The difference in the specimen length was so dramatic that it can clearly be discerned in the photograph appearing as Figure 5.4. Actual data from the evaluation are shown in Table 5.3.

5.1.3.2 COLD TEMPERATURES

Excessively low temperature levels will significantly delay the set and hardening of grout. Further, if it is allowed to freeze before attaining an initial strength of about 500 psi (3.4 MPa), both low strength and poor durability will result. Aside from cement selection, the use of heated water is especially beneficial in raising grout temperature. In addition, the cement and any other ingredients should be stored in a heated enclosure or covered with heat blankets. Obviously, the mixing and pumping plant should be protected and heated as necessary and injection lines should be insulated. The use of accelerating admixtures such as calcium chloride should also be considered. Should the temperature of the formation to be injected be so low as to act as a heat sink, it may be best to start with a fairly hot mixture so that a final temperature well above freezing will prevail.

TABLE 5.3 Test Specimen Properties

Sample Number	Mixing Hours	Length After Demolding		Specimen Age	Compressive Strength	
		in.	mm		psi	MPa
B-1	0	1.8	45.7	13	2243	15.4
B-2	0	1.8	45.7	—	—	
B-3	0	1.8	45.7	13	1551	10.7
B-10	1	2.0	50.8	13	1800	12.4
B-11	1	2.0	50.8	13	1578	10.9
B-12	1	1.9	48.3	13	1855	12.8
B-19	2	2.5	63.5	13	1376	9.5
B-20	2	2.6	66.0	13	1318	9.1
B-21	2	2.6	66.0	13	1477	10.2
B-28	3	3.5	88.9	13	1303	9.0
B-29	3	3.6	91.4	13	1230	8.5
B-30	3	3.6	91.4	13	1591	10.9

5.1.4 Mix Water

The quality required for water used in grout mixtures can vary widely according to the type of grout and its end use. Where either good strength or durability of the grout is important, the water must be of reasonably good quality. Any water source that is potable for human consumption is usually adequate for grout. The use of nonpotable water should not be ruled out, however, as it can be readily available and entirely satisfactory for such use. In fact, most water that is found in rivers and canals and retained behind dams is entirely fit, and even recycled water from mining and other industrial operations can be adequate.

Impurities in mix water can have a deleterious effect on cement hydration, however, adversely affecting the properties of the grout. Principal contaminants and their influences on cementitious compositions, together with maximum allowable content, are displayed in Table 5.4. The levels indicated are for guidance only, and the effect will vary according to the origin. When the exact properties of nonpotable water contemplated for use are unknown, or the levels of contaminants approach the maximum levels indicated in Table 5.4, testing of trial mixes is recommended. Particularly important to grouting are the effects on setting time.

There are some grouting applications in which water quality is of little significance. An example is compaction grouting, in which the objective is simply to inject mass and the degree of cementation has little bearing. Moreover, many applications of fill grouting do not require a material that reaches structural strength, but rather one that will only fill the required space.

In the real world, almost all grout must contain an excessive amount of water in order to achieve the necessary flowability and resulting penetrability. Unfortunately, any water in excess of that which is chemically consumed has a negative effect on the hardened material. Where there are only small amounts of excess water, the effects will be limited to shrinkage and the development of small cracks. Where the quantity of excess water is large, however, as in virtually all highly penetrable cementitious suspensions, the resulting cracks can be large and closely spaced and even significant void areas can occur. These start as pockets of water that have exuded from the fresh grout. They become open voids when the water is lost to evaporation, as shown in Figure 5.5, where a grouted prestressing

TABLE 5.4 Allowable Contaminants in Mix Water and Their Effects

Compound	Maximum Acceptable Level (ppm)	Effect on Cement
Sulfates	1,000	Expansive reaction with cement
Carbonates and bicarbonates	3,000	Accelerated or delayed set, strength reduction
Organic matter	20	Accelerated or delayed set, strength reduction
Suspended solids < 5 micron	2,000	Strength reduction
Sucrose/sugar	500	Strong retardation of setting time and strength
Chlorides	20,000	Corrosion of encased ferrous metal

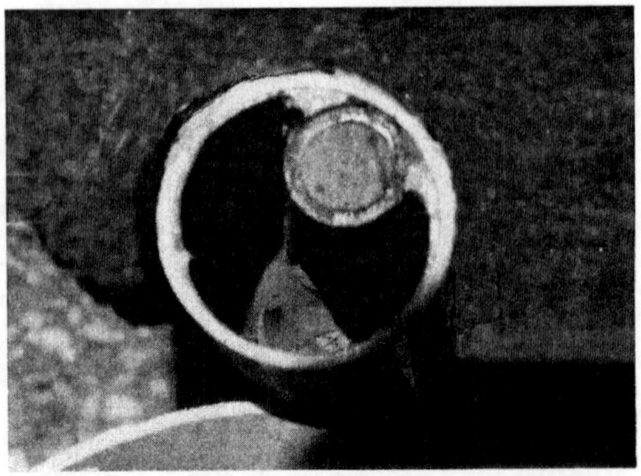

FIGURE 5.5 Void in the grout around a post-tensioning tendon resulting from sedimentation of the cement and accumulation of bleed water.

tendon has been removed and cut to expose the condition of the grout.

5.2 SPECIAL CEMENTS

A wide variety of cements are available that meet special performance requirements. With the exception of ultrafine cement, which is a fairly recent development, none have been developed for the express purpose of grouting. Although they are not often used in grouting, there are situations where the unique properties of these special products can be advantageous.

5.2.1 Expansive Cements

Expansive cements are portland cements that have been specially formulated so as to expand during the early hydration period. If a composition made with these cements is not restrained during hydration, significant expansion and cracking will result. If the compound is restrained once in place, however, it will self-stress through expansion at an early age. Drying shrinkage will subsequently occur, but the stress will simply relax, with the compound maintaining its full volume free of cracks.

The expansive mechanism is the production of *ettringite,* a substance that occupies more space than the volume of its constituents. There are three different reactive chemicals that can be interground in cement to provide ettringite. Each has been classified by ASTM under its Standard C 845, "Specification for Expansive Hydraulic Cements," as follows:

Type K
Type M
Type S

The only one now readily available in the United States is Type K, sulfoaluminate cement, which is made with a clinker that contains appropriate amounts of C_4A_3S and CS in addition to the normal ingredients. Grouts made with these cements tend to thicken and set rapidly, so they must be placed shortly after mixing. Although their rapid set complicates handling, they have been used in structural grouting, especially where low shrinkage is important.

5.2.2 Calcium Aluminate Cement

Calcium aluminate cement is also referred to as high alumina or fondu cement. Compounds made with this special cement tend to gain strength rapidly and will do so even at very low temperatures. Hardened grouts containing this cement exhibit excellent resistance to high temperatures and are highly resistant to sulfate, as well as most acids with pH levels to about 4. This cement does have one substantial limitation, however; it loses strength with time. The strength regression in concrete has been so serious as to require demolition of structures in which it was used. As a result, it is not widely available but can usually be found at suppliers to the refractory industry.

5.2.3 Plastic Cement

Plastic cement is made with either Type I or II cement, to which up to 12 percent, by volume, of plasticizing materials have been added during the grinding operation. This adds water retention, cohesion, and stickiness to grouts in which it is used. Although specifically made for masonry mortar and portland cement plaster, it provides good pumpability and is thus advantageously used in grout.

5.2.4 Masonry Cement

Similar to plastic cement, masonry cements also contain plasticizers and are designed especially for use in masonry mortar. There are three types of masonry cement, each with its own particular properties, but always directed for use in masonry mortars. Masonry cements are commonly available throughout the eastern half of the United States, but generally not in the western part, where plastic cement is used almost exclusively.

5.2.5 Rapid Setting and Hardening Cement

Within the last several decades, several rapid-setting cements have been manufactured, only one of which is widely available: Rapid Set, a product of CST Cement Manufacturing Company of Cypress, California. Rapid Set is calcium sulfoaluminate cement, more finely ground than common portland cement. As with most cementitious compositions, its setting time is closely related to temperature, which is usually too fast for most uses. Therefore, citric acid retarder is shipped along with the product to allow regulation of the set time. If the temperature is known, it is possible to formulate a mix that will provide virtually any desired set time, from minutes to hours, by varying the quantity of admixture. The rate of strength gain is directly linked to the set time, however, so excessive set delay should be avoided where early strength is required.

When used in normal concrete or mortar, a water : cement ratio of about 0.5 by weight is common. Because of its composition, the cement requires additional water to fully hydrate, so wet curing is required, and some of that water will be chemically consumed. Such cement used in grout will normally contain sufficient water, so curing becomes less important. Because the set occurs rapidly and the grains are quite small, little or no bleed occurs. This special cement is not widely used in grouts, but some structural applications can benefit from the rapid strength gain provided.

When dealing with a rapid-setting grout, premature hardening in the equipment and delivery system is always a risk. Trial batches should therefore be prepared at the site, under the same temperature conditions that will prevail during the production work, in order to develop and fine-tune the optimal mix. Obviously, pauses in pumping or other delays must be avoided when using these grouts.

5.2.6 Oil Well Cements

As previously discussed, the exact chemistry and properties of ordinary portland cement vary widely. Although this variability is acceptable for most cement usage, it is not tolerable for the special requirements of the oil well cementing industry. In this industry cement grouts are used to fill the annular voids around the well casings. These can be very deep, exposed to extreme pressures and very high temperatures. Working with pressures greater than 20,000 psi (138 bars) and temperatures of more than 312°F (156°C) are common requirements in the industry

To satisfy these needs, special oil well cements have been developed and standardized.

Nine different cements, known as Classes A through J, are specified by the American Petroleum Institute (API) under its Standard 10A (API, 1974). Most oil well cements start as ordinary portland cement, but with adjustment of the chemistry, especially the C_3A (tri-calcium aluminate), in order to provide the needed properties. In addition, a variety of special additives are blended into these cements, and some are ground much more coarsley than the more common cements. The properties of API oil well cement are summarized in Table 5.5.

These special properties can sometimes be advantageous to the grouter. An interesting example is discussed by Houlsby (1988), wherein reduced bleed for grout used in very large vertically oriented posttensioned tendon ducts was made with Type G oil well cement. It was found to be nearly equivalent in stability to grouts based on ultrafine cement, but at a substantially lower cost.

5.3 SUPPLEMENTARY CEMENTING MATERIALS

Supplementary cementing materials are generally *pozzolanic* materials, in that they will contribute to the hydration reaction of a cementitious composition but are generally not in themselves very reactive. These are siliceous or aluminosiliceous substances that react with free lime or another source of calcium in the presence of water. More precisely, they will become cementitious when in contact with calcium hydroxide, which is a normal product of cement hydration. They can be either natural or manmade, usually as by-products of other materials processing.

Natural pozzolanic minerals include:

- Volcanic glasses
- Volcanic tuff and pumice
- Clays and shales

TABLE 5.5 Types and Properties of Oil Well Cement

API Class	Special Properties	Intended Use
A	Same as ASTM Type I	Well depths of up to 6,000 ft (1,800 m) Temperatures of 80°–170°F (27°–77°C)
B	Similar to ASTM Type II, low C_3A, high sulfate resistance	Well depths of up to 6,000 ft (1,800 m) Temperatures of 80°–170°F (27°–77°C)
C	Similar to ASTM Type III, low C_3A, high sulfate resistance	Depths up to 6,000 ft (1,800 m) Temperatures of 80°–170°F (80°–170°C)
D	Low C_3A with set retarder	Depths of 6,000–12,000 ft (1,800–3,600 m) Temperatures of 170°–280°F (77°–138°C)
E	Low C_3A with set retarder	Depths of 6,000–14,000 ft (1,800–4,200 m) Temperatures of 170°–280°F (77°–138°C)
F	Low C_3A with set retarder	Depths of 10,000–16,000 ft (3,000–4,800 m) Temperatures of 230°–320°F (110°–160°C)
G and H	Coarse-grind ASTM Types II and IV	Temperatures of 80°–200°F (27°–93°C)
J	Essentially βC_2S and pulverized silica sand	Depths below 20,000 ft (6,000 m) Temperatures $>$ 350°F (177°C)

- Diatomite
- High-reactivity metakaolin

Synthetic pozzolans, which are generally by-products of the manufacture of other materials, include:

- Fly ash
- Silica fume
- Ground granulated blast furnace slag
- Rice husk ash

Substitution of pozzolan for a portion of the cement will lower the heat of hydration, which will extend the set time and slow the rate of strength development. The final strength will usually be higher, however, and the hardened material of lower permeability and thus more durable. Standard requirements for natural pozzolans and fly ash have been established by ASTM and are covered by its Standard C 618, "Standard Specification for Coal Fly Ash and Raw or Calcined Natural Pozzolan for Use as a Mineral Admixture in Concrete." Therein these materials have been divided into three classes, as follows:

Class N—Raw or natural pozzolans
Class F—Fly ash with pozzolanic properties only
Class C—Fly ash with both pozzolanic and cementitious properties

5.3.1 Natural Pozzolans

Class N natural pozzolans, which are based on volcanic glasses, are the result of very rapid cooling of volcanic magma, which is composed principally of aluminosilicates. These are formed following explosive eruptions wherein the magna is rapidly cooled such as by freezing conditions. The tuff minerals are basically of the same origin, but have been deposited as ash. Processing into commercial pozzolan usually includes grinding them to a fine powder, which is typically less than 45 microns in size, somewhat finer than portland cement. The basic mineralogy and grinding usually results in surface roughness of the grains.

Natural pozzolans based on clays and shales must be calcined at a temperature of 1200°–1800°F (650°–990°C) to convert the crystalline structure of their aluminosilicate minerals into an amorphous or disordered structure. This is usually done prior to final grinding. Diatomite consists of opaline or amorphous hydrated silica, which is derived from the skeletal remains of tiny single-celled marine growths know as diatoms. This material is pozzolanic when completely pure, but as found is usually contaminated; thus, it is also calcined prior to grinding. As with other natural pozzolans, the individual grains generally have a rough texture.

High-reactivity metakaolin is derived from kaolinite clay, which is composed primarily of hydrous aluminum silicate. It is similar to the other natural pozzolans based on clay, except for its manufacturing to a more stringent standard. The raw clay is water processed to remove impurities. It is then calcined and ground, but to a much finer powder, resulting in an average particle size of less than 2 microns. Because it is almost pure white in color, it is favored for use in architectural concrete; however, this should not be of much advantage in grouting.

5.3.2 Fly Ash

Fly ash is the most frequently used pozzolanic material in both concrete and grout. It consists of impurities released in the burning of coal, which escape and are carried away in exhaust gases of coal-burning furnaces. Because there is considerable variety in the chemistry of different coals, the properties of the resulting by-products are likewise variable. As the impurities cool, they develop into the form of tiny spherical particles composed primarily of silicate glass. The spheres vary in size from less than 1 micron to more than 100 microns; however, 70 to 90 percent are typically less than 45 microns, significantly finer than typical cement particles. Although some of the particles are hollow, most are solid, and they usually contain varying amounts of alumina, iron, and calcium, as well as very minor amounts of other constituents.

The wide spectrum of variation in the composition and quality of these by-products results in a considerable variation of reactivity. As previously discussed, the properties of fly ash are covered within the provisions of ASTM Standard C 618, wherein they are described as either Class C or Class F. Class C fly ash is produced from the burning of lignite or subbituminous coal and is composed of 10 to 30 percent calcium. It is therefore somewhat reactive on its own and will contribute to the hydration reaction of cement. Conversely, Class F fly ash, which results from the burning of anthracite or bituminous coal, contains only a minor amount of calcium, usually less than 10 percent, and is only slightly reactive. To qualify as either Class F or Class C fly ash, the carbon content must not exceed 2 percent of the total material.

Most fly ash will comply; however, some with carbon contents as high as about 5 percent do occur. A high carbon content will significantly increase the water demand of the resulting composition, producing a reduction in strength and durability. This must be avoided in cellular and structural grout; however, in some geotechnical and fill applications it can be of little consequence. Obviously, such "out of spec" fly ash can be obtained at a much lower cost, and it is not unusual to acquire it at no cost or, in some cases, even to be paid to haul it away. There is thus a strong cost advantage to its use, where appropriate. Because of their almost perfect spherical shape, which can be observed in the photomicrograph shown in Figure 5.6 (magnified 1000 times), the particles of fly ash act similarly to ball bearings when included in cementitious compositions, greatly enhancing pumpability and flow properties.

5.3.3 Silica Fume

Silica fume is an extremely finely divided pozzolan. It is a by-product that develops in the exhaust of electric arc furnaces used for the manufacture of silicon alloys. Silica fume first rises as vapor from the very hot furnaces, then condenses and is collected for disposal. It is composed of extremely fine spherical particles, generally less than 1 micron in size. This is about 100 times smaller than the particles of common cements, and in the lowest size range of the clay mineral.

At first thought, the extreme fineness of the powder can appear to be an advantage in grouting, especially in tight formations. This is not always the case, however, because compositions that include it tend to be excessively sticky. Silica fume is so fine, and its bulk density so low, that spills tend to drift off, as would cigarette smoke, especially if the wind is blowing, which

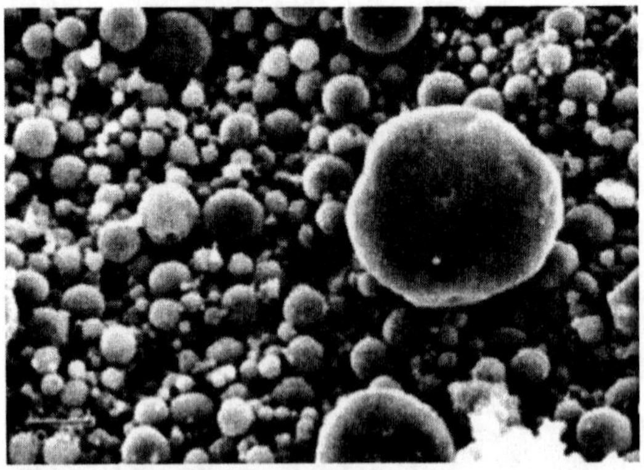

FIGURE 5.6 Photomicrograph showing perfectly spherical particles of fly ash.

renders it extremely difficult to handle in its raw state. It is thus commonly available only in a fluid slurry or densified (agglomerated) form.

Addition of silica fume to grout will greatly enhance its cohesion and reduce its susceptibility to washout. It will also lower permeability and increase the strength of the hardened mass. Because of its perfectly spherical particles, it can aid flowability of the grout, but considerable pressure may be required to initiate movement. This is due to its thixotropic nature and extraordinary fineness. The amount of silica fume used is thus commonly limited to a maximum of about 10 percent, by weight, of the cement.

5.3.4 Ground Granulated Blast Furnace Slag

Ground granulated blast furnace slag is a byproduct of the steelmaking industry and results from impurities in the original iron ore. It is produced with molten iron and rapidly cooled, such as by deposition into water, breaks up to form sandlike granules. These are dried and ground to a fine powder, with a grain size generally less than 45 microns, somewhat finer than that of common cement. The resulting powder is pozzolanic, although it can possess some cementing qualities as well. Other pozzolans are available as separate materials that can be included in a grout mix, but slag is usually combined with other ingredients to form a blended cement. The polar activity of slag is very low, so it does not tend to agglomerate.

5.3.5 Rice Husk Ash

Rice husks are shell-like materials that are removed from rice in order to prepare it for the market. In some parts of the world, large quantities of this waste product are created, which are usually reduced to ash by burning. With ordinary burning, this ash is composed of nonreactive silica, but when burned under controlled combustion, a highly reactive ash containing silica in a noncrystalline form results. Because there is a great abundance of rice husks in many parts of the world and disposal is a problem, much research has been conducted to convert this waste product to a useful pozzolanic material. According to Mehta (1991), production is now under way, so it may be available.

5.4 BLENDED HYDRAULIC CEMENTS

Blended cements are composed of normal portland cement and a pozzolanic material, which are combined during manufacture. The most frequently used pozzolanic materials are granulated blast furnace slag and fly ash, although any pozzolanic material can be used, as can be hydrated lime and even cement kiln dust. Blended cements usually set and gain strength more slowly than common portland cement, but often provide greater resistance to common deteriorating mechanisms, such as sulfate attack. These cements are controlled by the requirements of ASTM C 595, "Specification for Blended Hydraulic Cements." They are divided into six types, as follows:

Type IS	Portland–blast furnace slag cement
Type 1P	Portland-pozzolan cement
Type P	Portland-pozzolan cement (slow strength gain)
Type 1(PM)	Pozzolan-modified portland cement
Type S	Slag cement
Type 1(SM)	Slag-modified portland cement

Type 1S cement is produced by mixing 25–70 percent, by weight, of a granulated blast furnace slag with an ordinary portland cement. The slag can be previously ground or it can be interground with the ordinary cement.

Type 1P is produced by intergrinding or mixing 15–40 percent appropriate pozzolan with ordinary portland cement. The pozzolan must be selected so as to not retard the set or strength of the resulting mix excessively, although a slight reduction of strength gain for the first 28 days can occur.

Type P cement is very similar to Type 1P, except the restrictions on set time and hardening rate are less severe, which allows use of a broader range of pozzolanic materials. This cement typically produces less heat during hydration and is thus frequently used for massive placements of concrete, such as in dam construction. It is also useful for grouting applications that involve large volumes, or are installed in hot environments, and do not require rapid set or hardening. A guide to the set delay expected from different quantities of cement replaced with fly ash is shown in Figure 5.7.

Type 1(PM) cement is composed of portland cement or portland blast furnace cement with a pozzolan. The combination can be of preground components or produced through intergrinding. Pozzolan content is to be less than 15 percent, by weight, of the finished product.

Type S is a slag cement that contains not less than 70 percent ground granulated blast furnace slag. The other components can be portland cement, hydrated lime, or a combination thereof. This cement is not intended to be used alone, but rather in combination with a common portland cement.

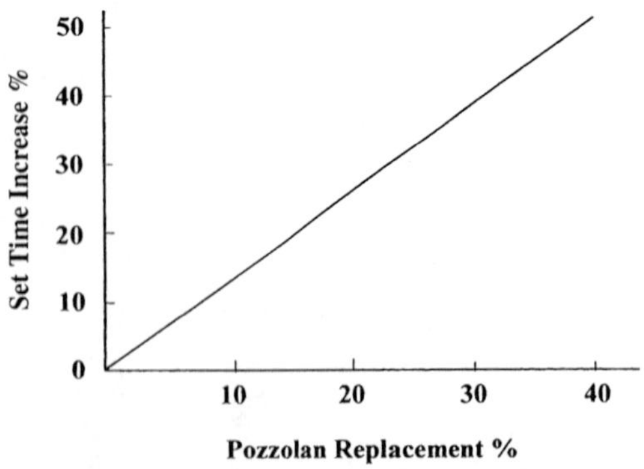

FIGURE 5.7 Effect of pozzolan replacement on setting time.

Type 1(SM), slag-modified cement, is a combination of portland cement and finely ground blast furnace slag. The slag, which can be either interground or separately combined, must be less than 25 percent of the finished cement.

5.5 ULTRAFINE CEMENT

Ultrafine cements of different origin and manufacture are readily available in the United States and most other countries. These cements are ground much more finely than ordinary portland cement. Figure 5.8 provides the grain size distribution for several ultrafine cements, as well as a typical Type I portland. Grain size data have been taken from technical information provided by the various manufacturers. For products this finely ground, virtually all manufacturers use the same test method, which involves atomic absorption, so the data are believed to accurately portray the fineness of the respective cements.

Because of the significantly smaller grain size of the suspensions they form, these grouts have far more penetrability than those containing ordinary cement. In fact, some can be blended so as to result in a waterlike suspension, even at relatively low water : cement ratios. High-range water reducing/plasticizers are commonly used with these products and are, in many cases, supplied by the manufacturers with the cement. The combination of the smaller, and thus much lighter, particles and a low water : cement ratio results in grout with a minimal propensity to bleed.

Such suspension grouts with water : cement ratios as low as 1 are common. Although conventional thinking might expect the finest-ground cement to have the greatest penetrability, this is not the case among the ultrafine products.

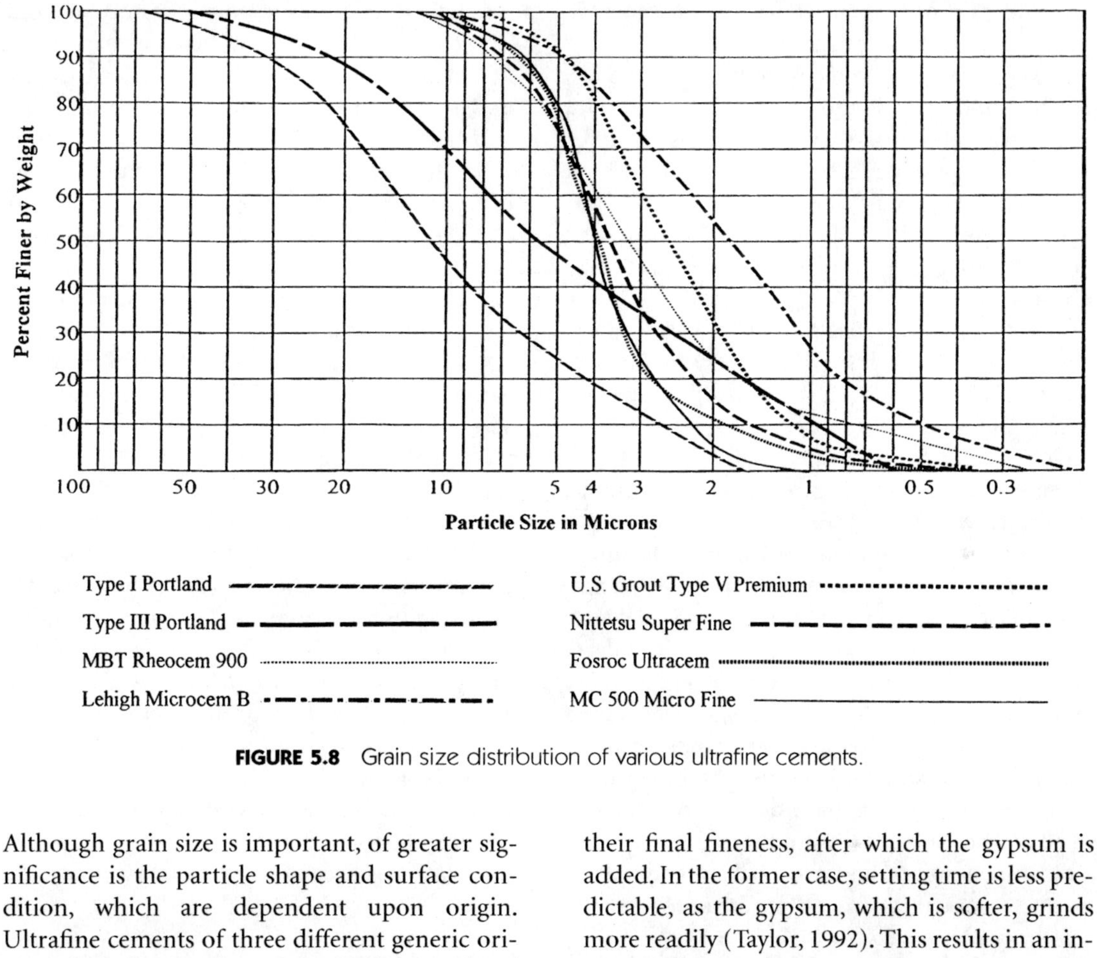

FIGURE 5.8 Grain size distribution of various ultrafine cements.

Although grain size is important, of greater significance is the particle shape and surface condition, which are dependent upon origin. Ultrafine cements of three different generic origins, including ordinary portland, granulated blast furnace slag, and blended cements, are commonly available. There are fundamental differences in the final product derived from each, however.

As previously discussed, portland cement consists of various natural materials that are combined, crushed, pulverized, and burned in a rotary kiln. Because the original minerals vary, the proportion of gypsum required for set control will vary. Some ultrafine cements are made by further grinding an already produced portland cement to which the gypsum has been added, whereas other ultrafines are ground to their final fineness, after which the gypsum is added. In the former case, setting time is less predictable, as the gypsum, which is softer, grinds more readily (Taylor, 1992). This results in an increase of the gypsum surface area. The reaction rate of the cement will thus vary, depending on the amount of gypsum in the beginning cement and the extent of increased surface area resulting from the additional grinding, but it will always be excessive.

For this reason, retarders such as citric acid are sometimes included. Portland cement–citric acid combinations are extremely sensitive to temperature changes, so the proportions of the additive must be matched to the temperature of both the working environment and the medium being injected. Problems of flash set and difficulties with both the setting time and the rate of strength gain have been reported with cements of this origin.

Where portland cement clinker is ground to its final fineness prior to the gypsum addition, care can be taken so as to match the gypsum content to the specific reactivity and actual surface area of the particular cement. Uncontrollable setting time is thus not as great a problem, as the reactive surface area of gypsum can be precisely controlled.

Pozzolanic materials such as slag and pumice contribute little if any early hydration activity. Therefore, cements containing appreciable amounts of these materials set more slowly. Such pozzolanic materials are available separately and are sometimes included as a grout component; however, they can also be interground into the cement during manufacture. Whereas ultrafine cements based on portland cement can have very fast set times, especially in hot environments, this is not the case with those containing appreciable amounts of pozzolanic material. Long setting times are usually quite beneficial for grouting in very fine voids, such as in soil or rock, so retardation is seldom a problem and can, in fact, be a distinct advantage. Well-established accelerating admixtures are available, however, and their behavior is quite predictable, so if faster set times are needed, they can be readily achieved.

Slag-based cement is, of course, pozzolanic, so it requires a source of calcium that is typically provided by inclusion of a minor proportion of ordinary cement. The reaction and set time depend on the amount of calcium included and is normally controlled to be slower than that of ordinary cement. Slag-based products typically promote slower hydration activity than common portland cements.

There are also substantial differences in the polar strength and surface condition of the various ultrafine products. This has become especially evident through a series of large-scale tests, conducted to evaluate penetrability of ultrafine cement grouts into the pore structure of sand. These tests have been conducted as part of the field demonstration at the Annual Short Course on Grouting offered by the University of Florida and held in Denver, Colorado, each year from 1997 through 2003. A further demonstration/test was held during the American Society of Civil Engineers Geo-Institute's GeoDenver conference, also held in 2000 in Denver.

In these tests, grouts prepared from a variety of different ultrafine cements were injected into identical columns, carefully filled with uniformly graded sand, as illustrated in Figure 5.9. The columns consisted of transparent plastic tubes with an inside diameter of 7-1/2 in. (190 mm) and a height of 5 ft (1.5 m). They were equipped with top and bottom plates bolted to flanges on each end. A 3/4 in. (19 mm) diameter threaded fitting for the grout inlet was attached to the bottom plate. The tubes were filled from the top with a kiln-dried, commercially processed quartz sand, which was graded so that 100 percent passed a No. 30 mesh sieve and 40 percent passed a No. 40 sieve. A specific amount of sand was placed into the top of the tubes while they were subjected to vibration by way of a pneumatic vibrator at their base. The removable top plate was then secured to restrain any up-

FIGURE 5.9 Sand-filled columns used for evaluation of ultrafine cement grouts.

ward movement of the sand resulting from grout pressure. Several small holes penetrated the top plate to allow venting of air and grout, should it permeate that far.

Although the scope and control of the tests have expanded and improved through the years, all operations from the very beginning have been performed by well-experienced grouting professionals, and the level of performance is the highest that could reasonably be expected on a production grouting job. Full-scale mixing and pumping equipment from three of the major manufacturers, Chemgrout, Colcrete, and Hany, have been used, including both plunger and progressing cavitation (Moyno) pumps.

Grouts composed of Type I cement, as well as a variety of different ultrafine cements, have been mixed and pumped into the columns. Ultrafine cements from seven different manufacturers have been used at least once, and several of them have been employed for the each of the seven years the event has occurred. The grout is injected into the bottom connection at a uniform pressure of 10 psi (0.7 bar) for a period of not more than 20 minutes. A standard circulating injection system is used, with the pressure controlled by manipulation of suitable valves at the header, as shown in Figure 5.10. All quality-control tests that might be employed in full-scale grouting have been used to monitor the work. These include tests for grout temperature, density, Marsh funnel and ASTM C 939 flow cone efflux times, and cohesion.

FIGURE 5.10 Injection pressure is controlled through manipulation of valves at the grout header.

Although temperature and other environmental conditions have varied over the time period when the tests have been conducted, the trends in regard to penetrability are surprisingly clear and the slag-based ultrafines have shown superior performance. The relative performance of these grouts is provided in Table 5.6. In the table, the indicated penetration distance, if less than 60 in. (1.5 m), is for the 10 psi (0.7 bars) pressure being held a full 20 minutes. These data appropriately apply only to injection of similar grouts into similar sand materials.

Simply because a particular cement performed relatively poorly in these tests does not

TABLE 5.6 Penetration of Various Ultrafine Cement Grouts into Experimental Sand Columns

Cement Type	1999 Penetration—in. (mm)			2000 Penetration—in. (mm)		
Type II portland	4	(102)		11	(279)	
Slag base #1	60	(1524)	7 minutes	60	(1524)	4 minutes
Slag base #2				60	(1584)	2 minutes
Slag base #3				60	(1524)	8 minutes
Portland-pozzolan	57	(1448)		10	(254)	
Portland base #1	9	(229)		24	(610)	
Portland base #2				32	(813)	

mean that it is inferior for all uses. As discussed later, the best grout properties for a given application are not necessarily the best for all applications, and a particular mix that does very well in one case may perform quite poorly in another. Interestingly, however, quite consistently, the grouts made with slag-based ultrafine cement have exhibited the best performance even though they are not the most finely ground. This behavior is consistent with real-world experience on many projects where ultrafine cement grout has been injected into granular soil.

FIGURE 5.11 Bulk supply tanks of liquid admixture with proportioning pumps to facilitate dosing.

5.6 ADMIXTURES FOR CEMENTITIOUS GROUTS

Admixtures are compounds other than water, cementing materials, aggregate, or fillers, that are included in cementitious grout to modify its properties in either the as-mixed or hardened state. The reaction of most admixtures is with the cement only, so the amount and type of aggregate, if any, will usually not influence its effect. Relatively minor admixture doses can provide dramatic changes in the properties and behavior of a grout mixture. These compounds have been commercially available and used in concrete for more than 50 years, and many grouters, especially those outside the United States, have used them routinely. Yet the grouting industry as a whole has been late in recognizing the benefits of their use. This is especially the case in the grouting of dam foundations in the United States, where many "old-timers" still prefer to use simple cement-water grouts.

Modern admixtures are available in both liquid and dry forms. Where large quantities of liquids are required, they are conveniently delivered to the job site in bulk tanks mounted on pallets, as illustrated in Figure 5.11. These can be easily handled with a forklift. Manufacturers often furnish proportioning pumps to provide the proper admixture dose for each grout batch, which can also be seen in Figure 5.11. Where supplied in the dry form, they are best prepackaged with the correct amount for each batch of grout. Some manufacturers use soluble packaging material so that the package and its contents can be deposited in the mixer. This is especially convenient, as it speeds up the work and ensures the correct dosage.

Specific grout properties that can be altered and/or imparted through judicious use of admixtures include:

- Mix-water reduction, absent any change in consistency
- Retardation of the setting time
- Acceleration of the setting time
- Hydration delay
- Increase in cohesion (antiwashout)
- Enhancement of pumpability
- Entrainment of air or gas
- Increase in grout bond strength

5.6.1 Water-Reducing Admixtures

Water-reducing admixtures are available that can allow the preparation of flowable grout with much less water than would otherwise be required. This can lessen the amount of bleed and significantly improve both strength and durability of the hardened product. The science of water reduction is extraordinarily advanced, and many different types of water reducers are readily available. Some will accelerate the setting time, others will delay it, and still others will have no influence at all. Standards have been developed for water reducers; these are covered in ASTM C 494, where they are classified as Types A through F, according to their water-reducing ability and effect on setting time, as follows:

Type A	Water reducing
Type B	Retarding
Type C	Accelerating
Type D	Water reducing and retarding
Type E	Water reducing and accelerating
Type F	Water reducing, high range
Type G	Water reducing, high range, and retarding

Water-reducing admixtures are marketed as normal or mid-range products, which provide water reduction of 5–12 percent, or high range products, sometimes referred to as *superplasticizers,* which can reduce the required water by as much as 30 percent without changing the mix consistency. High-range admixtures are obviously the most effective, but are also the most chemically complex and the most costly. These function by encapsulating the cement particles with a negative charge that causes them to repel one another. This effectively prevents agglomeration and greatly improves dispersion of the cement. High-range water reducers will decrease grout cohesion, improving penetrability and reducing the required pumping pressure.

The change in consistency resulting from the inclusion of such products can be dramatic, as illustrated in Figure 5.12, which shows a 0.6 water-to-cement ratio grout with and without the admixture. They can be used in combination with other admixtures, including normal and mid-range water reducers. The complexity of their chemistry can clash with some cement and chemical combinations, however, so trial mixes should be evaluated to confirm the propriety of any particular combination prior to production use. Because portland cement has a rather variable chemistry, a reevaluation of the mix design and admixture performance should be made any time the cement supply is changed.

FIGURE 5.12 Grout of 0.6 water:cement ratio, with and without a high-range water-reducing admixture.

Lignosulfonates, which are by-products of the pulp and paper industry, are some of the longest-used water reducers. Although at normal dosages they provide only about a 5 percent reduction of mix water, they become much more effective at higher dosages. The producers do not recommend higher dosages, as they will impart excessive retardation; however, this is often not a problem in grouting. Prior to development of the more efficient water reducers, massive doses of lignosulfonate water reducers were sometimes used to great advantage where rapid setting was not required. They remain available, and because of their ease of use and considerably lower cost, should be considered where appropriate.

5.6.2 Set-Retarding Admixtures

The rate of set and hardening of all cementitious compositions is temperature dependent. Normal setting time is based on a mix temperature of about 70°F (21°C). Setting time becomes shorter as the temperature increases, and delayed as it lowers. Induction of hydration occurs as soon as water contacts the cement, and because that reaction is exothermic, it will generate heat and cause the temperature of the grout to rise, even when held at rest or circulating in the system. This is a particular problem where large batches are mixed and/or very slow grout takes are experienced. Mixing and circulation within the system exert considerable energy into the grout, raising its temperature even more.

The effects of high temperature are well understood in both the oil well cementing and concrete industries. Elevated temperatures not only affect the setting and hardening time, but also influence the strength and durability of the final composition. Consideration must be given to both the temperature of the grout and that of the formation into which it is being injected. Where high temperatures, a slow pumping rate, or other factors that would benefit from a delayed set prevail, the use of a set-retarding admixture should be considered.

Most set retarders are also water reducers, and among the best of these are the lignin-based materials previously discussed. There is virtually no grout that will not benefit from a reduction of the mix water. This makes these admixtures especially attractive where a delayed set or additional working time is needed. Set retarders will always delay both the initial and the final set, although the level of that delay will be temperature dependent.

5.6.3 Set Accelerators

Just as high temperatures shorten the setting time of cementitious compositions, cold temperatures will extend them, and at temperatures much below 40°F (4°C) the setting time of ordinary cements will be extremely delayed. Even where temperatures are not low, there are situations, especially in water control work, where rapid set is desired. A very inexpensive and easy-to-use accelerator is calcium chloride. It is available in either liquid or dry form. If the dry form is used, it should be prehydrated, that is, mixed with water to form a fluid, before it is added to the grout. It goes into solution quite readily, so hand mixing in buckets or an open-top barrel is effective, although a mechanical mixer is preferred.

Normally, a maximum of about 2 percent, by weight, of the cement is used, although in grouting, greater amounts have been employed. As with all cementitious compositions, the set time is still strongly influenced by temperature. This influence will be even more pronounced with the inclusion of an accelerator such as calcium chloride. At high ambient temperature conditions, and depending on the chemistry of the cement, the grout can flash set at high dosages, so trial mixes should be cautiously made prior to production usage. The accelerating performance of different levels of calcium chloride dosage will vary with both different cements and grout temperatures. Figure 5.13 shows the typical performance that may be expected at a moderate temperature of about 70°F (21°C).

Calcium chloride in dry form is least expensive and available in two different forms. Regular flake is the most common and least costly and contains a minimum of 77 percent calcium chloride. Also available, however, are concentrated flake, pellet, and granular forms, which contain not less than 94 percent calcium chloride. One must thus be aware of which concentration is being used in order to accurately determine the dosage rate.

A great amount of damage to structural reinforced concrete, resulting from the corrosion of embedded reinforcing steel or other ferrous metals, has been due to calcium chloride in the concrete. Thus, this chemical is very much

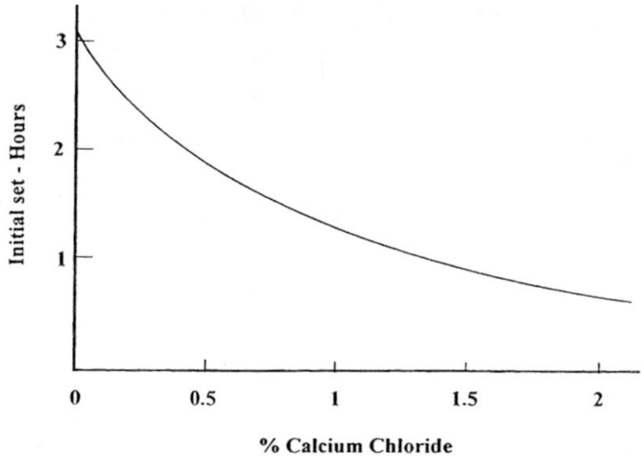

FIGURE 5.13 Effect of various dosages of calcium chloride on setting time.

frowned upon for use with concrete in general, and some material suppliers and even ready-mix plants no longer stock it. Special accelerating admixtures that do not contain chlorides are available, but are generally less efficient and considerably more costly. In grouting, embedded metals are seldom encountered, so this is not usually a problem. Calcium chloride is therefore quite acceptable and is by far the most economical and simple admixture to use. In structural applications where ferrous metals might be present however, any accelerator that contains chloride should be avoided.

5.6.4 Hydration Control Admixtures

Recent admixture technology has provided the ability to delay initial set for up to about 72 hours. Such admixtures form a protective barrier around the cementitious particles, preventing the initiation of hydration. When it is desired to bring the grout back to life, another admixture is introduced, which overcomes the barrier and allows normal hydration to proceed. By correct proportioning of the re-activator, the setting time of the grout can be controlled within a range of several minutes to several hours. As is common with all cementitious mixtures, temperature has an effect on the performance of these admixtures and must be considered. The re-activator must be physically mixed into the grout, so provisions for remixing must be made. These admixtures are very useful in projects where very long pumping distances exist or interruption of the injection is likely to occur.

5.6.5 Antiwashout/Viscosity Modifiers

All cementitious compositions, including concrete and grout, are subject to dilution and separation upon contact with moving water, wherein cement is washed out and separated from the mass. Inclusion of *viscosity-modifying* admixtures, most commonly known as *antiwashout* admixtures, provides extraordinary resistance to such separation. They alter the liquid phase (water and other liquid amendments) of the grout, which is changed to a viscous and cohesive state. This is illustrated in Figure 5.14, in which cementitious suspension grouts, with and without the admixture, are being poured into water. Note that the grout on the left settles into a dense state on the bottom of the jar, whereas that on the right is diluted and mixed with the water immediately upon contact.

FIGURE 5.14 Identical grouts with and without viscosity-modifying antiwashout admixture.

Antiwashout admixtures were originally developed to prevent the dilution of concrete placed under water, but have found extensive use in grouting. Just as they resist washout of the cement, they can reduce or completely eliminate sedimentation of the solids and the resulting bleed of suspension grouts. These are specialty admixtures, which are not widely stocked and thus usually require ordering well in advance. Other than at manufacturers' warehouses, they are not readily found, but may be on hand with some large ready-mix concrete producers.

5.6.6 Pumping Aids

Pumping aids provide water retention and/or lubricity to the grout in order to reduce or eliminate water loss under pressure, which is known as *pressure filtration.* This is usually accomplished by increasing the cohesion of the grout with materials similar to the viscosity modifiers. The sole objective in using pumping aids is enhancement of concrete pumpability. They can cause fluid behavior of grout in the ground, when used in even stiff mortarlike grout and so are often inappropriate for compaction grouting of soil. A more thorough discussion appears in Chapters 5 and 11.

5.6.7 Air- or Gas-Generating Admixtures

Some admixtures react with cement so as to generate a network of either gas or air bubbles. Such amendments satisfy two distinctly different objectives.

5.6.7.1 EXPANSIVE GROUT

Production of a grout that expands slightly to offset any shrinkage is effected with the use of air- or gas-generating admixtures. For this purpose, they are usually supplied in dry form and added to the grout during mixing. Either hydrogen or nitrogen bubbles are generated in the grout. The most common reactive component is finely divided aluminum powder, which reacts with the alkalis in the cement. Only a few grams are required to affect an entire bag of cement, so the gas promoter is usually supplied as part of a plasticizer, which also acts as a dispersing agent. There are a number of different types and grinds of aluminum powder, and their reactivity with cement varies. For these reasons, the addition of just any raw aluminum powder to grout must be avoided.

The amount of gaseous expansion will vary according to the exact chemistry of the cement, which can itself be quite variable. It is thus good practice to prepare test batches of grout, composed of the same cement that will be used in the production work, to determine the optimal dosage. These compounds are sometimes used in structural grouts to ensure complete filling of the intended spaces. They are also used in grouts for preplaced aggregate concrete to ensure intimate contact of the grout with the underside of the large aggregate. Because of the somewhat unpredictable performance of gas-producing admixtures, they are being replaced by more user-friendly shrinkage-compensating compounds but are still widely available in some areas.

Unrestrained expansion will result in marked reduction of both strength and durability of the grout, so these admixtures should be used only where the grout will be completely injected prior to the development of significant expansion and fully restrained until hardened. The working time will vary with temperature, but can be as little as about 20 minutes at high temperatures.

5.6.7.2 LOW-DENSITY GROUTS

A very different purpose for the use of foam generators is to lessen the weight of the grout used for filling applications where a low density is required. The technology for production of *controlled low-strength material* (CSLM), sometimes referred to as *cellular* concrete or grout, is highly developed. Committee 229 of the Amer-

FIGURE 5.15 CLSM grouts can be nearly self-leveling.

ican Concrete Institute has produced several publications on its technology.

Most of these admixtures also promote fluidity of the grout so as to render it nearly self-leveling, as can be observed in Figure 5.15, where it is being applied for an open fill. This makes them ideal compositions where the grout must flow a considerable distance. They are especially beneficial in grouting of annular spaces of horizontal linings and conduits where flotation of the inner lining may occur.

There are two forms of these admixtures, the first being a compound that is added directly to the premixed grout. This type can produce an air content of up to about 35 percent. Because it is formed by simple addition of the foam-generating admixture into the grout mixer, it is easy to use with virtually any size batchtype mixer.

The other form is a highly concentrated foaming agent that is first diluted in water and then combined with compressed air to form a prefoam by means of a foam generator. The prefoam is then carefully introduced into the grout after all of the other ingredients have been premixed. Care must be exercised in handling and pumping the final mix, as the foam network is somewhat fragile, especially at the lower range of grout densities. Although lower weight can be achieved, most grouting is accomplished with a final density on the order of about 40 lb per ft^3 (640 kg/m^3).

Because of the two-step development and introduction of the latter type, it is not very suitable for small batches of grout and is thus typically used in large-volume applications. Accordingly, both high-capacity stationary mixers and ready-mix trucks are commonly employed. Foamed grouts with as much as 80 percent air and densities as low as about 18 lb (8 kg) per ft^3 (28 L) can be made. As with all cementitious compositions, strength is directly linked to density. Table 5.7 provides data taken from ACI

TABLE 5.7 Typical Strength Properties of Low-Density CSLM Based on Density

	IN-SERVICE DENSITY		MINIMUM COMPRESSIVE STRENGTH	
Class	**lb/ft³**	**(kg/m³)**	**psi**	**(KPa)**
I	18–24	(288–384)	10	(69)
II	24–30	(384–480)	40	(276)
III	30–36	(576–672)	80	(552)
IV	36–42	(576–672)	120	(828)
V	42–50	(672–800)	160	(1104)
VI	50–80	(800–1280)	320	(2208)
VII	80–120	(1280–1920)	500	(3450)

229R-99 (1999), which classifies CLSM materials by density and provides density/strength relationships.

5.6.8 Bond Strength Enhancers

The bond strength of cementitious grouts is only modest. Although it is usually adequate for geotechnical grouting, in structural work higher strengths are often required. The bond of cementitious grout can be increased by inclusion of water-soluble latex bonding admixtures. These are dispersions of an organic polymer in water; they are usually supplied in liquid form, although some are also available as spray-dried powders. They are manufactured by a process known as *emulsion polymerization*, and the resulting milky whitish solutions, are often referred to as emulsions.

These admixtures can be derived from several polymer origins, but all appear similar and perform about the same, with one exception. After drying, some will reemulsify if kept constantly wet, whereas others will not soften once they are cured. This is important, as the drying time of these bond enhancers is usually short, often only 15 or 20 minutes, especially at high grout temperatures. Once drying starts and they become skinned over, they act as bond breakers. Grouts containing these agents must thus be promptly injected.

In cases where the grout batch cannot be fully injected before the latex skins over, only formulations that reemulsify should be used. Be aware, however, that these may not remain durable if subject to long-term rewetting. Although they can be easier to handle, those products provided in a dry form do not develop a bond strength quite as high as the liquids, inasmuch as there is some loss in properties resulting from the spray-drying process. Many preblended grouts and repair mortars contain admixtures of this type.

5.6.9 Air-Entraining Admixtures

Air-entraining admixtures create a network of very closely spaced microscopic air bubbles in the grout. They are widely used in geographic areas where new concrete may be exposed to freeze-thaw conditions while saturated, as exposed concrete allowed to freeze while critically saturated will be damaged by water expanding as it turns to ice. The network of entrained air bubbles provides a relief area for the expanding water and thus prevents such deterioration. Except in cold weather applications, freezing of water in grout is not much of a problem, and air entertainers are therefore seldom required.

There are by-products of the air entrainment, however, that can be advantageous in grouting. The minute bubbles can significantly reduce sedimentation of the solids and thus reduce bleeding. Further, they act as a lubricant, which can greatly improve pumpability. For these reasons, air-entraining admixture is frequently used in ready-mixed grout used for fill grouting. Vinsol resin, the most common air-entraining admixture, is widely available at ready-mix batch plants except in geographical areas not subject to freezing weather. It is inexpensive and very easy to use.

5.6.10 Special Combinations of Admixtures

Multiple admixtures can be especially proportioned, blended, and packaged in batch-size amounts to facilitate use. Such prepackaging will result in improved accuracy of dosage and can greatly improve mixing efficiency. This simplicity is especially beneficial on projects that lack well-experienced operating personnel. Packaging of this type can also be useful for stockpiling admixtures for either long-term storage or possible emergency use.

For ease of batching, most concrete admixtures are marketed as solutions or pumpable slurries, even though they may start as dry powders. For example, both welan gum and cellulose-based antiwashout admixtures start as fine powder. They are typically sold, however, as fluid suspensions. Unfortunately, these are subject to sedimentation, and thus their shelf life is fairly short, usually no greater than a year. Although they can be batched dry, their extreme fineness requires distribution within a dispersing agent.

An admixture combination was created for an agency that maintains several hundred miles of concrete-lined canals. Leaks occasionally develop in or under embankments, which occur in many places where the water level is above the surrounding terrain. These leaks can be substantial and, when discovered, require rapid plugging to prevent catastrophic failure. Emergency treatment is usually accomplished by rapidly pumping ready-mixed concrete into the leak source. Washout of the cement is a major problem, and inclusion of an antiwashout admixture crucial, but the available concrete suppliers do not routinely stock such admixtures, and thus cannot supply them on short notice as required. Also desirable for such work is a set accelerator to cause the plug to become immobile quickly. Because repairs are made only to meet emergencies, there is no idea of how long the concrete will remain in the mixer, or for that matter, whether it will even be used, so prebatching an accelerator can be risky.

A special cocktail consisting of dry accelerators, a welan gum antiwashout/viscosity modifier, a superplasticizer, and dispersing agent was developed. It was specially prepared by a custom blender and packaged in 5 gal (19 L) plastic pails for maximum protection and long-term storage. The pails were filled so that two contain the correct amount of admixture for 9 to 10 yds^3 (6.9 to 8 m^3) of concrete, which represents the capacity of ready-mix trucks commonly used in the area. The compound is stored at maintenance facilities along the route of the canals so that it can be quickly taken to the site of any leaks. This allows simple pump mix to be ordered from the nearest ready-mix supplier. Once at the job site, the admixture can be quickly and accurately added immediately prior to pumping.

5.7 FILLERS

The principal purpose of fillers is to lower the cost of a grout without compromising its strength and durability, but they are occasionally used to limit the travel of the grout as well. More common cementitious compositions such as concrete and mortar are composed of the cementing materials and water, which form the cement paste. The paste does just as its name implies; it fills the interparticle pore spaces and pastes or glues the individual units of filler material together. To minimize shrinkage and maximize strength and economy, a fundamental requirement of concrete mix design is to minimize the amount of cement paste, and thus both the cement and water of which it is composed.

The pore volume, and thus the water requirement, of small particles is greater than that of larger particles. The amount of fines in the aggregate, usually considered particles finer than a No. 200 sieve, is thus severely restricted for concrete. Moreover, to minimize the interspatial pore space, a well-graded aggregate ranging from the smallest particle to the largest is desirable, as illustrated in the saw-cut specimen shown in Figure 5.16.

Just as sand and graded rock constitute the normal filler materials in concrete, they can be used in grout, although the inclusion of rock is not frequent; where it is used, it is limited to grout filling of large or massive cavities. There are fundamental differences between the gradation and shape of aggregates used in concrete or mortar and those appropriate for grout. Aggregates that have clean, hard surfaces and are

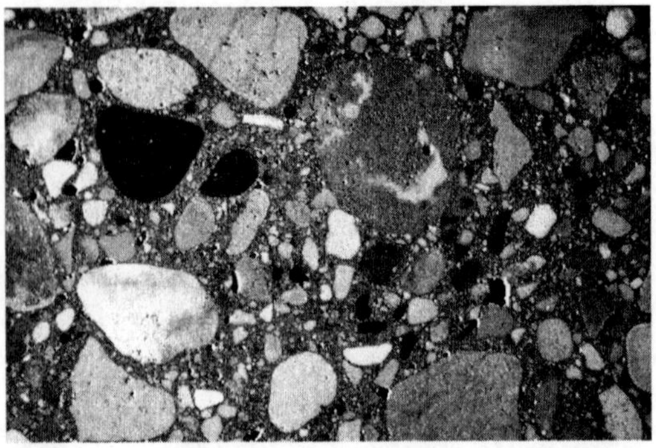

FIGURE 5.16 Sawed specimen shows inclusion of well-graded aggregate.

essentially void of fines are required for concrete, where strength is the primary requisite. In grouts, however, pumpability and penetrability are the most important properties and strength is usually of less importance. Accordingly, sand that is round grained, fine to very fine, and containing appreciable fines, (minus No. 200 sieve) is usually desirable.

Because the voids into which grout is injected are commonly small, the primary requirement is good penetrability. Even in the case of structural grouts, the maximum particle size of sand is frequently required to pass a No. 30 or even No. 50 sieve, which is very fine by conventional thinking. Further, many applications, including compaction grouting in soil, do not require high grout strength. Such grouts do require appreciable fines in the aggregate, however, to provide the necessary pumpability. As a result, most commercially produced sands are not very suitable for grouting work.

In addition to gradation, shape, and cleanliness of the particles, the *fineness modulus* (FM) is considered an important property of sand. This is an index value of fineness, irrespective of grain size distribution. Materials of different particle size distribution can thus have the same FM. It is determined by adding the cumulative percentages by weight that is retained on a specified series of sieves, and dividing the total by 100. Sieve sizes include 150 μm (No. 100), 300 μm (No. 50), 600 μm (No. 30), 1.18 mm (No. 16), 2.36 mm (No. 8), 4.75 mm (No. 4), and 9.5 mm (3/8 in.). The FM increases as an aggregate becomes coarser.

ASTM has established two different standards for sands. For guidance in selecting aggregate for concrete, ASTM C 33, "Specification for Concrete Aggregates," is used. Virtually all aggregate used in ready-mix concrete conforms to this standard, and such sand is widely available. Unfortunately, it is far too coarse for most grouting work.

Table 5.8 provides the grain size distribution called for in C 33. As shown, it starts effectively at plus 1/4 in. (6 mm) on the high end and descends to the No. 100 sieve. Fines, which pass the No. 200 sieve, although not specified, are commonly limited to no more than 5 percent and often even less. As can be observed, large variation in the actual distribution of the grain sizes is permitted. The C 33 standard allows variation of the FM all the way from 3.1 down to 2.3.

Alternatively, ASTM C 144, "Standard Specification for Aggregate for Masonry Mortar," provides for a somewhat finer product and allows the grains to be either naturally rounded or "manufactured" (crushed), which are angular in

TABLE 5.8 Fine Aggregate Grading as per Requirements of ASTM C 33

Sieve Size	Percent Passing by Weight
3/8 in. (9.5 mm)	100
No. 4 (4.75 mm)	95–100
No. 8 (2.36 mm)	80–100
No. 16 (1.18 mm)	50–85
No. 30 (600 μm)	25–60
No. 50 (300 μm)	10–30
No. 100 (150 μm)	2–10

TABLE 5.9 Aggregate Grading as per Requirements of ASTM C 144

	PERCENT PASSING BY WEIGHT	
Sieve Size	**Natural Sand**	**Manufactured Sand**
No. 4 (4.75 mm)	100	100
No. 8 (2.36 mm)	95–100	95–100
No. 16 (1.18 mm)	70–100	70–100
No. 30 (600 μm)	40–75	40–75
No. 50 (300 μm)	10–25	20–40
No. 100 (150 μm)	2–15	10–25
No. 200 (75 μm)	0–5	0–10

shape. No FM is specified, although it is noted that it should not vary by more than 0.20 "from the value assumed in selecting proportions for the mortar." Further, the aggregate "shall not have more than 50 percent retained between any two consecutive sieves . . . nor more than 25% between 300-μm (No. 50) and the 150-μm (No. 100) sieve." The gradation requirements of C 144 are shown in Table 5.9.

Available in many geographical areas is unwashed, natural sand that does not conform to any standard but is preferentially used where allowed. It is sometimes referred to as *dry*, or *unwashed*, either *plaster* or *masonry* sand and normally contains ample minus-200 sieve (75 μm) fines. If it does to not contain excessively large particle sizes, this can be a good sand for use in grouts, as the high fines content provides good pumpability and flow. Sand suppliers often have screening equipment and the ability to scalp out the larger-size particles if required.

As with all aggregates, the fines will detract from the strength of the final grout, and thus their inclusion may be restricted in structural installations. A number of special sands are available from specialty sand suppliers, such as those used in the foundry and abrasive blasting industries. These are usually kiln dried and available in a variety of gradations. They are typically packaged in paper bags, each containing about 88 lb (40 kg), although they are often available in bulk.

Clays are sometimes used as cheap fillers for grout, especially in Europe. They retain the mix water and are thus easy to pump, as well as highly penetrable. When they are used in cementitious grouts, however, a significant reduction in strength and durability will occur. Except in very special situations, their use is not recommended.

5.8 BLOCKING AGENTS

Blocking agents are used as an aid in stopping water moving through a system of relatively small passages. They become caught and jam up in, or over, the passages so as to reduce water flow. A number of different types of ingredients have been used for this purpose. Among the more effective and most readily available are the following:

- Lost circulation additives
- Synthetic fibers
- Short plastic tabs or punchings
- Feed grains

5.8.1 Lost Circulation Additives

A variety of proprietary materials are available for the express purpose of reducing or eliminating loss of circulation flush during drilling. These are used for all types of drilling, but particularly drilling for oil wells. Although the exact contents of the individual formulations are usually not disclosed, they generally contain a variety of waste materials, including sawdust, wood shavings, almond husks, ground walnut shells, old newspaper bits, and plastic scraps.

Lost circulation aids are available in an assortment of grades, which are generally based on the maximum size of the largest component. They usually contain flexible, flat pieces, such as shown in Figure 5.17, and can be about 1/4–3/4 in. (6–19 mm) in size. Suppliers to the drilling industry usually provide these materials, and virtually all major oil well service firms employ them routinely.

5.8.2 Synthetic Fibers

Synthetic fibers are commonly included in concrete mixtures to reduce plastic shrinkage and provide secondary reinforcement. They are thus readily available and are stock items at many ready-mix plants and concrete material dealers. There are several types, including those made of steel as well as a variety of synthetic polymers. Of the synthetic fibers, polypropylene is the most common and is widely available, although several others, including nylon, polyethylene, and polyolefin, are made. They come in a wide variety of lengths, sizes, shapes, aspect ratios, and end configurations. For use in concrete, these parameters can be of some significance, but to act as blocking agents in grouting they become less important. Here, the only real considerations are pumpability and their effectiveness in forming a plug.

FIGURE 5.17 Commercially available lost circulation aids make fine blocking materials for grout.

For this purpose, the polypropylene fibers are entirely adequate, and they are certainly the most readily available in a variety of lengths, the longest of which is 2 in. (50 mm). Standard lengths up to 2-1/2 in. (63 mm) are manufactured; however, and any length can be obtained through special order. On first thought, the use of the longest practical length seems advantageous, but there are two contrary schools of thought on the length issue.

Generally speaking, it is recognized that on a pound-for-pound basis, the amount of fibers that can be incorporated in concrete, without excessively reducing its workability and pumpability, is dependent on their length. Short fibers can be included in larger amounts than long fibers. A given number of short fibers may overlap each other, resulting in a greater bridging ability than that provided by a smaller number of long fibers. Because the short fibers allow a greater dosage, the space between them in the mix will be less, which can have a greater blocking effect. In any event, the length of the fiber should not exceed the diameter of the delivery line through which it is to be pumped.

Polypropylene fibers are available in two configurations, *monofilament* and *fibrillated.* Monofilament fibers come as individual pieces of fiber that maintain their cross section throughout the mixing operation. Fibrillated fibers, however, di-

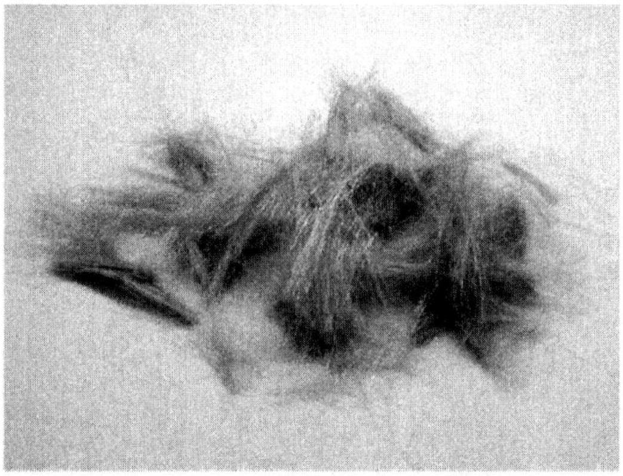

FIGURE 5.18 Fibrillated fibers form an interconnected random network.

vide into an interconnected, randomly attached network, as illustrated in Figure 5.18, usually with an individual fiber diameter not unlike that of a spider web. These are the mostly frequently used and are quite satisfactory as blocking agents.

For normal use in concrete, dosage rates are no more than about 0.05 percent of the volume, or about 7.5 lb per yd^3 (4.5 kg per m^3) of concrete, and often less than this quantity. When fibers are used as a blocking agent, however, it is desirable to include the highest practicable dosage that is still pumpable. Because this will vary with the length and type of fiber, the nature and proportions of the other mix ingredients, the injection line size, and the particular mixing and pumping equipment used, no single dosage is valid and it is best to determine the optimal quantity on the basis of the individual mix. The dosage is usually on the order of 1 to 3 percent, by volume, of the grout. The slump of the mix will tend to lower as the fiber quantity increases, so additional water will be required.

Synthetic fibers are usually packaged in convenient quantities, usually in cartons or bags of 10 lb (4.5 kg) or less. Some manufacturers package them in soluble bags that can simply be thrown into the mixer. Not only do these have the advantage of not requiring opening and emptying, but because the bags are consumed, no waste remains to be handled and disposed of.

5.8.3 Plastic Tabs and Punchings

A variety of plastic film pieces, such as tabs and punchings, as shown in Figure 5.19, are produced as a by-product of the manufacture of plastic bags and similar goods. These mix readily with grout and form excellent blocking agents, especially when they encounter networks of cracks or joints, such as in rock. Being composed of a synthetic, they do not appreciably detract from the durability of the grout. Because they are by-products, there is no established market in which to acquire them. Ready supplies have been found at little or no cost, however, directly from the producers of film products. As waste products, one must accept pieces of the thickness, size, shape, and configuration available at a particular time.

Fabricators of plastic film products will often custom produce small tabs of virtually any shape or size. Although this will result in a greater cost, uniformity of the product will be ensured. In addition, by having the pieces custom

FIGURE 5.19 Plastic film punchings provide a good blocking material.

made, the thickness and any other desired properties can be specified.

5.8.4 Feed Grains

Feed grains such as wheat and oats tend to swell as they absorb moisture from the grout mix water. This swelling does not occur immediately, but rather slowly over an extended time period, so that the grout does not reach its greatest volume for some time following the initial mixing. Feed grains are readily available, either in bulk or in burlap bags containing 50–100 lb (23–45 kg). They have a substantial disadvantage, in that they are organic and will decompose with time, so grout in which they are used will be of poor durability. Although feed grains were widely used in earlier times, the synthetic materials have largely replaced them.

CHAPTER 6

Cementitious Grout Mixtures

6.1 BLEED WATER

6.2 WATER-TO-CEMENT RATIO

6.3 MIX DESIGN

6.3.1 Cementitious Suspensions
6.3.2 Pourable Slurries and Pastes
6.3.3 Expansive and Nonshrink Grout
6.3.4 Limited-Mobility Grouts
6.3.5 Ready-Mixed Mortar and Concrete
6.3.6 Low-Density Cellular Mortar and Concrete
6.3.7 Low-Slump Limited-Mobility Grout
6.3.7.1 Compaction Grout

As previously discussed, by far the most widely used grouts are based upon cement. They can be formulated to have a broad range of consistencies varying from watery fluid to very stiff and mortar-like; they can consist of only cement and water or can include a variety of chemical admixtures, fillers, and blocking agents. Cementitious grouts can thus be formulated to fill a wide range of grouting requirements. Historically, however, simple cement-water combinations, often referred to as *neat* cement grout, have been the most frequently used. These continue to be widely employed, although modern technology allows compounding of far superior mixtures, usually containing a variety of modifying constituents.

Neat cement grout can be prepared with a nearly infinite number of different water : cement ratios. When the proportion of water to cement is less than about 0.4, the mix will be of a paste consistency; water : cement ratios greater than about 0.6 will result in viscous fluids. Obviously, the greater the ratio of water to cement, the lower the resulting viscosity will be. And as the water : cement ratio increases, the strength and durability of the resulting grout is reduced.

6.1 BLEED WATER

All fluid grouts are subject to sedimentation of the solids, known as *bleed,* in which clear water accumulates on the top surface. With time, the water will evaporate and the space it formerly occupied will become air filled, and thus a void. Bleed of suspension grouts has historically been, and continues to be, one of the greatest shortcomings of grouting practice. An arbitrary amount of 5 percent bleed has become fairly well established as the maximum reasonable amount, and grouts that experience greater quantities are referred to as *unstable.*

Although, historically, little effort to minimize bleed has been made in the United States, this has not been the case elsewhere. European grouters in particular have long promoted the use of *stable* suspension grouts. The stability of these grouts has typically been achieved through inclusion of a small amount of bentonite in the grout mix. With present grout formulation technology, stability can be rather easily achieved in a number of ways, usually involving chemical admixtures. Although advanced mix designs are becoming common throughout the world, the use of unstable suspension grouts, unfortunately, continues.

Application techniques are available that can reduce the amount of bleed water that develops. Foremost is quality mixing of the initial grout so that all of the particles are well dispersed. As discussed earlier, common cement is subject to high polar activity, which causes the individual grains to be attracted to one another and form clumps or agglomerations. Being heavier than individual grains, the clumps tend to settle out of suspension more readily, but this can be minimized through high shear mixing.

The amount of bleed that occurs will be directly proportional to the quantity of mix water in the beginning grout, as illustrated in Figure 6.1, reproduced here from Houlsby (1985). As can be observed, even thick suspensions, such as those with a water:cement ratio of 1, which are the thickest typically used in rock grouting, will promote considerable bleed. For thinner mixes, which are still commonly used, the amount of bleed water can be greater than the volume of the cured grout. For this reason it is imperative to formulate grouts with the least amount of water practicable for the individual application. Far better is to use stable grouts, which can be easily prepared using well-established mix design technology.

Bleed water that does accumulate can often be removed through a *bleeder* valve attached to the grout header (Figure 6.2). Because surplus water tends to gather at the top of a grout mass, especially when it is moving very slowly or at rest, bleeding is usually performed during the later periods of injection, when grout take is very slow. The bleed valve is opened while grout pressure is maintained. The fresh grout, being of higher density than the accumulated water and the dilute grout, will tend to descend, while the latter exits the bleeder valve.

Bleeding should continue until good-quality grout appears. This is normally done at intervals of about 15 minutes toward the end of injection. Unfortunately, bleeding is not very practical where packers which are discussed in Chapter 37 are being used, because of the generally small injection tube inherent to the process. It also complicates the injection and thus requires more competent personnel. And because the bleed occurs slowly, it requires more time to complete the work.

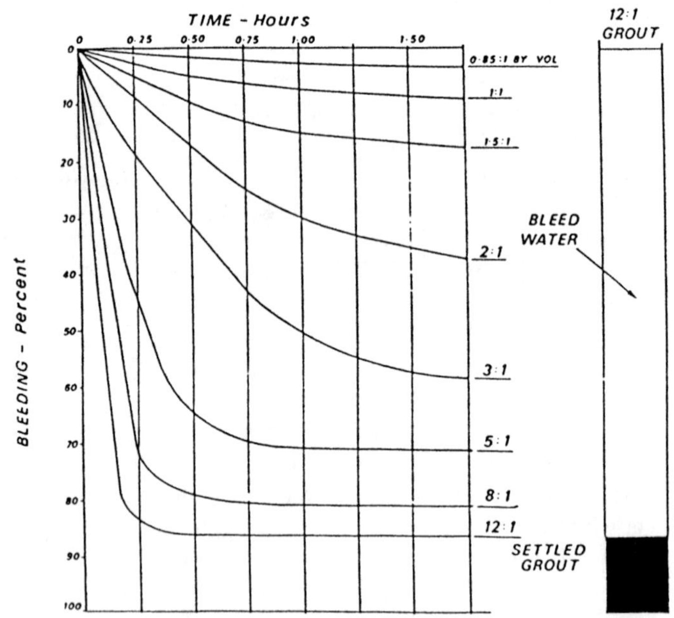

FIGURE 6.1 Percentage of bleed water rising from grout at different water:cement ratios. (From Houlsby, 1985.)

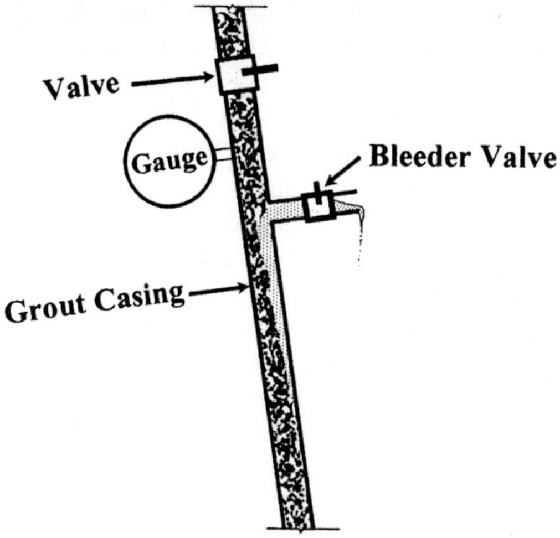

FIGURE 6.2 Bleeder valve on a grout header.

6.2 WATER-TO-CEMENT RATIO

The thickness of a grout mixture, and to a large degree its injectability, is dependent on the proportion of the fluid water to the dry cement. This is commonly referred to as the water-to-cement ratio, water: cement, or simply w:c. This seems simple enough, but, unfortunately, it is not so simple, as the amounts of both the water and the cement can be based on either weight or volume. To add more confusion, volume amounts can be either absolute or bulk, and should bulk volume be used, as is common in the United States, how much bulking is assumed? The water: cement ratio is a fundamental property in all cementitious mixtures, including not only those for grout, but those for both concrete and mortar as well. This important parameter dictates the consistency of the final mixture and, more important, its achievable strength and long-term durability. Standards for the water: cement ratio of concrete are well established and widely recognized. Unfortunately, the same cannot be said for typical grouts.

In the United States, a standard bag of cement contains 94 lb (42.7 kg) of mass and its bulk volume is considered to be 1 ft^3 (28.3 L). The absolute volume of that bag however, is only 0.48 ft^3 (13.6 L), and the precise amount of bulk volume can vary quite considerably. For example, soon after entering the grouting field, I placed a standard 94 lb (42.7 kg) bag of portland cement into a 1 ft^3 (28 L) container. The cement was poured slowly into the container so that it would result in the loosest packing. Interestingly, the container was completely filled before the bag was empty.

The cement surface settled, however, when the side of the container was tapped with a steel rod, such that the full 94 lb bag could be emptied into it. Continued tapping resulted in more settlement, and further investigation revealed that considerable variation in both specific gravity and bulk density of common cement is well recognized. A publication of the Portland Cement Association (Kosmatka and Panarese, 1990), reports that the bulk density of portland cement can vary from 52 to 103 lb (23.5 to 46.7 kg) per cubic foot. Thus, proportioning cementitious grout mixes by volume, with the assumed volume of a bag of cement at 1 ft^3 (28.3 L), is not very accurate at best.

The reason for the traditional use of volume proportioning is easy to understand, as in years past most grouting in the United States and some other countries was done with bagged cement. Assuming a volume of 1 ft^3 (28.3 L) for each bag was simple and a convenience on-site. Although it may still be convenient to use volume measurements, doing so can add great confusion when using cement packaged in other than 94 lb (42.7 kg) bags, which is common in many geographic areas outside the United States and sometimes even in this country.

During a shortage of domestic cement in the 1970s, imported cement was being used in a grouting project. It was packaged in 88 lb (40 kg) bags, and no adjustment had been made in the water used per bag. Thus, problems arose. More recently, in an investigation of a problem where both common portland cement and ultrafine cement grout were being used, it was found that two bags of ultrafine were substituted for one 94 lb (42.7 kg) bag of portland cement. The ultrafine cement was packaged in 20 kg (44 lb) bags, so two bags had a weight of 88 lb (40 kg). Again, the problem was batching by bags without adjusting for the specific gravity or volume of the bulk cement therein.

To more easily understand the difference in units, assume that a 94 lb (42.7 kg) bag of cement contains 1 ft^3 (28.4 L), as is common.

- To prepare a grout with a water : cement ratio of 1, by volume, would require 7.5 gal (28.4 L) or 62.4 lb (28.4 kg) of water, which is equal to 1 ft^3 (28.4 L).
- If the grout were proportioned by weight, 94 lb of water (42.7 kg), or 11.3 gal (43.7 L), would be required, which would raise the water : cement ratio, by volume, to 1.5.

The figures used here are very close to precise but are derived from values that have been rounded off. In this regard, there are conversions that are useful in grouting, also rounded off but certainly precise enough for the work:

Water
1 gal = 8.3 lbs = 3.8 L
1 ft^3 = 7.5 gal = 62.4 lb = 28.4 L

Portland cement (traditional bag)
1 bag = 94 lb = 42.7 kg = 1 ft^3 = 28.4 L (bulked)
= 0.48 ft^3 = 13.6 L (absolute)
Specific gravity of portland cement ~ 3.15

Conversion of w : c ratio
Weight to volume, multiply by 1.5
Volume to weight, multiply by 0.66

Common portland cement is available in both 47 and 94 lb (21.4 and 42.7 kg) bags. In addition, the use of both imported portland and ultrafine cement is now common, and these are virtually always packaged in masses other than 94 lb (42.7 kg). In fact, specialized cements such as the ultrafines, which are finding ever-increasing use in grouting, come in various size bags, and virtually no standard as to size exists. In recent research, six different, although commonly available, ultrafine cements were used. These were manufactured in the United States, Germany, and Japan, and the weight of the individual bags varied from 30 to 72.5 lb (13.6 to 33 kg). As if this did not produce enough confusion, the specific gravity of the cement itself varied from 2.9 to 3.3.

6.3 MIX DESIGN

Mix design and batching of grout by weight is not unusual. In fact, it is common in many countries of the world and is seeing increased use in the United States. In addition, virtually all concrete worldwide is proportioned by weight, and it is nearly always batched with bulk materials. With advances in grouting technology, the ecological impact, and the greater expense of supplying and disposing of paper bags, an ever-growing amount of grouting now employs bulk cement, which is, of course, batched by weight. Some old-timers and "granddaddy grouters" continue to prefer proportioning by volume, but the tendency is to convert fully to the use of weight or mass.

Cementitious grouts can be formulated to provide an endless range of different properties and are conveniently placed in the following categories:

- Suspensions
- Pourable paste or slurry
- Plastic consistency
- Low-density cellular paste or slurry
- No-slump mortarlike low mobility

Although, historically, most grouts were simply composed of cement and water, today innumerable admixtures, supplementary cementing materials, and fillers are used. It is not unusual to see a grout mix that contains several different ingredients. These may be added and combined while mixing on the job site, or the dry ingredients may be preblended and delivered to the job in bags or other containers. Whereas grouting has traditionally been dominated by simple mixtures with few ingredients, the tendency now is toward the use of much more sophisticated mix designs containing many components. For this reason, the use of preblended grout mixtures, to which only water or perhaps fluid admixtures need be added, is becoming more common. These are usually available in standard paper bags or in large bin bags, as illustrated in Figure 6.3, which are lifted with a forklift or small crane and emptied into the mixer.

6.3.1 Cementitious Suspensions

Cementitious suspensions were the very first grouts used and are still widely employed, especially for grouting in rock. In their simplest form, they are mixtures of cement and water without any other ingredients. Water: cement ratios can vary from about 0.3:1 by weight, which is about

FIGURE 6.3 Bulk bin bags of preblended grout mix.

TABLE 6.1 Conversion of Mix Proportions from Volume to Weight

w:c by Volume	Equals	w:c by Weight
0.4:1		0.26:1
0.5:1		0.33:1
0.6:1		0.40:1
0.8:1		0.53:1
1:1		0.66:1
2:1		1.32:1
3:1		1.98:1
4:1		2.64:1
5:1		3.30:1

the thickest that can be handled by equipment in common use, to more than 10:1, although the durability of unmodified mixtures with a w:c ratio greater than about 3:1 is questionable. Hereafter in this handbook all references to water: cement ratio are based on weight unless specifically stated otherwise. The conversion of the water: cement ratio from volume to weight for common grout mixes is shown in Table 6.1.

In the United States, grout mixtures as thin as 10:1 by volume have been used historically, in grouting rock for dam foundations. Many grouters consider such thin grouts to be little more than dirty water, and extensive testing and research (Houlsby, 1982; Weaver, 1991) has shown them to provide little durability.

Of particular significance is the work of Houlsby (1985), wherein large-scale tests were conducted to better understand the relative performance of different water: cement ratios, as well as the different mixing and injection equipment and techniques. In these tests, 3 ft (0.9 m) square concrete slabs were paired so as to provide a void opening of varying width. One edge of the slabs was clamped tight, with zero opening, whereas the opposite side was shimmed to provide a gap of either 1/16 or 1/8 in. (1.5 or

3 mm). This resulted in openings ranging from zero to either 1/16 or 1/8 in. (1.5 or 3 mm) in width. The slabs were securely bolted and clamped, together. A single injection hole was provided in the center of the top slab, such that injected grout could travel radially in all directions. Gaps of less than 1/16 in. (1.5 mm) were left open, but to prevent excessive leakage, any larger spaces were sealed with foam rubber. Air and water could, however, still escape through the smaller gaps. In all, 22 separate pairs were injected, using a variety of mixes, mixing, and pumping equipment.

A standard grout header, containing a bleeder valve, was secured to the top slab for injection. The gaps were first filled with water, which was allowed to drain off immediately prior to the grout injection in all but two of the tests (numbers 20 and 21). Those slabs were injected while moving water under a 15 ft (4.5 m) head traveled between them. Injection used full-scale field equipment and procedures. A pressure of 15 psi (1 bar) was used for the first 15 minutes, after which it was raised to 30 psi (2 bar). The pressure was maintained until 30 minutes after grout take ceased. After 24 hours, repeated injection was attempted on most of the specimens. The slabs that formed specimen 22 were separated and reused several times. This was done in an effort to assess the performance of grouts with different water: cement ratios in an identical void. The effort was not very successful, however, probably because of difficulty in reassembling with the same exact void width. A summary of the salient features of the test program and the individual test particulars are provided in Table 6.2.

The objectives were to evaluate:

- Effectiveness of different water: cement ratios
- Grout quality with paddle vs. high shear mixing
- The grout takes with piston vs. Moyno pumps
- Effect of thickening the mix during grouting
- Effect of multiple grout applications
- Influence of the roughness of the crack walls
- Influence of the crack size
- Effect of additives in the grout

After the grout had hardened, the slabs were separated and an assessment of the amount of filling was made. Because of considerable variation in the mode of filling and the final geometry of the grout, as illustrated in Figures 6.4 and 6.5, the estimates were somewhat subjective. They were, however, the result of careful visual inspection by three different engineers, including Houlsby. A summary of the evaluations, including the estimated degree of the filling, is provided in Table 6.3.

From the tests and observations of actual grout performance in curtains at several dams, Houlsby concluded that a water: cement ratio of 3 would provide the most complete filling of cracks less than 1/16 in. (1.5 mm) thick. Further, because grout with a water: cement ratio of 1 penetrated only about 1 ft (0.3 m), it was not satisfactory for acceptable filling of thin cracks. In cracks of greater thickness, however, he concluded that a water: cement ratio of 1, by volume, would be optimal. He also concluded that grouts thicker than those with a water: cement ratio of 3:1 by volume (which is about 2:1 by weight) were of good quality, whereas those with a water: cement ratio of 5:1 by volume (3.3:1 by weight) or greater were of questionable effectiveness.

A further conclusion was that use of high shear mixing and a Moyno pump resulted in better filling than use of a paddle mixer and a piston grout pump. Unfortunately, Houlsby did not use other combinations, such as a paddle mixer with a Moyno pump, or high shear mixing with a piston pump. Thus, it remains questionable whether the better performance was the result of superior mixing, or the steady output of the Moyno pump, or perhaps both, as he concluded. It is my experience that high shear mixing of common cementitious grouts does provide

TABLE 6.2 Injection Features of Test Program

Test No.	Gap Size	Starting Mix	Mixer	Pump	Comments	TAKE IN BAGS First	TAKE IN BAGS Second	TIME TO REFUSAL (MIN.) First	TIME TO REFUSAL (MIN.) Second
1	0–1/16	8:1	Paddle	Piston	Perimeter caulked	0.04	3.47	15	45
1-A	0–1/16	8:1	Paddle	Piston		2.15	0.01	60	15
1-B	0–1/16	8:1	High Shear	Moyno		1.32	0.01	15	15
2	0–1/16	5:1	Paddle	Piston		0.31	0.02	30	15
3	0–1/16	5:1	High Shear	Moyno	Roughened surfaces	2.46	0.02	30	15
4	0–1/16	3:1	Paddle	Piston		16.43	0.03	45	15
5	0–1/16	3:1	High Shear	Moyno		0.40	0.03	15	15
6	0–1/16	1:1	Paddle	Piston		0.20	0.07	15	15
7	0–1/16	1:1	High Shear	Moyno		0.20	0.07	15	15
8	0–1/8	8:1	Paddle	Piston		1.42	0.02	30	15
9	0–1/8	5:1	Paddle	Piston		13.05	0.02	90	15
10	0–1/8	3:1	Paddle	Piston		7.34	0.03	45	15
11	0–1/8	3:1	High Shear	Moyno		2.88	0.03	30	15
12	0–1/8	1:1	Paddle	Piston		117.88	0.08	210	15
13	0–1/8	1:1	High Shear	Moyno		11.46	0.07	30	15
14	0–1/16	1:1	High Shear	Moyno	Lignin water reducer used	0.47		15	
15	0–1/16	1:1	High Shear	Moyno	Lignin water reducer used	1.07		30	
16	0–1/16	1:1	High Shear	Moyno	Lignin water reducer used	0.40		15	
17	0–1/8	1:1	High Shear	Moyno		9.73		15	
18	0–1/16	5:1	Paddle	Piston		1.45		45	
19	0–1/8	5:1	High Shear	Moyno		61.13		240	
20	0–1/16	5:1	High Shear	Moyno	Water flow at 15 ft head	0.40		15	
21	0–1/16	3:1	High Shear	Moyno	Water flow at 15 ft head	0.46		15	
22	0–1/16	Various	High Shear	Moyno		Various			

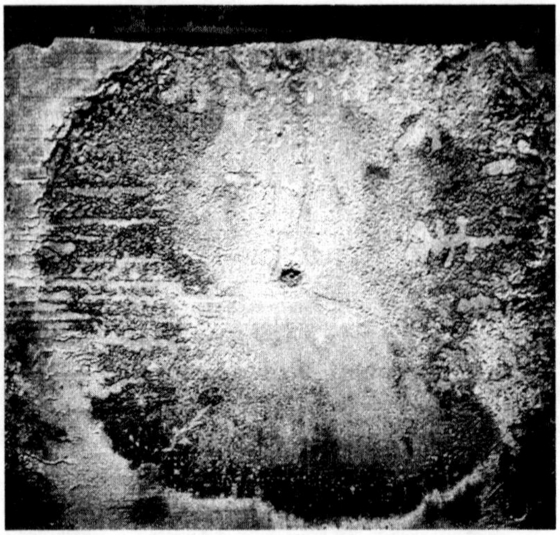

FIGURE 6.4 Test No. 4 grout dispersion.

superior grout and better penetrability. However, I question whether the steady flow of a Moyno pump provides better penetration than the pulsing flow of a piston pump. Although this has been the subject of considerable debate among grouters, documentation of valid comparisons are yet to be made.

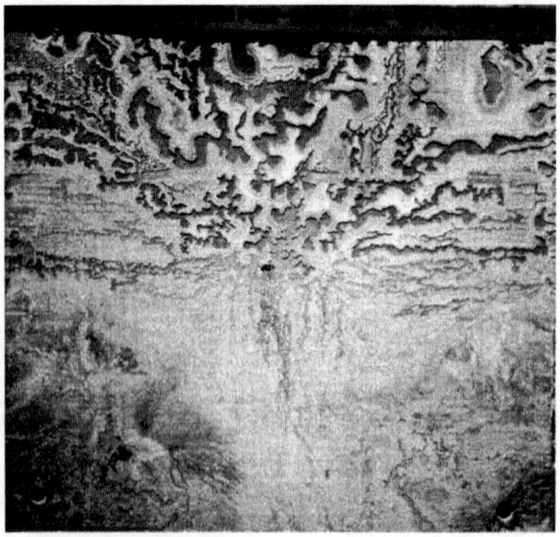

FIGURE 6.5 Test No. 5 grout dispersion.

Current thinking suggests that thicker mixtures (maximum w:c = 2:1 by weight) are far more appropriate for most applications. In part citing the results of the aforementioned research, Houlsby (1982) presented a well-documented discussion of water-to-cement ratios, concluding that a starting water:cement ratio of 2 by volume, which is 1.3 by weight, had been used on thousands of projects and was the optimal for general use.

Previously stated several times but so important that repetition is warranted, unmodified common cement-water suspensions suffer from excessive settlement of the solids and separation of mix water, as illustrated in Figure 6.6 which is a close-up of Figure 6.5. This subject was thoroughly discussed by (Kravetz 1959). The essence of his conclusions, as reported by (Houlsby 1990), are as follows:

1. Cement grains, when mixed with water, tend to aggregate and form clumps. This slows the wetting process, as does air attached to the grains. The effect of high-speed shearing or laminating, plus the centrifugal effect, is to thoroughly break up the clumps and to separate air bubbles. As a result, each individual grain is rapidly and thoroughly wetted and put into suspension.
2. During cement hydration, needle-like or springlike elements of hydrates form on the superficial layer of each wetted grain of cement. In a high-speed mixer, the laminating effect and high-speed rotation keeps breaking these hydrates away from the grain of cement, thus exposing new areas to the water and consequently bringing the formation of new elements. These hydrate elements are of colloidal size, and as the amount of these elements in the mixture increases, the grout becomes colloidal in character.

Colloidal particles are so small and lightweight that they will not settle, but will, rather,

TABLE 6.3 Evaluation of Test Program Effectiveness

	THINNER SIDE		THICKER SIDE		
Test No.	% of Void Volume Filled	% Area of Roof Contact	% of Void Volume Filled	% Area of Roof Contact	Comments
1	90	40	40	10	
1A	70	50	20	15	Soft powder on surface
1B	30	5	25	20	Soft powder on surface
2	50	5	20	5	Soft powder on surface
3	30	5	20	10	Much weak, soft powder
4	50	15	50	30	
5	90	10	80	15	
6	30	25	60	40	Partial penetration only
7	60	40	85	70	Excellent bond to upper slab
8	30	5	40	5	
9	60	5	60	5	
10	40	5	60	15	
11	15	5	20	10	
12	95	50	95	10	
13	30	25	75	25	
14	85	75	75	60	
15	85	75	75	60	
16	85	75	75	60	
17	80	25	75	15	
18	50	30	40	35	
19	70	30	85	75	
20	—	—	—	—	Mostly washed away
21	—	—	—	—	Mostly washed away
22	—	—	—	—	Purposely washed away

float in water. In reality, virtually no common cementitious suspension grout is truly colloidal, as even well-dispersed cement grains will settle unless appropriate admixtures are used. High shear mixing will greatly reduce bleeding, however. In some places, but generally not in the United States, a grout that exhibits little or no bleed is regarded as "stable." Such grouts are usually considered to gain no more than 5 percent bleed. European practice commonly calls for "stable" grouts, which typically include a few percent bentonite or other modifier to act as a suspension agent. Although bentonite will lessen cement grain settlement and the resulting bleed,

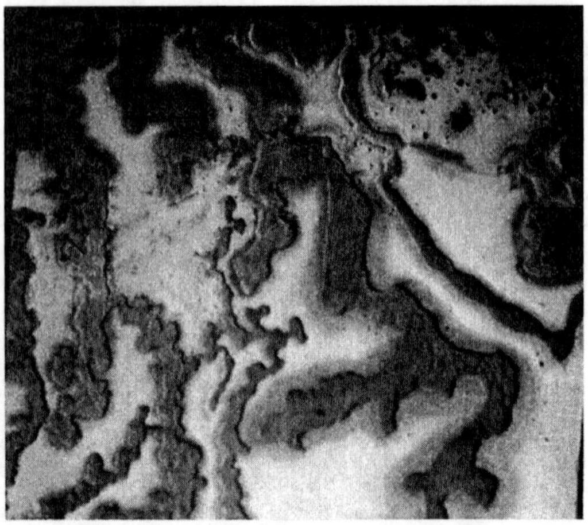

FIGURE 6.6 Closeup view of upper left Test No. 5.

there are significant disadvantages to its use. Modern admixture technology offers a variety of compounds that can minimize bleed without the undesirable aspects of bentonite.

In more recent times, there has been considerable discussion by a number of investigators relative to the filling of very fine rock fractures and soil pore spaces with particulate grouts. The discussion has been concerned mainly with mixtures containing ultrafine cement. The emphasis has been on theoretical ratios of the grout particle size to the width of a defect or the grain size, and thus the porosity, of a given soil into which a particular cement grout can be injected. There has not been a great deal of agreement as to the appropriate ratios, and especially whether they should be based on the mean or maximum particle size of the cement. Although grain size and particle dispersion have been emphasized, it has been recognized that though significant, these are not the only important parameters. The grain shape, surface condition, and polar strength, as well as resistance to pressure filtration, are equally important.

Suspension grouts can be injected under fairly high pressures, and pressure filtration—that is, loss of water from the solids due to the injection pressure—must not be allowed to occur. This is best prevented by minimizing the amount of water in the mix through the use of water-reducing admixtures in the grout, combined with high shear mixing to maximize dispersion, and uniform wetting of all the cement particles. Inclusion of viscosity-modifying admixtures will also reduce the susceptibility to pressure filtration; however, this may increase cohesion, thus requiring higher injection pressure.

The grain size distribution and maximum particle size are of varying importance, as previously discussed in Section 4.2.5.2. In filling fissures or joints in rock, the size of the largest particles is relatively unimportant, as long as there are not too many of them, they are smaller than the void to be penetrated, and are well disbursed. Conversely, when injection is into a granular mass, the size of the largest grains is of crucial importance, as these are the first to plug the soil pore structure in development of a filter cake, followed by progressively smaller particles until permeation all but stops.

Ultrafine cement suspension grouts are highly penetrable and can, in some cases, form nearly colloidal mixtures. These are seeing ever-increasing use for the solidification of sandy soils. Shimoda and Ohmori (1982) described the development of a new ultrafine cement in Japan and reviewed the laboratory research, as well as two significant application case histories. Suspension grouts, made with the then new cement, were injected into fine to medium sands.

Although fluid suspension grouts for soil and rock injection nearly always contain only neat cementitious materials and, perhaps, admixtures, such grouts used for the production of preplaced aggregate concrete and other structural grouting include fine sand. Sand and other fillers are also included in grouts used for the filling of unusually large joints or other defects in rock. Where included in suspension grouts, the proportion of sand is seldom more than about twice that of the cementitious material and is more typically no greater than equal to the proportion of the cement.

Suspension grouts all suffer from bleed, as the larger and heavier particles tend to settle out of the suspension. Therefore, the sands are usually significantly finer than those used in conventional concrete and mortar. Such sands are usually graded such that 100 percent will pass a No. 16 sieve, with a fineness modulus (FM) of less than about 2.0. The FM for sand commonly used in concrete and mortar ranges from about 2.3 to 3.1, with the higher number being the coarsest. Ideally, sand used in grout will contain ample fines and will preferably be uniformly graded all the way down to the No. 270 sieve size. It may even contain finer particles, which would be classified as silt. Unless a considerable proportion of the sand is finer than the No. 200 sieve size, the resulting grout will likely require admixtures and/or supplementary cementing materials to be resistant to sand separation.

6.3.2 Pourable Slurries and Pastes

As grouts of 0.5 water : cement ratio are about the thickest that can be readily mixed with the mixers commonly used in grouting, this is considered the lower end of pourable cementitious suspensions. Grouts with lower water : cement ratios result in a mixture with a paste-like consistency, generally referred to as a slurry. Slurry consistency grouts are sometimes used for filling voids within a range of about 0.3–3 in. (7.6–76 mm) in both geotechnical and structural grouting, as well as for the creation and filling of controlled fractures in soil. They have also been used for filling large voids, such as abandoned mines (although in my opinion, this is an inappropriate use).

Slurries are generally too thick to be mixed in commonly available high shear mixers. They are thus most often prepared in horizontal shaft paddle mixers, such as those used to mix plaster and mortar or in ready-mix trucks. Because these mixers do not provide high shear mixing, the individual particles are often not well disbursed. This can increase their tendency to settle within the mix and the grout's propensity to bleed. Slurry consistency grouts are frequently filled with fine sand or other fillers. Lightweight cellular mixtures and controlled low-strength materials are often mixed to a slurry consistency.

6.3.3 Expansive and Nonshrink Grout

Structural grouting applications often call for grout that will be completely free of shrinkage, and there are many suppliers of prepackaged grout mixtures marketed to provide this characteristic. In the real world in which we must perform, however, it is virtually impossible to formulate a cementitious composition that has the necessary flowability to be pumped and yet will not undergo at least some shrinkage. The closest we can come to a truly nonshrink material is to formulate a mix that will expand upon curing. Admixtures that promote expansion, as already discussed, as well as preblended expansive grout mixtures, are available. Even these will shrink, however, upon final drying, but after expansion has occurred. If everything works as planned, the final drying shrinkage will be slightly less than the prior expansion so as to reach the original as-cast volume, but this seldom occurs in the real world of grouting.

In reporting on the evaluation of volume change of 46 different preblended repair mortars, several of which were promoted to be *nonshrink* or shrinkage *compensated*, Carter and Gurjar (1987) found that only 7 experienced shrinkage that would be considered less than that of conventional portland cement mortar, and 1 of those experienced permanent expansion, as illustrated in Figure 6.7. None attained their original volume upon drying. Standard

ASTM C 157 length change tests were conducted on the different materials by a single commercial testing laboratory.

In a prior effort in the 1960s, I evaluated some 16 different grout mixes, all of which were designed to be expansive or nonshrinking. Six separate preblended products from four manufacturers were included. The rest were experimental mix designs that contained chemical admixtures to limit or eliminate shrinkage and/or promote expansion. In this test, two heavy steel boxes, 4 ft (1.2 m) square and equipped with removable top plates, were used. One was 3 in. (76 mm) and the other 12 in. (305 mm) deep.

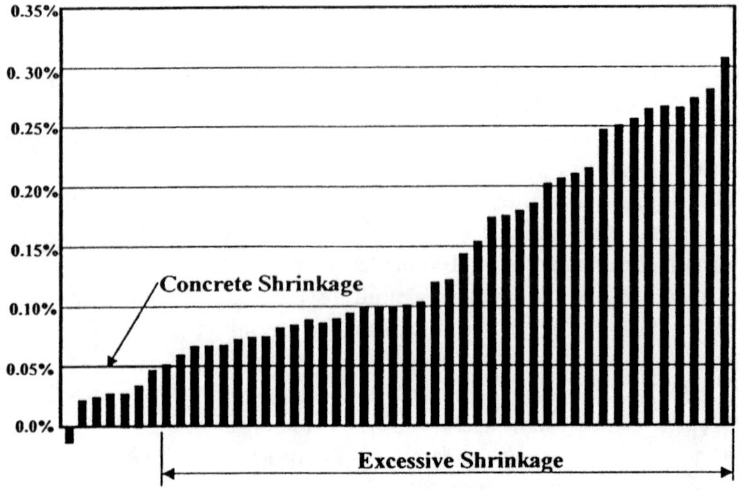

FIGURE 6.7 Length change of specimens prepared with 46 different commercially available cementitious mixes.

Each was filled with a particular grout, injected into one corner at the bottom, using full-scale grout mixing and pumping equipment, as shown in Figure 6.8. Appropriate vent holes were provided, and the grout progression, as well as the pumping pressure, was carefully monitored in accordance with the best practice that would be expected in the field. Once a box was filled, a positive pressure of 10 psi (0.69 bars) was maintained for ten minutes. Because of the constant expulsion of bleed water, occasional pump cycling was required to maintain the 10 psi (0.69 bar) pressure level.

When the grout hardened, the top plates were removed, exposing the grout surface for visual inspection, as shown in Figure 6.9. No specimens were completely free of at least some voiding under the plate. As part of the evaluation, a white powder (casting plaster) was rubbed into the voids to highlight them, and a string grid placed over the specimen to facilitate determination of the extent of void area, as illustrated in Figure 6.10. It varied from a minimum of about 2 percent to more than 90 percent of the total area. There was, however, wide variation in size, depth, and configuration of the voids among the different grout mixtures.

Both core specimens and saw-cut cubes were removed for strength evaluation and further testing. The strength and, presumably, durability of the hardened grout in the top of the

FIGURE 6.8 Test injections of grout into enclosed steel boxes.

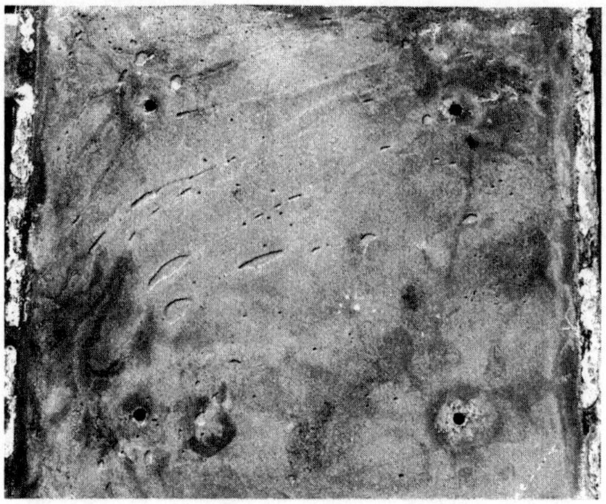

FIGURE 6.9 Top plates removed to allow visual inspection of any remaining voids.

12 in. (305 mm) deep box was found to be significantly lower than that in the bottom. Regardless of the shrinkage minimizing effort or material used, none had nearly as great an effect as did minimizing mix water and water : cement ratio. In fact, the very best performance resulted from several mixes that contained a massive dosage (12 oz per bag of cement, which was about three times the manufacturer's recommendations) of a lignin-type water-reducing admixture, in combination with either methyl cellulose or short-fiber asbestos.

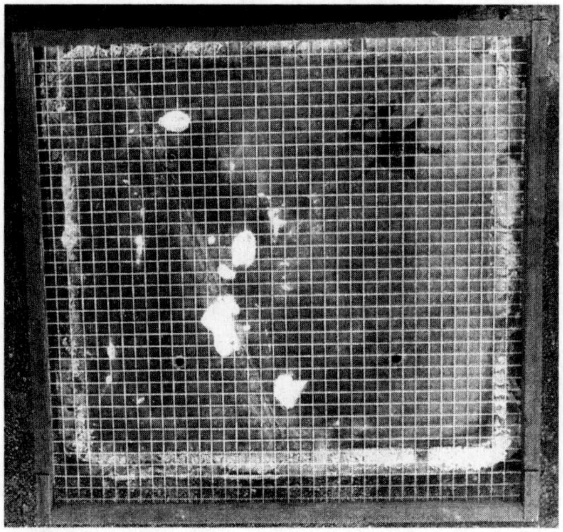

FIGURE 6.10 A string grid aids enumeration of void areas.

Lignin water reducers are based on purified or refined salts of lignosulfonic acid. These are a by-product of the pulp and paper industry, and as such the chemistry is not well controlled. They all contain at least some sugar and thus retard the set and hardening time. As with most admixtures today, they are commonly supplied as a liquid to simplify addition to the mixer. The particular admix used in the tests, however, was obtained in a powder form and classified as low sugar (LS). Although it was more expensive than common lignin products, it resulted in less set retardation and proved to be well worth the additional cost. Nonetheless, the initial setting time was greatly delayed and the rate of strength gain was quite slow, especially at low temperatures.

Although the performance of grouts containing gas-producing expansive admixtures appeared generally good in relatively small test cylinders, their overall performance in the larger volume was very poor. A large amount of surface void area resulted, with voids in excess of 1 in. (25 mm) deep. Performance was particularly poor in grouts with the higher water : cement ratio, and for this reason I generally avoid their use.

Three different mix designs emerged from this research and were successfully used on literally hundreds of projects. It should be noted, however, that the setting and hardening times are greatly affected by temperature. The mix designs and other particulars of the mixtures are shown in Table 6.4. The strengths shown are those obtained at average temperatures above 70° F (21° C) during casting and curing of the grouts.

The sands were kiln-dried products that had been processed with some precision to their particular gradation and were commercially available from a major aggregate supplier in the Los Angeles area. They were supplied in 100 lb

TABLE 6.4 Mix Designs for Low-Shrinkage Grouts

	Mix No. 1	Mix No. 2	Mix No. 3
TYPE I CEMENT	94 lb (42.7 kg)	94 lb (42.7 kg)	94 lb (42.7 kg)
S 8-100 SAND	300 lb (136.2 kg)		
S 16-50 SAND		200 lb (90.8 kg)	
SN 20			100 lb (45.4 kg)
METHYL CELLULOSE	1.5 lb (0.68 kg)	1.25 lb (0.57 kg)	1 lb (0.45 kg)
LS LIGNIN WATER REDUCER	12 oz (0.34 kg)	12 oz (0.34 kg)	12 oz (0.34 kg)
WATER	7 gal (26.5 l)	4.5 gal (17 l)	3 gal (11.4 l)
STRENGTH AT 28 DAYS	1630 psi (11.2 MPa)	2200 psi (15.2) MPa)	3150 psi (21.7 MPa)
STRENGTH AT 56 DAYS	3140 psi (21.6 MPa)	4180 psi (28.8 MPa)	7000 psi (48.3 MPa)

(45 kg) paper bags, which enabled easy and convenient batching. The specific gradation and other properties are shown in Table 6.5.

The research was performed in the 1960s, admittedly a long time ago, but the importance of minimizing the water content of grout has not changed. Modern water reducers and viscosity-modifying admixtures will result in equivalent or even better properties. With the present highly advanced technology for water reduction, as well as control of the viscosity of the liquid phase of grout mixtures, composition of grout of very low shrinkage is practicable and not at all difficult. Clearly, minimizing the amount of water in all cementitious grouts is the best way to ensure good strength and durability. Although important in all grouts, it is of particular significance for those grouts compounded for structural applications. Because of the variation of cement and other components, where low shrinkage and high strength and durability are required, experimental batches should be made

TABLE 6.5 Gradation and Properties of Sand

Sieve Designation, U.S. Standard Square Mesh	CUMULATIVE PERCENTAGE, BY WEIGHT, PASSING		
	S 8-100	S 16-50	SN20
4	100		
8	98	100	100
16	88	95	99
30	64	60	90
50	27	25	44
100	6	5	13
200	3	2	4
Fineness modulus	2.17	1.85	1.52
pH	8	8	8

and evaluated. These should employ the exact materials that will be used in production work.

6.3.4 Limited Mobility Grouts

Historically, all grouts were pourable fluids, and even to the present time it is not uncommon to hear a grouter observe, "If it isn't pourable, it isn't pumpable." Such opinions not withstanding, plastic consistency grouts that are of limited mobility have been effectively employed in a wide range of grouting projects for more than 50 years. Thick grouts see extensive use in the filling of massive voids, as well as in many structural applications. Not all limited-mobility grouts behave alike, however. Even though they may be very thick and of low slump, they can behave as fluids if injected into soil, resulting in hydraulic fracturing of the formation. This is especially likely with concrete and mortar mixtures containing clay, some pumping aids, and water-retaining admixtures. With the exception of mixes used for compaction grouting, plastic consistency grouts can be batched and mixed on the job site or batched at a ready-mix plant if large quantities are required.

Compaction grouting, which sees extensive use for strengthening soil through densification, requires grout that will behave as a growing solid, rather than a liquid, in soil. It must be able to withstand high sustained pressure and possess sufficient internal friction so as not to hydraulically fracture the formation. These requirements can be met only through the use of an appropriately graded aggregate that contains considerable silt size material, as discussed further in Section 6.3.7.1.

6.3.5 Ready-Mixed Mortar and Concrete

Where massive voids require filling, they are best filled as rapidly as possible with the most economical material available. This often is simply ready-mixed concrete, which is, obviously, delivered in ready-mix trucks and is widely available from conventional batch plants. One must, however, consider the different requirements of conventional concrete and grout, which limits the available component adjustments to provide the desired properties. And these adjustments must be doable within the normal workings of a conventional concrete batch plant.

The objective of normal concrete mix design is to obtain the required strength at the least cost, which means minimizing the cement content, along with using the largest-size aggregate that can be handled and including the maximum practical amount thereof. The reason for this practice is that the aggregate is relatively inexpensive as compared with cement, and in the mix it acts only as a filler and so will not contribute to shrinkage. As strength always rules in concrete production, plants are set up to use high-quality clean sands and large aggregates. They also have a limited number of silos and bins for the supply of mix ingredients, so it is often not possible to batch mixtures that contain nonstandard materials. Unfortunately, this restriction can conflict with the needs of grouting work, wherein flow properties and pumpability are paramount and strength is often not very important.

The pumpability and flow of a composition can decrease as the size of the aggregate increases. Further, grout mixes usually benefit by the inclusion of ample fines (that is, smaller than No. 200 sieve size) in the sand; however, this is not likely available at a batch plant, because it would be unacceptable in most concrete. Further, truck mixers depend on the large aggregate's rolling off the mixing fins, carrying the mortar and contributing to the mixing. Accordingly, their mixing efficiency goes down as the mean size of the ingredients decreases, and this is particularly so in mixtures with a high fines content.

There are, fortunately, many things that can be done within the confines of such sources to enhance the properties of a grout. As detailed in Chapter 5, the grain size distribution of sand for concrete is called out in ASTM Standard Specification C 33 and is provided in Table 5.8. Such sand is generally quite coarse, and the largest grains often approach 1/4 in. (6 mm). Fines that pass a No. 200 sieve are commonly limited to no more than 5 percent. As shown in Table 5.8, large variation in the grain size distribution of concrete sand is permitted. The C 33 standard allows variation of the FM (fineness modulus) all the way from 3.1 down to 2.3.

Large aggregate lowers the cost of the grout, but if it is used, it should usually be limited in both size and quantity in order to avoid an unacceptable loss of flowability. For most work, aggregate no larger than 1/2 in. (13 mm) is preferable and its proportion should be limited to no more than about 35 percent of the weight of the sand. If still larger aggregate is desired, its size should be additional to the $<1/2$ in. (13 mm) size and normally no larger than 1 in. (25 mm). The aggregate in a grout mix should be well graded from the largest to the smallest particles, and thus where larger-size aggregates are used, they should be in addition to the intermediate $<1/2$ in. (13 mm) portion and no more than about 40 percent of the weight.

The shape of the aggregate is also important, as it affects the pumpability of the grout and thus the amount of aggregate that can be used. Most natural aggregates are round grained, as shown Figure 6.11, whereas crushed aggregates (Figure 6.12) are generally angular. Compositions made with angular aggregates are appreciably more difficult to pump, and this becomes especially problematic when a considerable portion of the aggregate consists of relatively long, thin particles. Common criteria for mix proportions are based on average-shaped aggregates, which do not contain an appreciable amount of elongated flat particles.

FIGURE 6.11 Natural round-grained gravel.

These are often combinations of crushed and natural stone. Should the shape of large aggregate be angular, the proportions might best be reduced, whereas they can be increased in the case of natural, rounded gravel.

The mix rheology is also affected by the overall fineness of the sand. Greater amounts of large aggregate can be used when the sand is well graded, with an ample portion on the fine side (fineness modulus of less than about 2.5). The presence of large aggregate has relatively little influence on the strength obtained by the grout, as it is the cement paste and paste-sand mortar that prevail. Table 6.6 provides guidance as to the quantities of cement, sand, and water required

FIGURE 6.12 Angular crushed aggregate.

TABLE 6.6 Mix Quantities for 1 yd^3 of 4 in. Slump Cement-Sand Mortar, Containing Standard Concrete Sand

	CEMENT			CONCRETE SAND SSD[a]		WATER			28-DAY STRENGTH	
Mix	**Bags**	**lb**	**kg**	**lb**	**kg**	**lb**	**kg**	**W:C Ratio**	**psi**	**MPa**
1:1	17.9	1683	765	1794	815.5	478.9	217.7	0.28	7100	48.9
1:1.5	14.2	1335	607	2120	963.6	468.1	212.8	0.35	6100	42.0
1:1.75	12.8	1203	547	2235	1015.9	467.3	212.4	0.39	5600	38.6
1:2	11.7	1100	242	2334	1015.5	463.1	210.5	0.42	5200	35.8
1:2.25	10.8	1015	461	2417	1098.6	459.8	209.0	0.45	4900	33.8
1:2.5	10.0	940	291	2485	1129.5	458.1	208.2	0.49	4500	31.0
1:2.75	9.3	874	397	2536	1152.7	460.7	209.4	0.53	4050	27.9
1:3	8.7	818	372	2574	1170.0	464.0	210.9	0.57	3700	25.5
1:3.5	7.6	714	325	2626	1193.6	477.3	217.0	0.67	2900	20.0
1:4	6.7	630	286	2661	1209.5	491.4	223.4	0.78	2250	15.5
1:4.5	6.0	564	256	2676	1216.4	506.3	230.1	0.90	1700	11.7
1:5	5.4	508	231	2692	1223.6	518.8	235.8	1.02	1200	8.3
1:6	4.6	432	196	2715	1234.1	533.7	242.6	1.24	800	5.5
1:7	4.0	376	171	2747	1248.6	540.3	245.6	1.41	450	3.1
1:10	2.8	263	120	2797	1271.4	557.8	253.5	2.12	150	0.9

[a] Saturated surface dry

for a variety of mortar mixtures and the approximate strength that can be expected from each. These data are based on the aggregates that were common in Southern California in the 1970s and conformed with ASTM C 33. Aggregates from other areas should provide similar results, subject only to their conformance to ASTM Standard C 33. The water content shown will result in a grout with about a 4 in. (100 mm) slump, using the ASTM C 143 test. The particulars of this test are discussed later in Chapter 8.

The flow properties and pumpability of grout are also significantly influenced by the type and proportion of the cementitious material used. As cement is typically the most costly component, it is usually advantageous to limit its quantity. A minimum amount is required, however, to provide the necessary flow properties and achieve the required strength. Most concrete batch plants have on hand either a combination portland-pozzolan cement or separate fly ash or natural pozzolan.

Because the particles of the supplementary cementing materials are spherical, they lower friction within the mass and greatly improve flow properties. These materials are most often used in a proportion of 20–35 percent of the weight of the cement, although they can be included in a proportion up to the weight of the cement or even more. Their inclusion will provide significant improvement in both pumpability and the flow of the grout. Be aware, however, that although pozzolanic materials contribute to the hydration reaction, their contribution is much slower than that of portland cement, so that strength gain takes longer than it would with an equivalent amount of cement.

One of the most useful amendments for the grouter is an antiwashout admixture, which will make the mix very cohesive and resistant to separation from running water. Unfortunately, this is normally used only in concrete that is to be placed under water and thus is not often on hand at ready-mix plants, but requires advance ordering. If water-control grouting is being done in an emergency, with insufficient lead time to obtain the admixture, the inclusion of fly ash or a natural pozzolan, though not nearly as powerful as an antiwashout admixture, will nonetheless provide some additional cohesion. Silica fume, which is sometimes on hand at batch plants, will provide even greater cohesion, although it tends to make the mix quite sticky. Because of the stickiness it imparts, silica fume is usually not used in proportions greater than 12–15 percent, by weight, of the cement.

Where either strength or durability is an important criterion, the amount of mix water should be minimized. This is facilitated through use of water-reducing admixtures. Because they are frequently used in concrete, they are commonly on hand at ready-mix plants. Set accelerators and retarders are also normally on hand. Air-entraining admixtures are widely used, especially in cold climates, and are usually available at ready-mix plants. They generate tiny air bubbles in the mixed grout and can significantly improve its pumpability. These have perhaps the lowest cost of all admixtures, and it is often possible, without sacrificing pumpability, to replace up to a bag of cement per cubic yard of grout, with about 10 percent entrained air in the mixture.

6.3.6 Low-Density Cellular Mortar and Concrete

As previously mentioned, low-density air-foamed grout, which is often referred to as *controlled low-strength material* (CLSM), is available in many different controlled densities and can be as light as about 25 lb per ft^3 (400 kg/m^3). Such compositions can be very flowable and nearly self-leveling, as shown in Figure 5.15. These properties render CSLM mixtures especially useful where long linear voids must be filled. These materials also find extensive use as lightweight backfill of all types.

The strength of hardened cellular grout is variable and dependent on the original starting mix and resulting density. Initially, the mix can be either neat cement and water or a sanded grout and can include pozzolans. The resulting fill of the lower-density mixes can be readily excavated in the future if desired. Because a substantial portion of the final material is air, the quantity of solid components required to produce a given amount of grout is greatly reduced. This can be a substantial advantage in remote areas and in situations where working space is limited or material transportation costs are high. Typical mixes, including their approximate weights and strengths, are delineated in Table 6.7.

There are two types of foam-generating substances, the first being an admixture that is usually in a dry form and added directly to the mixer after the other constituents of the grout have first been blended. This can produce grout with an air content of up to about 35 percent. Because it is formed by simple addition of the dry foam-generating admixture directly into the mixer, it is easy to use with virtually any size batch-type mixer. It is, however, the more expensive option.

The other is a highly concentrated, usually liquid, foaming agent that is first diluted with water and combined with compressed air by means of a foam generator (Figure 6.13) to form a very stable prefoam (Figure 6.14). This prefoam is carefully introduced into the mixer after the other ingredients have been premixed. Because of the two-step development and introduction of this type of foaming agent, and the need for a foam generator, it is not very suitable

TABLE 6.7 Typical Density and Strength of Cellular Grout Mixtures

Type Mix	Mix	DRY DENSITY		CEMENT		Water:Cement by Weight	28-DAY STRENGTH	
		lb/ft³	kg/m³	lb/yd³	kg/m³		psi	MPa
	—	39	625	884	524	0.57	350	2.41
Neat Cement	—	34	544	790	468	0.56	210	1.45
	—	28	448	668	396	0.57	130	0.896
	—	23	368	535	317	0.65	50	0.346
Cement-Sand	1:1	58	928	724	429	0.40	460	3.17
	1:2	78	1248	630	374	0.41	820	5.6
	1:3	100	1600	602	357	0.51	2190	15.0

for small batches of grout and is thus most often used in large-volume applications. Ready-mix trucks and other high-capacity mixers are commonly used when a considerable amount of such material is required. With this system, it is possible to produce foamed grouts composed of as much as 70 percent air, with densities as low as about 25 lb per ft³ (400 kg per m³), as shown in Table 6.7. As with all cementitious compositions, strength is directly linked to density, so it will be noted that very low density compositions develop only modest strengths.

FIGURE 6.13 Foam generator used to develop prefoam, which will be fed into a mixer.

6.3.7 Low-Slump Limited-Mobility Grout

Low-mobility grouts were first used in the United States in the mid-1950s in the then-developing technology for structural grouting. Other low-slump, limited-mobility grouts have found extensive use in the filling of solution cavities in rock, sinkholes, and other large voids, as well as in constructing displacement piles. Although such applications do require grout of low slump that will not readily travel far from its point of deposition, the more rigid requirements for use in compaction grouting are not applicable.

FIGURE 6.14 Prefoam being placed in a mixer.

Limited-mobility grouts of the same slump are not necessarily equal in mobility under pressure, either in the delivery system or in the ground. Whereas slump alone may be a sufficiently accurate measure for conventional concrete or mortar, and such mixes are quite satisfactory for fill grouting of large spaces, the mix requirements for compaction grouting are much more demanding. There is an enormous amount confusion and misunderstanding within the grouting profession about this fact, and it cannot be overstressed.

Adding to the confusion, many technical papers have been published describing the use of *compaction grouting* where no compaction occurred and, in some cases, where the grouting was not even performed in soil. Graf (1992) lamented, "Much work that is called 'compaction grouting' by specialty contractors does not meet the current basic definition." In an effort to correct this situation, Byle (1997) proposed the term *limited-mobility displacement* (LMD) *grout* to apply to all low-mobility grouts.

Other than the LMD grouts used in compaction grouting, most are supplied in ready-mix trucks and used for filling-type applications. They are generally concrete or mortar mixtures, not much different from those discussed in Section 6.3.4, except for being of low slump. However, in order to make them pumpable, additives are used to provide water retention and lubricity. These may be in the form of pumping aids, viscosity modifiers/antiwashout admixtures, bentonite, or pozzolanic material. Concrete mix design is highly developed, and established ready-mix suppliers should be able to provide appropriate mix designs for such use.

6.3.7.1 COMPACTION GROUT

Compaction grouting involves the in-place controlled densification of soil at depth. It requires a grout mixture that will behave as a growing solid, under pressure, in the soil to be improved. When an appropriate mix is used, a reasonably regular mass of grout, as illustrated in Figures 2.6 and 2.10, will result, and this will provide reasonably uniform compaction. The importance of the grout mix rheology, based on the observed shape of numerous exhumed injections, has been made clear in many publications, the most complete of which are Warner (1992) and Warner et al. (1992).

Performance requirements dictate the use of a rather narrow range of acceptable grout compositions. Satisfactory mixes require the use of a closely controlled aggregate gradation, as shown in the acceptable gradation envelope displayed in Figure 6.15. As can be seen, although a considerable amount of fines, including silt-size particles, are required to provide the necessary pumpability, coarse sand and, preferably, small gravel are also required. This provides internal friction to control the grout behavior as a growing solid. As with all pumped mixes, round-shaped aggregate is highly preferable.

Although not necessary, compaction grout normally contains between 6 and 12 percent cement mixed with the aggregate, and just enough water to form an essentially zero-slump mixture. Because injection is made at fairly low pumping rates and required quantities are usually not predictable, compaction grout is nearly always mixed on-site. This is most often accomplished with auger-type mixers, which are amply discussed in Chapter 33.

To provide greater economy, some grouters attempt to use ready-mixed grout for this work; however this is not advisable. Although it is possible to mix very stiff compositions in ready-mix trucks, as discussed earlier, such compositions seldom perform as required in the soil, precluding controlled densification. Concrete batch plants have no normal use for the special aggregates that are required, and they are thus not readily available. Moreover, because of the high fines content, these aggregates tend to pack and to not flow well in the weigh hoppers and bins in such plants. The fines also tend to make

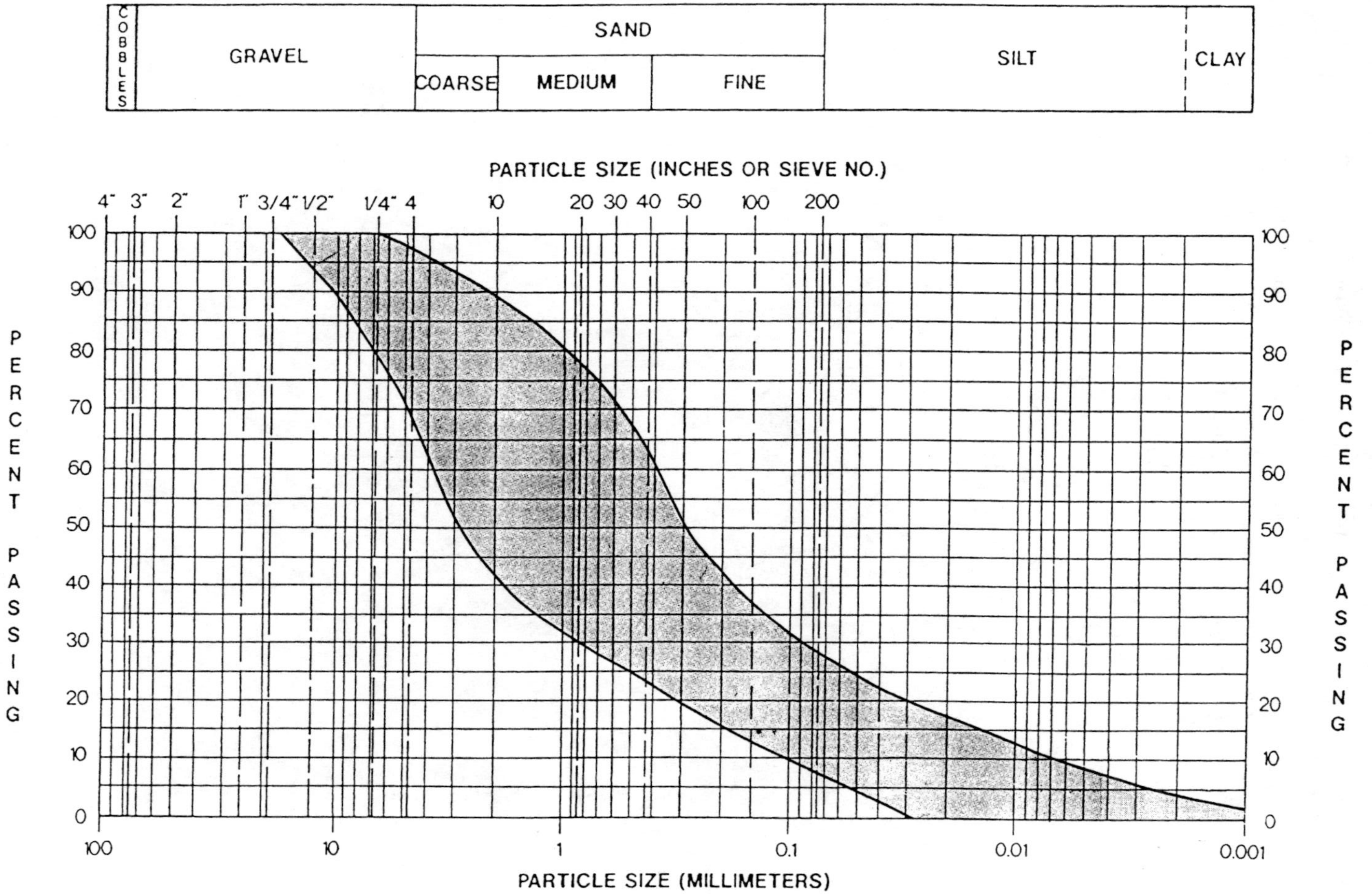

FIGURE 6.15 Envelope of aggregate gradation preferred for compaction grout.

the final grout mixture sticky, so that it is not readily mixed in a truck mixer. Some significant compaction grouting projects employing ready-mixed grout have been reported in the literature. This is unfortunate, as follow-up reveals that many have failed to perform as intended, which casts a negative light on the procedure, rather than directing attention to the faulty mixtures and injection procedures that were used.

CHAPTER 7

Noncementitious Grouts

7.1 CHEMICAL SOLUTION GROUTS

7.2 CHEMICAL SOLUTION GROUTS FOR STRENGTHENING
- **7.2.1** Sodium Silicate
- **7.2.2** Sodium Silicate Reactant Systems
  - **7.2.2.1** Group 1 Reactants
  - **7.2.2.2** Group 2 Reactants
  - **7.2.2.3** Group 3 Reactants
- **7.2.3** Durability of Chemical Solution Grouts
  - **7.2.3.1** Effects of Continuous Loading

7.3 CHEMICAL SOLUTION GROUTS FOR WATER CONTROL
- **7.3.1** Polymers
- **7.3.2** Polyurethane Grouts
  - **7.3.2.1** Hydrophobic Urethanes
  - **7.3.2.2** Hydrophilic Urethanes
  - **7.3.2.3** Chemical Origin
- **7.3.3** Water-Soluble Acrylic Polymers
  - **7.3.3.1** Acrylamide Grouts
  - **7.3.3.2** Acrylate Grout
  - **7.3.3.3** N-Methylolacrylamide Grouts

7.4 RESINOUS GROUTS
- **7.4.1** Epoxies
- **7.4.2** Urethanes
- **7.4.3** Phenolic Foams
- **7.4.4** Polyesters

7.5 MISCELLANEOUS GROUT MATERIALS
- **7.5.1** Asphalts/Bitumen
  - **7.5.1.1** Hot Asphalt
  - **7.5.1.2** Asphalt Emulsion
- **7.5.2** Clay Grouts

ALTHOUGH CEMENTITIOUS GROUTS are by far the most widely used grouts, there are a number of grouting objectives for which they are not the best and many for which they are simply not suitable. This is particularly so for water control and structural grouting. Very fast gel times and the ability to resist dilution or washout are common requirements for water control, especially where moving water is involved. High bond strength is often required in structural grouting. For these applications, noncementitious grouts will provide superior results.

7.1 CHEMICAL SOLUTION GROUTS

Chemical solution grouts are used to effect two different, but equally important, functions. The first is stoppage of flowing water through soil, rock, or structures wherein resistance to dilution and rapid setting are often requisite. In addition, in seepage control, especially in connection with structures, some flexibility of the cured grout is advantageous to accommodate slight movements in the host structure as a result of temperature variations and other events.

Another large area of use is solidification of

soils in order to increase their strength, which is accomplished, essentially, by gluing the grains together. In such applications it is usually most beneficial to obtain a hard, rigid cured gel. Thus, the same type of solution grout is seldom optimal for both uses, although it is not uncommon for special interests to promote their products as equally applicable to all potential situations. In the wide world of grouting, when dealing with noncementitious grouts, it is important to recognize that there simply is no one formulation that is equally applicable to all applications.

Chemical solution grouts are composed of two or more components, which are either premixed and pumped into the formation as a single solution, or stream mixed, whereby the various components come together and are blended as they enter the grout hole. On rare occasions a static or dynamic mixing head is used to thoroughly blend the components. At some point in time, usually predictable, the fluid will set in a gel or foam. Such substances may be hard and rigid or, depending on the formulation, soft and flexible.

In earlier times separate injections of the components were made, with the intention of mixing them in the ground. This was done to accommodate very rapid reaction times, which precluded prior mixing. Lack of complete blending, however, often resulted in unreacted chemicals and less than satisfactory results. Virtually all chemical solution grouts presently used are premixed prior to injection.

Some solution grouts are thin, waterlike fluids, and others can have considerable resistance to flow. Most grout manufacturers and many grouting professionals refer to the *viscosity* of a particular grout, implying that injectability is related to only this property. As discussed earlier in Section 4.2.5, this is not correct, as viscosity is only one property that contributes to a grout's injectability. Surface tension and affinity to the formation are every bit as important and must also be considered.

The injectability properties of a proposed solution grout must be related to the grain size, and permeability of the formation to be grouted. Although high penetrability is often desired, there are applications where this may not be advantageous, and even some situations in which it would be undesirable. For example, in a formation of high permeability above the water table, a grout that is excessively penetrable can "run away" from its intended location, whereas a less penetrable grout would remain where injected, as illustrated in Figure 7.1.

A search of the chemical grouting literature can uncover much discussion as to the effect of soil permeability on the viscosity of a grout to be used. This can be quite misleading, however, as such discussions seldom consider pumping rates in their conclusions. For example, in one case, several chemical grouts with widely differing viscosities were injected in a full-scale field evaluation. The test involved a deposit of fine to medium sand containing trace amounts of

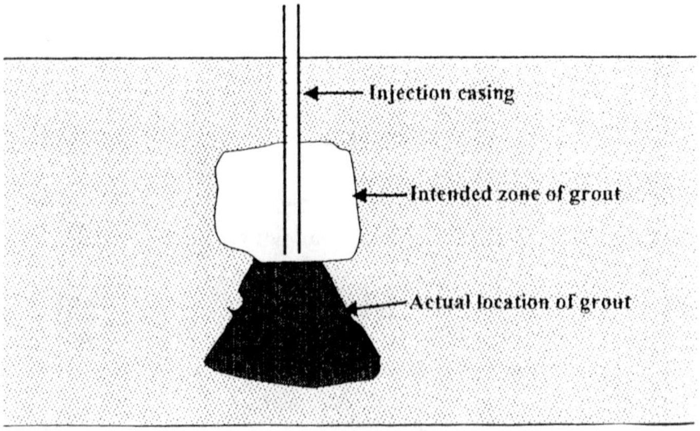

FIGURE 7.1 An excessively penetrable grout can run away from its intended location in a very permeable formation.

silt. It was interesting to note that whereas an acrylamide grout with a viscosity of about 1 centipoise, which is similar to water and is about the most penetrable of all grouts, could be injected at a rate of about 7 gpm (26.5 L/min), a much more viscous sodium silicate–based formulation, with a viscosity of about 8 centipoise, could still be injected into the formation, although it required a slower pumping rate of only about 3 gpm (11.4 L/min).

Set times of solution grouts can range from instantaneous to several hours, although for most work it is desirable for set to occur soon after injection. This is especially so when working below the water table, in order to minimize dilution of the grout. Grouts that resist dilution are available and should be used where such risk exists. Some can be formulated so as to offer extremely short or even zero set times, and in the case of some urethanes, instantaneous foaming upon the grout's first coming in contact with water. Because these more specialized grout formulations tend to be quite expensive, their use is sometimes avoided. Such thinking is flawed, however. Although they can be quite expensive on a unit basis, their significantly greater effectiveness often offsets the higher cost.

It is important to understand the setting behavior of grouts, especially where dilution is likely to occur. Some formulations will remain at or near their initial penetrability until shortly before setting, whereas others will start to thicken immediately upon mixing until they finally set. Either of these behaviors can be advantageous, and may even be required, depending on the particulars of the individual application.

Some grouts are hydrophobic, that is, they repel water, and others are hydrophilic and attract it. Hydrophilic grouts can take up any available water and thus be diluted. In some cases, they will continue to take up water even after fully curing. Because water will be lost upon drying, such grouts may not be appropriate for installations where this is likely, such as in arid environments. Once cured, hydrophobic grouts are usually dimensionally stable, unlike hydrophilic formulations, which tend to shrink as they dry. Most hydrophilics that have undergone shrinkage, however, will expand upon rewetting. Yet there will always be a time lag between the rewetting and return to greater volume. In many cases including complete resaturation, the original volume will not be recovered.

Some of these grouts will immediately gel or foam upon contact with water. The particular properties a grout should possess to provide optimal results will depend on the purpose of the injection. The ability of a grout to resist dilution and/or extrusion within the host formation, or syneresis (a squeezing of the water from the gel), are also important criteria. No single noncementitious grout is equally suitable for all applications.

7.2 CHEMICAL SOLUTION GROUTS FOR STRENGTHENING

For strengthening soil, the most important selection parameters are safety, strength, and cost. Many of the historically used grouts are toxic and have thus been removed from the market. Hard, rigid gels are usually desired for strengthening applications, as they provide the greatest stiffness as well as strength. Cost can also be an important factor, because this type of work often requires very large quantities of grout.

7.2.1 Sodium Silicate

Grouts based on sodium silicate are the most frequently used for strengthening applications. Appropriately, the technical literature is filled with publications relative to such grout. Much of it is quite academic, and some downright misleading, however. For example, several publications present strength data for *sodium silicate grouts,* but make no reference whatever to the reactant system. Sodium silicate may constitute the largest portion of a chemical solution grout, but unless it is able to completely dry out, it requires a reactant component to produce the hardened end product. The particular reactant system and the proportions thereof will directly determine the attained strength as well as the durability of the grout.

Sodium silicate is a composition of inorganic, amorphous glasses in solution. Such silicates are soluble by way of a high-alkaline sodium oxide content, which maintains a pH level sufficiently high to dissolve the silica. Sodium silicate is produced by fusing silica sand with either soda ash or potash under a very high temperature. Once cooled, the resulting glass is processed into either liquid or powder form with a number of different properties. Pure sodium silicate solution is a good-quality adhesive on its own and will form strong bonds if allowed to dry. However, these bonds will soften if exposed to water and, with long-term exposure can be completely lost. Sodium silicate solutions are commonly referred to as *water glass,* which is an accurate description, considering that, strictly speaking, it is dissolved glass in water.

If the pH of the liquid silica descends below about 10.7, solubility is reduced and it will start to gel as the silicate species polymerize. This is accomplished by mixing it with a reactant solution that lowers the pH. Although acidic materials abound, the reaction must be delayed sufficiently to allow injection, and this greatly limits the possible reagents. Depending on the particular system used, the bond can be quite strong and durable. As discussed in Section 7.2.2, the reactant system may consist of one or more chemicals, and the particular choice will have a great effect on the developed strength and durability of the grouted formation. In the rather rare case of injection into acidic soil, it is possible to obtain reaction without the use of a separate component. If the pH is very much below 7, however, the reaction can be very fast, inhibiting proper saturation of the soil. The strength of the chemical bond is not as great as that obtained with drying, however.

Sodium silicate solutions are available in a variety of densities and viscosities. The density of the initial silicate concentration has a significant effect on the properties of the grout, including long-term durability. Weight ratios of soluble silicate are expressed by their density in degrees Baume. This determines the solubility of solids and reactivity of the silicate, as well as physical properties such as pH and viscosity. Typical silicate ratios vary from about 1.6 to 3.2. Strength is increased with increasing ratio, whereas both viscosity and pH are lowered. The sodium silicate most used for grouting has a weight ratio of 3.2 and a density of about 41 degrees Baume. It weighs 11.6 lb per gal at 68°F (20°C) and is highly alkaline, with a pH of 11.3. The consistency is similar to light syrup, with a viscosity of about 180 centipoise. Viscosity will vary with temperature, however, as shown in Figure 7.2.

The largest supplier of sodium silicate in the United States is the PQ Corporation; it markets

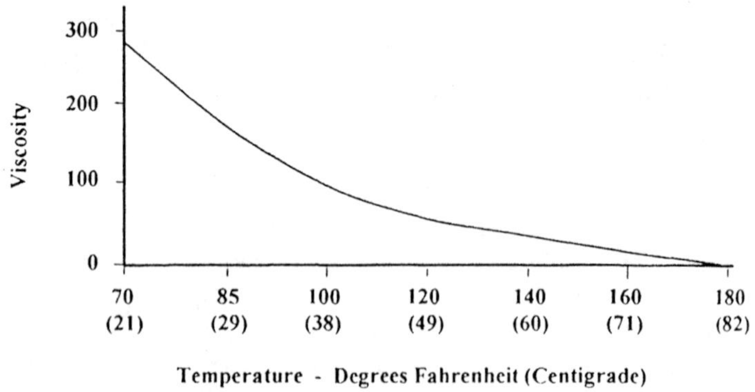

FIGURE 7.2 Temperature effects on sodium silicate.

a silicate denoted as *Grade N*, which possesses the aforementioned properties. The viscosity of the solution will be reduced with the addition of water, as shown in Table 7.1. As can be observed, a starting viscosity of about 180 centipoise drops off rapidly with solutions below about 60 percent. A 50 percent solution is generally less than 10 centipoise, and a solution of this strength is the most frequently used where good strength of the grouted medium is required. Where especially high strengths are needed, however, a 60 percent solution can be used, but because it is considerably thicker, a slower injection rate will be required. Solutions stronger than 60 percent are seldom considered in grouting.

The sodium silicate base solution usually makes up 30–60 percent of the total grout volume. To ensure good strength and long-term durability for permanent applications, a minimum solution strength of 50 percent sodium silicate base should be used. Although such grouts have a fluid viscosity somewhat greater than other widely touted *low-viscosity* grouts, they are nonetheless quite penetrable. Penetrability can also be enhanced by the inclusion of a non-ionic surfactant or wetting agent, which can effectively reduce the surface tension between the grout and the host formation. Only a small amount of surfactant is needed to greatly improve penetrability; surfactants that can be used advantageously include:

- Alkylated diphenyl oxide disulfonates (ADPODS), distributed by Dow Chemical Company under the name DOWFAX 201
- Sodium 2 ethylhexyl sulfate, distributed by Niacet Chemical Company under the name Niaproof Type 8
- Polyoxyethylene alkyl esters, available from several chemical distributors

I discovered the benefits of the polyoxyethylene surfactant in the early 1960s. Its use lowered the surface tension of the grout so effectively that the pumping rate could be increased about 40 percent without changing the injection pressure. The first two surfactants listed are of more recent development. Some sodium silicate manufacturers recommend them as particularly beneficial, however.

In the past, some grouters have opined that silicate solutions stronger than about 30 percent are too thick to pump, as an excuse for their poor performance. This is incorrect, as the viscosity does not start to rise significantly until solution strength greater than about 60 percent is reached, as shown in Table 7.1. In fact, I have been directly involved with the injection of literally hundreds

TABLE 7.1 Properties of Different Solution Strengths of Grade N, 3.2 Weight Ratio, Sodium Silicate

Percent Solution by Volume	Viscosity, cp at 68°F (20°C)	DENSITY			
		lb/gal	(kg/L)	lb/ft^3	°Baume
100	180	11.6	1.38	86.8	40.8
70	92	10.6	1.27	79.6	31.5
60	14	10.3	1.23	77.2	27.9
50	9	10.0	1.20	74.7	24.1
40	5	9.6	1.15	72.3	20.0
30	4	9.3	1.12	69.8	15.6

of thousands of gallons of grout that was of either 50 or 60 percent N grade sodium silicate.

7.2.2 Sodium Silicate Reactant Systems

There are many different reactant systems, which consist of one or more components. Strength and other properties of the resulting grout will vary greatly with the different reactant compounds. The grout produced with some reactant systems will break down and lose strength over time, whereas others have proven durable for the long term. Reactants commonly used in grouting can be classified in three generic groups: (1) inorganic salts, (2) organic/aliphatic esters or amides, and (3) stabilizers.

These compounds can be used alone or, in some cases, combined with each other. The dosage and proportions of the different components, when more than one is used, can have a dramatic effect on both the set time and the strength of the resulting grouted mass.

The reactant groups are as follows:

Group 1—Inorganic Salts
Calcium chloride
Sodium bicarbonate
Sodium aluminate
Calcium sulfate

Group 2—Organic/Aliphatic Esters and Amides
Dibasic esters
Acetates/acetins
Formamide
Glyoxal

Group 3—Stabilizers
Portland cement
Slag-based cement
Class C fly ash

There are many other chemical compounds that can be used; however, those listed are the most effective and the only ones that are in common use.

7.2.2.1 GROUP 1 REACTANTS

Of the salts, calcium chloride is the fastest acting and produces the strongest gels. If the silicate is combined with a sufficiently strong solution of calcium chloride (greater than about 40 percent), the resulting grout would provide a strong, permanently solidified mass. Unfortunately, it is not possible to mix such a combination under the conditions common to grouting, because of an excessively rapid gel time. Interestingly, however, the Joosten process patented in the United States in 1925, which was one of the first proprietary solution grout systems, involved only these two components. They were injected in separate adjacent grout holes, so that the actual mixing occurred in the soil upon injection. When the two components did satisfactorily combine, a very strong mass resulted. Unfortunately, however, the combined solutions precipitated an almost instant reaction, such that only those portions that were very near the reactant injection hole were properly blended and much of the silicate remained unreacted.

It has been found that fairly small inclusions of calcium chloride, in combination with reactants from the organic/aliphatic esters and amides group, can greatly improve strength. Whereas the gels resulting from the use of these materials alone will not be permanent, their use in combination with calcium chloride can be long-lasting. Relatively small amounts of these less permanent reactants can provide a buffering effect, which allows a greater proportion of calcium chloride to be used at a given gel time. This results in a higher-strength grouted medium. A variety of example mixtures are discussed in detail shortly.

The other members of the inorganic salts group have not proven to provide such high strengths or to be as permanent. They are, however, easy to use and should be considered where neither high strength nor long-term durability is required. Sodium bicarbonate is particularly easy to use, and in reasonably clean dry sands can provide a grouted mass with unconfined compressive strength in excess of 150 psi (10.4 kPa) and a usable strength, not subject to creep, of about 40 psi (28 kPa). The effect of creep on the usable strength is discussed in Section 7.2.3.1. Although the surface of such a mass will start to deteriorate shortly after exposure to air, as illustrated by the growing white "whiskers" shown in Figure 7.3, the deterioration propagates quite slowly after exposure of the face. In fact, on one project the solidification was performed in lieu of shoring, which was intended to remain in place for a period of only a few weeks. Because of changes in plan, however, it was open for a period of more than five months without measurable deterioration and, overall, performed very well.

It consisted of two parts, by volume, of a 60 percent solution of N grade sodium silicate, combined with one part, by volume, of a 6.7 percent (by weight) solution of sodium bicarbonate. This is about the strongest solution that is practical with these components, and 6.7 percent is about the greatest concentration of sodium bicarbonate that will stay in solution, even with continuous agitation. This grout had a working time of 30–60 min, and the unconfined compressive strength of the resulting mass exceeded 150 psi (10.4 kPa).

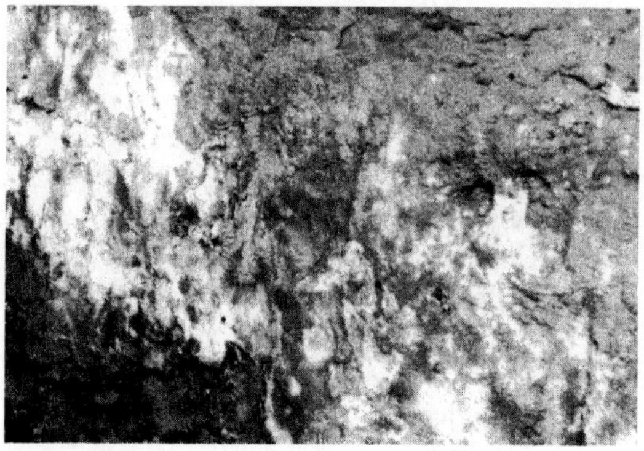

FIGURE 7.3 White "whiskers" on cut face are the result of grout deterioration.

As with most grouts, the reaction time generally becomes shorter as the temperature rises, and increases with a temperature reduction. Bicarbonate reactant grouts are especially sensitive to temperature, as detailed in Table 7.2 . The data shown are for the aforementioned mixture. One must be mindful that it is the grout temperature that counts, and this can be much different than either the ambient temperature or that of the soil. Because a lower grout temperature allows a greater amount of reactant to be included at a given gel time, thus providing a stronger grouted mass, it is preferable to keep it as low as practicable. Shading the mixing and storage areas from direct sunlight is recommended. In addition, crushed ice can be substituted for some of the mix water in hot weather.

A word of caution should be heeded in the use of bicarbonate reactants: Silicate-bicarbonate grouts do not develop good bond to grains of soils that are saturated or submerged. For this condition, sodium aluminate reactants perform better; however, they tend to have shorter gel times at equivalent solution strengths and are more difficult to handle. The moisture content of the soil has a dramatic effect on the final properties of the grouted mass. Generally speaking, the maximum strength of a grouted mass injected and cured under water will be significantly lower than that of one that is allowed to dry at

TABLE 7.2 Gel Time of Bicarbonate Reactant Grouts at Different Temperatures

TEMPERATURE		
Fahrenheit	**(Celsius)**	**Gel Time**
50	(10)	24 hours
60	(16)	6 hours
70	(21)	2 hours
80	(26)	30 minutes
90	(33)	5 minutes

least somewhat. Grouted masses that dry completely will benefit not only from the lowering of their pH, but also from the natural adhesive properties of the drying silicate, regardless of the reactant used.

Combination reactants, which include both calcium chloride and one of the organic/aliphatic esters or amides, continue to be used and are certainly dependable. They do require more knowledgeable personnel and sophisticated proportioning pump and processing equipment, however. To obtain the best properties of such multicomponent grout, the different parts must be intimately combined. This is accomplished through high shear mixing by a rapidly rotating element. Figure 7.4 shows such a mixer, in which the grout components are directed into a squirrel cage blower rotating at high speed on the vertical shaft extending from the air motor on top of the tank.

7.2.2.2 GROUP 2 REACTANTS

Because of their simplicity and flexibility in gelling behavior, Group 2 reactants are by far the most frequently employed, especially in large-volume applications. As a group, however, they tend to yield lower strengths and are generally more expensive than the Group 1 compounds. *Acetins* and *dibasic esters* are especially popular because of their ease of handling. These can be more readily combined with sodium silicate, and are simpler to compound and use, than the multicomponent mixtures.

FIGURE 7.4 Grout ingredients are rapidly combined by the mixer in a "flash" tank.

Dimethyl esters, which are the most commonly used and economical of these reactants, are a by-product of the manufacture of nylon. As recovered, they contain a combination of *dimethyl succinate, dimethyl gluterate,* and *dimethyl adipate.* Because dimethyl esters are a by-product, the proportion of these three subcomponents is not constant and can vary widely, depending on the time and the source of a given specimen of the compound. Although for many uses the exact proportioning is of little consequence, such is not the case with sodium silicate grout. Both gelling time and strength of the injected mass are affected by the proportions of the three components. To obtain consistent set, as well as uniform strength, the proportions should be reasonably consistent.

The amount of basic ester is usually within a range of 5 to 10 percent of the total grout solution, although for strong and durable applications the higher value should be used. The rest of the grout is water. A typical mixture contains a 50 percent sodium silicate, 7–8 percent dibasic ester, and has a viscosity of about 10 centipoise. In the simplest operation, equal parts of full-strength N grade sodium silicate and water, along with the appropriate amount of ester, are pumped separately to the injection header, where they are combined.

Basic esters are solvents and not highly soluble in water, especially at low temperatures. Although they can be combined at temperatures above about 70° F (21° C), such combination becomes less efficient at cooler temperatures. It can thus be advantageous to heat the mix water by way of in-line heaters prior to mixing. Alternately, these reactants can be made easily miscible at lower temperatures through the addition of non-ionic surfactants, such as discussed in Section 7.2.1. These emulsify aqueous solutions and are sometimes referred to as emulsifiers. Although not necessary, blending the solutions with high shear mixing immediately ahead of the header will ensure complete integration and through mixing.

The mechanism by which silicate grout hardens is reduction of the silicate pH. In the presence of water, the esters hydrolyse to acid, which then converts the silicate to a solid. The rate of hydrolysis of the three different ester types, and thus the grout gel time, varies greatly. Dimethyl succinate hydrolyzes in minutes, gluterate in hours, and adipate in days. The proportions of these in the reactant solution will dictate the gel time of the grout. Although grouts employing acetins and basic ester reactants are widely used, I am unaware of any documented data relative to long-term durability and suggest that they not be used for permanent solidification until suitable performance data become available.

Both glyoxal and formamide can be used alone; however, this is much more costly than when they are used in combination with a salt. Further, when used alone, both strength and durability are lowered and control of the set time can be difficult, especially at low temperatures. Ammonia gas is produced as a by-product of the reaction of formamide with sodium silicate. This can present a problem in enclosed spaces and has, in fact, previously resulted in the closure of facilities until the uncontrolled gas could be vented.

7.2.2.3 GROUP 3 REACTANTS

Of the Group 3 reactants, portland cement is the most frequently employed. It is not generally used for routine strengthening of soil; rather, it is commonly used for applications that require control of moving water, either with or without strengthening. Sodium silicate acts as a powerful set accelerator for cement, as illustrated in Table 7.3, so combinations of the two can provide fast setting and be highly resistant to dissolution or washout. Because the primary reaction is hydration of the cement, the resulting grout is of good strength and durability. But because the cement grains are too large to penetrate most soils, these grouts are not used for ordinary strengthening work.

Cement-sodium silicate grouts are very effective where water is moving through the host deposit. It is not uncommon for the soil fines to have been previously washed away, and the existence of eroded passages is common. Thus, there are often spaces that easily accept the cement grains, so both rapid set and good strength are advantageous. Grouts of this type are used where rapid set is required. They are best injected by use of separate pumps for the cementitious suspension and the sodium silicate. Combinations of equal parts of N grade sodium silicate and a cementitious suspension grout

TABLE 7.3 Set Time of Cement-Sodium Silicate Mixtures

Cement-Silicate Ratio by Volume	Time of Set	Typical Usage
1:1	1 to 10 seconds	Controlling flowing water
10:1	1 to 10 minutes	Water control/limiting grout mobility
30:1	10 to 20 minutes	Compounding rapid-setting cement grout
50:1	30 to 60 minutes	

with a water:cement ratio of about 1, work well. Generally, only the very rapid setting mixtures have been used, and they are extremely effective in cutting off flowing water in relatively small passages. The reactions resulting from slag base cements or Class C fly ash will be slower, so, in general, there is no real advantage in their use over that of portland cement.

7.2.3 Durability of Chemical Solution Grouts

The American Society for Testing and Materials (ASTM) Standard D 4219, "Standard Test Method for Unconfined Compressive Strength Index of Chemical-Grouted Soils" (1993), applies to chemical solution grouts. It addresses only *short-term* strength, which certainly precludes application in permanent installations. Unfortunately, *short-term* is not defined, although it has been found that virtually any quality silicate base grout will retain significant strength for a period of at least a month or more, subject only to containing at least 50 percent sodium silicate. This will, no doubt, allow sufficient time for many temporary applications.

Long-term strength can, however, be obtained with sodium silicate base grouts. In research reported by Graf, Clough, and Warner (1982), several different silicate grout compositions proved to be durable for more than ten years. Upon testing of chemically grouted soil specimens aged 9–11 years, they concluded, "Except for the very weakest samples, the strengths of the stabilized soils tested with no environmental change, showed no change or a modest increase over those measured after one to two years' aging." Included in the report was a graph, reproduced here as Figure 7.5, summarizing the long-term strength history of grouted sand specimens of five different grout compositions.

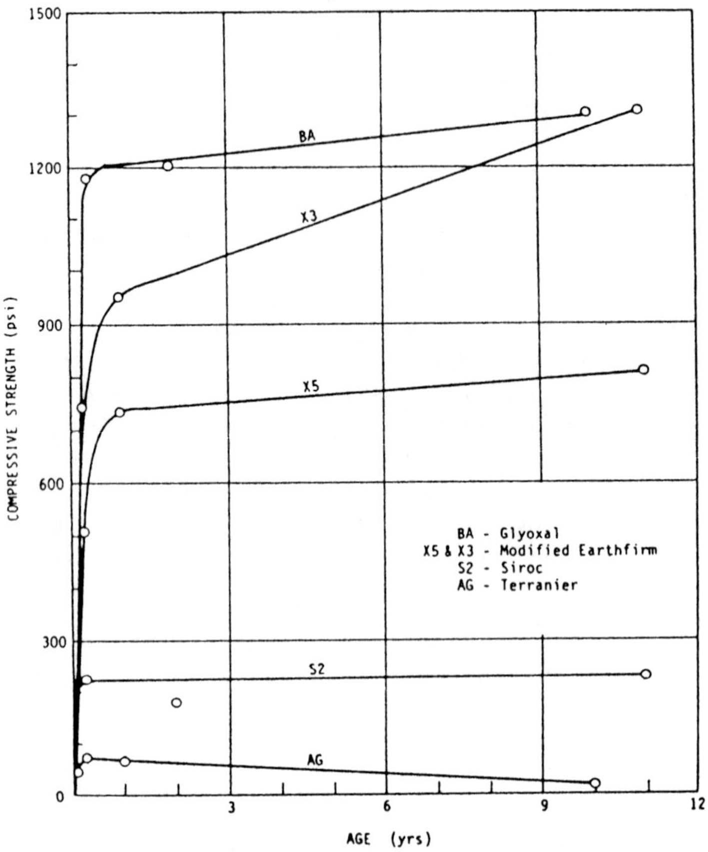

FIGURE 7.5 Strength gain over time for various grouts. (From Graf, Clough, and Warner, 1982.)

Results of research that had been ongoing for several years prior to 1971 were also reported (Warner, 1972). That work involved the preparation and testing of more than 2500 laboratory specimens of chemically solidified soil under a variety of test and environmental conditions. The specimens were made with eight different

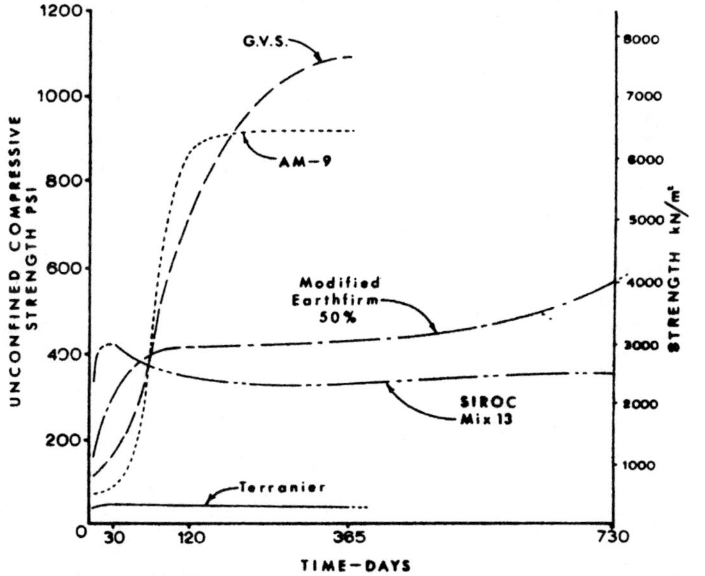

FIGURE 7.6 Strength of normal cured solidified specimens.

chemical grout systems and cured in three different environments: under water; contained in a waterproof specimen mold in a controlled laboratory environment, referred to as *normal* cure; and oven dry. Sodium silicate was the base material in four of the systems. Also reported were the strengths of about 100 specimens from sodium silicate grouted masses, procured from a variety of actual field installations made between 1965 and 1971, including several permanent applications. These correlated well with the laboratory results.

Although that work was completed a long time ago, nothing comparable has since been presented and most of the results remain pertinent. Several factors were found to affect the indicated strength of the specimens. Some, such as the rate of loading, have a dramatic effect on the indicated strength. Faster loading will produce significantly higher indicated strengths. Thus, comparisons of reported strengths are not valid unless the loading rate is known and is identical for all specimens.

The moisture level within a specimen also has a substantial influence on the indicated strength. Dry specimens always indicate a substantially higher strength than otherwise identical moist ones. It is thus imperative that design strength be based on reliable test or performance data, under the same moisture conditions that will exist in the proposed work. To illustrate the effect of the curing moisture condition, Figures 7.6 and 7.7 show the strength of identical specimens of various grouts under both normal and submerged cure, respectively. Many sodium silicate grout formulations that are entirely effective above the water table will not perform adequately when submerged and

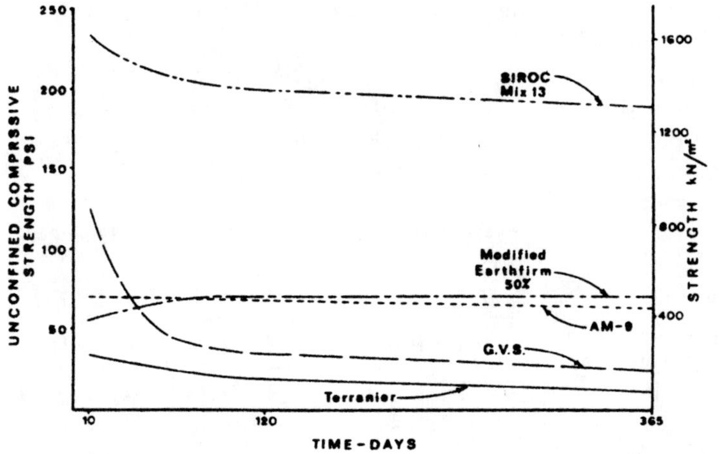

FIGURE 7.7 Strength of wet-cured solidified specimens.

should not be so used. The most significant finding was significant creep of the solidified masses under sustained load. The strength under continuous load was but a small portion of that indicated by the relatively fast loading of standard laboratory compression tests.

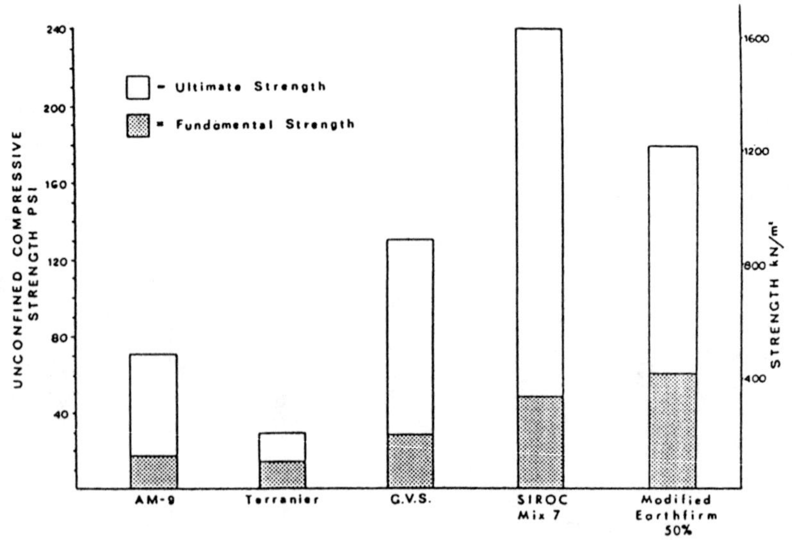

FIGURE 7.8 Ultimate and fundamental strength of masses injected with different grouts. (From Warner, 1972.)

7.2.3.1 EFFECTS OF CONTINUOUS LOADING

The strength obtained under long-term loads, that is, the *fundamental strength,* was within a range of only 20–80 percent of the *ultimate* strength, indicated by the standard loading rate. Noteworthy also was the large variation in the strengths of specimens that were identical in all respects except for the reactant system. The loading rate was 20 psi (0.14 MPa) per second, as per the provisions of ASTM Standard D 1663. The ultimate and fundamental strengths of several of the grouts evaluated at five days are shown in Figure 7.8, which is reproduced from Warner (1972). Therein, G.V.S., SIROC, and Modified Earthfirm all contain sodium silicate as the base component. I have had extensive experience with Modified Earthfirm and have found the long-term fundamental strength to be about 70 percent of ultimate. Mix proportions for that grout are provided in Table 7.4.

Another finding was a large variation of strain at rupture. The poorest performing was AM-9 acrylamide grout, which had a strain of 19.7 percent. Although not now easily available and virtually never recommended for strengthening applications, at the time of the research it was so recommended. Obviously, unless the solidified masses were fully restrained, less than satisfactory performance could be expected with such great strain deformations. The strain behavior of several grouts is shown in Figure 7.9. Many of the formulations were proprietary at the time of publication in 1972. Such rights have now expired, and the mix contents of the three main silicate base formulations are provided in Tables 7.4 through 7.6.

Another research effort was reported by Clough, Kuck, and Kasali (1979). Therein 150 specimens of silicate grout compounded with

TABLE 7.4 Mix Constituents for Modified Earthfirm, 50 percent

Ingredient	Laboratory Batch	Field Batch
Sodium silicate	350 ml	50 gal (189 L)
Calcium chloride	6.72 g	8 lb (3.6 kg)
Ethyl acetate	14 ml	2 gal (8 L)
Water	336 ml	48 gal (181 L)

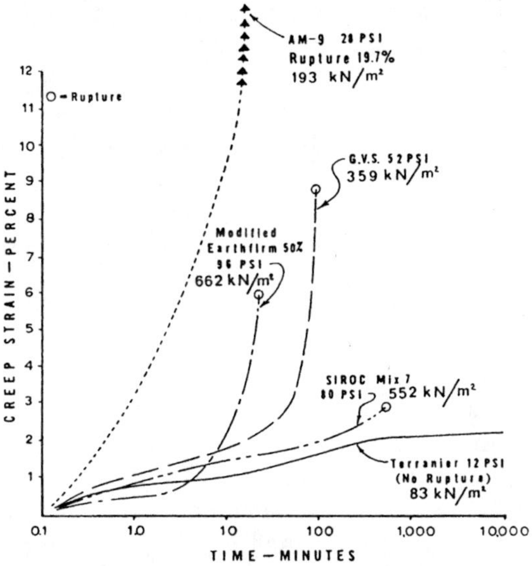

FIGURE 7.9 Strain behavior of soils solidified with different chemical grouts. (From Warner, 1972.)

varying amounts of formamid were evaluated. The findings of these investigators were in general agreement with those of Warner (1972), with one exception: Greater strengths were obtained in finer sands. Although more recent strength data have been reported, important criteria such as loading rate, curing regime, and the like, either varied or were not reported, so meaningful comparisons are impossible. ASTM Standard D4219 was not developed until 1993.

TABLE 7.5 Mix Constituents for G.V.S. Chemical Solution Grout

Ingredient	Laboratory Batch	Field Batch
Sodium silicate	350 ml	50 gal (189 L)
Calcium chloride	4.5 g	6.5 lb (2.9 kg)
Glyoxal	35 ml	5 gal (19 L)
Water	315 ml	45 gal (170 L)

TABLE 7.6 Mix Constituents for SIROC Mix 7

Ingredient	Laboratory Batch	Field Batch
Sodium silicate	350 ml	50 gal (189 L)
Formamide	63 ml	9 gal (34 L)
Sodium aluminate	4.3 g	6 lb (2.7 kg)
Water	287 ml	41 gal (155 L)

Further, it applies to only *short-term* unconfined compressive strength and, unfortunately, did not adopt the loading rate of any of the previous work, but provides only for a constant but unspecified rate of load. The only limiting requirements are that failure is not to occur in less than 2 minutes and that the maximum strain rate is not to exceed 1 percent per minute.

Such wide latitude in test protocol precludes meaningful comparison of different results, even though evaluated in conformance with the standard, unless identical values were used and noted. Another limitation of the standard is that reported strength is to be obtained at specimen rupture or at an accumulated strain of 20 percent. This is a very large strain, and it is unlikely that many structures partially supported by such grout would tolerate the resulting deformations.

There has been much discussion, as to the appropriateness of unconfined tests for such evaluations, and it has been suggested that triaxial tests, which provide lateral confinement of the specimen, would be more suitable. This view has some validity if the solidified mass of the real application is to be continuously confined. Most grout solidification, however, is performed for the retention of excavated soil faces, as illustrated in Figure 7.10, and lateral confinement does not exist. Even in situations where the mass is completely confined, extensive laboratory studies by Ata and Vipulanandan (1999) have established little, if any, effect of the confining pressure on either the shear strength or the modulus of the

FIGURE 7.10 Chemical solution grout is often used for the unrestrained temporary retention of soils during construction.

grouted soil. Use of the much easier and less costly unconfined test is thus certainly justified.

Although properly injected silicate grouted soils have proven adequate for many years, there has been less-than-acceptable performance in some cases, and catastrophic failure in a few. The use of a sufficiently strong solution of sodium silicate is essential, and of even greater importance is recognition of the creep properties. The strength under continuous loading will always be much less than that indicated by the rather fast loading of typical unconfined compression tests. The widespread belief that grouts containing more than about 30 percent silicate will be too thick to inject or will result in gelling too rapidly, is a further problem. And then there is, unfortunately, always a strong incentive to minimize the amount of expensive chemical by simply increasing the proportion of water in the grout.

Poor results are virtually always caused by faulty mix design and are not indicative of the behavior of properly compounded formulations. The vast majority of present chemical grout strengthening applications involve sodium silicate–base grouts. Although in earlier times many other grouts were used, most contenders have fallen because of environmental, economic, or performance limitations, and so are not discussed further here.

7.3 CHEMICAL SOLUTION GROUTS FOR WATER CONTROL

Although grouts that set to hard rigid gels are preferred for soil strengthening, some flexibility is often desirable for water control applications. Similarly, although long gel times allow extended pumping at any one location and are usually advantageous for strengthening, this is not the case in most water control. Here, resistance to dilution or displacement by moving water is the most important criterion, and very rapid set times are often required. When the gel time is necessarily very short, occasional gelling in the pump or delivery lines is likely. A grout that reacts first as a soft gel will allow pumping of the young gel out of the system before it stiffens excessively.

In strengthening work, individual grout constituents are often purchased separately and combined by the contractor, but in water control efforts, because the chemistry of the grouts is more sophisticated, they are usually purchased as grout systems from a single supplier. There are three main generic classes of chemical grouts that are particularly suited for water control and are used in most of the work performed:

1. Acrylic polymers
2. Polyurethane
3. Sodium silicate (combined with cement)

7.3.1 Polymers

With the exception of cement/silicate combinations, the grouts listed here fall under the classification of *polymers*, as do many other grouts used in structural work. What are polymers? To answer that question let us first start with monomers, which are the units of which a polymer is constructed. Monomers are small simple molecules, which can be chemically linked together so as to form a continuous molecular chain. Polymers are chains of many molecules and typically contain several thousand monomer units. Think of a monomer as a single bead. If we link several of the beads in a continuous string, as shown in Figure 7.11, we will have constructed a polymer. If the string of beads includes monomers of different types, the result will be a *copolymer.* The chemical process of linking monomers is known as *polymerization.* From a practical standpoint, this is the curing of a polymer-based grout, which may occur rapidly or can require an extended period of time. The reaction time of most polymers is strongly influenced by temperature, and many polymers require an external source of heat to cure. Those materials used in grouting, however, virtually all employ a *promoter,* which is intermixed into the compound, allowing polymerization to occur at ambient temperature.

Polymerization can occur in several different ways. *Addition* reactions occur when monomers link to one another to form a polymer. The molecular weight of that polymer will be the sum total of all of the monomers of which it is composed. Thus, the terms *long chain* and *molecular weight,* often provided in technical data, refer to

FIGURE 7.11 A polymer is many monomers linked together, like a string of beads.

the size and structure of the polymer. Polymers can also be formed by *condensation* reactions, wherein successive monomer units combine by the disposition of very small molecules from within.

The polymerization process can be altered through the use of a number of different compounds. These are classified as follows:

- Initiators
- Promoters/accelerators
- Inhibitors
- Miscellaneous fillers and additives

Initiators are chemicals that are added to monomers to initiate, or start, the polymerization process. They are often referred to as *catalysts,* although this term is incorrect, as catalysts are substances that *increase* or *decrease* the rate of a chemical reaction, not start it. Heat is often necessary for polymerization to occur, but it is not needed if *promoters,* or *accelerators,* are included. These are chemicals that initiate polymerization without the need for external heat. By variation of the amount of promoter, the rate of the reaction can be controlled. Promoters can also be used to accelerate polymerization and/or allow it to take place at relatively low temperatures. For this reason, most polymer-base grouts are supplied in two or more parts that are mixed immediately prior to or during injection.

Miscellaneous additives are used to vary the final properties of a polymer compound. *Diluents* are used to improve the penetrability of the polymer, whereas *fillers* will thicken it to paste-like nonsag or mortar consistency. They will also extend the volume and thus reduce the cost of a formulation. Some fillers are mixed into the basic polymer by the manufacturer, and others can be mixed into the compound when it is prepared on-site. One of the most common fillers is ordinary sand. Where it is used, however, it must be of the correct gradation recommended by the manufacturer and scrupulously clean, in order to attain the required bond to the fluid component. Fillers can greatly reduce shrinkage, which is a serious shortcoming of some polymers.

Inhibitors are compounds added to monomers to postpone polymerization or prevent premature reaction. Other additives can change the stiffness of the final composition, enhance the curing process, or provide resistance to ultraviolet light. The overall quality of a polymer is directly influenced by the inclusion of such amendments, and, of course, cost will be likewise affected. One must thus be especially wary of low-cost polymer grouts, as they are often of inferior quality, lacking the most optimal modifying additives and/or unreasonably extended with low-cost fillers or diluents.

Polymer compounds can be either *thermosetting* or *thermoplastic.* Most polymers used in grouting are thermosetting, in that their chemical transition from liquid to solid forms an irreversible polymer solid upon curing. Thermoplastic polymers, however, can be reprocessed by subsequent heating and reuse. These are not nearly as stable or durable as their thermosetting cousins. In grouting, the term *resin* is often used interchangeably with *polymer* to describe such materials. Yet strictly speaking, *resin* refers to the cured or polymerized monomers, which usually attain a solid or semisolid state.

7.3.2 Polyurethane Grouts

Among the chemical solution grouts employed for water control, those based on polyurethane chemistry are by far the most extensively used, especially in the correction of leakage into structures. These are highly versatile compositions and are of many different types and supplied in various forms. They can provide a wide range of properties in both the fluid and hardened states. Some are furnished as single solutions that need only be combined with water, whereas others consist of two separate solutions that are premixed, the reaction being completed upon exposure to moisture. Different formulations will react to form gels, solids, or foams.

The strength of the gel, or density of the foam, is dependent on the particular formulation and the amount of water with which it has combined. As with most compositions involving chemical reaction, temperature has a significant influence on both the viscosity of the individual components and the mixed grout. Time to gelling and/or curing of the mixed resins is also sensitive to temperature. Urethane grouts are generally very stable once in place. They provide high resistance to acids, but only fair resistance to strong alkalis, and are subject to degradation upon exposure to ultraviolet light.

Some formulations react completely upon coming in contact with moisture, but others require thorough mixing with a certain minimum amount of water. Single-component urethanes are prepolymers; that is, initial polymerization has already taken place, but the process cannot be completed without their coming in contact with moisture. The amount of water required for complete polymerization in the required time frame will vary and is dependent on the individual formulation. Because the stability and quality of the reacted material are totally dependent on the inclusion of a sufficient amount of water, a knowledge of that required, and positive assurance that it is provided, are crucial. Many cases of less-than-acceptable grout performance have been attributed to lack of sufficient moisture.

7.3.2.1 HYDROPHOBIC URETHANES

Urethane grouts can be divided into two main classes, *hydrophobic* and *hydrophilic.* Hydrophobic formulations react with water, but only a very small amount is required for their proper cure. Once that amount has been combined, they repel further moisture to which they are exposed. Because only a limited amount of water is allowed to contact the polymer during the reaction, gas is trapped in the polymerized matrix, which develops as a foam in an unrestrained form.

Hydrophobic formulations tend to produce relatively rigid foams and are available in a wide variation of densities. The actual density results from the particular chemical formulation and, of course, exposure to sufficient moisture. To enhance the wetting of contacted surfaces and stimulate the water–grout reaction, surfactants are frequently used, resulting in an increased chemical bond.

Because they repel excess water, the ability of hydrophobic formulations to bond to wet surfaces is not great; however, this is not a limitation in soil pore space or random cracks with rough surfaces. It can become a problem in generally smooth walled joints in rock or concrete or in a void in soil immediately adjacent to a smooth interface. Some hydrophobic formulations will immediately (within 15–20 seconds) foam on contact with water, as shown in Figure 7.12. This can be beneficial in stopping flows of moving water.

Because reacted hydrophobic urethanes take in little outside water in the cured state, they are generally free of shrinkage, even when allowed to dry. They can also be formulated to be of very low viscosity, which allows penetration into the finest cracks and defects and into virtually any permeable soil. Even though initial contact with water is required for reaction, once it is started, the grout will be resistant to further dilution. Hydrophobic resins can be formulated to have a very high expansion potential, and grouts with unrestrained expansion of nearly 30 times their original volume have been reported. Although the expansion within a fine fissure or soil pore space will be considerably reduced because of frictional constraint, a high expansion rate will lower the cost of material for a given amount of work.

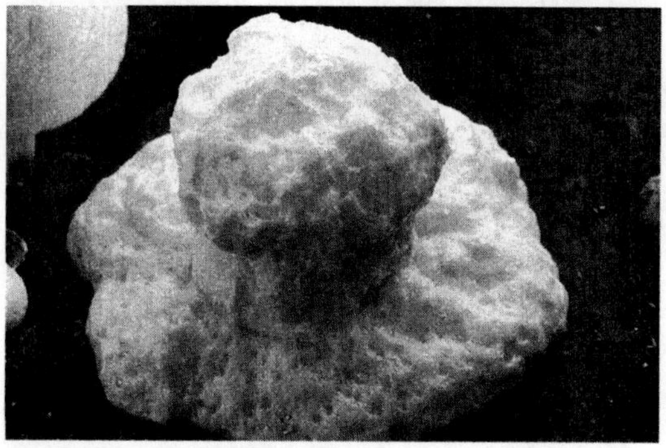

FIGURE 7.12 A few drops of water on a hydrophobic urethane results in immediate forming.

7.3.2.2 HYDROPHILIC URETHANES

Hydrophilic formulations also react with water, but many continue to attract it after completion of the initial reaction. These formulations often continue to expand if possible, forcing the resulting gel into pore spaces and voids beyond those originally filled. Because they attract moisture, they tend to be drawn into water-filled pore space and microcracks on the surfaces of the filled defects. In addition, they develop a much greater bond to those surfaces as long as they are not excessively diluted.

With small amounts of mix water, most hydrophilics will form a gel. Once formed, however, the acceptance of further water results in outgassing, creating foam. The foam density is dependent on the original chemical formulation and the amount of water that has been combined. If additional moisture is supplied very slowly, much of the foaming gas will be absorbed, or will escape into the atmosphere, resulting in a solid piece of urethane rubber. If ample water is supplied immediately, however, large amounts of carbon dioxide gas will be created and the foam will be of much lower density. The density of the grout can become very low and excessively weakened if excessive amounts of water are involved in the reaction.

The quality of the cured resin is totally dependent on the amount of water it contains, so it is important to supply only that which is required for the desired gel or foam. In the repair of wet cracks, many contractors simply inject the raw resin, relying on the moisture therein to initiate the cure. In dry cracks, injection holes can be drilled with water or the cracks can be flushed prior to injection of pure resin.

Because hydrophilic formulations attract and hold large quantities of water, they are especially prone to significant shrinkage if allowed to dry. These formulations are thus best avoided for use in areas that are likely to dry periodically. Because some hydrophilic formulations require a minimum amount of water to react fully, good practice dictates that the grout is stream-mixed with water during injection with the use of an appropriate dual pump proportioning system.

Small amounts of water will provide the strongest gels, and they will become progressively weaker as the amount of water is increased. Once the water provided has reached a given amount, which is dependent on the exact chemical formulation, foaming will start. The density of the foam is directly proportional to the amount of further water added, decreasing as the amount of water increases. There is a point beyond which further water will be included in the cured product; however, that will generally not occur until an extremely weak foam has developed. Although most of the readily available hydrophobic urethanes will behave as previously described, it is possible to formulate resins that will not foam. The gel produced from these materials, however, will become increasingly weak as further water is included.

7.3.2.3 CHEMICAL ORIGIN

Urethane grouts are of two different categories chemically, those based on toluene diisocyanate (TDI) and those based on diphenylmethane diisocyanate (MDI). The designations TDI and MDI are universally used and well-established terms and are often used alone for reference. TDI-based formulations, which are usually hydrophilic, will combine with all available water during initial cure. The result is a gel or flexible foam, the strength of which depends on the proportion of the combined water. These grouts stick tenaciously to wet surfaces and require effort to remove or clean up if they are deposited on an exposed surface. Spills of hydrophilic urethane grout are quite unsightly, as illustrated in Figure 7.13.

MDI formulations are generally hydrophobic, and although requiring water to react, they will accommodate only a specific amount of water. These usually do not develop a good bond to wet surfaces, as they tend to reject surface water. They can be formulated to provide flexible or rigid foams or solid masses. Because they can be compounded to be of very low viscosity (50 centipoise or less) and to promote extraordinary expansion, they are preferred for injection into granular soil.

Although the TDI liquids are generally safe to handle, their vapor pressures can exceed established threshold limits. Full protective precautions are thus required with their use. MDI mixtures, however, have lower vapor pressures and are less hazardous. With either type, airborne droplets, such as may result from a leaking delivery line, are hazardous, although once

FIGURE 7.13 Spills of hydrophilic urethane are unsightly and stick tenaciously to any surface.

cured, the resulting gels are not. The vapor pressure of either type will rise with temperature and become excessive at temperatures greater than about 150°F (66°C). The safety aspects of working with urethane grouts are thoroughly covered in "Recommendations for the Handling of Aromatic Isocyanates" (1976).

7.3.3 Water-Soluble Acrylic Polymers

Mixed acrylic polymer grouts, sometimes referred to as *hydrogels,* are waterlike solutions that are highly penetrable into both soil and cracks in rock, concrete, and like substances. These will react into flexible gels, which tend to swell with continued exposure to water. Unfortunately, they also shrink if allowed to dry, and although they expand upon rewetting, the original gel properties are never completely recovered. These grouts are prepared by first mixing two different stock solutions, which are then stream fed for final mixing at the injection collar. The two components do not ordinarily require physical mixing, but are self-blending when combined. Because they require substantial amounts of mix water, preparation to react in a 1:1 proportion is usually possible.

7.3.3.1 ACRYLAMIDE GROUTS

Acrylamide grouts are waterlike aqueous solutions, which are highly penetrable. The base solution consists of two basic monomers, acrylamide and methylene-bis-acrylamide, which are proportioned by the grout manufacturer to provide the desired gel. The optimal proportions are about a 95:5 ratio, which will produce a clear, firm gel, but the ratio can vary either way. Increasing it to 97:3 will result in the production of a very elastic, sticky gel with relatively low strength, whereas reducing the ratio to about 90:10 will produce a stronger, opaque white gel (Karol, 1990).

Solution grouts of these resins can have solids contents up to about 20 percent and yet maintain low viscosity, similar to water. The cost of the grout is directly proportional to the solids content, and most work employs a solution containing about 10 percent solids. This will provide an adequate strength for most applications. Some of these compounds will promote gels at concentrations as low as about 3 percent, however, and they can be used at strengths as great as about 20 percent. Obviously, stronger gels result from higher solids content. Because they are easily diluted, higher solids contents should be used to reduce the effects of dilution when injection is into moving water.

The grout consists of a base solution combined with two or more reaction components. The reaction system is typically an aqueous solution containing an initiator, commonly ammonium persulfate, and an organic activator. The gel time of the grout is strongly affected by temperature, though varying the makeup of the reactant system and the proportion mixed into the base solution renders it controllable. A benefit of this grout is its ability to be used when gel time must be very short, often only seconds. The reaction time can be lengthened up to several hours by the inclusion of an inhibitor in the reactant solution.

Once mixed, the solutions are quite stable; however, they are subject to gelation upon exposure to ultraviolet rays and must be shielded from sunlight. Exceptional and precise control of the gel time is facilitated through variation in the makeup of the stock mixes, including the addition of appropriate chemical amendments, and adjustment of the final mix proportions. The gel is soft upon initial setting, enabling it to be pumped out of the delivery system if necessary. This facilitates the use of very short gel times without risk of the grout's hardening within the system.

Although these are benefits from the standpoint of handling, they can be significant limitations to effectiveness. Because the gel is weak in the early stages, it is subject to extrusion from significant voids or large defects within the formation. Common to water control work is an increase in the driving force of the water once the "safety valve" effect of the larger leaks has been lost due to their stoppage during initial grout injection.

The gels can be stiffened by inclusion of filler materials such as cement, pozzolan, clay, and other finely divided materials. Although manufacturers recommend the use of filled grout where additional strength is required, these are probably not the best formulations to be used for such requirements. If the void network being filled will accept particles of these fillers, there is no logical reason to use this much more penetrable (and expensive) grout. If it is used, be aware that inclusion of the fillers will result in a dramatic reduction in the grout's penetrability.

Acrylamide grouts were among the first commercially available chemical solution grout systems. They were heavily marketed by the American Cyanamid Company under the trade name AM-9. As part of the marketing effort, comprehensive research into their properties and application was performed. First available in 1955, these grouts were extensively used for water control applications until 1978, when they were suddenly removed from the market. This was due to a toxicity problem in Japan, which resulted in a ban on use in that country, although their production continued for export. The cured grouts are innocuous, but the individual components are toxic, so unmixed solutions can present a serious risk.

Though not widely promoted, acrylamide grouts are still available in the United States and continue to see extensive use for the stoppage of leaks in sewers and pipelines. They are generally sold only to firms that are properly equipped and have personnel especially trained in their safe use. The base material is available in either a dry granular form or a fluid slurry. The slurry is considered less hazardous than the dry form. The individual components are proportioned and impelled by separate synchronized pumps to the point of injection, where they are stream mixed.

A variable proportioning pump system is required to enable variation of the mix ratio during injection as required for gel time control. Changes can also be required to accommodate temperature changes or variation in the leakage. The reactant portion of the grout is highly corrosive, so mixers, pumps, and other appurtenances on that side of the pumping system must be of plastic or stainless steel. Moreover, the reactant's contact with iron will accelerate gel time, so it should be avoided.

The acrylamide monomer is a neurotoxin, which dictates careful handling and avoidance of direct contact or inhalation of the vapors. The reaction is nonreversible, and the gelled grout chemically stable, inert, and nontoxic. Most work now performed with these grouts involves sophisticated pumping systems designed to minimize physical handling and exposure to both unmixed ingredients and the resulting grout. They are commonly used in pipelines, where they are injected from remotely controlled vehicles, used in combination with closed-circuit TV observation.

7.3.3.2 ACRYLATE GROUT

Acrylate grout was made available and promoted as an exact replacement for acrylamide grout when AM-9 was taken off the market. Although the chemistries of the gelled grouts are similar, the premixed origins of these materials are different. Acrylate grout uses the same catalyst system, initiators, and promoters and offers many of the advantages of acrylamide, although it is not as penetrable. Further, it does not develop as great a strength and the reacted gel will absorb available water up to about 400 percent of its volume if unrestrained. This results in lower strength, so use should be limited to fine defects that provide restraint to volume expansion.

Acrylate, which forms the base solution, is a prepolymer and is thus not neurotoxic, as is acrylamide monomer. The mixing and pumping equipment requirements are essentially the same for both grout types. Acrylates are generally not as stable in solution as the acrylamides, however, and their shelf life, even in unopened original containers, is quite short. Some emit a very foul odor and they can be unfriendly to handle, so although marketed as a replacement for acrylamide, they are very different.

Both acrylamide and acrylate gels are highly elastic, so soils with which they are solidified are subject to substantial creep when under a continuous load. Krizek et al. (1992) provide creep data, shown here as Figure 7.14, illustrating this property very well. Their work agrees with that reported by Warner (1972), wherein AM-9 acrylamide grout exhibited nearly 20 percent creep. For this reason, the use of these grouts should be limited to water control work, as their cured properties are not satisfactory for most strengthening.

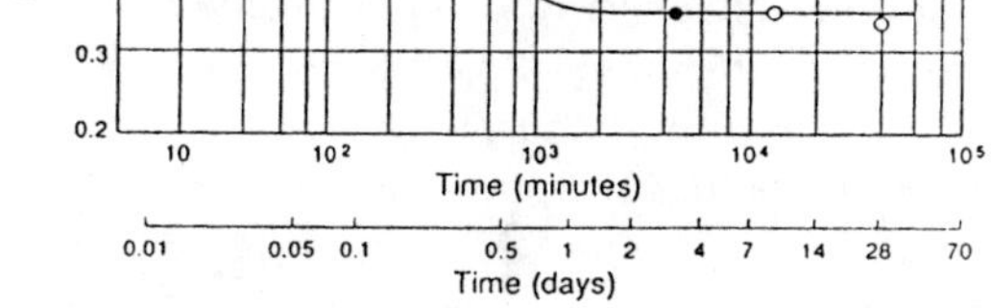

FIGURE 7.14 Creep performance with acrylate solidified sands. (From Krizek et al., 1992.)

7.3.3.3 N-METHYLOLACRYLAMIDE GROUTS

The N-methylolacrylamide grouts are somewhat in a middle ground between the other acrylics in composition and performance. Their gel time can range from a few seconds to an hour or more, and the result is a flexible gel. Once fully reacted, the gel will be stable and irreversible. It will, however, absorb water up to about 200 percent of its original volume. Normally, the use of these grouts should be restricted to applications that prevent it from swelling.

7.4 RESINOUS GROUTS

The term *resin* is loosely used to describe polymers that, once cured, tend to flow when under continuous stress and/or elevated temperature. Resins usually soften, undergoing a significant reduction in elastic modulus at high temperatures, but stiffen and become brittle as temperature lowers. Resinous compounds are used in the grouting of structures and have also seen some use in grouting rock, where both high strength and durability are required.

Resinous compounds used in grouting, include, in order of descending usage:

- Epoxy
- Urethane
- Phenolic foams
- Polyester

7.4.1 Epoxies

True epoxies are always supplied as two component materials, a basic epoxy and a hardener. These resin materials are available in a wide variety of types and consistencies and can be formulated to provide a nearly infinite range of cured physical properties. They develop into cured masses that vary from soft/rubbery to hard/

rigid and even brittle compositions. Whereas some formulations are for dry environments only, others are moisture compatible and many will even cure under water. Epoxies are very sensitive to temperature; however, formulations are available for use in both very hot and very cold environments, including below freezing.

The word *epoxy* is descriptive of a chemical reaction—specifically, the linking of the ethelyne oxide ring with a reactant material (Lee and Neville, 1967). There are virtually an infinite number of chemical formulations that can properly be considered epoxies. Accordingly, it is not possible to describe properties that are typical of all potential formulations. However, those commonly used in grouting become brittle when cold and significantly soften as temperature rises, although the magnitude of variation differs between formulations.

Freshly mixed, unreacted epoxy also has a range of consistencies, which vary from near waterlike fluids to very stiff nonsag pastes. Regardless of consistency, they all tend to be sticky in the uncured state. Ideally, the different constituents are supplied in two distinct colors, as illustrated in Figure 7.15, so that a third color will result from mixing. The uniformity of that final color and the absence of color striations will confirm that the required thorough mixing has occurred.

FIGURE 7.15 Different colored components will result in a uniform third color when properly mixed.

Epoxies constitute an extremely complex family of chemicals. When properly used, they are advantageous for a variety of applications, but can be difficult to apply and are extremely sensitive to misuse. The basic epoxy component is characterized by the epoxy equivalent, which is the weight of resin that contains 1 molecular unit of epoxy groups. These are sometimes referred to as epoxide sites and are the elements responsible for the eventual polymerization. The hardener is specifically proportioned to provide an equal number of reactive units so as to maintain *stoichiometry*, which is the numerical matching of reactive groups in both the resin and the curing agent. The two components must thus be precisely proportioned so that an equal number of reactive units is supplied in each, because every unit of hardener must connect with its epoxy counterpart.

The chemistry of the basic epoxy is always the same, but hardeners can vary widely. Interestingly, the two most often used hardeners have opposite pH values. Those based on amines, which are the most frequently encountered, are acidic, and those based on mercaptan are alkaline. It is through composition and manipulation of the hardener that the epoxy chemist is able to constitute systems offering cured resins with a wide range of properties.

Basic epoxy is sticky, whereas the hardener systems can vary from thin, waterlike fluids to sticky pastes. To complicate matters, both are dramatically affected by temperature change and the flow characteristics will change with variations therein. For example, a significant increase in temperature, such as that caused by the sun rising so as to shine on the injection hoses, can result in a substantial decrease of the base component's resistance to flow, but have little, if any, effect on the hardener. A good understanding of these properties is essential if one is to avoid substantial problems in their use.

As mentioned earlier, the two components must be combined in precise proportion and thoroughly mixed to ensure proper polymerization. Although the particulars of different formulations vary, even slight variation in the proportioning of some formulations can result in a substantial change in the performance of the cured resin. In the case of batch mixing, each component must be carefully measured, which is difficult because of the sticky resin, that tends to cling to both the mixing element and the container. For this reason, positive displacement proportioning pump systems are commonly used.

Even with the best pumping equipment, however, the differential properties of the components can be a problem if subject to varying temperature. The efficiency of practically all pumps will vary according to the pressure head and the viscosity of the material being displaced. Because pressure variations are inherent in most injection and flow properties of the components subject to substantial variation with temperature change, the use of high-efficiency, positive displacement pumps, along with frequent checks of the used quantity of each component, is mandatory. Rotary-type pumps, which virtually all result in some "slip" under pressure, should be avoided.

Epoxy grouts can be composed of 100 percent solids and/or can contain one or more amendments, extenders, diluents, or finely divided fillers. To enhance penetrability, diluents are often included so as to reduce the viscosity and/or surface tension of the mixture. Where these are used, they should be reactive, that is, contain reactive groups, which become an integral part of the cured resin. Although some formulations contain fillers as supplied, suitable filler materials can also be intermixed immediately after the two resin components have been thoroughly blended.

Most fillers used during the original manufacture of a resin system are finely divided solids, usually finer than the No. 200 (75 μm) sieve. Both the cleanliness and moisture content of a filler are of great importance to ensure a proper bond with the resin. For this reason, many manufacturers recommend that only their own supplied fillers be used, and some will sell their products only in kit form, including the filler. Separately obtained fillers can provide proper results, however, but careful attention must be given to matching the properties the resin manufacturer recommends, especially those for moisture content and gradation.

The proportions of epoxy components can vary from 1:1 to 100:1 or greater. Most systems commonly used in grouting involve ratios of 1:1, 1-1/2:1, or 2:1. Obviously, mixes of equal parts are preferable in that they will be the least sensitive to proportioning inaccuracies. Mix ratios of 20:1 are not uncommon, however, especially with formulations for use in very cold or otherwise special environments. Many available systems will cure at temperatures below freezing, but other factors such as condensation moisture or the existence of ice often control the work in such conditions.

In general, epoxies provide extremely high bond strength, and for this reason they are often used in the repair of concrete and masonry structures. The cured resins have high resistance to most chemical exposures, and especially to strong alkalis. Most formulations will provide good compressive and tensile strength, although they are typically subject to high creep, especially at elevated temperatures. Commonly used formulations nearly always have strength characteristics greater than those of the formations into which they are injected. As with other properties, however, the strength potentials of epoxy can vary greatly, especially with temperature variation. Typical values for formulations commonly used in grouting fall within the ranges shown in Table 7.7.

An important property of epoxy is the *heat-deflection temperature* (HDT) of the cured resin.

TABLE 7.7 Typical Strength of Epoxy Grouts

	psi	MPa
Compressive	4,000–15,000	27.6–103.4
Tensile	3,000–8,000	20.7–55.2
Bond	2,000–4,000	13.8–27.6

This is determined in accordance with ASTM Standard D 648. In essence, the HDT is the temperature at which the cured resin changes from a solid to a plastic. It is usually within a range of 120° to 130°F (49° to 55°C). Even a slight change in the temperature will result in a change in the elastic modulus, or stiffness, of virtually all cured epoxies. The elastic modulus is subject to change among different formulations and, in fact, will vary tremendously even within a given formulation. The elastic modulus will always increase as temperature decreases, and lower as the temperature increases. Virtually all epoxies will noticeably soften at temperatures above about 110°F (43°C). The magnitude of variation and the exact behavior differs among various formulations, however. Variation in stiffness can be reduced by filling the system with sand or another inert filler.

The pot life or open time of epoxies are strongly influenced by both temperature and mass. Epoxies will always set faster at elevated temperatures, and dwell time will be extended as the temperature descends. The pot life of mixed resin will vary according to the formulation, but is usually less than one hour, and very often only five or ten minutes. It can be extended through composition of the hardener, and formulations with an open time of up to twelve hours or more are available.

Epoxies are exothermic; that is, they produce heat as a by-product of the chemical reaction of their cure. The exothermic heat of most formulations is fairly high, and this has led to many problems in the field. The gel time provided in the manufacturer's literature is usually based on mixing a fairly small mass, seldom more than a few hundred grams. With larger amounts, the heat builds on itself and set can be greatly accelerated. In large masses, say several quarts (liters), of the neat resin, boiling can occur, resulting in a composition with the appearance of caramel popcorn.

The amount of heat generated by a particular formulation is dependent not only on the amount of the mass, but also on the ambient temperature as well as that of the formation into which it is injected. It is therefore important to select an epoxy formulation that possesses exothermic properties compatible with the factors anticipated in each individual application. Accordingly, where large masses of material are involved, especially if the ambient or formation temperature is high, only low-exothermic formulations should be used. Conversely, in the case of small quantities or thin films and a cold ambient or substrate temperature, the set time will be greatly extended, so a high exothermic (hot) formulation will probably be best.

Many epoxy resin applications have been unsatisfactory because of the resin's boiling as a result of excessive exothermic heat. When selecting epoxy, careful attention must thus be given to the heat that will be generated, the heat absorption potential of the formation into which it will be placed, and any environmental temperature conditions that will affect the work. Where large masses are used, intermixing with sand, or preplacing larger aggregate in the void into which the resin is injected, should be considered. This not only reduces the total amount of heat-producing resin, but also increases the available heat absorption media. Either the substrate, admixed material, or both can be artificially cooled to increase their ability to absorb heat. Although epoxy resins thicken and become harder to pump when cold, in warm environments it can be helpful to cool them, as in the freezer illustrated in Figure 7.16, prior to mixing and injection. In

FIGURE 7.16 A household chest freezer is used to cool epoxy components in warm weather.

extreme cases, pipes or tubing, through which cool air or chilled fluid can be circulated during the exothermic period, can be installed within the mass.

The significance of temperature to the behavior and curing of epoxy is well recognized. In fact, ASTM C 881, "Standard Specification for Epoxy-Resin–Base Bonding Systems for Concrete," addressees the subject by dividing epoxy formulations according to their temperature behavior. The various formulations are classified and separated into groups based on the recommended application temperature. Six classes, A through F, are intended for use in different temperatures, as follows:

Class A—For use below 40°F (4.5°C), the lowest allowable temperature to be defined by the manufacturer of the product

Class B—For use between 40° and 60°F (4.5° and 15.5°C)

Class C—For use above 60°F (15.5°C), the highest allowable temperature to be defined by the manufacturer of the product

Class D—For use between 40° and 65°F (4.5° and 18.0°C)

Class E—For use between 60° and 80°F (15.5° and 26.5°C)

Class E—For use between 75° and 90°F (24.0° and 32.0°C)

Any epoxy used where moisture is present must be formulated to be resistant to that moisture. In addition, special provisions may be required to prevent entrapment of condensation resulting from exothermic heat developed during the curing reaction. Condensation entrapment can result in severely reduced bond strength. Many field problems with epoxy have been related to a combination of exothermic heat and moisture.

Epoxy base systems can also be formulated to develop lightweight closed-cell foams. Although foam products are more common to other resins, especially those based on polyurethane, epoxy foams are distinguished by their especially tenacious bond strength. Their bond strength will always be greater than their tensile strength, so it is the latter that prevails; however, the tensile strength of epoxy foam tends to be higher than that of other resins. Both compressive and tensile strengths are related to the density of the final product and are variable. Resins have been noted with free foaming ratios up to about 20:1. The foaming will be reduced, however, when the material is injected into semirestricted areas, such as cracks or voids, as a result of friction between the expanding resin and the void surface.

7.4.2 Urethanes

Urethanes can be formulated to provide a nearly endless array of different forms and properties, which has led to their use in a variety of applications. As foams, they are among the most efficient thermal insulating materials known; as solids, they provide high-quality protective coatings. As previously discussed, they are used in grouting primarily for seepage control. Although the propensity of the resin to attract water can be an advantage in such work, it can be a serious limitation in working in areas of high humidity. On a project in an underground utility tunnel, leaking steam resulted in a very humid environment. The containers of hydrophilic polyurethane started to react soon after their containers were opened and created such a problem that this material had to be replaced with another grout.

Cured polyurethanes are especially resistant to high temperatures, and some formulations can produce compositions that will resist up to about 250°F (121°C). They can be compounded to have good resistance to both acids and alkalis and, in fact, are frequently used for high-durability coatings. The urethane compositions most often used in grouting are for the purpose of water control and are thus designed to be flexible and often of limited strength. It is possible, however, to formulate urethanes with strengths approaching those of epoxies.

7.4.3 Phenolic Foams

Phenolic resin foams are two-component systems that develop into crusted, rigid foam masses upon cure. They act and perform very much like the hydrophobic urethane grouts, but are distinguished by their high strength-to-weight ratio and their self-extinguishing inability to support flame. As with similar products, their strength potential is proportional to density, although foams of these materials offer some of the highest strength-to-weight ratios available, as shown in Figure 7.17, reproduced from Randolph (1960). When load failure occurs, it is in a brittle mode and happens at rather low total strains of 2–5 percent. These grouts promote good bonding to rock, concrete, masonry, and similar materials.

Phenolic foams are rather specialized, and their use is primarily in the mining industry, where their properties make them particularly

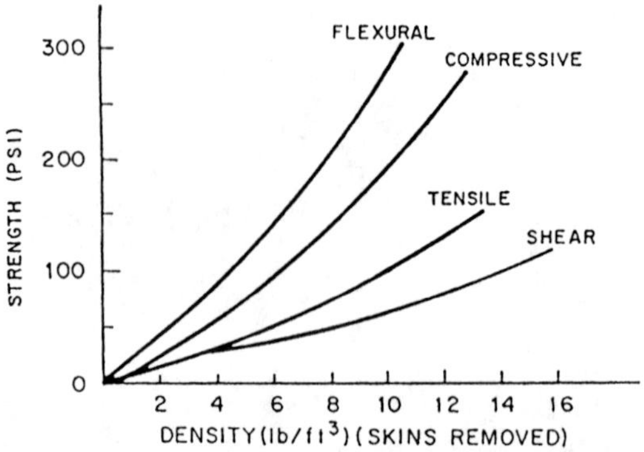

FIGURE 7.17 Properties of foam grouts.

advantageous. In mining, flammability presents risks that are unacceptable. Phenolic foams are resistant to temperatures up to 250°F (121°C) or more, will not support flame, and are self-extinguishing. Their high strength-to-mass ratio is especially important, as the unfoamed resin materials require relatively little space for storage in the usually limited confines of mines.

7.4.4 Polyesters

Polyester resins can appear much like epoxy, but they have several fundamental differences. They are subject to substantial shrinkage during hardening, and this results in reduced bond strength. To minimize potential shrinkage, they are usually filled to the maximum practical extent.

They do offer some advantages, however, being very forgiving and easy to use as well. as relatively unaffected by temperature. Their set time and curing rate are controlled by the ratio of base polyester to hardener component and can be adjusted from minutes to hours. Polyesters are very forgiving of poor proportioning or lack of proper mixing and will cure to their full properties even though poorly prepared. They provide good chemical resistance to most common substances. These grouts are not widely employed, although they have seen some use for filling voids in both rock and masonry.

7.5 MISCELLANEOUS GROUT MATERIALS

There are a few other materials used in grouting that do not fit into any of the aforementioned classifications. These generally see relatively little use, but they can be advantageous in some instances.

7.5.1 Asphalts/Bitumen

Asphalt materials, similar to those widely employed for roofing and asphalt paving, can also be used for grouting. These materials, which are derived from the residues of petroleum distillation, come in two forms: hard asphalt solids, which must be heated to become flowable, and *cationic asphalt* emulsions. Although not frequently used for grouting, asphalts in both forms have been injected almost as long as grouting has been practiced and are widely available. They are sometimes employed in connection with water control in geomaterials, where their special properties can be useful.

7.5.1.1 HOT ASPHALT

Asphalts supplied in solid form must be heated to become flowable; the asphalt commonly used in the roofing industry is referred to as *roofing grade* asphalt. Asphalts may be obtained as solids, usually molded in cylindrical forms with a weight of about 50 lb (22.5 kg), which are referred to as *kegs*, and are melted on the job site in portable kettles. Alternatively, hot fluid bitumen can be obtained directly from an asphalt plant. Although the precise temperature behavior varies according to the particular asphalt, those frequently employed in roofing and grouting are normally maintained at a temperature of about 300°F (149°C). The asphalt thickens and becomes very viscous as the temperature drops much below about 270°F (132°C), and a minimal threshold temperature of 275°F (135°C) is thus typically maintained.

Because these materials are so widely used for roofing, they are readily available as heated liquid in virtually all large population centers. They are routinely delivered in insulated, and usually heated, tank trucks and trailers and are commonly pumped to the roofs of structures, with rotary pumps integral to the melting kettle or delivery tank. Although these materials are readily available with suitable pumps in populated areas, considerable time and expense is required for constructing appropriate melting facilities in remote locations. To maintain the temperature of the asphalt, the lines through which it is pumped should be insulated. When time comes to shut down the injection, the pumps can be reversed to return any remaining material to the tank, or heated oil that will remain fluid upon cooling can be circulated.

My first experience with hot asphalt was in the *subsealing* of concrete highway paving, which was a fairly common practice before 1955. Holes were drilled through the pavement, and the asphalt was injected to create a moisture barrier on the underside of the concrete. The asphalt was melted in a trailer-mounted butane-fueled asphalt kettle, from which it was pumped into the holes. Working with hot asphalt was somewhat risky. As with all grouting, squirting leaks and spills were bound to occur, often resulting in burns experienced by the workers. Although very popular at the time, such work is seldom done now.

Today most hot asphalt is used in connection with difficult projects in which rapidly moving water must be controlled. The asphalt is highly flowable when hot, but rapidly thickens and becomes relatively immobile upon cooling. When injected into water, it tends to form stringers, the aspect ratio of which will vary, depending on the temperature and flow rate of the water. The stringers tend to gather and "bunch" together so as to plug the flow. Even when cold, asphalts do not have much strength, however, so it is prudent to follow up with injection of cementitious grout. Cement grout can also be simultaneously injected with asphalt, either through use of a coaxial header, as shown in Figure 7.18, or through adjacent grout casings, extending to the target area.

7.5.1.2 ASPHALT EMULSION

Asphalt is commonly available from pavement plants as a *cationic* emulsion, in which all the particles are positively charged. Because most mineral surfaces are negatively charged, the asphalt globules are attracted by and attach to most soil and rock surfaces. About 35 percent of a typical asphalt emulsion is water. When it is allowed to evaporate, the emulsion is said to *break,* leaving only the asphalt solid. Asphalt emulsions are very sensitive to change of pH, and exposure to a strong alkali will cause them to break. As the resulting asphalt falls out of suspension, it forms sticky, globular masses. In moving water, the globules will become elongated. Because they are positively charged, they will be attracted and tend to stick to adjacent soil or rock surfaces.

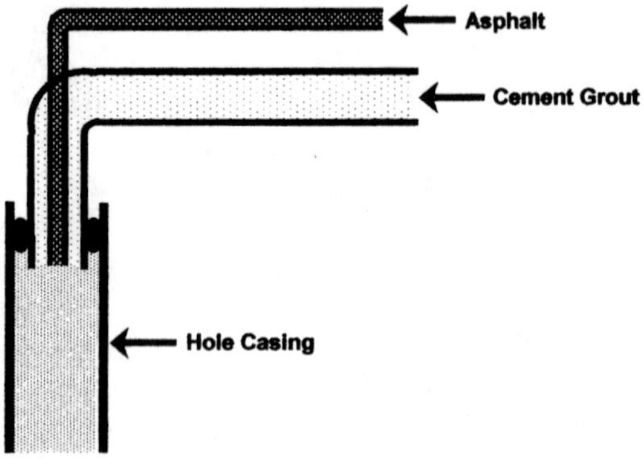

FIGURE 7.18 Coaxial injection header for plural-component grouts.

The time required to break and the size of the resulting globs are dependent on the ratio of the alkali to the asphalt.

An efficient source of alkali is hydrated lime. This can be conveniently made into a stock mixture by combining two gallons of water for each pound of lime. When injected at a 1:1 ratio with the asphalt emulsion, the break time will be on the order of ten minutes. As with most grouts, the reaction time is also dependent on, and will vary with, the prevailing temperature. To shorten the break time, the lime mixture can be strengthened or the proportion to be pumped increased. Conversely, the break time can be increased by reducing the proportion of lime.

7.5.2 Clay Grouts

Historically, clay materials have been used in grouting especially for the control of water. Clays in general, and bentonite especially, will swell once wetted. They are quite inexpensive in their dry state, and the economy becomes even greater when they are hydrated. They will expand up to several times their dry volume and produce grout that is equal to many times their original volume; thus, they are, without doubt, the least expensive materials available for grouting. There have been some reports of injection of unmodified pure clay mixtures, but these often exhibited poor durability and performance in the grouted work. In most reports, the clays have been combined with portland cement, which provides some minimum strength and yet is still quite economical.

When used for grouting, clay is typically prehydrated, that is, mixed with water and allowed to swell to its maximum volume, before being combined with the cement. To minimize the amount of clay, it is preferable to use a type that has the greatest amount of expansion upon hydration. Of the three principal types, *kaolinite, illite,* and *montmorillonite,* the last is by far the most active. For this reason, the most frequently used clay is *bentonite,* which is composed primarily of the montmorillonite mineral. Montmorillonite consists of extremely fine, flake-like microscopic particles, which have a laminated crystalline structure. These are usually smaller than 2 microns and are thus colloidal, in that water on their surface becomes a significant part of the structure. In addition to the water that coats the surfaces, the particles can also absorb substantial amounts of water (molecules) directly into their laminations.

Both the physical properties and amount of expansion of montmorillonite vary, depending on the nature of the substances contained in the adsorbed layers. Sodium bentonite gives the best performance, and will swell from 12 to 18 times its original volume. The highest-quality sodium bentonite originates in the state of Wyoming and is often referred to as *Wyoming bentonite,* or sometimes as *wy-o-gel,* in that it will form an expansive gel. Calcium bentonite is much less active, with an expansion of about 3 to 7 times its original volume. It does not have extensive use in the United States, because of the abundance of the more active sodium bentonite, but is commonly used in Europe where sodium bentonite is not widely available.

It is quite common for small amounts of bentonite (usually only 1 or 2 percent, by weight, of the cement) to be used as an admixture for cementitious suspension grouts. The intent of such use is to minimize settlement of the cementitious solids and the resulting bleed, and this material is not considered to be clay grout.

Grouts with significant clay contents have been used for both water control and strengthening of granular soils. As reported in AFTES (1991), "Grout proportions vary depending on the result required. Grouts used for waterproofing will include much clay and little cement, whereas grouts used in consolidation works will contain much cement and little clay." Unfortunately, this standard fails to give any definitive information on the actual proportions, other than to state that the clay content will depend on its "quality . . . and pre-hydration time."

A widely held opinion is that grouts that contain significant amounts of clay have not proven to be very durable. They have seen considerable use in the former Soviet Union, although very little has been published as to the durability of the work. In most of the cases reported, the clay was combined with portland cement to constitute a final mixture of about two parts clay to one part cement. The clays that were used were generally bentonites or other locally available materials that were primarily of sodium or calcium montmorillonite origin. These materials are now seldom used, but they do occasionally reappear, usually on projects that are being done on a very low budget.

CHAPTER 8

Tests for Evaluation of Grouts and Grouted Masses

8.1 FLOW CONES

8.2 SPECIFIC GRAVITY

8.3 EVALUATION OF BLEED

8.4 PRESSURE FILTRATION

8.5 SLUMP

8.6 STRENGTH TESTS
- 8.6.1 Special Provisions for Testing Expansive Grouts
- 8.6.2 Strength of Permeation Grouted Masses
- 8.6.3 Bond Strength

SIMPLE FIELD TEST METHODS to measure and ensure the quality and propriety of a grout are unquestionably among the greatest needs in grouting. Unfortunately, there are few tests that have been designed specifically for grouting; thus, in many cases, the tests that are used were originally intended for compositions other than grout. Even with these, few evaluation methods are available that can be easily executed in the field for either very thin or very thick grouts. A number of tests are available, however, for mid-range grout consistencies, which are generally fluid suspensions.

8.1 FLOW CONES

Several different configurations of funnels have been proposed or used to evaluate the flow properties of grout. The time required, in seconds, for a given amount of grout to flow out of a funnel, known as *efflux time*, is determined with these devices. Two different types of funnels have become somewhat commonly used in grouting. The "flow cone" (Figure 8.1), originally developed in the early 1940s, has been used for several decades by the U.S. Army Corps of Engineers, under the designation CRD-C611, for both research and field quality assurance. It was adopted by ASTM as C 939, "Standard Test Method for Flow of Grout for Preplaced-Aggregate Concrete (Flow Cone Method)," in 1988.

The top of the funnel is 7 in. (178 mm) in diameter, and it is 7-1/2 in. (190 mm) deep, with a 3 in. (75 mm) high cylinder on the top. A 1/2 in. × 1-1/2 in. (12.7 × 38.1 mm) orifice, which is usually removable, is on the bottom. The volume of the cone is 1725 ml (1.82 qt), which is to be within an accuracy of plus or minus 5 ml. Although not always with the cone, but required by the ASTM Standard, is a 3/16 in. (5 mm) diameter pointed level indicator rod, which is mounted normal to the funnel. The point of the level indicator is positioned so as to be precisely at the top of the funnel flare.

FIGURE 8.1 ASTM C 939 flow cone used to evaluate the flow properties of grout.

FIGURE 8.2 Marsh funnel.

In use, the cone is rinsed with water one minute before filling to the proper level with grout. A plug or finger is held over the orifice until the cone is filled, after which it is released and the time until the first break in the continuous flow of grout is noted. This must be timed and noted to within 0.2 seconds with an appropriately calibrated stopwatch. The efflux time for water is 8 seconds. Typical times for grout can vary up to 2 minutes, although the appropriateness of the test is questionable with efflux times much over 1 minute. Not provided for by ASTM, but of interest, is the availability from at least one manufacturer of a 3/4 × 1-1/2 in. (19 × 38.1 mm) orifice that can be used interchangeably with the standard one.

The Marsh funnel (Figure 8.2), long used to evaluate the flow properties of drilling fluids, is also often used for grouts. The top of the cone is 6 in. (152.4 mm) in diameter, with a height of 12 in. (304.8 mm) and an outlet orifice 3/16 in. (4.76 mm) in diameter by 2 in. (50.8 mm) long. The cone has a volume of about 1.6 qt (1500 ml), but the test result is properly based on the passing of 1 qt exactly. Correctly used, the cone is filled to near its top while a finger is held over the outlet. A 1 qt (946 ml) container is held under the orifice, the finger is removed, and the time, in seconds, that is required to completely fill the 1 qt container is noted.

There has been considerable confusion and misuse of this funnel. Although it has a capacity of about 1.6 qt (1500 ml), some technicians fill it only to 1 L (0.95 qt), and others to 1 qt (946 ml), and then wait for it to completely empty, often without a proper receiving container. In a particular instance, the funnel was filled all the way to a screen on the top, so that about a full 1.6 qt (1500 ml) was used, and the time for it to completely empty was recorded. The results from such practices can be very misleading. Not only are the beginning grout quantities subject to question, but some of the material is bound to stick to the funnel walls, which can compound the error. The efflux time for water should be 26 sec, and for grout it will vary from about 26 to more than 60 seconds. Grouts requiring more than about 75 seconds are not appropriate for use of this test.

The C 939 cone is typically made of cast aluminum with a removable stainless steel orifice. The Marsh funnel is usually in the form of a single continuous piece of molded plastic. Neither material can have wettability properties at all similar to those of the wide variety of geomaterials into which a grout may be injected. The cones do, however, provide confirmation of the uniformity of grout batches and are widely used in quality assurance testing.

8.2 SPECIFIC GRAVITY

Another device originally developed for the evaluation of drilling fluids is the Baroid Mud Balance (Figure 8.3). This is a simple and very rugged device for determining the density of drilling mud or grout. In application, the cup and beam are removed from the fulcrum, and the cup is dipped into the grout until completely filled. Excess grout on the beam or outside the cup is removed. Alternatively, to minimize the amount of grout that must be cleaned from the beam and cup, the cup can be filled by pouring grout into it, which has been obtained with a separate dipper. The beam and cup are then placed on the fulcrum, and a rider weight slid along the beam until the bubble of an attached spirit level is centered. The density is then read directly from a scale on the beam, on the fulcrum side of the rider weight. The Mud Balance is a particularly useful device for quality assurance testing of cement-water suspensions, in that the water-to-cement ratio can be easily and accurately determined.

FIGURE 8.3 Baroid mud balance used to determine the specific gravity of grout.

8.3 EVALUATION OF BLEED

As previously discussed, controlling the bleed of suspension grouts is important, especially when slow pumping rates are used or large voids are being filled. The bleed potential of a suspension grout can change quite markedly with changes of water-to-cement ratio, shear mixing energy, or changes in the properties of the cement or other constituents. Bleed evaluation is easily accomplished by filling a transparent tube or jar with the grout and, after about two hours, observing the amount of clear water that collects on top. Where accurate determinations are desired, 1000 ml graduated cylinders, as shown in Figure 8.4, are used. Such use is outlined in ASTM C 940, "Standard Test Method for Expansion and

FIGURE 8.4 Evaluation of the bleed of two cementitious suspension grouts with different water : cement ratios.

Bleeding of Freshly Mixed Grouts for Preplaced-Aggregate Concrete in the Laboratory."

In the C 940 test, the cylinder is placed on a level surface free of vibrations. It is then filled with 800 ml of grout, covered so as to preclude evaporation of the contents, and the time is noted. Readings are taken of the amount of bleed water on top of the grout at 15 minute intervals for the first hour, and one hour intervals thereafter. Generally, the test runs for no more than three hours, but it should properly extend to at least the set time of the particular grout being evaluated. The amount of bleed is expressed as a percentage of the original 800 ml and is determined as follows:

$$\text{Bleeding \%} = \frac{V_w}{V_1} \times 100$$

Where: V_1 = volume of sample at beginning of test (800 ml)

V_w = volume of decanted bleed water, ml

Although this test procedure is useful for laboratory evaluations, the required settlement time usually renders it impractical for field use.

8.4 PRESSURE FILTRATION

The result of pressure filtration is essentially that the bleed water is forced out, ahead of the grout, because of the pumping pressure imposed on it. The bleed potential can be evaluated for common cement grouts with a standard American Petroleum Institute (API) filter press or a Gelman pressure filter. The API instrument consists of a grout reservoir that is mounted within a rigid frame. The reservoir has a top cap that is fitted with a pressure inlet, and a bottom base containing a screen that retains the actual filter paper. The filter area is 7.1 in.^{2} (45.8 cm^2), and the normal working pressure is 100 psi (6.9 bars). Pressure is usually applied to the top of the reservoir by means of bottled nitrogen gas. Any liquid that escapes out of the filter when the chamber is pressurized is collected in a graduated cylinder placed under the reservoir.

The Gelman filter device is similar, although somewhat smaller. It has a filter 47 mm (1.85 in.) in diameter, which provides 17.34 cm^2 (2.68 in.^{2}) of filter area. As with the API chamber, pressure is applied to the top by way of nitrogen gas, as shown in Figure 8.5. A variety of filters are available; however, a disposable glass fiber filter works well with grouts based on common cement.

With either system, the expelled fluid is collected into a graduated cylinder placed under the chamber. It is not unusual, however, to have virtually all the free water pushed out of the grout specimen so that only gas continues to escape. This can occur in as little as a few minutes with unmodified cementitious suspension grouts. Subject only to the maximum pressure capacity of the grout reservoirs, any desired pressure can be applied to the grout. Ideally, this will be the highest pressure level planned to be used with the particular grout being evaluated. Although the rated pressure on the API filter press is only 100 psi (6.9 bars), API does have an alternate

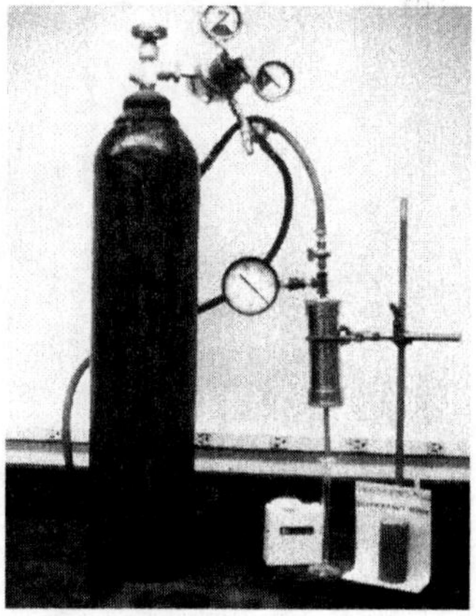

FIGURE 8.5 Pressure filtration evaluation using a Gelman pressure filter.

apparatus for use at both high pressures and high temperatures.

8.5 SLUMP

The ASTM C 143 slump test is very well established and is routinely used for evaluation of the consistency of concrete. Because it is so well known, there is a tendency by some to refer to or specify its use for all cementitious mixtures. This is inappropriate, however. The test involves the careful filling of a truncated conical mold with the composition to be evaluated, and then carefully pulling the mold off the specimen. As the restraint of the mold is released, the height of a plastic consistency mix will subside, as shown in Figure 8.6. The value of this subsidence, which is taken at the center of the specimen, denotes the material's slump. The C 143 mold is 4 in. (102 mm) in diameter on top, 8 in. (203 mm) at the bottom, and 12 in. (305 mm) high.

There are several other established slump tests, however, and they all involve molds of different sizes and configurations. The most common is the *half-size slump test.* It involves a mold that is 2 in. (51 mm) in diameter on top, 4 in. (102 mm) at the bottom, and 6 in. (152 mm) high. Although this test mold is not covered by any widely established authority such as ASTM, it is referenced in several codes of practice and widely used for evaluation of portland cement plaster and masonry mortar. Both of these are void of any large aggregate, as are most grout compositions. Although it is probably more appropriate to use the smaller mold for grouts, this is not commonly done, presumably because virtually every material testing laboratory has and regularly uses the larger C 143 mold.

FIGURE 8.6 ASTM C 143 slump test.

As previously discussed, ASTM Committee C-9 (1978) stated in its Special Publication 169 C that its standard slump test is not appropriate for either "very wet" or "very dry" mixtures, which, unfortunately, include most commonly used grout compositions. Further, the most important properties for grouts are penetrability and flowability, so they are typically rich in fine particles, and even sand, when used, is relatively fine. The mixtures are thus quite cohesive and often sticky and thixotropic, so that the level of shear stress required to start movement may exceed the force of gravity, on which slump is based. When acted on by the greater force of a grout pump, however, many of these mixtures will flow readily and, in some cases, actually behave like fluids. Thus, even though a mixture displays low slump in the test, its actual behavior when pumped can be quite different.

Another limitation of the C 143 test with grouts is the requirement for filling and compaction of the mold in three separate layers, with the volume of each being about one-third that of the mold. So far, there is no problem, but then each of these layers is required to be rodded 25 times with a 5/8 in. (15 mm) diameter steel rod in order to provide appropriate compaction. In very cohesive mixtures, this tends to leave holes in the grout, as its consistency does not allow self-closing as the rod is pulled out. It is thus often difficult to fill the mold, and virtually im-

FIGURE 8.7 Voids and tears in lower portions of test specimen.

possible to compact it according to requirements of the standard. An example of these problems is shown in Figure 8.7, in which the bottom portion of the specimen has not been completely filled. Further, torn areas caused by grout sticking to the mold are apparent.

There are other shortcomings of the test even when it is used with concrete, for which it was developed, and there are many caveats as to its applicability within the body of the C 143 Standard:

> This test method was originally developed to provide a technique to monitor the consistency of unhardened concrete. Under laboratory conditions, with strict control of all concrete materials, the slump is generally found to increase proportionally with the water content of a given concrete mixture, and thus to be inversely related to concrete strength. Under field conditions, however, such a strength relationship is not clearly and consistently shown. Care should therefore be taken in relating slump results obtained under field conditions to strength.
>
> If two consecutive tests on a sample of concrete show a falling away or shearing off of a portion of the concrete from the mass of the specimen, the concrete probably lacks necessary plasticity and cohesiveness of the slump test to be applicable.

Grouting specifications calling for a given slump or range of acceptable slumps, without calling out the dimensions of the mold to be used, and/or mandating use of the C 143 mold for grouting, have resulted in many claims and much litigation. Thus, except for the rather rare instances of *fill grouting,* in which standard concrete that is neither *very wet* nor *very dry* is being injected, the test should not be specified or used in grouting.

In those instances where some type of slump test is desired for a grout, the dimensions and other particulars of the intended mold should be specifically called out. Where the C 143 mold is to be used, it should be explicitly specified, but the method of filling and compaction should be otherwise called out. Where slump tests are used, the best method for filling is to pump directly into the mold. By forcing the grout in under pressure, it will become fairly well compacted. Tapping the mold can also be helpful to provide complete filling. Even with these procedures, however, it must be recognized that the test is not well suited for most grouts, especially those that are sticky or highly thixotropic. Thus, except in rare grouting applications, slump tests should *not* be specified.

8.6 STRENGTH TESTS

In most cases of geotechnical grouting, strength is of little concern as long as it is at least as great as that of the soil or rock being treated. In fact, a common error is specification of cementitious grout strength based on typical concrete or mortar standards. Specifying an unnecessarily high strength for grout is seldom beneficial, can present difficulties for the contractor, and almost always results in increased costs. In those cases where strength is important, either standard unconfined tests or triaxial compression tests can be made. Because grouts seldom contain any large aggregate, specimens are usually standard 2 × 4 in. (51 × 102 mm) cylinders or 2 in. (51 mm) cubes.

Cylindrical specimens are usually cast according to the provisions of ASTM C 31, "Practice for Making and Curing Concrete Test Specimens in the Field." They must have a diameter at least three times the dimension of the largest aggregate in the mix. The height should be twice the diameter. The widely used 6 × 12 in. (152 × 305 mm) molds commonly used for concrete are excessively large for most grouts, however, and are seldom needed or justified. According to Standard C 31, cylinder molds are to be filled in three layers, each being rodded with 25 strokes of a 5/8 in. (15.6 mm) diameter steel rod with a hemispherical tip similar to that used in the slump test. As with the slump test, this method of compaction does not work very well with the often sticky grout mixtures and may require modification.

The preparation of cube specimens is addressed by ASTM C 109, "Standard Test Method for Compressive Strength of Hydraulic Cement Mortars (Using 2-in. or 50-mm Cube Specimens)." The individual molds are to be filled in two layers. Each layer is to be tamped 32 times with a tamper made from a smooth nonabsorptive material and having a square tamping face 1/2 × 1 in. (13 × 25 mm), which is perpendicular to the length of the tamper. Unlike the round rod called for in the compaction of cylinders, this larger and flat surface, combined with the shallow depth of the cube mold, works satisfactorily with grout. Cube molds commonly contain three compartments and are made of metal.

8.6.1 Special Provisions for Testing Expansive Grouts

Expansive grouts, if left unrestrained, will expand so as to excessively reduce their density. The strength of all cementitious compositions is directly affected by density, so this will result in a lower indicated strength. Filled specimens should thus be capped with sufficient weight to restrain any expansive force, as shown in Figure 8.8. In the evaluation of expansive grouts, absorption of grout moisture into the substrates should be considered. A useful test developed to provide for both restraint and absorption is illustrated in Figure 8.9. In this test a 16 in. (406 mm) long piece of 8 in. (203 mm) clay sewer pipe is clamped to a steel base plate fitted with a grout inlet. The pipe is capped with a 2 in. (51 mm) thick precast concrete cap. Clamping rods on either side of the mold are tightened so as to prevent any heave of the grout as it tries to expand. Any rising water can escape out the minute space between the end caps and the clay pipe. Once the grout has hardened, the clamps are removed. Bond to the top concrete cap, if any, can be determined by placing lateral and uplift pressure on it. As further confirmation of ac-

FIGURE 8.8 Specimen molds restrained with steel weight.

FIGURE 8.9 Test setup for expansive grouts.

ceptable bonding to the overlying slab, cores can be drilled and broken off well into the sample.

8.6.2 Strength of Permeation Grouted Masses

In the case of permeation grouted geomaterials, it is not the strength of the grout that prevails, but rather, the strength of the solidified composite. To verify field strength, specimens are ideally carved from the actual grouted mass. This will require excavating to the grouted zone for access. Core drills can be used to obtain specimens of most geomaterials that have been grouted with cementitious grouts. They can sometimes be effective in soils that have been solidified with higher-strength chemical solution grouts, provided that the soil has been completely permeated and essentially all pore space filled. Grouted soil is usually of much lower strength than grouted rock, and some erosion and disturbance of the core surface often results from the drilling operation. It is thus useful to procure a core of greater diameter than required and then carve out acceptable specimens.

Very often, however, it will not be possible to obtain a proper core specimen of grouted soil. This is especially difficult where weaker grouts are used or the soil pore space is not completely filled, as is often the case. The strength of many grouts, and virtually all chemical solution grouted masses, are significantly affected by their moisture condition. It is thus crucial to preserve the moisture existing in the formation in all specimens until they are tested. This often precludes use of core drilling, as common core bits require water to cool them and expel the cuttings, which will wet otherwise dry specimens.

Where laboratory evaluation is employed to establish the effects of a particular grout on a confined mass of a given soil, complete saturation with grout is required. ASTM Standard D 4320, "Test Method for Laboratory Preparation of Chemically Grouted Soil Specimens for Obtaining Design Strength Parameters," is applicable to the preparation of such specimens. In this test the soil is compacted into vertical tubular molds to the desired density. Water is then injected from the bottom, at a pressure of not more than 5 psi (3.5 Pa), until the specimen is uniformly saturated. The grout is then injected from the bottom until fresh undiluted grout comes out through a vent on top. Once it has cured, the mass is removed and can be tested immediately; it may also be sealed within a plastic membrane, or placed in a moist cabinet with a temperature of 73.4 ± 2°F (23 ± 1.1°C) and a relative humidity of not less than 96 percent, until testing.

These sample preparation parameters were used, essentially, in the early stages of an extensive research program conducted in the 1960s and reported in the *Journal of the Soil Mechanics and Foundations Division of the American Society of Civil Engineers* (Warner, 1972). That program involved evaluation of more than 2500 individual laboratory-prepared specimens, as well as 40 field specimens gathered from a variety of actual field applications. Because a large number of test specimens were to be prepared and the injection was quite slow, even with a four-chamber test apparatus, more efficient preparation methods were investigated.

It was found that quality specimens could be prepared by slowly pouring the soil into molds that had been partially filled with the solution grout. The molds started with more grout than required, and the soil was manually pressed into it with thumbs until the densest possible condition was obtained. The excess grout was displaced and wasted, and the specimens screeded off to provide a smooth surface. It was found that uniformity of the resulting samples correlated well with the injected specimens. As this sample preparation method was much faster and more efficient than injecting into tubes, per the subsequently adopted ASTM standard, it was used for most of this effort. It is suggested that this much faster procedure can produce high-quality test specimens and is therefore fully satisfactory.

Regardless of the preparation method, virtually all laboratory tests rely on remolded specimens. Because the positioning and structure of the individual soil grains will likely differ from those in the natural deposit, the indicated strengths may not compare closely to those of the in-place grouted soils; however, the indicated strength should give a good idea of the strength levels that can be reasonably expected.

The horizontal surfaces of all specimens must be perfectly smooth and parallel when placed in the compression machine. It is possible to cast or cut a specimen of the proper profile, but it is much easier to correct any deviations by providing a proper cap. Although concrete cores require a very strong cap, grout specimens seldom have strengths of even 500 psi (3.4 MPa), and thus capping with more easily used material is sufficient. An entirely satisfactory method is to place a daub of casting plaster on the ends of the core and then press a metal plate onto it so that the plaster is spread completely over the core surface and the plates perpendicular to the axis. It has been found that 3 × 3 in. (76 × 76 mm) squares of 1/8 in. (3.2 mm) thick steel plate work well for 2 in. (51 mm) cubes or cylinders.

Because chemically grouted masses are subject to significant creep, and the *fundamental strength*—that is, the strength under sustained loading—is often significantly less than indicated by the rather rapid loading of standard tests, it is preferable to employ variable stress compression apparatus, such as shown in Figure 8.10. The simple device illustrated operates on compressed air, and the imposed load level is easily controlled by varying the air pressure with a regulator, such as shown at the bottom right. A long-term, constant force is applied to the specimen until all strain (creep) ceases. This equipment is both rugged and simple and can be used on the job site, facilitating testing immediately upon procurement of samples.

FIGURE 8.10 Constant stress compression test apparatus.

8.6.3 Bond Strength

In most geotechnical grouting, the required bond strength of the grout to the host formation is not great, and the commonly used grout materials usually provide considerably more strength than required. Testing for bond strength is thus not a frequent requirement in most grouting. Yet there are times when bonding to the substrate is required. Such is the case in applications involving concrete or masonry repair and, occasionally, in the grouting of rock, especially in connection with rock bolts or anchors.

Although not widely recognized, the bond strength of compositions based on portland cement is generally high, provided that the compound does not shrink excessively. Unfortunately, to aid pumpability and improve penetrability, most cement grouts contain so much excess water that high shrinkage occurs. Shrinkage is proportional to the water : cement ratio, resulting in an ever-increasing reduction of bond as the amount of water increases. Bonding fails when shrinkage stresses become greater than the bond strength. Grouts with a water : cement ratio less than about 0.35 will not undergo much shrinkage and thus provide good bond strength, whereas those with higher ratios will experience significant shrinkage and thus lower bond strength.

Resinous grouts, however, virtually all provide high bond strength. This is especially true of epoxies, which exhibit some of the highest bond values obtainable. The primary objective of most grout injection with epoxy is to integrate the different surfaces and/or pieces of the host formation into a competent monolithic mass. Required bond strength is usually a value exceeding the tensile strength of the host material, which can be quite high, especially in the case of very competent concrete or rock. The tensile strength of such materials is on the order of one-tenth the compressive strength, and the bond is usually required to be at least as great. Values up to or even greater than about 500 psi (3.4 MPa) are not uncommon. Although these can be easily procured with resins, cementitious grouts will require bond-enhancing admixtures, as well as very low water : cement ratios, to comply.

Laboratory tests can establish the bond strength of a particular grout and are often used to prequalify different grouts for a given requirement. The *slant shear* test is the most commonly used. It involves a normal concrete test cylinder that is diagonally saw-cut longitudinally. Most typically, 3×6 in. (76×152 mm) specimens are used, but larger sizes can be employed. Alternatively, two pieces of the correct configuration and dimensions can be precast. In the case of existing concrete or rock, a core of appropriate dimensions is used. Once cut, the two halves are banded together, the joint sealed (usually with a paste epoxy), and the subject grout injected so as to adhere the two pieces. The specimen is then placed in a compression machine and loaded to failure. The bond strength of the grout is determined by dividing the load imposed at failure by the area of the bonded surface. Obviously, if failure is outside the bond plane, the grout exceeds the tensile strength of the formation.

This test can be conducted as set out in ASTM Standard C 882 "Standard Test Method for Bond Strength of Epoxy-Resin Systems Used with Concrete by Slant Shear." It specifies the use of precast sections for the test core. Although it does not limit the strength of the precast sections, it does require a minimum of 4500 psi (31 MPa) at seven days. Presumably, the precast sections could be of such great strength that the grouted bond would fail before the concrete. The ASTM Standard does not provide for job-procured cores, but is directed to material producers that need to establish the blanket strength of their products.

Another test that is performed in a standard compression testing machine is the splitting tensile test as outlined in ASTM C 496, "Standard Test Method for Splitting Tensile Strength of

Cylindrical Concrete Specimens." It is employed to evaluate the bond of a single repaired fracture, wherein a core specimen, with the grouted defect parallel to the axis of loading, and very near the center of the cross section, is required. The core is placed horizontally in the compression machine with the repaired defect in a vertical orientation, as illustrated in Figure 8.11. Alternately, the test may be used to determine the strength of several different joints or fissures within a core, such as would be the case in masonry. In all cases, plywood bearing strips are placed between the core and both the upper and lower platens of the testing machine. These are typically 1/8 in. (3 mm) thick, about 1 in. (25 mm) wide, and at least as long as the test specimen. The two bearing strips of the core must be perfectly parallel and a uniform distance apart, which sometimes requires preparatory grinding of any high areas.

The specimen must be precisely centered in the machine so that the compressive force is applied exactly in the center of the circular cross section. A continuous load is applied at a constant rate within a range of 100 to 200 psi/min (689 to 1380 kPa/min) until splitting failure occurs. The splitting tensile strength is then determined by calculation, as follows:

$$T = 2P/\pi l d$$

Where: T = splitting tensile strength, psi (kPa)
P = maximum applied load indicated by the testing machine, lbf (kN)
l = length, in. (m)
d = diameter, in. (m)

The splitting tensile test is extremely sensitive to proper positioning of the specimen, which must be precisely centered in order to return valid data. The ASTM Standard goes into great detail as to the proper methods for both locating the bearing lines and precise positioning in the machine.

Another test, which can be used in either laboratory evaluations or in the field in situations where the bond line is parallel to the face and at a reasonably shallow depth, is the ACI 503 direct tensile test, part of a guideline reported by ACI Committee 503 (1989), "Use of Epoxy Compounds with Concrete." For this test, a core drill is used to penetrate perpendicular to and beyond the bond line to be tested, usually with a 2 in. (51 mm) core. A threaded pipe cap, or a steel disk containing a threaded hole in the center, is epoxy adhered to the top of the core and allowed to cure. The core is pulled out in direct tension by means of a pulling device that will ensure concentric loading and axial tension. Pullout mechanisms conforming to the requirements of ASTM C 900, "Standard Test Method for Pull-out Strength of Hardened Concrete," are adequate and commonly used, as illustrated in Figure 8.12. The unit tensile strength is determined by simply dividing the maximum pullout force by the specimen area.

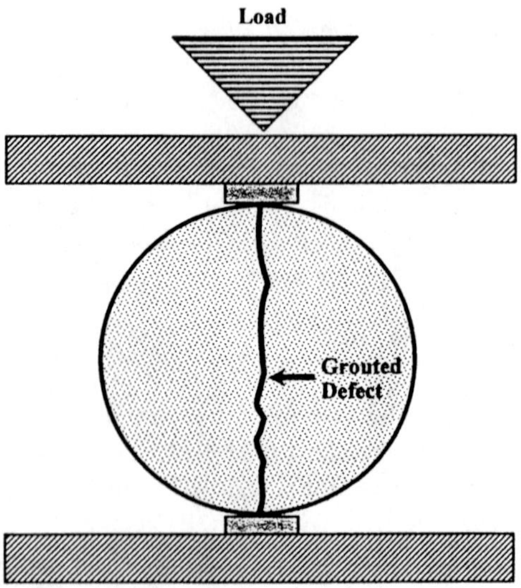

FIGURE 8.11 Horizontally placed core with repaired defect in a vertical position near center.

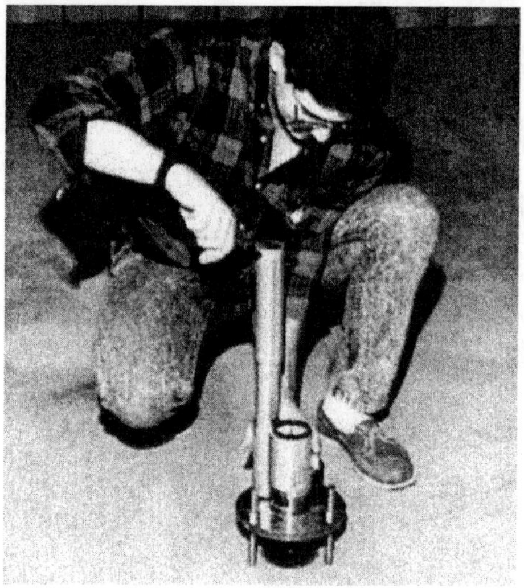

FIGURE 8.12 Typical pullout device for direct tension test.

Many have attempted to perform this test by using a center-hole hydraulic jack or another simple loading source. This usually is not satisfactory, however, as the accuracy of the results is completely dependent on a perfectly concentric pulling force. Unless such a force is provided, which is difficult with simpler tools, the specimen tends to break off from lateral forces and the indicated value can be significantly lower than the actual bond strength.

Direct shear tests can be conducted on specimens of different grout materials or grouted formations. There are no clear standards for such tests, and special appliances are often used in their performance. Shear testing requires the specimens to be of a size and configuration appropriate for the anticipated test methods and appliances, which often requires diamond sawing in their preparation. Shear testing is usually limited to highly specialized applications that require exceptional input as to the realized strength. These tests are therefore best performed by highly experienced engineers and technicians who not only possess the required knowledge of material science, but also have the ability and resources to fabricate the required test appliances.

CHAPTER 9

Injection Fundamentals

9.1 PRINCIPLES OF FLOW

9.2 EFFECTS OF GROUT RHEOLOGY

9.3 GROUT PENETRATION

9.4 INJECTION PRESSURE
- 9.4.1 Pressure–Volume Relationship
- 9.4.2 Appropriate Pressure Levels

9.5 PUMPING RATE

9.6 PRESSURE BEHAVIOR

9.7 VACUUM

REGARDLESS OF THE TYPE of grout being used or the purpose of its injection, there are several important parameters that will always be present and must be considered. During injection, the pore structure of the formation into which grout is injected comes under increased pressure and is often subjected to mechanical disruption. The magnitude of such disruption can be effectively controlled through adjustment of the grout properties and control of the injection parameters. A good understanding of these factors and their relationships are thus fundamental to the proper design and field performance of grouting. It is essential to understand these factors and, with such knowledge, to make prompt adjustments of the grout composition or injection parameters.

9.1 PRINCIPLES OF FLOW

Flow is described as either *laminar* or *turbulent.* In laminar flow, all portions of the cross section of flowing material remain in their respective positions as they are being propelled forward. Because of surface friction, flow confined in a delivery line or linear void will move at different velocities, depending on its location in the cross section. The velocity will be slowest at the periphery of the contained grout, because of friction with the conduit surface, and will become faster as the distance from the periphery increases, as illustrated in Figure 9.1. Obviously, the fastest progress will be in the center of the conduit cross section, as illustrated.

In turbulent flow, the various portions of the cross section of flowing media constantly change position relative to each other, resulting in eddy currents within the stream. There is thus constant change in both the velocity and direction of flow within the conduit, although the general direction of movement is forward (Figure 9.2). Although the principal influences of flow generally do not begin until the material enters the delivery conduit, the pump action can also have an effect. Piston pumps, for example,

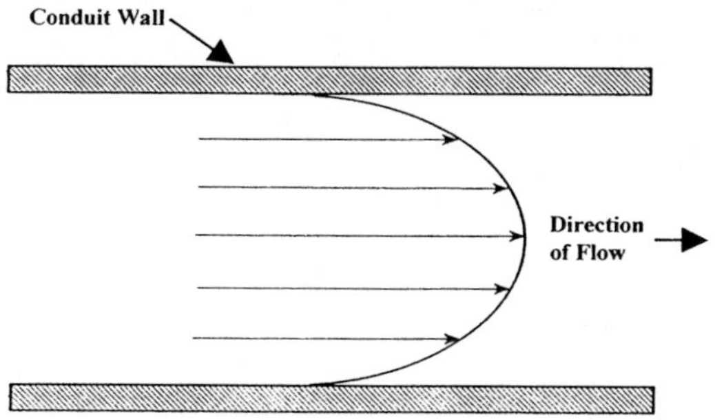

FIGURE 9.1 Laminar flow—velocity is fastest at core of flow.

result in pulsations in the pumped media, and this can continue into the delivery line as turbulence. Conversely, most rotary pumps produce a constant output and usually do not contribute turbulence.

When grout, or for that matter, any material, flows through a delivery system or void network, resistance to flow movement occurs. The flow of grout, and, in fact, of all fluids, including water, is affected by both the frictional resistance with the wall of the delivery line and turbulence-producing irregularities within the injection system or the formation being grouted. The effect is usually a pressure loss, which increases progressively with distance from the grout pump. This is commonly referred to as *head loss* and is represented as a portion of the pumping pressure.

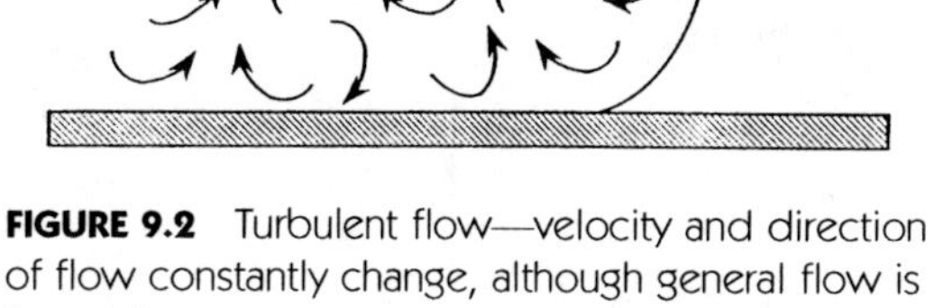

FIGURE 9.2 Turbulent flow—velocity and direction of flow constantly change, although general flow is forward.

The type of flow and the resulting effects are also influenced by the properties of the grout. For example, turbulent flow of a true solution grout would likely be of little consequence other than to slightly increase the required pumping pressure. But should the grout contain solid particles, such as the fluid suspensions commonly used in rock grouting, even though the pressure head loss would increase, turbulence would likely be advantageous in that it could help keep the solids in suspension, thus reducing separation. It can also cause intermixing of fines or other substances that are picked up by the advancing grout. Conversely, should the grout be of a thick plastic consistency, flowing as a solid extrusion, turbulence could result in separation of the ingredients and plugging of the injection system.

In the case of grouting deep vertical holes, or where the grout pump is located at a higher elevation than the hole collar, the force resulting from the static weight of the grout may be greater than the cumulative head loss resulting from friction. The density, or weight, of the grout can be significant, and it varies widely. Table 9.1 provides data for various cementitious grouts occupying 1 in.^{2} (645 mm^3) of vertically positioned line at different depths. Note that even neat cement-water suspensions, identical except for the water : cement ratio, will exert different pressure heads.

It is thus possible to develop positive *head gain* at the point of grout deposition, simply due to the weight of the grout itself, with no additional pump pressure. This often occurs when pumping downhill or down a vertical shaft. As a general rule, the value of head gain will increase with the increasing diameter of the grout line. In such cases it may be worthwhile to place a section of tightly coiled hose or another

TABLE 9.1 Approximate Static Head Weight of 1 in.^{2} (645 mm^2) of Various Grouts in a Vertical Delivery Line

Height of Column		Static Head Weight per Square Inch of Line in Pounds (kg)							
ft	m	Water	0.5:1 Grout	1:1 Grout	2:1 Grout	3:1 Grout	4:1 Grout	5:1 Grout	4 in. Slump Cement-Sand
1	0.3	0.43 (0.20)	0.89 (0.4)	0.73 (0.32)	0.61 (0.27)	0.56 (0.25)	0.53 (0.24)	0.51 (0.23)	0.88 (0.40)
10	3	4.3 (1.9)	8.9 (4.0)	7.3 (3.32)	6.1 (2.7)	5.6 (2.5)	5.3 (2.4)	5.1 (2.3)	8.8 (4.0)
20	6	8.6 (3.9)	17.8 (8.1)	14.6 (6.63)	12.2 (5.4)	11.2 (5.1)	10.6 (4.8)	10.2 (4.6)	17.6 (8.0)
30	9	12.9 (5.9)	26.6 (12.1)	21.9 (9.95)	18.3 (8.1)	16.8 (7.6)	15.9 (7.2)	15.3 (7.0)	26.4 (12.0)
40	12	17.3 (7.9)	35.6 (16.2)	29.2 (13.3)	24.4 (11.1)	22.4 (10.0)	21.2 (9.6)	20.4 (9.3)	35.2 (16.0)
50	15	21.7 (9.9)	44.5 (20.2)	36.5 (16.6)	30.5 (13.9)	28.0 (12.7)	26.5 (12.0)	25.5 (11.6)	44.4 (20.0)
60	18	26.0 (11.8)	53.4 (24.3)	43.8 (19.9)	36.6 (16.2)	33.6 (15.3)	31.8 (14.5)	30.6 (13.9)	52.8 (24.0)
80	24	34.6 (15.7)	71.2 (32.4)	58.4 (26.7)	48.8 (22.2)	44.8 (20.4)	42.4 (19.3)	40.8 (18.5)	70.4 (32.0)
100	30	43.4 (19.7)	89.2 (40.5)	73.0 (33.2)	61.4 (27.7)	56.2 (25.5)	53.0 (24.1)	51.0 (23.3)	88.8 (40.0)
150	45	65.0 (29.5)	133 (66.7)	109 (49.8)	91.5 (41.6)	84.4 (38.4)	79.5 (36.1)	76.5 (34.8)	123 (60.0)
200	60	86.8 (39.4)	178 (81.1)	146 (66.4)	123 (55.8)	112 (51.0)	106 (48.2)	102 (46.4)	177 (80.0)

restriction at the bottom of the vertical run to reduce the magnitude of head gain.

These various losses and gains comprise two distinct elements, commonly recognized as the *static* and *dynamic* heads. The static head within a system represents the pressure or resistance to flow, which is present regardless of whether the grout is moving or still. It is primarily dependent on the *unit weight* of the grout. Conversely, the dynamic head is the result of *frictional resistance* of the moving material as it passes the surfaces against which it is being propelled. The value of the dynamic head is thus sensitive to the size, type, and condition of the delivery line, as well as the surface condition of the formation being injected.

The frictional resistance of pumped material is influenced by the rate of movement over the particular surface. Thus, the dynamic head of a particular grout will vary in a straight-line relationship with its velocity. The same is true of the relationship of delivery line size to the pumping rate. The velocity of grout through a delivery system affects not only the level of resulting head loss, but also the amount of wear of the system components. Table 9.2 provides the velocity of material through delivery lines of different sizes at varying pumping rates.

TABLE 9.2 Approximate Grout Flow and Velocity in Lines of Various Sizes

	FLOW			VELOCITY	
ft^3/minute	gal/minute	L/minute	ft/minute	m/minute
1/2 IN. (12.7 MM) LINE				
	0.5	1.9	0.8	14.6
0.1	1	3.8	1.6	29.3
0.3	2	7.6	3.3	60.4
0.4	3	11.4	4.9	89.6
0.5	4	15.1	6.5	118.9
3/4 IN. (19 MM) LINE				
0.1	1	3.8	0.7	12.8
0.3	2	7.6	1.4	25.6
0.4	3	11.4	2.2	40.2
0.5	4	15.1	2.9	53.0
0.7	5	18.9	3.6	65.8
0.8	6	22.7	4.4	80.5
0.9	7	26.5	5.1	93.3
1.1	8	30.1	5.8	106.0
1 IN. (25.4 MM) LINE				
0.3	2	7.6	0.8	14.6
0.5	4	15.1	1.6	29.3
0.8	6	22.7	2.5	45.7
1.1	8	30.1	3.3	60.4
1.3	10	37.9	4.1	75.0
1.6	12	45.4	4.9	89.6
1.9	14	53.0	5.7	104.3
1-1/4 IN. (31.7 MM) LINE				
0.5	3.8	14.4	0.7	11.0
1.0	7.5	28.4	1.6	29.3
1.5	11.3	42.8	1.8	36.6
2.0	15.0	56.8	2.4	42.1
3.0	22.5	85.2	3.5	86.0
4.0	30.0	113.6	4.7	117.0
1-1/2 IN. (38 MM) LINE				
0.5	3.8	14.1	0.6	11.0
1.0	7.5	28.4	1.2	21.9
1.5	11.3	42.8	1.8	32.9
2.0	15.0	56.8	2.4	43.9
3.0	22.5	85.2	3.5	64.0
4.0	30.0	113.6	4.7	86.0
2 IN. (50.8 MM) LINE				
1.0	7.5	28.4	0.7	12.8
1.5	11.3	42.8	1.0	18.3
2.0	15.0	56.8	1.4	25.6
3.0	22.5	85.2	2.2	40.2
4.0	30.0	113.6	2.9	53.0
5.0	38.0	143.8	3.6	65.8
6.0	45.0	170.3	4.4	80.5

Good practice in grouting requires sizing delivery lines to provide sufficient grout velocity to prevent excessive aging before reaching the grout header. In addition, the flow rate through the delivery system must be sufficiently rapid that no setting or gelation of the grout occurs during injection. Conversely, the delivery lines must be large enough to prevent excessive dynamic head. Knowledge of the range of pumping rates that will be used on a given project is thus required in order to select an appropriately sized delivery system. The relative resistance contributed by the formation or void system into which the grout is pumped must also be considered.

Because of the nearly infinite variation in the properties of grout and the surfaces against which it will flow, the establishment of standard values for dynamic head is not practicable. However, the frictional properties of water through pipe are well understood and often used as a baseline to estimate frictional losses. Table 9.3 provides the approximate value of head loss for water flowing through 100 linear ft (30 m) of pipes of different sizes at a variety of flow rates.

The values shown in Table 9.3 are based on the pipes' being in a continuous, straight run and horizontal configuration. When a conduit is tipped upward from horizontal, has fittings put on it, or bends are put in a hose, additional frictional losses are experienced. Again, in regard to water, the head losses resulting from various standard pipe fittings are shown in Table 9.4. It follows, then, that head loss within a delivery system can be greatly reduced by minimizing the number of fittings to the greatest extent practicable. Substantial reduction can also be achieved through the use of wide sweep bends, such as those typically used with electrical conduit, rather than ordinary pipe fittings.

Substantial turbulence can also occur at the couplings in a delivery line. In hose, it is common for the cross-sectional area to be reduced at a coupling as a result of the smaller inside diameter of the hose barb. Where rigid pipes are used, the pipe ends are often slightly separated at the couplings. This can occur with standard threaded couplings as well as with the clamp-type couplings often used with the larger line sizes, as illustrated in Figure 9.3. Although the frictional effect of pipeline couplings on water is fairly minor, this is not the case with cementitious suspension grouts. These grouts tend to fill such openings first and then slowly build up. Given enough time, the buildup can nearly block the openings.

Valves are another source of increased turbulence. They have various cross-sectional geometries, and the flow area is often less than that of the line. Ball valves that provide full-size bore openings are readily available and are certainly advantageous to use.

When grout enters a formation from small holes or nozzles, as are common in sleeve port pipes (tube à manchette) and special drive needles, it is important to consider the pressure loss effect. Table 9.5 provides information on the effect of water passage. Although this can give some guidance, it must be kept in mind that virtually all grouts are much less penetrable than water. Cementitious suspensions are especially affected when impelled through small openings, and these should be no less than about 0.25 in. (6 mm).

In a particular instance, the contractor on a project was convinced that higher grout pressures were required. He was using grout needles made of 3/4 in. (19 mm) steel pipe. A pointed plug was welded onto the bottom, and three 1/8 in. (3.2 mm) holes were provided for the grout to exit. Upon investigation, the only problem found was the small holes. When the grout was pumped into a needle laid on the ground with no restriction to the exit nozzles, about 90 psi (6 bar) pressure was required to simply pump the grout at 3 gal (11.4 L) per minute. Even then, the holes completely plugged within a few minutes of pumping. Once they were

TABLE 9.3 Approximate Head Loss of Water per 100 ft (30 m) of Pipes of Various Sizes

FLOW RATE		PIPE SIZE, IN. (MM)															
		1/2 (12.7)		3/4 IN. (19)		1 (25.4)		1-1/4 (31.7)		1-1/2 (38.1)		2 (50.8)		2-1/2 (62.5)		3 (76.2)	
gpm	L/min	psi	kPa	psi	kPa	psi	kPa	psi	kPa	psi	kPa	psi	kPa	psi	kPa	psi	kPa
2	7.6	2.1	14.5	0.5	3.5												
3	11.3	4.3	29.7	1.1	7.6												
4	15.2	7.4	51.1	1.8	12.4	0.5	3.3										
5	18.9	11.2	77.3	2.7	18.6	0.8	5.5										
6	22.7	15.8	109.0	3.9	26.9	1.2	8.3										
7	25.8	21.1	145.6	5.1	35.2	1.6	11.4										
8	30.2	27.2	187.7	6.5	44.9	2.0	13.8										
9	34.0	34.0	234.6	8.2	56.6	2.5	17.1	0.5	3.5								
10	37.8	41.6	287.0	10.0	69.0	3.0	20.7	0.8	5.5								
12	45.4	50.2	346.2	14.1	97.3	4.2	29.0	1.1	7.6	0.5	3.5						
14	52.9					5.6	38.6	1.4	9.7	0.7	4.8						
16	60.5					7.2	49.7	1.8	12.4	0.9	6.2						
18	68.0					8.9	61.4	2.3	15.9	1.1	7.6						
20	75.6					10.9	75.2	2.8	19.3	1.3	9.0						
22	83.2					13.1	90.4	3.4	23.5	1.5	10.4	0.4	2.8				
24	90.7					15.5	107.0	4.0	27.6	1.8	12.4	0.5	3.5				
26	98.3					18.0	124.2	4.6	31.7	2.1	14.5	0.6	4.1				
28	105.8					20.9	144.2	5.2	35.9	2.4	16.6	0.7	4.8				
30	113.4					23.7	163.5	5.9	40.7	2.8	19.3	0.8	5.5				
35	132.3					31.8	219.4	7.9	54.5	3.6	24.9	1.0	6.9	0.5	3.5		
40	151.2					41.2	284.3	10.2	70.4	4.7	32.4	1.3	9.0	0.7	4.8		
45	170.1					54.0	372.6	12.8	88.3	5.9	40.7	1.7	11.7	0.8	5.5		
50	189.0							15.6	107	7.1	49.0	2.0	13.8	0.9	6.2	0.5	3.5
60	226.8							22.1	152	10.1	69.7	2.9	20.0	1.5	10.4	0.6	4.1
70	264.6							29.9	206.3	13.6	93.8	3.9	26.9	2.0	13.8	1.0	6.9
80	302.4							38.7	267.0	17.6	121.4	5.0	34.5	2.8	19.3	1.3	9.0
90	340.2							48.6	335.3	22.1	152.5	6.2	42.8	3.3	22.8	1.6	11.0
100	378.0							59.9	413.3	27.0	186.3	7.6	52.4	4.1	28.3	1.9	13.1
120	453.6									38.3	264.3	10.7	73.8	4.9	33.8	2.7	18.6
140	529.2									51.6	356 .0	14.4	99.4	6.0	41.4	3.5	24.2

TABLE 9.4 Approximate Friction Loss Created in Fittings—Equivalent Length of Straight Pipe Head Loss

	PIPE SIZE, IN. (MM)															
	1/2 (12.7)		3/4 (19)		1 (25.4)		1-1/4 (31.7)		1-1/2 (38.1)		2 (50.8)		2-1/2 (63.5)		3 (76.2)	
	ft	m	ft	m	ft	m	ft	m	ft	m	ft	m	ft	m	ft	m
90° ELL	3.6	0.1	4.4	1.3	5.2	1.6	6.6	2.0	7.4	2.3	8.5	2.6	9.3	2.8	11.1	3.4
45° ELL	0.71	0.2	0.92	0.3	1.3	0.4	1.7	0.5	2.1	0.6	2.7	0.8	3.2	1.0	4.0	1.3
T (THROUGH LINE)	1.7	0.5	2.4	0.78	3.2	1.3	4.6	1.4	5.6	1.7	7.7	2.3	9.8	2.8	12.0	3.7
T (BRANCH)	4.2	1.3	5.3	1.6	6.6	2.0	8.7	2.7	9.9	3.0	12.0	3.7	13.0	4.09	17.0	5.2
COUPLING	0.21	0.06	0.24	0.07	0.29	0.08	0.36	0.01	0.39	0.01	0.45	0.1	0.47	0.01	0.53	0.02

FIGURE 9.3 A gap in the flow surface often occurs at couplings.

enlarged, the perceived problem disappeared and no pressure adjustment was required.

Although it is difficult to control the bends and radii in a typical run of flexible hose, it is important to recognize the head loss effects that occur. It is always preferable to minimize the number of bends and their tightness. To this end, stretching out the hose in a reasonably straight orientation and positioning it so that the bends are relatively wide is beneficial. These are things that experienced grouters do instinctively, often without exact knowledge of the mechanics involved.

Both head loss and gain within the delivery system have an effect on the grout pressure in the formation. To minimize this effect, grout pressure should be monitored and controlled as near the area of deposition as possible, which is usually at the hole collar. Probable effects of head loss or gain should be considered, and appropriate adjustments of injection pressure made. On sensitive applications, especially where deep grout holes are involved and limitation of pressure within the formation is important, trial injections through a casing identical to that proposed for the production work, or other mock-ups that simulate production conditions, should be considered.

In setting up such trials, it is important to understand that the orientation of the conduit will affect frictional resistance. For instance, resistance through a horizontal casing will virtually always be different than when the same casing is inclined or in a vertical position. The reason is that only frictional head loss is experienced in the level orientation, whereas both frictional and static head change will occur when the conduit is inclined or vertical.

The performance of flexible hose, which is most often used in grouting, is similar for the most part, but there are some differences. As mentioned earlier, resistance within a length of tightly coiled hose will be much greater than in the same hose stretched out in a straight line. In addition, hose is subject to kinking, which results in severe restriction and/or turbulence. The condition of the hose wall also has an influence on both frictional resistance and the amount of turbulence. Flexible hose nearly always entails greater resistance than relatively smooth walls of a steel delivery line. For this reason, experienced grouters routinely use rigid line for especially long runs. Both rigid line and sweep fittings rated for high pumping pressures are commonly available in diameters of 1-1/2 in. (13 mm) and greater from suppliers to the concrete pumping industry.

9.2 EFFECTS OF GROUT RHEOLOGY

The properties of a grout have an effect on the flow through the injection system, as well as into the formation. Both the static and the dynamic heads will be affected. For example, very sticky grout with a high affinity to the injection line surfaces will develop significant resistance and thus dynamic head loss. Conversely, a similar grout of poor affinity to the line will not develop as much resistance to flow.

TABLE 9.5 Theoretical Discharge of Nozzles

HEAD PRESSURE		DISCHARGE VELOCITY		DIAMETER OF NOZZLE, IN. (MM)																			
				1/16	(1.6)	1/8	(3.2)	3/16	(4.8)	1/4	(6.4)	3/8	(9.5)	1/2	(12.8)	5/8	(15.7)	3/4	(19.1)	7/8	(22.2)	1	(25.4)
psi	kPa	ft/min	cm/s	g/m	L/m	g/m	L/m	g/m	L/m	g/m	L/m	g/m	L/m	g/m	L/m	g/m	L/m	g/ m	L/m	g/m	L/m	g/m	L/m
10	69	38.6	19.6	0.37	1.40	1.48	5.61	3.32	12.58	5.91	22.5	13.3	50.4	23.6	89.4	36.9	139.9	63.1	239.1	72.4	274	94.5	358
15	104	47.3	24.1	0.45	1.71	1.81	6.86	4.06	15.39	7.24	27.4	16.3	61.8	28.9	109.5	45.2	171.3	65.0	246.4	88.5	335	116	440
20	138	54.6	27.8	0.52	1.97	2.09	7.92	4.69	17.78	8.35	31.6	18.8	71.3	33.4	126.6	52.2	197.8	75.1	284.6	102	387	134	507
25	119	61.0	31.1	0.58	2.20	2.34	8.87	5.25	19.90	9.34	35.4	21.0	79.6	37.3	141.4	58.3	221.0	84.0	418.4	114	432	149	565
30	207	66.9	34.1	0.64	2.43	2.56	9.70	5.75	21.79	10.2	38.7	23.0	87.2	40.9	155.0	63.9	242.2	92.0	348.7	125	474	164	622
35	242	72.2	36.8	0.69	2.62	2.77	10.50	6.21	23.54	11.1	42.1	24.8	94.0	44.2	167.5	69.0	261.5	99.5	377.1	135	512	177	671
40	276	77.2	39.4	0.74	2.80	2.96	11.22	6.64	25.17	11.8	44.7	26.6	100.8	47.3	179.3	73.8	279.7	106	401.7	145	550	188	713
45	311	81.8	41.7	0.78	3.00	3.13	11.86	7.03	26.64	12.5	47.4	28.2	106.9	50.1	189.9	78.2	296.4	113	428.3	153	580	200	758
50	345	86.3	44.0	0.83	3.15	3.30	12.51	7.41	28.08	13.2	50.0	29.7	112.6	52.8	202.1	82.5	312.8	119	451.0	162	614	211	800
55	380	90.4	46.1	0.87	3.30	3.46	13.11	7.77	29.45	13.8	52.3	31.1	117.9	55.3	209.6	86.4	327.5	125	473.8	169	641	221	838
60	414	94.5	48.2	0.90	3.41	3.62	13.72	8.12	30.77	14.5	55.0	32.5	123.2	57.8	219.1	90.4	342.6	130	492.7	177	671	231	875
65	449	98.3	50.1	0.94	3.56	3.77	14.29	8.45	32.03	15.1	57.2	33.8	128.1	60.2	228.2	94.0	356.3	136	515.4	184	697	241	913
70	483	102.1	52.1	0.98	3.71	3.91	14.82	8.78	33.28	15.7	59.5	35.2	133.4	62.5	236.9	97.7	370.3	141	534.4	191	724	250	948
75	518	105.7	53.9	1.01	3.83	4.05	15.35	9.08	34.41	16.2	61.4	36.4	138.0	64.7	254.2	101	382.8	146	553.3	198	750	259	982
80	552	109.1	55.6	1.05	3.98	4.18	15.84	9.36	35.59	16.7	63.3	37.6	142.5	66.8	253.2	104	394.1	150	568.5	206	781	267	1012
85	587	112.5	57.4	1.08	4.09	4.31	16.33	9.67	36.65	17.3	65.6	38.8	147.1	68.9	261.1	108	409.3	155	587.5	211	800	276	1046
90	621	115.8	59.1	1.11	4.21	4.43	16.97	9.95	37.71	17.7	67.1	39.9	151.2	70.8	268.3	111	420.8	160	606.4	217	822	284	1076
95	656	119.0	60.7	1.14	4.32	4.56	17.28	10.2	38.66	18.2	69.0	41.0	155.4	72.8	275.9	114	432.1	164	621.7	223	845	292	1107
100	690	122.0	62.2	1.17	4.43	4.67	17.70	10.5	39.80	18.7	70.9	42.1	159.6	74.7	283.1	117	443.4	168	636.7	229	868	299	1133
105	725	125.0	63.8	1.20	4.55	4.79	18.15	10.8	40.93	19.2	72.8	43.1	163.3	76.5	289.9	120	454.8	172	651.9	234	887	306	1160
110	759	128.0	65.3	1.23	4.66	4.90	18.57	11.0	41.69	19.6	74.3	44.1	167.1	78.4	297.1	122	462.4	176	667.0	240	910	314	1190
115	794	130.9	66.8	1.25	4.74	5.01	18.99	11.2	42.45	20.0	75.8	45.1	170.9	80.1	303.6	125	473.8	180	682.2	246	929	320	1213
120	828	133.7	68.2	1.28	4.85	5.12	19.40	11.5	43.59	20.5	77.6	46.0	174.3	81.8	310.0	128	485.1	184	697.4	251	951	327	1239
125	863	136.4	69.6	1.31	4.96	5.22	19.78	11.7	44.34	20.9	79.2	47.0	178.1	83.5	316.5	130	492.7	188	712.5	256	970	334	1266
130	879	139.1	70.9	1.33	5.04	5.33	20.20	12.0	45.48	21.3	80.7	48.0	181.9	85.2	322.9	133	504.1	192	727.7	261	989	341	1292
135	932	141.8	72.3	1.36	5.15	5.43	20.58	12.2	46.24	21.7	82.2	48.9	185.3	86.7	328.6	136	515.4	195	739.1	266	1008	347	1315
140	966	144.3	73.6	1.38	5.23	5.53	20.96	12.4	47.00	22.1	83.8	49.8	188.7	88.4	335.0	138	523.0	199	754.2	271	1027	354	1342
145	1000	146.9	74.9	1.41	5.34	6.62	21.30	12.6	47.75	22.5	85.3	50.6	141.8	89.9	336.3	140	530.6	202	765.6	276	1046	360	1364
150	1035	149.5	76.2	1.43	5.42	5.72	21.68	12.9	48.89	22.9	86.8	51.5	145.2	91.5	346.8	143	542.0	206	780.7	280	1061	366	1387
175	1208	161.4	82.3	1.55	5.87	6.18	23.42	13.9	52.68	24.7	93.6	55.6	210.7	98.8	374.5	154	583.7	222	841.4	302	1145	395	1497
200	1380	172.6	88.0	1.65	6.25	6.61	25.05	14.8	56.09	26.4	100	59.5	225.5	106	401.7	165	625.4	238	902.0	323	1224	423	1603

Similarly, very dense cementitious grout will have a greater effect on static head loss than a light, perhaps air-entrained, mixture. Many cementitious grouts contain air unintentionally entrained during mixing and handling, or sometimes purposely entrained through admixtures. Air in grout can be greatly reduced, however, as a result of pumping pressure destroying air bubbles. Although pressure filtration, wherein water is forced out and ahead of the grout, can occur, high pumping pressure can cause mix water to be forced into the grout constituents as absorbed moisture.

Particularly difficult to pump are compositions that contain especially heavy aggregate such as ferrous metal filings, which are a component of many preblended machine base grouts. Lead pellets are sometimes included with other heavy media in grouts, which will serve as radiation shielding. Such compositions result in particularly high static head gain or loss, especially where the grout line is inclined or vertical. Admittedly, these are unusual applications that many grouters will never see. They do occur, however, and one must be careful to recognize the potential difficulties.

Another consideration is the degree of saturation of solids in the grout mixture. Most particulate grouts start as fully saturated compositions. They are subject to pressure filtration, however, and with the exertion of sufficient pumping pressure can become only partially saturated. This influences the frictional resistance and, to a lesser degree, causes static head variation. Unfortunately, such changes usually occur only when the grout is at higher pumping pressures, and thus the particulars are not easily measured or controlled. The degree of saturation loss, if any, will vary according to the different levels of injection pressure, which further complicates achieving a reasonable understanding. Resistance of a particular grout to pressure filtration can be determined prior to injection, as discussed in Section 8.4.

A trend in the formulation of preblended grouts is inclusion of short fibers that are intended to both limit shrinkage and provide increased flexural, tensile, and shear strength. In dealing with such a material containing an especially large proportion of fibers, high frictional loss was experienced. In fact, to achieve practicality in pumping the material even at a low pumping rate of only 1 gal (3.8 L) per minute, it was necessary to use 1 in. (25 mm) diameter Teflon-lined hose. Even with the resulting low grout velocity of only 0.4 ft (122 mm) per minute, head loss of 3 to 4 psi (21–28 kPa) per foot of hose occurred.

9.3 GROUT PENETRATION

Most grouts are thixotropic, whereby they require a greater pressure to start flow than to maintain movement once initiated. Thus, sufficiently high pressure must be allowed in order to start grout movement, although it can be lowered once continuous flow occurs. Historically, most grouters thought of grout as either thick or thin and were generally unaware of the properties beyond these points. Injection was started with thin grout, and if it permeated readily, sequential thickening through reduction of the water would follow. The logic for this well-established practice, which continues to the present, must be carefully considered.

In the real world of grouting, larger defects intersected by a hole will be filled first. They are filled more readily and at lower pressures than smaller defects. Only after such filling does grout attempt to enter the smaller defects. Because they are smaller, however, both thinner grout and higher pressure will be required. The use of thick grout, either as a starting mix or as a result of intentional thickening during injection, must be carefully considered, as the thicker grout will block the intrusion of more penetrable mixtures that may follow.

9.4 INJECTION PRESSURE

A consideration of injection behavior must begin with an understanding of the nomenclature used in discussing pressure. In the United States, pressure is usually stated in pounds per square inch (psi). In most other areas of the world some form of metric unit is used. The most common of such units is the bar, which is equal to 14.5 psi. Pressure gauges are available with dual scales, including both psi and bars, as illustrated in Figure 9.4, and are especially advantageous when personnel of different backgrounds are on-site. Other metric units of pressure are also sometimes used. Table 9.6 provides typical conversions.

Although often ignored, the most important influence on pressure is the pumping rate. Developed pressure is directly proportional to pumping rate. It will rise as the pumping rate increases, and fall as it descends. For this reason, notation of the maximum pressure reached during injection is meaningless unless the pumping rate is also stated. Both pumping rate and pressure should therefore be monitored and recorded. Although the pumping rate is often not metered, it can be readily calculated if current time is noted at each pressure entry. Optimally, recording of *both* pressure and grout *quantity* should be done at every interval level, and inclusion of the current time should be *mandatory* for all record entries.

FIGURE 9.4 Gauges that display dual scales in both psi and bars are useful.

TABLE 9.6 Conversion Factors for Grout Pressure

1 psi	= 0.07 bar	= 0.07 kg/cm^2	= 6.9 kPa
1 bar	= 14.5 psi	= 0.98 kg/cm^2	= 100 kPa
1 kg/cm^2	= 14.2 psi	= 0.98 bars	= 98 kPa

For economy, use of the highest safe rate of injection is desirable. Where circulating injection systems are used, the rate will be ever variable through manipulation of the control valves to maintain whatever pressure is desired. Where direct injection lines are used, the flow must be controlled by varying the pump output. This is most often accomplished by varying the pumping rate, but can also be effected by bypassing some of the grout immediately as it exits the pump. The maximum pressure output of some pumps driven by hydraulic power transmission can be preset by restricting the pressure of the hydraulic oil within the system. Machines so equipped typically have a simple rotary knob to make the desired adjustments. Likewise, the maximum output pressure of pneumatically powered pumps can be established by control of the input air pressure, which is accomplished via a common pressure regulator.

The optimal pressure level will be variable and is dependent on many factors, including the type and rheology of the grout, the purpose of the injection, and the hole or void orientation, geometry, and volume. Of special significance are depth of injection, specifics of formation strength, stratigraphy, and void network. The optimal injection pressure for use on most projects will vary, not only between different grout holes, but often within a given hole. Obviously, careful monitoring of an injection is required to provide valid information needed to guide such variation. However, variation will always be based on the *pressure behavior,* that is, the relative rates of increase and loss of pressure at various injection rates. This very important issue is discussed in greater detail shortly.

Pressure used in grouting can be categorized in two distinct modes, that resulting from friction and any restriction within the delivery system (pump output to lower end of casing) and that resulting from resistance to permeation or displacement of the formation or other medium being filled. Where injection is through a void of uniform cross section, such as a pipe or injection hose, the total pumping pressure will be dependent on the length of that void and will be the product of the unit length resistance times the number of units of length. Unless a void is filled with water or isolated so that air cannot escape, pressure at the leading edge of the grout will always be zero. Where injection is made into a porous mass such as soil, or into voids of varying size and configuration, the pressure distribution within the grout will be ever variable. It will, however, always be greatest at the point of injection, which is typically the lower end of the injection casing, and will usually decrease with distance from that point.

Because effective pressure on the formation will invariably start at the casing outlet, it is important to consider the pressure differential between that point and the top of the casing, where pressure is typically observed. This value may be either positive or negative, depending on the particulars of the grout and the delivery system. For example, on a significant compaction grouting project, the contractor proposed use of a 6 in. (150 mm) diameter casing for some very deep holes. Because of concern that the static head pressure might exceed the frictional resistance, a trial injection with the 6 in. (150 mm) casing was made. A pressure cell (Figure 9.5) was lowered into the hole to measure the actual pressure at depth. A *positive* pressure of about 0.45 psi (0.03 bars) per vertical foot (0.33 m) of

FIGURE 9.5 A pressure cell that was lowered into the grout hole to evaluate downhole pressure.

casing was measured. Because the resulting pressure at the bottom of the hole would be excessive, a smaller 3 in. (75 mm) casing was used for the production work. It resulted in a head *loss* of about 0.45 psi (0.03 bars) per vertical foot.

Through experimentation and calculation, the approximate pressure differential between the point of injection and casing outlet can be determined. The term *approximate* is used because pressure within a delivery line is dependent on many variables. Major contributors, in addition to resistance within the host formation, are rheology of the grout being pumped, pumping rate, size of the delivery lines and resulting grout velocity, and frictional resistance at the grout conduit wall interface, which is a function of the wall's smoothness and its affinity for the particular grout being used. Once the grout is in the formation, both the size of the individual fractures or pores and their surface condition are major factors. Obviously, the rheology of the grout and its affinity to the formation surfaces are also significant.

9.4.1 Pressure–Volume Relationship

The quantity of grout pumped at any one location, and the shape it forms in the host material, have a significant effect on the formation reaction. A given pressure on a very large mass of grout will affect a greater area and thus exert a much larger *total* force within the formation, than will the same pressure on a small mass, as illustrated in Figure 9.6. Depending on the particulars of the individual injection, this will affect the pressure level that can be safely used. Where large quantities of grout are placed, the pressure for initial injection may be too high once a significant mass of grout has developed. Therefore, as the quantity grows, reduction of pumping rate, and thus pressure, is often prudent.

The shape and orientation of the injected grout mass are also important. For example, if a grout were to form an essentially horizontal lens, as illustrated in Figure 9.7, a large horizontal area would result. This could translate into high total force in a vertical orientation, jacking the overlying formation upward. However, if the injection were into a vertical fracture, the primary effect of the pressurized grout would be horizontal, resulting only in a beneficial tightening of the formation and little risk of jacking the overburden.

As previously discussed, initiation of *controlled* groutjacking is often accomplished with highly mobile grout, but once started, control of lifting usually dictates the use of a lower-mobility grout. Jacking can occur with any grout, however, if the injection rate exceeds that which the formation is able to accept. Much damage has been done as a result of *uncontrolled* jacking resulting from excessive pressure caused by a too rapid pumping rate. This is a particular risk when injection is into large or adversely oriented defects.

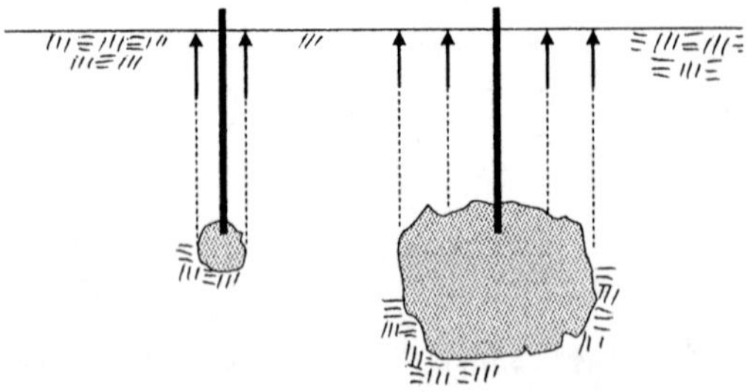

FIGURE 9.6 The total uplift force of a large mass of grout will be much greater than that of small mass.

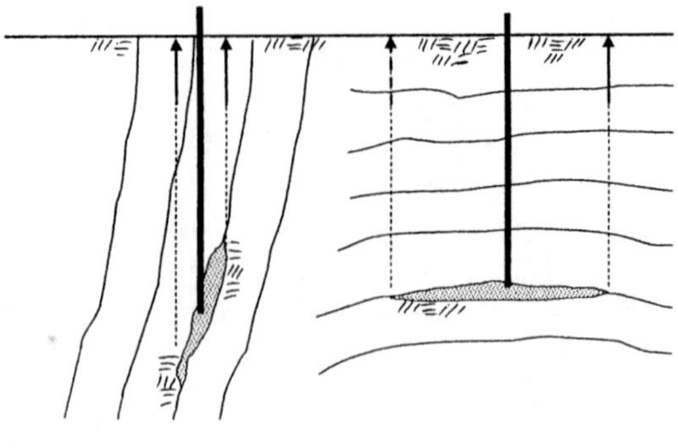

FIGURE 9.7 A horizontally oriented grout lens will provide greater jacking force than a vertically oriented one.

9.4.2 Appropriate Pressure Levels

Excessive pressure can result in damage to the formation and must generally be avoided. Yet the realities of the real world of grouting must be faced, wherein determination of the maximum safe injection pressure is usually not possible before at least some actual injection. Even then, determination usually involves exceeding a safe pressure occasionally, which can result in minor damage to the formation. A sudden drop of injection pressure at a constant pumping rate almost always indicates the use of excessive pressure. One of the primary qualities needed in grouting engineers and technicians is an ability to quickly recognize such pressure behavior. When such a drop occurs, prompt corrective action must be taken, which usually involves lowering the pumping rate or the cessation of injection.

The extent of damage resulting, if any, will vary with the type of grouting and the nature of the formation. In many cases, the damage will be inconsequential. For example, solidification of soil by permeation grouting is done at the highest possible pumping rate to achieve the greatest economy. However, the pressure will increase as the pumping rate increases, until hydraulic fracture of the formation occurs. Although occasional minor fracture is usually acceptable, its prompt detection and control are required, as grout flowing in a fracture will not be deposited in its intended location. Because the properties of different deposits vary, the value of acceptable pressure is different for each project.

The optimal pumping rate may be determined by slowly increasing it until the pressure undergoes a sudden drop, indicating initiation of a fracture. A production rate slightly lower than that at which the pressure drop occurred is then adopted. As there is often variation, even within soil deposits on the same site, the initially adopted rate may prove to be excessive as the work progresses. Accordingly, to detect any further fracturing, the pressure must be monitored continuously, and immediate remedial measures taken upon occurrence of further sudden losses. In grouting soil, as long as fracturing is not frequent and is promptly detected and remedied, damage to the formation is minor and of little consequence.

However, a sudden pressure drop when grouting in rigid materials such as rock or in voids within a masonry structure, may indicate that jacking of the formation is occurring. In such cases, the structure of the formation can be badly damaged, even with minimal disruption. Selection of the maximum pressure must thus be made conservatively and must include the application of appropriate safety factors. Interestingly, a number of "rules of thumb" apply to the grouting of rock, but there is enormous variation in the acceptable pressure levels.

Establishing the maximum allowable injection pressure is perhaps the most contentious issue in rock grouting. A widely recognized rule of thumb in the United States dictates that maximum pressure in psi should be limited to the injection depth in feet. Well-documented experience in the United States and other countries, however, has proven this idea faulty. It is always preferable to use the highest *safe* pressure, as this will force the grout into the smallest defects and cause it to permeate farthest. It will also cause the work to proceed faster and thus be more economical. The pressure is, however, usually restricted to a level that is *perceived* to avoid undesirable hydraulic fracturing, jacking, or upheaval of the host material.

Interestingly, some European practitioners, especially those in France, intentionally jack the formation in the belief that it is the only way to ensure complete filling of all defects. They reason that the volume of injected grout will shrink as a result of water loss from the mix. Thus, by placing a larger amount of beginning grout, which results in jacking of the foundation, the defect will remain filled as the formation relaxes. Although I do not agree with this philosophy except in very special situations, it does have merit and has been extensively discussed in the literature (Nonveiller, 1968; Rigny, 1974; Cambefort, 1977).

Excessive pressures can lead to disastrous consequences even in nonrigid formations. For example, in compaction grouting, pressures used satisfactorily in a totally restrained formation can be excessive when the work is being done near an unrestrained downslope or retaining wall. Unfortunately, displacement of slopes and damage to improvements occur far too often and are simply the result of too high a pressure as a result of excessive pumping rates.

It is important to realize that unless a void is filled with water or compressed gas, the pressure at the leading edge of the advancing grout is always zero. Moreover, damage caused by excessive pressure is the result of total force, as previously discussed and illustrated in Figures 9.6 and 9.7. The resulting force is grout pressure multiplied by the effective area of the deposited mass. Substantially higher pressures are appropriate for use when only small quantities of grout have been injected. Conversely, extreme care must be exercised when the grout quantity is large. Where maximum allowable grout pressures are mandated, inclusion of a range of permissible pressures based on the relative volumes injected are recommended.

Criteria for establishing maximum allowable pressures continue to be among the most contentious issues in the pressure grouting industry, especially for geotechnical applications. It has been my experience however, that while the various players are often completely inflexible, and totally adamant as to the correctness of their opinions, they seldom have documented data or rational reasons therefore. The establishment of rational pumping pressures must include consideration of the size and orientation of the defect to be filled, as well as the amount of mobile grout within the formation at any given time. Only when these factors are considered can an even approximately rational pressure be determined. In most work a suite of several permissible pressure values, based on the volume of the injected mass and other specifics, will be appropriate. Of fundamental importance: Never forget that the developed pressure is directly related to the injection rate.

9.5 PUMPING RATE

An increase in pumping rate will *always* result in higher pressure, and a reduction will lower the pressure. This principle is well illustrated in Figure 9.8, which is an actual printout of a computer-generated record of pressure behavior on a project where the pump malfunctioned. The piston speed varied so that one piston was traveling nearly twice as rapidly as the other. The actual pump strokes can be observed by their pressure differentials, and it can be noted that pressure on the short stroke (higher pumping rate) was more than 100 psi (6.9 bars) greater than on the slower long stroke. It is also note-

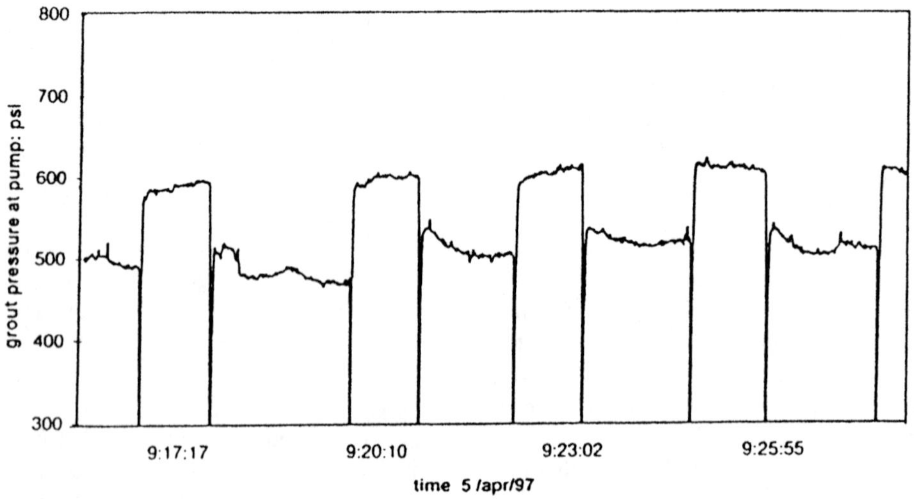

FIGURE 9.8 Grout pressure responding to the varying pumping rate of a malfunctioning pump.

worthy that the pressure increased slightly during each of the short strokes, indicating a generally optimal pumping rate, whereas there was a slight decay during the long strokes, indicating a slower than optimal rate.

Resistance developed within a formation is beyond the grouter's control; therefore, the injection rate must be limited so that excessive pressure does not develop. Because wide variation in formation resistance is common, the use of adjustable output pumps or circulating systems is required. In these systems, grout is continuously circulated, as previously shown in Figure 2.1. The amount of grout entering the hole, and thus the injection rate, are controlled by "throttling" the valves. Because economy dictates the use of the highest practical rate, the valves are adjusted to allow the fastest possible flow that will not exceed the maximum allowable pressure.

Single-line direct delivery systems, in which the grout flow is directly from the pump to the grout hole, should be used only with variable output pumps. In this system, the speed of the pump, and thus the rate of injection, are regulated so as not to exceed the allowable pressure. Variation of injection rates in discrete increments facilitates observation of the affects of even slight pressure changes, allowing optimal evaluation of the conditions within the formation.

Grout pumps and delivery systems should be sized for reasonable injection rates in order to preclude the development of excessive pressures during injection. All else being equal, the system pressure will usually diminish as the pipe or hose size increases. The delivery system should not be so large, however, as to preclude complete emptying within a reasonable time. Many grout materials, especially cementitious mixes, tend to build up on the walls of the delivery system with time if sufficient grout velocity is not maintained. This results in an ever-decreasing opening for the grout to travel, increasing both grout velocity and the resulting pressure. This subject was discussed earlier and illustrated in Table 9.2, which shows the velocity of grout through various size delivery lines.

The importance of the injection pressure/pumping rate relationship cannot be overemphasized. Many practitioners have the erroneous impression that maximum grout pressure alone indicates the degree of improvement or filling obtained. Many instances of faulty grouting have occurred for which available records show only the hole number, grouting depth, and *maximum* pressure reached. These are usually next to useless in understanding what actually happened in the ground. The pumping rate/pressure relationship is so important that it warrants special attention. The specifying or recording of injection pressure is of little use unless the injection rate is also provided.

9.6 PRESSURE BEHAVIOR

Except in those rare cases with previously installed geotechnical instrumentation, the only parameter providing timely information for control of grouting is the injection pressure and, more important, its variation during injection. Although a given pressure level in itself is of little value, pressure variations at a constant pumping rate reveal much about the constantly changing conditions in the host formation. In addition to its beinig affected by the frictional resistance within the delivery system, the pressure is directly affected by both the grout rheology and the pumping rate. Whereas changing grout rheology usually requires a time delay, the injection rate can be adjusted instantly, provided that proper equipment is being used. Continuous observation of the behavior of the grout pressure and controlling it through appropriate variation of the flow rate are necessary for optimal performance.

A basic law of physics is that at a constant pumping rate, an increase in pressure *always* indicates greater resistance to flow and a reduction indicates less resistance. Although somewhat subjective, knowing the value and nature of the pressure movement allows a person with experience to make an informed prediction as to cause. For example, if it is assumed that a pressure of 100 psi (6.9 bars) is required to pump a grout through a given length of hose, then should a connection located in the middle of the length become broken, the pressure would sharply drop to approximately half of its original value. Similarly, if a restriction were to be placed at the outlet end, a sudden increase in pressure would result. Applying this understanding to the real world of grout injection, it is widely recognized that a sudden reduction in pressure indicates a likely disruption in the mass being grouted (Houlsby, 1990; Warner and Brown, 1974).

Experience indicates that sudden pressure changes or variations of injection rate at a constant pressure *always* indicate the occurrence of a significant event. In most cases this leads to negative results, which can be greatly minimized, and in many cases completely averted, through quick response, usually the immediate lowering of the pumping rate or complete cessation of the injection.

Typical events that are accompanied by a loss of pressure include hydraulic fracturing, displacement or jacking of the formation, grout loss into a concealed pipe or other substructure, outward displacement of a downslope or retaining wall, grout entering much larger fractures or voids or encountering a softer or more permeable formation, thinning of the grout or other rheological change that increases mobility, leakage of grout, and pump malfunction.

An interesting example was observed during the injection of a very deep compaction grouting hole. A sudden pressure drop was noted. Consideration of the value of the loss, combined with the depth of injection and injection rate, ruled out hydraulic fracturing as the cause. The particular circumstances of the injection led to the conclusion that some sort of break in the injection line had occurred. It was assumed that a casing joint at a fairly shallow depth had failed. After withdrawal of 40 ft (12 m) of casing, however, a vertical split, as shown in Figure 9.9 was found. There were no underground pipes in the area of the injection; if there had been, the pressure behavior could have indicated grout leakage into such pipes.

A case in point involved several hundred feet of 8 in. (203 mm) diameter sewer pipe that had been abandoned for many years and was to be filled with grout. All connections between adjacent buildings and the pipe were to have been plugged when a replacement sewer was installed,

FIGURE 9.9 Vertical split in casing was first detected by a sudden drop in pressure.

but it was possible that one or more open connections might remain. Hence, the injection pressure was continuously monitored. In two different instances, distinct reductions in the rate of pressure increase were detected. The amount of pipe that would have been filled, based on the volume of grout that had been injected up to the point of the pressure change, was calculated.

An excavation was then performed to expose the pipe at the calculated locations. In each instance, open connections were found within about 6 ft (1.8 m) of the expected point. Had injection continued, damage would have occurred, inasmuch as one of the connections led directly into the basement of an adjacent building.

During construction of a facility for the U.S. Air Force, several hundred feet of 1/2 in. (12.7 mm) internal diameter instrumentation conduit were to be filled with a polyester resin grout. There were several connections to the conduit, which were to have been plugged during original installation. Because fast-track construction was employed, there was concern that some plugs might not have been made. As it was important that grout not enter the connections, continuous pressure records were obtained with a chart recorder. One defective plug was found to exist, as clearly evident by a blip on the record, followed by a slight lowering of the rate of pressure increase. The detection allowed corrective action to be taken before unacceptable damage occurred.

Consider another example: Epoxy adhesive was injected into several thousand linear feet of cracks in a structure to remedy earthquake damage. As the resin possessed physical properties much different from those of the concrete substrate, it was not to be placed in defects greater than about 1/4 in. (6 mm) in width. During injection a large quantity of the resin was introduced at a single port location, without the expected pressure increase. Injection was suspended and a test core removed from the area, disclosing a zone of shattered concrete with openings as great as 1 in. (25 mm). Completion of the injection was then delayed until the unsatisfactory concrete was replaced.

It is important to monitor pressure behavior in groutjacking. In such a process, two distinct events are always of concern: filling a hidden substructure and "popping" the improvement to a higher elevation than planned, which occurs all too often in lifting slabs. In regard to filling a substructure, a pressure drop will always precede such an event, dictating immediate corrective action. Sudden unwanted movements of slabs are virtually always caused by their becoming wedged or bound, due to containment of surrounding improvements. The onset of this condition is *always* preceded by a substantial, rapid pressure increase. This behavior is well recognized and was reported in the *Journal of the Geotechnical Engineering Division of ASCE* in 1977 (Committee on Grouting, 1977). It is thus difficult to understand why so much slabjacking continues to be performed without the benefit of a pressure gauge, all too often resulting in damage from uncontrolled raising.

In a report by Meyers (1994), the author stated, "You can tell when binding occurs when . . . you hear the pump laboring due to the strain of the bonded slab." Obviously, if the pump is laboring, it is operating very near maximum pressure capability. In this instance, it is likely that pressure gauges were not being used. Would it not be better to more quickly and positively detect such conditions by observing the pressure? It is interesting that in the same publication, the author states, "If it seems you are using an abnormal amount of grout and obtaining little or no lift, stop immediately and find out where the grout is going. Every slabjacker can tell stories of sewer lines, drain tiles, crawl spaces, and pools he has filled with grout."

I cannot agree with such statements, as they are contrary to my personal experience, in which pressure was routinely monitored and potential problems were detected and corrected before any damage occurred. In addition, I have encountered many competent contractors who do the appropriate monitoring and complete their work without unanticipated damage. It is unfortunate however, that some contractors continue to work without the benefit of a pressure gauge. Such was the case when a large drainage culvert was severely damaged by grout intrusion during slabjacking at the Detroit Metro Airport. As shown in Figure 9.10, the metal culvert was severely distorted and a huge amount of grout deposited therein.

In addition to the binding of slabs and structures during groutjacking, events that are preceded by a pressure increase include plugging or restriction within the injection system or formation, thickening or lowering the mobility of the grout, and completion of the filling of a fracture or void.

FIGURE 9.10 This culvert was badly distorted and contained a massive amount of grout.

In grouting soil, the compaction grouting procedure employs relatively high pressures even at shallow depths. This results in large pressure variations, which are especially instructive and indicate changing conditions within the zone of influence of the growing grout mass. When defective soil is of generally uniform, though inadequate, condition, such as, for example, soil containing wind-deposited granular materials, pressure buildup will generally be constant, as illustrated at the left side of Figure 9.11.

Often, however, the deficiency is caused by buried brush, nested boulders in or under a fill, uncompacted soil resulting from initial clearing or development of haul roads, or insufficient removal of slopewash or other weak surface materials, before fill is placed. Both the existence and the approximate volume of such defects can be determined through continuous monitoring of pressure behavior. A nonuniform pressure buildup, as illustrated at the right side of Figure 9.11, will be experienced. The approximate volume of the apparent defect can be ascertained by subtracting the amount of grout injected at the time of initial pressure drop, from the quantity when that pressure level is reestablished.

As grouting is frequently performed within, under, or in close proximity to existing structures, underground piping and other substructures are common. There is thus, inherent in the procedure, a risk of grout leakage into such utilities. However, a significant drop of pressure will usually precede such an occurrence, so corrective action can be taken. Pressure drops can also indicate other events, but when grout has entered an underground pipe, a slow but steady increase of the initial entrance pressure will typically occur.

Although significant pressure movement or change in injection rate will accompany major events, only minor variations will accompany more subtle occurrences. Unless real-time com-

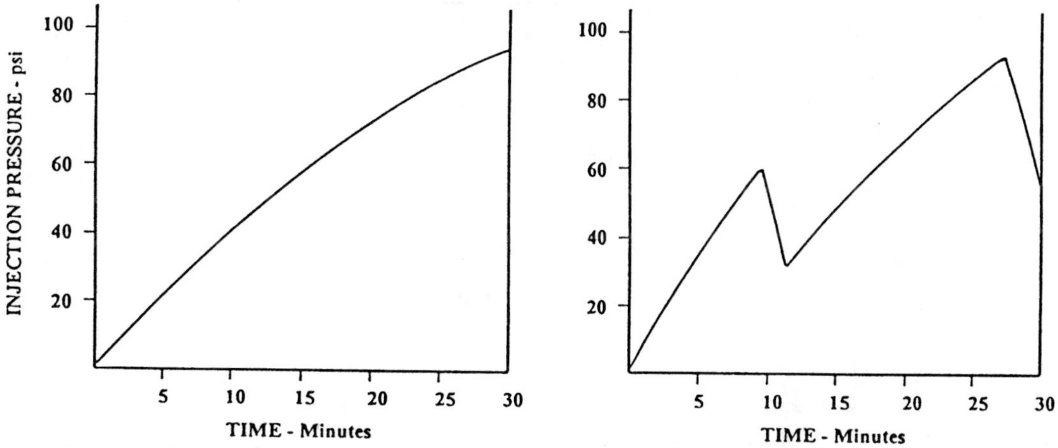

FIGURE 9.11 Constant pressure buildup (left) indicates uniform soil, whereas uneven buildup (right) suggests varying conditions.

puter monitoring is being used, these will not be easily observed in circulating injection systems or in those subject to pulsations caused by pump stroking. Thus, where detection of minor events is important, the use of a constant output pump, free of any significant pulsation, combined with careful and continuous pressure monitoring, is recommended. Piston pumps operating at rates greater than about 100 strokes per minute, combined with a minimum hose length of about 100 ft (30 m), have been found satisfactory because, at high rates, the stroking causes only a flutter of the gauge needle and dampening occurs in the flexible hose.

Pressure pulsations resulting from stroking of piston or diaphragm-type pumps can also be lessened through the use of a hydraulic accumulator near the pump outlet and/or an increase of at least 50 percent in the diameter of the injection hose for a distance of about 25 ft (7.5 m), starting at the pump outlet. Conversely, as illustrated in Figure 9.12, the actual pressure of individual pump strokes can be observed in the case of piston pumps operating at very slow stroke rates. The pump strokes are clearly delineated by the sudden negative pressure spikes.

Pressure behavior can be noted through continuous gauge readings and manual entries on an appropriate form, with fully automated equipment employing either a continuous disk or strip chart recorder, or continuous computer data acquisition and processing. Regardless of the recording method, the actual time of each entry should be included. When manual monitoring is performed, it is good practice to record entries at predetermined uniform increments of pressure. Alternatively, entries can be made at regular time intervals. The magnitude of the pressure increments or time intervals will vary according to the individual application and type of grout being used, but should be of sufficient frequency to allow plotting of curves that accurately display all significant changes.

Although disk recorders, which are commonly used, especially in Europe, are satisfactory, continuous stripe chart recorders can include many injection parameters displayed in a overlying horizontal configuration similar to the computer-generated record shown in Figure 9.12. This enables instant comparison of the various parameters. Regardless of the method, records should include real time, ideally formatted as

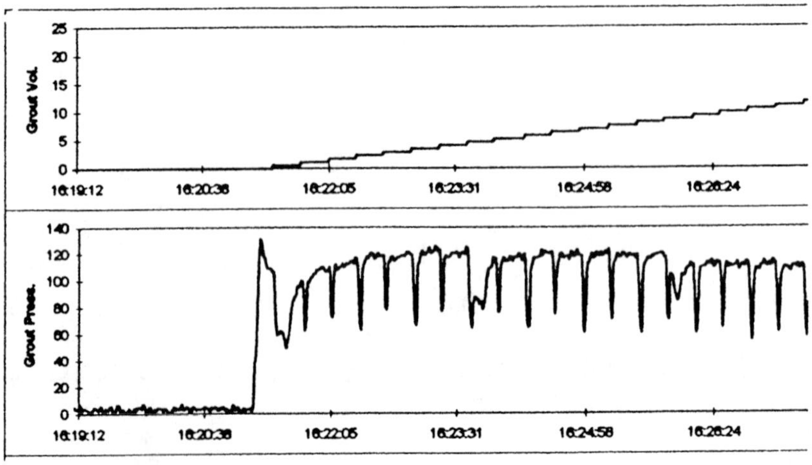

FIGURE 9.12 Each stroke is clearly evident in the downward spikes in the pressure plot.

shown in Figure 9.12 below the plots. They must facilitate immediate observation and interpretation, as well as provide permanent records for later reference.

Computer monitoring systems are also available, which allow instant readout of all parameters. Although there are two different philosophies concerning their location, they all work in a similar manner. Some commercial systems are built into rugged enclosures located at the grout plant, as shown in Figure 9.13. This allows the operator to view the record in real time and make appropriate adjustments as the work progresses. Alternately, the rather sensitive computer equipment can be located in gentler surroundings, with only sensors at the grout header and the computers housed in portable trailers or control cabins.

In either case, signals are taken at a rate of several times per minute and are usually routed into a format such as Microsoft Excel. Computer equipment has the advantage of providing a permanent record on disk and the ability to print hard copies as desired. In traditional practice employing circulating systems, the constant variation of the injection rate virtually precludes accurate manual recording. The use of real-time, continuous computer monitoring is thus especially beneficial.

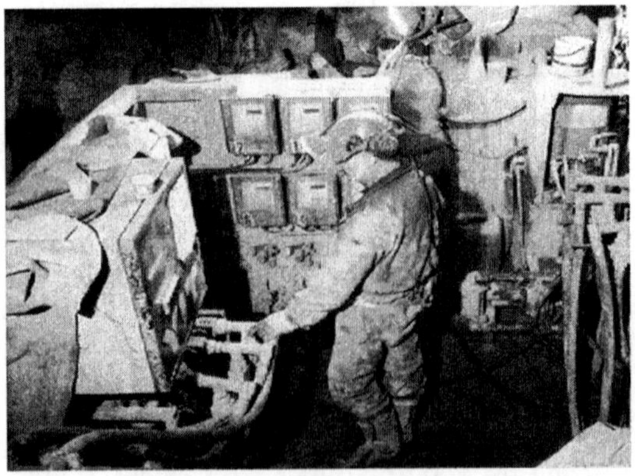

FIGURE 9.13 Computer monitoring equipment at the grout plant allows the operator immediate observation of the injection parameters, facilitating prompt adjustment.

9.7 VACUUM

The air we live in has weight. Because it surrounds us at all times and the weight is not great, we are not very aware of its existence. Should we go to a very high elevation, however, we will perhaps become lightheaded because of the lesser pressure surrounding our bodies. The air at higher elevations is often said to be *thin,* but what actually affects us is lower overlying weight. In addition to elevation, the weight of the surrounding air varies with both temperature and the prevailing weather. The *barometric* pressure given in weather reports is based on the height of a column of mercury that is raised as a result of the prevailing pressure. It is thus a measure of the actual weight of the air at a particular location and time which can be converted to the more familiar units of psi or bars, as shown in Table 9.7. At sea level, this pressure can vary from about 28 in. of mercury, or 13.75 psi (0.95 bar), to about 31.25 in. of mercury, or 15.43 psi (1.06 bar).

TABLE 9.7 Conversion of Barometric Pressure to psi and bars

Barometric Pressure	psi	bars
28.00	13.75	0.947
28.25	13.87	0.956
28.50	13.99	0.964
28.75	14.12	0.973
29.00	14.24	0.981
29.25	14.36	0.989
29.50	14.48	0.998
29.75	14.61	1.006
30.00	14.73	1.015
30.25	14.85	1.023
30.50	14.98	1.032
30.75	15.10	1.040
31.00	15.22	1.049
31.25	15.34	1.057

Expressed as atmospheric pressure, the weight of air is about 1-1/4 oz (35 g) per ft^3 (28 L). The term *normal atmospheric pressure* refers to the average pressure exerted at sea level and a temperature of 59°F (15°C). It is equivalent to a column of mercury 30 in. (76 cm) high or a 34 ft (10.2 m) high column of water. This is expressed as *1 atmosphere* and is equal to 14.7 psi, which is, of course, equal to the frequently used grouting term for pressure of 1 bar (barometer). Whereas temperature and atmospheric conditions have a rather minor effect on the weight of air, and thus pressure, elevation is of major significance and contributes a considerable effect.

A vacuum occurs when the weight of the air inside an evacuated space is less than the prevailing atmospheric pressure. The value of the vacuum is expressed as the difference between the atmospheric pressure that bears down on the space and the pressure level remaining in the space, as illustrated in Figure 9.14. If there is a

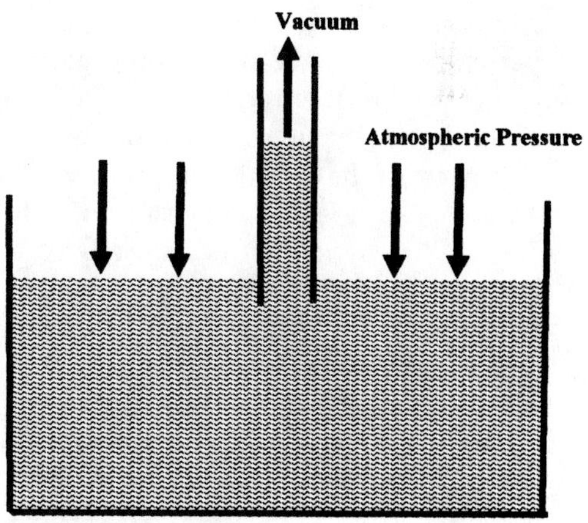

FIGURE 9.14 Atmospheric pressure pushes down on a fluid, causing it to rise when overlain by a lower pressure (vacuum).

TABLE 9.8 Relative Atmospheric Pressure at Different Elevations

ELEVATION		BAROMETRIC PRESSURE		
m	ft	Inches of Mercury	psi	bars
−500	−1,640	31.74	15.59	1.074
0	0	29.93	14.70	1.013
250	820	29.03	14.26	0.983
500	1,640	28.20	13.85	0.954
750	2,460	27.36	13.44	0.926
1000	3,280	26.53	13.03	0.898
1250	4,101	25.74	12.64	0.871
1500	4,922	24.96	12.26	0.845

100 percent vacuum inside a chamber at sea level, it has to resist pressure of 14.7 psi (1 bar) and must be capable of lifting a 34 ft (10.2 m) column of water. As the elevation of the chamber is increased, the atmospheric pressure acting against it will decrease because of a reduction of the height of overlying air. For example, at 10,000 ft (3000 m), the pressure will be about 10.1 psi (0.7 bar) and a complete vacuum will lift only a 23 ft (6.9 m) column of water. In reality, it is nearly impossible to create a complete 100 percent vacuum, but in vacuum grouting a more intense effort to create vacuum will be required at higher elevations than at sea level.

The effect of the elevation at which work is to be accomplished is thus important to understand, as it directly affects the intensity of the vacuum that can be drawn. This fundamental principle applies to many everyday activities, such as sucking a drink through a soda straw and using a vacuum cleaner, and to many types of water pumps as well. This is also the reason that the proportion of fuel going into internal combustion engines must be reduced at higher elevations; a greater proportion of air is required, as its density is less than that at lower elevations. The effects of different elevations on prevailing atmospheric pressure are provided in Table 9.8.

Boiling occurs when the pressure of a liquid becomes less that that of the confining pressure. Consequently, the temperature at which various fluids will boil decreases with higher elevation. Should a vacuum pump be used to remove air in a void completely isolated from the environment, any fluid in the void will boil, which will assist in drawing down the pressure (increasing the vacuum) inside the void. Because all liquids will boil more readily at high temperatures than at low, the ease with which boiling occurs will increase as the temperature rises, and decrease as it descends. Because boiling plays an important part in the drawing of a vacuum, both elevation and temperature must be considered.

CHAPTER 10

Grout Holes

10.1 ESTABLISHING THE HOLES
10.1.1 Hole Spacing and Layout
10.1.1.1 Split Spacing
10.1.1.2 Holes for Permeation Grouting of Granular Formations
10.1.1.3 Grout Holes in Structures and Rock
10.1.1.4 Grout Holes for Discrete Voids
10.1.2 Hole Identification
10.1.3 Hole Size

10.2 HOLE DRILLING
10.2.1 Drilling Methods
10.2.1.1 Rotary Percussion Drilling
10.2.1.2 Dual-Head Duplex Rotary
10.2.1.3 Rotary Drilling
10.2.1.4 Diamond Core Drilling
10.2.1.5 Rotary Sonic Drilling
10.2.1.6 Drive Drilling
10.2.1.7 Special Drive Needles
10.2.1.8 Becker Hammer Drilling
10.2.2 Cutting Removal
10.2.2.1 Circulation Flush
10.2.3 Grout Hole Alignment and Deviation
10.2.3.1 Monitoring Hole Deviations
10.2.4 Protection of Drilled Holes

10.3 HOLE CLOSURE AND ABANDONMENT

10.1 ESTABLISHING THE HOLES

Most grouting, and virtually all that is performed in soil and rock, requires the drilling of holes to provide a route for the grout to access the voids and defects. Obviously, a sufficient number of holes must be provided and spaced so as to reach all of the formation to be improved. Whereas vertical grout holes are easier and usually less costly to drill, it is important for the holes to intersect and penetrate any defects that are to be filled, and this sometimes requires the placement of inclined holes.

In structural grouting, the space to be filled is often known, or at least partially visible, as in the case of cracks. Surface-set ports, as shown in Figure 10.1, can thus be used. Where cracks or voids are hidden, however, grout holes must be drilled. Even where cracks are visible, it may be advantageous to drill holes that intercept them at some depth in order to optimize placement of the grout.

10.1.1 Hole Spacing and Layout

Proper location of the grout holes is obviously fundamental to the effectiveness of a grout injection. Although there are variations in the methods and materials of grouting, as well as in the establishment of grout holes, there are many fundamentals that apply equally to all work. In

FIGURE 10.1 Surface-set injection port over a crack.

the layout of grout holes, the effective distance of grout flow or permeation must be considered. This distance will obviously vary, according to the properties of the host formation and the size, configuration, number of connections, and other attributes of the void or pore system.

When the grout is to fill large volumes of the host formation, a network of many holes will be required. These will usually be spaced so that the zones of injection will slightly overlap to ensure complete filling of the defined mass to be improved. The amount of such overlap and the resulting thoroughness of the injection will vary according to the individual project requirements.

There is a tendency by many grouters to space the holes excessively so as to minimize the amount of drilling. Although this will reduce the drilling cost, it is not always in the best interest of the work. The amount of control and the degree of thoroughness of the injection are directly related to the hole spacing. As a general rule, closer spacing results in more orderly injection and more uniform filling of the void network. It is much better for holes to be spaced too closely than for them to be spaced too widely.

10.1.1.1 SPLIT SPACING

The *split spacing* principle of hole layout is used in virtually all geotechnical grouting. In practice, the initial injection is into *primary* holes, which are spaced at a distance at least twice that of the anticipated final spacing. The spaces between primary holes will be split equally for the placement of *secondary* holes, as shown in Figure 10.2. In some situations the spaces will again be split for tertiary holes, and then quaternary, and, in rare instances, continued with quinary, sextary, septenary, and so on.

By starting with the widely spaced primary holes, the largest defects will be filled, but there should be little, if any, interconnection of grout between those holes. Holes of a given hierarchy should be completed prior to any drilling in between. As the spacing is split, succeeding holes

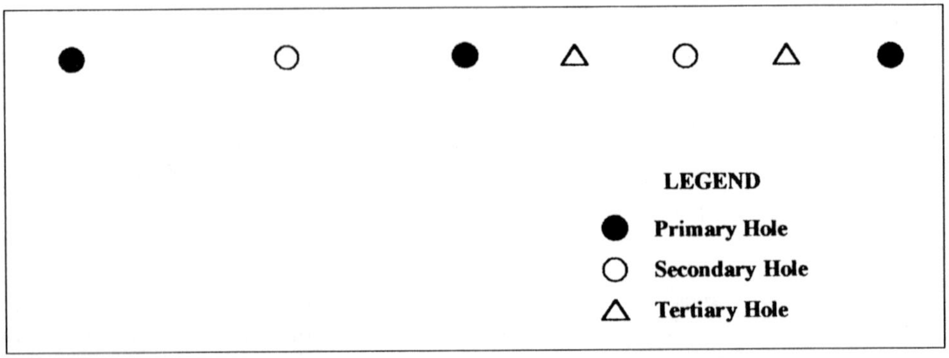

FIGURE 10.2 The space between the two primary holes is split equally by split spacing.

should have progressively lower grout takes. Should this not occur, excessive spacing is indicated. In grouting the pore network of soils for strengthening, a strictly primary-secondary pattern is typically used. Conversely, for the creation of impermeable curtains in either soil or rock, the use of tertiary and even quaternary holes is not unusual.

10.1.1.2 HOLES FOR PERMEATION GROUTING OF GRANULAR FORMATIONS

For permeation grouting in substrates such as soil, which has a relatively consistent pore void system, each hole should have similar attributes of injection rate, grout take, and effective radius. The grout holes should therefore be equally spaced over the area to be treated. The exact spacing is dependent on the effective distance of grout permeation in the particular soil to be improved, at the pressures and injection rates to be used. Such spacing will normally be within a range of 2 to 4 ft (0.6 to 1.2 m).

Alternate primary holes should be drilled and grouted first, followed by the secondary holes. The grout *take* in the secondary holes should be slightly less than in the primary holes. If the take is greater than that in the primary holes, the spacing is too great and should be reduced. Conversely, if the grout take in the secondary holes is a great deal less than that in the primaries, the spacing is likely too tight and can probably be increased. Where a high level of solidification is required, the hole spacing should be tightened so as to provide greater overlap; the exact amount is based on the degree of perfection required. For permeation grouting in porous formations, split spacing beyond the secondary holes is rare.

When multiple rows of holes are used, the alternate rows should be offset so as to provide the most efficient grout filling, as illustrated in Figure 10.3. The exact spacing to be used is dependent on the effective distance of grout permeation in the particular soil. To optimize the

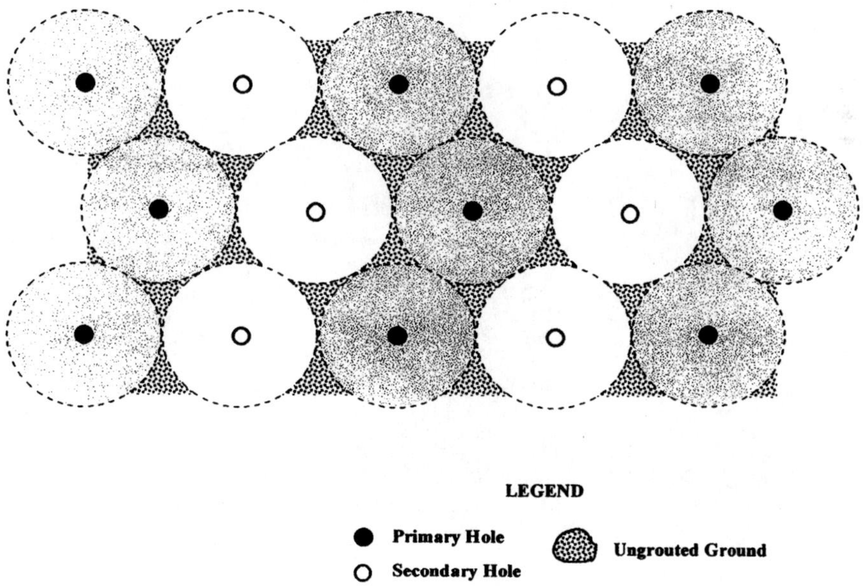

FIGURE 10.3 Intermediate rows are offset to provide the most efficient grout permeation.

hole layout, it is essential to maintain good records of the injection parameters. The hole spacing is often adjusted upon analysis of the injection data during the early stages of grouting. Such data will provide guidance as to any appropriate adjustments that are warranted.

When the purpose of the grouting is to block water flow through soil, the spacing of the holes must be sufficiently close to fill essentially 100 percent of the pore space. This will require considerable overlapping of the injections, as illustrated in Figure 10.4. As can be observed, the amount of grout injected into the primary holes, which are the first to be filled, will be considerably greater than that required for the subsequently injected intermediate holes. The creation of an effective watertight curtain in soil usually requires a minimum of two rows of holes, although three-row curtains are common.

Where the purpose of the grouting is to increase the strength of the formation, filling of 100 percent of the soil voids is usually not required, and minor amounts of untreated soil, as illustrated in Figure 10.3, are acceptable, which will allow an increased hole spacing. In grouting for water control, the effectiveness is obvious by a reduction of leakage during injection, but in grouting for strengthening applications, thoroughness is not so evident. Thus, test excavations are often used to confirm the effective injection radius and optimize the hole spacing.

It is not unusual for the grout to pick up fines in the soil as it progresses away from the injection point. Similarly, fine particles such as ultrafine cement that are in the grout can be filtered out, especially when penetrating fine granular deposits. Such action obviously becomes more intense with increases in the distance of the advancing grout from the hole. Although it is possible to permeate most soils for a distance of 2 ft (0.6 m) or more, it has been found that varying levels of solidification often result with

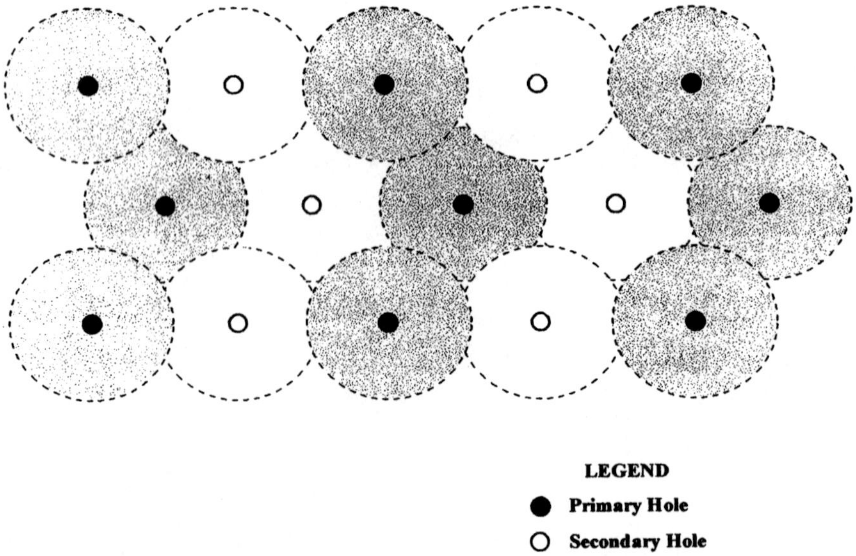

FIGURE 10.4 Considerable overlap of the grout will occur with the close hole spacing required for 100 percent saturation of the formation.

such travel distances, and thus a tighter hole spacing may be advantageous, especially when a high degree of grout saturation is required. Such spacing is usually on the order of not more than about 2 ft (0.6 m).

10.1.1.3 GROUT HOLES IN STRUCTURES AND ROCK

Unlike the voids in a soil, those in many structures and rock are not consistent, but vary in size and distribution. This results in large variations of both the appropriate injection rate and the amount of grout injected into different holes. Determination of the best spacing for grout holes, as well as their orientation, is thus greatly complicated. As illustrated in Figure 10.5, holes that intersect significant geologic defects can require very large quantities of grout, whereas adjacent holes may not intersect significant defects or, possibly, any defects at all. Further, the orientation of joints and other defects in rock can vary almost infinitely. For example, the major defects in rock of sedimentary origin will be normal to the dip or strike of the bedding plane, which is explained in Chapter 21, and which can be steeply inclined, inhibiting intersection by vertical holes, as illustrated in Figure 10.6. Inclined holes must thus be used.

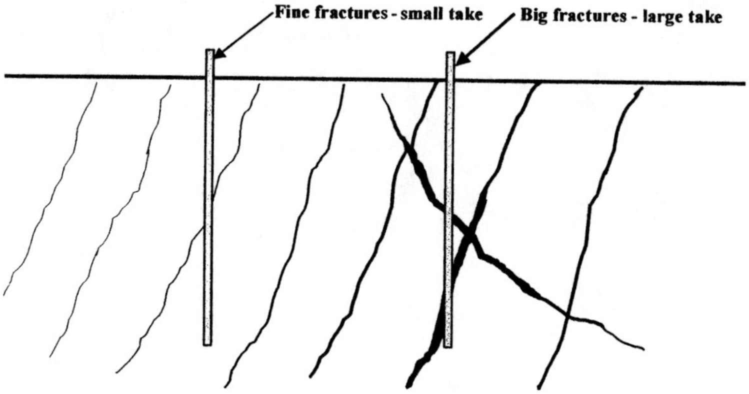

FIGURE 10.5 The quantity of grout injected in rock can vary greatly.

To ensure the optimal hole inclination, good geologic mapping is required during the planning of a grouting program. In addition, continuous monitoring to detect any changes in the anticipated void orientation should be provided during drilling. It is preferable to obtain the greatest number of possible intersections of a given defect with the grout holes. This often requires them to be inclined in complex geology, and several different inclinations may be required, as illustrated in Figure 10.7.

It is usually more difficult and expensive to drill inclined holes and, especially, to control their alignment. Thus, they are not inclined more than about 30 degrees off vertical, although flatter orientations may be required in rock with very complex jointing. Note that the inclination of grout holes is commonly stated as the degrees off vertical. This is contrary to geologists' common reference to the *dip* of rock, which is the degree of tilt off horizontal. The geological formation of rock and its influence on grouting is discussed further in Chapter 21.

10.1.1.4 GROUT HOLES FOR DISCRETE VOIDS

When injection is into a discrete void or other defect, only a limited number of holes are required. Where isolated large voids are being filled, both injection and vent holes should be provided at the most distant and opposite extremities, which are usually at opposite ends, as shown in Figure 10.8. The grout is typically pumped into a hole at the bottom of the void, with venting provided on top at the most distant location. Where there are pockets in the top of the void that might entrap air, venting must be provided for. This can be accomplished either by provision of additional holes intersecting the

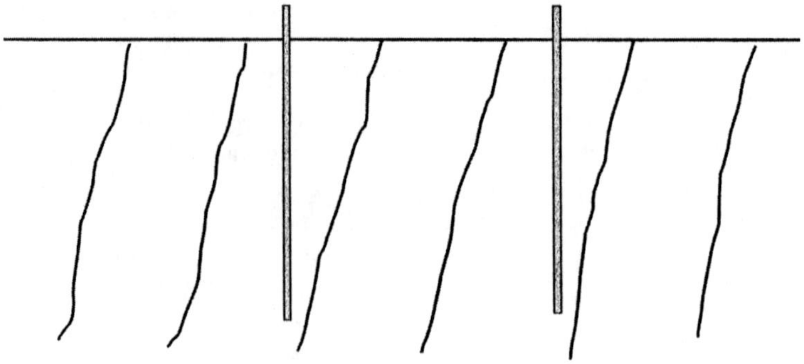

FIGURE 10.6 Vertical grout holes can completely miss rock joints that are strongly inclined.

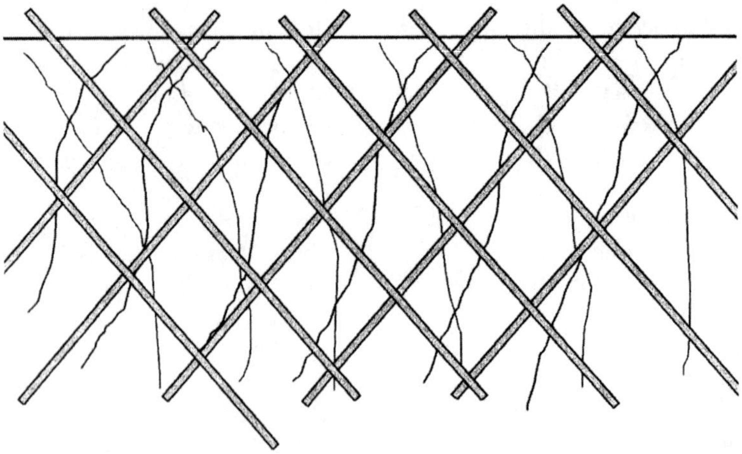

FIGURE 10.7 Several different hole inclinations may be required in rock with complex geology.

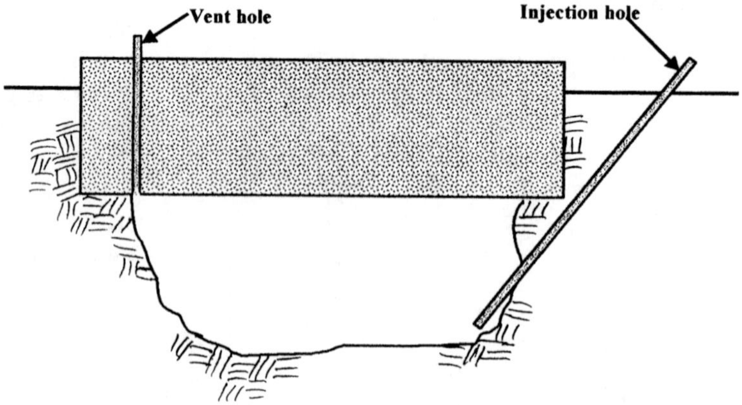

FIGURE 10.8 Injection and vent holes are located at the opposite extremities of large voids.

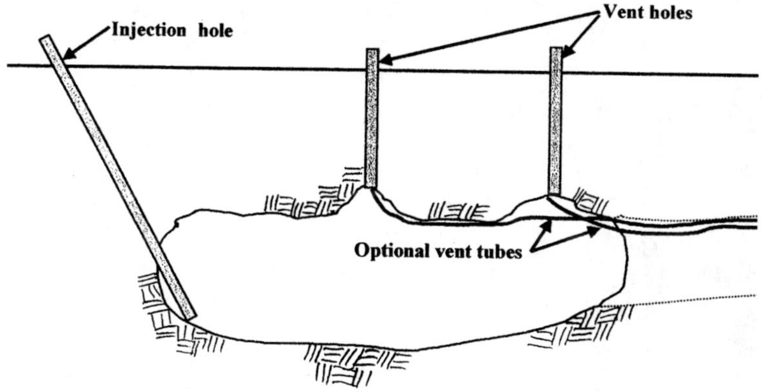

FIGURE 10.9 Small tubing can be run from isolated pockets in the roof area of voids to provide the required venting.

pockets or placement of small vent tubes leading from the pockets, as illustrated in Figure 10.9.

Except when relatively small voids are being filled, the provision of additional regularly spaced holes is advisable. These will provide further venting, and should the initial injection hole become obstructed, the grout injection can be moved to the next open hole once the advancing front of grout reaches it. Holes that are not needed for further injection should be plugged once good-quality grout exits. Although most grout develops considerable friction with the surfaces it contacts, increasing the required injection pressure, this is not always the case in filling large voids. In these applications, holes can usually be spaced much farther apart.

10.1.2 Hole Identification

The grout holes should be accurately laid out and durably marked on the grout casing and/or formation surface, including the hole number or other identification, as illustrated in Figure 10.10. This can be best be accomplished with a can of spray paint. It is very easy to become confused during the work, and misidentification of grout holes is a common problem. Thus, it is good practice to use different colors to mark the hierarchy of the various holes—perhaps red for primaries, yellow for secondaries, and so forth. The importance of durable marking cannot be overstressed. As grouting progresses, dragging hoses, spilling grout, and other actions can obliterate even the best of markings. The grouting crew should have a can of each color of paint used at all times in order to easily refresh the markings that become covered or otherwise damaged.

In situations where markings are particularly prone to obliteration, it is a good idea to provide offset markings or other external control points that can be used to relocate lost holes. It is also beneficial to mark each hole upon completion of grout injection. This can be readily accomplished by inserting a surveyor's flag, as shown in Figure 10.11. Note the comment "Bad Boy" on the flag, in addition to the hole number and depth. This was the crew's way of identifying erratic injection and issuing a caution to be alert during adjacent work.

FIGURE 10.10 Hole layout is expeditiously performed with pressurized cans of spray paint.

FIGURE 10.11 Surveyor's flag used to denote a completed grout hole.

A convenient way to differentiate between holes of different hierarchy is to place a large washer or "donut" of oversize pipe or casing around each hole of a given level. In some cases, short rings of standard pipe of various lengths—3 in. (75 mm) for primary holes, 2-1/2 in. (63 mm) for secondaries, and 2 in (50 mm) for tertiary holes—have been used. Although there is a nominal cost involved, it is inexpensive insurance against misidentification and erroneous injection, which occurs all too often.

10.1.3 Hole Size

The size of grout holes has been the subject of much debate among grouters. Some contend that a larger hole will intersect greater lengths of potential defects, reducing required pumping pressure. Although larger-size holes will not adversely affect the injection process and may even benefit it slightly, the additional cost is usually not justified. For grouting in soil and rock, the most common hole sizes range from 2 to 3 in. (50 to 75 mm).

Where very deep holes, or holes drilled to a close tolerance, are required, larger sizes are justified and should be considered. Moreover, a larger-size hole can be advantageous in the penetration of particularly difficult soil, especially one that contains rocks or boulders, which tend to cause small drill systems to drift off course. Although rather rare, holes of 6 in. (150 mm) diameter or even greater have been used in special grouting applications.

Small-size holes are often the most favorable in structural applications. Because the depth of such holes is usually not very great, the grouts often used quite fluid, and the pumping rates low, injection into very small holes is appropriate. Further, the chance of a small hole encountering embedded reinforcing steel is much less than that of a larger hole. The selection of size is thus often based on use of the smallest diameter for which a sufficiently long drill bit is available. The issue of hole size is discussed further as appropriate to the different types of grouting.

10.2 HOLE DRILLING

Most grout holes are formed by drilling, but in soil they may also be placed by driving a temporarily plugged casing or a special injection needle. To simplify the subject, *driving* is treated here as a *drilling method,* although, by strict definition, it is quite different. The small holes typically used are generally capable of remaining open for the normally short time periods required prior to grout injection. Where it is likely that holes will cave in however, it will be necessary to install casing during drilling. Larger-diameter holes are sometimes used for depths of more than 100 ft (30 m) or where drilling to close tolerances is required. They are also warranted where a fast pumping rate may be appropriate, such as in fill grouting of massive voids.

There is a wide variety of types of drilling equipment, although there are similarities common to all methods, as shown in Figure 10.12. In operation, a *drill head* rotates a *string* of drill *rods,* to which the drill *bit* is attached. The drill

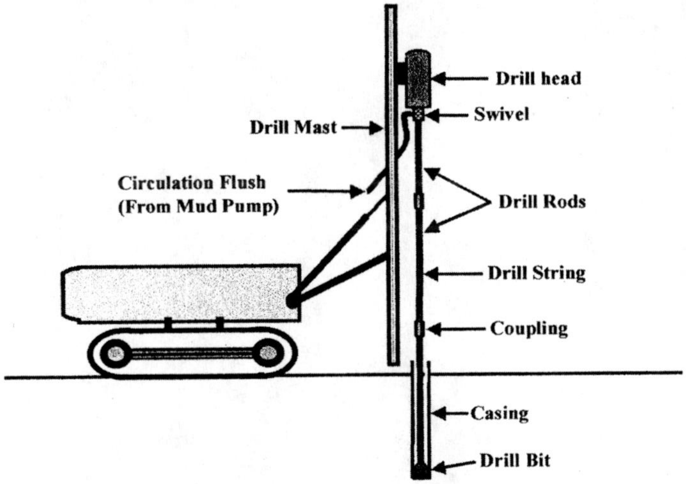

FIGURE 10.12 Basic elements of a drill rig.

head is most often mounted on a *mast,* or *feed,* which is part of a drill *rig.* Rotary and rotary percussion drill rigs are similar in this respect; however, the equipment used with percussion drilling must be heavier and more robust than that used with rotary drilling.

The drill rods are hollow, and a *swivel* is provided so that during rotation air or fluid can be forced through the drill string to the bit. The air or fluid is referred to as *circulation* or *flush.* The circulation flush is forced, under pressure, through the hollow stem of the drill rod to the bit, where it picks up the cuttings and returns them to the surface, as illustrated in Figure 10.13. The makeup and type of the circulation flush and the manner in which it is circulated is crucial to providing a clean hole, as discussed later.

In operation, the drill head is raised so that a section of rod, with the bit attached to the lower end, can be joined to it. The rod and bit are then simultaneously rotated and forced into the ground to form the hole. Most drill rigs provide a positive downward *thrust,* such that uniform pressure can be maintained on the bit to facilitate its penetration. When the entire length of the drill rod has penetrated the formation, the head is disconnected and raised and an additional section of rod added. This operation is repeated until the desired depth has been reached.

As previously mentioned, a wide range of drill rig types are used in grouting. The trend is toward use of self-contained fully hydraulic drill rigs powered by diesel engines, as shown in Figure 10.14. These machines are available in a range of sizes and capacities, and most offer the ability to adjust the drill mast to an unlimited number of positions under power, as well as to extend a considerable distance from the base. The only disadvantage is limitation to access on very steep terrain, as their diesel engines must not be angled so much as to interfere with the circulation of the lubricating oil.

For work on slopes or other difficult terrain, the long established *air track* drill is ideally suited. It is powered by a remotely located air

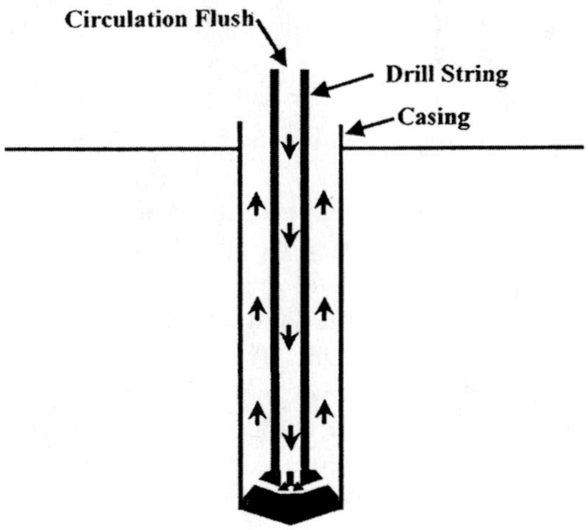

FIGURE 10.13 The drill cuttings are removed by the flow of the circulation flush.

FIGURE 10.14 Self-contained fully hydraulic drill rig.

compressor, which feeds a separate independently controlled and reversible air motor for each track and can thus "turn on a dime." Because it has a very low center of gravity, it is inherently stable regardless of its position. These machines can operate in virtually any attitude, and air tugger winches are often mounted on them to assist their traversing on steep slopes. In extreme cases they can be suspended from above with a cable. They feature masts that can reach far from their tracks and be placed in virtually any inclination or other position under power, as illustrated in Figure 10.15. Although air tracks were originally used primarily for percussive drilling of blast holes in rock, they can be fitted with either rotary or combination rotary-percussive drill heads.

Although the aforementioned drill types are the most frequently used for drilling grout holes, many different types of specialty drills, particularly suited to accessing and working in confined locations, are available. These range from the handheld wet head bore motor, shown in Figure 10.16, to very powerful independently powered hydraulic drills, such as shown in Figure 10.17 working in the gallery of a dam. In between are a myriad of both self-powered and independently powered drills and drill attachments for virtually any demand.

FIGURE 10.15 Remotely powered pneumatic "air track" drills can be placed in the most difficult positions.

The actual drilling equipment used often depends on the preferences of the individual contractor and is strongly influenced by the type of equipment a contractor may have available for a particular job. There are, however, certain fundamentals that must be considered, the most relevant being the particular makeup of the formation into which the holes are to be advanced. For example, deep holes can be readily drilled in soils that are free of rocks or boulders with the use of handheld drills, as illustrated in Figure 10.16.

Should similar soils contain embedded rocks or boulders, however, a much stiffer drill string would be required as well as a heavier drill rig to drive it. Some of the most difficult drilling conditions involve penetrating large boulders in soil, especially where several rocks are nested together and/or the density of the encasing soil is not great. Drilling forces can cause the boulders to be-

FIGURE 10.16 Handheld pneumatic bore motors can drill in soils to 100 ft (30 m) or more.

come dislodged and their exact position changed. This will result in a misalignment of the hole, which can cause the bit to become wedged under or between boulders. Both high torque and a large upward withdrawal thrust are required under such conditions, and this obviously means a powerful and heavy drill rig. There are a number of drilling techniques and available equipment for such difficult conditions. The most common of these are sonic drilling, dual rotary drilling, and Becker hammer drilling, which are discussed in the following sections. The particulars of drilling equipment are discussed in Chapter 30.

FIGURE 10.17 A powerful independently powered hydraulic drill fills this dam gallery.

10.2.1 Drilling Methods

Virtually all drilling of grout holes can be grouped in one of three general classifications, *rotary, rotary percussion, driving,* or combinations thereof. Rotary drilling is the most frequently used for work in soil. It is also used to procure cores of rock or other hard materials by the use of diamond core bits. Although rotary drilling can be used to drill ordinary holes in rock, its use is not common for the small diameters typically used in grouting. Rotary percussion is usually faster and is thus the most commonly used method. The employment of drive drilling is obviously limited to penetration of soil.

Regardless of the method, drilling in soil, rock, or other hard materials always involves the rotation of a drill *bit* that is simultaneously thrust into the bottom of the hole. There are many different types of bits, as described in Chapter 30, but they all work according to the same principle, which is to loosen or fragment the formation. As the formation is loosened by the action of the bit, the cuttings are expelled out the annulus surrounding the drill rod by the upward flow of the circulation flush, as illustrated in Figure 10.13.

Although rotary drilling involves only rotation of the bit under the thrust force, rotary percussion provides simultaneous rotation and percussion of the bit. It is thus faster and more economical for drilling hard materials. Simple rotary drilling can be performed with very light equipment and is fairly quiet, but heavier equipment and loud noise are common to percussion drilling. In addition, an oily mist is sometimes discharged from the drill head, which soils anything with which it comes in contact.

10.2.1.1 ROTARY PERCUSSION DRILLING

Rotary percussion drilling is the fastest and most economical method for establishing grout holes in virtually all rock, as well as in concrete and masonry. Although soils can usually be efficiently drilled without the high energy of percussive equipment, there are instances where such equipment is advantageous. Penetration of very hard or strongly cemented soils, such as caliche, or advancing holes through boulders is often best done with percussive equipment. Modern drill rigs can be fitted with either rotary or percussive drill heads or combination rotary-percussive heads.

Percussive drilling involves rapid pounding on the drill bit as it is simultaneously rotated. This is accomplished with a hammer on the drill head, along with rotation of the rod. The simplest form of such drilling is the common handheld pneumatic sinker drill (Figure 10.18) which is often incorrectly dubbed a "jackhammer." For much drilling, more powerful hammers are required, which is why they are mounted on some type of carrier, as shown in Figures 10.14 and 10.15.

FIGURE 10.18 Handheld pneumatic sinker drill.

Because considerable impact energy is absorbed in the drill string, drilling efficiency decreases as hole depth increases. Thus, the maximum effective depth of top-head hammer drills is on the order of 100 to 200 ft (30 to 60 m). Where greater depths are involved, it is most expedient to use a *down-the-hole* hammer. These mount directly to the drill bit and provide the percussive pounding while the rotation is from the top head by way of the drill rod. Down-the-hole hammers usually increase production and provide straighter holes, but they require a minimum hole size of about 3 in. (75 mm).

Although, historically, percussive hammers worked on compressed air, fully hydraulic drills, as shown in Figure 10.14, are now common. Hydraulic power is capable of exerting an especially high impact and is thus well suited to power percussive drills. It does, however, present one limitation when used in connection with independently powered equipment. The inherently stiff hose lines filled with hydraulic oil are much heavier and harder to handle than air lines.

There are two distinctly different types of drill bits used with percussive drilling, chisel and button. Chisel bits tend to produce larger cuttings than button bits, and these require more energy to expel from the hole. They are less likely to block small defects that need grouting, however. The trend is to use button-type bits because they do not wear as rapidly and are thus more efficient. Most chisel tooth bits contain four blades, although two- and six-blade bits are also manufactured.

Of particular advantage in drilling small holes for structural grouting are single-chisel-tooth bits of carbide that are integral with a drill shank. These are available in standard lengths up to 26 in. (660 mm) and in hole diameters ranging from 1/2 in. (13 mm) to 1-1/2 in. (38 mm). Longer lengths are available on special order. They are very efficient, and when used in small pneumatic handheld drills, as illustrated in Fig-

FIGURE 10.19 Drilling with a lightweight pneumatic sinker drill with an integral single chisel-blade bit.

ure 10.19, will produce relatively large-size cuttings that are efficiently removed by the blast of air circulating from the chisel blade. Such removal of large cuttings from shallow holes is one area in which air circulation has a real advantage, and it results in exceptionally clean holes.

10.2.1.2 DUAL-HEAD DUPLEX ROTARY

Most grout holes are simply drilled with a common rotating drill string to produce an otherwise open hole. In soil, however, it is sometimes necessary to case the hole either partially or to full depth. Where required, the casing is often simply pushed in place by the available thrust pressure of the drill rig. If it is necessary to penetrate rock as well, drilling to the full outside diameter of the casing will be required. This can be accomplished by simultaneous rotation of both the drill string and the casing, which has been fitted with a separate hollow *crown* bit, or simultaneous underreaming of the casing through use of a special eccentric bit that extends beneath it.

Simultaneous rotation of both the drill string and the casing in the same direction can be achieved with ordinary drill rigs as long as sufficient torque is available; however, this tends to displace the intended hole trajectory. By rotating the bit and casing in opposing directions, penetration is faster and the hole tends to be straighter. Such drilling requires rigs that provide the necessary opposed independent rotation. This is generally limited to medium to large and heavy machines. Such drilling can be performed with either rotary or percussive drills.

10.2.1.3 ROTARY DRILLING

Rotary drilling is the most frequently used type of drilling to produce grout holes in soil. Drilling in soil requires much less energy than drilling of rock and other hard formations. Accordingly, much lighter equipment can be used. Some contractors, however, tend to adapt their percussive equipment to work in soil, which may not be completely rational. Lighter equipment is more maneuverable and economical and is fully adequate for penetrating most soils. In fact, many tens of thousands of grout holes, with depths to 100 ft (30 m) or more, have been drilled with handheld equipment such as that illustrated in Figure 10.16. This type of drilling can be used in the most inaccessible of locations, such as in the culvert depicted in Figure 10.20.

Where handheld drills are used, it is prudent to limit the length of the drill rod to 3–5 ft

FIGURE 10.20 Drilling in a culvert with a handheld pneumatically powered drill.

(1 to 1.5 m), which enables the workers to remain standing at ground level. For hand drilling, the weight of the drill rod becomes significant, especially when long lengths of drill string are required for deep holes. Custom, nonstandard lightweight rod is thus often used with these drills. Although the lightweight tools do not drill as quickly and efficiently as larger rig-mounted drills, their light weight and ability to move quickly from one hole to the next makes them extremely productive.

A great deal of soil grouting is performed around and within structures that provide limited access and frequently cramped conditions. The use of larger machines can increase the actual drilling production, but the time required for moves and setup will be much greater. In addition, larger equipment often damages the surfaces on which it operates, and in some instances it simply cannot fit in congested areas of limited access.

There have been instances in which holes have been poked into the ceiling of a building to enable a drill mast to be raised. This practice results in unnecessary damage, greatly extends the time required to complete the work, and is always more expensive. Further, repair of the damage will entail increased time and costs once the grouting is completed. Light equipment is therefore preferred for drilling in soil around and under existing structures. Such drilling may be done with handheld drills such as shown in Figure 10.16 or with a remotely powered drill head mounted on a lightweight frame with a winch to raise it, as illustrated in Figure 10.21. The only limitation to this type of rig is the lack of positive down thrust, although weights can be hung from the drill carriage if needed.

FIGURE 10.21 Pneumatically powered drill head mounted on lightweight frame.

10.2.1.4 DIAMOND CORE DRILLING

Rotary methods are also used to core through rock and reinforced concrete where penetration of the reinforcing steel is required. There are two distinctly different categories of core drilling equipment and procedures. The first involves the use of relatively thin-wall core bits of limited length, which is often referred to as *construction* core drilling. The equipment and procedures, originally developed in the 1950s, were directed to the drilling of relatively short holes through concrete and masonry, which remains their main use. Although there are no strict standards for this equipment, most manufacturers use similar patterns that are interchangeable.

Diamond bits with thin walls have a distinct advantage in that they remove only a thin kerf thickness of material. This can be readily observed in Figure 10.22, where the core barrel is only slightly larger than the diameter of the core being removed. These bits cut faster and require less power to operate than the more traditional *geotechnical* diamond core bits, which cut a much wider kerf. Because of the thin cut, this equipment is especially useful for penetrating reinforcing steel and other metal embedments.

Another area in which this equipment is especially beneficial includes situations where grout holes must be placed through hard-to-match

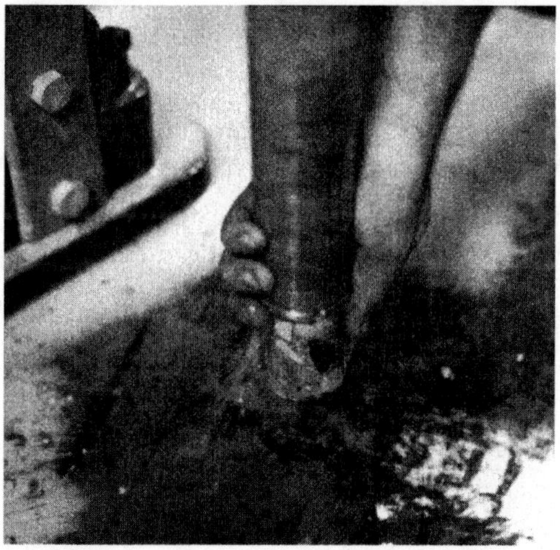

FIGURE 10.22 Only a thin kerf section is removed with thin-wall diamond-core bits.

finished surfaces. In such cases, an oversize core can be carefully removed from the surface, as illustrated in Figure 10.23. It should be marked with the hole number, so as to indicate the location from which it was removed, and its exact orientation within that hole. The grout hole can then be drilled and injected as usual. Upon completion of the grouting, the original core can be replaced in its proper orientation. The thin annulus, when filled with a blended mortar, will be hardly noticeable, especially with the passing of time. Even immediately after placement it is hardly noticeable, as shown in Figure 10.24.

Although considerable thrust is required, these drills are amazingly small and highly portable. To provide the necessary thrust, they are typically anchored temporarily to the surface from which the core is being withdrawn. This is illustrated in Figure 10.25, where the column supporting the drill head is anchored to the concrete with a single wedge-type expansion anchor. Where marring of the surface cannot be tolerated, vacuum pads in combination with a vacuum pump can secure the rig. In either case, the operator provides thrust to the drill bit manually through a rack and pinion gear drive mechanism.

Standard thin-cut core bits are 14 in. (355 mm) long for core diameters of up to 12. in (305 mm). They are able to penetrate only about 12 in. (305 mm) into the formation, however, as the end cap that connects to the power

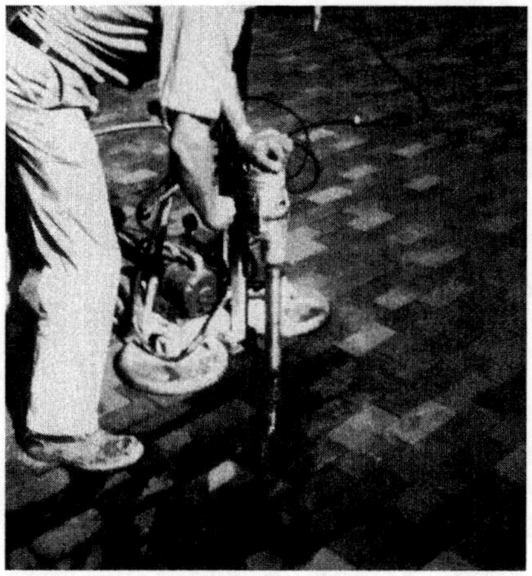

FIGURE 10.23 Removal of a core of the ceramic surface, which will be replaced upon completion of the grouting.

FIGURE 10.24 The surface is restored, with the hole barely visible in the center of the photo.

FIGURE 10.25 Highly portable drills are temporarily anchored to surface being drilled.

source bottoms out on the core. Bits for diameters greater than 12 in. (305 mm) are usually 24 in. (610 mm) long and have a maximum drilling depth of about 22 in. (559 mm). The wall thickness of these bits is only about 1/8 in. (3 mm), so it is not possible to make flush-coupled extensions as is common in more traditional geotechnical coring. Longer thin-wall core bits are available on special order, like that shown in Figure 10.25.

The standard procedure for drilling greater depth holes is to withdraw the bit once it bottoms out, break off and remove the core, and then, with an extension rod on the bit, continue to drill, break off, and so forth, until the desired depth has been reached. With this procedure there is no theoretical maximum length that can be drilled. Withdrawal of the bit and, especially, repeated break off and extraction of the core, are time-consuming and become progressively so with depth. Practicality thus limits drilling lengths to about 10 ft (3 m) or less.

This type of core drilling is most often done by specialty contractors who do both diamond drilling and sawing. This is, no doubt, due to the high cost of the diamond tools, the longevity of which is dependent on the operator's skill. A substantial financial benefit thus accrues through employment of experienced full-time workers.

Core bits for holes of up to 36 in. (0.9 m) in diameter are commonly available, and significantly larger bits are offered on special order. For example, in one instance a core bit was observed cutting 12 ft (3.6 m) diameter cores out of 24 in. (610 mm) thick concrete airport paving. Costs increase with the diameter of the bit, especially for special orders. Very large bits are thus obtained only for unusual applications that can justify their high cost.

Unlike those used for thin-wall construction diamond drilling, the equipment and procedures used for the more traditional geotechnical exploratory drilling have been long established and practiced. In fact, the first reported use of the method was in 1863, when Leschot, a Swiss engineer, used a rotating diamond-tipped hollow tube to scour rock, while washing the cuttings to the surface with water (Acker, 1974). In 1929, several manufacturers of specialty equipment for core drilling organized the Diamond Core Drill Manufacturers Association (DCDMA) to develop standards so as to make the products of different manufacturers compatible.

DCDMA generally limited its activities to so-called *down-the-hole* equipment, such as drill rods, casing, core barrels, and bits, although it also set standards for the power factors that drive the drill string. These include the brake horsepower and torque capabilities of gasoline and diesel engines, electric motors, and pneumatic motors, all of which are used as power sources on drill rigs. DCDMA was organized in the United States, and its provisions were based on the Imperial (inch) system of measurement. These were quickly accepted in the United States, Canada, Australia, England, and South Africa and are now recognized worldwide. Metric equivalents have more recently been added to the DCDMA standards.

Concurrently, European manufacturers developed a system based on the metric system of measurement, which has now been recognized as an alternate to the DCDMA standards. The original DCDMA inch-system equipment covers a much wider range of sizes and is by far the most widely used worldwide. In general, it provides for greater clearances between the casing and the wall of a hole, and the components are of somewhat more robust construction. It is thus more durable and can be used for drilling deeper holes. The European equipment is generally tighter fitting and of lighter construction. Its design has been directed toward the drilling of shallower holes, but in generally harder rock.

This type of drilling is most widely used for deep exploratory work in rock. Although, historically, it was used extensively for grout holes, the method has been largely replaced by rotary percussion drilling due to its slower production and greater cost. For projects with complex geology or where a good knowledge of the subsurface defects is lacking, it has the advantage of core recovery, which can add greatly to an understanding of the actual subsurface conditions.

Standard hole and core size for DCDMA equipment in the inch and metric systems are shown in Tables 10.1 and 10.2.

There are two basic systems of core drilling equipment, standard and wire line. With the standard system, a core bit is attached to a core *barrel.* The core barrel contains a core *lifter* immediately above the bit. The lifter tightly grips the bottom of the core when an upward force is exerted. As the core bit penetrates rock, the core barrel encases the specimen. Once the bit has descended a distance equal to the length of the core barrel, an upward force on the rod causes the core to break off. The drill string is then withdrawn, the core recovered, and the process repeated until the final hole depth has been reached.

The advanced *wire line* technique is widely practiced, especially in the coring of soft rock and in very deep holes. Herein, the drill rod is essentially a thin tube, the outside diameter being almost that of the drill hole. The equipment for down-the-hole tooling is similar to that of the standard system except for the drill rod, which has flush wall threads and is usually identified by the suffix "WL" or "Q" after its size designation. Once drilling has progressed by the length of the core barrel, an *overshot* tool

TABLE 10.1 Standard Diamond-Drilled Holes and Cores in the Inch System

Designation	Hole Diameter in in. (mm)	Core Diameter in in. (mm)
RW	1.160 (29.4)	0.735 (18.6)
EW	1.470 (37.3)	0.905 (22.9)
AW	1.875 (47.6)	1.281 (32.5)
BW	2.345 (59.5)	1.750 (44.4)
NW	2.965 (75.3)	2.313 (58.7)
HW	3.890 (98.8)	3.185 (98.8)
2.75 × 3.87	3.840 (97.5)	2.690 (80.8)
4 × 5.50	5.435 (138.0)	3.970 (100.8)
6 × 7.75	7.655 (194.4)	5.970 (151.6)

TABLE 10.2 Standard Diamond-Drilled Holes and Cores in Metric System

Designation	Hole Diameter in mm (in.)	Core Diameter in mm (in.)
36 mm	36.0 (1.42)	21.7 (0.85)
46 mm	46.0 (1.81)	31.7 (1.25)
56 mm	56.0 (2.20)	41.7 (1.64)
66 mm	66.0 (2.60)	51.7 (2.04)
76 mm	76.0 (2.99)	61.7 (2.43)
86 mm	86.0 (3.38)	71.7 (2.82)
101 mm	101.0 (3.98)	86.7 (3.41)
116 mm	116.0 (4.57)	101.7 (4.00)
131 mm	131.0 (5.16)	116.7 (4.59)
146 mm	146.0 (5.75)	131.7 (5.18)

attached to a wire line is lowered into the hole. The overshot device grips an inner core barrel, which is then raised to the surface for core removal. The wire line typically consists of flexible wire rope with a diameter on the order of 3/16 in. (4 mm). It operates at a speed of about 100 ft (30 m) per minute, so core withdrawal is fairly rapid.

Each system has its own relative merits and limitations. The width of the drill kerf of standard bits is substantially less than that of a wire line bit. For example, although the thickness of the cut varies according to the hole size, standard bits usually cut a kerf on the order of 3/8 in. (9 mm) wide, whereas the thickness of cut of wire line bits will be closer to 1/2 in. (13 mm). Wire line drilling thus involves a greater amount of diamond contact on the rock surface, which both slows progress and increases the cost of diamond usage.

However, it is not necessary to withdraw the entire drill string for core recovery, so this system is faster, especially in deep holes. Regular inspection of the bit condition, which is otherwise routine at each core withdrawal, cannot be done, because it remains in the hole while the core is removed. In the drilling of very hard rock, it is necessary to renew the bit at regular intervals, and this would normally be done when the drill string is removed for core retrieval. With wire line work, complete withdrawal will be required any time the bit needs attention. The system thus has its greatest advantage in the drilling of deep holes in relatively soft rock, in which bit life will be relatively long.

10.2.1.5 ROTARY SONIC DRILLING

Rotary *sonic* drilling involves high-frequency vibration of the entire drill string as it is slowly rotated, with sufficient thrust to advance the cutting bit of a dual tubular drill string. The frequency of the vibration is adjusted to attain harmony with the particular medium being penetrated. Because the natural frequencies of different formation materials vary, the frequency of most modern sonic drilling equipment can be adjusted to the exact resonant frequency required. Typical frequency rates fall within a range of about 50 to 150 Hz, or cycles per second, which is at the lower range of sound vibration detectable by the human ear.

Sonic drilling does not involve standard drill rod, but rather a combination of continuous flush-threaded casing and *core barrel.* The resonate energy of the drill string and bit simultaneously imparts fracturing, shear, and displacement forces on the formation. When sharply different geological materials are encountered, such as a hard boulder in relatively soft soil, an appropriate adjustment of frequency is made to maintain resonance at the bit. This results in near liquefaction of the formational materials in contact with the bit, which tend to move away from the drill string. Very few drill cuttings are thus returned to the surface, and a continuous core of the penetrated geomaterials is obtained.

With the bit in resonance with the material being penetrated, it is possible to remove an almost perfect continuous specimen of virtually any geomaterial. In fact, samples have been removed from large boulders encased in soft residual soils. Because the system does not involving circulation fluids, there is no static head applied to the hole and only minor amounts of cuttings removed. This makes the method one of the safest for drilling within water-retaining embankments. The only real shortcoming for grouting is the minimum outside diameter of readily available sonic drill strings, which is on the order of 6 in. (150 mm).

10.2.1.6 DRIVE DRILLING

Drive drilling involves the physical driving of a casing into the ground. The bottom of the casing is typically closed with an expendable plug during the driving. For small casings, of less than about 1.5 in. (38 mm), a standard round-head bolt or rivet can be used. On larger casings, it is best to have a pointed probe, which will tend to keep the hole centered. Once the casing has reached the planned depth, it is withdrawn a short distance and the plug is knocked out by hammering with an inserted rod or a rope-suspended weight. To facilitate elimination of the plug, it should fit loosely into the casing tip. To prevent its falling out of the casing prior to and at the beginning of driving, it can be temporarily secured with duct tape.

FIGURE 10.26 Large equipment-mounted hydraulic driver.

The driving force is usually provided by some type of pneumatic or hydraulic hammer. Where sufficient access allows, equipment mounting as shown in Figure 10.26 can be advantageous. Such drivers are best contained in a vertical lead to maintain an accurate trajectory, because constant hammering and inaccurate alignment, as shown in Figure 10.26, accelerates wear and negatively affects alignment. The hammering and the use of high-frequency pneumatic tools, such as shown in Figure 10.27, are especially hard on the threaded couplings typically used. Drivers that supply high-energy, low-frequency pounding,

FIGURE 10.27 Standard pneumatic hammer used for driving casing.

FIGURE 10.28 Pipe rammers provide high-energy, low-frequency pounding.

such as the pipe rammer shown in Figure 10.28, have been found to be much less damaging.

10.2.1.7 SPECIAL DRIVE NEEDLES

In soil that is reasonably free of rocks, injection needles can be simultaneously driven as fluid or slurry grout is being injected. This procedure is effective to a depth of about 30 ft (9 m), which is the length of the longest practical needle. In this procedure the grout is injected from the top down, which is an ideal sequence. The needle is driven to the top of the zone to be injected. A precalculated amount of grout is then pumped, followed by driving to the next stage, injection, and so on, until the final depth is reached. When injection is near the ground surface, it is usually best to stop pumping when driving to the next level; however, once a depth of more than about 6 ft (1.8 m) has been reached, injection can usually be continued during driving.

The drive needles must be pointed, with grout exit holes positioned so that they are not easily clogged during penetration. Moreover, the outside diameter of the main shaft should be slightly larger than that of the grout exit location so as to resist leakage up and around the shaft. Such needles have been found to be especially useful for producing high-quality work economically and rapidly. Because they are not available commercially, optimal criteria for their fabrication is provided in Chapter 30. These needles are generally of a small diameter, on the order of only 1-1/4 in. (31 mm), so they do not require powerful drivers. For lengths up to about 6 ft (1.8 m), a small handheld pneumatic driver can be used. For greater depths, the driving hammer is best mounted on a support frame with a winch to raise it, as shown in Figure 10.29.

10.2.1.8 BECKER HAMMER DRILLING

The Becker hammer drill employs a diesel pile hammer to drive a special dual wall casing into the ground. The two separate tubes of the casing act as a very stiff, single unit, with high resistance to deviation or bending. This provides the strength to displace even large rocks and boulders without suffering deflection. It is fitted with tapered threads to facilitate the makeup of the flush joint. The casing is commonly available in two sizes, 5.5 in. (140 mm) OD by 3 in. (76 mm) ID and 6.63 in. (168 mm) OD by 4 in. (102 mm) ID. Normally, a toothed bit is used to break up the geomaterial as it is impelled into the formation. It can be either hollow or

FIGURE 10.29 Light frame with winch for handling drive needles.

plugged. Where desired to obtain specimens of the material penetrated, a reverse circulation method is used. With this method, compressed air is forced down the annulus between the two tubes and returns, with any cuttings, up the center passage.

The standard diesel pile hammer provides variable speeds, with up to 95 blows per minute at a force up to 8000 foot-pounds (1106 kg-m) per blow. The drill has operated to a depth of about 400 ft (120 m) and offers one of the fastest drilling methods available. It was specifically designed for exploratory drilling in sands, gravel, and boulders. Although such rigs are large and heavy, they are particularly efficient for penetrating rocky soils, which are especially challenging for the more common drilling methods.

10.2.2 Cutting Removal

In any drilling method, with the exception of drive drilling, a fundamental requirement is complete removal of the drill cuttings and preventing their entering and clogging any voids or defects that the hole traverses. This is best accomplished with a drilling method that produces relatively large cuttings and then expels them from the hole quickly. The optimal methods vary, however, according to the particular grouting requirement, and are discussed further with the various grouting applications.

10.2.2.1 CIRCULATION FLUSH

The process of *circulation* with a fluid, referred to as a *flush*, removes the cuttings. In common usage, either of these terms can be used individually, or they can be used together. The circulation flush can be water, air, or air-water foam. Drilling mud, which is routinely used for circulation flush in the boring of exploratory holes, generally should not be used for grout holes. The mud will tend to seal the surfaces of the hole, impairing grout permeation, and in the case of compaction grouting, any mud remaining in the hole can cause hydraulic fracturing of the soil when pressurized during initiation of grout injection.

Water is the most commonly used medium for circulation flush. It can be continuously circulated through the drilling system with a *mud pump* or impelled by normal tap water pressure. Where tap water is used, it is usually not recirculated, but drained to waste upon return to the surface with the drill cuttings. Depending on the area of use and any pertinent environmental restrictions, this may require separation of the cuttings, which is commonly facilitated with settling tanks, as shown in Figure 10.30.

Sufficient circulation flow must be maintained for efficient washing of the cuttings to the surface. The velocity must not be so great, however, as to excessively erode the walls of the hole. Although the required volume of flush will vary according to the size of the cuttings, which is dependent on the nature of the formation, type of bit and the thrust thereon, a rule of thumb calls for a starting velocity of about 150 ft (45 m) per minute. The approximate flow injection rates required to maintain this velocity when using the standard DCDMA drill rod for the respective hole sizes are provided in Table 10.3.

FIGURE 10.30 Settling tanks used to aid separation of drill cuttings.

TABLE 10.3 Approximate Circulation Flow Rates Required to Maintain a Circulation Flush Velocity of 150 ft (45 m) per Minute in Various Size Holes

Hole Size	Diameter in. (mm)	gal/minute (L/minute)
EW	1.470 (37.3)	1.5 (5.7)
AW	1.875 (47.6)	2.7 (10.2)
BW	2.345 (59.5)	5.9 (22.3)
NW	2.965 (75.3)	11.6 (43.9)
HW	3.890 (98.8)	17.7 (67.0)

Compressed air is commonly used for circulation in drilling blast holes or other common holes in rock and concrete. It should generally not be used for grout holes, as it tends to impel the cuttings into existing joints and fractures, inhibiting optimal grout penetration. In addition, if moisture is present, the drill cuttings, and especially soil fines, will tend to become sticky and ball up, becoming difficult to expel from the hole. In extreme cases, the resulting mud can completely plug the hole, resulting in the pressurized air being forced into the formation. This must be avoided, as high-pressure air tends to permeate into both rock and soil much more readily than water or grout. It can provoke plugging of the defects, jacking displacement of the formation, and in the case of soil, rearrangement of the grains and erosion of the hole.

The efficiency of cutting removal can sometimes be increased with the use of a foaming agent to form an air-water foam. The foam weighs much less than water alone, which considerably reduces the static head pressure exerted at the bottom of the hole. This can be significant when drilling deep holes, and outright hazardous when those holes are within a water-retaining embankment.

During drilling it is possible for circulation to be partially or totally lost; that is, the circulation flush does not return to the top of the hole, but runs into a void or particularly permeable zones at some depth below the surface. Continuing to drill when this occurs, which is referred to as *drilling blind,* is risky at best and can result in the drill string becoming bound in the hole because of cuttings settling over the bit. Of greater significance, however, is the risk of drill cuttings being washed into defects and thus blocking them. Unless circulation can be regained within a few feet of the point of loss, it is best to terminate drilling and inject grout, which will seal the hole so that drilling with proper circulation can resume. Although total loss of circulation can readily be discerned, the drillers must also be alert for any partial loss, which occurs gradually over an extended period of time. If there is a significant loss of flow volume, drilling should be terminated and the hole grouted.

A "T" fitting, or another appropriate device, as shown in Figure 10.31, should be attached to the top of the casing as required to direct the circulation flush into a open-top *mud tank,* or into a hose that can transport it to a satisfactory disposal location, such as the settling tanks shown in Figure 10.30. A *mud pump* (Figure 10.31) is used to either circulate the flush from the mud tank, through the drill string, and back to the tank, or to direct it from the mud tank through a hose to a disposal location. The mud pump shown in Figure 10.31 is a very compact air diaphragm pump. This is only one type that can be used for circulation; virtually all others are much heavier, often requiring lifting equipment to handle. The different types of pumps that can be used to maintain circulation are discussed in Chapter 30.

Rather than make the effort to set a proper standpipe and control the spent flush fluid, some drillers simply let it run out of an otherwise open hole, as shown in Figure 10.32. This not only makes for a messy site, but more important, can cause serious contamination if the flush runs into a previously drilled hole. Capture and con-

FIGURE 10.31 Collection device used to direct circulation return to mud tank.

trol of the circulation flush should thus be a requirement in most grouting work.

Most large rotary drill rigs, especially those used for exploratory drilling, have self-contained mud pumps mounted on them, and use relatively large mud tanks, fitted with baffles to promote settlement of the cuttings. Because these rigs are heavy, they can require considerable time to set up and move. Their handling and use is difficult, if not impossible, for holes located on steep inclines or where access is limited, such as within a structure. Smaller and more maneuverable drilling equipment, including much smaller mud tanks and circulation pumps, are thus common in grouting, as shown in Figure 10.31. The air diaphragm pump shown in Figure 10.31 can develop sufficient fluid circulation pressures for drilling to depths in excess of 100 ft (30 m). Alternately, it can be used to pump the fluid from the mud tank directly to a disposal area. Because of their light weight and small dimensions, the pump and tank can be quickly moved and set up by hand labor.

FIGURE 10.32 Failure to use a wash T or other device leads to a messy operation.

10.2.3 Grout Hole Alignment and Deviation

Although it is ideal for all grout holes to be drilled perfectly straight, such accuracy seldom occurs. It is possible to monitor the deviation of holes, and the amount of deviation is usually controllable; however, drilling to a very fine level of tolerance requires a heavier drill string and much greater care in the operation. It is thus significantly more time-consuming and expensive. In addition, considerable time and expense will be required to conduct a deviation survey, which can result in increasing the cost of drilling by a factor of 2 or more. Designers must thus use care in specifying unnecessarily stringent deviation requirements.

The drilling method and the specific equipment used can have a substantial influence on the straightness of a hole. Generally, rotary drilling will provide a straighter hole than percussive drilling. Where percussive equipment is used, down-the-hole hammers will produce straighter holes than top hammers. Further, the use of button bits will produce straighter holes than the use of chisel bits. In all cases, larger and stiffer drill rods will provide greater hole accuracy, and even closer tolerances can be achieved through the use of drill string centralizers and guide rods.

Unfortunately, all of these techniques come at a cost. For example, in hard formations rotary-drilled holes virtually always require more time and greater expense to effect than percussion-drilled holes. Down-the-hole hammers require larger hole diameters than top hammers. This can result in greater expense for both the drilling and casing, as well as additional grout required to fill the hole. Further, the hammers that operate on air exhaust it at the bottom of the hole, which can damage the formation. Chisel bits produce larger cuttings that are less likely to enter and block fissures that the hole has traversed. And although larger and stiffer drill strings will reduce deviation, they will also result in a smaller annulus, inhibiting the removal of the cuttings. Inclusion of centralizers will result in even greater interference in the orderly removal of cuttings.

All decisions relating to drilling methods and equipment thus involve some compromise, at best. The greatest influence on the straightness of a hole, however, and over which we have absolutely no control, is the composition of the formation through which the drilling must be done. For example, should the drill bit strike the side of a boulder or particularly hard inclined rock layer, as illustrated in Figure 10.33, it will tend to drift away from its intended trajectory. Holes drilled in formations of varying consistency and hardness will similarly deviate more than those in a consistent formation.

However, the accuracy of the holes largely depends on the care exercised by the drill crew. Most grout holes are drilled through a casing or standpipe that is embedded a minimum of about 2 ft (0.6 m) into the surface rock or other substrate. It is either cemented into an oversized drill hole (Figure 10.34) or embedded in a concrete grout cap, as illustrated in Figure 10.35. In either case, it is essential that this element be precisely set to the proper hole alignment. A convenient method of sealing the starting casing in oversized holes is to use casting plaster (plaster of paris). This cementing material is supplied as a powder, which is simply mixed with water into a thin slurry. It is commonly available with different setting times, typically 10, 20, and 30 minutes. The set casings should be initially positioned to a perfect alignment, rechecked as the material stiffens, and adjusted as required before it hardens.

Also of immense importance is maintaining the proper alignment of the drill string at the start of drilling. The drill rig must be firmly situated and the mast precisely positioned. Figure 10.36 shows a driller plumbing the mast in preparation for starting a new hole. Drilling technique, especially the use of excessive thrust on the drill bit, can cause the bit to wander off course; however, the effects of driller technique

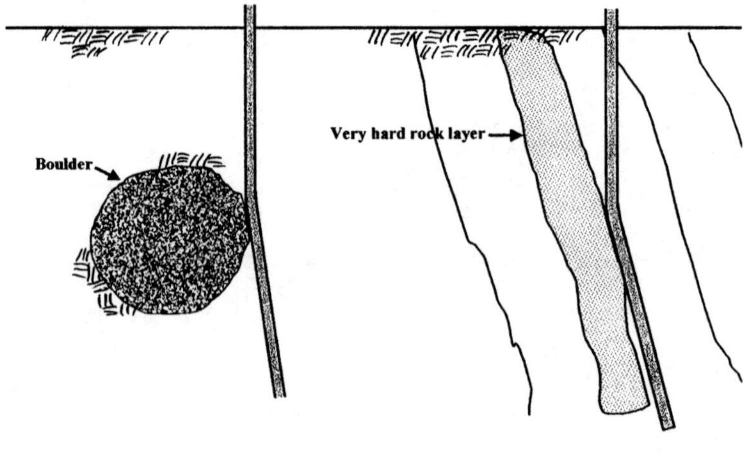

FIGURE 10.33 Striking a boulder or hard inclined surface will cause the drill string to drift away from its planned trajectory.

FIGURE 10.34 Encased standpipes must be set precisely to ensure proper trajectory of the drill hole.

FIGURE 10.36 Driller using a spirit level to plumb the drill mast prior to production drilling.

on hole deviation, once the hole is being advanced, are minor as compared with those resulting from poor initial positioning of the drill and/or starting casing, and/or adverse geology.

In relatively homogeneous formations and with reasonable care in setup and drilling, the holes should remain within a tolerance of approximately 3 degrees to a depth of 100 ft (30 m) and 5 degrees to 250 ft (75 m), which is about as deep as grout holes commonly descend. Conversely, where the formation is of varying hardness or the holes penetrate adversely oriented geologic structures, unless extraordinary efforts are taken, deviations up to five times greater can be expected. The direction of deviation will be random for vertical holes, but will usually be in the direction of inclination for holes battered more than a few degrees off vertical.

FIGURE 10.35 Standpipes are sometimes embedded in a concrete grout cap.

In summary, the precision with which grout holes are typically drilled is not very great, and in many instances it is downright poor. Where accuracy of hole trajectory is critical, very close tolerances can be achieved, but this is accomplished at a generous additional cost of both time and money. A variety of established methods are available to determine the exact grout hole route when required. Because of the considerable expense of drilling to tight tolerances as well as the required surveys, in most grouting it is best to simply recognize that the hole locations at depth will vary. With this in mind, redundancy through closer hole spacing and/or provision of additional rows of holes is probably the most practical approach. This is undoubtedly the reason that most grout curtains contain multiple rows of holes.

10.2.3.1 MONITORING HOLE DEVIATIONS

Where excessive hole deviation cannot be tolerated, the alignment can be surveyed by a number of different methods, all of which involve *tripping* (inserting) some sort of instrument down the hole, as illustrated in Figure 10.37. Although some evaluation tools send data to the surface in real time, either through a cable or by telegraphy, others store it in the instrument itself and it is downloaded once the tool returns to the surface.

The simplest survey method involves evaluation of downhole magnetic fields relative to the axis and position of the instrument. Such monitors are typically lowered into the hole on a nonmagnetic wire line. They are stopped at each elevation at which a reading is desired. Typical instruments contain magnetometers to determine magnetic north, and multiple solid-state tilt meters and/or accelerometers to measure dip. Thus, dip and azimuth are defined for each individual reading.

These data, including a correction from magnetic to geographic north, are typically processed to define the exact borehole path via special software, which also interpolates between the reading points. The resulting data, including 3-D coordinates, dip, and azimuth, can be displayed on either a laptop computer or a special handheld computer supplied with the instrument. Depending on the exact instrument and software, the data can usually be processed with output of color plots or sections. Obviously, the precision of the survey is subject to the distance between readings, and closer readings will improve the accuracy. Because these instruments operate on magnetism, they cannot be used inside steel-cased holes and should be isolated for at least 10 ft (3 m) from any magnetic source. They are usually about 4 ft (1.3 m) long, and many that are commonly available will fit inside a 2 in. (50 mm) diameter hole.

FIGURE 10.37 Inserting deviation survey instrument into a grout hole.

Some of the longest-established methods for borehole monitoring use gyroscopic instruments to establish azimuth when in the hole. Their orientation to geographic north must be established prior to tripping down the hole, but once set, they are not affected by magnetism. Most rely on multiple accelerometers or tiltmeters to define the relative position off vertical. They can be used to virtually any depth and provide both deviations from verticality as well as azimuth attitude. Gyroscope-based instruments are not influenced by naturally occurring or induced magnetic forces, so they can be used inside drill casing or in other locations exposed to magnetic flux. Most modern gyroscopic instruments are based on *surface reading gyroscope* (SRG) technology, which allows reading in real time on the ground surface.

Perhaps the most versatile instruments currently available operate on the basis of comparative optical scans. A light source on one end of a long, rigid tube is aimed at a target on the opposite end of an interior flexible tube. As the tube departs from a true vertical orientation, the point of light impingement on the target varies. The exact points are recorded with optical sensors, and the resulting data are saved in an on-board memory. These are downloaded into a handheld computer, once the instrument is withdrawn from the hole. The instruments are preferentially used in casing, as it provides the most uniform section for the insertion and return trip. The readings are obviously not affected by magnetic influence.

What these instruments assess is the relative differential in trajectory at each measuring interval, typically at a spacing of 10 ft (3 m). These values are then correlated to the known orientation of the instrument at the hole collar as the probe was initiated. The relative hole curvatures, once processed, provide a continuous log of the trajectory. To provide accurate information, the instruments need minimal clearance with the wall of the hole, so as to pass freely. Shims and/or hole centralizers should be used where the clearance exceeds a few millimeters. Available instruments will fit into a 1.5 in. (38 mm) hole. They can be used to virtually any depth.

Another evaluation method involves the lowering of a sensor head down the hole, attached to rigid single-axis hinged rods. The rigid rods, which are usually rectangular in section, are maintained to the correct azimuth at the hole collar. Precision inclinometers or tiltmeters on the rods evaluate the relative verticality of the hole at each rod interval, and, of course, the azimuth is controlled by the attitude of the hinged rods. These values are then processed to provide an appropriate plotted log. Because such instruments are not affected by magnetism, they can be used in cased holes, but they are generally limited to vertical or steeply inclined holes with depths of as much as 300 ft (90 m).

10.2.4 Protection of Drilled Holes

The cleanliness of a hole when grout is being injected is of paramount importance. Therefore, a hole should usually be flushed out with water until it runs clear and free of any cuttings or other foreign matter. This can be accomplished with a normal circulation flush of water through the drill string upon completion of drilling or by a separate flushing hose, as illustrated in Figure 10.38.

FIGURE 10.38 Flushing a grout hole until only clear fluid returns.

The drilled holes must then be protected during the interval between drilling and injection. Of particular importance is to not allow drill spoils or circulation flush to enter nearby holes. This is not much of a problem where an injection casing or standpipe has been cemented in place, but holes drilled directly into rock or concrete, especially in depressions or drainage paths, must be protected. Even where casings have been used, it is good practice to plug or cap the holes after drilling. Standard pipe caps are handy where threaded casing is used, and tapered wooden plugs can be used where threads are lacking.

10.3 HOLE CLOSURE AND ABANDONMENT

Once the grouting has been completed, it is important to properly abandon the holes. Virtually all cementitious suspension grouts will suffer from bleed, whereby the solids settle out of the mix, leaving water in the upper portions. As this evaporates, an open segment of hole will result. If the hole is cased and the casing is to be removed, it is preferable to "grout the casing out," that is, to perform continuous injection of a thick grout as the casing is removed. If the hole is not cased, a tremie tube or pipe should be inserted full depth, through which a thick grout can be pumped. Even with these procedures further settlement of the grout can occur, leaving a void at the top, so multiple treatments may be required.

CHAPTER 11

Grouting in Soil

11.1 COMPACTION GROUTING
- **11.1.1** The Basics
  - **11.1.1.1** Grout Holes
  - **11.1.1.2** The Grout Mixture
  - **11.1.1.3** The Grout Injection
  - **11.1.1.4** Monitoring and Control of the Injection
- **11.1.2** Improving Deficient Soil
  - **11.1.2.1** Densification of Loose Soil, Fill, and Backfill
  - **11.1.2.2** Mitigation of the Liquefaction Potential
  - **11.1.2.3** Improving Water-Retaining Embankments
  - **11.1.2.4** Densification of Submerged Soil
- **11.1.3** Settlement Correction
  - **11.1.3.1** Common Structures
  - **11.1.3.2** Pile-Supported Structures
- **11.1.4** Mitigation of the Settlement Trough in Tunneling

11.2 PERMEATION GROUTING IN SOIL
- **11.2.1** The Basics
  - **11.2.1.1** Degree of Solidification
  - **11.2.1.2** Injection Probe Layout
  - **11.2.1.3** Required Grout Quantity
- **11.2.2** Injection Holes/Probes
  - **11.2.2.1** Top Down with Special Drive Needles
  - **11.2.2.2** Top Down by Jetting
  - **11.2.2.3** Bottom-Up Staging
  - **11.2.2.4** Sleeve Port Pipes
- **11.2.3** Monitoring and Control of Injection
- **11.2.4** Permeation Grouting Examples
  - **11.2.4.1** Short-Term Strengthening
  - **11.2.4.2** Permanent Strengthening Applications
  - **11.2.4.3** Controlling Water Flow
  - **11.2.4.4** Strengthening and Water Control
  - **11.2.4.5** Altering Soil Dynamic Problems

11.3 FRACTURE (CLAQUAGE) GROUTING
- **11.3.1** Lime Injection
- **11.3.2** Cementitious Grout Injection

11.4 JET GROUTING (MIXING)

FOR THE GROUTING of all mediums except soil, there are only two types of grouting, permeation and fill, both of which result in the filling of open void space. In dealing with soil, however, a full range of grouting methods is applicable and many choices are available. The mechanism by which each grouting method affects the soil is different, however, as illustrated in Figure 11.1. It is thus important to select a method or combination of methods that will result in the desired outcomes. Often the preferred result can be achieved by two or more methods, in which case cost and availability become the deciding factor in the selection process. In many instances, however, only a single method will provide the required mechanism of improvement and thus be satisfactory. Each of the four methods for grouting in soil are examined in general, and selected case histories will be presented to provide an understanding of how things come together in actual projects.

Compaction grouting does just as the name implies: It compacts or densifies the soil, which significantly increases its bearing strength. Although the individual grains are forced into a tighter packing, they achieve little additional cohesion and improvement in shear strength is usually not great. In the United States, this is by far the most frequently used method for the repair of settlement damage, mitigation of the potential for liquefaction, and increasing the density of in-place soils. Because the grout behaves as a growing solid in the ground, the risk of hydraulic fracturing or other damage resulting from out-of-control grout is minimal. Depending on the particulars of the soil, a small to moderate decrease in its permeability will result from compaction grouting.

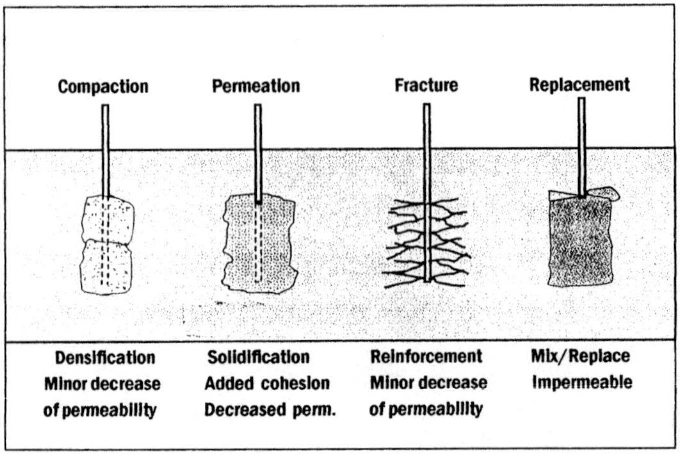

FIGURE 11.1 Each of the four grouting methods for soil results in a different improvement mechanism.

Compaction grouting can be applied to a wider range of soil types than any other grouting method. It is applicable to the full range of sands and silts including clayey materials, as long as the permeability is sufficient to allow expulsion of the pore water. Based on the cost per unit of improved soil, compaction grouting is usually the least expensive means of soil improvement. In addition, it is readily performed in areas of poor access or other restriction and can result in the least disruption or messiness. It is thus particularly advantageous for use under or around existing structures.

Permeation grouting fills the interparticular soil pore system, essentially gluing the individual particles together. This greatly increases both the shear and bearing strength. It will also result in a significant decrease in the soil permeability and when thoroughly applied, will completely block the flow of water. Its use is limited to those soils that possess sufficient permeability to allow thorough penetration of the grout. From a practical standpoint, this means sands and gravels, although such materials containing minor amounts of silt-size particles are also treatable. Because permeation grouts behave as fluids in the ground, the risk of loosing control, including leakage and hydraulic fracturing of the soil, is high. Based on the unit of soil treated, the cost of permeation grouting is relatively high.

Fracture (claquage) grouting involves the intentional fracturing of the soil in a presumably controlled manner. It is usually performed with fluid-consistency particulate suspension grouts at sufficient pumping rates to cause high pressure and the resulting hydraulic fractures in the soil. Because the direction and configuration of the fractures cannot be controlled, the quantity of grout injected at any one location should be strictly limited. Likewise, an injection procedure that clearly defines the location of the individual injections must be employed. The principal mechanism of improvement is believed to be a general reinforcement of the soil as a result of the interconnected network of relatively strong grout lenses, although some densification of the soil also occurs. Fracture grouting will result in a reduction of the soil's permeability. The level of such reduction will vary considerably according to the structure of the grout lenses and is virtually impossible to predict or control. Because of the difficulty in controlling the grout deposition, the process is best reserved for improvement of soils that are not treatable by other grouting methods, such as compaction grouting. The procedure is thus most often used to improve clayey formations.

Jet grouting involves turbulent mixing of the soil particles with a fluid-consistency cementitious grout to form "soilcrete." It can also be performed so as to replace virtually all of the affected soil with grout. Its mechanism is thus replacement of the original soil with a new mixed composition. It can be performed in virtually all soil types. The presence of large rocks and boulders will make the procedure much more difficult to apply, however, and can result in areas of untreated soil due to the shadow effect of the obstructions.

The strength of the resulting "soilcrete," which will vary according to the original soil composition as well as the proportion of grout mixed with it, will be high, relative to that resulting from other grouting methods. Strengths can exceed 2000 psi (13.8 MPa) in coarse-grained soils, whereas those obtained in clay will be significantly lower. The range of obtainable strength as a function of soil type is illustrated in Figure 11.2. The soilcrete will be essentially impermeable, and completely watertight walls can be constructed with sufficient overlapping of the injection probes.

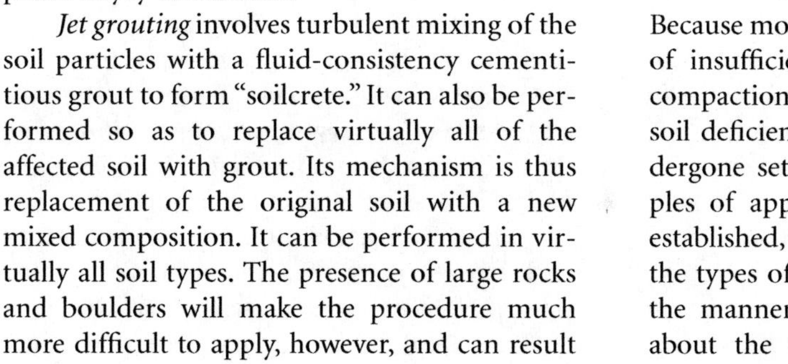

FIGURE 11.2 Typical strengths obtained of the masses resulting from jet grouting vary with soil type.

11.1 COMPACTION GROUTING

Because most foundation problems are the result of insufficient density of the underlying soil, compaction grouting is widely used to remediate soil deficiencies under structures that have undergone settlement. Although the basic principles of application are quite simple and well established, there is widespread variation in both the types of projects that are accomplished and the manner in which different contractors go about the work. The basic requirements for proper performance, are examined here, followed by a review of several case histories to reveal the flexibility of the procedure and the many conditions under which it can be effectively used.

11.1.1 The Basics

As discussed in Chapter 2, injections can proceed either from the bottom up or the top down. Bottom-up staging is simpler, faster, and thus more economical. It is generally not effective at shallow depths of less than 15 ft (4.5 m), however, so work in this zone should be reserved for top-down performance. Where injection is to be made from deep to shallow elevations, a combination of top-down, followed by bottom-up, staging can be used. Further, if the top of the grout-densified zone will be at depth, such as in the case of improving soil underlying the tip elevation of end bearing piles, a few stages should be performed from the top down prior to treatment of the underlying soil from the bottom up. This will increase confinement of the lower grout and prevent its escape to the higher elevations, where it could react negatively on the piles.

As discussed in Chapter 6, the rheology of the grout must be such that it will behave as a growing solid in the ground. This point cannot be overstressed, because grout that has insufficient internal friction and thus behaves as a fluid can cause hydraulic fracturing regardless of content or consistency, precluding proper densification. Such out-of-control grout can result in serious damage, not only to the soil, but to adjacent structures and substructure as well.

Although the actual injection is quite simple, the mixer, pump, delivery lines, and fittings must be able to accommodate the rather harsh grout that is used. Of extreme importance is that continuous careful monitoring of the ground surface and any adjacent structures or substructure be provided. Structural movement often occurs and is generally a signal to cease injection at the particular location in a given grout hole. Unfortunately, much damage has occurred on many projects where less than sufficient and appropriate monitoring was provided. Such damage and poor performance casts a bad light on the procedure itself and is inexcusable.

11.1.1.1 GROUT HOLES

Grout holes are generally placed on a grid with a spacing of 6–12 ft (1.8–3.6 m). Although the spacing is soil specific, 10–12 ft has proven to be an optimal spacing when the pumping rate is within a range of 1 to 2 ft^3 (28 to 57 L) per minute. In one test application, soil deformation was actually measured 12 ft (3.6 m) away from the point of injection. The effectiveness of the grouting is, obviously, directly related to the quantity of grout injected into the ground. It is well established that greater amounts of grout will be injected at slow pumping rates than at faster rates. Thus, the hole spacing must take into account the pumping rate to be used. The employment of excessively high pumping rates is common in compaction grouting, and this mandates closer than otherwise necessary hole spacing.

Although the holes are typically vertical, they can be inclined in virtually any configuration. Holes inclined more than 20 degrees off vertical will, however, have an excessive horizontal "roof area," as discussed in Chapter 2 and illustrated in Figure 2.20. This can cause jacking of the surface at a relatively low injection pressure, so hole inclination far off of vertical is usually reserved for very special applications, with the realization that the amount of injected grout, and thus its effectiveness, will be limited.

Most compaction grouting is accomplished with 2 in. (50 mm) inside diameter casing. It has been performed with casing as small as 1.5 in. ((38 mm), but the use of such small casing is not recommended, as initiation of the grout bulb can be difficult at the start of pumping. Larger-size casings are acceptable, but except in the case of very deep holes or those that present difficult drilling, such as in the penetration of boulders, their use is questionable. Not only are larger casings heavy to handle and more costly, but if they are larger than about 4 in. (102 mm) in diameter, they will cause positive pressure at the bottom as a result of the static head of grout. In contrast, a head loss will result in casings 3 in. (75 mm) or less in diameter. The head losses/gains of typical compaction grout in different sizes of vertical casings, as measured in test evaluations, are provided in Table 11.1.

Because most holes for compaction grouting are in clean soil and often in areas difficult to access, the use of large drill rigs is neither needed nor sensible. Although some contractors prefer to use standard drill rigs, those with

TABLE 11.1 Head Gain/Loss per Linear Foot of Different Sizes of Vertically Oriented Grout Casing

Casing ID, in. (mm)	1.5 (38)	2 (51)	3 (76)	6 (152)
Head loss (gain), psi	0.75	1.0	0.48	+(0.48)
Head loss (gain), kPa	0.52	6.9	3.3	+(3.3)

extensive experience have adopted a variety of lightweight drills and drivers. Lightweight drills, such as those that were discussed in Chapter 10 and illustrated in Figure 10.21, are widely used for compaction grouting holes. Handheld bore motors, as illustrated in Figure 10.16, are likewise often employed, especially inside or under structures or in other confined or difficult-to-access locations.

For bottom-up applications, driving can be the most expedient method to establish the grout holes. Although casing alignment is easier to maintain with mast- or equipment-mounted drivers, handheld units such as those shown in Figure 10.28 are often used. Regardless of placement method, it is essential that the casing fit tightly in the hole and that no annular space exists around it. Should the casing not fit tightly enough, leakage of the grout around it to the surface is likely and effective compaction of the soil will be precluded. This problem is especially common with bottom-up injections at shallow depths.

With bottom-up grouting, the removal of casing joints requires a pause in the injection. It is important to minimize such interruptions in pumping to the greatest extent practicable. Typical compaction grout is somewhat thixotropic, as well as chemically active when it contains cement. As the delay time increases, a higher initial pressure will be required to start the grout moving upon resumption of pumping. Excessive delay can result in complete blockage of the delivery line. Furthermore, delays in pumping extend the time required to complete a given job and thus add expense. A good rule to follow is, "Minimize pumping delays to the greatest extent practicable."

Thus, it is prudent to use the lightest casing system that is of adequate strength and will perform satisfactorily. In this regard, the weight of the grout in the casing section being removed becomes significant. The weight of the grout in different sizes of casing is shown in Table 11.2. The length of the individual sections is also of great significance. Not only do long lengths contain more grout, and thus weight, but they require more time to remove because of poorer access to the fittings on top. With casing lengths of 5 ft (1.5 m) or less, the workers can make the necessary removals while standing on the ground or perhaps on a low box or other stand. It is for this reason that most experienced contractors choose such short casing lengths for their work. Yet longer lengths are occasionally used, most often 10 ft (3 m). These require the erection of a scaffold or other means of access, or worse, the employment of workers with monkey-like ability, as illustrated in Figure 11.3.

TABLE 11.2 Weight of Grout per Vertical ft (0.3 m) of Different Diameters

Casing ID, in. (mm)	1.5 (38)	2 (51)	3 (76)	6 (152)
Weight in pounds	1.9	3.3	7.4	29
Weight in kilograms	0.86	1.5	3.4	13

FIGURE 11.3 Acrobatic skills are required of workers when no means of reaching connections are provided.

Another factor that influences casing makeup and break time is the design of the joint thread. In general, coarser threads are easier to make up and break, as long as they have sufficient strength. Because the large end of a taper telescopes over the small end, taper threads require fewer revolutions to tighten than parallel threads. Further, a slight rotation of a taper thread will loosen it completely, so that very little effort is needed for further rotation. I have witnessed the workers of one contractor, using 3 ft (0.9 m) long joints of a proprietary double-taper threaded casing, break, reconnect, and resume pumping in less than one minute. Conversely, the breaking of 10 ft (3 m) long lengths of 3 in. (75 mm) casing typically requires more than ten minutes.

Obviously, the casing must be of sufficient strength and durability to withstand the forces of withdrawal and the constant makeup and breaking of the joints. Most compaction grouting is performed at depths of less than 50 ft (15 m), in soils relatively free of large rocks and not situated immediately adjacent to a hard boundary. For this most common situation, forces on the casing are not extraordinary, and such work is routinely accomplished with relatively light sections, including both standard diamond drilling casing and common steel pipe. Where depths become greater, these commonly used casings may not provide sufficient strength to withstand the pulling forces.

The greatest enemy of casing, however, is penetration through large rocks or boulders or placement in close proximity to a hard face, such as the wall of a rock face or concrete structure. Under these conditions large bending forces can develop, and it is not unusual to retrieve severely bent casing. Such deformed casing obviously requires a significantly higher than normal pulling force to extract. Bent casing in the ground can also exert often repeated bending stresses on the joints, which can become subject to fatigue failure. Under these severe conditions, the use of especially strong casing is required, and even then some bending should be anticipated.

Because casing failure during extraction usually occurs in the joints, experienced grouters often use welded joints for work in extreme conditions. These are then cut, either with a pipe cutter or a handheld portable saw (Figure 11.4), as the casing is withdrawn. Generally, lightweight

FIGURE 11.4 Casing with welded joints during placement will require physical cutting during withdrawal.

and small casing is entirely satisfactory for most routine work, but the use of large-size casing with heavy walls and extra-strong joints will probably be necessary in situations where boulders are present. This will often require the use of large equipment for handling and withdrawal. In such situations, it is not uncommon to use a relatively heavy drill rig for both placing and withdrawing the casing. This, of course, requires that the drill rig be maintained over the hole throughout the injection and obviously increases the cost of the work considerably.

For top-down injection, especially starting at shallow depths, it is best to cement a starter casing into an oversized hole. This is easily accomplished with rapid-setting casting plaster (plaster of paris). Previous test applications have demonstrated the substantial advantage of cementing the casings in place. It allows higher pressures and up to nearly twice the amount of grout to be placed, all else being equal. For top-down work, the bottom of the casing is set to the upper limits of the grout injection or a minimum of about 5 ft (1.5 m) from the surface. The first stage is then drilled and grouted, followed by completion of successive stages through the same casing.

There is considerable variation in the casings preferred by different specialty contractors. Obviously, a casing must be compatible with both the drilling and withdrawal methods to be used. Several contractors have developed special systems, often involving custom-made casing, which work particularly well for them. The choice of the particular system to be used on a given project is thus best left to the contractor, subject only to the requirement that it be suitable to the conditions that will be encountered in the work and, possibly, the specification or limitation of the internal diameter.

11.1.1.2 THE GROUT MIXTURE

As discussed in Sections 1.2.2.1 and 6.3.7.1, suitable rheology of the grout used for compaction grouting is fundamental for a successful outcome. Regardless of the stiffness or slump of the grout, if it behaves as a fluid in the ground, hydraulic fracturing of the soil is likely and meaningful densification will cease. There are, unfortunately, no established tests to confirm the suitability of the mixed grout. A vast amount of experience has resulted in the ability to prescribe a grout mix that will perform satisfactorily, however. It bears repeating that even though a grout has a low slump, or for that matter, zero slump, it is not necessarily satisfactory for compaction grouting. Even very stiff compositions can behave like fluids when under pressure in the ground.

Gradation of the aggregate controls the suitability of the resulting grout. It must contain sufficient sand and gravel-size particles to provide sufficient "harshness" in order to behave as a growing solid in the ground, but sufficient fines to provide the necessary pumpability. Silt-size particles have proven over the years to be the best component to provide pumpability without any adverse effects on the resulting grout behavior. Although some concrete pumping aids and/or clay can provide pumpability, these have been consistently found not to be satisfactory and will likely result in the grout's behaving like a fluid when under pressure in the ground. Thus, grout aggregates should be essentially free of clay and none should be intentionally added to the grout mix. Ideally, the aggregate should consist of natural round-grained particles. Strongly angular or crushed particles, although beneficial to the effectiveness of the mix, will adversely affect the pumpability and should be avoided if possible.

In the early history of compaction grouting, a sandy aggregate with at least 95 percent passing a No. 8 sieve, as illustrated in Figure 2.7, was used almost exclusively. It has since been found that inclusion of larger-sized particles will minimize the risk of hydraulic fracture of the soil, and that even where a fracture occurs, gravel will tend to block its propagation, as discussed in

Section 2.2 and 6.3.7.1. This was made abundantly clear in 1997, when grout squeezed from a faulty seam of a casing during injection, as shown on the left side of Figure 11.5. Once the protruding grout had been washed off the casing surface, gravel in the grout was exposed and was found to be effectively bridging a portion of the split that was less than about 3/4 in (19 mm) in width, as illustrated in Figure 11.5 (right). Although a great deal of quality work is still performed, absent any gravel fraction in the aggregate, inclusion of the coarser material is definitely advantageous and should be required in any particularly sensitive work. A particle-sized envelope for the preferred aggregate gradation is shown in Figure 6.15.

Compaction grout usually contains portland cement in an amount equal to about 8 to 12 percent of the aggregate by volume. There are cases, however, in which hardening of the grout is not acceptable, such as adjacent to a flexible pipe or conduit or within a water-retaining embankment. Although the cement does provide pumpability, it is not necessary for either pumpability or performance of the resulting grout. In fact, many jobs have been successfully performed with grout that contained no cement whatever. To maintain pumpability, either a larger proportion of silt, or some other inert material, such as a natural pozzolan, can be used. Note that fly ash is a manufactured pozzolan with spherical particles, so it is generally not satisfactory.

FIGURE 11.5 Fin of grout forced out of split casing (left). Gravel forms a plug where split is less than about 3/4 in. (19 mm) (right).

Uniform consistency of the grout is fundamental to quality performance. The individual components must therefore be evenly batched and mixed. The mixer operator must be experienced and of proven ability to produce consistently uniform grout. The really good operators take pride not only in providing good-quality grout, but also in keeping their equipment clean. It is not uncommon to see an operator washing spilled and splattered grout off his machine from time to time. Although we must encourage such neatness, it is important that overflow water does not enter the pump hopper, as it will adversely affect both the consistency and uniformity of the grout.

Compaction grout is most often prepared with continuous auger-type mixers, but can be mixed in any equipment that will uniformly combine the various ingredients into the required very stiff mortarlike mass. Figure 11.6 (left) illustrates good-quality grout dropping off the mixing auger into the pump hopper and (right) shows grout breaking off as a hose is moved. Note the open texture of the grout surfaces and the absence of "slickensides," which are

FIGURE 11.6 Rough texture of good-quality grout falling from mixing auger (left). Extruded grout should break off with rough-textured face (right).

indications of good-quality grout. With experience, one can easily identify the grout quality by simply observing the material in the pump hopper or that which breaks off when the header is moved to a new hole. A granular texture should always be present.

There have been many attempts to batch and mix grout in ready-mix trucks. With few exceptions, this has not proven effective, however. Common batch plants do not normally have appropriate aggregate available. Such mixes usually contain large quantities of fly ash, which, as mentioned earlier, is composed of spherical particles that tend to add mobility to the grout, as do concrete pumping aids that are sometimes included. The fines in an appropriate aggregate tend to hang up in the storage and weigh hoppers and present problems with batching. Further, the somewhat sticky grout does not always mix well with the action of a ready-mix drum, which depends somewhat on large aggregate to carry the fine components of the mix off the mixing drum fins.

11.1.1.3 THE GROUT INJECTION

To facilitate compaction, the soil moisture content should be at optimum level or greater. Even where soil is on the dry side, its moisture is usually increased during drilling, assuming that water is used for the circulation flush, as is common practice. In the case of holes with insufficient moisture, water can simply be poured into the casing prior to injection hookup. Where very dry conditions occur, and especially in the case of fine-grained soil, a water injection hose can be connected and run for an extended period of time. Drive drilling, of course, does not leave an uncased hole and so does not allow permeation of water into the soil. Thus, it is best to restrict the drilling method to rotary drilling with water circulation flush in situations where the in situ soil is overly dry.

To prevent the casing from becoming lodged in deep holes, it is best to grout them shortly after drilling, preferably the same day. Where the holes have been drilled some time earlier, it is possible for rocks or other obstructions to have fallen in and become stuck at a depth short of the hole bottom. Even in the case of recently drilled holes, the full depth may not have been reached. Thus, it is always good practice to confirm the hole depth immediately before connecting the grout line. This can be accomplished either by stabbing the hole with a small pipe or rod, or assessing the depth with a weighted measuring tape.

Before connecting the delivery line to the grout header, it is a good idea to circulate grout through it, preferably back to the pump hopper. When the grout first goes through a dry line, it can lose moisture to the dry surface and stiffen. Furthermore, delivery lines are not always cleaned out as well as they should be, which may result in a thin coating of grout on the wall and/or grains of sand within. Such materials will tend to gather at the head of the advancing grout and can cause plugs. Beyond the initial filling, it is good practice to recirculate the grout when moving to new holes, especially when the prior injection has been at a slow rate and/or high pressure.

The quantity of grout injected is usually determined by the pump stroke count, but, unfortunately, pump cylinders often do not fill completely. It is thus a good idea to calibrate the pump output by filling a cubic foot box before beginning injection and at any time thereafter when it is thought that a change in output has occurred. It must be understood however, that this calibration, though useful, is not precise. First, the box is seldom filled with a precise number of full strokes, so interpolation of the grout quantity needed to fill the box from the final stroke must be made. Further, the pump will be more efficient with zero pressure at the outlet end of the hose while filling the box, than it will be when a positive pressure of some magnitude exists during grouting. It should be noted that pump efficiency is also affected by any internal expansion of the delivery hose, as well as by both the pumping rate and the pressure level. These variations can be significant, so knowing the average amount of grout displaced per pump stroke, as determined by a calibration test, gives only an approximation of the grout quantity that will be pumped during production. The indicated total amount of grout pumped in a given time period should thus be checked against the quantities of material used and appropriate adjustments made, especially where the quantity pumped is the basis for payment.

Low-mobility grout is delivered to a grout hole essentially as an extrusion. This requires the delivery line and all valves, gauge savers, and other fittings to be of full-flow design, so as to not excessively change the extrusion configuration. Only wide-sweep bends should be used. Figure 11.7 shows a typical compaction grouting header setup. The grout goes from the hose into a gauge saver on which a pressure gauge is mounted, from there through a full-size opening ball valve, then through a wide-sweep elbow, and into the grout hole.

Even though the pumping pressure may have been very high, a valve is not necessary to contain the grout in the hole upon disconnecting the header, as is the case with more conventional grouts. This is due to the high frictional resistance of the low-mobility grout within the delivery system and the soil. It is useful to provide a valve, however, to prevent grout dropping from the hose as it is moved from one hole to another. This greatly aids in keeping the job site clean. Both rigid pipe and flexible hose for the delivery line are readily available from suppliers to the concrete pumping industry.

FIGURE 11.7 Wide-sweep bend and full-diameter fittings used in compaction grouting.

Protection of the pressure gauges from contamination with grout is essential and must be provided by some type of gauge protector. As discussed in detail in Chapter 36, commercially available gauge protectors tend to be very heavy and awkward to handle and/or contain small passages that can become impacted with grout, so most contractors have developed their own, and considerable variation exists. It is a good idea to ascertain the accuracy of the gauges at the beginning of the work and at any time thereafter that accuracy is questioned. Be rational however; compaction grouting is not a precise science, and, a high level of accuracy is not needed in determining the correct pressure. As discussed earlier, it is not the pressure level itself that is of importance, but the relative pressures within a given hole. Gauges that read to within 5 percent of the actual pressure should be acceptable.

The sequence in which the holes are grouted is of great importance, though often ignored by the novice and sometimes even by experienced grouters who should know better. Holes are generally injected in a primary-secondary manner, with the secondaries splitting the space between the primaries. Where more than two rows of holes are used, those around the periphery are usually completed prior to those on the interior. In this way, the zone of improvement is somewhat contained. Where injection is made near a downslope or retaining wall, the work should be started in the holes nearest to that feature so as to restrain the grout when injecting the interior holes. Figure 11.8 illustrates the sequence of in-

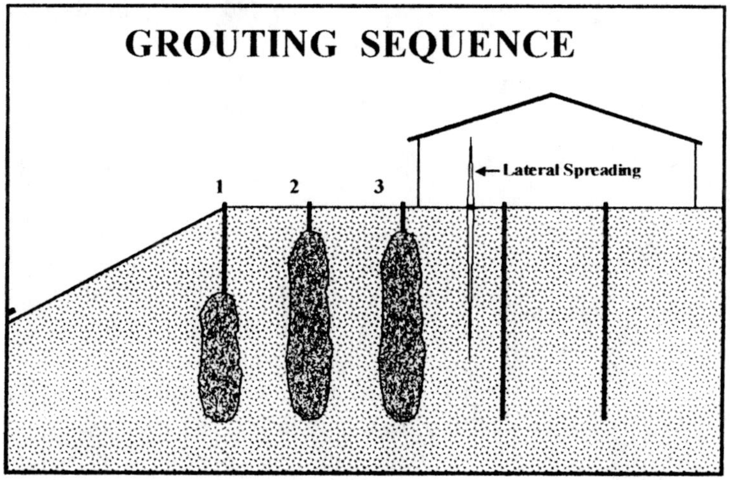

FIGURE 11.8 The injection sequence should always progress from holes nearest a downslope or retaining wall.

jection, starting with row 1 and continuing with rows 2, 3, and so on.

Injection is continued in a given hole stage until one of the following criteria are met. These are listed in the order of most common experience:

- A minute heave of the ground surface or of an improvement thereon
- A predetermined quantity limitation
- Obtaining a pressure limitation at a specified pumping rate

Establishing final cutoff criteria in advance of some actual grouting or completion of a test section is difficult at best. Thus, rather conservative criteria are generally used, and for large or particularly critical applications, a test section is strongly recommended. As a practical matter, a surface heave or other disturbance will be the cutoff criterion when the grouting is being done within about 25 ft (7.5 m) of the ground surface. A volume-limit criterion generally should be reserved for those cases where the in situ soil is fairly consistent.

Pressure limits are generally employed in situations in which widely varying grout takes will be required to bring the soil up to a somewhat uniform condition. This usually occurs in deeper applications where the soil conditions are variable or where significant voids or weak areas exist. It has been noted that the root cause of most *apparent* fill failures has repeatedly been found at the bottom of the fill or in the underlying natural materials. Aside from weak soil in this zone, piles of vegetation and clusters of buried boulders have been found as well. Accordingly, this zone commonly requires a much greater volume of grout than the overlying fill. Setting uniform volume limitations in such situations is thus not recommended.

Damage to the site can result from excessive heaving of the ground surface. Moreover, little further densification occurs once surface heave has initiated unless it is well restrained, such as by a heavily loaded foundation. It is thus mandatory that the ground surface, as well as any improvements within about a 30 ft (9 m) radius of the injection point, be diligently monitored continuously throughout the period of injection. Obviously, grout injection should normally be stopped immediately when surface movement is noted.

Where the existing soil is fairly uniform but of insufficient density, the volume of grout required to provide an acceptable increase in density can be determined and an appropriate volume limitation established. When this is done, it is important to provide a reasonable factor of safety, keeping in mind that the volume actually in place may not be as great as that

TABLE 11.3 Equivalent Diameter of Grout Column per ft (0.3 m) of Hole for Different Quantities of Injected Grout

Grout Quantity, ft^3 (L)	Column Diameter, feet	Column Diameter, meters
1 (28.3)	1.2	0.37
2 (56.6)	1.6	0.48
3 (84.9)	2.0	0.61
4 (113)	2.3	0.70
5 (141)	2.5	0.76
10 (283)	3.6	1.10
15 (425)	4.4	1.30
20 (566)	5.0	1.50

indicated by the pump stroke count, as previously discussed. Although a truly uniform column of grout virtually never occurs, it is a good idea to keep in mind the approximate amount of space that will be filled with different volume limitations, as shown in Table 11.3.

When pressure limitations are used, it is important to ensure that the pumping rate is uniform. As previously discussed in Section 9.4, the grout pressure is directly related to the pumping rate. As the pumping rate is increased, the pressure will also increase. Conversely, the pressure will lower with a reduction of the pumping rate. Therefore, in order for pressure to be used as a valid cutoff criterion, the cutoff pressure must always be taken at the same pumping rate.

11.1.1.4 MONITORING AND CONTROL OF THE INJECTION

When appropriately monitored, each grout hole can act as an exploration hole, providing valuable information as to the actual conditions existing prior to the grout injection. This is particularly applicable in those cases where there is considerable variation of the subsurface conditions. Quality performance thus dictates that the injection parameters be closely monitored and the observed data recorded. Ideally, the data will be automatically acquired with direct input to a computer. Although this is routinely done in many parts of the world, U.S. contractors have not been keen on initiating the use of such technology. Thus, most monitoring is visual and the data hand entered.

Recording data is of little importance, however, unless it is promptly evaluated so as to become useful in directing the course of the work. This becomes particularly important in situations where there is considerable variability in the project soil. In such instances, knowing the behavior of the earlier grouted holes can be immensely useful in going about the subsequent work. Even though the original data may have been taken manually, it is convenient to enter it into a standard spreadsheet program, such Microsoft Excel, to facilitate its being easily displayed.

On projects with a large number of holes or that are otherwise important, it is useful to display the data graphically, as shown in Figure 11.9 (top). In this illustration, records of the individual holes have been posted on the wall of the office trailer in the same relative positions in which they exist in the field. The spaces between some of the records on the bottom row represent holes that are yet to be grouted. The horizontal bars on the individual records represent grout take, and the depths of significant take can be readily seen in even this small photo. Figure 11.9 (bottom) shows a simple plan view layout on the same project. Here, holes have been color coded according to take, and those that had significant behavior have been highlighted or otherwise marked. In most instances, the behavior of a single grout hole is not nearly as significant as the trends that develop within a network of many holes. Plotting and posting the particulars of a grouting program, as illustrated, allows one to stand back and see how the various conditions are interrelated. In grouting it is nearly always valuable to stand back and observe the entire forest, rather than analyze in detail only single trees.

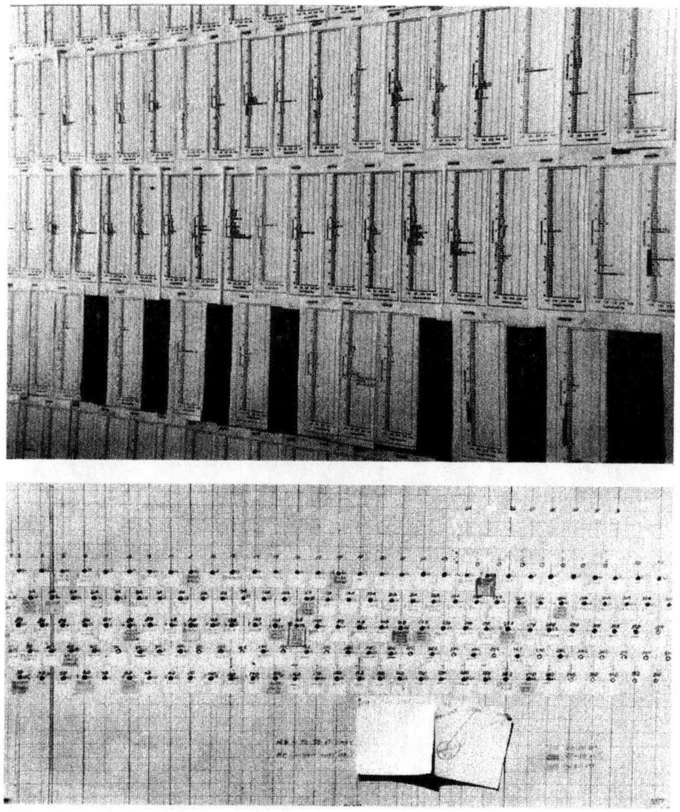

FIGURE 11.9 Injection trends can best be interpreted when records of the individual holes are posted according to the actual layout in the field (top). A plan view of the hole layout, marked up and color coded, is helpful in recognizing trends (bottom).

There is always a need to evaluate the effectiveness of grouting, and though really good methods are not always available, some form of postinjection testing is often performed. An analysis of good continuous records of the actual injection parameters is, however, the single most important source of data for verifying the quality of the work. Knowing the grout quantities of the various stages and holes, and especially the injection rate and pressure behavior during injection, can provide an excellent portrayal of the pregrouting conditions. This subject is discussed further in Section 26.3.

A word of advice is in order here. Upon examination, a large number of records generated by a number of contractors often provided less than complete knowledge of what actually happened during injection. The contractor's most important activity is getting grout in the ground, and keeping records is usually not high on the list of things to do. It is thus suggested that the best party to monitor and record the grouting parameters is the engineer or the owner's representative. It is of great importance to remember that the noting of pressure *alone* is of no usefulness whatever. The pressure developed is always related to the injection rate, and these two parameters should always be shown when either is entered into the record.

Monitoring the grouting site and any structures on it is also an important function. It should be done by personnel who are well experienced and thoroughly familiar with the typical signs of movement. Unless the point of control for the grout pump is at the header or within an easy line of sight and hearing, some type of communication that will allow an instant message to the pump operator must be provided.

Surface heave should be detected as soon as it begins and prior to any measurable amount. This is especially important on work that is being performed from the bottom up, as the increments of heave accumulate and the total heave can be considerable, especially when injecting into deep holes. Although most bottom-up grouting is performed in 1 ft (0.3 m) stages, the use of longer stages should be considered in situations where controlling the total heave is important. Stage lengths of 3 ft (0.9 m) are not unreasonable and have been used on many projects.

Accumulative heave is not much of a problem when the work is carried out from the top down. With this approach, stage lengths are typically 3–6 ft (0.9–1.8 m). Because there is pause for a minimum of several hours between injection of adjacent stages, any heave that has occurred tends to recover before the next stage is injected. Obviously, because of the longer stage lengths, there are not as many stages from which incremental heave can accumulate.

Monitoring for surface movement should include both vigilant observations for any visual or audible abnormalities and the use of one or more types of observational instruments. The methods used can range from simply watching the bubble of a spirit level placed over a pavement joint, as shown in Figure 11.10 to using surveying instruments to continually scan the surface area around the injection hole. Some instruments will both detect and record the magnitude of any movement, others will simply signal that movement has occurred. Although the latter group can be very useful in sounding an alarm, such instruments should always be backed up by a system capable of providing an accurate measure of the actual amount of any heave or other movement.

Where surveying instruments are used, multiple targets should be distributed over the area of interest in such a manner that the instrument operator can quickly scan the full array of targets. This can be easily accomplished by placing a short line on the surface of existing walls or columns, or on rods either supported on stands or driven into an earth surface. Alternately, a short ruler can be attached with masking tape to the various surfaces, as illustrated in Figure 11.11. The ruler is usually set so that the original level, before any grout injection, will fall on an even inch or other appropriate primary graduation.

A simple and very effective means of monitoring is the common liquid level, scientifically referred to as a manometer. Because a fluid will always seek its own level, a change in vertical position will be indicated if the position of the liquid in a transparent tube changes while the other end remains stable, as illustrated in Figure 11.12. One advantage of using such instruments is that virtually any number of individual tubes can be run off a single reservoir located remotely from

FIGURE 11.10 Differential movement across a crack or joint can easily be observed by the movement of the bubble in a carpenter's level.

FIGURE 11.11 Calibrated targets facilitate evaluation of the amount of movement.

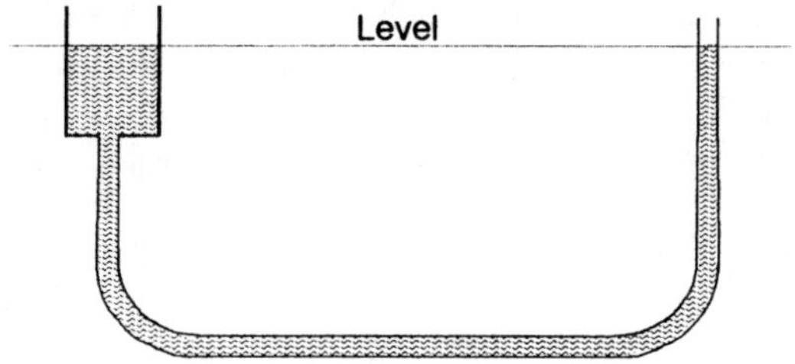

FIGURE 11.12 Water will always seek its own level.

the area being observed. Where multiple lines are used, however, the reservoir surface area should be considerably larger than that of the sum of all of the tubes. Thus, its level will not be unduly affected if the terminal ends of any one tube (or a group of tubes) drops, which will effectively lengthen that tube and increase the amount of fluid it holds.

In the most common application, multiple tubes are attached to a common reservoir, as illustrated in Figure 11.13 (left). The individual tubes are then distributed as desired around the work area. Figure 11.13 (right) illustrates a readout terminal simply taped to the wall of a building, again in combination with a ruler so that the exact measure of any fluid level change can be readily observed. Alternately, individual reservoirs can be mounted at the locations that are being observed, as shown in Figure 11.14 (top), with attached single tubes running to a remotely mounted readout point. This is especially helpful when the relative elevations of several different locations must be read from one central point. The terminal lines can be mounted together on a wall or other convenient surface, as illustrated in Figure 11.14 (bottom).

In addition to the advantages of monitoring injection with manometers, these devices offer an easy way to make a survey and/or contour the relative elevations of a near horizontal surface, either before or after the grouting. In this application, a single tube is run from the reservoir to a rod, to which it is attached securely, as

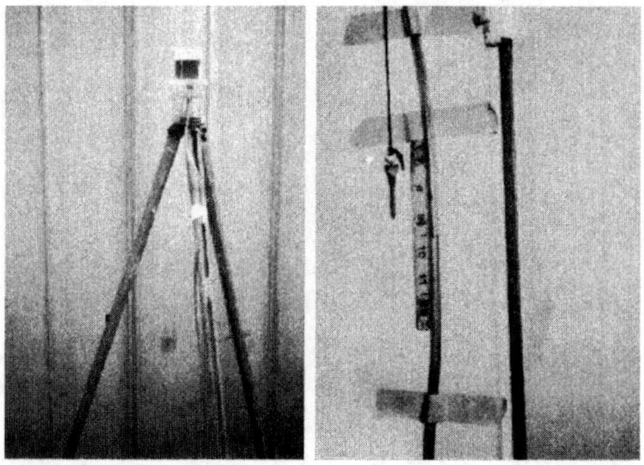

FIGURE 11.13 Manometer reservoir mounted on a standard surveyor's tripod (left). Readout simply taped to the wall of a building, with accompanying ruler to assess the amount of movement (right).

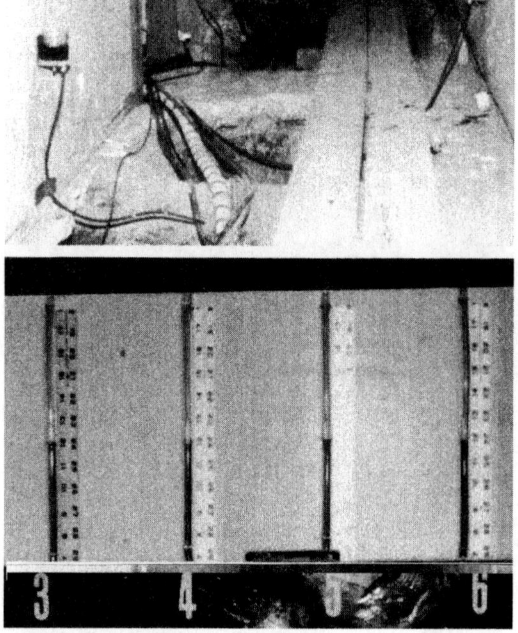

FIGURE 11.14 Individual manometer reservoirs attached to wall being monitored (top). Multiple manometer terminals placed at a single readout area (bottom).

FIGURE 11.15 A manometer terminal on a rod allows rapid evaluation of floor elevations. Sharpened steel point on base can be inserted in tight areas or poked through carpet (inset).

illustrated in Figure 11.15. To allow it to be inserted into tight areas or poked through carpeting to reach a slab, the rod can be fitted with a steel pointed end, as shown in the inset. With this arrangement, a single technician can run contours of several thousand square feet per day, regardless of the existence of walls or other obstructions. Because a line of sight is not required, the need to move the instrument frequently, as in traditional surveying, is eliminated.

Rotating laser beacons are also sometimes used. Properly set up, these shoot out a continuous laser signal on a perfectly level trajectory. Separate receivers, commonly referred to as *targets,* are attached to the structure, or set on stands as required, and adjusted vertically so as to be on the exact plane of the laser signal. The targets will trigger an audible beep upon a change of elevation relative to the signal plane. Virtually any number of targets can be used with a single rotating laser instrument, but to operate within the design tolerance, the receiving targets must be matched to the particular laser beacon.

Unfortunately, these instruments do not provide a measure of the change in elevation detected, and because standard practice is to readjust a target's height to the signal elevation once it has indicated a change, an accumulation of total movement at any one target location is not readily obtained. It is thus necessary to provide a backup system capable of acquiring an accurate measure of the total difference in elevation when these instruments are used. A conventional surveyor's instrument is most often used for this purpose.

There is a considerable difference in the operating tolerance of available rotating laser instruments. Some of the lower-cost models will barely provide accuracies on the order of 1/4 in. per 100 ft (6 mm per 30 m), whereas the higher-quality units can operate to a tolerance of less than 1/32 in. (0.79 mm) per 100 ft. Because of the importance for immediate alert of even a minute surface heave, none but the highest-quality laser instruments are usually acceptable. Virtually all available laser instruments will operate to a range greater than 100 ft (30 m), which is more than that required for most grouting work.

No instrument or group of instruments, however, can replace the work of vigilant, skilled monitoring personnel. It is not unusual to detect fine cracks radiating from an injection casing as upward movement begins, and grout injection should normally be terminated immediately upon the initial detection of any such surface disturbance. Figure 11.16 shows a very competent technician sweeping the ground surface with a whisk broom to allow him to more carefully observe any micro-movements of the ground surface. So skilled were the personnel on this particular project that they often discerned

FIGURE 11.16 A worker sweeps the ground so as to be able to see any micro-cracks that develop.

movement before it showed up on the fairly high precision instrumentation also used. The importance of the demonstrated qualifications of monitoring personnel cannot be overstated.

On many projects, the only monitoring required is immediately adjacent to the grout header, so it can be performed by the crew members who are already there. However, when there are improvements within the zone of influence of a given grout hole, either above or below the ground surface, they must be continuously observed during injection. Where walls isolate areas, or improvements are underground or otherwise isolated from the injection header, qualified monitoring personnel must be dedicated to these areas full time. In such instances, the maintenance of good communication, such as by radio or phone, is also mandatory.

Small-diameter sewer and drain pipes within the area of injection must also be monitored. This is best accomplished by accessing the conduit both upstream and downstream of the grouting and running clear water through the line. The water is directed into the upstream access and is observed as it passes downstream. Any leakage of grout into the line will be indicated by a change in color produced by the grout. Upstream access is usually readily obtainable with the use of easily accessible plumbing fixtures or cleanouts. Manholes or cleanouts are often located downstream, but if these are not available, excavations should be made to the pipe and an inspection hole cut into the pipe. When any grout leakage is detected early on, the conduits can be promptly flushed out, precluding any permanent damage.

11.1.2 Improving Deficient Soil

Although there are many different methods to improve deficient soil in place, compaction grouting is often the only method that can be easily and safely performed around or inside existing structures or other areas of limited access. Further, the grout injection may be limited to only those zones or horizons that require improvement. This can be an important consideration, especially where the improvement is being performed under defective end bearing piles or other deep foundations. Because compaction grout remains in a controlled mass at the point of its injection, the defective zone requiring improvement can be treated with a reasonable degree of precision. Moreover, damage resulting from uncontrolled wandering grout can be easily avoided.

Compaction grouting has thus proven to be an effective procedure for the solution of a wide range of foundation deficiencies and has been effective in the solution of many other types of problems. To illustrate the diversity of proven use, several actual case histories follow.

11.1.2.1 DENSIFICATION OF LOOSE SOIL, FILL, AND BACKFILL

Compaction grouting is especially well suited for the densification of backfill material that was not properly compacted when originally placed. Many projects have been completed where faulty fills behind retaining walls, or around and over buried pipelines, have been improved. Figure 11.17 illustrates such a case, where settlement of

FIGURE 11.17 Loose backfill behind the basement wall in a shopping center being grouted.

the fill behind the basement wall of a shopping mall was being densified. As can be seen, the operation was fairly orderly, and there was relatively little interference with the shoppers as the work was being done while the stores remained open.

Shortly after the original opening of the Japanese Pearl Divers attraction at Sea World in San Diego, California, unacceptable leakage of a man-made lagoon threatened to close the facility. On examination, the problem was found to be the result of insufficient compaction of backfill materials adjacent to the underwater viewing areas, during original construction. The owners number one criterion for correction of the deficiency was that normal business not be interrupted in any way.

Because it could be accomplished with minimal intrusion on normal operations, compaction grouting was chosen to densify the faulty soils (Figure 11.18). Most of the work was under raised decks and shops over the lagoon, with only about 4 ft (1.2 m) of overhead clearance. Holes of 2 in. (50 mm) diameter were drilled through the wood decks to allow the drilling of grout holes to proceed from above. Those holes extended for a depth of 20 ft (6 m) below the lagoon floor and were placed with the use of handheld drills. Once the grout hole at a given location was completed, wood plugs were placed to restore the deck.

FIGURE 11.18 Compaction grouting of backfill to correct the settlement of a pond liner.

The grout mixer and pump were located outside the park about 600 ft (180 m) distant from the work, which was accomplished at night and in the early morning hours prior to opening of the attraction. The grout delivery hose was picked up and stowed beneath the facility at the end of each work shift, as were all tools and other equipment. Thus, except for the absence of water in the lagoon, there was no indication of a problem or ongoing repair work to the visitors of the attraction, which remained open and fully operational throughout the work. The project was monitored for several years following completion, and no further leakage or other distress was noted.

The 1991 failure of a large sewer conduit in Houston, Texas, resulted in the formation of a sinkhole in the overlying road. Subsequent investigation disclosed that the backfill around and overlying some 8000 ft (2400 m) of reinforced concrete pipe was of low and inadequate density. The silty and clayey sand backfill was found to have Standard Penetration Test (SPT) "N" values averaging 12, with many below 10. Compaction grout was injected through rows of holes on each side of the pipe, at 10 ft (3 m) centers, as illustrated in Figure 11.19. The injection casings were driven to full depth, and grout injection was made from the bottom up in 1 ft (0.3 m) stages. The interior of the pipe was monitored continuously with a caliper during injection. The specification requirements for improvement to an average SPT "N" value of 20, with no test below 15, were easily met and verified. One hundred thirty-four postgrouting SPT tests, drilled midway between the grout holes by the same drill rig and operator, produced "N" values averaging 22.3.

In a somewhat similar but more sensitive case, compaction grouting was used to densify loose backfill around and overlying several miles of distressed storm drain pipe. In this case the grout had to be very carefully injected, as the relatively flexible corrugated metal pipe (CMP) could be easily deformed or crushed. The backfill soils consisted of fine to medium sand and silty sand, with silt and clayey silt lenses. The drains were under the pavement of a major freeway and had resulted in some localized settlement, prompting a geotechnical investigation. Initial cone penetrometer testing CPT disclosed cone tip resistance (Qc) to be generally less than 30 tons/ft^2 (29 kg/cm^2), which indicated very loose conditions. It was concluded that a cone tip resistance on the order of at least 50 tons/ft^2 (49 kg/cm^2) would be required to ensure against further settlement.

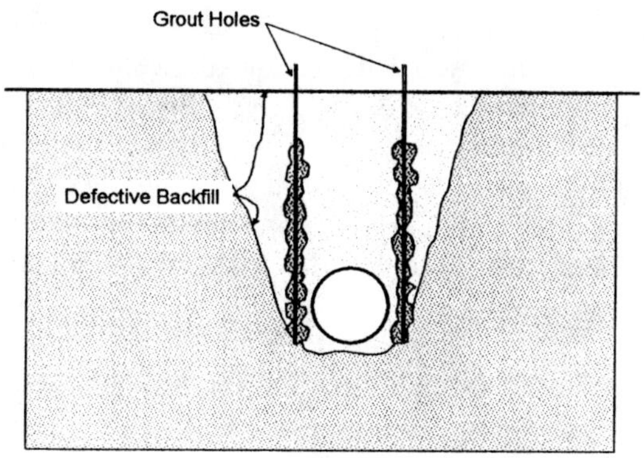

FIGURE 11.19 A row of casings was grouted on each side of the pipe.

A test section in which to evaluate the effectiveness of compaction grouting was developed, and a full-scale test application made. Because of the relative fragility of the corrugated metal pipe (CMP), the 32 injection points were spaced at a conservative 7 ft (2.1 m) on center. The grout was then injected from the bottom up in 1 ft (0.3 m) stages. A total of 318 ft^3 (9 m^3) of grout was injected, using a maximum pressure of 400 psi (28 bars). Details as to the aggregate gradation and the rate of pumping, which was done on an emergency basis, are not known. However, after completion, the minimum (CPT) values of 50 tons/ft^2 (49 kg/cm^2) were generally met, and in most cases greatly exceeded.

As a result of the successful test application, a contract was let to improve the soil around several miles of the conduits. In this production grouting, 3 ft (0.9 m) lengths of bottom-plugged, 2 in. (50 mm) proprietary flush wall casing were driven in a single row on each side of the drainpipes. They were spaced on 7 ft (2.1 m) centers and were driven with a hydraulic hammer mounted on a standard rubber-tired backhoe (Figure 11.20). Prior to injection, the casing was raised about 1 ft (0.3 m), and the plug knocked out. Grout was then pumped at a rate of less than 1.5 ft^3 (42 L) per minute. The interiors of the conduits were constantly monitored for deflection, and in several instances injection had to be cut short because of the initiation of deformation. For verification of the grout's effectiveness, another set of CPT tests were performed after completion of the work. The required degree of soil improvement was easily achieved.

FIGURE 11.20 A hydraulic casing driver mounted on a standard backhoe.

Compaction grouting is not often performed in high-plasticity clays, because their low permeability does not allow contained water to be driven out rapidly enough. In simple terms, grout cannot be injected into a soil any faster than the water can be expelled. This means extremely slow pumping rates in plastic clays, and as the degree of plasticity increases, the pumping rate must be evermore lowered. Further, the soil pore pressure must be continuously monitored during grouting to ensure that damaging pore pressures are not reached. Because the pumping rates that can be safely used are so low, the cost rapidly escalates so that the work is usually not economically justified.

There are, however, exceptions, as in the case of a building in Long Beach, California, where uneven settlement was occurring. The building was constructed on conventional spread footings bearing directly on a highly plastic clay soil layer about 15 ft (4.5 m) thick. The structure had a continuous footing around the periphery, with six columns supported on individual 12 ft (3.6 m) square concrete pads. Settlement of up to 10 in. (254 mm) was anticipated in the original design.

Shortly after construction was completed, significant uneven settlement of the columns was noted. This prompted a program of elevation monitoring for the entire foundation system. It was found that the outside walls of the structure were settling at less than half the predicted rate, whereas the rate of column settlement was nearly twice as much as that of the walls. This nonuniform settlement was adversely affecting the reinforced concrete structural system, which was undergoing considerable distress.

This situation presented a problem for which there was no really good solution, and although it would be painfully slow and abnormally costly, a very carefully orchestrated compaction grouting program was formulated to slow the rate of the columns' descent. Because the key to success was drainage of the clay and limitation of increased pore pressures, an array of piezometers were installed around the work area prior to any injection. They were monitored for about two weeks to establish threshold piezometric levels. A very slow and cautious injection program was then initiated.

The individual 2 ft (0.6 m) thick foundation pads were 12 ft (3.6 m) square and were based about 5 ft (1.5 m) below the top of the concrete slab floor of the ground-level parking garage. Four vertical grout holes, 9 ft (2.7 m) apart and spaced in a square pattern, were established concentrically around each column. Pilot holes of 3 in. (76 mm) diameter, which penetrated about 6 in. (150 mm) into the 2 ft (0.6 m) thick concrete foundation pad, were first drilled. Standard pipe casings of 2 in. (50 mm) ID were then inserted and cemented into the concrete to form a tight seal. Working through the installed casings, the rest of the foundation was then penetrated and the hole extended 3 ft (0.9 m) for injection of the first stage of grout.

Because it would be difficult to remove the cemented casings from the footing, they were composed of two pipe sections joined by a coupling. The coupling was placed so that it was about 1 ft (0.3 m) below the floor surface but above the top of the footing, such that the top piece of casing could simply be threaded out for abandonment at the conclusion of the work. Following injection of these four primary holes, an additional hole was drilled immediately adjacent to the column and inclined so as to exit the bottom of the concrete footing exactly centered under the column, as shown in Figure 10.21.

Initially, grout was injected at a relatively slow rate of 1/2 ft^3 (14 L) per minute; but this injection rate was found to be too fast, as revealed by an excessively rapid increase in pore pressure. The pumping rate was then adjusted to between 0.1 and 0.2 ft^3 (2.8 and 5.6 L) per minute, which was about as slowly as the pump could operate. Even at this very slow pumping rate, on many occasions it was necessary to cease work for a period of several days to allow the pore pressure to relax. The outside four holes were first grouted to somewhat stiffen the confining soil prior to injection of the last hole, which was directly under the column.

As the pore pressures increased, minute upward movement of the 6 in. (15 cm) thick reinforced concrete floor slab was detected. Because the slab as designed and built was quite stiff, steel shores were placed between it and the overhead concrete floor slab to help prevent any measurable heave. This, in combination with the very slow pumping rate, was effective, and maximum measurable heave of the floor upon completion of the work was limited to about 1/8 in. (3 mm). Interestingly, with great caution and some trepidation on the part of the participants, localized pore pressures near the hole being injected were allowed to exceed the overburden pressure. For this to occur, the piezometer tubes had to be extended with plastic tubing (shown in Figure 11.21) hanging from the overhead utilities. The injection parameters were directly dictated by the pore pressure behavior, as interpreted by the geotechnical engineer, Ray Maurseth, seen at the left in Figure 11.21. He was the engineer on the fateful swimming pool project in which the compaction grouting mechanism was first discovered in 1957, as explained in Section 1.1 "History of Grouting."

FIGURE 11.21 Piezometer tubes hanging from the ceiling indicate pore pressures greater than the overburden pressure.

11.1.2.2 MITIGATION OF THE LIQUEFACTION POTENTIAL

Liquefaction occurs while loose submerged granular soils are shaken. In this process the soil goes into a quick condition that can allow overlying structures to literally tilt or sink. Compaction grouting is often used to raise the density of such soils in order to eliminate these risks. In 1983 the first use of the process for this purpose involved improvement of the site for an addition to the existing pile-supported Kaiser Hospital building in South San Francisco, California. The project is of considerable interest, as it was conceived, designed, and the resulting performance evaluated with a high level of engineering control. This included an initial full-scale test program and evaluation of several application variations, including extensive testing as the work progressed. Further, the completed project was tested subsequently by an actual major earthquake, the Loma Prieta, in 1989.

The site was composed of uncontrolled fill that included primarily sands and gravels, as well as some clay and construction debris, to a depth of 6–8 ft (1.8–2.4 m). This material was underlain by a layer of loose to medium-dense hydraulically placed fine- to medium-grained sand fill. It varied in thickness from about 2 to 30 ft (0.66 to 9 m) and was underlain by competent natural materials. Analysis determined that to be resistant to liquefaction resulting from the ground shaking of a significant earthquake, the hydraulic fill zone would have to be densified to a minimum of 70 percent relative density.

The work was immediately adjacent to the existing structure, which remained fully operational throughout construction. Excessive noise or vibration could thus not be tolerated. Of the several candidate methods evaluated for the soil improvement, compaction grouting was the only one that met all of the requirements. Not only would it be free of any vibration, but it was reasonably quiet to perform. There were, however, no well-documented data available wherein the procedure satisfactorily densified similar soils. A large-scale test program was thus formulated to qualify the method.

Casings of 2 in. (50 mm) ID were driven from the ground surface to the top of the hydraulically placed fill in an offset grid of 8 ft (2.4 m). The holes were then extended and grouted in three to 4 ft (1.2 m) stages from the top down. This continued to the bottom of the liquefiable sand layer, which was at a depth of about 34 ft (10.2 m). Grout injection was continued until surface movement was noted or a pressure of 600 psi (41 bars) developed at a pumping rate of approximately 2 ft^3 (56.7 L) per minute. The total ground heave upon completion was slightly less than a 1/2 in. (13 mm) and represented about 10 percent of the total grout injected. Both PT and CPT test probes were used before and after the grouting, disclosing that a more than adequate increase in density was achieved.

Unable to meet the project schedule for subsequent production grouting, the contractor requested and was granted a change in the injection procedure. This provided for injecting from the top down in a single continuous operation. The casing was driven to the top of the treatment layer, followed by grout injection. It was then alternately driven and injected in 1 ft (0.3 m) intervals until the bottom of the treatment zone was reached. Considerable difficulty was experienced in starting the grout injection following the casing advance, apparently caused by plugging of the casing during the intermittent driving. Accordingly, this method was abandoned.

A second alternate procedure was then evaluated, namely, traditional bottom-up staging. The casing was driven to the bottom of the treatment zone and then pulled 3 ft (0.9 m) for the first-stage injection. Alternate withdrawal and injection were then carried out in 3 ft (0.9 m) stages until the upper limit of the improvement zone was reached. Although grout takes at the lower stages were similar to those of the top-down holes of the initial test section, they were noticeably lower in the upper portion of the holes. Subsequent testing indicated that the densities obtained with this method were generally adequate below a depth of about 16 ft (4.8 m), but deficient at higher levels. This was not unexpected and confirmed previous experiences of the inability of bottom-up staging, to provide adequate improvement at shallow depths.

For the final production work, grout holes were established in a staggered grid pattern of 8 ft (2.4 m) over the entire site of the addition, with a peripheral row 5 ft (1.5 m) beyond. To provide greater restraint for the shallower injection, the upper level of the soils, from 7 to 17 ft (2.1 to 5.1 m) in depth, was first grouted from the top down in 3 ft (0.9 m) stages. This was followed by treatment of the deeper remaining soils, starting at the base of the liquefiable sands at a depth of about 34 ft (10.2 m). Injection stages of 3 ft (0.9 m) were then accomplished from the bottom up, until the 17 ft (5.1 m) depth was reached. Additional grout holes were required in several areas that failed to meet the acceptance criteria, resulting in a final spacing of 4 ft (1.2 m).

The grout was composed of silty sand, cement, and water mixed to a stiff consistency, with a slump of 1 to 2 in. (25 to 50 mm). Average grout take was 2.7 ft^3 (76 L) for each 1 ft (0.3 m) of grout hole. Injection pressure was typically between 400 and 500 psi (28 and 34 bars). In the early stages of work, insufficient monitoring and a tendency to pump at excessively high injection rates resulted in total surface heaving of as much as 11 in. (280 mm). Stringent monitoring and more careful control of the pumping rates were then initiated, and the total heave was brought down to a maximum of about 2 in. (50 mm). Although the excessive heave should not have occurred, and probably reduced the amount of compaction obtainable, postinjection testing indicated an increase in the relative density of the in-place soils of about 20 percent. The overall improvement, as determined by posttreatment boring and extensive penetration testing, was even greater, suggesting that compaction grouting also has a significant influence on the stress state of the soil. Results of the postgrouting penetration tests are shown in Figure 11.22.

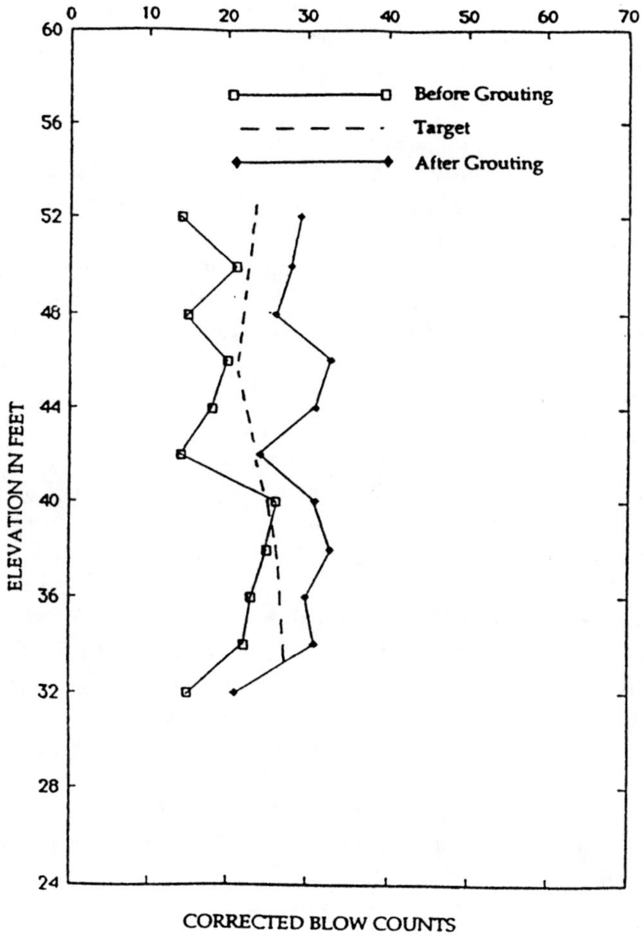

FIGURE 11.22 Degree of densification is shown by corrected blow counts.

This project represented the first documented use of compaction grouting for general site improvement and mitigation of liquefaction. Because it survived the Loma Prieta earthquake of 1989 unscathed, in spite of heavy damage in the immediate surroundings, it has been of special interest to earthquake engineering professionals, and the performance of the site has been well documented. Upon careful inspection following the 1989 event, the building addition showed no damage nor was there any indication of adjacent liquefaction. A through account of the grouting program and resulting soil condition was presented by Donavan, Becker, and Lau (1984), and a report on the project's performance during the Loma Prieta earthquake has been provided by Mitchell and Wentz (1991).

In 1988 compaction grouting was used to densify foundation soils underlying Chessman Dam near Helena, Montana. This was the first of many similar projects in which compaction grouting has been employed to upgrade the seismic resistance of dams and other large civil works. A similar project was carried out on Croton Dam on the Muskegon River in Michigan in 1998. The culprit soils consisted of fairly uniform loose sands and gravels, which had been placed as hydraulic fill during the original dam construction. These low-density materials overlaid very dense original ground.

Grout holes were laid out in a grid of 8 ft (2.4 m) over the treatment zone, which covered an area in the general form of a trapezoid with dimensions of about 140 by 220 ft (42 by 66 m). Treatment was required to depths of about 50 ft (15 m). Injection continued until any one of three different events occurred: (1) upward heave of the ground surface, (2) injection of a volume of 5 ft^3 (141 L) of grout per foot of hole, or (3) reaching a maximum pressure, which varied with the depth of injection from 400 psi (28 bars) at depths of less than 10 ft (3 m) to 800 psi (55 bars) at depths greater than 26 ft (7.8 m).

To mitigate the risk of liquefaction from seismic activity, a massive project was started in 1999 to improve soil underlying the Haneda Airport, which serves Tokyo, Japan. The effort involves one of the largest undertakings to date in terms of both the amount of grout used and the volume of soil treated. Grout holes are laid out in a grid pattern on a spacing of 10 ft (3 m) each way. A 2 in (50 mm) diameter proprietary casing with a double-taper flush joint and 1/4 in (6 mm) thick wall is driven to the bottom of the stabilization zone.

Most of the activity is under busy taxiways (Figure 11.23), and these could not be shut down for the work for more than several hours at any one time. Thus, one of the key requirements in selecting an improvement method was that the work could be done at night, with the work area open each day for normal airport operations. Compaction grouting was the only technique that could comply with this requirement, as neither large equipment nor any invasive removals would be required. Further, no damage to the existing concrete pavement would occur. Most of the work is being performed at night, with the casings being completely removed and the work area cleared and cleaned up so that normal activity can resume in the early morning hours.

FIGURE 11.23 Compaction grouting under active runways and taxiways.

The grout is proportioned from a standard portable volumetric batching plant and mixed with an attached auger-type mixer. It is fed directly into a 4 in. (102 mm) Putzmeister swing tube pump. The batch plants and pumps are set up to the side of otherwise active areas, where they can remain during the daytime airport operations. Rigid pipelines are then run to the areas of injection, which are often several hundred feet distant. The pipeline is removed as required for normal airport operations. The grout plants are normally not moved until grouting of the surrounding area is completed. They are resupplied with cement and aggregate as needed.

Grout is injected from the bottom up in 1 ft (0.3 m) stages. The casing is supplied in 3.3 ft (1 m) lengths, allowing easy manual handling as it is withdrawn. Both the grout quantity and pressure, as well as any surface heave, are monitored electronically in real time. All data are transmitted to a portable microprocessor station, from where a technician controls the grout pump (Figure 11.24).

FIGURE 11.24 Data are transmitted to a pump from a portable control station.

11.1.2.3 IMPROVING WATER-RETAINING EMBANKMENTS

Perhaps the most sensitive of any geotechnical remedial applications involves work in water-retaining embankments. Should excessive pore pressures develop, uplift leading to complete failure of the embankment may occur. Further, should hydraulic fracturing occur, leakage through the embankment could develop and also lead to complete failure. Nonetheless, compaction grouting has been satisfactorily performed to improve the soil in a number of water-retaining earth embankments.

Because compaction grouting allows close control of the grout deposition location, it is, when properly performed, one of the safest remedial procedures for correcting defects in such embankments. To achieve maximum protection against the occurrence of hydraulic fractures, careful design and rigid control of the grout mix is of crucial importance. In this regard, a minimum of 20 percent of the grout aggregate should be retained on a No. 4 sieve. The combined aggregate should otherwise comply with the particle size distribution given in the envelope of acceptable aggregate gradation provided in Figure 6.15. In addition the aggregate should be completely free of any clay component, which will require confirmation by standard hydrometer testing.

The California Aqueduct is one of the largest man-made water-conveying facilities in the world. Much of its route winds along the base of the Coast Range of hills, and its course runs alternately through the cut of traversing ridges and the fill of the valleys. There are substantial earth embankments, as illustrated in Figure 11.25, on the downhill sides of many of the valleys. Investigation following leakage at the base of an embankment, which had been controlled by an earlier emergency response, found both the natural and the fill soils in the base, as well as under the embankment, weakened because of dissolution of the underlying gypsiferous foundation material. It was further believed that in addition to the weakened soil, some open voids might exist.

A method to "find" and remediate any such weak areas or voids was needed. Compaction grouting was selected as the safest method, because it would provide for the greatest control of the grout deposition area and, properly performed, would not result in hydraulic fracturing of the embankment. Fracturing was of particular concern, because bagged bentonite had been placed in the embankment during the earlier emergency response. Neat bentonite gel is known to act as a fluid when under pressure in soil and

FIGURE 11.25 Typical embankment on the California Aqueduct.

would likely initiate a hydraulic fracture if subjected to the pressure that could result from grout injection.

A very carefully controlled compaction grouting program was thus adopted. Injection was performed in stages of 1 ft (0.3 m) from the bottom up. Initially, a single row of grout holes, spaced at 16 ft (4.8 m) increments, was established adjacent to the top of the channel for the 300 ft (90 m) segment that was suspect. Following injection of these primary holes, secondary holes were placed midway between for a final spacing of 8 ft (2.4 m). Based on data gained from early injection, additional rows were established, as the problem was found to be far more extensive than originally anticipated. The initial spacing of 8 ft (2.4 m) was maintained for all additional holes, which were offset to provide a staggered, triangular pattern.

Drilling was accomplished by rotary wash methods, using a special double-taper thread joint, flush-walled casing furnished by the contractor in 3 ft (0.9 m) lengths. The base of the bottom casing section was notched so as to mate to an expendable bit, which was knocked out prior to the start of injection. As illustrated in Figure 11.26, employment of these relatively short casing sections provided for easy breaking and removal as the grouting progressed and allowed the workers to remain on the ground surface at all times. In fact, there was a friendly rivalry between the workers to see how fast the casing could be broken, the injection hose reconnected, and the pumping resumed. Amazingly, these operations were frequently accomplished in less than a minute.

Grout was prepared with the use of aggregate of the preferable gradation, illustrated in Figure 6.15, containing the maximum amount of 25 percent gravel and 10 percent cement. It was proportioned from a self-contained truck-mounted volumetric batching plant and mixed with an attached auger-type mixer. Injection was performed with a swing-valve-type concrete pump fitted with 4 in. (102 mm) pistons and swing tube. Staging progressed from the bottom up, in 1 ft (0.3 m) intervals, at a conservatively low pumping rate of 1/2 ft^3 (14 L) per minute. All holes were grouted using split spacing, with a primary-secondary order of injection. A general view of the drilling and grouting operation is shown in Figure 11.27.

As the work progressed, the extent and severity of the deficiencies were amplified, and by the time it was finished, five rows of holes had been completed. The original length of treat-

FIGURE 11.26 Casing lengths of 3 ft (0.9 m) are easy to remove from the ground level during withdrawal.

FIGURE 11.27 Overall view of the work area. Canal operations are maintained but at a reduced level.

ment had also increased by more than 100 ft (30 m) and, in fact, extended beyond the embankment into the natural soils of the southern ridge. A total of 4523 ft^3 (128 m^3) of grout was injected into 281 holes. Although most of the embankment soil was found to be quite competent, several pockets of very loose soil were encountered in the bottom of the fill and, most frequently, in the underlying natural deposit.

Because of the sensitivity of the work, which was done with the aqueduct in constant operation albeit at a reduced level, continuous monitoring of the embankment was performed. In addition to careful observation and recording of the grout injection parameters, surface monitoring with multistation manometers and a continual roving optical survey were used. Carriage bolts, 12 in. (0.3 m) long, were driven into the ground surface at the same spacing as the grout holes, but staggered so as to be midway between them. These served as convenient benchmark monuments for the repetitive scanning survey that was continued during the work.

Some urgency was placed on better assessing the subsurface conditions, so the grouting program proceeded 24 hours a day until the work was essentially completed. Rapid mobilization precluded the use of continuous real-time computer monitoring, which would have been preferred. Comprehensive records of the grout take and injection behavior were kept, however, and promptly reviewed. They confirmed that the faulty soil was much more extensive than originally thought, and that indeed the deficiency involved the original formational materials underlying the embankment. A graphical presentation of the performance of each grout hole was prepared; these were assembled, according to the actual positions of the holes in the embankment, on the wall of the office trailer to facilitate comprehension of the overall project and the interrelation of different holes. In these presentations, grout take was indicated by a horizontal bar at each respective stage depth so that the actual configuration of the low-density soils could be readily visualized.

Concurrently with the start of grouting, a subsurface geological investigation, which eventually included 12 carefully logged bore holes and one test pit, was carried out. It was determined that the embankment had been founded on soil materials underlain by a gypsiferous formation. The deficiencies were determined to most likely be the result of dissolution of the gypsiferous materials caused by water permeating through an overlying sand strata underlying the embankment fill. This conclusion was in line with the results of the grouting program.

The work was a success in that it not only improved the embankment, the stability of which was found to have been seriously compromised, but also led to a much more complete understanding of the existing condition than would geological exploration alone. Further, it dramatically demonstrated the seriousness of the overall situation, which led to a decision to make an orderly shutdown of the aqueduct to allow a more comprehensive repair, consisting of a continuous membrane protected by an overlay of shotcrete. Of greatest significance, however, it allowed the aqueduct to function for several months until a planned shutdown could be made.

In 1996 a sinkhole was discovered in the core of WAC Bennett Dam in the northern part of British Columbia, Canada. Bennett is one of the largest and highest earth embankment dams in the world. During investigation of the problem, a second sinkhole was discovered. In a broad study of possible remedial methods, Garner and colleagues (2000) concluded that compaction grouting would be the most advantageous. This method was chosen because of the minimal amount of invasive activity required and the ability to prevent large pore pressure increases in the embankment. Further, it would require a minimum amount of large equipment on the limited area of the dam crest during the work.

Compaction grouting had not previously been accomplished to such depths, however, or under such sensitive conditions within the core of a major dam. Should the embankment be breeched during repair, the losses would be enormous and catastrophic, so a very conservative approach was mandatory. Further, in order to complete the work in time to allow the spring freshet to be impounded, it had to be accomplished entirely in severe cold weather during the winter months, which would greatly complicate the application. This was a critical problem, which warranted the very best analysis, design, and remedial implementation.

An extensive investigation determined the disturbed zone to be generally less than 20 ft (6 m) across, but about 400 ft (120 m) deep. It was not in the form of a uniform vertical orientation, but rather varied in cross section and wandered randomly in different directions. The undisturbed core material was found to be hard and very competent, whereas the soils within the sinkhole were variable but generally very loose. Precision drilling to a very tight tolerance would obviously be required if the grout holes were to remain within the loose sinkhole soils. And even then, it was not unlikely that some holes might enter an "overhang" of undamaged embankment, only to break through to further loose material.

The wandering configuration of the loose sinkhole debris could present other problems as well. Unlike most compaction grouting, in which the soils to be treated extend over a considerable area, in this case the extremely loose sinkhole debris was surrounded by a boundary of very dense core soil. With such a contrast of conditions, the drill casing would have to be stiff enough to maintain a straight course when entering and leaving the relatively stiff embankment. Further, if it were to strike the hard interface at an oblique angle, it would tend to follow down that hard face. Not only would this present a problem for the drilling, but for the grout injection as well. These factors suggested the advisability of using a relatively large and stiff drill casing, which is unusual in compaction grouting.

As previously discussed, when the injection casing is in near proximity to a hard boundary, it tends to bend from it during grout injection, especially if injection is in loose materials and at high pressure. Although it was not known what magnitude of pressure would be required at the great depths, a generally high pressure would obviously be required. The great depth also caused concern that pressure from a static head of either drilling fluid or grout could create excessive pore pressure within the embankment, which could not be allowed. Obvious requirements would be precision drilling to ensure that the holes were correctly placed, design of a highly stable grout mix, and control of the injection parameters so as to preclude the risk of hydraulic fracturing or other damage to the embankment. To follow the grout behavior and provide timely response, extensive instrumentation was installed to enable constant knowledge of the in-place conditions during injection.

To qualify the proposed methods and criteria for the work, a full-depth test hole was drilled and grouted. This hole was in an area just south of Vancouver that was already instrumented and was composed of similar soils to those of the dam core. One of the objectives of the trial was to fully acquaint the staff who would be responsible for monitoring the production work, with the full range of injection behavior that might be experienced. In this regard, part of the plan was to intentionally create a hydraulic fracture of the soil with a sudden increase in pumping rate and/or addition of clay to the grout. This would allow the working personnel to visually observe and become acquainted with the sudden pressure drop signaling such an event. Another objective was to confirm pumpability of the proposed grout mix, which was to contain a minimum of 25 percent of aggregate retained on a No. 4 sieve, but *no* cement.

It had been decided to employ continuous electronic monitoring of both the drilling and grouting parameters, and the effectiveness of this could be determined, as well as the staff become familiar with its operation, during the trial. In addition to the grouting parameters, drilling thrust and torque were continually recorded and converted into a numerical value of specific drilling energy. All objectives of the test hole were realized and additional shortcomings of the contractor's operation identified. Although it represented a considerable cost, the test proved to be extremely valuable to the overall operation and is credited with enabling the successful completion of the work.

Drilling the holes to within 1 percent of verticality was required. To meet this provision, the drilling contractor, Foundex Explorations, used a heavy and very powerful Foremost DR24 dual rotary drill rig (Figure 11.28). To achieve the required tolerances, a relatively stiff 6 in. (152 mm) diameter 0.219 in. (5.56 mm) wall-welded seam casing was used. This was combined with a variety of heavier lead sections. The casing joints, which occurred every 20 ft (6 m), were carefully aligned and welded as the hole descended. This required cutting with a pipe cutter, as required during retrieval, as illustrated in Figure 11.29. To prevent the static head of drilling fluid from contacting the hole bottom, a 6 ft (1.8 m) length of tightly fitting flight auger was used at the base of the lead casing so as to effect a soil spoil plug between the fluid and the hole bottom. Upon completion of drilling, a gyroscopic logging instrument was run the full depth to confirm that the hole was within the required verticality tolerance. The drilling method and casing system, combined with skilled operators, were found capable of easily meeting the requirements.

FIGURE 11.28 Powerful dual rotary drill rig working on test hole.

FIGURE 11.29 With rotation supplied by a drill rig, a pipe cutter is used to break joints during withdrawal.

It was originally planned to establish seven grout holes for each sinkhole. Six equally spaced holes in a 6 ft (1.8 m) diameter circle were to be placed around a central hole. Because of conflicts with instrumentation and lost drill rod and casing from prior drilling, as well as better investigative definition of the areas needing treatment, significant changes in the hole layout were made as the work progressed. The holes extended to depths as great as 380 ft (115 m) from the surface.

The initial sinkhole had been hastily filled with a mixture of sand, gravel, and cobbles that extended to a depth of about 50 ft (15 m). To work through this fill, an 8 in. (203 mm) casing was first placed to that depth. The hole was then continued to final depth using the 6 in. (152 mm) diameter casing. Initially, a 0.219 in. (5.6 mm) wall-welded seam casing was used, which was combined with a variety of heavier lead sections. Following problems with this casing bending and breaking in the first production grout hole on the dam, a heavier 0.432 in. (11 mm) wall, seamless steel casing was addopted. This proved to be quite adequate and was employed for the rest of the project without further difficulty. With but one exception, in which the drill is believed to have encountered an abandoned drill string, the required verticality tolerance was readily achieved.

The contractor's proposal was to inject the grout directly through the 6 in. (152 mm) casing. There was concern, however, that unacceptable pressures could develop at the bottom as a result of the static head of such a large column of standing grout. Most compaction grouting has been performed through 2 in. (51 mm) casing. A pressure head/loss of about 1 psi/ft (0.23 bar/m) has been established for this size. No such criteria were available for the much larger 6 in. (152 mm) casing, however. Thus, an evaluation of the grout behavior in the larger casing was made during injection of the test hole, utilizing a down-the-hole pressure cell. A head *gain* of 0.45 psi/ft (0.03 bar/m) of depth was experienced with the larger casing. As this could not be tolerated, a smaller 3 in. (76 mm) grout casing was inserted down the 6 in. (152 mm) casing upon completion of drilling. It resulted in a head *loss* of about 0.45 psi /ft (0.03 bar/m). This smaller casing was sealed at the bottom with a mechanical interlock so that grout could not run up the annulus. In production grouting, 20 ft (6 m) long sections of both casings were removed as the work progressed.

From a geotechnical standpoint, it was desirable to match the properties of the grout as closely as possible to those of the core material. The inclusion of cement was undesirable and was thus not included in the grout. As previously discussed, maximizing the coarse grain proportion, while providing just enough silt-size particles in the aggregate to allow pumping, will result in the greatest control during injection and minimize the likelihood of hydraulic fracturing. Because the work was to be done in the winter under freezing conditions, any moisture in the finer aggregate fractions would be subject to freezing and, obviously, it would be extremely difficult to excavate and handle such frozen material. Further, no natural deposits of satisfactory aggregate could be located within a reasonable hauling distance from the dam.

Accordingly, a special site-blended mixture was designed. It consisted of a gap-graded concrete sand and round-grained pea gravel, both of which remained in stockpiles from the original dam construction, combined with a naturally occurring silt that existed adjacent to the dam. To preclude freezing of the silt, it was excavated during good weather, and stored in an underground access tunnel leading to the powerhouse for later use. These components were precisely proportioned so as to result in a stable grout that provided the maximum control and the necessary pumpability, without the need for inclusion of cement. Moreover, this mixture provided a grout that very closely matched the gradation of the core material, shown in Figure 11.30. This aggregate blend was combined only with water to supply a very stiff, essentially zero-slump, grout.

A fundamental project requirement was to prevent damage to the core of the dam during the grout injection, as it would remain under high reservoir head throughout the work. Parameters were thus established to ensure forces of injection would be reasonably gentle to the soil. In this regard, an early decision was made to inject generally small quantities of grout into many closely spaced holes, rather than more typical quantities into a small number of more widely placed holes. On this basis, an initial volume limit of 4 ft^3 (113 L) per foot (0.3 m) of hole was established. As the work progressed, based on analysis of the retrieved data, the volume limit

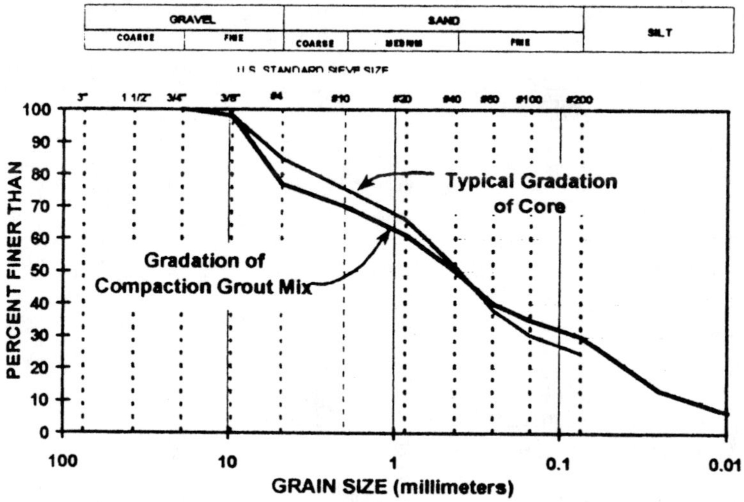

FIGURE 11.30 The grain size distribution of the grout aggregate was matched to that of the embankment core.

was increased to 7 ft^3 (198 L) per foot (0.3 m) of hole. Hole spacing was typically about 3 ft (0.9 m).

Further requirements included avoiding the development of excessive pore pressures and ensuring continuous drained behavior of the core soils during injection. Simply stated, the grout could not be injected faster than the pore water could escape from the area under densification. The embankment's pore pressures were continuously monitored throughout the grouting with an array of vibrating wire piezometers combined with monitoring wells. Readings of the instruments were taken at two-minute intervals and fed into an automated data acquisition system for real-time immediate display.

Because of the low permeability of the core soil, an extremely slow initial injection rate of 0.25 ft^3 (7 L) per minute was selected. This resulted in essentially drained behavior of the core soil as required. During the early stages of work, the contractor used a grout pump with a 6 in. (152 mm) cylinder. It was found to be virtually impossible to control at the very low pumping rates, and there was considerable variation of pump output within a range of 0.2 to 0.6 ft^3 (5.7 to 17 L). During this early work, several sporadic excessive head gains in the piezometers dictated a pause in the injection. Replacement of the original pumps with better-controlled machines containing smaller 4 in. (102 mm) cylinders enabled more uniform control of the injection rate.

Once the new pumps were operational, the optimal injection rate was established experimentally through careful manipulation within a range of 0.25 to 0.5 ft^3 (7.1 to 14.2 L). In the early stages of the work, the bottom portion of one grout hole was injected at a rate of 0.25 ft^3 (7.1 L) per minute without incident. As the injection level came to within 15 ft (4.5 m) of adjacent piezometers, the pumping rate was increased to 0.5 ft^3 (14.2 L) per minute. Piezometer head increases, of as much as 43 ft (12.9 m) were recorded at this higher pumping rate. A production rate of 0.38 ft^3 (10.7 L) per minute was then adopted for the remaining work, which continued without incident.

Because there had been no prior experience with compaction grouting to such depths, the pressure levels that would be required were not positively understood. Because it is known that grout pressure is directly related to the injection rate, the use of an injection rate, and thus a grout pressure, that would not cause excess pore pressure, as discussed, would result in safe injection. An initial pressure limitation of 1000 psi (69 bars) was proposed, with the requirement that the grout pump be capable of operating at up to 2000 psi (138 bars) should it become necessary. Actual experienced pressure at the grout header was on the order of 1200 psi (83 bars) in the lower portions of the holes, descending to about 500 psi (34 bars) in the upper parts. Figure 11.31 shows the drilling and injection parameters experienced in a typical grout hole.

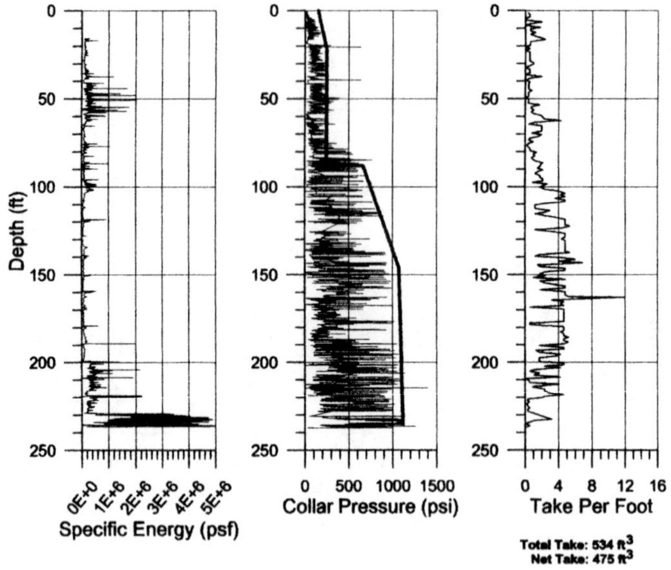

FIGURE 11.31 Typical values of the drilling and grouting parameters.

Unlike their positioning at the test site, the grout casings in the actual work were sometimes very close to the hard boundaries of the sinkholes. Moreover, in the beginning of the work, the contractor was unable to pump grout at the time the first hole was drilled, requiring the casing to remain in place for a period of several days. To prevent it becoming "frozen" in the hole, the drillers periodically rotated it. This activity, combined with the close boundaries and the high grout pressure, resulted in bending the very stiff casing. To overcome these problems, an even heavier 0.432 in. (11 mm) wall seamless steel casing was adopted and used on the rest of the job without further difficulty.

As previously mentioned, the work had to be accomplished during the winter to allow the reservoir to rise with the spring freshet. This required virtually all work to be performed within heated shelters, as illustrated in Figure 11.32. The mast of the drill rig was enclosed in hording, which extended out about 25 ft (7.5 m) to provide for the 20 ft (6 m) long joints of drill rod and casing. An additional tent was provided for the mixing and pumping equipment. Within this tent, a stockpile of aggregate, supplied to the grout mixer by a skid steer loader (Figure 11.33), was also enclosed. A separate trailer was provided for the computers and control operations. The setup illustrated in Figure 11.32 is for sinkhole No. 1. A similar setup was made at sinkhole No. 2, where work was carried out at the same time in order to meet the required completion deadline.

In addition, to detect any soil deformation within the embankment, during grouting several inclinometer casings were installed. The inclinometer was run periodically, and in some instances significant deformations within the

FIGURE 11.32 Heated shelters were required to withstand the cold winter conditions.

FIGURE 11.33 Aggregate was supplied to the mixer by skid steer loader in the shelter.

sinkhole debris were noted. In addition, arrays of survey prisms were mounted on both faces of the dam, as shown in Figure 11.34. These were constantly scanned during the project in order to detect any surface movement. Survey instruments for this work were protected from the weather in heated cabins, which were constructed just over the edge of the crest and provided with windows, as required to see the prisms. One of these shelters can be seen just to the right of the drill enclosure at the bottom of Figure 11.32. Preset monuments in the crest were also continuously scanned with a high-precision laser level throughout the work. No measurable surface movements were observed, with the exception of very minor heave immediately around the injection casings during grouting at shallow depths.

A dirt ramp had to be built to provide access to sinkhole No. 2, which was at the top of the upstream face. This required the importation of considerable soil to form a bench. As part of the final remedial work, the soil cap was excavated down to the top of the injected grout. That grout was found to exist in a perfect columnar mass, as illustrated in Figure 11.35. The success of this project proved the engineering validity of a properly designed and controlled compaction grouting program, even in very sensitive situations.

11.1.2.4 DENSIFICATION OF SUBMERGED SOIL

In 1978 a newly constructed cantilevered concrete sheet-pile seal wall began to tip outward as backfilling was initiated behind it. The wall was composed of pretensioned interlocking concrete piles, 12 × 48 in. (0.3 × 1.2 m) in cross section. They had been jetted into the ground underlying the water, utilizing a barge-mounted crane. The wall was to function as a continuous cantilever

FIGURE 11.34 Arrays of prisms were constantly scanned to detect any surface movement.

FIGURE 11.35 Perfectly columnar mass of exposed grout.

structure, as illustrated in Figure 11.36. Investigation determined that a silt layer under the sea floor had been badly disturbed, during the jetting operation, along the entire length of the 1.5 mi (2.4 km) wall. To prevent corrosion of the prestressing tendons within the piles due to exposure to salt water, it was imperative that the remedial work did not result in cracking of the piles.

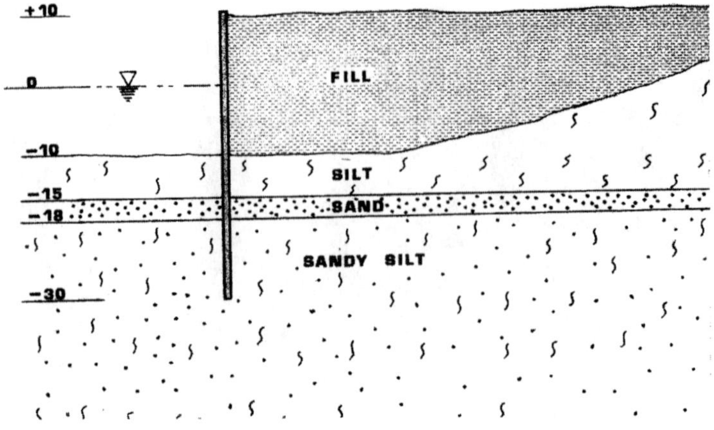

FIGURE 11.36 The wall was to function as a cantilever from the underlying soil.

Because of the serious consequences of imparting excessive forces on the sheet piles, a full-scale test was performed on a 100 ft (30 m) long section of the wall to qualify compaction grouting as the production method for repair. Eight inch (203 mm) aluminum whalers were secured to each side of the wall at the top. Load cells bearing on the whalers and reacting against brackets that were attached to 10 in. (254 mm) wide flange steel beams, anchored to land side deadmen, were then installed, as illustrated in Figure 11.37 (left). These were used to monitor the forces on the wall during grout injection. The forces were found to be well within acceptable levels, the greatest being 300 lb (136 kg) per linear foot (0.3 m) of wall.

Because the work was under water and any cracks or displacements that were to develop on the sea floor could not be readily seen, a very conservative grouting program was formulated. Holes were located adjacent to the existing bulkhead faces at a spacing of 8 ft (2.4 m). They were established by first drilling 3 in. (76 mm) holes about 4 ft (1.2 m) into the submerged surface. This was accomplished by workers with handheld pneumatic bore motors, working from the top of the wall and assisted by scaffolding that had been positioned on each side of it, as shown in Figure 11.37 (right).

With the assistance of a diver, 2 in. (50 mm) standard Schedule 40 steel pipe casings, which extended from the top of the wall to the bottom of the submerged holes, were inserted. One by one, the casings were raised a few inches, and a rapid-setting cementitious grout pumped into them until it completely filled the underwater annulus, as confirmed by the diver. Prior to final set, the fast-setting grout was drilled out of all but about the last 3 in. (76 mm) at the bottom of the casing. The production drilling then progressed through these casings.

FIGURE 11.37 Test setup with load cells (left). Grout holes drilled with handheld drills (right).

The grout mixer and pump remained on the adjoining grade, with the delivery line supplying a standard header. Grout with aggregate conforming to the criteria illustrated in Figure 2.7 was injected from the top down at a maximum rate of about 1 ft^3 (28 L) per minute. The first two stages were only about 2 ft (0.6 m) in length. The lengths of the stages were then increased with depth, to a final length of 4 ft (1.2 m). Grout pressures varied from 50 to 150 psi (3.4 to 10.3 bars) but were most often within a range of 50 to 70 psi (3.4 to 4.8 bars), especially in the upper stages. Approximately 4 ft^3 (113 L) of grout was injected for 1 ft (0.3 m) of the bulkhead wall.

Going into the work, it was well recognized that upward heave of the underwater surface would be the controlling cutoff criterion for the injection. This would require special procedures, as the underwater surface, obviously, could not be visually scanned. Subsurface heave indicators were placed on each side of the wall adjacent to the grout holes. These simply consisted of 5 gal (19 L) pails filled with concrete, into which a vertical piece of 3/4 in. (19 mm) pipe had been cast. Extension pipes were then added so as to rise above the top of the wall. Targets consisting of penciled marks were placed on the pipes, oriented so as to be visible from a surveyor's level located on the adjacent land.

In addition, slightly more sophisticated, but nonetheless simple, instruments, in the form of inductance monitors, were placed in the grout holes as well as in jetting holes that penetrated the length of the sheet piles. These were positioned very slightly above the water surface near the grout injection location. Underwater stresses that caused a rise of the water would trigger a loud horn activated by the affected instrument. A diver was also in the water at all times of grout injection, but because of limited visibility, his effectiveness in first spotting a heave was limited. Careful monitoring of the grout pressure behavior, as well as the instrumentation, did make it possible, however, to identify even a minor uplift as it developed. The diver could then be directed to the area indicating heave to provide firsthand information of any significant activity. Quite remarkably, considering the difficulty of monitoring the submerged surface, the greatest amount of total uplift was on the order of only 1.5 in. (38 mm).

To assess grouting effectiveness, cone penetrometer probes were made both prior to and following the grouting. Two rows of penetrometer probes were placed on the water side of the wall, where dense soil was required for the cantilever to function. The first row was approximately 5 ft (1.5 m) from the face of the bulkhead, with probes being made every 15 ft (4.5 m). The second row was 15 ft (4.5 m) off the bulkhead, with probes at 25 ft (7.5 m) intervals. This work was accomplished with the use of a 10 ton (0.9 t) capacity penetrometer secured to a floating barge, and the total penetration force included that portion of the weight of the barge lifted by the force acting on the penetrometer rods. When the rods could no longer penetrate, the barge was lifted partly out of the water, which established refusal.

Cone tip resistance prior to grouting was less than 41 tons/ft^2 (40 kg/cm^2) at a depth of 23 ft (7 m), and decreased to about 20 tons/ft^2 (20 kg/cm^2) above that depth in the first row. In the second row, beginning values were less than 5 tons/ft^2 (5 kg/cm^2) to a depth of 23 ft (7 m), and about 20 tons/ft^2 (20 kg/cm^2) below that depth. The postgrouting value increased with depth in both rows to nearly 51 tons/ft^2 (50 kg/cm^2) at about 20 ft (6 m), below which refusal was reached at about 82 tons/ft^2 (80 kg/cm^2) at a depth of 26 ft (7.8 m).

The test section provided satisfactory evidence that compaction grouting was a suitable method for remediation, as the soils were improved to acceptable values. Further, it did not require the use of heavy equipment at the wall location, which was surrounded by water. This resulted in its use to remediate the entire length of the wall, which was then backfilled and the planned construction of a commercial center begun. The site has been reported to be performing well, with no indication of the former distress.

Compaction grouting was also used to improve submerged soil underlying a pier structure in South Korea. Casings of 2 in. (50 mm), fitted on the bottom with tight-fitting expendable knock-off bits, were installed from the top of the pier with the use of standard rotary drilling, as shown in Figure 11.38. Once the final depth was reached, the casing was withdrawn 1 ft (0.3 m), and the bit knocked off with the force of a dropped weight. The weight was attached to a rope so it could be recovered when the bit broke loose. Injection was then made in 1 ft (0.3 m) stages from the bottom up. Extraction was accomplished with a common dual hydraulic ram withdrawal jack system. In addition to monitoring the grout injection parameters, the underwater submerged surface was continually scanned by divers. Because of limited space and weight restrictions on the pier structure, a portable batching and mixing plant and the grout pump were positioned on an neighboring rock jetty, as shown on Figure 11.39.

FIGURE 11.38 Drilling an underwater hole from a pier.

FIGURE 11.39 The grout plant located on an adjoining jetty.

11.1.3 Settlement Correction

Historically, and even to this day, the greatest proportion of all compaction grouting is used in the remediation of earth settlement. The procedure has a special advantage in this work, as it is possible to actually jack a settled structure back to its proper grade along with improving the soil. Although heaving of the ground surface is a frequent criterion for terminating grout injection, and preventing such heaving can be a problem where it is not wanted, it can be an advantage in dealing with settled structures. In such cases, incremental jacking is allowed to take place in a controlled manner with each grout stage. Groutjacking is discussed further in Chapter 14.

Faulty foundation soils and the resulting settlement of literally thousands of light residential structures have been corrected with compaction grouting. The great majority of these projects are typically on sites that have undergone significant grading and very often have downslopes composed of fill soils in near proximity to the settled structures, as shown in Figure 11.40. This type of project seldom involves working to depths greater than 30–40 ft (9–12 m)

FIGURE 11.40 Row of grout holes (left) adjacent to typical compacted fill downslope.

from the surface. The process has the advantage of not requiring large items of equipment near the structure, as the grout can be pumped with common procedures from a remote location, such as illustrated in Figure 11.41, as long as the distance is not more than about 500 ft (150 m).

As mentioned earlier, compaction grouting has also been used to correct settlement under a great number of very heavy structures, including large buildings, bridges, and equipment foundations. The work has been accomplished under virtually every type of foundation, such as large

FIGURE 11.41 A grout mixer and pump can be remotely located.

continuous rafts and deep pile-supported structures. The latter structures often involve working to depths of 100 ft (30 m) or more. Many successful applications have been made where settlement was so serious that it had caused structural damage, resulting in evacuation and, in some cases, condemnation of the affected structure.

11.1.3.1 COMMON STRUCTURES

The four-year-old building of Manitou Springs Junior High School in Colorado had to be vacated in 1980 because of severe structural distress resulting from several inches of differential settlement. Compaction grouting not only improved the underlying faulty soil, but also jacked the building back to proper level. A portion of a five-story wing of a concrete building in Rapid City, South Dakota, had settled several inches, resulting in serious structural distress. In 1986 faulty soils to a depth of 50 ft (15 m) were remediated under the shallow foundations, and the settled areas were raised to their original grade while the structure remained fully occupied. In 1991 the four-story Inage Welfare Center building in Chiba City, Japan, remained fully functional and occupied during the correction of nearly 3 in. (76 m) of differential settlement.

Severe settlement of some residences occurred at Mount La Jolla, an upscale retirement community built around a golf course near San Diego, California (Figure 11.42). The development was constructed on steep, hilly terrain and involved massive grading with significant cuts and fills of as much as 100 ft (30 m). Differential settlement of a few units started soon after they were occupied. The residents were obviously upset about the damage to their homes and demanded a repair that would not only cure the problem, but would not be unduly obtrusive or messy to perform. Compaction grouting was chosen as the best remedial method by the investigating engineer, who had a great deal of experience with this process.

FIGURE 11.42 Severe settlement occurred in many buildings in the Mount La Jolla development.

Accordingly, a contractor was retained to do the work. He was soon removed from the job, however, as his operation was excessively messy, which was unacceptable to the home owners. In the case of one of the structures, grout was splattered not only on the walls, but also on the roof. When the contractor was confronted with this problem, he is reported to have laughed and said, "So grouting is messy," His firm was replaced, as were two others that followed. These beginnings of what turned out to be a huge project are mentioned not to criticize, but to illustrate a common problem that has plagued the grouting industry for many years. Many professionals, both grouters and designers, seem to believe this falsehood. It is wrong to say that grouting work is inherently messy and, on that basis, to make no effort to operate in a more orderly manner. As very nearly occurred on this project, much potential grouting is not performed, because of the perception of owners and designers that such work must be very messy. And all too often such opinions are not based on perception, but on prior bad experience. As this case demonstrates, messiness is not necessary and grouting does not have to be excessively disruptive.

Although the property owners were disgusted with the idea of grouting, the engineer told of work he had supervised that was done in a very clean and orderly manner and strongly recommended that one last trial be made using a contractor who was known to run an especially clean operation. This suggestion was accepted, and now contractor number four mobilized to do the job with full knowledge of the disgruntled home owners with whom he would have to deal.

The interior surfaces of the first unit approched by the new contractor were covered with plastic sheeting, as shown in Figure 11.43. Knocked-down cardboard boxes were placed over the lower 4 ft (1.2 m) of the walls, as can be seen in Figure 11.44. A self-contained batcher mixing unit and connected trailer pump were simply parked on the street in front of the unit being worked on when required during the day, as shown in Figure 11.45. At the end of each work shift, the grout plant was removed from sight and the street washed down, as illustrated in Figure 11.46.

The home owners were happy with this contractor, but not with the developing problem. At

FIGURE 11.43 The walls were protected with plastic sheeting.

FIGURE 11.44 Knockdown cardboard cartons were used for additional protection of lower portions of the walls.

first, only a few units had been found to be settling, but many others were now showing signs of distress. What started as a fairly moderate undertaking developed into a huge project, which continued for more than two years. To keep the owners happy, the contractor continued his clean operation. His staging area, which included a silo for bulk cement, was hidden behind the landscaping in a remote fringe area of the development.

The actual work involved grout holes placed around the outsides of the structures, as well as on a 10 ft (3 m) grid in the interiors. The holes

FIGURE 11.45 The mixing plant and pump remained on the fronting street.

FIGURE 11.46 The work area was washed down at the end of each shift.

varied from very shallow to more than 100 ft (30 m) in depth. Rotary wash drilling methods were used, with rotation by handheld pneumatic bore motors and water for the circulation flush. Drilling employed a proprietary casing, which also served as the drill rod, in combination with an expendable bit that engaged grooves in the lead section. The returning flush water was directed into a small open-top tank or other containment to avoid a damaging mess on the concrete slab floors. Thus, the drilling did not create much of a mess, and absolutely no damage to the building itself.

Grout injection was from the bottom up, in 1 ft (0.3 m) stages. The casing was in 3 ft (0.9 m) lengths so they could easily be handled within the often restricted spaces inside the homes. A typical double hydraulic ram withdrawal system, similar to that illustrated in Figure 2.21, was employed for lifting the casing. The initial pumping rate was 1 ft^3 (28.3 L) per minute. As would be expected for this type of site, the injection quantities varied greatly, especially in the lower zones of fill and in the underlying natural materials. Pressure behavior indicated that there had been less than good removal of the original vegetation and slopewash and a lack of proper benching upon which to found the fill. It also suggested the nesting of large boulders within the fill. Unfortunately, these findings are quite consistent with apparent fill failures in general. Some grout was injected within the fill, but the greatest quantities were near the cut fill interface either in the bottom of the fill or the top of the original ground.

Although the primary operation was improvement of the underlying soils to arrest further consolidation and the resulting surface settlement, raising settled portions of the structure was also accomplished. A multistation manometer was used for elevation monitoring during the grout injection. The manometer system was supplemented with simple items such as string lines and carpenter's levels, which were all backed up with a surveyor's instrument.

Following completion of the grouting, necessary cosmetic work was performed and the homes returned to their owners. Although the special attention to cleanliness and protection of the existing construction did have a cost, much of that was recovered in cost savings in the cosmetic work that followed. The work was done in the 1980s and has performed well, with no further distress occurring.

A very different but equally serious settlement problem occurred in Ely, Nevada, in 2000. A highway and a single-track railroad were both benched into the wall of a steep canyon. The highway was at an elevation about 30 ft (9 m) higher than that of the railroad. To improve the highway, it was necessary to extend it out over the railroad for several hundred feet, which was to be accomplished by building a tunnel over the track, which would then be backfilled to support the new roadway, as illustrated in Figure 11.47. Geologically, this was a very complex site. There had been both natural and man-made filling activity in the canyon for a very long time. It was the primary drainage for an extremely large land area and had undergone flash flooding on many occasions. Some of this flooding had probably deposited mine wastes onto the site from a large open-pit operation about 5 miles (8 km) upstream. Further, other deposits had probably resulted from the drainage of an intersecting canyon.

Both mudflow and slide debris had been deposited, from the canyon wall where the bedding intersected the slope, and on which was located the highway and the railroad. The mudflow deposits, composed principally of fine soils and deposited as fully saturated, were probably quite dense. Conversely, the dry slide remnants, which were principally of rock debris, were expected to be in a very loose condition. And then there were the man-made fills, no doubt resulting from the original construction and subsequent improvements to the highway and railroad, as well as probable mine tailings from extensive mining in the area. All of these filling episodes had continued for a period of more than 100 years.

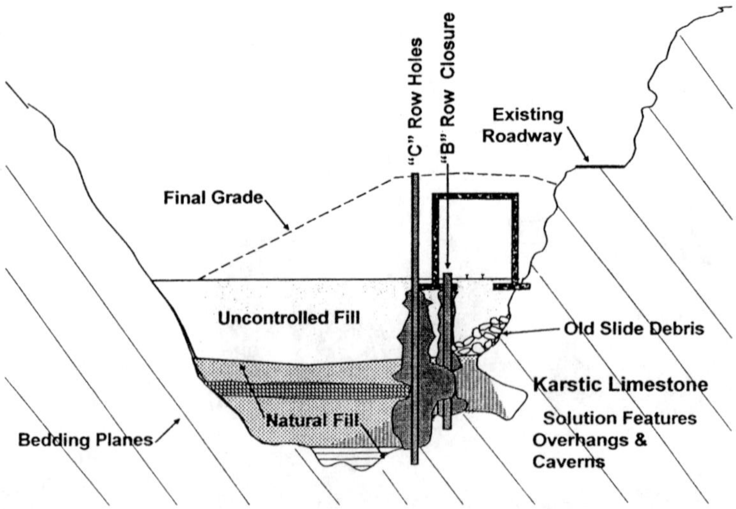

FIGURE 11.47 The tunnel was located on very complex geology.

To complicate things even further, there was considerable solution activity on the canyon side that was intersected by the bedding. This included many overhangs and much cavernous space. There were also areas that had been eroded at the various joints, especially those oriented in the direction of the strike of the rock. Where these features existed under the various fills, they might have been filled very loosely, and there could also be obvious voids that were not filled at all. Because the natural fills had been deposited over thousands or perhaps even millions of years, the site's origins and condition were bound to vary greatly. Because of the very long depositional period and the existence of such diverse content, it was virtually impossible to attain a good understanding of the conditions.

The tunnel structure and some adjacent retaining walls were built and backfilled during the construction season of 2000. As winter and cold weather set in, paving the roadway was put off until the following spring of 2001. As part of the construction, a new 10 ft (3 m) wide by 6.8 ft (2.1 m) high concrete box drainage culvert was to be built around the new tunnel. Unfortunately, this structure was not completed with the initial work so as to provide good management of the drainage. This factor, in addition to the massive changes of the historic drainage patterns, likely caused the saturation of some hydrocollapsible soils that existed in the various fills. It is possible that loose soils were also washed into existing voids in the adjacent rock face.

When the work was to be resumed the following spring, it was found that one side of the west end of the tunnel structure (the left side as viewed in Figure 11.48) had settled some 8 in. (0.2 m). The settlement lessened going toward the east and was virtually nonexistent on the east end. And there was virtually no settlement of the south side, which was only about 24 ft (7.2 m) away. Compaction grouting under the entire length of the affected northerly side of the structure was selected as the best remedial approach.

FIGURE 11.48 The left side of the tunnel settled as much as 8 in. (203 mm).

The fill soil was removed to about the roof of the tunnel, and a primary row of holes ("C" row) was placed immediately adjacent to the footing on the outside of the structure, as illustrated in Figure 11.47. It was anticipated that grouting of this row would confine the soil materials under the tunnel between it and the steep rock slope. Following completion of these "C" holes, the "B" row of holes was to be grouted, providing closure so as to properly support the tunnel wall. Because the wall footing had been poured immediately adjacent to the railroad bed, which had to be maintained during the work, core holes were drilled through the footing for the "B" holes, which were directly under it. A third row of "D" holes was contemplated to be placed to the north of the "C" row if it was found that those soils were deficient and the canyon wall continued to descend in that direction. Only one of these holes was grouted, as the conditions in that direction were found to be not as poor as in the terrain closer to the rock face.

Drilling of the grout holes was expected to be difficult at best. Embedment of large boulders was anticipated, which would have to be penetrated in the various fills. Moreover, to confirm that the holes were bearing on solid bedrock and not on an overhang or boulder, all primary holes were required to extend into rock a minimum of 10 ft (3 m). Rotary wash drilling was required by the specifications. This was to provide sufficient moisture to the collapsible soils to allow their maximum compaction during grouting. Early in the work, drilling was accomplished with a Davey DK500 rotary drill rig (Figure 11.49) in combination with a self-drilling casing, negating the requirement for a separate drill string. The lead section of casing was fitted with carbide teeth, as shown in the inset.

The contractor, however, chose to use the ODEX duplex rotary percussive drilling system for most of the work. In this system the bit contains an eccentric dog that rotates outward so as to undercut the casing as the bit is rotated in the drilling direction. The casing can thus follow the bit to the bottom of the hole. Once completed, rotation of the drill string is reversed, which causes the eccentric dog to retract so the entire drill string, including the bit, can be withdrawn. This drilling was accomplished with a pneumatic track drill. As specified, water was used for the circulation flush. For casing, standard NW flush-wall diamond drilling casing, which is 3 in. (76 mm), inside diameter, by 3.5 in. (88.9), outside diameter, was employed. Although this arrangement worked quite satisfactorily from a drilling standpoint, the casing proved to have insufficient strength, resulting in many difficulties during grouting, which are discussed shortly.

The grout consisted of a processed aggregate meeting the grain size distribution shown in the envelope of preferable gradation (shown in Figure 6.15) and between 10 and 15 percent cement. The specifications required a minimum cement content of 8 percent, but the contractor chose to increase this amount to improve pumpability. The grout was mixed in semifixed-location grout plants, initially set up at two places on the site but moved as the job progressed so that two setups were required for each plant. These plants consisted of a trailer-mounted proportioning batcher with an integral auger-type mixer that was positioned so as to empty into the hopper of a separate 4 in. (102 mm) swing tube concrete pump. The aggregate was supplied from an adjacent stockpile to the batcher, with a small skid steer loader, as illustrated in Figure 11.50. Bagged cement was used and fed to the cement hopper manually.

Alternate primary holes were injected first. The intermediate secondaries could be injected in any order as long as grouting of the adjacent primaries was complete. The grout was injected from the bottom up, using 1 ft (0.3 m) stages, with a pumping rate set at 1.5 ft (42 L) per

FIGURE 11.49 Rotary drill rig used for initial grout holes.

FIGURE 11.50 Mixer supplied with aggregate by a skid steer loader.

minute. This was the pumping rate on which the contractor's proposals were to be based. Its validity was confirmed once the work was started, as it was found to be the fastest rate that was reasonable for injection into the finer soil pockets, based on a requirement that a pressure increase of more than 8 psi (0.55 bars) per minute not be exceeded.

As previously discussed and illustrated in Figure 11.47, the soils were originally deposited by a hodgepodge of different mechanisms and were thus extremely variable in character. Although a faster pumping rate could have been used for much of the work, changing the rate while continuously pumping was not practicable with the pumping equipment that was used, so the slower injection rate, which was confirmed to be safe for all soil conditions to be encountered in any one hole, was used.

Because of the semifixed locations of the grout pumps, longer delivery lines were required than would otherwise be used with more portable grout plants. To minimize the frictional head loss within the line, the contractor elected to use a rigid steel delivery system up to within a short distance of the hole being grouted. This line was of 3 in. (76 mm) inside diameter, which, combined with the relatively slow pumping rate, resulted in a very low velocity of grout movement, thus contributing to frequent problems of plugging within the system.

Also used were 10 ft (3 m) long joints of casing. This required erection of scaffolds to gain access for detaching the delivery line and gauge saver, as illustrated in Figure 11.51, each time a joint was to be removed. Furthermore, because the casing was of a relatively large 3 in. (76 mm) internal diameter and filled with grout, it was heavy and presented a gargantuan task to the technicians to make the required removals. Because of this difficulty, as well as the time spent climbing up and down the scaffold, an injection delay of 5 to 15 minutes or more was common in removing these casing joints.

FIGURE 11.51 Access to the header required scaffolding because of 10 ft (3 m) casing lengths.

As mentioned earlier, the site was incredibly complex geologically. Moreover, prior to the work much of the terrain was steep and the access often restricted, which limited the optimal placing of test borings. The contour of the underlying rock surface had not been well defined, and the existence of rock outcroppings and overhangs, as well as large boulders embedded in the soil material, was probable. But the greatest problem was that the subsurface situation changed in such short distances that it was virtually impossible to obtain a complete and accurate understanding of the real conditions. Accordingly, unusually thorough monitoring of the drilling and grouting work was performed and the incoming data were promptly plotted and analyzed.

To establish the true contour of the foundation rock, all primary holes were drilled a minimum of 10 ft (3 m) into what was believed to be the underlying rock surface. This allowed confirmation of its elevation and assurance that the hole did not simply terminate in an overhang or boulder. The actual confirmed depths to rock were then plotted on a profile of the best estimate of its depth (Figure 11.52) so that the true conditions could more easily be visualized. It was disappointing that the actual rock profile

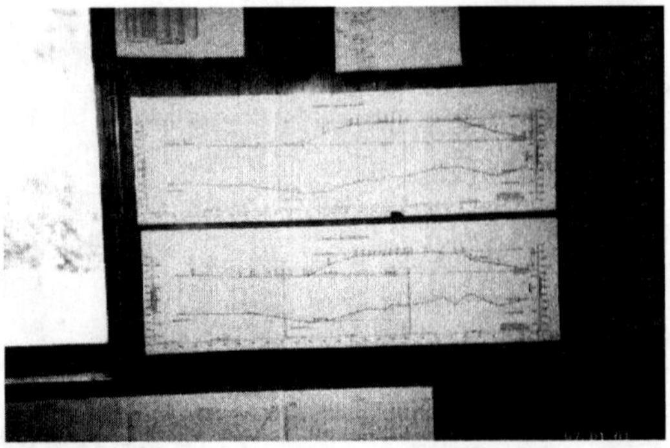

FIGURE 11.52 The plot of the underlying rock surface was updated on the profile.

was found to be significantly deeper than first thought at the beginning of the work. The starting elevation of the grout holes varied because of the topography, and, of course, as can be readily observed in Figure 11.47, the "B" row of holes started at the level of the railroad bed, whereas the ground surface for most of those in the "C" row was considerably higher. The "B" row holes varied in depth from about 17 to 72 ft (5.1 to 22 m), and the "C" row holes were somewhat deeper, varying from about 50 to 90 ft (15 to 27 m).

Although real-time computer monitoring of the injection parameters would be extremely beneficial on such a project with large variations in grout behavior, it was not available, so the initial data were acquired visually and entered manually into a field log. Pressure readings were entered at three-minute intervals. All significant events were noted, and the current time was indicated with all record entries. This involved an extraordinarily large number of entries, which then had to be manually loaded into the computer database. Once the numbers were entered, however, they could be easily processed and the output presented in a number of easy-to-understand formats.

Grout takes were plotted in a bar graph, and time/pressure/volume curves were prepared for those stages that exhibited significant behavior. As they were developed, these were posted on the wall of the office trailer, as shown in Figure 11.53 (top). The extreme variability of the grout takes can be readily seen in the bar graphs, which are at the top of the wall postings. Immediately under the graph appeared a tabular review of the parameters followed by the most noteworthy injection behavior curves, a sample of which is shown in the bottom of Figure 11.53. These were typically attached in several layers because of the limited wall space available.

Extremely large quantities of grout were injected throughout the project, but especially in the early work. Up to several hundred cubic feet were placed in some of the stages of the first holes completed. An evaluation of the early injections resulted in the adoption of a pressure limit of 450 psi (31 bars) at a pumping rate of 1.5 ft^3 (42 L) per minute. For the first several days of the work, pumping was continued until this pressure level was reached, which resulted in extremely large quantities of grout being injected. Upon analysis of this early injection, and in consideration of the resulting general tightening of the ground, a volume limit was also instituted. It was set at an ordinarily huge amount of 100 ft^3 (2.8 m^3) per stage in the "C" row. This appears to be, and is, an extraordinary quantity, but it was often reached at reasonably low pressures, in some cases less than 300 psi (20.7 bars), which further emphasized the uniqueness of the site. As injection continued in the "B" row, a volume limit of 50 ft^3 (1.4 m^3) was instituted. This quantity was very often not reached, and in general the "C" row holes had done their job, resulting in significantly lower takes in the "B" row.

The very large quantities of grout injected are attributed to the incredibly loose soils and probable voids in the adjacent karstic rock, as well as the forced collapse of the hydrocollapsible soils. Analysis of the pressure behavior did not indicate the filling of any outright voids, but the probability that very loose soils had been

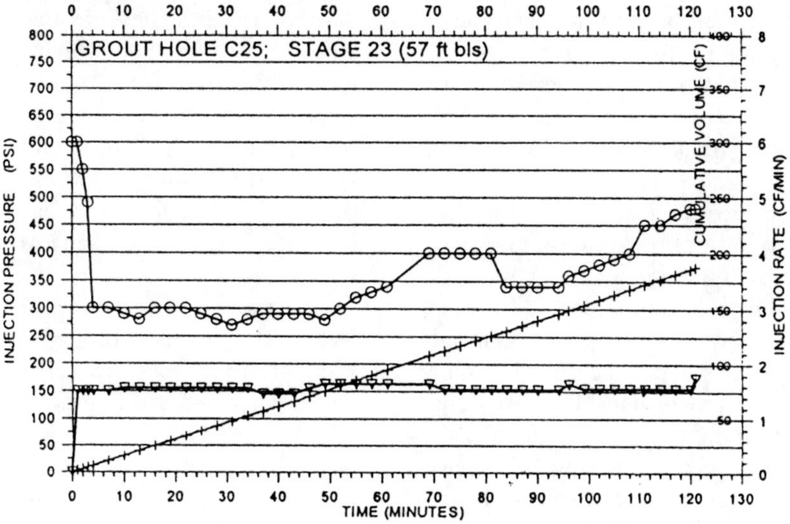

FIGURE 11.53 Grout records posted in proper order. Bar chart at top of photo provides grout take with depth with typical computer output plots for each stage below. On individual stage plots (bottom) curve (triangles) indicate pressure; top curve (circles), injection rate; sloped curve with ticks is the cumulative volume of grout.

compressed into cavernous voids was suggested on many occasions. There was, however, a huge difference in both the quantities of grout injected and the pressure behavior between different holes, as well as between the different stages of a given hole. The most extensively settled portions of the structures were groutjacked up several inches during the work.

The existing conditions proved to be too extreme for the standard NW drill casing, and numerous episodes of breakage, usually during extraction, were experienced. Further, the near proximity of the rock face was evidenced by the periodic withdrawal of bent casing. These problems again illustrated the importance of using a very tough casing system when working in such extreme conditions. In retrospect, employment of a thicker walled casing and perhaps even reverting to welding of the joints would have been a better option for this project. Although considerable expense is realized in the handling of a heavier casing, welding, and, even more, for the cutting during extraction, those costs could easily be offset by the cost of the lost casing, the rework involved, and the adverse effect on the final quality of the work.

The failure to recover all of the originally placed casing in bottom-up grouting presents a serious problem of quality assurance. Obviously, the length of the grout hole occupied by a lost casing has not received any grout. Thus, an immediately adjacent hole must be drilled to allow proper injection of that zone. Understanding where the lost casing is located, and thus the zone of soil that requires reworking, is not easy, however, and, in fact, often impossible. The occurrence of a break is often not known until the final piece of casing is withdrawn and found to be short. When more than one joint of casing is not recovered, there is no way to easily determine whether the shortage represents only one break or whether more than one may have occurred.

Sometimes, but not always, the incidence of a casing break is heard as a snapping sound during withdrawal and noted in an appropriate entry on the log. In such instances, if only one piece of casing is short upon complete withdrawal, the breakage depth is known and thus the ungrouted zone can be documented. In the case discussed here, however, it often happened that more than one length of casing was short upon final withdrawal, so it was not possible to ascertain whether all the casing was lost in a single break or whether there were other breaks not heard or noted. There were many breaks that were not heard at all, the only evidence of a problem being a shortage in the amount of withdrawn casing.

In such instances, the only way of ensuring that a hole has been properly grouted is to redrill and grout the entire depth of the hole. When a break occurs, continued grouting in the normal bottom up intervals deposits the grout at a higher elevation than the records would indicate and can be at a higher level than the grout is intended to be deposited, which was quite a problem on the project. Because the tunnel fill was sloped down to the track level on the ends, temporary benches had to be built to gain access. These often consisted of essentially uncompacted fill, and in many instances of casing breakage, such fill was inadvertently injected with the grout.

These problems had a terribly negative effect on the project's completion. The time lost resulted in the work extending into the winter, which presented a whole new set of problems, not the least of which involved the freezing of water and grout. And, of course, the much extended construction schedule resulted in a considerable increase in the owner's cost for inspection and administrative staff. Occasional bending or breakage of casing is simply one of the things that occurs in grouting and must be dwelt with. Extensive and continual breakage, such as happened on this project, is not acceptable, however. This experience with a contractor failing to correct such problems early on emphasizes that, in addition to requiring an adequate casing in the specifications, further authority must be vested in the owner's representative and/or engineer to order changes when a particular system has proven unreliable.

11.1.3.2 PILE-SUPPORTED STRUCTURES

The first significant use of compaction grouting on a pile-supported structure occurred in 1966 in Hollywood, California. Although it was done a long time ago, this particular job remains a landmark project and an outstanding example of advantageous use of the procedure for remediating failing end bearing piles. The 17 buildings of a luxury condominium complex, still under construction (Figure 11.54), were undergoing considerable settlement. The structures were sited on uncontrolled fill that had been formed over a period of many decades in a deep canyon. It was known to contain many large boulders in addition to an assortment of construction and other debris, along with a minor amount of organic waste.

The foundation consisted of reinforced concrete grade beams, supported by end-bearing, cast in drilled shaft piles that were to be socketed into bedrock. This was a high-profile project on which construction had been stopped for about a year because of a variety of problems.

FIGURE 11.54 Seventeen building development suffered settlement while under construction.

Upon resumption of work by the lender, it was found that differential settlement of several inches had occurred. The root of the problem was found to be that many of the piles were socketed into boulders rather than the underlying bedrock, as designed. The tip elevation of the piles varied in depth to more than 80 ft (24 m). Because these were end-bearing piles, it was considered imperative that no grout be deposited above the pile tip elevation, as it would act negatively and actually increase the load on the already failing piles.

A building permit for the remedial work was required, but because compaction grouting was not included in the prevailing building code, the method would have to be approved by a special Grading Consultants Board composed of some of the leading local engineers and geologists of the time. Nothing comparable had been done with compaction grouting up to that time, (and very little has been done since), which was of concern to the Board. It thus adopted several requirements for granting of a permit. First, two of the piles were to be grouted as proposed. Once this had been completed, 2 ft (0.6 m) diameter inspection shafts were to be drilled adjacent to them, so that a visual inspection could be made to confirm that the grout was under the tips where required and that none had migrated above the tip elevation. Assuming the work met this first requirement, the Board further called for the two piles to be load tested to 1.5 times their design capacity. Should the piles pass these tests, a permit would be granted, provided that the owner agreed to commission an annual inspection and survey to ensure that no further settlement had occurred.

Accordingly, prior to adoption of a full-scale remedial program, grouting was accomplished adjacent to the selected piles. The grout holes were spaced 6 ft (1.8 m) apart and placed in either a triangular pattern of three holes or a square pattern of four. Grout was injected from the top down, in vertical stages of 4 ft (1.2 m), starting at the level of the pile tip. Following the grout injection, the test borings were excavated, extending beyond the pile tips, thus allowing the required inspection. The photograph shown here as Figure 11.55 was taken from the bottom of one of the borings, which was about 80 ft (24 m) deep. As can be observed, grout was under the pile tip, as required, and extended above that elevation only less than 1 ft (0.3 m).

The core holes, which can be seen in the underlying grout, resulted from test specimens that were taken for strength evaluation. The grout strength as measured was more than twice the required 400 psi (2.7 MPa), to the surprise of everyone involved, thus helping to establish the fact that grout strength will always increase when the grout is exposed to elevated pressure under confinement for some period of time. This is believed to be the result of better wetting of the solids in the grout when it is under sustained pressure. The strength of grout specimens measured in their final place of deposition will virtually always be higher than the strength of the grout taken from the pump hopper. On this project, the strength of growth taken from the pump hopper was on the order of 450 psi (3.1 MPa), whereas that taken from grout under the piles approached 1000 psi (6.9 MPa).

The required load tests to 1.5 times the design capacity were then performed, showing no further settlement and even full recovery of the elastic deformation that had occurred, even though muddy conditions created by a rainstorm delayed removal of the load, illustrated here in Figure 11.56, by more than a week. In projects such as this, where the grout deposition zone must not extend above a given level, it is important to "seal off" the boundary by at least two stages of grout injected from the top down. This is especially important in such instances as this, where low-density soils and deleterious inclusions exist above the zone to be improved, as injected grout will naturally tend to grow into the weaker regions.

Here, a slow pumping rate was used and the pressure and grout quantity injected in the first stage were kept low, contributing to a limitation of the grout's travel. The second stage was then completed, using the maximum amount of

FIGURE 11.55 Inspection shaft shows grout under pile tip at depth of about 80 ft (24 m).

FIGURE 11.56 Lead weights being applied during one of the load tests.

grout that it would take. The work on this project was all completed working from the top down. Where there is considerable depth under the piles needing improvement, however, bottom-up procedures can be used once the upper boundaries have been established by at least two, or preferably three, top-down stages.

Four holes were placed in a 6 ft (1.8 m) square pattern around each pile for the production work. Virtually all of the holes were in confined or hard-to-reach locations. Those on the interior were in a subsurface parking garage subject to a height restriction of less than 8 ft (2.4 m), as shown in Figure 11.57, and those on the exterior were usually on sloping ground, as shown in Figure 11.58. Moreover, many of the holes were located in confined areas not accessible by vehicles or large equipment.

To penetrate to the pile tips for first-stage grouting, casing consisting of standard 2 in. (50 mm) Schedule 80 steel pipe was used. This was supplied in 5 ft (1.5 m) lengths and driven into place with a pneumatic hammer. To facilitate the driving and prevent breakage in the pipe joints, extra-long threads were placed on the pipe and straight-threaded electrical conduit couplings were used, so that the mating pipe sections could be tightly turned until the ends were in intimate contact with each another. In this manner, pounding stresses on the threads could be avoided. This subject is discussed further in Chapter 30.

FIGURE 11.57 Drilling in restricted headroom of subsurface parking garage.

FIGURE 11.58 Most of the exterior holes were on sloping ground with limited access.

In several instances, the casing encountered rocks, boulders, or other obstructions, which prevented penetration to the designated depth. When this occurred, an attempt was made to break or displace the obstruction by rotary percussive drilling, using a handheld rock drill equipped with standard 1 in. (25 mm) hex drill steel and a 2 in. (51 mm) four-bladed rock bit. This could be accomplished in a reasonable amount of time and worked satisfactorily so as to allow resumption of penetration by driving. Where such drilling failed, the previously placed casing was removed and the obstruction drilled through with an oversized bit. In a few instances, obstructions required relocation of the grout hole.

Once casings were set, the holes were advanced utilizing rotary drilling methods with water for the drill flush. The drilling was accomplished with lightweight man-handled drill rigs, as illustrated in Figure 11.57. These consisted of a pneumatic bore motor attached to a drill carriage supported by a light tubular frame. A pneumatically powered winch on the base was used to raise the drill head. Although the combined weight of the drill head and rod provided sufficient thrust for drilling in soil, it was insufficient for penetration of the frequently encountered boulders. To overcome this limitation, steel weights were hung on the drill carriage as required, or it was forced down by leverage from the overlying concrete slab when inside the structure. Standard "AW" drill rod was used and was fitted with a tri-bladed carbide insert bit. This equipment was much lighter than would be ideal for such difficult drilling; however, it was able to complete all of the required holes, many of which could not be accessed with heavier drill rigs.

The grout consisted of aggregate with a gradation falling within the acceptable envelope, as shown in Figure 2.7. It was mixed in a batch-type horizontal paddle mixer and contained about 12 percent cement. Because the work was spread out over such a wide area, the grout was manually batched off the bed of a truck, to which the pump was continuously connected. This allowed ready portability so that the pump could be quickly and easily moved without an excessive amount of effort or lost pumping time. The site was quite congested, with all the normal construction operations continuing during the grouting, as shown in Figure 11.59. This necessitated frequent moves, not only to facilitate the grouting, but to make way for other operations as well. The grout materials were replenished as needed with a front-end loader. Because of space restrictions, the truck often had to be moved for access.

FIGURE 11.59 The portable grout plant was batched off a flatbed truck to which it was hitched.

Upon completion of grout injection, the casings were generally withdrawn, utilizing either a skid-mounted pneumatically powered winch in combination with a mast, or a set of hydraulic jacks. In several instances the retrieved casing was severely bent. Where this occurred, perusal of the drilling records indicated penetration of large boulders, usually immediately before the tip elevation. A few instances of broken casing occurred, but this was quite limited. There were also a few instances in which the 80 kip combined force of the hydraulic jacks was insufficient to pull the casing, in which case it was cut off below grade and abandoned.

Depths to apparent bedrock were carefully logged for all of the holes. These were then plotted and compared to ensure that the indicated pattern was logical and geologically faithful. Regular inspections by the structural engineer and second-order optical surveys were made for the required ten years following completion and occupancy of the buildings. No structural distress or damage was found during that period, and the maximum vertical movement observed was less than 0.1 in. (2.5 mm).

A noteworthy example of the correction of settlement in a structure bearing on friction piles involves the repair of a courthouse in Westminster, California, in 1976. This project, also completed a long time ago, is especially pertinent in that the structure was so badly distressed that demolition had been previously ordered. Yet it has shown no further distress since the repair and remains in good condition to this day. As much as 5 in. (127 mm) of settlement had occurred in the westerly half of the West Orange County Municipal Courts Building. The structure is of contemporary design, basically a reinforced concrete rigid frame with both concrete and masonry infill walls. The floor and roof slabs are of reinforced concrete. Approximately two-thirds of the structure is one-story high with a slab-on-grade floor; the rest includes two structural floors above a basement. The foundation system consists of friction piles, cast in place, in driven corrugated steel shells.

The original geotechnical investigation provided for the piles to extend from the surface to depths of 25 to 40 ft (7.5 to 12 m). Following the original geotechnical investigation and prior to construction, the building was relocated approximately 100 ft (30 m) to the west on the same site. This was done arbitrarily, without further geotechnical exploration of the soils. At the time of construction, the groundwater level was approximately at the basement floor elevation and, in fact, an extensive underfloor drainage system was installed to prevent development of uplift pressures.

Distress was noted shortly after completion of construction. The building, nonetheless, remained in full service until a major reinforced concrete roof girder fractured eight years later. A forensic investigation disclosed 5 in. (127 mm) of differential settlement. Moreover, analysis concluded that the rigid structure was suffering from significant stress redistribution and that although some of the joints had already fractured, many structural members remained seriously overstressed. When the building had been moved from the originally planned location, the west half had inadvertently been placed over a wedge of very low density peat that existed only a few feet deeper than the pile tips, as illustrated in Figure 11.60. As part of the investigation, individual piles were disconnected from the grade beams and test loaded, in increments, up to twice their design capacity. Even with supporting these high imposed loads for five days, no significant movement of the piles was observed. It was concluded that the settlement involved not only the structure and its pile foundation, but also the entire block of soil above the peat layer.

Compaction grouting was thus performed in the peat layer underlying the pile tips. Two-inch (51 mm) ID casing was placed to the top of the layer, and grout injected in stages from the top down until good bearing soil was encountered. Interestingly, the building remained open and in full service during the work. The work was divided into five phases, each involving one courtroom and ancillary facilities. Each phase was completed prior to starting work on the next, so as to never have more than one courtroom out of service at any time. Access was through windows in the office areas adjacent to the courtrooms so that the work areas could be completely isolated from the continuing normal operations of the facility.

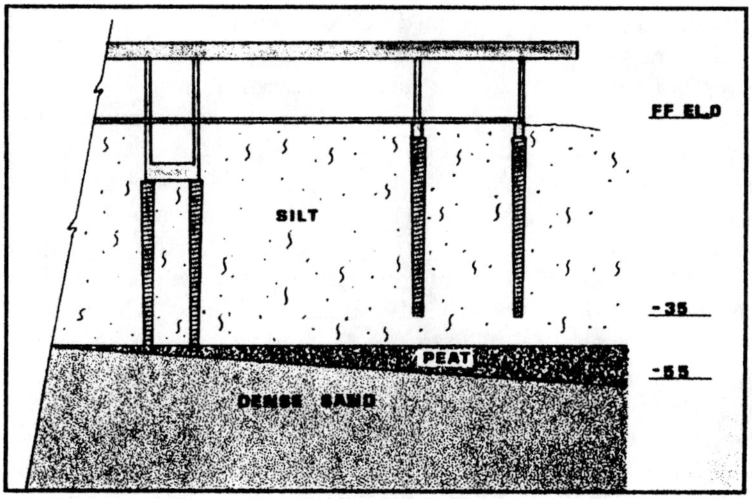

FIGURE 11.60 The pile tip elevation was just above a wedge of low-density peat.

In clearing the work areas of furniture and fixtures, two huge problems arose, one for the court clerk and another for the contractor. Each judge's chamber contained a full wall of bookshelves with a complete law library. Trustee labor was used to remove the furniture, but the court administrator was greatly concerned that if these workers were required to handle the books, they would get them mixed up and out of proper order. This problem was easily solved by the contractor, who said that the books could be left in the bookcase and would be protected. Simply wrapping the entire bookcase in protective plastic sheeting and a plywood cover provided more than adequate protection.

The contractor's problem involved the handling of very expensive built-in courtroom fixtures. These were of exquisite quality walnut veneer and included the judge's bench, the witness box, and the jury box, as well as miscellaneous divider rails, and so forth. Specifications required that these be removed and replaced following completion of the work, which would have required a substantial effort by highly skilled woodworkers. As with the bookcases, the solution was to first wrap them in plastic and then crate them in place with plywood, as shown in Figure 11.61. This greatly simplified the work and resulted in a large savings of both time and cost.

Grout holes were located around the piles on an 8 ft (2.4 m) spacing in either a square or a triangular pattern. The square pattern was used under continuous foundations, and the triangular arrangement was reserved for isolated piles. Drilling the holes, which were as deep as 65 ft (19.5 m), was done by handheld rotary wash equipment, as illustrated in Figure 10.16. Although this method was slightly slower than using a drill rig, the handheld equipment could be quickly moved and set up to facilitate work throughout the various areas, which often provided poor access or were otherwise restricted.

FIGURE 11.61 All surfaces were protected by plastic, and the main fixtures were crated in place with plywood.

The drill flush water was either recirculated or pumped off for disposal with the use of small mud tanks and pneumatic diaphragm pumps similar to those discussed in Chapter 10 and illustrated in Figure 10.31.

Grout batching, mixing, and pumping were performed outside the structure, and as much as 200 ft (60 m) of hose was used to reach the farthest holes. The average injection rate was slightly less than 1 ft^3 (28 L) per minute. Pumping pressures were on the order of 300 to 400 psi (20 to 28 bars), although pressures as great as 500 psi (34 bars) were not uncommon. Because of the very low density peat, rather large quantities of grout were injected, resulting in a large horizontal area of grout mass. The resulting uplift force was considerable, causing surface movement to be the most frequent criterion for terminating injection at any given stage.

Manual records of both the drilling and the grouting were maintained. The drilling logs enumerated the type and consistency of the material being penetrated, the drilling methods used, and any unusual conditions encountered, such as the presence of organic matter. In addition, frequent inspections of the drill cuttings were made, utilizing a small strainer to capture and separate organic material. The grout injection parameters were manually monitored at the header, including time, pumping rate, amount pumped at any particular time, and, of course, pressure. The injection foreman made a log entry at each change of the pumping rate and every pressure change of 10 psi (0.69 bars) or more. The records were collected immediately upon completion of any given stage for analysis and review. The project staff frequently reviewed these data so as to promptly address any changes in the injection program deemed appropriate.

An extensive multistation manometer system provided primary detection of any upward movement. It was backed up with traditional surveyor's instruments. In no instance was the total heave greater than about 1/8 in. (3 mm). Prior to initiation of the work, a system of vertical control points had been established on the flat concrete roof of the structure. It contained monuments placed in a grid generally corresponding to the pile layout and covering the entire building area. This system of control was surveyed weekly for some period of time before commencement of the repair work. The weekly readings continued throughout the repair period and for a few months thereafter. They were originally scheduled to continue on a monthly basis, but because virtually no elevation changes were detected in the months following completion, further surveys were eliminated.

A visual inspection in 1998, 22 years after the work was performed, found no evidence of distress or the major settlement damage that had occurred earlier. In fact, no occupant could be found who had been present when the work was done, and all expressed surprise at the idea that the building once had a problem so serious that it was considered for demolition.

11.1.4 Mitigation of the Settlement Trough in Tunneling

Another area in which compaction grouting has been found advantageous is in the prevention of surface damage resulting from soft ground tunneling. Such tunnels are commonly constructed within the protection of a tunnel "shield," which may best be compared to a giant cookie cutter that is forced through the earth. To minimize frictional resistance, an oversize cutter is placed at the head of the shield, which cuts a hole slightly larger than the body of the shield. This "tail shield" void then is able to propagate to the overlying surface.

Although the zone of disturbed soil starts directly over the tunnel, it extends up and out until it reaches the ground surface. This action can occur quickly, or it can continue over an extended time interval, depending on the depth of the tunnel and the particulars of the existing soil. The disturbed zone often encompasses buried utilities and can extend beyond the foundations of adjacent structures, so it can be damaging not only to the structures but also to the subsurface pipes and utilities within the zone of disturbance.

In 1980 a large evaluation program was conducted in the early stages of construction of the Bolton Hills tunnel, part of the Baltimore Metro project (Baker, Cording, and MacPherson, 1983). A trial application was made in a section of the actual tunnel where the adjacent structures were of minimal value. Compaction grouting was performed immediately over the tunnel shield as it progressed. This was accomplished through a single row of grout holes, which had been previously drilled from the surface of the street above, and extended to within 6 ft (1.8 m) of the tunnel crown, as illustrated in Figure 11.62.

A single stage of grout was injected immediately after the shield passed, in the tunnel below. The trial, which included extensive instrumentation and evaluation of the adjacent

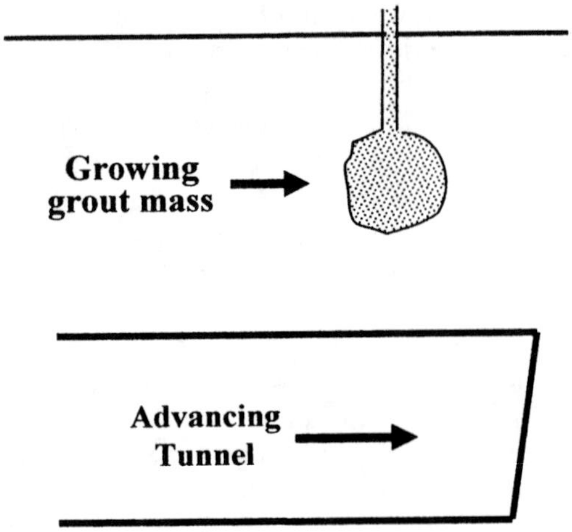

FIGURE 11.62 Grout is injected as the tunnel shield advances.

FIGURE 11.63 Grouting need not be messy. Here it is being performed adjacent to sensitive laboratory equipment.

soils, was so successful that the method was adopted for protection of many historical and irreplaceable structures on the rest of the tunnel route. Since that early experience, compaction grouting has been used in a similar manner on a number of other significant tunnels constructed in both North and South America. It was also specified for use in the construction of the Taipei Metro in Taipei, Taiwan, Republic of China.

As emphasized earlier, *compaction grouting need not be messy.* Figure 11.63 shows a grout hole casing within inches of sensitive laboratory equipment in the Spalding Research Laboratory at the California Institute of Technology in Pasadena, California. The grouting was being done to correct excessive vibration of the equipment that resulted from the insufficient density of a near surface layer of underlying granular soil. With reasonable care (note that the equipment is wrapped in plastic sheeting), compaction grouting can be successfully conducted in even the most sensitive environments. If numerous or deep holes were required, the equipment probably would have been further protected with plywood crating. As this job included only a single stage of grout in eight grout holes, the protective plastic sheeting, combined with special care on the part of the workers, was sufficient to prevent any damage.

11.2 PERMEATION GROUTING IN SOIL

As previously discussed, permeation grouting involves the filling of voids, and in the case of soil, this means the intergranular pore space. Obviously, such pore voids must be large enough to allow penetration of the particular grout. From a practical standpoint, this means that the procedure with common cement grout is limited to use in coarse-grained soils, primarily coarse sands and gravels although finer deposits can be treated with ultrafine cement or chemical solution grout. Permeation grouting is performed in soil for two fundamentally different purposes: one is to strengthen the soil, and the other is to block the flow of water through it. Of course, there are some applications in which both of these objectives must be attained.

11.2.1 The Basics

Where water control is required, the mechanism of improvement is the filling of all the voids. In this application, the strength of the grout is of relatively little importance as long as it completely fills the soil pore space and does not deteriorate or escape from the area in which it has been deposited. To completely block liquid flow, virtually 100 percent of all pore space must be filled. In strengthening operations, the primary mechanism of improvement is through adhesion, which is simply the gluing of the soil grains together. This results in a substantial increase in the cohesion properties of the soil, which will also increase the bearing strength. The degree of filling is relatively unimportant, however, as long as the grains are sufficiently bonded together. The degree of that bonding can vary widely, depending on the final strength requirement. Obviously, in applications that require both strengthening and water control, the pore space must be completely filled, and the strength of the grout becomes fundamental.

On the basis of a unit of improved ground, permeation grouting is generally much more costly than compaction grouting, so its use is typically limited to applications in which the primary requirement is either to block the flow of liquid or increase the cohesion of the soil, such as by solidification of a cohesionless running sand. Obviously, the amount of grout that must be injected, and thus cost of the work for water control, will be much greater than that required for most strengthening.

11.2.1.1 DEGREE OF SOLIDIFICATION

In water control essentially 100 percent of the pore void system must be filled, but in strengthening operations a lesser degree of filling is acceptable. The amount will vary, however, depending on the individual application requirements. Strengthening of granular soils is most often done to solidify running soils as an aid to the support of excavations. In some cases irregularity of the solidified surface is of little importance, as in the case illustrated in Figure 11.64 (top). In such instances, especially where the excavated wall can be laid back as shown, even a few windows in the solidification can be tolerated, so the completeness of the solidification is not of great importance.

However, where the grouted mass must be trimmed to a neat excavation line, such as to enable the placing of concrete against it, as shown in Figure 11.64 (bottom), a much more thorough filling of the pore space is required. Soil solidification is relatively expensive, and it is thus usually preferable to minimize the amount of

FIGURE 11.64 Uniformity of the grouted face can be of little importance (top). A flat excavation wall requires some precision of the grout injection and considerable overcut (bottom).

grout injected, and thus the volume of soil solidified. For this reason, the grouting is generally used only as part of an overall system where much of the load is supported by other means. Where the grouted mass is to serve as the sole support, a properly designed gravity structure is required, and this, of course, will result in significantly more grouted ground.

11.2.1.2 INJECTION PROBE LAYOUT

The solidification requirements, as discussed in the preceding paragraphs, will dictate both the probe layout and the amount of grout required. Where a flush trimmed surface is required on a grouted mass, considerable excess injection is required and much of the solidified ground is trimmed and wasted. Where only retention of the cut is required, the probes can be spaced so as to provide solidified columns that just touch each other, as shown at the top of Figure 11.64.

The spacing of the probes will also be dependent on the permeability of the soil and the penetrability of the particular grout to be used. It will also be influenced by the uniformity of the soil to be treated. Generally, wider spacing can be used in soils of uniform permeability than in those with widely varying properties. Because the volume of pore space in the different horizons of a nonhomogeneous soil will vary, the radius of travel or penetration of the grout will likewise vary, and this difference will obviously increase with the distance of travel.

Injection probe spacing usually falls within a range of 2 to 4 ft (0.6 to 1.2 m) or occasionally more. Experience shows, however, that closer spacing allows greater control of the work and generally results in a better finished product. It can also provide a significant reduction in the quantity of grout required. This is particularly so when the injection is made into soils that contain even trace amounts of silt or other fines. These tend to be picked up by the grout as it propagates, which reduces its penetrability and thus limits the effective radius.

11.2.1.3 REQUIRED GROUT QUANTITY

As discussed earlier, the quantity of grout used for a particular application depends on the thoroughness required of the work and the volume of the pore void system of the particular soil to be improved. The volume of the voids can vary greatly at a given density, depending on both the shape of the grains and their moisture content. Other than near the ground surface in an arid region, the surface of a coarse-grained sand above the water table is usually humid, whereas that of a fine sand can be moist to wet. A standard terminology to describe the moisture condition of sand has been established and is provided in Table 11.4

The volume of the voids is known as the *porosity* and stated as a percentage of the total volume of the particular soil. The porosity is significantly affected by the density as well as the grain shape and structure of packing of a given deposit. It can vary widely, as shown in Table 11.5. Porosity will be the greatest for clean, uniform, coarse grained materials and will be reduced as the materials become well graded and/or finer. Understanding a soil's porosity is fundamental to determining the amount of grout that will be required to treat a given volume of it. Keep in mind, however, that precise porosity is dependent on the grain shape and structure of the deposit, so the numbers in Table 11.5 should serve as guidelines only.

TABLE 11.4 Degree of Saturation of Sand at Various Moisture Contents

Description	% of Saturation
Dry	0
Humid	1–25
Damp	25–50
Moist	50–75
Wet	75–99
Saturated	100

TABLE 11.5 Dry Unit Weight and Porosity of Various Granular Soils

	POROSITY %		DRY UNIT WEIGHT, LB/FT^3 (KG/M^3)	
Soil Type	**Loose**	**Dense**	**Minimum**	**Maximum**
Uniform coarse sand	50	32	82 (1312)	115 (1840)
Uniform fine-to-medium sand	50	29	83 (1328)	118 (1888)
Well-graded sand	49	17	85 (1360)	138 (2208)
Well-graded silty sand and gravel	46	12	89 (1421)	145 (2320)

The amount of grouted ground that will be wasted as a result of trimming must also be considered. When a smooth flush cut is required, as shown at the bottom of Figure 11.64, the amount will be fairly large. This becomes a particularly significant issue when injections cannot be made normal to the plane of the surface because of the interference of overlying structures, as illustrated in Figure 11.65. In such instances it is necessary to inject variable amounts of grout to saturate from the area the hole to the final surface trim line. The overage of grout can be minimized by making injections at many different hole inclinations, as illustrated in Figure 11.66. There is a cost to this procedure, however, so the relative cost of additional wasted grout must be compared with that of the multiple injection planes.

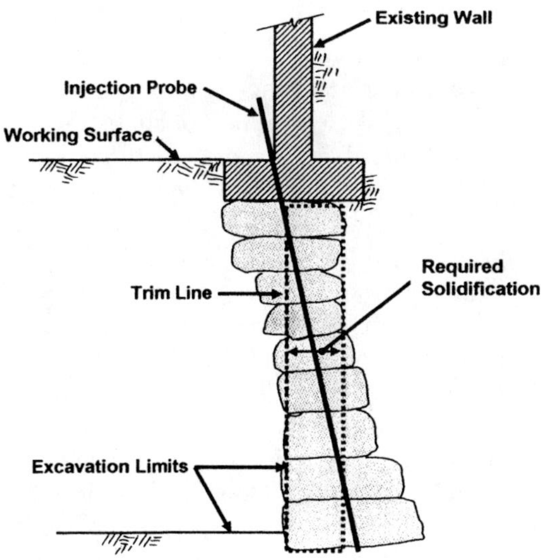

FIGURE 11.65 Excessive grout must be injected in inclined holes when the final trim line is under a foundation.

11.2.2 Injection Holes/Probes

Permeation grouting can be performed from the top down, from the bottom up, or at selected intervals with the use of a sleeve port pipe. Where the depth of treatment does not exceed 30 ft (9 m), the top-down method has been found to

FIGURE 11.66 Injection casings at many different angles to pinpoint grout deposition.

offer several advantages, which are discussed in the following paragraphs. Bottom-up staging should be only used in deposits that decrease in permeability with depth. The reason is that even though the injection pipe has been raised, the grout can continue to flow to the lower previously completed stages if they are in soil that takes the grout more readily. Where the soil permeability varies greatly, special care must be taken to ensure that the grout is actually deposited at the planned location. This can be accomplished by use of a special drive needle that positively restricts the location of grout deposition, or through the use of sleeve port pipes whereby the locations of grout deposition can be selectively chosen.

11.2.2.1 TOP DOWN WITH SPECIAL DRIVE NEEDLES

Where applicable, special drive needles have proven to offer the fastest and most economical means of downstage injection. Proper needles may not be available commercially, however, and may have to be specially fabricated; details are provided in Chapter 30. For grouting to depths of less than about 6 ft (1.8 m), the needles can be simply driven by handheld pneumatic drivers. Where greater depths are involved, however, an appropriate driver is best mounted on a mast, as illustrated in Figure 11.67, to allow raising of the driver to the required height and for guidance of the driver and thus injection needle, in the proper trajectory.

The needles are constructed of high-quality seamless tubing, but even so, constant driving will eventually result in a fatigue failure. To minimize such a possibility, it is best to use a driver that works on the basis of relative high amplitude and low frequency. Although most pneumatic hammers operate at a very high frequency, many of those based on hydraulic power transmission work more slowly and are thus more suitable. There are also readily available many standard hydraulic hammers that are equipped with adjustable frequency controllers. Because these probes take a great deal of abuse, they are best constructed in one piece without joints. The length is thus limited by the length of high-quality tubing that can be obtained, which is typically on the order of 30 ft (9 m).

FIGURE 11.67 Mast-mounted driver for placing injection needles.

In operation, a metered amount of grout is injected at each descended stage. When injection is near the ground surface, it is best to stop pumping while the needle is being driven deeper. However, once it has penetrated sufficiently deep to develop a good seal, it can be further advanced without the necessity of stopping the pump. This greatly increases productivity, and because the grout can flow only from the recess provided for it, close control of the zone of improvement is possible. The driving of needles from the top down works well in most soils as long as they are relatively free of rocks or other hard obstructions. It does not work well in gravelly soils, however.

11.2.2.2 TOP DOWN BY JETTING

For injections into clean sands that are not deeper than about 10 ft (3 m), the jetting of grout from an open-ended pipe is a reasonable approach. In practice, the jet pipe is typically of either 1/2 or 3/4 in. (13 or 19 mm) standard Schedule 40 steel pipe. The bottom end is usually cut on a bevel to lead the way and assist in penetration. With the grout flowing, the pipe is pushed and guided by hand as it works its way into the ground. At regular intervals, typically every 1 ft (0.3 m), the insertion pauses until the preset amount of grout has been injected. This continues until the planned depth has been reached.

Hand jetting takes some skill, and to be successful, must be performed with close control of the grout quantities and depths of injection. The grout flow must be rapid enough to erode ahead of the pipe tip, and, of course, the pipe must be guided so as to maintain the planned trajectory and resulting area of solidification. Hand jetting is thus best limited to soils that are sufficiently permeable to allow rapid penetration of the grout. The pumping rate must be controlled so as to prevent the grout from flowing up and around the injection pipe to the surface. Although it is a simple process, it is not a suitable injection method for the untrained. Considerable experience is required to ensure that proper amounts of grout will be accurately deposited at the intended stage locations.

11.2.2.3 BOTTOM-UP STAGING

Aside from hand jetting, the simplest method of injection is from the bottom up through predriven steel pipes. These can be placed, absent the need for any special equipment, with relatively little skill required. The injection pipes are most frequently standard 3/4 in. (19 mm) Schedule 40 steel pipe. A loose-fitting plug, often consisting of a rivet or a round headed bolt, is typically placed in the bottom to keep dirt from entering as the pipe is driven into the ground. The temporary plug is usually pushed out by the grout as injection starts. The pipe is then pulled out as injection continues in the intended stages.

At reasonably shallow depths and in clean soils free of rocks, the pipes can be placed reasonably accurately. However, if rocks are present or great depths must be reached, the pipes can easily bend and/or drift off course. The standard material of which these pipes are generally made does not have the strength of the special steel tubing used to fabricate the drive needles previously discussed. This limitation can be lessened through the use of a larger pipe or one with a heavier wall, such as Schedule 80 extra-heavy pipe.

Where the depth of treatment exceeds the length of available pipe, typically 21 ft (6.3 m), coupling of additional lengths will be required. Standard pipe couplings are supplied with a taper thread so that the butt ends of the pipe do not normally come in contact. Thus, the driving force is transmitted through the threads, which form a weak link and often break in their roots. This problem can be greatly minimized by use of electrical conduit couplings that are straight threaded. In fact, electrical conduit, which is similar to pipe and is furnished in 10 ft (3 m) lengths with a straight thread, can itself be used advantageously.

Withdrawal of the pipes can be accomplished with hydraulic jacks or by any other suitable means, such as with a forklift. Bottom up injection is a simple method, but it is often not very advantageous because of the frequent inability to keep the pipe on course, especially when rocks or other obstructions may be present. Its use should therefore be limited to depths no greater than about 20 ft (6 m) and only in soils that are reasonably free of rocks.

11.2.2.4 SLEEVE PORT PIPES

By far the most accurate way to place grout is with the use of a sleeve port pipe, also referred to as tube-à-manchette. This is simply a tube that has small port holes drilled at regularly spaced intervals, typically within a range of 1 to 2 ft (0.3 to 0.6 m). Such tubes most commonly consist of 1-1/4 or 1-1/2 in. (32 or 38 mm) plastic pipe, although they can be made from virtually any size or type of pipe. The ports, which are usually 1/4 to 3/8 in. (6 to 9 mm) in diameter, are regularly spaced so as to provide the desired grout stage length. They are typically covered with a short piece of thin rubber sleeve on the outside, to prevent any soil or encasement grout from entering the tube. Each end of a sleeve should be secured to the pipe, which is often accomplished by a few turns of tape.

Some high-quality sleeve port pipes are made with depressions, which have been milled from the outside of the pipe to contain the sleeve. This results in the sleeves being flush with the outside wall of the pipe, which is then of a uniform diameter. Although this feature is admirable, it is not really needed for most applications and it does run the cost up quite appreciably. Such a sleeve port pipe is shown in Figure 11.68, wherein the rubber sleeve has been folded back so the depression and the port holes can be readily viewed.

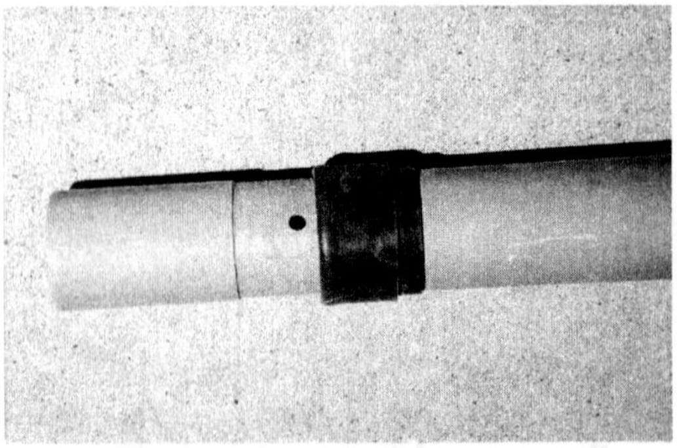

FIGURE 11.68 Closeup view of a port with the rubber sleeve folded back.

In use, the sleeve port pipe is most often placed in an oversize drilled hole. A weak grout, which is typically of cement-bentonite composition, is used to fill the annular space between the tube and the wall of the hole. For grout injection, a double packer is used to isolate any one of the ports for injection. Injection pressure breaks the weak confining grout and expands the rubber sleeve so that the grout is forced into the formation at that particular location. The packer can be placed over any selected port, and the different ports can be injected in any desired order. With sufficient pressure, the same ports can often be injected more than once. In soils that do not run excessively, steel sleeve port pipes can be driven or pushed into place so as to be in tight contact with the surrounding soil. When so used, the rubber sleeve is not required, as confining grout is not used.

Because only one grout stage is affected at any given time when sleeve port pipes are used, different amounts of grout can be injected into the various stages. In fact, completely different types of grout can be used if so desired. This is especially advantageous when treating soils that contain strata of significantly different void contents, as the exact grout dosage can be varied as appropriate.

Sleeve port pipes offer the only technique to reliably ensure the proper placement location of the grout when the injection is deeper than about 30 ft (9 m) from the surface or in rocky or gravelly soils. Because the oversized holes are drilled, their accuracy can be better controlled. In particularly sensitive applications where precise location of the grout deposition is important, it is possible to use holes large enough to allow the traversing of an instrument to log an actual hole bearing. As with any placement, sleeve port pipes placed in holes that require grouting of the annulus are usually limited to a downward vertical or inclined arrangement. Tight-fitting steel pipes can be placed in virtually any desired configuration.

11.2.3 Monitoring and Control of Injection

Permeation grouts all behave as fluids when under pressure in the ground. Should the injection rate exceed the rate at which the grout can permeate the soil, hydraulic fracturing of the formation will likely result. It is thus important to continuously monitor the grout pressure behavior to ensure that an excessive pumping rate is not causing such fracturing, which will always be indicated by a sudden drop in pressure. Although an occasional fracture can be tolerated, such an occurrence should not be frequent. To establish the optimal pumping rate for a given formation, it is common practice to slowly increase the pumping rate at the beginning of a job until the sudden pressure drop of a fracture signature is observed. A pumping rate for the remaining production injection is then set slightly lower than that at which the fracture occurred. Should a further fracture be experienced later in the work, the pumping rate should again be appropriately lowered.

Unlike compaction grouting, wherein many parameters require constant observation, permeation grouting involves a much simpler process for monitoring. Obviously, the proper quantity of grout must be placed in each stage in a controlled manner. This requires only that the volume and pressure be suitably controlled. There are but two adverse events that are likely to occur during such injection in soil. The first is hydraulic fracturing of the formation, which, as discussed earlier, is readily controlled and not likely to get out of hand with reasonable observation and control of the pumping parameters.

The other is not so innocent, and that is leakage into a substructure or a sudden explosive fracturing of the formation or an adjacent substructure. As with any grouting, caution is in order when there are pipes, conduits, or other substructure in the area of injection. There is always a risk of the more fluid permeation grouts finding their way into such a structure. Because they are so flowable, however, they can usually be diluted and flushed out as long as the leakage is discovered in a timely manner. Observation of the normal flow of all potentially affected conduits should thus be made from time to time.

When grouting in rock, a sudden pressure spike can result in severe damage, but such incidents are usually absorbed as minor hydraulic fractures when grouting in soil. These are, of course, readily corrected by immediately reducing the pumping rate while allowing the injection to continue. When grouting in soil, explosive fractures usually occur only at pumping rates and pressures that are significantly higher than reasonable. The development of such damaging events is readily observed by noting the pressure behavior, so they should not be of great concern as long as the pressure is being continuously monitored and the pump is promptly stopped on the occasion of any sudden changes.

11.2.4 Permeation Grouting Examples

There are many different requirements that can be fulfilled with permeation grout injection into soil. Obviously, the type of grout material and the procedures used for its injection will vary according to the requirements of the individual application. Both chemical solution and ultrafine cement grouts are commonly used. Common cement is not frequently used, as the grains are too large to penetrate any but the coarsest sand and gravel. Minor amounts of work have been accomplished with the hydrophobic urethane resin foams as well, but their high cost and inability to control the expansive pressure eliminates consideration of their use on most work.

11.2.4.1 SHORT-TERM STRENGTHENING

The largest single use of permeation grouting in soils is in short-term strengthening, which is most often used as a tool of construction during excavation to aid in the support of low-cohesion soils. Accordingly, the strength and durability of the grouted mass is pertinent only until the completion of construction. This is an important factor, as most chemical solution grouts do not provide long-term or permanent strength, as discussed in Chapter 7. Although it is possible to design and inject a self-supporting mass of grouted soil, it is usually more costly than providing supplementary support. Thus, the solidification is most often intended to provide cohesion only and is supplemented with more conventional shoring.

Historically, the solidification of sandy soils was done almost exclusively with chemical solution grouts, and, as discussed in Chapter 7, about the highest reliable strength that can be obtained with these materials is on the order of 100 psi (6.9 bars). Much higher strengths may be indicated in testing, but these strength levels will probably not be reliable, because the indicated strength is highly dependent on both the loading rate and the duration of the load. In addition, the moisture content of the specimen is fundamental to the indicated strength. Dry specimens will indicate significantly greater strengths than those that are cured under water.

As shown in Figure 7.8, chemical solution grouted masses are extremely sensitive to creep, and the load a mass will support continuously is but a small proportion of that indicated by normal test loading rates. It is thus imperative that these factors be considered and that the test method and, most important, the loading rate be mandated when specifying strength requirements. The strength attributes of different chemical solution grouts, and even of different mixtures of the same grout, can vary widely, so it is important to ensure uniformity of the composition. The specimens used for strength evaluation must obviously be prepared with the same grout type and mixture that is being used for the production work.

Ultrafine cement suspension grouts are experiencing ever-expanding use in the solidification of soil. Fortunately, these grouts are not subject to the strength variation extremes of their chemical base cousins. Further, the cementitious grouts do not loose strength with time, so they are appropriate for long-term installations. Moreover, solidified masses of much higher strength can be obtained. In fact, when injection is into clean, well-graded sand, unconfined compressive strengths of 2000 psi (13.8 MPa) or more are usually obtainable.

One must carefully consider the strength levels to be used, however, especially where the application will require trimming back to a given plane. Although high strength is desirable from a structural standpoint, it can be a hindrance for those who must later excavate and trim it. Furthermore, grouting in connection with temporary shoring is often performed in areas that may be excavated in the future. In such cases, dealing with a high-strength mass can be difficult and expensive.

Perhaps the largest single area of permeation grouting in construction is in increasing the cohesion of in situ soils prior to the excavation of soft ground tunnels. Huge quantities of grout have been used in such work throughout the world. The most common objective is to prevent soil runs during excavation, as well as to improve support for any structures lying adjacent to, or over, the tunnel alignment. Historically, the most commonly used materials for this type of work have been chemical solution grouts based on sodium silicate. As the grout improvement need last only long enough for the construction to take place, long-term durability of the treated formation is usually not a concern.

This type of work can be performed either from the tunnel heading sequentially, using fan-shaped arrays of grout pipes, as shown at the left of Figure 11.69, or from the surface, as illustrated on the right. Where practical, working from the surface is preferred because it can be done well ahead of the tunnel advance and thus not inter-

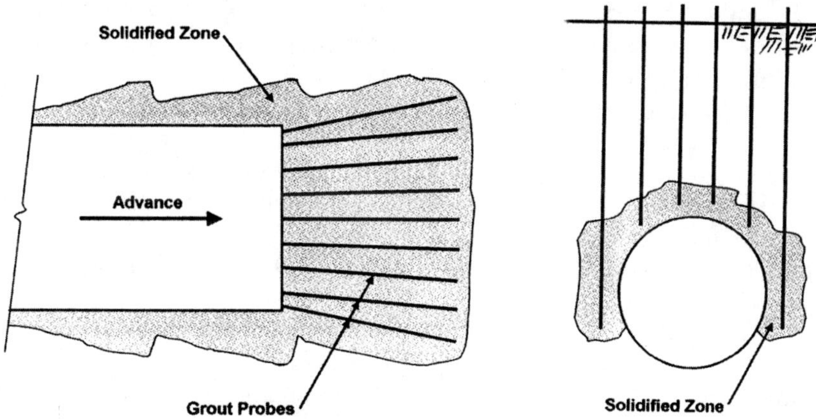

FIGURE 11.69 Grouting in sequential fan shaped hole arrays from the interior (left). Injection probes and grouted mass shown when working from the surface (right).

fere with the excavation and other tunneling operations. Although some cohesion is imparted to the soil within hours of injection, substantial strength usually does not develop for a longer period of time. Thus, a time lag is required between grout injection and the subsequent excavation, which is easy to provide when working from the top.

A typical surface application is shown in Figure 11.70. Here, the grouting is being accomplished from the bottom up and a set of withdrawal jacks is in place to raise the injection pipe as required. The grout being used is a sodium silicate base with a dibasic ester reactant system, as earlier discussed in Chapter 7. These components are supplied from separate compartments of a nearby tanker trailer. A pair of Moyno pumps, driven by the same pneumatic motor and sized so as to provide proper proportioning of the two components, are suspended from the underside of the tank.

FIGURE 11.70 Bottom-up grouting in progress for a transit tunnel.

In many instances, injection from the surface may not be feasible because of excessive tunnel depth or limitations of surface access. In these situations, where the work must proceed from the tunnel heading, the selection of a grout system that gains strength rapidly is essential. Such work is usually done on alternate shifts, with the grouting performed during a night shift so that excavation can take place the next day. Because the grout is literally only hours old when the excavation progresses, unless a rapid-strengthening formulation is used, thicker sections may be required. In either case, the cost will be substantially increased. Moreover, because the grout is being injected in a fan shape, there will be a greater amount of solidified ground that must be excavated and wasted.

Permeation grouting for temporary support during construction has also been used on a wide variety of excavations and structures. As part of the expansion of Los Angeles International Airport, it was necessary to underpin an existing concrete retaining wall, which was founded on wind-deposited fine dune sand. The wall was to be undermined and extended by a depth of 8 ft (2.4 m). A sodium silicate base grout was injected from the top down with the use of special injection needles that were advanced in 1 ft (0.3 m) increments.

Two rows of holes were used, with the lower portion of the grouted mass being injected from holes inclined about 10 degrees from vertical, and the upper portion from holes inclined at about 20 degrees. The quantity of grout injected at each level was adjusted so as to provide the required lateral saturation needed, considering the inclination of the injection needle. When injection must be performed by probes that are greatly inclined relative to the final plane of the excavated surface, much more grout must be used and a great deal of the resulting solidified mass will have to be removed to establish the excavation line, as illustrated in Figure 11.65.

The existing sands were improved to an unconfined compressive strength of about 150 psi (1 MPa) using the G.V.S. grout formulation shown in Table 7.5. The sodium silicate base material was supplied from a tanker truck that pulled the trailer-mounted grout plant. It provided for mixing of both the base material and the reactant system in individual batch-mixing tanks. These were then proportioned in the proper ratios with separate calibrated pumps, which fed the materials into a flash mixer, as described in Chapter 7 and illustrated in Figure 7.4. Excavation, which proceeded about two weeks after the grouting, required pneumatic tools because of the strength of the mass.

The rehabilitation of a failing side-hill structure supporting a roadway required the excavation of an immediately adjacent vertical face. The structure was founded on cohesionless dune sand. Because normal support was not possible, because of the existing steep downslope, permeation grouting was performed to solidify the soil under the structure. To make the shear vertical cut, which varied in height from 6 to 24 ft (1.8 to 7.2 m), it was necessary to actually remove a portion of the footing from the structure's supporting pillars, as shown in Figure 11.71.

A sodium silicate base permeation grout was injected into the wind-deposited fine sand. This was accomplished through injection of a single row of grout holes 15 in. (381 mm) behind the face of the structure and spaced 30 in.

FIGURE 11.71 A portion of the existing footing was removed, and pneumatic tools were required to cut the solidified mass.

(0.76 m) apart. Where the cut exceeded a 9 ft (2.7 m) height, a second row of holes was provided 3 ft (0.9 m) behind the primary row, so as to result in a solidified wall about 6 ft (1.8 m) thick. A quantity of grout that would solidify a theoretical 36 in. (0.9 m) diameter column, based on the soil's porosity, was injected into each hole in an alternate primary/secondary split spacing order. Holes in the second row were not grouted until the first row was complete.

Because the sand became less permeable with depth, uncontrolled downward loss of grout was not a problem. The injection pipes were thus driven to the final depth prior to injection, which was made in 1 ft (0.3 m) stages from the bottom up. As illustrated in Figure 11.71, pneumatic tools were required to trim back the solidified mass to the specified alignment. The flush cut was 240 ft (72 m) long and varied in height from 6 to 24 ft (2 to 7.2 m).

Seismic retrofit of the 70-year-old San Francisco Civic Center required undermining existing foundations by as much as 9 ft (2.7 m). The foundation soils were basically round-grained beach sands, with virtually no cohesion. The ground, which was to remain under the foundation elements, was solidified with an ultrafine cement grout. As shown in Figure 11.72, this allowed for otherwise unsupported excavation, as required for the new extended foundations. The specifications required the grouted mass to obtain a minimum unconfined compressive strength of 100 psi (0.69 MPa) and stipulated, "The grouted sands will extend from the bearing area beneath the footings to be stabilized, extending on a 2 to 1 H:V line, to 2 ft (0.6 m) below the bottom of the adjacent underpinning excavations." The means and methods to accomplish this were otherwise left to the contractor.

The grouting contractor's original plan was to place rows of primary grout holes at a spacing of 6 ft (1.8 m). After these were injected, the spacing was to be split to 3 ft (0.9 m) and secondary injections made. Based on a calculated soil porosity of 27 percent, an injection quantity of grout sufficient to form a 4 ft (1.2 m) diameter column was to be injected. To obtain the required strength, ultrafine cement grout would be mixed with a water : cement ratio of 6, by weight, of the cement. No other ingredients were to be included except for a high-range water reducer supplied by the manufacturer of the cement as an essential part of any mix for which it is used.

FIGURE 11.72 The existing wall was undercut to extend its depth.

A test section, was made and excavated to evaluate the propriety of these parameters. Insufficient overlapping of the grout columns was found. By comparing the actual diameter of the grout columns with the quantity of grout injected, the actual porosity was found to be about 35 percent. Moreover, the unconfined compressive strength was substantially higher than required and several of the specimens tested broke at well over 1100 psi (7.6 MPa). Although this was an added benefit from the standpoint of the engineer, it was a huge problem for the general contractor who had to perform the excavation. Of some significance was that the strength of the grouted mass was much greater near the injection hole than at some distance from it, indicating that the ultrafine particles were being filtered out as the distance of grout travel increased.

Hole spacing was then reduced to 2 ft (0.6 m) and the grout quantity injected, which was calculated to form a 3 ft (0.9 m) column, based on a porosity of 35 percent, and the water : cement ratio was also raised to 8. This was found to result in complete interlocking of the individual injections. Further, the strength was reduced to about 300 psi (2 MPa); this was still much greater than required, but it provided a good safety factor for the contractor and reduced the effort required in the following excavation.

The grout was mixed in a high-speed vertical paddle mixer, from which it was transferred to one of two adjacent agitators. Injection was from the bottom up through preplaced 3/4 in. (19 mm) injection pipes. These had been driven into place with a temporary plug in the leading end. The pipe was raised 1 ft (0.3 m) and the plug easily knocked off immediately prior to injection. In all, more than 120,000 gal (453,600 L) of grout was used. The ultrafine cement grout provided another significant advantage on the very congested construction site. Because it was supplied in paper bags and required only water to prepare, no space was required for storage tanks or tank trucks, which would be needed for chemical solution grouts.

When the terminal building at the Kansas City International Airport was originally built, the excavated space behind the basement retaining walls was backfilled with a very permeable, cohesionless granular fill sand. In 1997 a contractor building four access tunnels, connecting to the building, encountered this sand fill. It caved into his excavation, exposing utilities and undermining adjacent pavement. To mitigate the problem, ultrafine cement grout was used to solidify the cut face of the required excavations. The grout was injected from the top down through open-end jet pipes that were marked at 1 ft (0.3 m) intervals. The pipes were positioned at a 15 degree angle toward the proposed excavation. They were jetted with grout into the formation, with a pause at each 1 ft (0.3 m) interval, as required to place the predetermined amount of grout.

For most of the work, the holes were placed on 3 ft (0.9 m) centers. The calculated amount of grout to form a 4 ft (1.2 m) column was injected. Where the excavation was greater than 10 ft (3 m) deep, a second row of holes was provided, split spaced, 3 ft (0.9 m) behind the initial row. The grout was mixed in a high-shear colloidal mixer and transferred into an agitator tank that supplied a 2L8 Moyno pump, which was set for an injection rate of 2.6 ft^3 (74 L) of grout per minute.

Upon excavation, the sand fell freely from the injected column-shaped masses (Figure 11.73). They withstood the elements, including several cycles of freeze and thaw, for a period of several months. Both vibration and impact forces were created during demolition of the adjacent concrete wall, but had no detrimental effect on the solidified mass. The spacing of the injections would have been excessive if trimming

FIGURE 11.73 Loose backfill sand fell freely from grouted columns.

to a smooth plane had been required. However, it was entirely satisfactory being laid back at the 15 degree angle and needing only to prevent excessive movement of the retained sand.

In a most unusual project, permeation grouting was employed to arrest the retreat of a high bluff overlooking the Pacific Ocean near Dana Point, California, in 2003. Over time, the bluff retreated under the natural forces of wind and water erosion until it had taken a large area of the yard of an extremely expensive home located at the top. The owner retained a geotechnical engineer to devise a method to stop further erosion of the fine sand formation that existed. A permeation grouting program was adopted to solidify a section reaching from the bluff face back about 30 ft (9 m). This, of course, required treatment all the way to the surface of the very steep slope shown in the top photo of Figure 11.74.

Obviously, a long-term fix was desired, so the grout had to provide permanent durability. Ultrafine cement was selected; however, it was thought that it would not be sufficiently penetrable to treat the near surface soils. Accordingly, the plan was to solidify the main mass with ultrafine cement grout, but to use a sodium silicate solution grout, which was thought to be more penetrable, for the near surface zones. Grout holes were established on a 4 ft (1.2 m) offset grid on the slope as well as on the remaining somewhat flatter portion on top, as shown in the bottom photo of Figure 11.74.

One-inch (25mm) ID sleeve port pipes were placed in the holes, and grout injection was made at each port, starting from the bottom and working up. In the early stages of the work the original plan was followed, using the solution grout in the upper stages. As work progressed, however, it was found the cement grout could be used all the way to the surface at about the same pumping rate and pressure as the solution grout. As it provided greater strength and was permanent, the rest of the work was completed with the cement grout.

FIGURE 11.74 The steep slope required solidification all the way to the surface (top). Injection points were on a 4 ft (1.2 m) grid (bottom).

Permeation grouting was a crucial part of a project to increase the flow capacity of an existing 400 ft (120 m) long drainage tunnel in Oakland, California, in 2001. The tunnel served a large open channel known as Glen Echo Creek, which, like many such channels in California, was all but empty during the summer months but filled quickly with the winter rain. It formed a bottleneck, was not of sufficient capacity for rapid rainstorms, and studies indicated that flooding of the congested central city location it traversed would occur with 25-year storms.

Most of the length of the tunnel had apparently been conventionally mined and concrete lined with an arched cross section, as illustrated at the left of Figure 11.75. On the downstream end, however, was a 100 ft (30 m) extension that consisted of 84 in. (2.1 m) diameter corrugated metal pipe, as shown on the right side of Figure 11.75. This had apparently been placed by traditional open-cut methods. Soils in the bottom portion of the tunnel were free draining, consisting of fine to medium sand with some gravel. Much of the overlying soils, however, were cohesive silty clays.

To sufficiently increase capacity, the main tunnel section was being lowered by about 3 ft (0.9 m), as shown in Figure 11.75. The corrugated metal pipe (CMP) section was to be completely removed and replaced with a new larger-diameter reinforced shotcrete lining, as illustrated in Figure 11.75. Although all the work in both sections of the tunnel was to be completed in alternating sections of 6 ft (1.8 m), it was necessary to solidify the existing low-cohesion soils prior to any excavation. Further, control of groundwater in the lower portions of the cross section was required.

As with most tunnels, significant zones of weak soil and voids existed around the lining. Thus, contact grouting with a cementitious suspension grout was first performed. This was followed by the driving of specially made steel sleeve port pipes for the permeation injection. In the conventionally constructed section, they were limited to the bottom portion of the section, as illustrated in Figure 11.75, and the injection pipes were installed in a fan configuration at 6 ft (1.8 m) intervals along the longitudinal axis. In the CMP section however, the solidification had to be done for the full circular section, which was accomplished with regularly spaced radial pipes.

Initially, an injection of cement suspension grout was made to perform a basic tightening of the soil. This was followed with a solution grout composed of a sodium silicate base and dibasic ester reactant. The initial cementitious injection reduced the amount of the more costly solution grout required and provided greater strength to the areas penetrated. The subsequent injection of solution grout was able to permeate any remaining pore space, which provided additional strength and, in the case of the lower portions of the cross section, greater resistance to intrusion of groundwater.

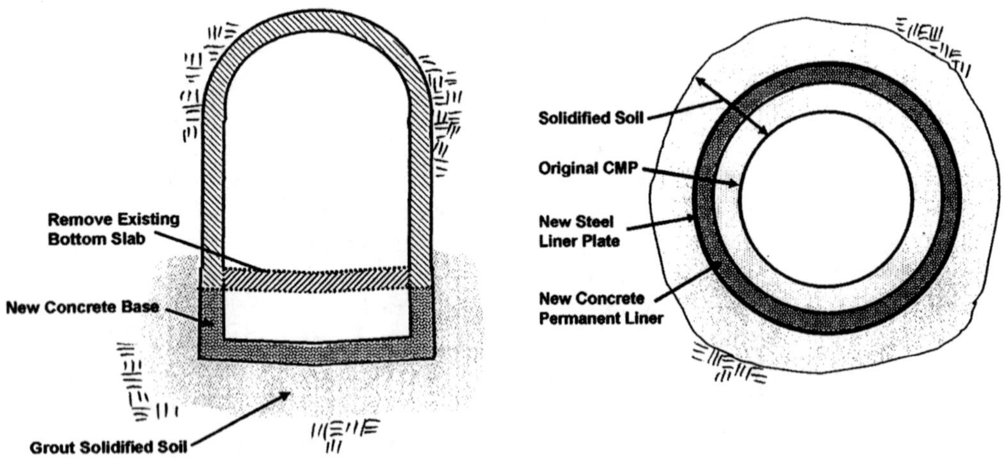

FIGURE 11.75 CMP was to be removed and the tunnel enlarged.

The work site was extremely congested. In addition, to meet a reasonable construction schedule, it was necessary to start the tunnel reconstruction work as soon as the grouting had progressed sufficiently. Thus, there was a great deal of activity in the rather cramped confines of the tunnel, which complicated traversing to the various work sites. Accordingly, the grout plants were stationed on timber platforms just outside the upstream tunnel portal, and rigid PVC grout lines, seen on the left side of Figure 11.76, were run the length of the tunnel.

The chemical solution grout plant was self-contained on skids, as shown in Figure 11.77. It was pneumatically powered and featured separate synchronized pumps to properly proportion the ingredients. Appropriate flow meters and pressure gauges were built into the unit at the control station. The cement grout was mixed in a high-shear mixer, from which it was transferred to a separate agitator. It was then fed into an open-throat pneumatically powered Moyno pump, just behind the illustrated solution plant. As it left the mixing area, the pumped grout passed through a chart recorder, where a continuous record of pumping rate and pressure was made.

FIGURE 11.77 Self-contained grout plants were located outside the portal.

The solidification was completely effective, as can be readily observed in Figure 11.78, which shows a segment in the CMP run immediately prior to installation of the liner plates that formed the initial lining. The new liner plates can be observed in the upper-left corner of Figure 11.78. The 6 ft (1.8 m) long excavated segment of grouted soil is standing freely up to the juncture with the smaller-diameter original CMP, which is yet to be removed. The grout supply lines are among the bundle of utilities that can be seen at the bottom of the photo.

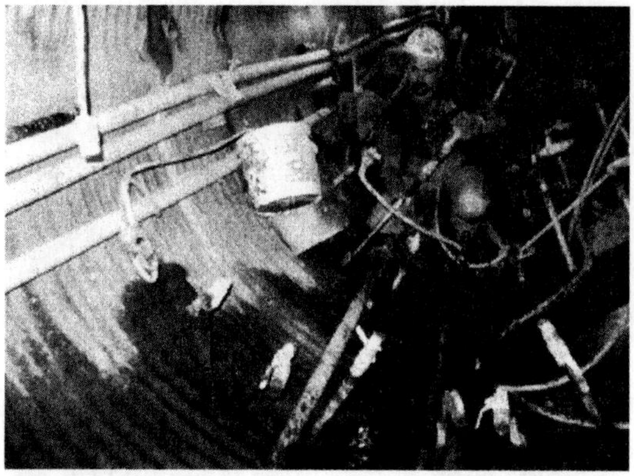

FIGURE 11.76 Grout lines were run the length of the tunnel and tapped at the injection area.

FIGURE 11.78 Six-foot (1.8 m) length of grouted tunnel stood unsupported.

11.2.4.2 PERMANENT STRENGTHENING APPLICATIONS

To save time and money, NASA used permeation grouted piles, rather than conventional piling, for the deep foundation of a 350 ton missile launch gantry (Figure 11.79) at Vandenberg Air Force Base in California. A tight schedule for the launch of a Physics Interplanetary Monitoring Platform satellite into orbit required very rapid construction of the launch facility. There simply was not sufficient time to construct a conventional pile foundation. The gantry rolled forward to extend out over the launch pit for assembly and fueling of a missile. It then was rolled back about 200 ft (60 m) so as not to be damaged during the blastoff.

The underlying soil consisted of about 25 to 30 ft (7.5 to 9 m) of fine to medium wind-deposited dune sand, underlain by shale bedrock. Groundwater was present at a depth of about 15 ft (4.5 m). The sand was not capable of taking the loads to be imposed, especially when the gantry approached the 30 ft (9 m) high retaining wall of the launch pit. Foundation loads were thus to be transferred, by piles, to the shale layer. The gantry traveled on sets of rails mounted to two parallel pile caps. These heavily reinforced concrete caps were, in turn, supported on two rows of piles, as shown in Figure 11.80. Pile design was based on a permanent, chemically solidified mass, with a minimum unconfined compressive strength of 100 psi (0.69 MPa). This was considered a very conservative design, as strengths of nearly twice that value were found to exist after seven years in a nearby installation that used the same grout formulation.

FIGURE 11.79 A launch gantry was erected simultaneously with the foundation construction.

Accordingly, the 188 piles were designed with a minimum diameter of 2.5 ft (0.76 m). The diameter was increased to 5 ft (1.5 m) for the portion below the water table, as chemically grouted masses are known to obtain lower strengths under water. The piles extended full depth to the underlying shale formational deposit, and because of the greater diameter on the tips, spread the load on it to relatively low unit values.

A distinct advantage of using injected piles was the ability to form them either before or after the pile cap had been cast. This greatly reduced the total time required for the foundation work. Initially, piles were injected prior to the casting of one of the parallel concrete pile caps while the other cap was being constructed. Sleeves of 1-1/2 in. (37 mm) pipe were cast into the second cap at each pile location. The remaining piles were then cast under the completed pile cap, using the pipe sleeves for access, while the first cap was being built.

A three-component sodium-silicate-based chemical solution grout was prepared and pumped from self-contained trailer-mounted solution grout plants located adjacent to the site. The grout plants contained two sets of mix tanks for each component, enabling continuous pumping without interruption. Sodium silicate, water, and a surfactant were combined on the base side. On the reactant side, the proper pro-

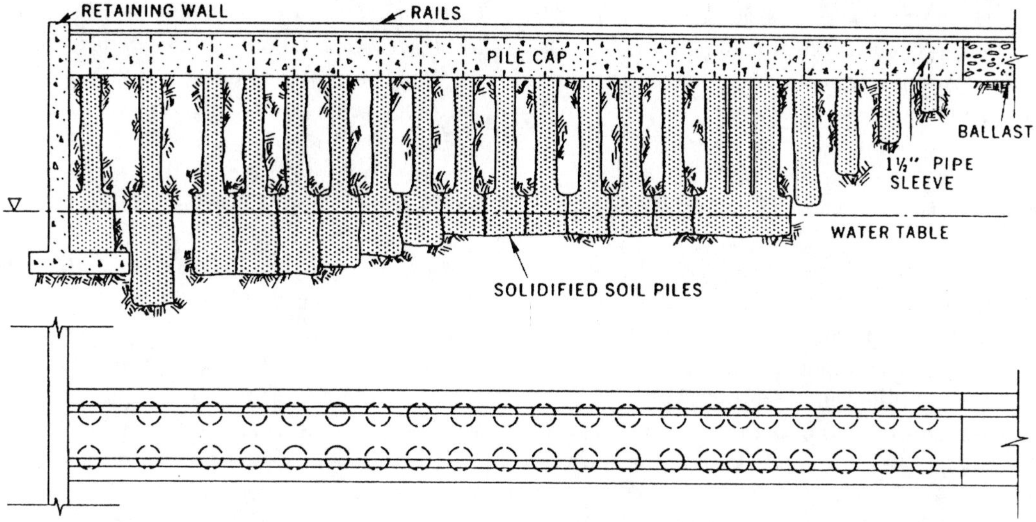

FIGURE 11.80 Longitudinal section of the foundation and solidified piles.

portions of both the liquid ethyl acetate and dry flake calcium chloride reactants were combined with water. The proper amount of sodium silicate was pumped directly from a tanker trailer to the mixer tank as required. The calcium chloride was delivered to the job site in batch-size preproportioned plastic bags, and the ethyl acetate was manually drawn from an adjacent truck-mounted tank, using a graduated container to ensure the correct dosage.

The two premixed grout components were accurately proportioned by way of properly calibrated metering pumps, which fed a "flash mixer" located immediately ahead of the main grout pump. The flash mixer was simply a 3 in. (76 mm) squirrel cage blower, enclosed in a small cone-bottom tank and rotating at about 1800 rpm. It was used because the injectability of the particular grout formulation is greatly improved through very high shear processing of the components immediately prior to injection. The final grout mixture contained 55 percent sodium silicate.

Injection was from the top down through special drive needles, as illustrated in Figure 11.81. The needles were advanced in 1 ft (0.3 m) increments, followed by injection of the proper dose of grout. A calculated amount of grout, sufficient to completely saturate a 36 in. (0.9 m)

FIGURE 11.81 Injection needles were driven with wheel-mounted driving rigs.

column, was injected for the 30 in. (0.75 m) sections, and 66 in. (1.65 m) for the 60 in. (1.5 m) sections. Additional soil was solidified adjacent to the end retaining wall where the depth of the rock strata dropped off quite severely.

The work continued around the clock for 11 days, using two separate crews and sets of equipment. In all, more than 130,000 gal (491,400 L) of grout was required. Tests carved from the solidified masses, both during the work and after completion, attained fundamental strengths somewhat greater than 200 psi (1.4 MPa). Simultaneous to the foundation work, the gantry tower was constructed on temporary rails set on conventional ballast and ties. It was moved onto the new foundation in less than a month after the start of construction. The project was hailed at completion for having the fastest turnaround time, from concept to completion, of any launch facility.

Ultrafine cement was not available at the time the work was performed, but such grout would be much more advantageous to use should a similar project be developed today. The reason is that cement grout is not adversely affected when under water, as is the solution grout, and its use would have significantly reduced the amount of grout required. Further, cement grout can be formulated so as to produce significantly higher-strength grouted masses, which would have further reduced the quantity needed.

In a rather unusual grouting application, it was necessary to solidify pea gravel backfill around the intake pit of a water pumping plant. The plant was adjacent to a reservoir, and a 24 in. (0.6 m) inlet pipe ran from the reservoir to the bottom of the pump pit, which was formed by a 60 in. (1.5 m) diameter vertically oriented corrugated metal pipe. The CMP was set into a concrete base slab, as shown in Figure 11.82. Gravel smaller than 3/8 in. (9.5 mm) had been used to backfill the space between the excavated soil and the wall of the vertical CMP during the original construction. A problem with maintenance of the pump impeller blades had developed over time, to the point that they had to be replaced frequently. Something hard was obviously striking them, but there should have been nothing in the water-filled pit to cause such damage.

Following the development of voids under the foundation of the enclosing pump house, it was obvious that something was radically wrong and corrective measures had to be taken. So that the actual conditions could be observed, the pumps were removed, a plug was placed in the intake pipe by a diver, and the pump pit was dewatered. The metal of the CMP pit lining was found to be corroded so severely that holes completely penetrated the wall of the pipe. This condition had allowed the cohesionless pea gravel to intrude, and several inches of it filled the bottom of the pit. The situation was so fragile that there was concern of immediate failure of the entire shaft, which could destroy both the pump house and the enclosed mechanical equipment. To make things worse, the only access was through

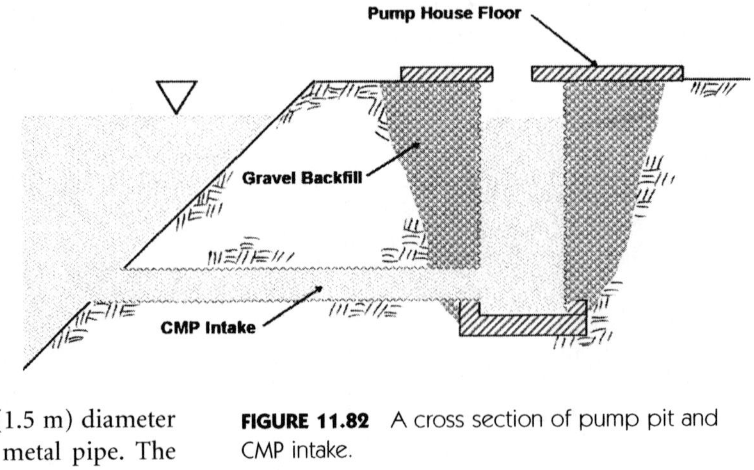

FIGURE 11.82 A cross section of pump pit and CMP intake.

the three small holes into which the pumps were mounted.

Something had to be done to solidify the loose gravel; however, the wall of the CMP pipe retaining it was so weakened that the disturbance caused by normal drilling and grouting would likely cause failure. It was thought to be unwise to try to advance holes in the gravel backfill, which had already run freely into the pump pit. The only other option was to drill through the CMP itself, which would present a problem if very large holes were required. In addition, it was not considered safe to place a very large quantity of unhardened grout at any one time, which would entail alternating between drilling, grout injection, and hardening over a considerable period of time. But extended time simply was not available, as there was no replacement supply of water available.

To fulfill these requirements, procedures that are normal to structural grouting but quite unusual in geotechnical applications were selected. The grout would be an expanding hydrophobic polyurethane foam, and it could be pumped through holes as small as 1/4 in (6 mm). This not only minimized the size of the required holes, but also greatly reduced the amount of grout that would have to be pumped, as the selected formulation would expand to about 12 times its original volume once in place. Further, it would be fully expanded and of considerable strength within minutes of injection. This type of resin grout is usually reserved for the control of water leaking into underground concrete structures and is routinely injected in such applications, albeit in generally much smaller quantities.

The first operation, however, was to stabilize the structure by filling the void existing under the pump house foundation. To allow inspection of the void area and facilitate placement of the grout with minimal pressure, relatively large 6 in. (152 mm) diameter holes were core drilled through the concrete floor of the pump house. The voids were substantial, so a ready-mixed flowable fill, designed to be virtually self-leveling, was used. It was simply pumped to the large core holes and allowed to run in, primarily under gravity pressure.

The fill grout was allowed to harden overnight. Preparation for injection of the urethane foam then continued. Starting at the bottom of the pit, one row of the required 3/8 in. (9.5 mm) holes was drilled through the wall of the CMP. These were spaced about 1 ft (0.3 m) apart and carefully positioned so as to initially avoid areas of serious corrosion to the greatest degree practicable.

The grout was then injected into one hole at a time, starting at that which appeared to be the most stable. Injection continued randomly, generally proceeding from the most competent conditions to those of lesser stability. Only enough grout to provide an expanded volume of about 0.5 ft^3 (14 L) was injected at any one location. The one component resin was pumped directly from the drums in which it was received, by way of a small air diaphragm pump. The pea gravel was generally wet, having been submerged until just a few days prior to the injection. The grout, being hydrophobic, immediately started to foam on contact with the moisture, and it attained substantial strength within minutes of injection. Only a very small area of the CMP wall thus felt pressure at any given time. The repair was in every way effective, and the pumps have been operating without incident since completion.

11.2.4.3 CONTROLLING WATER FLOW

The procedures used to create an impermeable barrier in soil are essentially the same as those used to inject grout in strengthening applications. The desirable properties of the cured grout, however, are different. Whereas hard, rigid grouts are desirable in strengthening applications, some flexibility is preferable in water control work. Because the grouted mass cannot be seen, as in those cases where the adjacent soil has been excavated, the degree to which the objectives have been achieved will usually be known only by way of the reduction of the flow achieved. This is usually observed in the drainage system, if there is one, or from pump drawdown tests conducted in drainage holes placed downstream of the curtain.

It is extremely difficult to effect an impermeable barrier with a single row of grout holes. Accordingly, two or more rows of holes should be placed, with split spacing. Where there is subsurface flow, the first row of holes to be grouted should always be the farthest downstream. Most grouters then inject the upstream row, reserving the middle row for final closure grouting. Where the flow is strong, however, some prefer to reserve the row farthest upstream of the flow for final injected closure holes.

Alternately, where an especially tight barrier is required, a hole layout with tertiary injections in the center row can be used. This will result in relatively small areas of remaining ungrouted soil in the center row, which is used for closure. With this sequence, higher grout pressures can often be used in the closure holes. In such cases, a quantity of grout that exceeds the theoretical quantity required should be placed, unless excessively high injection pressure develops, which would indicate that all available voids have been filled. As with most grouting, the individual holes in a given row should be injected in a alternate primary/secondary and perhaps tertiary sequence.

Stoppage of water leakage adjacent to a solid object such as the wall of a structure is somewhat easier, in that one boundary of the work is established. The grout will tend to follow along the hard boundary, in effect forming an impermeable membrane on it. This can usually be accomplished with a single row of grout holes placed adjacent to the wall. The radial distance from the wall should be only about 30 percent of the anticipated grout mass diameter so as to favor the grout's reaching the wall first.

Where outside access is difficult or otherwise undesirable, an alternate is to drill holes through the wall, on a regular spacing, followed by grout injection. Hole spacing for this option is typically about 2 ft (0.6 m) in each direction. The lowest row of holes should always be as close to the bottom of the wall as practicable, so as to seal off any flow coming from below the floor area. Such work is best done during periods of active leakage so that its effectiveness can be observed by way of a reduction in the flow.

Water was penetrating the concrete masonry unit wall of a subsurface equipment room. A hydrophobic urethane grout was injected in a single line of holes at 2 ft (0.6 m) intervals and about 8 in. (203 mm) outside the wall. Injection was started in the areas that had experienced the greatest amount of leakage. As shown in Figure 11.83, some of the grout actually penetrated the areas of leakage all the way through the wall, reacting on its surface. This confirmed that the cul-

FIGURE 11.83 The grout was forced through the former leaks and ran into the interior.

prit leakage paths were being plugged, although it also created a nasty cleanup job.

11.2.4.4 STRENGTHENING AND WATER CONTROL

Perhaps the greatest challenge in grouting soil is in situations that require both blocking the flow of water and increasing the soil strength. Because of the strengthening requirement, generally hard, rigid grouts must be used even though these are not necessarily the best for water control. As in any water control work, however, a relatively tight hole layout is in order. At least two rows of holes are generally required, and when a high degree of water blockage is required, a three-row curtain is often needed.

The installation of new rolling equipment in a steel mill required a foundation that would extend 10 ft (3 m) below the floor surface. This was about 6 ft (1.8 m) deeper than the prevailing groundwater. Moreover, the excavation was required to be within a few feet of a heavily loaded column footing supporting, among other things, a 50 ton (45 t) bridge crane that was to remain in service throughout construction. The soil was basically cohesionless sand, with some gravel and occasional cobbles. Because the new foundation would generally fill the excavation, there was no space for any internal bracing. This required solidification of a self-supporting watertight wall around the excavation. In addition, the bottom soils would have to be made essentially watertight to prevent seepage from below, and this bottom plug would require sufficient strength to withstand the resulting buoyant forces.

Chemical solution grout was used to form an impermeable solidified "bathtub" section for the excavation, as illustrated in Figure 11.84. The first operation involved the formation of a self-supporting perimeter wall. Three rows of grout holes, spaced in a conservatively tight pattern, were established. Because the new foundation was to be formed, the holes were inclined out of the excavation at about 5 degrees off vertical. They extended to a depth of 13 ft (3.9 m), which was 3 ft (0.9 m) below the bottom of the excavation. Holes were located on 2 ft (0.6 m) centers in the outer two rows, which were spaced 2.25 ft (0.68 m) apart. These were injected with a volume of grout calculated to form 2.5 ft (0.75 m) diameter solidified columns.

Because of its superior properties for strengthening and its lower cost, a sodium silicate base grout was used for injection in these two initial rows. After completion of their grouting, an acrylamide chemical solution grout, which is especially suited for permeation into tight formations, was injected in the center closure holes to provide a watertight wall. Upon completion of this work, a bottom plug was formed. It consisted of an injection of the sodium silicate grout from a depth of 10 to 14 ft (3 to 4.2 m), with the use of a 2 ft (0.6 m) offset grid throughout the interior of the tub. As in establishing the walls, this was followed by injection of acrylamide.

For the initial wall, the sodium silicate base grout was simply jetted in place with the use of an open-end 1/2 in. (13 mm) aluminum pipe. The formation was exceptionally permeable, allowing a rather rapid pumping rate

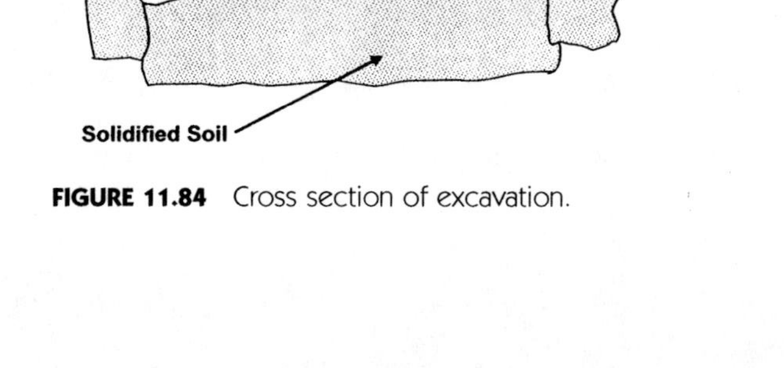

FIGURE 11.84 Cross section of excavation.

of about 12 gal (454 L) per minute. This promoted fast penetration of the jet pipe, which was provided with graduated markings at 1 ft (0.3 m) intervals along its length. Its penetration was paused at each 1 ft (0.3 m) of depth until the calculated quantity of grout had been injected.

The intent was to inject the acrylamide grout in the closure row using similar methods, but the jet pipe met resistance to penetration in about 30 percent of the holes, apparently resulting from solidification by silicate grout that had permeated to the injection location. These areas required predrilling for the acrylamide grout to reach final depth. Thus, a 2 in. (51 mm) diameter hole was drilled to the full depth of injection. It was then filled with 50 mesh sand, which was somewhat less permeable than the native materials. This ensured that further injection would not travel down the hole but would, rather, be deposited at the appropriate level. These variations allowed proper completion of the closure holes.

Once injection of the walls was completed, the bottom plug was formed. For the plug, 1/2 in. (13 mm) injection pipes were preplaced to a depth of 15 ft (4.5 m) for bottom-up grouting. This was 1 ft (0.3 m) deeper than had been used for the walls, and 5 ft (1.5 m) deeper than the bottom of the future excavation. Only the bottom 4 ft (1.2 m) of these holes was injected, however. This was done because there was no requirement to have the floor of the excavation solidified, nor any desire to excavate in the grouted soil. These holes were spaced in a grid of 2 ft (0.6 m) in each direction.

The thickness of the bottom plug was designed to resist the uplift forces and was thus much greater than normally required for the blockage of water alone. Therefore, the normal closure holes were not included. The alternate primary holes and the bottom stage of the secondary holes were injected with a quantity of grout that was calculated to form a column 3 ft (0.9 m) in diameter. Theoretically, this quantity of grout should have provided closure. The grout quantity was increased by about 20 percent for the upper 3 ft (0.9 m) of the secondary holes, which would theoretically fill the voids of significantly more soil than actually existed in the treated zone. The excess grout was expected to provide a good seal against infiltration, although it would penetrate to an elevation higher than the stage being injected. The injection therefore stopped 1 ft (0.3 m) short of the excavation bottom, as indicated earlier.

This "overtreatment" of the bottom plug was quite successful, and areas of solidified soil existed all the way to the subgrade elevation. Upon excavation, there were only a few areas of leakage, as illustrated in Figure 11.85. These were easily shut off by injection of a small quantity of the acrylamide grout. It was administered by driving short pieces of temporarily plugged 1/2 in. (13 mm) pipe as deep as possible and then injecting the grout. Most of these injection pipes encountered solidified soil within 1 ft (0.3 m) or so of the excavation floor. The leaks were quickly stopped, however, as a result of the ability to proportion the acrylamide grout to provide an almost instant reaction, as explained in Section 7.3.3.1.

Injection of the two-component acrylamide grout requires special proportioning equipment.

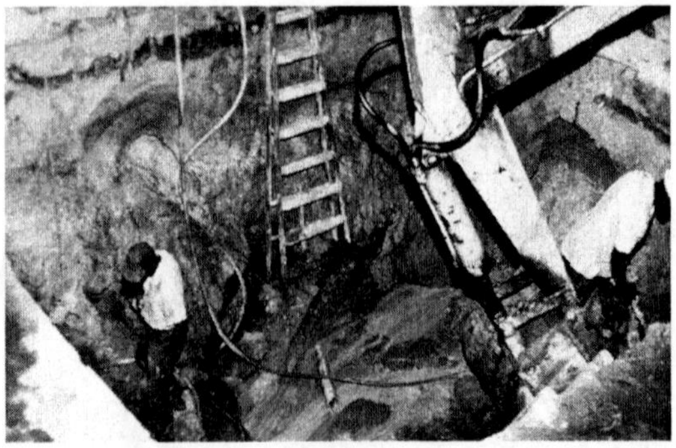

FIGURE 11.85 Only a few areas of minor leakage were found during excavation.

The reactant is highly corrosive, so the mixers, pumps, and all piping that will come in contact with it must be of stainless steel or a rigid plastic. The gel time of the grout is controlled by the ratio of the base and reactant portions, so a variable proportioning pump system is mandatory for proper injection. The two components are typically pumped to the injection site, where they are stream mixed. This was accomplished in a "mixing stand," where a flow meter and pressure gauges were located. A short whip hose was used between the injection pipe and the mixing stand.

Before the two components are mixed, they are hazardous, and for this reason special training is required for the technicians who will be handling them. Although acrylamide grout is not widely distributed, it is available and regularly used. The material is unique in that it is highly penetrable. But of greatest value is the ability to precisely control the gel time, decreasing it to as little as a few seconds if desired. It thus provides a distinct advantage for difficult water control applications.

11.2.4.5 ALTERING SOIL DYNAMIC PROPERTIES

The dynamic properties of soils vary every bit as much as their other properties. For most work, dynamic values are not of great concern, and thus most engineers are only peripherally aware of soil dynamics. An area in which these properties are crucial, however, is the design of foundations for equipment that will be required to absorb large dynamic forces. Should the natural frequency of the soil be in harmony with the output of the machine, serious vibration forces can develop.

An interesting case involves two large piston air compressors that were moved to a new location after the first two years of use. They worked flawlessly at the original location, but the vibration at the new location was so bad that florescent tubes were being shaken out of the overhead light fixtures. Yet the concrete foundation block and the surrounding building were identical to those at the first installation. In fact, the same set of plans was used at both locations. The only difference was the soil and, apparently, its dynamic properties. Investigation revealed that the foundation block was established within a deposit of silty fine sand. The sand had a natural frequency, which was in harmony with that of the compressors. It was concluded that an increase of stiffness of the sandy soil was required to control the problem.

Such soils would normally be considered too fine to permeate with a solution grout, but it was doubtful that the only viable alternative of compaction grouting would provide sufficient improvement. A rigid grout formulation containing 60 percent sodium silicate was used. Not only was the penetrability of the soil limited, but the penetrability of a grout containing such a high concentration of the base component is also significantly less than that of more conventional formulations. It was used, however, in order to provide the greatest degree of additional stiffness practicable.

With grout of such low penetrability, very slow pumping rates are necessary and the grout holes must be spaced closely. In addition to grout holes placed through the foundation, a single row was placed immediately adjacent to it with an additional peripheral row 1.5 ft (0.45 m) beyond, on all four sides. These holes were spaced 1.5 ft (0.45 m) apart and extended from the floor level to the minus 9.5 ft (2.9 m) elevation. An attempt was made to maintain the same spacing each way under the foundation block, but it could not be totally achieved because of obstruction of the installation.

The grout was injected from the top down through special drive needles, as illustrated in Figure 11.86. The fastest the grout could be injected was an unusually slow rate of only about 1/2 gal (1.9 L) per minute. Although the cost of injection at such a painfully low rate was very

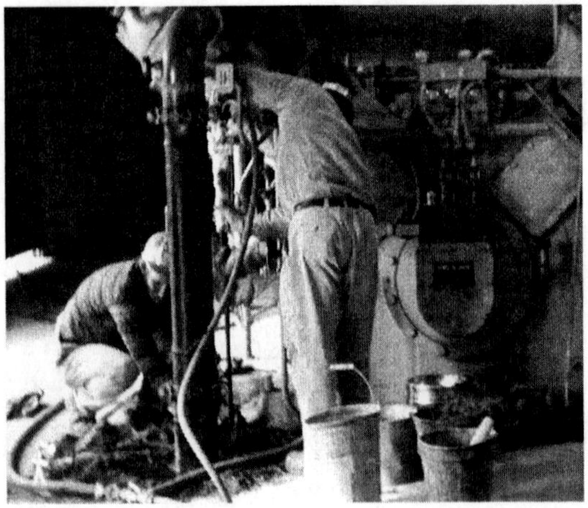

FIGURE 11.86 Injection from the top down, using a handheld pneumatic hammer to drive the injection needles.

high, in this instance it was warranted and the resulting benefits were very cost-effective. The grouting was completely successful, in that the excessive movements of the machines were completely eliminated.

An interesting example of the dynamic improvement of soil by grouting was presented by Partos, Woods, and Welsh (1982), wherein a die forging plant was relocated from Greenfield, Ohio, to Walterboro, South Carolina. The maximum peak particle displacements of the foundations existing at the original installation were on the order of 0.2 in (0.5 mm) at frequencies of 20 to 33 Hz, which is equivalent to about 1200 to 2000 cycles per minute. Although these vibrations were acceptable in the soils at the original installation, they could not be tolerated at the new plant location because its soils would consolidate under such vibration.

The new location was in an existing building that was founded on a deposit of loose fine sand that descended to a depth of more than 30 ft (9 m). To take the forces that the forging machines would impart, both the density and the dynamic modulus of the soil would require considerable improvement. The density requirements could be fulfilled through compaction grouting, but it was doubtful that this would satisfactorily improve the dynamic properties. A grouting program was thus formulated that involved both compaction grouting and permeation grouting with a chemical solution grout if required.

On the basis of analysis and extensive laboratory evaluation, the minimal acceptable properties for the soil were determined. These required improvement of the relative density to a minimum value of 70 percent. The lowest acceptable dynamic shear wave velocity was set at 623 ft (187 m) per second. A three-step ground improvement plan was developed, involving two phases of compaction grouting, followed by permeation solidification if required. The levels of improvement were evaluated after each phase of grouting.

The improvement in density was determined at exactly the same location after each grouting operation. This was accomplished with nucleonic depth density and moisture measurement so that the degree of change at any given location could be observed. The improvement in dynamic response was assessed by interval crosshole shear wave evaluation. The required levels of improvement were attained, although the process did require all three phases of grouting.

The compaction grouting did provide an ample increase in density, as expected, but it had only a modest effect on the dynamic properties. Of particular interest, however, was that the solution grouting resulted in a dramatic increase in the shear wave velocity, to more than double the required values. Partos, Woods, and Welsh (1982) provide an unusually thorough discussion of the studies that led to the remedial work. They describe extensive testing and analysis of both the dynamic output of the machines in their original installation, as well as the soil to be improved at the new site. Soil dynamics are not often discussed in detail, especially in the context of improvement by grouting, but are covered comprehensively by these authors.

Another example of excessive vibration for which multiple grouting technologies were used involves a very large horizontal piston-type air compressor. It was founded on an 18 ft (5.4 m) square, heavily reinforced 7 ft (2.1 m) thick concrete foundation block. This, in turn, was supported by six wide-flange steel piles driven into the underlying silty fine sand dredge fill. The compressor worked without any vibration problem for the first several years of its existence.

The problem for which the grouting was performed had developed over a period of more than ten years. It was initially a rather minor vibration, but with time it became a rocking type of movement. The displacement was in a linear direction parallel to the course of the piston stroke. The magnitude of this movement had been getting progressively greater, until it reached a frequency of about 200 cycles per minute, with an extraordinary amplitude of nearly 1/8 in. (3 mm), regardless of the operating load. At this point it was shaking a large area of the enclosing building and causing a significant maintenance problems. It was thus shut down until an appropriate repair could be made.

Investigation found that there was a void of about 2 in. (51 mm) under the foundation. It was postulated that the settlement resulted from the vibration, which was being carried to depth by way of the steel piles. Once sufficient settlement had occurred to prevent any lateral restraint between the soil and the base of the foundation, it apparently started rocking on the piles. Initially, this affected only the uppermost parts of the piles, but with time the base of the rocking movement deepened until a void developed on either side of the piles for an unknown but considerable depth.

A threefold approach was taken to correct the problem. Eight 2 in. (51 mm) diameter holes were core drilled through the foundation. These were placed about 1 ft (0.3 m) in from the face of the concrete at each corner and midway between. A conventional low-mobility cement mortar grout was then injected to fill the void. The injection was into one of the corner holes and continued until it had returned out of all those remaining, indicating that the void had been completely filled. Although after the fact it was known that the grout would flow in a logical manner, filling from one corner to the opposite corner, it is not wise to expect this to happen in all cases. Thus, with a series of holes as used, the exact flow of the grout can be followed, ensuring complete filling of the void.

Following the fill grouting, the holes were extended a distance of 10 ft (3 m) below the bottom of the foundation and compaction grouting was performed. The purpose of this operation was to generally improve the density, and in particular, to reestablish intimate contact of the soil against the steel H-piles. This was followed by permeation grouting of a 10 ft (3 m) thick layer of soil immediately under the foundation. The solidification continued for 3 ft (0.9 m) beyond the limits of the foundation block and was carried all the way to the underside of the floor in that area of the periphery.

Additional grout holes were required for this latter injection. Although obstruction by the machine precluded the drilling of an orderly grid of holes, an effort was made to establish them on a basis of 2 ft (0.6 m) each way. This was accomplished with vertical holes for at least two rows on the outside limits of the work. For the interior area, however, inclined holes were required as the uniform spacing could not be maintained. Additional quantities of grout were injected, as practicable, in those holes that had excessive spacing.

Figure 11.87 illustrates the activities at the grout header during the permeation injection. The technician is shown procuring a sample of the grout to confirm its gel time. This is good practice with all resin- and chemical-based grouts that are proportioned within the pumping

FIGURE 11.87 Taking a grout sample at the mixing stand during injection.

system. The gel time of such grouts is largely dependent on proper proportioning, so such checks can ensure that the pumped volumes are correct. Gel time is also influenced by temperature, so any resulting changes can also be evaluated in this manner, and appropriate changes made in proportioning.

11.3 FRACTURE (CLAQUAGE) GROUTING

Fracture grouting has its roots in France and is thus referred to by the French term for *fracture*, which is *claquage*. It involves intentional hydraulic fracturing of the soil with a fluid suspension grout, resulting in what is hoped to be an interwoven network of grout-filled fractures. Although some densification occurs, the primary improvement mechanism is reinforcement of the soil by the network of relatively strong interconnected lenses. Cementitious suspension grouts are most frequently used. These provide good stiffness and compressive strength, but exhibit only limited tensile capacity. Very short polymer fibers have been included in the grout mixture to improve the tensile strength of the hardened mass. Although this is probably of some benefit, the exact performance is unknown.

Unfortunately, it is virtually impossible to control the direction, interconnection, or extent of fractures, so the results are somewhat unpredictable, especially at the micro-level. Thus, exceptional care must be taken to ensure that the injection is at the intended location. Sleeve port pipes are thus generally used for the grout injection, with the injection ports spaced at fairly small intervals, typically no more than about 1 ft (0.3 m). To preclude possible damage, such as uncontrolled hydrofracture resulting from runaway takes, the quantity of grout placed at any one location is usually limited and not great.

This is not a widely practiced grouting technology, although some significant projects have been performed with its use, especially in Europe. It has often been used in connection with *compensation grouting*, that is, compensating for ground lost in tunneling or other underground construction, prior to the resulting voids reaching the ground surface. Several contractors in the United States have promoted this type of grouting, using a variety of terms, often protected by trademarks, to define it. Such terms as *Soilfrac*, *Frac*, and *Lens* are often seen in this regard.

Such verbiage may be of benefit to marketing efforts, but it is not altogether descriptive of any special improvements achieved in the formation. A number of excavations have been made to evaluate the results of such work. In most instances, the fractures were not closely spaced, nor were they frequently interconnected. In practice, compaction grouting will give more predictable results, and generally at a significantly lower cost in most soils. There are soils, however, in which compaction grouting is not appropriate, most notably in plastic clays, especially those that are highly expansive.

In these soils there are two very different improvement mechanisms that are applicable. The first, of course, is reinforcement, as already mentioned; the other is a cation exchange, which alters the fundamental chemistry of the soil and results in a tighter binding of the clay platelets to each other. The consequence is a much more dimensionally stable formation, not nearly as sensitive to moisture change as the beginning soil. The process is thus most often used in unstable plastic clays and usually those that are exhibiting uncontrolled expansion. Cation exchange is one of the very few available and frequently used techniques for control of expansive soils.

11.3.1 Lime Injection

The most frequently used grout material for lime injection is high-calcium hydrated lime slurry, which is usually applied from the top down by way of special injection needles. Injection may be performed with a single needle that is hand-thrust into the soil for shallow treatment, as shown in Figure 11.88. Although this is a simple procedure that involves very little investment in equipment, it is difficult to get the penetration started and is thus quite messy; a large amount of slurry often leaks to the ground surface, as illustrated. Alternately, a needle can be driven with a small rig so that the injection need not be started until it has penetrated a few feet. This provides much better confinement of the grout and thus less leakage to the surface. Equipment-mounted multiprobe injection is the most frequently utilized method for deep or large-scale injection, however.

FIGURE 11.88 Manual injection of lime slurry can be very messy.

In 1967 three large dormitories at the Hollygrove Orphanage in Hollywood, California, required remediation of damage to the slab on grade floor. An investigation disclosed that the floor slab had experienced upward movement of as much as 1.5 in. (38 mm). An additional heave of 2.75 in. (70 mm) was predicted. The cause of the upheaval was determined to be a layer of highly expansive clay underlying the floor at a depth of 2 to 7 ft (0.6 to 2.1 m). Lime injection was considered somewhat experimental at the time, but because funding was restricted and moving the occupants out of the structure long-term was not possible, it was adopted as the most feasible remedial method.

As mentioned earlier, lime injection can be quite messy, so great care was taken to perform the work with as little mess as possible. Injection holes were located on a 3 ft (0.9m) grid, normal to the module of the existing 12 in. (0.3 m) square floor-covering tiles. The work was accomplished with only minor disturbance, and the structures were occupied throughout the work. Limiting replacement of the floor covering to only those tiles at the hole locations and waxing the entire floor upon completion of the work resulted in a large cost savings to the owner. The replacement tiles were of bright colors to provide an accent, which actually improved the aesthetics of the original construction.

The slurry consisted of high-calcium hydrated lime mixed with water, at the rate of about 4.2 lb (1.9 kg) of lime per gal (3.8 L) of water, to make a nearly 50 percent mixture by weight. The injection rate was held to a relatively low value of about 0.6 ft^3 (17 L) per minute in order to minimize leakage to the surface. Injection pressure varied between 50 and 250 psi (3.4 and 17.2 bars). A report prepared after completion of the work read as follows:

> The slurry was injected from the top down through special injection needles made especially for this work. The needle was driven downward from the surface until the clay layer was "felt," after which a metered quantity of 0.25 (7 L) cubic feet of slurry or 7.5 pounds (3.4 K) of lime was injected. The needle was then continued downward in 6 in. (152 mm) increments, the metered amount of slurry injected at each, until the bottom of the clay layer was "felt" or minus 7 ft (2.1 m) was reached. In some cases it was not possible to inject the full measure of slurry due to uplift of the structure walls or floor.

Several elevation surveys had been taken prior to the work. Others were taken immediately following completion and at 30, 90, 180, and 360 days thereafter. Immediately following the injection, an additional upward heave of about 0.25 in. (6 mm) was recorded. No further movement was noted thereafter, and, in fact, after one year a portion of the original 0.25 in. (6 mm) of heave had recovered.

An application of much greater scale occurred on the Chicago and Northwestern rail line between Des Moines and Kansas City. The tracks traversed a 1000 ft (300 m) long embankment of plastic clay. The soil consisted of about 30 ft (9 m) of clay fill on top of a 10 ft (3 m) deep clay layer that was underlain by generally competent sands. In spite of previous remedial work, including widening the embankment with ballast, installation of drainage pipes, and buttressing the toe of the fill with riprap, maintaining the track profile had become a major undertaking. In fact, settlement of the tracks had become such a huge maintenance problem that train speed was restricted to no more than 10 miles per hour. As there were no alternative methods to stabilize the track bed without creating long delays in rail service, fracture grouting with a lime slurry was chosen to remedy the continuing dilemma.

The entire clay fill was treated to a maximum depth of 40 ft (12 m). This was accomplished with special rigs, each containing a large tank to serve as a reservoir for the lime slurry and three individual 40 ft (12 m) high masts, each supporting an injection needle. These were mounted on ten-wheel trucks that had been fitted with rail wheels so as to ride the existing rails, as illustrated in Figure 11.89. Problems with excessive leakage of the slurry through the rail ballast stone were experienced during the early stages of the work. This was explained by the railroad's records, which indicated that some 200 railcars of ballast had been added under the rails over the previous seven years. The slurry freely permeated and leaked from areas of ballast rock, which were 8 to 10 ft (2.4 to 3 m) deep in some sections.

An initial injection phase was adopted that placed the three simultaneous probes every 10 ft (3 m) along the track. These were limited to a depth of 14 ft (4.2 m). After a three-to-four day curing period, similar injections were made

FIGURE 11.89 Large-scale lime injection on the railroad bed.

midway between the initial ones for a final spacing of 5 ft (1.5 m). These also were limited to a 14 ft (4.2 m) depth. In addition to leakage of lime slurry, large amounts of free water were forced out of the embankment during this initial injection. The final phase of work involved injection all the way to the bottom of the fill, with one hole located just inside each rail, at 5 ft (1.5 m) intervals.

Large amounts of lime were required for the work, and up to a ton of lime was injected into some of the 40 ft (12 m) deep holes. Upon completion of the work, a total of more than 3000 tons of lime had been applied. This, of course, required the ability to supply large quantities of slurry, which requires prehydration. Accordingly, large mixing and holding tanks, which in a single operation could prepare an entire semi bulk hauler full of the dry lime, were used. So successful was the work that within a month after its completion, the trains were able to return to the timetable rate of speed, 49 miles per hour.

11.3.2 Cementitious Grout Injection

Likely the largest application of fracture grouting in the United States was used to halt settlement of a large bulk mail handling facility in Clarksburg, West Virginia. The structure, originally built in 1989, was founded on the former site of an open pit mine that had been backfilled with surplus mine waste through the years. This material was supposed to have been replaced with a properly controlled fill prior to the building's construction, but up to 6 in. (152 mm) of differential settlement had nonetheless occurred. The foundation system consisted of conventional shallow spread footings with a slab-on-grade concrete floor. The single-story steel framed building had undergone considerable structural distress, which was also presenting serious operational problems.

The owner had great concern about continuing regular operations during the remedial work and did not want any activity whatever in the interior of the building. Therefore, a system was devised whereby the injections would be made from horizontal sleeve port pipes installed from four pits on the exterior of the building. The pits extended from the ground surface to a depth of 25 ft (7.5 m), which was the bottom of the treatment zone. They were 20 ft (6 m) in diameter and lined with shotcrete, as illustrated in Figure 11.90. Horizontal injection pipes were then placed in fanlike arrays under the structure, as shown in Figure 11.91.

The probes were 3 in. (76 mm) closed-end steel sleeve port pipes specially fabricated for the operation. Because of concern that conventional drilling could initiate greater settlement of the existing soft silty clay soils, the probes were hydraulically jacked into place using the back of the shaft for reaction. The pipes were placed in two horizontal layers at depths of 10 ft (3 m) and 15 ft (4.5 m) below the underside of the floor slab. Two additional rows were inclined from the 15 ft (4.5 m) level to deposit grout at the 20 and 25 ft (6.0 and 7.5 m) levels. To ensure complete coverage of the treatment area, every pipe was surveyed to map its exact trajectory prior to grout injection.

FIGURE 11.90 Sleeve port pipes were jacked into place from a large shaft.

A stable cementitious suspension grout was then injected sequentially through the ports of each pipe. The grout pressure and injection rate were continuously monitored in real time throughout the injection. In addition, the structure had been instrumented and was also continuously computer monitored. Thus, all of the data pertaining to both grout behavior and building response could be assessed at any given time from the same computer monitor. Moreover, a well-experienced technician continuously observed conditions above the area of injection on the interior of the building.

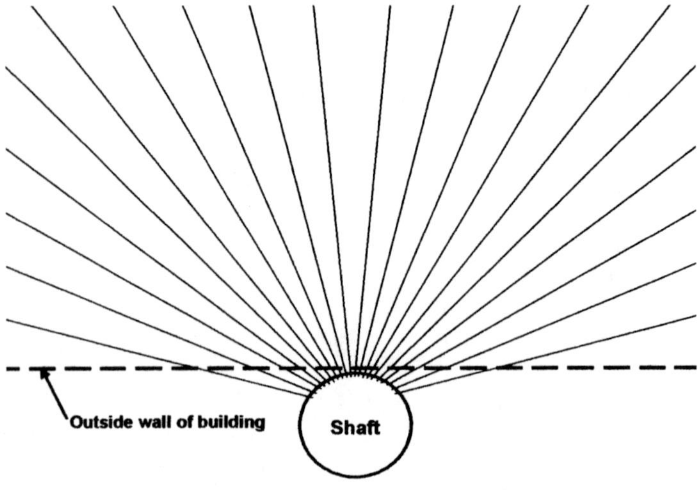

FIGURE 11.91 The injection probes radiated out from the shaft.

11.4 JET GROUTING (MIXING)

As discussed in Chapter 2, jet grouting involves the mixing of in situ soils with a cementitious suspension grout so as to produce new, relatively homogenous, mixed masses, as illustrated in Figure 11.92. The mixing is accomplished through erosion of the soil by a high-pressure stream of water, or grout, or air-enshrouded water or grout, emitting from a jet monitor that is mounted at the tip of a drill stem. The soil is first eroded and then mixed via the turbulence caused by the high-pressure jetting together with rotation or other movement of the nozzle. There are, as discussed in Section 2.4.1, three basic procedures used; their radii of effectiveness differ but they produce similar mixed products.

Although jet grouting can be used in any type of soil, it is most effective in granular formations, especially fairly clean sands and gravels. In these ideal materials the erosion effect is greatest, resulting in larger effective distances than achieved in cohesive soils. Moreover, the strength of the final mixed product is the greatest and can reach concretelike levels. Jetting in clay requires more input energy, which usually results in slower progress of the work and a smaller effective radius of the jet. In addition, the strength of the resulting mass is significantly lower in clays than in granular formations. Whereas unconfined compressive strengths up to about 2000 psi (13.8 MPa) can be achieved in clean sands and gravels, those obtained in clay seldom reach 500 psi (3.4 MPa) and are most often within a range of 100 to 300 psi (0.69 to 2 MPa). Generally, the resulting strength increases with increasing grain size and cleanliness

FIGURE 11.92 Jet grouting results in concrete-like masses of mixed material.

of the host soil, and drops as the soil becomes finer.

For each unit volume of grout injected, one or more units of spoil will result. The spoil exits through the annular space between the drill rod and the wall of the pilot hole. If for any reason this passage should become blocked, very high pressures of the jet will be induced into the soil, which will result in fracturing, heaving of the surface, and likely displacement or breaking into any existing substructure. This can be a problem in all soils, but becomes a significant risk with increasing soil cohesion and depth of injection, especially when working in saturated or submerged deposits. In submerged conditions a blocked spoil return will result in immediate over pressurization of the groundwater, which will cause hydraulic fracturing sufficient to relieve the excessive pressure. Thus, well qualified operators constantly observe the returning spoil and promptly act when any change occurs.

FIGURE 11.93 Some contractors are very messy, spilling jet grouting spoil widely (left) while others are very tidy, such as by collecting into a small pit and pumping to a disposal area (right).

The process can be quite messy unless the spoils that are produced are conscientiously controlled. In this, some contractors take great care as illustrated in Figure 11.93 (right) while others are literal slobs, with the worksite covered in mud (Figure 11.93 (left). Good practice calls for directing the spoil to an excavated pit, usually behind the drill from which it is pumped to a disposal lagoon or settling basin. Alternatively, it is removed by a vacuum truck for disposal. The spoils always contain a significant amount of grout and, given enough time, will dry to a solid state. The strength of that dried waste, of course, varies with both the characteristics of the beginning soil and the proportion of grout that has been intermixed, but will usually be quite low. The volume of spoils will usually approach that of the grout quantity injected, and can develop to considerably more than 100%.

Obtainable strength of the soilcrete is primarily determined by the soil type, but also by the amount of cement intermixed, which is a function of the water:cement ratio and the jetting parameters. Typical grout mixes contain only cement and water, the water:cement ratio being within a range of about 0.6 to 1.2, by weight. The strongest masses are created with the one-tube system as the compressed air that is used with the other systems tends be become entrained into the soilcrete. As with any cementitious composition, the strength is directly related to the density and, of course, this is reduced with entrained air.

Injection parameters vary with the particular system used as well as contractor preference and equipment capability. They will normally be within the ranges shown in Table 2.1, and are closely related so that change in one will result in a compensating change in another. The exact parameters to be used are typically determined

FIGURE 11.94 Large and usually high-mast drill rigs are required.

FIGURE 11.95 Typical setup, with cement storage silos over containerized grout mixing plant.

by trial and error at the beginning of a project. Once determined, they are usually entered into a computer on board the drill rig, which will virtually control the subsequent work.

Because of the considerably large and heavy equipment needed, and the time and effort required for setup, jet grouting is usually reserved for projects that will continue for at least several weeks. Heavy drill rigs with high masts, as shown in Figure 11.94 are typically used. But it is the high-performance pumps and material-handling and mixing equipment that require extensive setup. Very large and high-powered grout pumps are required to provide the extremely high pressures and flow rates commonly used. To ensure a continuous supply of grout, bulk silos of cement typically feed large mixers, which are often totally automated.

Figure 11.95 shows a typical setup in which the grout mixers are within the container directly under the cement silos. The pump is self-contained in another container slightly to the right of the silo on the right. The entire arrangement has been placed on an specially built concrete slab and includes supporting generators and compressors so as to be self-supporting. Obviously, such a setup cannot be moved without much time and great expense, so it will typically be centrally located to the injection. Pipelines for grout supply and utilities are then run the length of the work. Smaller equipment can be used in combination with the one-fluid system, although the jet grouted elements will not be as large. It is thus used primarily in granular soils and can be advantageous for smaller or congested sites. Although few contractors so perform, there are noteworthy exceptions in both the United States and Japan.

An additional requirement, which is unique to jet grouting, is an absolute necessity for the grout to be completely free of any large particles or foreign matter. Such contaminants will result in rapid wear of the jet nozzles, and serious damage to the equipment may occur if the passages become completely plugged. For this reason, it is common to use in-line strainers between the grout pump and jetting monitor. Obviously, all components of the delivery line must be capable of withstanding the very high pressure that is commonly used.

As is apparent, jet grouting is a rather specialized procedure that requires a high initial investment in equipment. Furthermore, the skills required to accomplish the work are somewhat different from those necessary for the more common grouting applications. Thus, jet grouting is performed by a relatively small number of specialty grouting contractors. Accordingly, the best source of information is generally the various contractors who regularly perform the work. Be aware, however, that not all such contractors use the same jetting methodologies operate with similar orderliness or produce work of equivalent quality.

The particular jet grouting system to be used is often influenced by the contractor. There are, however, individual project requirements that clearly favor one system over another. Use of the three-fluid or Jumbo/Super Jet systems should probably be avoided in near proximity to improvements that are to be preserved. In this respect, use of either one- or two-fluid systems will result in considerably smaller grouted elements, so that both the amount of time required for completion and the cost will be noticeably greater. The additional cost versus that of repair of any damage which might result from the more aggressive systems must thus be weighed. When work is performed through existing pavement or other areas where spoil must not gather, the one-fluid system will generate less spoil requiring handling.

Jet grouting has become very widely used, especially on large projects. The use has not been limited to vertical applications, but it is also used for earth restraint in tunneling where fan shaped arrays of elements are placed sequentially from within the bore (Figure 11.96). It is particularly advantageous for temporary earth support in cohesionless granular deposits. While a few contractors are biased toward use of the more aggressive and economical systems, there is much justification for both the one- and two-fluid systems, especially in small or sensitive applications.

The one-tube system was used for temporary earth retention during trenching for a new sewer in Pacific Coast Highway immediately adjacent to the Pacific ocean in Santa Monica, California. The soil was cohesionless beach sand with the groundwater level only a few feet below the surface. Depth of treatment varied from about 15 to 25 ft (4.5 to 7.5 m) and the work was executed through the existing pavement of a blocked off lane, immediately adjacent to regular traffic. The single-fluid system was used and produced columns about 3 ft (0.9 m) in diameter. Because of the immediately adjacent traffic, rapid collection and disposal of the spoil was required. This was easily handled by vacuum trucks. The suction hose was connected to a small prefabricated reservoir around the individual grout holes.

FIGURE 11.96 Jet grouting in a tunnel using special drill rig.

In another project, the single-fluid system was used to construct underpinning of an existing building at Yale University in New Haven, Connecticut. A new addition immediately adjacent to the existing wall would undermine the foundation by about 16 ft (4.8 m) in cohesionless sand and gravel. The job was not large; only about 80 linear feet (24 m) of existing wall required treatment, but access was very poor and the site extremely confined. A drill head was installed on a skid steer loader in order to access all of the required areas. The compact grout pump was specially built by the contractor. It consisted of ten standard 4:1 ratio air piston pumps mounted in series, so as to achieve a total grout pressure of 6,000 psi (414 bars). Solidified masses as great a 2.8 ft (0.85 m) in diameter were constructed which allowed a shear

FIGURE 11.97 Jet-grouted wall supports undermined foundation during construction.

excavation to be made in the nearly cohesionless soil, as shown in Figure 11.97. Jet grouting on this site would not have been possible with the more common large equipment.

Both one- and two-tube system were used to solidify loose soil under the submerged Webster and Posey tubes which carry traffic from Oakland to Alameda Island, California, as part of a seismic retrofit. The objective of the work was to reduce the liquefaction potential of the soil, thus inhibiting floating of the tube in the event of an earthquake. It was also intended to prevent settlement of the end sections of the tube which were attached to pile-supported portal structures, once they were cut loose and replaced with flexible joints, as part of the overall retrofit. A gentle approach was warranted to prevent any damage to the tubes which were to remain in service throughout the work.

Initially the one-fluid system was used at the Webster tube, as displacement of the tube sections immediately adjacent to the jetting, could not be tolerated. The cementitious suspension grout had a water : cement ratio of 1 and was injected with a flow rate of 74 gal/min (280 L/min) at 5700 psi (393 bars) pressure. Monitor rotation was at 14-20 RPM with a withdrawal rate of 2–4 in. (5–10 cm) per minute. This produced columns about 31 in. (79 cm) in diameter. They extended from the bottom of the existing foundations to a depth of about 90 ft (27 m).

Although the work was satisfactory, the progress was slow because of the large number of elements required due to their small diameter. Consequently, the two-tube system was adopted for work at the Posey tube. This formed much larger columns with an average diameter of about 72 in. (183 cm), as revealed by a test pit. The same grout mix was used, but it was injected at a flow rate of 76 gal/min (287 L/min) at 6100 psi (420 bars) pressure. Monitor rotation was 9 RPM with a withdrawal rate of 7.9 in. (20 cm) per minute. The grouted elements extended from the bottom of the existing foundations to a depth of about 95 ft (28.5 m).

With this more aggressive operation, extreme care was taken to prevent any damage. All grouting parameters were automatically recorded in real time and instrumentation which had been previously installed in the tube structure as well as the adjoining soil, was continuously monitored. It included ground settlement devices in the area of the actual jetting, as well as precision tile meters at the tube joint.

To reduce the risk of liquefaction and provide stability for a 100 year flood, the banks of

FIGURE 11.98 Drill rig required frequent moves to isolated areas, but grout plant and all utilities were located remotely and delivered by temporary pipeline.

the lower American River in Sacramento California, received slurry walls constructed to depths of 60 to 80 ft (18 to 24 m). Slurry wall construction was not amenable for sections with street or utility crossings however, so these were effected by jet grouting. Considering the depth and amount of work to be completed, the 3-fluid system was selected.

The work was based on constructing jet grouted elements 8.3 ft (2.5 m) in diameter on a spacing of 6.7 ft. (2 m), so as to form a continuous secant pile wall. Grout with a water: cement ratio of 1 was injected at a rate of 90 gal (340 L) per minute and 2320 psi (160 bar) pressure. Initial erosion of the soil was accomplished with water pressurized to 6525 psi (450 bar) and flowing at 53 gal (200 L) per minute. To enshroud both the cutting water and grout, about 200 CFM (5.7 m^3) of air was supplied at a pressure between 100 and 130 psi (7–9 bar). The jet monitor was turned at 2 RPM and withdrawn at a rate of 3.6 in. (9 cm) per minute.

The individual gaps of jet grouting were spread out over more than a mile of the river route, requiring frequent moves of the drill rig, Figure 11.98. A central plant was set up for mixing and pumping the grout as well as basing the air and water utilities. Temporary pipelines were run the length of the project for distribution of the grout and utilities. The soil was run to a small pit behind the drill rig, Figure 11.93 (right), from where it was pumped to a central settling pond. Although construction of the temporary pipelines was obviously costly, the setup did allow the work to be done in an urban setting with only minor inconvenience to the public.

12 CHAPTER

Grouting in Rock

12.1 KNOW THE ROCK PROPERTIES

12.2 GROUTING FOR WATER CONTROL
- **12.2.1** How Tight? The Improvement Standard
- **12.2.2** Water Tests
  - **12.2.2.1** The Lugeon Unit
- **12.2.3** Rock Grouting Design
  - **12.2.3.1** Blanket Grouting
  - **12.2.3.2** Consolidation Grouting
  - **12.2.3.3** Curtain Grouting Planning
- **12.2.4** Grouting Methodologies
  - **12.2.4.1** Conventional Technology
  - **12.2.4.2** Grouting Intensity Number (GIN)
  - **12.2.4.3** Amenability Theory

12.3 GROUTING ROCK FOR STRENGTHENING

12.4 GROUT HOLES
- **12.4.1** Hole Size and Alignment
- **12.4.2** Casing/Collar Standpipe
- **12.4.3** Surface Treatment
- **12.4.4** Grout Caps
- **12.4.5** Grouting Galleries
- **12.4.6** Drilling
  - **12.4.6.1** Drill Access
  - **12.4.6.2** Cleaning the Holes
  - **12.4.6.3** Logging of Drill Holes

12.5 THE GROUT MIX
- **12.5.1** Grout Mix Properties
- **12.5.2** Grain Size/Chemistry
- **12.5.3** Optimal Mix Consistency

12.6 INJECTION CONSIDERATIONS
- **12.6.1** Injection Sequencing
- **12.6.2** Hole Staging
  - **12.6.2.1** Stage Length
  - **12.6.2.2** Injection Downstage from Collar
  - **12.6.2.3** Downstage with Packer
  - **12.6.2.4** Grouting Upstage with a Packer
  - **12.6.2.5** Circuit Grouting
  - **12.6.2.6** Multiple Packer Sleeve Port Pipe (MPSP)
- **12.6.3** Before Hookup
- **12.6.4** Grout Plant
- **12.6.5** The Injection
  - **12.6.5.1** Grout Supply
  - **12.6.5.2** Monitoring the Injection
  - **12.6.5.3** Injection Pressure
  - **12.6.5.4** Connections Between Holes
  - **12.6.5.5** Handling Surface Leaks
- **12.6.6** Refusal Criteria

THE PRIMARY OBJECTIVE of rock grouting is typically to reduce or stop the movement of water, strengthen the formation, or both. Most grouting in rock is done in connection with the construction and maintenance of dams and tunnels, although it is also used in shaft sinking and mining applications, as well as other types of underground construction. A minor amount is performed strictly for the strengthening of weak rock to increase its bearing capacity or, in the case of rock anchors, its pull-out resistance. Cavernous voids in soluble rock such as limestone are sometimes filled either to prevent or to remediate the intrusion of overlying soil. This type of work, which is frequently accomplished with standard concrete pumps and ready-mixed concrete or mortar, is referred to here simply as *fill grouting*, and it is not discussed further in this chapter.

In underground construction, grouting is often performed for the purpose of reducing and controlling water infiltration, although strengthening of the rock is also a frequent requirement. Although applications for shallow tunnels can be done from the overlying ground surface if accessable, underground grouting is most often performed from within the tunnel or excavation. It is commonly performed by miners as part of their regular activities, although specialist grouting contractors are sometimes employed, mainly for unusual or challenging situations.

The primary purpose of grouting in connection with dams is usually to minimize seepage through the foundation rock, but here, also, strengthening of poor-quality rock is sometimes desired. As in the case of earth embankments, control of leakage is required to prevent excessive uplift forces, minimize unacceptable leakage, and prevent internal erosion and resulting piping voids. Although the formation of water-resistant curtains has historically been the most common use for grouting, such work now accounts for only a small portion of such work. Most of the available grouting literature and, indeed, most "guidelines," however, are directed strictly to dam applications

All rock grouting involves the filling of existing cracks, joints, and other defects. This is nearly always accomplished with cementitious suspension grout, although both chemical solution and resinous grouts are occasionally employed. The size and distribution of typical joints and defects in rock vary enormously. Whereas wide open features are usually readily filled, the effective treatment of fine fissures can be more than challenging. And of greatest difficulty is the filling of small defects that occur in combination with larger ones. A good understanding of rock properties and, in particular, the likely joint and defect geometry of a given formation, is fundamental to planning rock grouting. For such work to be effective, all defects and apertures must be accessable from at least one grout hole.

12.1 KNOW THE ROCK PROPERTIES

The properties of rock, especially the size, spacing, and orientation of the joints and apertures, vary widely. All rock contains some cracks and openings; however, the extent of these vary, depending on the origin and stress history of the particular formation. A basic understanding of geology is thus fundamental to effective grouting of rock and is discussed more thoroughly in Chapter 21, "Understanding Geology."

Sedimentary rock is derived from the consolidation of loose sediment that has accumulated in layers or strata over very long periods of time. It contains bedding planes that visibly separate each successive layer of the stratified rock. Depending on the past geologic activity, these planes—which were more or less horizontal when formed—can now be oriented at virtually any inclination, including completely overturned. Such joints, parallel to the bedding plane and remaining essentially horizontal, are usually quite tight and seldom need grouting.

Historical uplifting and folding of strata can cause slipping along the bedding planes, however, resulting in joint sets that are inclined and can open to varying degrees. Uplifting and folding can result in significant fractures or partings in a rock. Mechanical weathering, whereby the rock breaks down or comes apart along well-defined joint planes, can create large open joint systems that intersect the primary strata, as illustrated in Figure 12.1. Such open fissures and interconnected joint networks can require considerable quantities of grout to fill.

Igneous rock, formed from the molten magma of volcanoes, can result in very complex joints and fissures, as well as tubular passages. Magma often cracks upon cooling, especially if it comes in contact with a very cold environment such as snow or ice on the ground surface at the time of eruption. Because these formations have usually resulted from numerous eruptions over long periods of time, they are often interspersed with sedimentary deposits. It is not unusual to find a large assortment of non-connected defects and joint sets in such rock, which will obviously require special attention and generally close grout hole spacing.

Fundamental to the proper design of rock grouting is a competently prepared geological assessment of the formation to be treated. This will typically include core samples of the rock to be encountered. Downhole fiber-optic or TV inspection of core holes passing through shattered zones and/or where less than 100 percent core recovery occurred will often provide important additional information. Although a good geological report can provide the basis for grouting design, practicality and the complexity of many rock formations virtually always preclude a complete understanding of the existing conditions. Continuous assessment of grout behavior and constant reevaluation of the properties of the host rock during drilling and grouting are therefore essential.

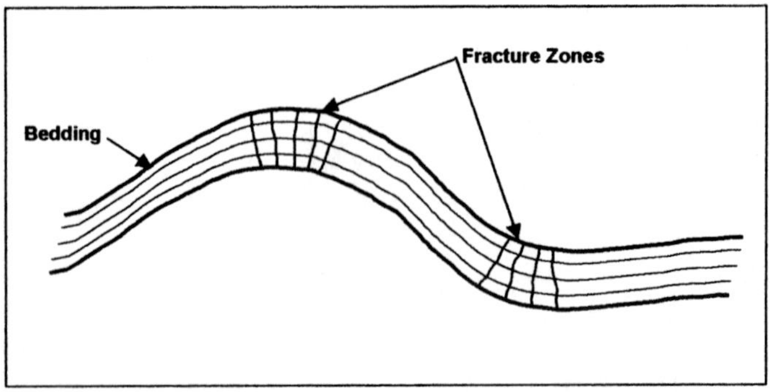

FIGURE 12.1 Large open joint sets can develop due to geologic folding.

12.2 GROUTING FOR WATER CONTROL

Work being performed to restrict flow under a dam involves the creation of a vertical plane mass, known as a *grout curtain*. Where the rock immediately under a dam structure is highly fractured, the entire surface layer up to a depth of about 20 ft (6 m) is sometimes treated. This is referred to as *blanket* or *consolidation* grouting. Where similar work is performed around the entire section of a tunnel or shaft, it is called *encapsulation*, and the sealing of the top portion of a tunnel section only is known as *cover grouting*. In contrast to the many different methods used for grouting soil, only a single method is used in rock and that is permeation grouting which fills existing voids.

12.2.1 How Tight? The Improvement Standard

Unique to the reduction of water flow is the relative effort required to produce a given decrease in flow. Large continuous defects that facilitate the flow of large amounts of water can usually be readily filled with a minimum number of relatively widely spaced grout holes. Conversely, tight apertures and crevices, although capable of transmitting water, albeit at a low rate, can be very difficult to heal and typically require a large number of closely spaced holes. Simple common cement-water grouts can be readily injected into the larger defects, but smaller ones will require much slower injection rates and often more expensive grout constituents, such as ultrafine cement or perhaps a chemical solution grout.

The cost of the work is not proportional to the leakage improvement. The number of holes and the injection effort required to block the last 10 percent of leakage can easily surpass that required to stop the first 50 percent. One must thus consider both the nature of the defects and the amount of reduction reasonably required. Calling for a greater reduction in seepage than is rationally required will virtually always come at a high price.

Where grouting is performed simply to reduce the amount of leakage to a manageable level, such as in tunnel construction, it is usual that only significant flows require stoppage, the remaining seepage being readily handled by canalization and pumping. In connection with dams, there is virtually always some amount of leakage. In either case, cost is directly related to the amount of seepage permitted to continue. Thus, in the case of dams and other water-retaining embankments, the value of the water must be considered. Where it is especially precious, a high standard of grouting is justified, whereas a lesser effort will often be adequate where water is plentiful and of lesser value.

In the case of water-retaining embankments, it is important to consider the effects of the leakage volume and the pressure by which it is driven. Should these be excessive, erosion along the flow path can occur. This will usually present a serious safety issue and, unless remedied, can cause catastrophic failure. Internal erosion is known as *piping* and is the most frequent cause of distress and failure in water-retaining embankments. A through discussion of the relative amounts of leakage tolerable has been presented by Houlsby (1990) and is highly recommended for those with an interest in the subject.

12.2.2 Water Tests

The ability to measure the amount of seepage following the completion of grouting is requisite to conformance with any particular standard. Ideally, one could establish the total amount of leakage through a given area of a formation precisely, but this is not achievable in the real world of grouting. Most dams are provided with a downstream drainage system, and observation of such a system can disclose total leakage, although the origin and precise flow route are usually beyond detection.

Water can be injected into upstream drill holes at known flow conditions, as illustrated in Figure 12.2, wherein a gauge and a water meter are being observed. The volume injected, and that exiting at downstream drill holes or an existing drainage system, are compared. Although this will give an approximation of the formation conductivity, the actual route the water has taken is usually not determinable. Much of the injected water can run to areas outside the influence of the downstream monitoring holes.

In situations where there are many suspected leakage origins, dye can be added to the suspected source water. Dyes are commercially available that will color water at concentrations of less than 2 parts per million. Red, green,

FIGURE 12.2 Typical water test in progress.

yellow, and violet colors can be especially potent at low concentrations. Where leakage is suspected at several different sources, separate colors can be used to permit identification of the source by the outflow color. Although blue used alone will not show up as readily as the colors noted, it is useful in combination with yellow to identify separate sources. By placing blue dye in one source and yellow in another, the exit color will be green if the two have indeed combined. The color shade will indicate which source is providing a significantly greater portion of the flow.

An approximate measure of formation conditions can be ascertained by the introduction of three successive water injections, all at the same pressure and of five minutes duration. In most instances, the amount of water take in all of the runs will be about equal, signifying that passages in the rock are continuous and stable. Should the water take decrease on successive runs, it can be assumed that cracks and other defects are isolated and are being filled with water. Conversely, should the quantity increase during the successive injections, scouring or erosion of soft rock, or washout of infilled material, is probably occurring. At high pressures an increasing water take can also indicate an opening or jacking of the rock. Although the latter behaviors are the exception, where they do occur, they are important to understand.

Routine water testing of grout holes is a contentious issue. Some grouters are of the opinion that every stage of every hole should be tested prior to grout injection. Others consider such testing to be a waste of time and of little value. Ewert (1985) presents several chapters on the subject of the geologic variations of different rock types and their effect on water pressure testing. He concludes,

> The WPT rates are unsuitable for deciding whether to grout, because they consider neither the specific permeability behavior of the rock type, nor the local parameters belonging to the water economy or geology. Moreover, the conventional WPT rates do not allow prediction of a certain correlation between water takes and solids.

Having downplayed the pertinence of water tests, he then laments,

> In spite of the limited ability to supply evidence, the water pressure test is the only practical possibility of learning something about the permeability of rock. Therefore, it cannot be abandoned. On the contrary, the attempt must be made to consider the conditions of the local specific percolation behavior and to perform and evaluate the test carefully. A better interpretation of the permeability can then be achieved, in spite of the limitations.

Water testing is a useful tool, but as with all tools, its use is best subject to judgment on the basis of the individual project or grout hole. The ease of flow, or resistance to it, during a test can provide a rough measure of the permeability of the rock to be injected. Although this can provide guidance in selection of the starting grout mix, it must be recognized that the rheologies of water and grout are very different, and simply because the water flows easily is not assurance that the always thicker grout will behave similarly.

In drilling grout holes in rock, water is usually employed as a circulation flush to remove the cuttings. Further, all holes should be washed out with water under sufficient pressure to thoroughly clean the walls and expel any drill cuttings or other deleterious matter upon completion of drilling, as illustrated in Figure 12.3. It is not unusual to loose part or all of the water into the formation during these activities. Should a large portion or all of the injected water fail to return to the surface, as is not uncommon,

FIGURE 12.3 Washing out the hole to thoroughly clean it of all cuttings or other deleterious matter.

it is apparent that significant defects exist. With this behavior, considerable grout take will obviously occur and the potential benefit of formal water testing is not great. Conversely, no water loss during these operations can suggest a generally tight hole, and in this situation a water test may provide valuable information. Sometimes a water test may disclose conditions so tight that no grout injection is warranted.

The consistency of grout take from one hole to another, and even between different stages of a hole, will vary. The amount of variation depends on the origin and resulting properties of the particular rock. Behavior patterns will be established with time, providing a basis for decisions regarding subsequent water tests. Unless the holes have taken large amounts of water during drilling and subsequent cleaning, such tests are best made in the early phases of most projects. The thoroughness, and indeed the necessity of continuing such tests, can then be evaluated on the basis of the results of the early performance.

Likewise, water tests should be conducted in all closure holes toward the end of grouting. This is not done to only determine groutability, but rather to confirm the effectiveness of the work and that seepage has been reduced to no more than the maximum level allowed. All stages of these latter holes should be individually evaluated and the results documented.

12.2.2.1 THE LUGEON UNIT

A simple evaluation system was developed in 1933 by Swiss geologist Maurice Lugeon and has since become used throughout the world in determining seepage. It measures the quantity of water that can be forced out of a given length of drill hole in one minute, under a set pressure. That amount of water is defined as a Lugeon unit. One Lugeon unit is equal to 1 L (0.26 gal) of lost water, per meter (3 ft 4 in.) length of hole, per minute, at a pressure of 10 bars (145 psi).

The test was originally developed for determining inflow of water wells drilled in competent rock, for which the 10 bar (145 psi) pressure was apparently appropriate. This relatively high pressure will, however, be excessive for many grout holes, wherein lower test pressures will be required. In such cases, the standard 10 bar (145 psi) pressure can be divided by the actual pressure used. This assumes a straight-line pressure relationship, which may not be exactly correct but should be well within the accuracy limitations of such evaluations.

12.2.3 Rock Grouting Design

In most instances the final design of grouting will not be completed until the actual work is finished. This is due to the great variance of most sites, which dictates constant evaluation and modification of the work as it progresses. The first and usually most important criterion is establishment of the standard to which the work must be done. It is dependent on the value of the water and the particulars of the formation, especially the propensity for erosion of the rock or any contained inclusions. Although some seepage will virtually always remain, both flow rates and velocities must be reduced to acceptable levels.

It is common to provide a suite of different mix designs for a any given job. In the simplest form, this entails changing the grout consistency by varying the water content of neat cement grout, but can also involve many sophisticated mixes, including different admixtures, supplemental cementing materials, and other additives. The design of rock grouting is amply discussed in Chapter 24.

12.2.3.1 BLANKET GROUTING

Blanket grouting is performed in the near surface rock to strengthen the formation and reduce permeability, usually under and prior to construction of a dam. It typically involves injecting a grid of uniformly spaced holes over the area to be improved, although additional holes may be included to treat geological anomalies. Blanket grouting is nearly always completed ahead of curtain grouting, and is often laid out and executed as a separate operation. It is suggested, however, that blanket grouting holes be laid out so as to be parallel to and in harmony with the future curtain grouting holes. Blanket grouting is limited to the near surface rock and seldom extends more than 20 ft (6 m) from the surface.

Most blanket grouting is performed with vertical holes, although inclined holes may be considered where defect features are likely to be missed otherwise. Holes are usually established with spacing of about 10 ft (3 m) each way. Individual rows are usually staggered to result in a triangular pattern. As with most grouting, alternate rows of holes are usually grouted first, as are alternate holes in any given row. Once these primary holes are completed, those remaining are injected in a *split spacing* progression, as illustrated in Figure 12.4.

12.2.3.2 CONSOLIDATION GROUTING

Similar to and often considered synonymous with blanket grouting, consolidation grouting is performed in weak or highly fractured rock near the ground surface. The primary goal is to cement the surficial rock into a more monolithic mass, capable of providing greater restraint for grouting the subsequent curtain. In the case of embankment dams, it is used to close any near surface defects that may receive eroded material from the overlying fill. It is also intended to strengthen foundation rock to support dams and other structures.

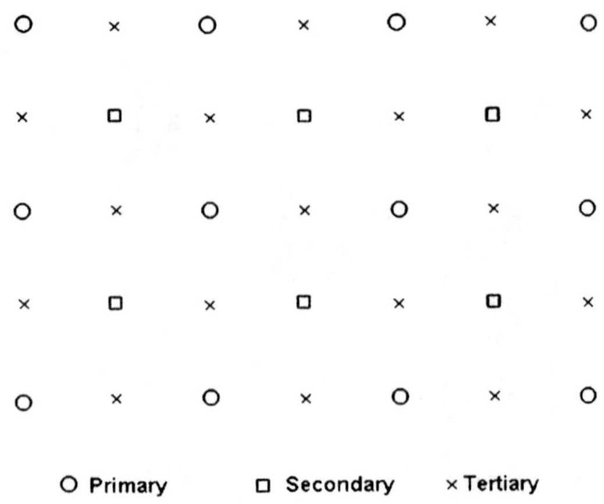

FIGURE 12.4 Typical layout of blanket grouting holes.

12.2.3.3 CURTAIN GROUTING PLANNING

There are a number of criteria for determining grout curtain depth. Most criteria are based on a depth equal to the height of the dam plus some constant, commonly about one-third the height. Although this may effectively increase the flow length of seepage, it does not rationally consider the nature of the rock or its permeability at depth. Thus, a detailed geologic investigation on which to base the design is necessary. Should the lower portions of the primary holes exhibit low takes, it is common practice is to reduce the depth of lower-order holes.

The grout holes must be laid out and oriented so as to intersect the maximum number of cracks and fissures. Grout can flow freely in large defects, but travel will be limited in narrower fissures. One must be aware of subsurface conditions to ensure an optimal hole layout with a maximum number of intersections with defects.

In curtain grouting, common practice is to first inject *primary* holes, which are widely spaced. Major defects are thereby filled for a considerable distance, but penetration of finer fissures will not be great. Once primary holes are completed, the space between them will be equally split and *secondary* holes drilled and grouted. This split spacing continues until preinjection water tests and/or the grout takes indicate permeability has been adequately reduced. Although any desired starting spacing can be used for the primary holes, 40 ft (12 m) is common. This space will then be split to 20 ft (6 m) for secondary holes, 10 ft (3 m) for *tertiary* holes, and perhaps 5 ft (1.5 m) for *quaternary* holes. Although seldom employed, the space can be further split with *quinary, sextary,* and so forth, holes.

Holes of the final order are the last injected and are known as *closure* holes. If all has gone as planned, the gout taken will decrease in the holes of each descending order. Closure holes should take considerably less grout than those of higher order. Because the larger defects will be filled from the primary holes, the take of these holes can be substantially greater than the takes of those that follow. Voids encountered by primary holes can be very large, resulting in the grout easily permeating a much greater distance than required. Volume limitations are thus sometimes assigned to the primary holes, thus preventing unreasonable *runaway* takes, which are wasteful and serve no rational need.

Generally, all stages of closure holes should be water tested prior to injection. The water loss should not exceed the maximum limitation of the target seepage. Further grout take should be significantly less than in the next higher-order holes. When to declare closure is of critical importance and can be a difficult call to make. Here, the importance of good records becomes patently clear. One must analyze those records, noting the ratio of grout take in the holes of the various orders. The nature of remaining defects should become reasonably clear from the prior grout behavior.

One must exercise common sense regarding the issue of very close hole spacing, however. Grout holes are typically small, usually within a range of 2 to 3 in. (50 to 76 mm) in diameter. It is unlikely that many achieve accurate alignment when drilled. The lateral distance of deflection will obviously increase with depth. Thus, holes that start at a spacing of, say 5 ft (1.5 m), could easily be 10 ft (3 m) apart at depth. Where excess leakage remains as the distance between holes becomes smaller, it may be best to offset additional holes a few feet, probably in an upstream direction. Moreover, where closely spaced holes have not provided the anticipated reduction in seepage, one must question the inclination and consider changing the orientation for any additional holes.

During injection, grout will travel freely and for a distance in wide, open fractures, but its penetration will require a much greater effort and will be significantly slower in fine cracks and fissures. Small fissures, even though connected to larger ones, may become sealed off as the grout travels freely while filling larger defects. The grout can also pick up loose debris or rock fragments as it permeates. Although this is minimized by thoroughly washing the holes prior to injection, additional debris can be created by slight movements of the rock during grout injection. This is a particular problem in highly fractured soft rock.

Although vertical holes are the easiest to drill, they may not be satisfactory where the defects are in a predominately vertical orientation. Satisfactory treatment will require inclination of the holes, as illustrated in Figure 12.5. Even with this approach, significant fissures may not be penetrated, so a second row of inclined holes

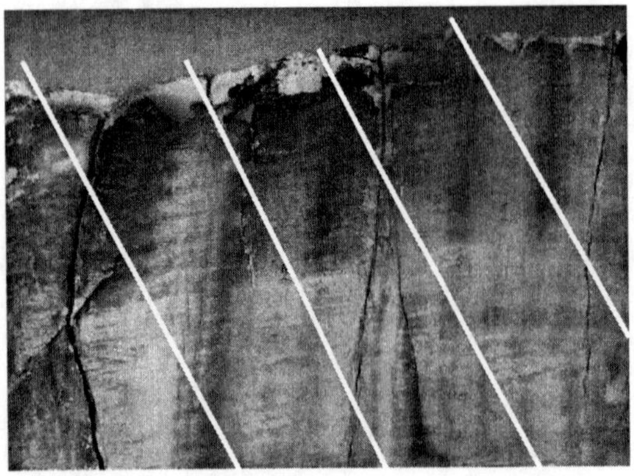

FIGURE 12.5 Inclined holes are required to intersect vertical joints.

oriented to intersect both the defects and the plane of the initial holes may be required. Although they are rare, very complex joint patterns may require multiple rows of holes at a number of different inclinations.

Weaver (1991) suggests that a two-row curtain with the holes drilled at opposing angles, as illustrated in Figure 10.7, "tend to reduce the likelihood for large ungrouted windows to be left in the foundation." Where multiple rows are used, they should be oriented to interact with each other. The typical distance between rows is generally no more than 3 to 5 ft (0.9 to 1.5 m). The downstream row is usually injected first. If three rows are used, the upstream row is injected next, with the middle closure row grouted last.

Both achievable and required standards of seepage reduction have been comprehensively discussed by Houlsby (1990). Seepage is examined from two perspectives: first, economics, based on the value of the water; and second, safety, as related to uplift pressures, erosion, and piping within embankments. Pertaining to the economic value of the water, he has presented a flowchart, reproduced here as Figure 12.6.

Houlsby cautions that the values shown are only for guidance, but in extensive discussion provides convincing evidence for their applicability, starting with the first work of Maurice Lugeon, who concluded in 1933 that grouting was required for dams in the Swiss Alps when seepage exceeded 1 Lugeon for dams more than 100 ft (30 m) high, and 3 Lugeons for those of lesser height. In practice, reducing seepage to no more than 4 or 5 lugeons is relatively easy. Conversely, reducing seepage further requires progressively greater effort, especially in rock containing numerous fine fissures.

Regarding dam safety, a second chart, reproduced here as Figure 12.7, provides acceptable levels of seepage under or through dams. Here the type of dam must be considered, as well as the thickness of the core in the case of earth embankments. Thinner cores will experience steeper hydraulic gradients and thus require grouting to a higher standard. Of even greater importance, however, are the nature of the rock and, in the case of earth embankments, the nature of the various fill materials. Experience has shown that over time, even minor seepage can cause piping, which can become severe in easily eroded formations

12.2.4 Grouting Methodologies

There are three established methodologies for constructing impermeable grout curtains, all of which have seen significant use in the past. Although there is considerable contention among leading grouters as to the best method, the fact is that each has its own advantages and limitations. Properly executed, however, any one can produce good-quality work.

12.2.4.1 CONVENTIONAL TECHNOLOGY

Conventional injection technology uses a circulating injection system, single pressure level, and usually a simple neat cement-water grout. Although the use of grouts with water : cement ratios of 10 and above has been reported, current thinking generally agrees that mixes thinner than

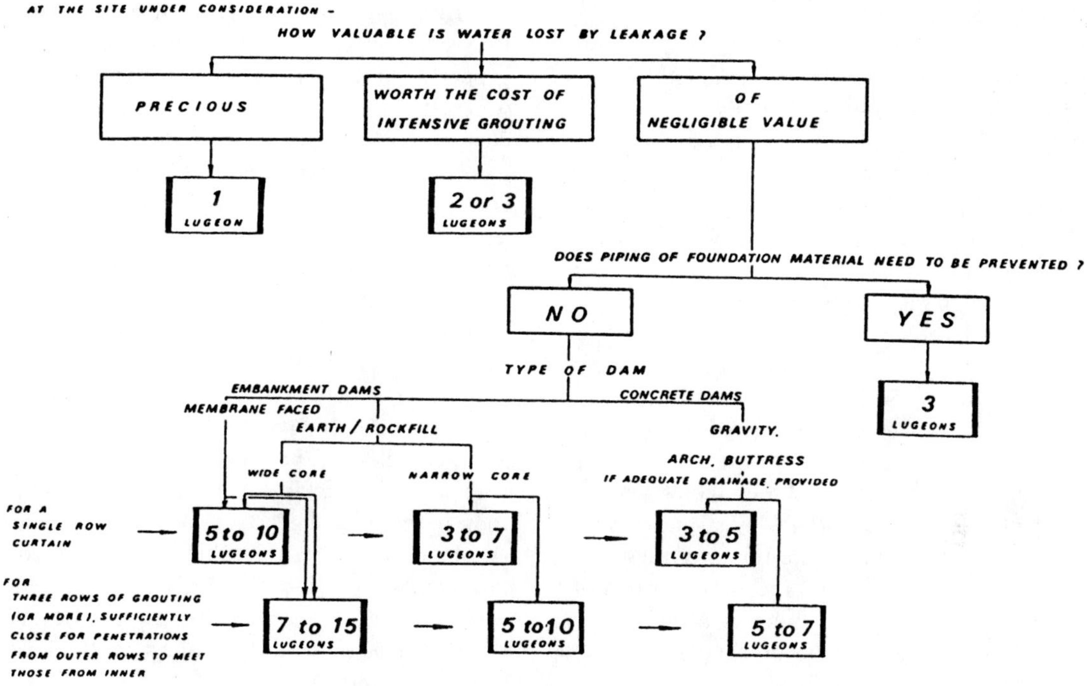

FIGURE 12.6 Allowable seepage based on economics. (From Houlsby, 1990.)

5 : 1 should be avoided. Such grouts, which are typically absent any admixtures, are unstable and will undergo considerable bleed. This effect must be mitigated during injection to achieve satisfactory results.

Typically, the thinnest grout contemplated is used as a starting mix and is initially run into the hole slowly. The injection rate is steadily increased until the predetermined maximum pressure is reached. This pressure is maintained

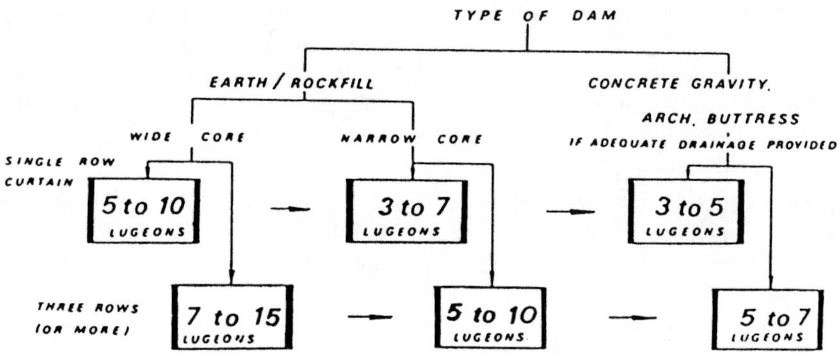

FIGURE 12.7 Allowable seepage based on dam safety. (From Houlsby, 1990.)

throughout injection, first by varying the injection rate of the beginning grout mix, and then by thickening the mix where warranted. Thickening adjustments are based on injection rate, as indicated by manual assessments of the grout pumped out of the agitator with the use of a calibrated *dipstick.* Mix adjustment involves increasing or reducing the water:cement ratio incrementally.

Although such adjustment sounds simple, it is not. First, the quantity of grout remaining in the agitator and delivery system must be ascertained to determine the amount of additional cement needed to achieve the desired consistency. This is cumbersome and difficult to calculate quickly in the field, so, ideally, tables will be available for reference in making such determinations. Illustrating the importance of these adjustments, Houlsby (1990) includes 45 pages of tables for thickening and thinning grout.

Another disadvantage of having to make such a change is interruption of the injection. In practice, the valve leading to the grout hole is closed and the grout is continuously circulated until the new mix has appeared in the return. The previously placed grout in the hole and formation will, of course, be motionless while this is accomplished, usually for a interval of at least several minutes. During pumping, there is turbulence in the grout as it travels, which aids suspension of the solids. Conversely, bleed occurs when the grout is at rest, especially in the hole and the larger defects.

Unfortunately, considerable time can pass between observing the need for mix modification and actually injecting the modified grout. This results from both delays in dipstick reading and circulation through the delivery system. Even with the best practice, dipsticks are read only every few minutes. It is common for experienced and conscientious personnel to constantly observe the rate of grout lowering in the agitator, but such attention is often lacking in the real world and excessive delays can occur.

Although very thin grout will permeate a formation faster, it is fallacy to think that it is properly filling defects, as it contains mostly water. It is the cement solids that remain upon curing, that counts not the water, which dissipates. Some grouters are of the opinion that viscous grout cannot be as readily injected, but this is not necessarily so. The fact is that grout penetrability depends more on grain size and absence of flocculated particles than on the amount of water in which they are carried. Injection of a lesser amount of grout carrying a considerable amount of cement is superior to placing a larger quantity composed mostly of water.

Regardless of the water:cement ratio, bleed will occur in unstable grouts, especially when they are moving slowly or at rest, as is common as a hole stage nears refusal. As this occurs, thin grout and water will gather at the top of the hole. This content should be bled off periodically through a *bleeder* valve installed on the grout header, as illustrated in Figure 6.2. The valve should be installed as high as practicable, as water and thin gout tend to rise. Where holes are inclined, the bleeder should be on the high side.

The use of unstable grouts is contentious at best, and their application is questioned by many. If they are used, periodic bleeding is essential for quality performance. Unfortunately, this is an area field crews often neglect and for which they fail to provide the meticulous attention required. Perhaps the greatest proponent and prime authority for such activity is Houlsby (1990). Although he provides extensive discussion of the importance of bleeding, he also acknowledges that "it tends to be skimped or overlooked unless operators are suitably enthused as to its importance." Within my experience, motivating operators to become "suitably enthused" about bleeding is difficult if not impossible.

12.2.4.2 GROUTING INTENSITY NUMBER (GIN)

The concept of grout *intensity* was first proposed by Lombardi and Deere (1993). They recognized that the grout pressure that can be safely used without jacking or damaging the formation is directly related to the area of defect surface in contact with the fresh grout, as discussed in Section 9.4.1. Knowing exactly the size of the area affected by the grout at any particular time is not possible, but they reckoned it would correlate roughly with the quantity of grout injected. The grouting intensity number (GIN) was thus set as the product of the injected volume times the pressure at a given time.

To simplify the process, injection employs a single stable grout, injected at a uniform rate and conforming to a GIN limitation. Recognizing the poor correlation between water test Lugeon values and groutability, GIN eliminates such testing, further simplifying the process. Instead, computer-generated graphs showing the grouting parameters of each hole stage in real time are continuously evaluated, and the GIN number limitation is not exceeded. Furthermore, use of a constant injection rate allows use of a simple direct injection system. The use of GIN involves:

- A single stable grout mix
- A constant, uniform pumping rate
- A continuous evaluation of the grouting parameters during injection
- Limitation of grout volume at low pressures
- Low volumes at high pressures
- Progressively higher pressure as the rock tightens
- Prevention of damaging high pressure *combined with* high volume

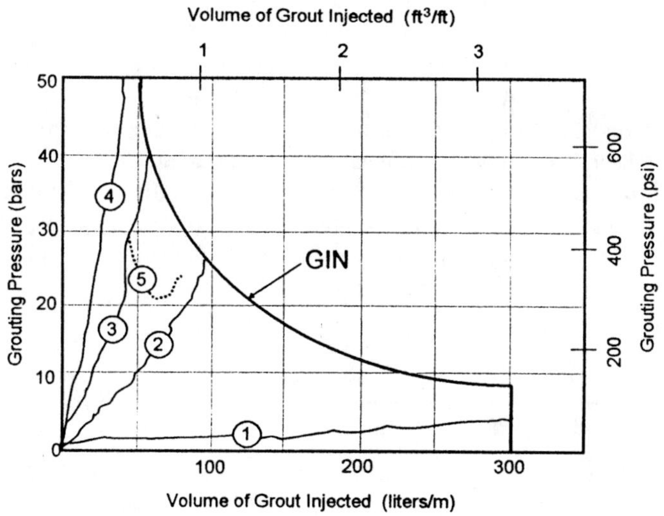

FIGURE 12.8 Parabolic GIN curve with typical pressure volume relationships: 1, wide-open defects; 2, common fissures; 3, fine fissures; 4, very tight fissures; 5, hydraulic fracture.

Once the limiting volume and pressure are chosen, the GIN curve is constructed. It will always be parabolic, as illustrated in Figure 12.8. Shown here are a range of pressure-volume curves that may reasonably be expected to occur during typical injections. Note that when filling fractures in the early injection phases, the volume will be high and the pressure low. Conversely, higher pressures will be experienced in the later phases, but the volume will be significantly less. The magnitude of the GIN curve must be carefully chosen, based on the condition of the particular rock to be treated and the required standard of seepage reduction. An example of typical values first presented by Lombardi and Deere appears here as Figure 12.9.

For most grouting a moderate GIN, on the order of curve 3, will be applicable. As the seepage reduction standard increases, a higher GIN will be required, and vice versa. Similarly, a higher GIN is warranted for grouting good rock at considerable depth than would be wise for poor rock at a shallow depth. It is not necessary to limit all grouting on a project to any one GIN value. For example, injection along the base of a dam may incorporate the fairly high GIN of curve 2, whereas that on steep treading abutments where the pool is shallower, is treated at a lower intensity.

Table 12.1 provides the typical values of maximum pressure and volume for a variety of common GIN values, as shown in Figure 12.9. Particularly sensitive injection, such as into essentially horizontal near-surface fissures or overconsolidated rock, will require a low-intensity number, such as delineated by curve 5 or even lower. The curves shown in Figure 12.9 should be considered for guidance only. Values either greater or lower may be found optimal for a given requirement.

Any combination of grout pressure and take can be used, as long as their product falls under the selected GIN curve. Thus, in the early phase of grouting when larger defects are filling, GIN will limit the distance of travel by controlling volume. The common problem of runaway takes, wherein grout travels a great distance beyond that required, will be largely mitigated. Conversely, in later injection, larger defects will have already been filled, as will a good portion of those that are smaller. Because the grout volume, and thus the affected area of defects, is now small, a significantly higher pressure can be safely used. This will more effectively impel grout into the smaller apertures, speeding up the work and meeting a higher-quality seepage standard.

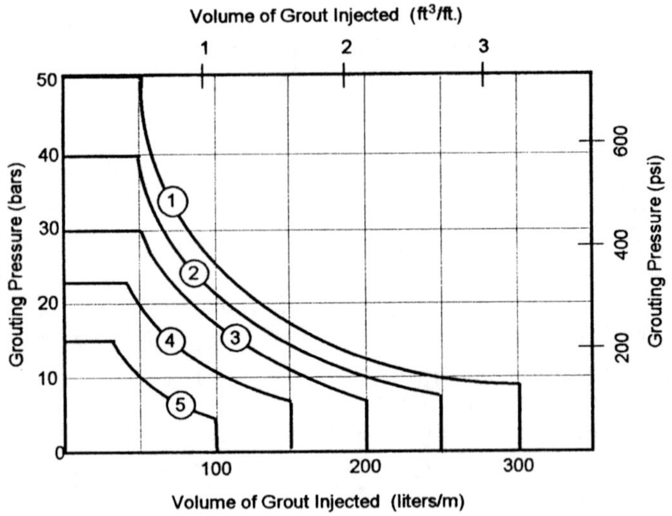

FIGURE 12.9 Typical GIN curves. (From International Water Power & Dam Construction Magazine. Copyright Wilmington Media, 1993.)

12.2.4.3 AMENABILITY THEORY

Acknowledging that the penetrability of water into in a given formation can be very different from that of grout, Naudts (1995) proposed that the *amenability* of a particular grout to a given rock is important. Because a given grout is easily driven into one formation does not ensure that it is amenable to all formations. Thus, the particulars of the geologic formation and those of the grout must be matched. Because there is little we can do to change the formation, we must

TABLE 12.1 Typical Values for Grouting Parameters in Using GIN

		GIN		MAXIMUM VOLUME		MAXIMUM PRESSURE	
No.	**Intensity**	**Bars × L/m**	**(psi × ft³/ft)**	**L/m**	**ft³/ft**	**bars**	**psi**
1	Very high	>2500	388	300	3.2	50.0	725
2	High	2000	310	250	2.7	40.0	580
3	Moderate	1500	233	200	2.1	30.0	435
4	Low	1000	155	150	1.6	22.5	326
5	Very low	<500	76	100	1.1	15.0	217

tailor the grout mix to provide properties most amenable to penetration into the particular rock.

Amenability is the ability of a particular grout to penetrate joints and other defects permeated with water. It is defined by the amenability coefficient (A_c) of the grout, which is expressed as follows:

$$A_c = \frac{Lu_{gr}}{Lu_{wa}}$$

Where: A_c = amenability coefficient
Lu_{gr} = Lugeon permeability of the grout
Lu_{wa} = Lugeon permeability of water

The grout rheology is to be adjusted so as to maintain as high an amenability coefficient as possible. This should generally be greater than 75 percent, and preferably higher. Real-time computer monitoring is required to constantly evaluate ongoing amenability. If the coefficient descends to an unacceptable level, a change to a more amenable mix is made.

Naudts places great emphasis on grout rheology and calls for *balanced stable mixes*, which he considers to be optimal for the work. Six or more components may be included, along with one or more viscosity-modifying admixtures as a mainstay. The mixtures are evaluated by thorough testing, including, at a minimum, Marsh funnel efflux times and pressure filtration properties. Typically, a suite of different mixes will be available and may even be used on the same hole. Should none of these attain sufficiently high amenability upon injection, the need for continuing the work is deemed questionable.

It is reasoned that the percentage of water permeability above that of the grout discloses ungrouted defects, resulting in secondary permeability. Although this may seem rational, there is a possibility that many of the apertures originally open to seepage have become isolated or blocked by grout, even though not completely filled. It is common to find larger defects connected to very fine fissures that are penetrated by water. Such small apertures can easily become blocked, preventing water access, as a result of the grouting. Further, having many different grout mixes available presents complications in performing the work. Where reasonably required, this can be dealt with, but it is counterproductive to complicate a job more than necessary.

Because the use of stable grout mixtures and continuous computer monitoring are advantageous and usually cost-effective, this approach should be more widely utilized. However, the enormous variability of rock into which injection is made must be recognized, and the use of "cookbook" solutions to all grouting must therefore be questioned. Improved technology is certainly needed, and knowledge of the type, history, and mechanics of the particular rock being treated is requisite. Considered judgment will continue to be an essential part of all grouting, however.

12.3 GROUTING ROCK FOR STRENGTHENING

The procedures used for the purpose of strengthening rock are not really much different from those used for water control, except that grout strength becomes important. Conversely, there is seldom a need to penetrate the very smallest of defects, so the number of grout holes and the effort required are usually less. Cementitious grouts are used almost exclusively for this work, although resinous grout may be used on rare occasions when high bond strength is required. In such situations, it is important to first ensure that the rock surfaces to be bonded are sound and clean.

The stability of the grout is important in strengthening applications, especially where defects are thick and/or oriented horizontally or nearly so. Bleed water will always migrate to the top of the grout, resulting in weak laitance or a void, which prevents bond to the top of the void. Although cementitious grouts will bond well to wet surfaces, many resins, especially those in the epoxy family, will not. It is thus important that any resinous grout considered for the work be compatible with moisture. Also keep in mind that even though a particular grout is moisture compatible, if it is strongly exothermic, heat will be produced upon polymerization. In large masses, high exothermic heat can cause boiling and moisture condensation, which adversely affects bond, compressive, and tensile strength. Highly exothermic resins should thus not be used in large masses and should never be deposited into standing water.

Although water testing is usually not requisite in strengthening applications, rock that is particularly absorptive and/or dry should be prewetted so that it does not pull an excessive amount of water out of the advancing grout. Most strengthening of rock is at fairly shallow depths and is thus often in dry rock. If water circulation flush has been used for drilling, sufficient moisture will usually remain unless there has been a long delay between drilling and grout injection.

12.4 GROUT HOLES

Until very recently, the method for drilling grout holes was a contentious issue at best. For grouting dam curtains, one large government agency in the United States proclaimed that any type of percussive drilling would cause the joints and fissures to somehow become clogged, such that subsequent grout injection would be impaired. This thinking is still ingrained in the minds of many old-timers, but its fallacy has been repeatedly exposed by a vast amount of successful work using rotary percussion drilling. Although both rotary and rotary percussion drilling can provide satisfactory grout holes, because the latter is significantly faster and more economical, it is the preferred method for most work.

The method used for removal of the drill cuttings is important to the cleanliness of the hole, however. For holes more than about 10 ft (3 m) deep, standard rotary percussive buttonhead bits are the most commonly used. These literally pulverize the rock, producing cuttings that are very fine. Many drillers prefer to blow the cuttings to the surface with compressed air, but this has not been satisfactory for grout holes. If the drilling is in moist or wet conditions, as is common, the cuttings will combine with moisture to form a fine slurry, which can readily block any cracks or crevices it contacts. Although contamination of defects with at least some cuttings is unavoidable, water has been found to be the most innocuous flush medium for use in rock. In deep holes, where good removal is difficult or high static head pressures are forbidden, an airfoam/water flush can be used, as discussed in Section 10.2.2.1.

12.4.1 Hole Size and Alignment

As discussed earlier, grout holes in rock are usually 2–3 in. (51–76 mm) in diameter. Maintaining good alignment of such small holes when drilling to considerable depths is difficult at best. It becomes nearly impossible when the drill string encounters a near-vertical bedding of rock of variable hardness. In this situation the bit will follow the bedding when it encounters a hard layer, as illustrated in Figure 12.10. It can also become a problem in rock that is heterogeneous and contains many hard and soft areas.

For holes to be drilled to good accuracy, much more time will be required and the cost will be high. Performing alignment surveys will present a further cost and take even more time. Working to rigid alignment requirements is thus

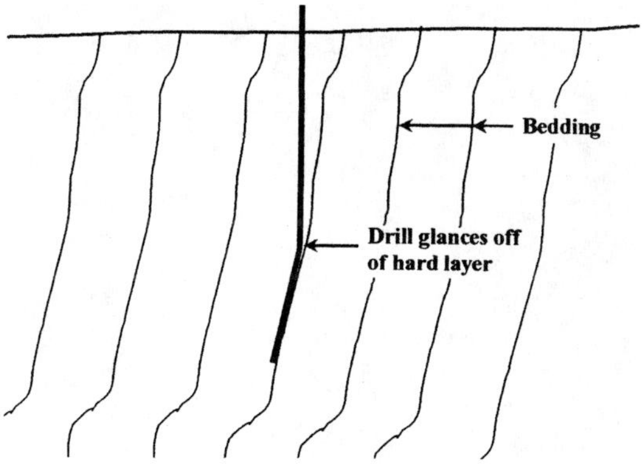

FIGURE 12.10 The drill will tend to follow the direction of steeply inclined bedding.

seldom warranted for grout holes. It should therefore be expected that such holes will stray off course and miss some defects. This is the reason that water pressure tests should be conducted in all closure holes, some of which will likely display excessive seepage. It is nearly always more economical to drill and grout additional holes in such situations than to achieve perfection in hole alignment.

12.4.2 Casing/Collar Standpipe

To provide a suitable connection for the grout header, or a good surface to seat a packer, some form of casing or pipe must be secured into the formation at each hole location. For most rock grouting, this is a short piece of pipe, usually cemented into an initially drilled oversized hole. Because such standpipes can be subject to much abuse, they must be sufficiently strong and embedded firmly with a good-quality grout or mortar. They are typically made of steel pipe, a size larger than that of the drill bit to be used; standard Schedule 40 galvanized pipes are most often employed. These are usually cut into lengths of about 3 ft (0.9 m), about two-thirds of which is embedded into the rock. The top is usually threaded to accept injection fittings and/or a suitable cap.

Standpipes should be used even if subsequent injection is done with downhole packers, as they will prevent waste water or other deleterious materials from entering the hole. When a pipe is not in use, a suitable cap or plug should be loosely fixed to the top end for additional protection. The pipes must be set to precise alignment of the grout hole. This usually requires checking and adjustment as the embedding grout stiffens, immediately before final set. The embedment grout must harden sufficiently to hold the pipes firmly in place prior to drilling. Where rapid hardening is desired, fast-setting casting plaster (plaster of paris) can be used.

12.4.3 Surface Treatment

For curtain grouting of new dams, excavation of a *key trench,* upon which the core will be founded, is often completed prior to grouting. It is important that the surface be clean and sound. On surfaces that are even and free of large depressions, but of weak or soft friable rock, or where open cracks and fissures exist, *slush grouting* is sometimes used. In this process, a cement-sand grout or mortar that has been mixed into a slurry consistency is simply applied to the surface. It is worked into any cracks or other defects with stiff brooms. The thickness usually varies from 1/8 to 1/2 in. (3 to 13 mm), although it will fill in any isolated depressions to greater depths. Because it will run into any substantial cracks and seal over the smaller ones, it helps greatly in avoiding grout leaks when injection is at shallow depths.

One step up from slush grouting is the use of shotcrete to form a solid membrane of about 1 in. (25 mm) to several inches (centimeters) in thickness. There are two methods used to spray shotcrete, *dry* mix and *wet* mix. In the dry mix process, a premixed dry cement and aggregate mixture are blown through a hose to the nozzle, where water is added through the radial jets of a *water ring*. Dry-process shotcrete is usually absent large aggregate, containing only sand. In the wet mix process, a mortar or concrete is premixed and pumped through a hose to the nozzle, where a blast of compressed air propels it onto the receiving surface. Wet-process material is usually supplied in ready-mix trucks and pumped with common small concrete pumps. The wet mix process is most often used to place membranes over rock, as it is faster and more economical.

Because many types of rock cannot be cut uniformly, the base of the key trench is often rough, as illustrated in Figure 12.11. Such a surface is difficult to work on. In addition, waste water and spilled grout will run into and fill the depressions with weak material that will later require removal. Such surfaces can be treated by filling depressions with *dental* concrete, sometimes referred to as *regularizing* concrete, so as to leave a relatively smooth surface, as illustrated in Figure 12.12. Regularizing concrete is typically delivered in ready-mix trucks and pumped or sprayed into place. It is usually the most economical mix that provides sufficient strength and easy placement.

FIGURE 12.11 Very rough finished surfaces are common with some rock types.

FIGURE 12.12 Depressions in the rock can be filled with dental concrete to provide a relatively smooth surface.

12.4.4 Grout Caps

In situations where surface rock is heavily fractured or weak, the provision of a concrete *grout cap* can be advantageous. In its simplest form this may consist of a simple trench filled with concrete. It is important, however, that the trench walls be well roughened and the width at the bottom be no less, and preferably more, than that at the top, so as to provide good restraint against upward movement during injection. Caps are typically 2–4 ft (0.6–1.2 m) deep. They should extend a minimum of 2 ft (0.6 m), both up- and downstream from the nearest row of grout holes.

Grout caps can be reinforced and tied down to the rock with steel anchors, as shown in Figure 12.13. Note that the standpipes have been installed so as to allow for drilling following the placement of the concrete. Where this is done, it is imperative that the pipes are properly spaced and well secured at the correct alignment. In situations where the concrete is lightly reinforced or not reinforced at all, it is probably best to wait until it has hardened and then drill for the standpipes, in order to simplify their being correctly aligned. In the case shown in Figure 12.13, the

FIGURE 12.13 Grouted anchor rods secure this reinforced grout cap to the rock surface.

reinforcing steel is so dense that it would be virtually impossible for drilling to proceed without frequently encountering it; thus, preplacement is required.

12.4.5 Grouting Galleries

Where grouting must be carried out on a surface of severely flawed or badly shattered rock, leakage to the surface is a huge problem. Likewise, because there is little vertical restraint, jacking of the surface is a constant risk. And, of course, only minimal grout pressure can be used, which limits the area of effect of individual holes. In such cases, it is advantageous to construct at least the bottom portion of the dam before grouting. In such cases, a *gallery* is formed in the lower portion of the dam, from which the grouting can be performed at any time.

There are many advantages in using a grouting gallery. First, the completed portions of the structure offer greatly increased restraint to the rock, allowing higher grout pressures and virtually eliminating the problems associated with grout leakage and surface uplift. Further, they allow the grouting to be performed concurrently with the other construction operations offering little interference to them, and to continue in the event of inclement weather. And, of course, they facilitate any future remedial injection following completion of the dam or reservoir filling. Because such further grouting can be readily performed with minimal effort or disturbance, a less thorough and conservative initial grouting program can be justified.

With all these advantages, one might ask, why not use grouting galleries for all new construction? Unfortunately, there are disadvantages to their use, which are substantial. Grout galleries are typically small tunnels cast into the structure, most often only about 5 ft (1.5 m) wide and seldom more than about 10 ft (3 m) high. All supplies and equipment must be manhandled and carried into the work area. Further, the linear and very confined working area precludes the use of optimal drilling equipment and restricts the length of the drill rod used.

Overall efficiency is also adversely affected, as movement for one operation past another is severely restricted. Furthermore, long delivery and return lines must be run to the grout plant, which is typically on the outside. This complicates changing the grout mix and restricts cleaning the line or header, as any grout waste must be manually carried out. And then there is the noise, especially when percussive drilling is used, making verbal communication nearly impossible.

The use of grout galleries is not limited to concrete dams. They can be constructed under or within earth and rock embankments. In several cases they have been formed into much more extensive cutoff structures under large embankment dams. Whether the advantages outweigh the disadvantages is a subject of frequent debate among dam designers and builders. Where galleries are employed, they are often restricted to the lower portions of a structure, with higher sections of the abutments being grouted in a conventional manner from the surface.

The combination of methods used for grouting during construction of Olivenhain Dam in Southern California makes an interesting example. Here, injection into the valley of the dam was effected from a grout gallery, whereas that on the upper portions of the abutments was performed from the surface. In a large hollowed area on the left abutment, however, it was performed from a massive concrete "shaping block," all as illustrated in Figure 12.14.

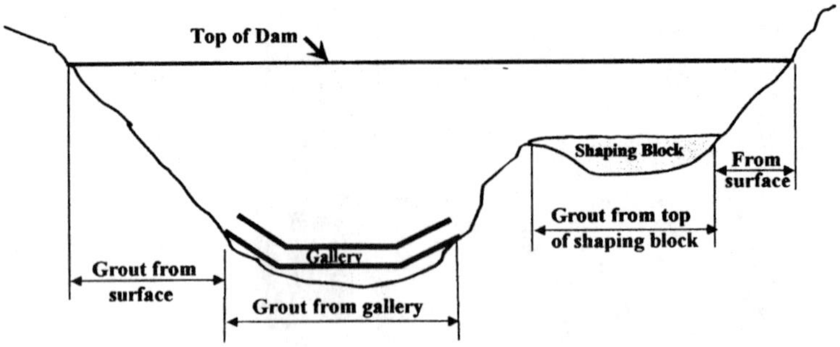

FIGURE 12.14 Grout started under different surface conditions on Olivehain Dam.

12.4.6 Drilling

As discussed earlier, the most frequent method for drilling grout holes is rotary percussion with water for circulation flush. For reasonably accessible areas, self-contained hydraulic drill rigs are used. When drilling is to be accomplished on steep inclinations or in areas of poor access, pneumatically powered track drills are usually employed. Either self-contained hydraulic drill rigs or individual drifter drills mounted on a carrier or *jumbo* are commonly employed for work in tunnels or other underground construction. These carriers are usually self-propelled, mounted on rails or rubber tires, and often contain two or more drills. Handheld pneumatic rock drills can, of course, be used for shallow holes.

12.4.6.1 DRILL ACCESS

Access for drilling grout holes is often less than easy and can be a significant challenge for the driller. For steep inclined surfaces such as dam abutments, relatively lightweight air track drills are common, often with a supplemental winch to aid in traversing steep slopes. These have a low center of gravity, so they are resistant to tipping should one track become lower than the other.

On sites where a large number of holes or very deep holes are required, a drill rig can be mounted on a wheeled carrier that runs on some sort of fixed track. Because construction of such facilities can be a major undertaking in itself, these arrangements are usually reserved for situations where access is very poor and a large amount of drilling must be accomplished. Figure 12.15 shows tracks on continuous formed concrete guide walls for the drilling of holes up the long abutment of the Portugues Dam in Puerto Rico. Such an installation is very expensive to construct, but in this case allowed a large and powerful self-contained drill rig (Figure 12.16) to be used. The cost here was more than justified by the significantly more productive drilling of many deep grout holes.

FIGURE 12.15 Tracks placed on continuous guide walls up a dam abutment slope.

FIGURE 12.16 Powerful drill rig is able to operate from rail-wheel-mounted platform.

12.4.6.2 CLEANING THE HOLES

As previously mentioned, thorough cleaning of grout holes is essential. When drilling is in hard stable rock, free of impacted joints and fissures, one must circulate the flush water only until it appears clear. Drilling in shale or other rock that contains clay, or in weathered rock, can produce very fine cuttings that tend to seal small defects and coat the hole wall. Obviously, such deposits must be removed, as feasible before grout injection. In some rock the holes may penetrate crevices infilled with soft material. Where existing, this also should be removed to the greatest extent practicable.

Cleaning the hole walls of clayey substances takes considerable effort. It is best performed with a jetting pipe fitted with horizontal orifices and plugged on the bottom. This is easily made by simply drilling a number of small radial holes at the base of a plugged piece of pipe or drill rod. It is then worked up and down the hole and simultaneously rotated so that the water jets impinge on the entire wall surface. The amount of effort required will vary with the particular rock, but the cleaning action must continue until clear water runs from the hole when the jet is at the bottom.

The clearing of infilled fractures, although performed in a similar manner, requires significantly greater effort, and even with such greater effort, is not always completely successful. To be effective in such applications, much higher pressures are required than are commonly used for normal circulation flush. Commonly available pumps used for high-pressure washing and/or hydrodemolition are better for this application. These are available in two categories—high pressure, which refers to pressures up to 20,000 psi (1380 bar), and ultrahigh pressure, which exceeds 20,000 psi (1380 bar) and is available with a pressure capability up to about 40,000 psi (2759 bar).

Care must be exercised in the use of such equipment, as the jets can scour even good-quality rock for a considerable distance. Should this occur, seating a packer may become impossible in the scour area, and, of course, a good deal of grout will be wasted. The use of such equipment should thus be limited to the intervals of a hole that need be scoured out. If infilled fractures are intersected by multiple grout holes, jetting can sometimes be performed progressively so as to wash the deleterious materials from one hole to the next.

It is advisable to follow such jetting with normal washing from the bottom of the hole, working sequentially from one hole to the next until clear water returns. Where large defects occur, they are best filled prior to the general grout injection. When they are continuous, extending through several holes, it will be necessary to place packers over the defect in the holes adjacent to that being grouted, in order to preclude their being filled with grout. Be aware, however, that it is extremely difficult to thoroughly clean and fill infilled defects.

12.4.6.3 LOGGING OF DRILL HOLES

The cost of drilling grout holes is considerable. Thus, the comparatively small additional cost for continuous logging, which is recording of drilling data, can be money well spent in most instances. Regardless of the thoroughness of the original geological investigation, much usually remains to be known about the actual conditions, which are often greatly variable. Although detailed logging can slow the work and be excessively costly, simple notation of such parameters as drill penetration rate, drill string *drops* that are indicative of open voids, the general nature of the rock as evaluated by examination of drill cuttings, the percentage of lost circulation, and other salient features of the work can provide valuable input.

A reduction in the drill flush return provides a good idea of the rock permeability. As a general rule, where a significant loss occurs, the drilling should be stopped and the hole grouted before further advancement. This will prevent an excessive amount of drill cuttings from being washed into the defects, which would interfere with subsequent grout filling. One might ask, What is a "significant" loss? Unfortunately, there is no single, always correct answer.

Making such evaluation requires experienced judgment. Before commencement of drilling, a trained professional should have a general idea of the conditions expected, based on a review of geologic reports. The actual conditions will become more clearly known upon drilling and grouting the early holes. In most instances, the experienced behavior will be in line with that reasonably expected. Where this is not the case, however, special attention must be given to carefully recording all parameters, including those of any grouting that is performed to correct circulation loss.

In rock that contains many small fractures, a loss of 25–30 percent may result in significant plugging of the apertures, whereas a 75 or 80 percent loss may not cause significant damage where running into a single wide defect. Obviously, leakage into massive voids that may exist in soluble rock is not of great concern. From a practical standpoint, however, drilling *blind,* that is, without cutting return, is risky at best. Unseen cuttings can become packed in the hole above the bit, making it difficult or impossible to withdraw. This will often require drilling of adjacent replacement holes, which is a cost to the contractor, but, more significant, will result in variation of the hole spacing and a disruption in the normal work progress. Furthermore, when injecting replacement holes, grout will likely make its way into the original hole, which cannot be properly staged because of interference of the stuck drill rod.

Software programs are available that enable continuous monitoring of the drilling parameters in real time. These are often used by European practitioners and have been employed on significant jobs within the United States, as well as in other countries. Drilling parameters such as torque, thrust, rotation speed, and rate of advance can be entered so as to provide the relative drilling energy. Although this is of greater significance in larger holes, and especially those in soil, such data can also assist in developing a good understanding of the downhole rock conditions.

Generally, at the very least, manual records should be maintained of any significant behavior during the drilling of every grout hole. These are best noted on forms specially prepared for the particular needs of the individual project. The required entries should include, but not be limited to:

Date and start time
Hole identification
Hole location, including row and exact position.
Name of driller
Penetration rates for each increment of depth
Drill string drops or significant voids encountered
Zones of caving
Nature of drill cuttings from return
Estimated percentage of lost circulation
Unusual drill behavior

And, most important, the current time should be entered at *every* data entry.

DRILLING LOG Date__________ **Hole No.**__________

XYZ Dam Project **Prepared By**__________ **Page**______**of** ______

Type Drill__________ **Flush** __________**Location**__________ **Surface Elevation**__________

Time	Depth Interval	Rock Type	Circ. Loss %	Comments
	to			
	to			
	to			
	to			
	to			
	to			
	to			
	to			
	to			
	to			
	to			
	to			
	to			
	to			
	to			
	to			
	to			
	to			
	to			
	to			
	to			
	to			
	to			
	to			
	to			
	to			
	to			

FIGURE 12.17 A typical form used to log the drilling of grout holes.

Every individual project has it own special character and nuances. Some of these will be known prior to the work, but many will become evident only as the effort progresses. Record forms must be kept simple and designed with the goal of minimizing required writing, as the primary job of field technicians is to observe what is happening, not to write books. It is thus better to provide boxes to receive checkmarks or perhaps numbers, rather than to expect monitors to write in repetitive comments. A typical Drilling Record form is shown in Figure 12.17. Should new categories of needed information develop during execution of the work, new forms should be prepared to include such data. Field personnel will maintain good records if it is reasonably easy to do so.

12.5 THE GROUT MIX

The subject of grout materials and mixtures has been fully discussed in Chapters 3 through 8, so it will only be summarized here in regard to injection into rock. To penetrate fine cracks and fissures, highly penetrable grout is required. Traditionally, grout was composed of water and cement only. To thin such mixtures, water was added, resulting in very high water: cement ratios. Such grouts are unstable and undergo excessive settlement of the solids and resulting bleed. The amount of bleed increases as the quantity of the water increases.

Although they have been used extensively historically, and in some cases to this day, unstable grouts will only partially fill joints and other defects. The space initially filled with bleed water will become an air-filled void upon evaporation of the water. This is a particular problem in filling defects in an essentially horizontal configuration, as the resulting voids can be continuous for long distances. Conversely, in vertical channels, the grout will normally settle so that the lower portions of the defect are tightly filled and the top portion contains little if any grout. It is thus preferable to use stable grout that has been designed to exhibit little, if any, bleed. Although some may disagree, it is my opinion that, with the technology readily available today, there is no valid reason for continued use of unstable grouts.

12.5.1 Grout Mix Properties

Stable grouts are easily compounded and mixed. The first consideration is minimizing the mix water without changing the grout consistency, which is accomplished with readily available water-reducing admixtures. Obviously, as the amount of water in the grout is reduced, less of it will be available to rise as bleed. For water reduction, a variety of admixtures with different capabilities are available. The *high range* water reducers, sometimes referred to as *superplasticizers*, provide the greatest reduction, but at a high price. Similar water reduction can be achieved with very large dosage of a less expensive lignin-based water reducer. Large doses of these will retard set and significantly lengthen the time of strength gain, however long-term strength and durability are not affected.

Viscosity-modifying agents will thicken the liquid phase of the grout, decreasing the propensity of solid grains settling. Both the proportion and type of viscosity modifier must be carefully selected, however, as these agents will increase the grout's cohesion, reducing its penetrability. Historically, stable grouts have been achieved through the inclusion of bentonite at a rate of about 2 percent, by weight, of the cement. Bentonite, however, will greatly increase the water requirement, negatively affect the strength of a grout, and possibly its durability, and markedly increase cohesion, so it is often better to use a viscosity modifier designed for use with concrete.

12.5.2 Grain Size/Chemistry

When very fine apertures must be filled, both the mean and the maximum size of the cement grains must be considered. In general, Type III cement is ground finer than the more commonly used Types I and II. Where still greater fineness is required, ultrafine cement can be used. In addition to the benefit of providing greater penetrability, the finer grains also have less propensity for sedimentation and the resulting bleed. The relative grain size distribution of various cements is presented in Figure 5.8.

Further consideration must be given to the origin and chemistry of the particular cement, as discussed in Section 5.5. Some ultrafine cements have a high affinity for a particular rock and are thus drawn to it, whereas the affinity of others is not so strong. In general, the cost of ultrafine cement is greater than the costs of other types, but the resulting grouts are more penetrable, allowing faster injection and better filling of the defects.

12.5.3 Optimal Mix Consistency

According to conventional thinking, a given hole or stage should be started with a relatively thin grout, which is thickened during injection if the take is considerable or the pressure rise is slow.

The justification for this approach is that a thicker starting mix would tend to block off finer defects prematurely. Although this may seem logical, and the thinking well ingrained in present grouting technology, a reconsideration may be in order.

When injection is started, grout goes to the area of least resistance, which is always the larger or thicker defects, although it may enter the finer ones for a limited distance. As this occurs, individual grains of cement are deposited on the walls of the passage and form a coating. Such coatings are often seen on the interior of grout hoses, which are much smoother and more resistant to such buildup. Although grout contacts the wall of a hose for a longer time than the typical duration of injection of a hole, there can be no question that this duration is sufficient for some blockage of the finer apertures to occur.

When excessively thin grouts are used, they travel an excessive distance in the larger defects, likely increasing cost, without benefit. Furthermore, because they always contain much more water than thicker grouts, bleed is increased and strength and durability lessened. Although these effects will be less dramatic in rock that is relatively tight, the deleterious influence can be substantial where large defects exist. Under such conditions, thicker grouts may best be used in the earlier holes and the thinner mixes reserved for the higher-order holes or perhaps the later grouted rows.

12.6 INJECTION CONSIDERATIONS

Once the holes have been drilled and cleaned and the grout mix established, injection can begin. In practice, these operations are all done simultaneously; however, the sequence must be carefully planned. In general, no two adjacent holes should have either drilling or grouting in progress at the same time, because of risk of interconnection. In highly fractured rock, with significant fissures likely to interconnect several holes, a greater distance between active holes is in order.

12.6.1 Injection Sequencing

Judgment must be exercised in hole sequencing. Where rock is generally sound and the risk of interconnection small, drilling the upper portion of a hole may be appropriate while grouting is performed in lower portions of an adjacent hole. In general, the spacing of activities can be closer in relatively sound rock, but it must be spread out in highly fractured rock, where interconnections are likely. There is always pressure to complete the work quickly, and this sometimes demands closer working distances, but one must carefully weigh whether this is prudent. The extra time required to correct improperly grouted connecting holes can be significant, as can the additional cost. Regardless of the activity intervals chosen, a split-space order of injection should always be respected.

12.6.2 Hole Staging

Not only must the sequence be considered, but also the injection, which will usually proceed in stages; that is, only a portion of a hole will be injected at any one time. Staging can proceed from the top down (*downstage* or *descending stage*) or from the bottom up (*upstage* or *ascending stage*). Downstage progression can be made from a connection at the hole collar or through use of downhole packers. In unsound or heavily fractured rock, *circuit* grouting, which is described shortly, may be used. Thoughtful selection of the length and progression of individual stages is important, as it affects both the cost and the quality of the work.

12.6.2.1 STAGE LENGTH

The lengths of individual stages can vary widely, although optimal selection is crucial to good performance. Shorter stage lengths will always provide higher quality, whereas longer stages are more economical. In some projects, stage lengths of more than 50 ft (15 m) have been used, which is excessive for most work. As a general rule, longer stage lengths are appropriate as the depth increases and/or the rock quality improves.

A length of 5–10 ft (1.5–3 m) is usually appropriate in near-surface stages, which may be increased to 10–20 ft (3–6 m) at depth. Where outstanding features are encountered, or in especially critical zones such as the contact of a dam base, very short lengths of only a few feet (0.6 m) should be considered. These are only guide numbers, however, for there are many factors to be considered in choosing the best length. Although shorter stages improve quality, economy usually pushes for longer lengths, sometimes pushing so hard that quality (and often economy as well) is adversely affected.

The allowable pressure increases with depth, but must be based on the highest elevation of a stage. Thus, the lower portions of long stages will receive less pressure than would be the case were shorter stages used. Furthermore, long stages expose defects with greater size variability. The grout will tend to fill larger voids while initially sealing over smaller ones, as illustrated in Figure 12.18. The effectiveness of the injection will thus be compromised, and seepage reduction may not be satisfactory upon completion of the originally planned holes. This will require additional holes, especially if seepage must be reduced to a high standard. Thus, what initially appears as savings can well develop into a hefty cost increase.

Short stages, however, will encounter fewer defects at any one time and thus not be subject to as much variability of aperture size. This will result in more thorough filling. Where large defects are known to exist however, unless previously filled during drilling due to lost circulation, they should be grouted off prior to the remainder of the hole. And, of course, as previously mentioned, any large cavernous voids should be filled ahead of the general grouting, for the same reason.

The selection of optimal stage lengths is one of those parts of grouting that depend largely on experience. They are also probably best determined as the work progresses, and the best stage length will likely vary, not only between different holes, but within a given hole as well. In summary, shorter stage lengths should be used where the rock quality is poor, in particularly sensitive areas, and nearer the surface than at depth, as well as in holes with a wide variety of aperture openings. Longer stages are applicable to deeper injection, in holes with more consistent defect openings, and in better-quality rock.

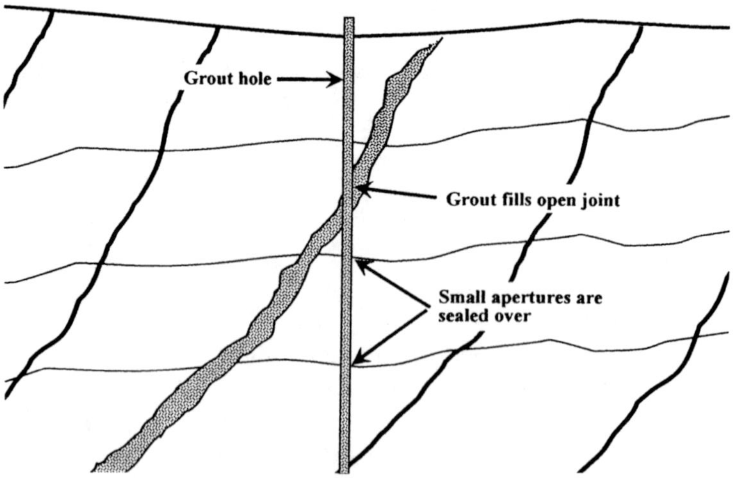

FIGURE 12.18 The smaller defects are likely to get sealed over while the grout goes to larger ones where the hole crosses both.

12.6.2.2 INJECTION DOWNSTAGE FROM COLLAR

Downstage grouting involves drilling to the preselected base elevation of the stage. A connection is then made at the hole collar, and grout is injected. Once this grout has attained initial set, the hole is generally washed out so as to minimize the required drilling for the next stage. When the grout has hardened, the hole is redrilled and extended to the depth of the second stage. The grout line is then reconnected to the collar, and the second stage is injected. The hole is washed out, and so forth, and the work progresses until the entire depth of the hole has been treated. In downstage work, similar depth stages of all holes in the particular vicinity are usually completed prior to work on the next lower stage. This procedure is required because connecting at the collar for the deeper holes will expose the previously completed stages to the generally higher pressures used at depth.

Working from the top down is advantageous in highly fractured rock or where caving is likely, as it lessens the risk of rock fragments being displaced, falling, and entrapping the drill bit. Moreover, as the drill cuttings pass only through the portion of hole to be immediately grouted, none will be washed into later treated defects. It also mitigates the adverse effects of lost circulation or the fairly rare case of artesian flow. In addition, making a connection to the collar is fast and easy. However, the continual washing of grout out of the holes requires time and is messy. Diluted grout is apt to settle into depressed areas, from which it must eventually be removed. More work is required to reset the drill for each stage, and because a given stage must be completed in many holes before advancing the adjacent holes, more time and effort are required for moving and setup. Furthermore, the upper grouted rock is exposed to the higher pressure used at depth. Although considered a limitation by many, this is actually beneficial, as exposing the previously completed work to the greater pressures provides a good test of upper stage quality and completeness. Working downstage from the collar nonetheless requires more time and is usually more costly.

12.6.2.3 DOWNSTAGE WITH PACKER

Working downstage with a packer is quite similar to using a collar connection, except that the grout is placed through a packer that is set at the top of the stage being grouted. This avoids exposing the upper portions of the hole to the higher pressures used at depth. It does, however, require additional effort to constant trip the packer in and out of the hole and set it for each grout stage.

Because the packer is often suspended from drill rods or other rigid tubes, the stage length should be coordinated with the rod length, which is typically 5 or 10 ft (1.5 or 3 m), or more. Packers can be problematic, especially in small holes or in zones of highly fractured or unstable rock. They can leak or become plugged, dislodged, or stuck in the hole. Grout can flow in defects so as to bypass the packer seal. This was more of a problem in earlier times when contractors fabricated their own packers; however, high-quality packers are now available commercially and tend to be less trouble prone. From an operational standpoint, however, it may be wise to avoid the use of packers whenever possible, except perhaps in the case of very good quality rock.

12.6.2.4 GROUTING UPSTAGE WITH A PACKER

In rock that is generally sound, a hole can be drilled to its final depth and grouted in stages, starting at the bottom and working upward. In sufficiently massive rock with defects generally uniform in width, this is unquestionably the fastest and most efficient method. However, this ideal condition seldom exists in the real world of grouting, so the method is often problem prone, especially where fractured rock zones are encountered. The initially perceived savings can therefore rapidly decrease.

There are also substantial technical shortcomings to this progression. Because the hole is initially drilled to full depth, circulation flush, will flow the entire distance to the top. This will result in the higher joints and defects being exposed to up-flowing flush for an extended period of time. Obviously, a greater amount of drill cuttings will find their way into the higher joints and defects. Significantly, this is usually the zone containing the largest number of defects and where the highest quality of work is needed because of contact with the dam body.

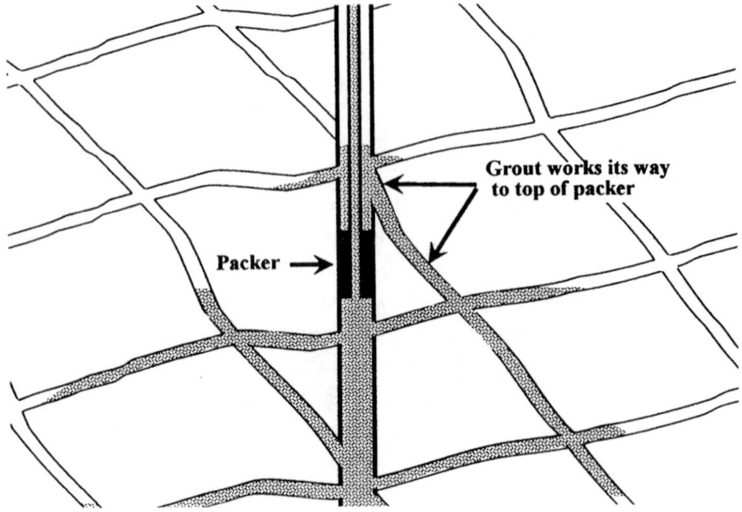

FIGURE 12.19 The grout can bypass a packer when working from the bottom up.

The seating of a packer at depth, to be leak free, can be difficult at best. This is a particular problem in holes passing through significant fracture zones, or in those that are not perfectly round as a result of drill chatter, which is common when they are passing through changing rock conditions. There is also a high risk of grout bypassing the packer through fractures in the rock, as illustrated in Figure 12.19. This not only contaminates higher portions of the hole, but can also cement the packer in place. The grouter is usually not aware of such an episode until grout returns out the hole collar, as can be seen in Figure 12.20. Much time, effort, and expense have been exhausted in attempts to retrieve stuck packers. When they cannot be freed, replacement holes must be drilled.

A further significant limitation is that grout used at the end of a given stage is often thicker than appropriate for beginning the next stage. This thicker grout will require removal, so the packer must be tripped out and flushed clean. This takes time and results in considerable expense. Because the upper rock zones have not yet been strengthened by injection, pressure will be limited at depth, restricting the penetration distance. Furthermore, the risk of jacking or dislodging the upper rock is increased. With these substantial limitations, this method should probably be reserved for grouting in very competent rock.

FIGURE 12.20 Grout bypassing the packer sometimes is not discovered until it exits at the hole collar.

12.6.2.5 CIRCUIT GROUTING

Circuit grouting can be used for integrating badly shattered rock and is sometimes employed in holes subject to caving or rockfall. The procedure consists of running a drill string to the full depth of a hole and injecting grout, which flows up the annulus to the top, where it is captured and returned to the agitator. A screen must be supplied at the agitator to remove any oversized cuttings that have been picked up by the rising grout. Grout is continually circulated until no further take occurs. Pressure is typically limited to the static head to the ground surface, but can be increased by running grout through a riser pipe firmly secured to the standpipe.

The only advantage is the ability to place badly shattered rock under a constant, albeit low, pressure for an extended period of time. Obviously, such grouting treats the entire depth of the hole at one time. Because this is considered a last resort practice, it is reserved for only the most difficult fracture conditions. Thus, downstage sequencing is the only realistic method, and the stage lengths are usually quite short so as to minimize the effect of rocks falling on top of the drill bit, locking it in place.

Where this risk occurs, knowledgeable drillers grind a bevel on any flat areas on the back of the bit so as to reduce the wedging action of the spoil that does drop onto it. In addition, on sites where binding of the drill string becomes a problem, it is a good idea to weld teeth to the back of the bit to facilitate *drilling it out,* should this be required. These teeth should be set in slightly so that they do not contact the rock during regular drilling, which may result in out-of-round holes or additional debris being dragged into the hole. Although not common, special drill bits are commercially available that allow drilling in either direction.

Circuit grouting is limited to downstage sequencing, either with a direct connection to the collar or with a packer. A larger hole will be required when using a packer, in order to accept both the supply and return grout lines. The previous comments relative to packers are even more applicable here because of the additional grout line. Many grouters experienced with the method have vowed to never be involved with it again. It is, however, one of the few procedures available to handle badly caving conditions and can tighten things enough to allow more conventional operations in subsequent holes. It is thus a useful tool to be aware of.

12.6.2.6 MULTIPLE PACKER SLEEVE PORT PIPE (MPSP)

Another way to handle badly broken or collapsing rock is use of the multiple packer sleeve port pipe (MPSP) system, which consists of a standard sleeve port pipe around which fabric bags have been fixed at regular intervals. In application, duplex drilling is used to produce fully cased holes. The specially fitted sleeve port pipe is then placed in the hole and the casing removed, allowing any caving material to collapse against the pipe and the enveloping bags. The bags are then inflated, using grout injected through the appropriate sleeve port. The grout is expected to expand the bags and form seals at predetermined depths, thus isolating the intended grout stages.

Although the procedure sounds easy, this is seldom the case. First, rock debris can fall against the bags before they are filled, resulting in unequal restraint to inflation, and should the rock have sharp edges, the bag can be cut or otherwise damaged. Extreme care must be exercised in filling the bags so as not to exceed their volume, which can result in bursting. At first glance, this may seem simple, as the original volume of the bags is known, but their resulting volume can be reduced greatly by fallen rock debris.

A further limitation is that the inflated bags can seal any cracks and voids with which they come in contact, isolating those defects from orderly penetration of the subsequently injected grout. Furthermore, should substantial voids exist in the caved rubble around the isolation bags, subsequent grout could easily bypass them. This is another tool for the grouter to have available, but to be used only as a last choice.

12.6.3 Before Hookup

It is important to be sure that all equipment is in proper working order and that a continuous supply of fresh grout will be available before hookup of the delivery line. The cost of drilling grout holes and the time required are considerable, and loss of a hole due to lack of proper injection must be avoided to the greatest extent possible. Make sure that sufficient material is on hand to prepare the maximum amount of grout likely to be required for a hole. Enough spare hoses and fittings should also be available to facilitate immediate replacement if necessary. If surface leakage is likely, a sufficient supply of wood wedges and caulking materials should be on hand, along with tools for their installation.

Before the first hookup of a given shift, it is advisable to run water through the system prior to filling with grout. This will wet the surfaces to prevent water from being drawn out of the grout and will also confirm that proper connections have been made. When hookup is made following completion of a previous hole, grout should be circulated in the system until uniform fresh grout returns to the agitator. This is especially important when the prior hole has had a slow take or the grout has otherwise been in the line a long time.

This is also a good time to clean the screen through which the return line empties into the agitator. Finally, immediately before hookup, open the valve leading to the header to ensure that good grout is running freely. This is important, as partially set grout may have developed within the header fittings, especially if the previous hole was a slow taker. It is also good practice to partially close the return valve so as to build up pressure in the system, allowing confirmation of proper operation of the pressure gauges.

12.6.4 Grout Plant

The particulars of the grout mixing and pumping plant will vary, depending on the general layout of the work. Some grouters prefer to establish a single central plant to supply the entire job. This has the advantage of easy access, facilitating optimal equipment arrangement and the use of bulk cement and other grout ingredients. Yet, very long grout supply and return lines can be required, resulting in the grout's aging excessively within the delivery system. This is particularly a problem in warm environments, where setting time is relatively fast, and becomes especially critical when grout takes are slow, resulting in the return of most of the pumped grout to the agitator.

Another disadvantage is that the delivery lines are always in the way and can greatly impede movement around the site. Other operations are typically performed simultaneously, requiring reasonably free movement of vehicles and other equipment. Traffic should not be allowed to cross over the grout lines unless they are protected from direct contact, as this can cause a pressure spike and breakage is likely. A line break, at the minimum, causes a pause in injection, and if spare lines are not readily available, a delay of considerable length.

For these reasons, many grouters prefer to use small, highly portable plants that can be easily moved around the project site and near to the area of use, as shown in Figure 12.21. This allows the use of shorter lines, thus preventing grout from aging in the system. Because the plant operator can visually observe what is happening at the header, communication is also improved. When these plants are frequently moved, the less-efficient bagged cement will generally be required because of the effort required to relocate bulk silos and the frequent limitation to the often restricted sites of the plants. Because of poor access for grouting steep abutments, fixed grout plants can be established at the bottom of a slope or other accessible location.

FIGURE 12.21 Highly portable skid-mounted grout plant.

Underground work is usually performed on alternate shifts or an on-demand basis, so the grout plant must be frequently moved out of the excavation. Thus, it is usually mounted on some sort of vehicle. Cement for such work is nearly always supplied in bags. The logistics of underground grouting can be difficult, and such operations must be carefully planned. This subject is covered more thoroughly in Chapter 16.

12.6.5 The Injection

Once the system is proved to be working satisfactorily, hookup is made and injection initiated. The valve leading to a hole is slowly opened to the point that injection starts at a relatively low pressure. Personnel should be vigilant and constantly scanning the surface around the hole for any leakage, communication with other holes, or surface jacking. This is especially critical when injection is at shallow depths or a direct connection to the hole collar is being used. If all appears well, the flow can be uniformly increased to the maximum allowed pressure. The return valve will be partially closed to choke off the grout return if necessary to develop the desired pressure.

12.6.5.1 GROUT SUPPLY

Once the flow is established, the grout hole valve is manipulated as required to maintain the appropriate pressure. It is important that the injection be continuous and without interruption. The restriction of grout flow, and thus pressure, may be necessary if the grout is being accepted too freely and more rapidly than new grout can be mixed. This is important, because it may not be possible to reinitiate permeation of smaller defects if an interruption in the injection occurs. Personnel must remain alert to any leakage or disruption of the rock surface as injection continues.

If the hole continues to take grout freely for more than about five minutes, thickening of the mix may be considered. Once a few holes on a new site have been completed, the crew should develop a "feel" of what to expect in the various stages of the work. Changing the grout mix takes time and is subject to errors, so it should be done only when clearly needed. Starting mixes should thus be as thick as practicable, usually no thinner than 3 : 1 by volume.

12.6.5.2 MONITORING THE INJECTION

As with all grouting, the best information as to behavior in the formation and the quality of the work is obtained by closely monitoring the injection parameters, especially the injection rate and resulting pressure. The most common procedure in rock grouting relies on the use of a preset maximum pressure, with constant variation of the injection rate to maintain that level. Because traditional practice depends on dipstick measurements of the grout level in the agitator, the amount of grout pumped and subtle changes (and even some that are substantial) in the injection rate are often missed.

The ability to monitor grouting parameters in real time with computers is now well developed. Even the most minor variation of grout behavior will be apparent, which can allow time to make suitable adjustments and optimize the injection, thus greatly improving efficiency and the final quality. It is my opinion that real-time computer monitoring should be mandatory for every significant project.

Where jacking of the rock is likely to occur, continuous monitoring of both pressure behavior and the formation surface is essential. Although periodic surface surveys are useful, the time delay in receiving data usually prevents sufficiently prompt corrective action. This is not the case with continuous monitoring, however, and corrective measures can be promptly taken, in the same manner used in grouting soil, described in Section 11.1.1.4. The injection parameters usually give advance notice of any impending jacking, providing notice that the elevation monitoring instruments need be read.

12.6.5.3 INJECTION PRESSURE

Perhaps no element of rock grouting is more contentious than selection of the injection pressure. In fact, if one were to make a survey of the pressure criteria used around the world, an extraordinary variation would be found, as illustrated in Figure 12.22. Virtually all the values used are based, at least somewhat, on the depth of the injection below the surface. Although many take into account the general condition of the rock, almost none consider either the grout quantity previously injected or the predominant attitude of the defects. These, of course, have an important influence on the propensity to jack overlying rock, as discussed in Sections 9.4.1 and 9.4.2, and illustrated in Figures 9.6 and 9.7.

When one considers the large variation of pressures used around the world, and the significantly higher levels routinely employed successfully for many years, it becomes clear that limiting pressure to unduly low levels is not justified. Both the number of holes required and the speed with they are injected depend on the allowed injection pressure. Likewise, the time required to complete the grouting and its cost are also directly influenced. It is thus advantageous to use the highest pressure that can be employed without undue risk of formation damage.

In earlier times, available monitoring methodologies did not facilitate rapid acquisition of the injection parameters nor allow rapid comparisons of them. With modern computer technology now available, such comparisons can be made quickly and appropriate injection adjustments can be initiated. Therefore, with reasonable care and attention, the risks of excessively jacking the rock or causing other damage have been largely mitigated. For this reason, it may be that significantly higher pressures can be used than are now common, particularly in the United States, where inordinately low limits have been the norm.

12.6.5.4 CONNECTIONS BETWEEN HOLES

Although the permeation of grout for a good distance in a hole is desirable, it sometimes connects with other open holes, which is not good.

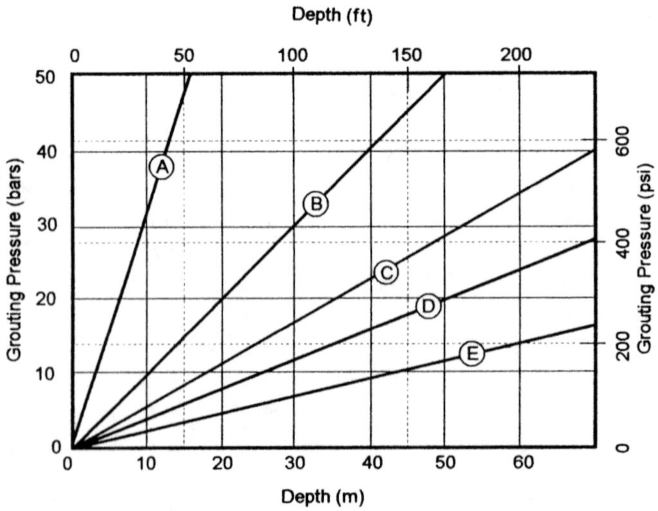

FIGURE 12.22 Variety of allowable grout pressures from different sources: A, Good rock—Swedish prcatice; B, European rule of thumb—1 bar/m of depth; C, normal rock—Swedish practice; D, Swedish prctice—weak rock; E, U.S. rule of thumb—1 psi/ft of depth.

Connections usually occur early in a program during injection of the primary holes inasmuch as major defects have yet to be filled. When this occurs, the first consideration should be an increase of the distance between open holes, either by greater spacing of the layout or sequencing primary holes alternatively.

A connection is usually first noted when grout returns out of a hole near the hole being injected and is often preceded by the return of water or diluted grout. When this occurs, anything other than good grout should be allowed to completely exhaust and should be directed to waste. Occasionally such leaks block themselves off prior to the return of good grout. Unless this occurs, the return flow should be continued until a steady flow of good-quality grout appears.

Once the good grout appears, a second set of injection fittings should be installed on the leaking hole and connected to a second T fitting in the circulation line so that both holes are injected simultaneously. It will be necessary to stop pumping to make the hookup, but such an interruption must be minimized to the greatest extent possible. Where hole connections are likely, additional header fittings and hose should be laid out and preassembled so they can be added to the system quickly if required. When flow from the leaking hole is considerable, there is always a temptation to simply block it at the collar. This is not good practice, however, as the flowing grout has likely picked up rock fragments and other debris during its travel and thus does not have the same degree of penetrability as that in the circulation lines.

12.6.5.5 HANDLING SURFACE LEAKS

When fractures and defects extend to the rock surface, grout leakage is bound to occur, especially when near-surface stages are being treated. The likelihood of such leakage should be considered prior to injection. Isolated large defects can often be filled by simply pouring grout into them under gravity pressure. Smaller apertures are more difficult to treat, and where a high propensity for leakage exists, sealing the entire surface, as discussed in Sections 12.4.3 and 12.4.4, is advisable.

Minor seeps from very tight cracks are difficult to treat during injection, and as long as the amount of grout lost is less than that being injected, may best be allowed to continue. Such seeps often start with the expulsion of water, followed by thin grout, and so on, and will often block themselves off prior to or shortly after a return of good-quality grout. Where the surface rock is hard and competent, leaks that appear in larger seams can often be controlled by caulking with a flexible absorbent material such as burlap or *dry* oakum.

The caulking medium is typically pushed ahead and into the cracks by wood wedges, firmly driven into place, as illustrated in Figure 12.23. Although wedges and caulking act in concert to retard the flow, thickening of the grout as water is lost to the caulking is also a major factor. This is the reason for emphasizing use of *dry* oakum, as it comes in two forms, oiled and dry. Oiled oakum is more readily available, but does not act as an absorbent and so is far less effective.

To control leaks, many grouters subscribe to either thickening the mix or stopping the injection periodically for a pause of up to several minutes. They reason that this gives the grout a chance to firm up on its own and thus self-heal the leak. Although this can occur, what may be overlooked is the fact that self-healing will also happen at depth, sealing over the defects. Hence, effective permeation often ends when such actions are taken. Blocking the flow of grout to the surface once it has started is extremely difficult. The best action is often no action other than cessation of the injection when the return quantity approaches that of the injected quantity or the leakage spoil becomes excessively difficult to handle. Depending on the quantity of the grout

FIGURE 12.23 Surface leaks can sometimes be stopped by caulking with oakum and wedges.

and its behavior prior to leakage, the establishment of additional adjacent grout holes may be necessary to satisfactorily complete the work.

12.6.6 Refusal Criteria

Agreement on what constitutes sufficient injection, and thus completion of grouting a given stage, is almost as contentious as the selection of the pressure required. Some grouters think that refusal should be considered to have occurred when the injection rate decreases to some fixed value, usually on the order of 1 ft^3 (28.3 L) in a period of 5 to 15 minutes. Others argue that the grout line should remain connected and pressurized for some amount of time after all measurable flow ceases.

Houlsby (1990) recommends holding full pressure for a period of 15 minutes after refusal, "in terms of whatever definition is used," has occurred. He further reasons that this will hold the grout "firmly in place" until thixotropic stiffening occurs and the grout becomes resistant to flowing water. Weaver (1991) states that the California Department of Water Resources uses comparable refusal criteria. He further quotes W. H. Bussey (1973): "It is very important to keep pumping as long as there is measurable 'take' in any ten minute interval." Both the U.S. Bureau of Reclamation and the Army Corps of Engineers take a somewhat relaxed view, generally considering refusal as a take of less than some given amount of grout, often 1 ft^3 (28.3 L), in a given period of time, usually varying within a range of 5 to 20 minutes.

Note, however, that all of these authorities have traditionally used unstable grouts, which are subject to settlement of the solids and substantial bleed. By maintaining pressure for a time period after apparent permeation has ceased, much of the bleed water will rise and can be bled off, as emphasized by Houlsby (1990) and discussed in Section 12.2.4.1. The space it formerly occupied will then be filled with fresh grout, somewhat compensating for the unstable behavior. This could very well be the action Houlsby refers to when mentioning the grout being held "firmly in place."

Extensive testing has shown that about 60 percent of total bleed of unstable grouts occurs within the first 15 minutes when at rest. Although there is considerable scatter in the data, no doubt due to the particular grout mix and energy used in its mixing, the trend is clear that significant bleed occurs rapidly. Accordingly, there can be no doubt as to the validity of maintaining pressure for some extended time, combined with appropriate bleeding, when unstable grouts are used.

As previously discussed, I consider unstable mixtures to be outdated and their further employment inappropriate, and thus suggest that with stable grouts, refusal should be considered any point during injection at which no measurable grout passes in a fairly short time period, perhaps 1 to 5 minutes. Where grout quantity is determined by dipstick, the 5-minute period should be used. Conversely, where computer monitoring is employed, an immediate and accurate measure of flow is displayed, so the cessation of flow for only a minute or two should be sufficient.

CHAPTER 13

Grouting in Concrete and Masonry Structures

13.1 **FILLING STRUCTURAL CAVITIES AND VOIDS**
- 13.1.1 Grout Material
  - 13.1.1.1 Cementitious Grouts
  - 13.1.1.2 Resinous Grouts
  - 13.1.1.3 Resinous Foams
- 13.1.2 Grout Holes
- 13.1.3 Grouting in Concrete Structures
- 13.1.4 Grouting in Masonry
  - 13.1.4.1 Injection into Masonry
  - 13.1.4.2 Dealing with Leaks

13.2 **FILLING CRACKS AND JOINTS**
- 13.2.1 Injection Materials
  - 13.2.1.1 Epoxy
  - 13.2.1.2 Polyester Resin
  - 13.2.1.3 Cementitious Grout
  - 13.2.1.4 Urethanes
- 13.2.2 Crack Repair
  - 13.2.2.1 Epoxy Crack Injection
  - 13.2.2.2 Construction Joints in Mass Concrete Structures
  - 13.2.2.3 Joints Between Precast Elements

13.3 **CONTACT GROUTING**

13.4 **POSTTENSION TENDON GROUTING**
- 13.4.1 Rock and Soil Anchors
- 13.4.2 Repairing Defective Tendon Grouting

13.5 **GROUTING BEDPLATES**

13.6 **ABANDONING CONDUITS AND TANKS**

13.7 **FILLING ANNULAR SPACE/BACKFILLING**

13.8 **FILLING FABRIC FORMS**

13.9 **PREPLACED AGGREGATE CONCRETE**
- 13.9.1 Coarse Aggregate
- 13.9.2 Formwork
- 13.9.3 Injection Ports and Pipes
- 13.9.4 Placing the Aggregate
- 13.9.5 The Grout Mixture
- 13.9.6 Injecting the Grout
- 13.9.7 Quality Control

13.10 **AN UNUSUAL STRUCTURAL APPLICATION**

STRUCTURES SUFFER from the same or similar problems as soil and rock, but the sections of structures are always much smaller than the nearly infinite range of geomaterials. Furthermore, their strength requirements are almost always greater, although the volumes of the defects are typically smaller and the amount of material to be injected thus significantly less. Injection rates for structures are usually much lower in order to preclude displacement. Unique to structural grouting is the frequent additional requirement of aesthetically preserving a structure's exposed surfaces.

Injection holes for structural grouting are usually much smaller, and the pumping rate slower, than in geomaterials. This can be an advantage, as difficult access and confined working areas are frequently encountered and are much more readily handled with the use of smaller equipment. Of extreme importance in grouting into structures is the weight of the added grout. Existing foundations often have little, if any, reserve capacity to support additional loads.

13.1 FILLING STRUCTURAL CAVITIES AND VOIDS

Deleterious voids in concrete and masonry usually occur as part of the original construction. There are two major types found in concrete, *rock pockets* or *honeycomb* and empty spaces resulting from a lack of proper placement and/or compaction. The latter are typically in areas that were shadowed by large conduits or other blockouts during construction. The purpose of filling these defects in concrete may be to gain the intended strength or to prevent corrosion of the reinforcing steel by blocking access to deleterious agents.

Void areas are common in masonry construction. Normally, these occur during construction as a result of incomplete filling between the units with mortar or lack of filling all cavities. Unless excessive, voids between the units are not usually of great concern; however, they may need filling as part of a repair or for strengthening the structure. Furthermore, cavities that contain reinforcing must be properly filled for the structural design to perform. Many older masonry structures were built without reinforcing steel. In very old construction, it is common to find walls with thicknesses of several feet constructed of stone, brick, or a combination. Reinforcing steel grouted into predrilled holes is often added to such structures as part of a rehabilitation effort.

13.1.1 Grout Material

Many different materials are used in structural grouting. The hardened properties of cementitious grout are similar to those of the substrate, which is advantageous. Such grouts are usually the simplest and most economical to use, yet because they harden to a dense mass, they will add considerable weight to the structure. Resinous materials, although generally more difficult to work with and always more expensive, have distinct advantages. They can provide better properties, especially in terms of bond strength and low weight, and are extensively used where these performance characteristics are needed.

13.1.1.1 CEMENTITIOUS GROUTS

Cementitious grouts, very similar to those employed in rock, are also used in structures. Unlike the situation in most geotechnical grouting, splatters or spills of grout can be a major problem in much structural work, as the finished surfaces must be preserved. Further, leaks, such as illustrated in Figure 13.1, are a constant prob-

FIGURE 13.1 Leaks are a constant problem when grouting into structures.

lem that must be dealt with. Thus, the ease with which a grout spill can be cleaned off a surface becomes an important consideration.

Because strength is also usually important, settlement of the solids and the resulting bleed are especially detrimental, so only stable grouts should be used. Unlike the voids in rock, those in structures are often large and open, so sanded grouts are common. To provide good penetrability, however, the sand is usually on the fine side; No. 16 mesh is a common maximum size, although coarser sand can be used in larger defects.

Both compressive and bond strength are usually important. Adequate compressive strength is generally easy to attain, but this is not the case with bond strength. Because virtually all injectable grouts will experience at least some shrinkage, they are dimensionally unstable, which causes damaging stresses at the bond line. Such stresses can be reduced by the judicious use of water-reducing and other admixtures. As the cement content is typically high and the sand gradation on the fine side, these grouts will nearly always exhibit more shrinkage than typical concrete mixes.

Inclusion of a latex-type bonding material will improve the bond strength. Be aware, however, that latex material can act as a bond breaker once it has skinned over, which occurs quite rapidly. It should thus be the last ingredient added during mixing, and the grout must be injected promptly. Open time will vary and is dependent on temperature, but will usually be within a range of 10 to 30 minutes. It will shorten as the temperature increases, and become longer as it lowers. Accordingly, it may be useful to cool the grout ingredients and/or use chilled mixing water or crushed ice in latex-modified grout mixtures. The as-mixed temperature of the grout should not drop below about 40°F (5°C), however.

Fly ash or a natural pozzolanic material is often included in structural grouts, as it will result in improvement of both pumpability and penetrability. It will also contribute to the final strength of the grout, which is usually important in structural applications. In addition to water reducers, viscosity-modifying admixtures approved for use in structural concrete should be included to provide grout stability. Bentonite and other clays should be avoided, however, as they will result in a significant lowering of the strength and can also result in increased shrinkage and thus poorer bond strength.

13.1.1.2 RESINOUS GROUTS

Resinous grouts can provide properties not available with cementitious compositions. Resins such as epoxy and some urethanes can provide bond strength greater than the tensile strength of either the grout or the substrate, which can be important. They can be formulated to contain 100 percent solids, so they will exhibit little, if any, shrinkage. Moreover, such materials usually cure and gain strength rapidly and are highly resistant to deterioration.

Like all materials, however, they do have disadvantages and limitations. They are significantly more trouble prone and difficult to use than their cementitious cousins. In addition, their cost is several times greater than even the very best cementitious mixtures. Extremely sensitive to temperature, their open time can be very short (seconds), especially when mixed in large quantities and/or at high temperatures. When resinous grouts are used in larger masses, boiling can occur that virtually destroys their strength properties. In this regard, the particular formulation used should be matched to the prevailing temperature. Epoxies are especially subject to creep, so they should not be used in large masses. Where very high bond strengths are required, however, they may well be the only option available.

13.1.1.3 RESINOUS FOAMS

Foaming resinous grouts have a great advantage in structural work, in that they can be of minimal density and still offer good strength properties.

Furthermore, because they typically expand to several times their original volume, much less material must be handled than needed with more conventional nonexpansive grouts. And, of course, the expansion will result in more complete filling of the injected cavities, for which the size and configuration are usually not known.

The foaming resins tend to have the same characteristics as the ordinary resinous materials, but are even more sensitive to temperature and generally much more difficult to work with. Because they tend to foam shortly after mixing, and sometimes instantly, special proportioning pump, in head mixing and injection equipment is usually required for their handling. This results in a relatively heavy mixing head that is fed by multiple hoses, as illustrated in Figure 13.2. Obviously, the mixing head can be somewhat restricting and may require considerable effort to maneuver. Some resin foams, especially those

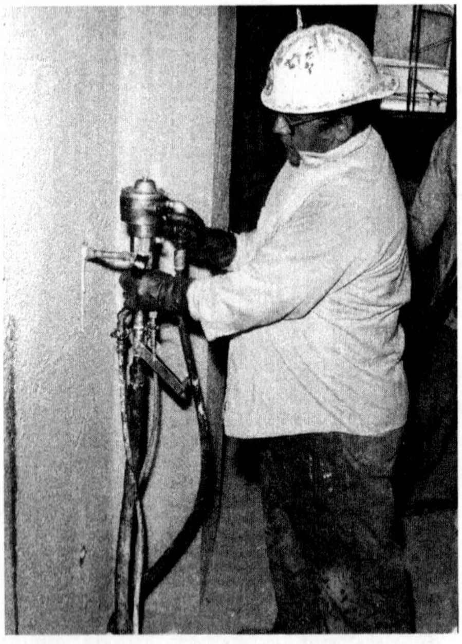

FIGURE 13.2 Dynamic mixing head for multiple component polymer foams.

based on urethane technology, require heating as well, with the further requirement that the hose bundles must also be insulated.

Although an expansive grout may be advantageous, the pressure of expansion can be a major liability in its use. Because the volume of the individual defects is not known in advance, it is not possible to limit the amount injected so as to just provide complete filling upon full expansion. Thus, in nearly all instances, more of the unfoamed resin that is actually required will be injected. Once it has expanded so as to completely fill the cavity, the remaining expansive pressure can be sufficient to displace and/or burst the substrate section.

There is a considerable difference in the expansive energy of various formulations, and it is imperative to select one with relatively low expansive pressure for most injection work. Expansive grouts are available based on both epoxy and urethane chemistry. The expansive pressure potential of different formulations varies considerably. A number of low-pressure formulations are available, however, with the lowest restrained expansion products exerting an expansive force of about five psi (.34 bars).

13.1.2 Grout Holes

Cementitious grouts usually require larger holes and nipples for injection than do resinous materials. They are thus more likely to be spilled on the surface, and such spills are more difficult to protect against. In the case of resinous grouts, small holes often suffice, so it is practical to wrap the finished surface in plastic, through which larger holes than are needed for the grout port are cut and then firmly attached to the substrate with duct tape. The smaller port hole is then drilled through the protective tape, and injection is made. In masonry, this should be done only where the joints have been sealed and are tight, however. Even then, it is important to use only

FIGURE 13.3 Pipe nipple with grout header secured into hole.

clear sheeting so that any leaks that develop will be easily observed and promptly cleaned up.

Injection holes for cementitious grouts are usually on the order of 1-1/8 in. (28 mm), which accompanies a standard 3/4 in. pipe nipple. The nipples can be either caulked in the holes with lead wool or oakum, or simply fixed with wood wedges, as shown in Figure 13.3.

As an alternate to pipe nipples, special plastic ports fitted with one-way valves, allowing the grout to be injected but restraining it from leaking out following injection, can be used. One example is molded from rubber with a corklike shape and a hole through the body to accept the injection nozzle. The interior end has been cut so as to leave a solid flap about 1/8 in. (3 mm) thick. The flap will open into the hole during injection, but will be forced shut by back pressure once injection has ended. Although this sort of device can be used for virtually any grouting, it is usually reserved for applications in which a measurable pressure will remain after injection is completed.

13.1.3 Grouting in Concrete Structures

The solid components in concrete consist of cement, sand, and large aggregate or gravel. Sufficient cement paste, composed of cement and water, must be present to thoroughly coat all the sand particles. Likewise, there must be sufficient mortar, including sand, to fill all the pore voids of the large aggregate. It is preferable for the concrete to contain the greatest proportion of large aggregate practicable, as it is a relatively inexpensive filler and, more important, because it will not shrink, its inclusion lessens concrete shrinkage.

Because only enough mortar to completely encapsulate the large aggregate is included, it must be uniformly distributed for the concrete to harden to a uniform dense mass. For this to occur, the fresh concrete is compacted, usually by vibration during placement. The large aggregate can become segregated into masses that are deficient in mortar during transportation and placement, however, if less than good practices are used. Such mortar-deficient areas are referred to as *honeycomb* or *rock pockets.* It is quite common to see these on a surface once the forms have been removed, as illustrated in Figure 13.4.

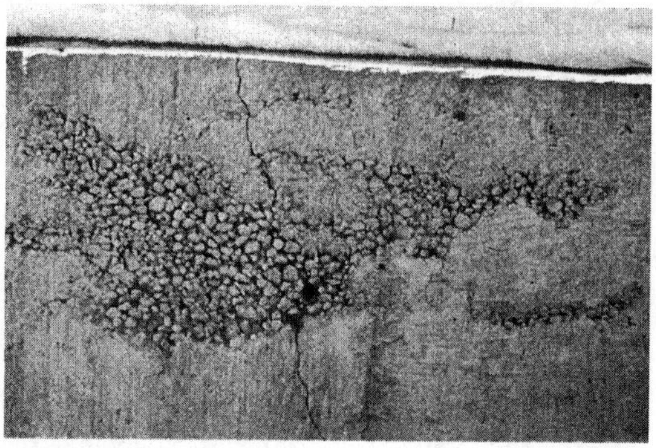

FIGURE 13.4 Mortar-deficient honeycomb on a formed concrete surface.

They are routinely repaired by chipping out and patching with repair mortar.

Yet rock pockets can occur on the interior of the mass, which are thus neither optically visible nor readily repaired. This condition is usually detected by some form of ultrasonic pulse testing, which usually is called for only when excessive surface defects are observed. Both surface and interior honeycomb can be repaired through injection of grout consisting of cement and fine sand. For interior defects, small holes, usually on the order of 1 in. (25 mm), are drilled so as to intersect the defect at a minimum of two different locations on the periphery. One hole is for injection, and the other is to allow venting and the escape of any water or diluted grout. As is typical for any void filling, the holes are best located at opposite extremes of the defect. In the rather rare cases in which only one hole is able to penetrate the void, vacuum grouting, as described in Section 2.6, can be used.

Although traditional repair of surface honeycomb has involved removal and patching from the surface, grouting is an ideal method, as the grout will fill isolated pockets that are not visible. For such repair, tightly fitting plywood forms are placed over the honeycombed surface and ports are drilled through the form as needed. Injection should start at the lowest elevation of the defect, with venting at the higher points. Injection connections can be readily established by screwing pipe nipples into snugly drilled holes through the form. Generally, 3/4 in. nipples screwed into one in. (25 mm) diameter holes will provide good fixing for the ports. Vent holes do not require the fixing of a nipple. They are typically 1/2 in. (13 mm) diameter or smaller and are simply stopped with a cork once good-quality grout appears.

Larger open voids can be filled in a similar manner, except that the grout may contain coarser sand and, in extreme cases, even large aggregate. Cementitious grouts are preferable for virtually all void filling in concrete, as they set up and harden to similar physical properties as the host material. Some marketers of resinous systems, especially epoxy, strongly promote these materials for such work, usually citing superior properties such as the very high bond strength. This is not needed for most concrete repairs, and the fact is that resinous formulations all cure to properties much different from those of the parent concrete. As discussed in Section 7.4, there is no way the proprieties of these fundamentally different materials can be made similar.

As with all void filling, it is important to provide injection holes or ports at the lower extremities of the defect, with vents provided at the higher areas. For simple applications, only one injection hole will be necessary, but for large or geometrically complex areas, two or more will be required. Obviously, in either case, sufficient vent holes must be provided to preclude entrapment of air, water, or diluted grout. For sanded grouts, 3/4 in. (19 mm) injection holes are adequate; however, larger holes and delivery lines will be required for grouts containing large aggregate.

13.1.4 Grouting in Masonry

Voids in masonry are far more common, tend to be more continuous, and represent a larger amount of a section's volume than those in concrete. Injection of significant quantities of grout is thus common in these structures, so the further weight added often becomes critical and must be carefully evaluated. There are several reasons for grouting in a masonry structure. Perhaps the most frequent is to improve the fixity of the individual units in order to reduce the risk of their becoming loose and falling during earthquakes or other dynamic loading events.

Older masonry was traditionally laid up with lime base mortars, which tend to be quite weak. The individual units were thus not highly bonded together originally and so could have become further loosened due to age, settlement, or prior seismic activity. In such cases, they can be reintegrated by grouting. Both cement- and resin-based grouts have been extensively used for this purpose. Here, also, cementitious grouts have a greater affinity to the parent materials and so are preferable for most applications.

Although the polymer resin grouts will provide substantially greater bond strength, this is not usually needed, so the higher costs and greater effort required for injection is seldom justified or cost-effective. Should voids represent a large portion of the host section so that substantial grout quantities will be required, the additional weight of a cementitious grout may be excessive. Thus, when the ability of the existing foundations to support the added mass is questionable, a lighter-weight resinous foam grout should be considered.

During original construction, cladding materials such as cut stone and marble are often set at a small distance out from the substrate with fixing hardware. This leaves a continuous cavity, usually on the order of 1 in. (25 mm) thick. Historic activities, as well as corrosion of the hardware, can cause these elements to become loosened. At the same time, more stringent current requirements can call for greater fixity than would exist even if the hardware were in first-class condition, which is seldom the case.

Again, either a cementitious or resinous grout can be injected to fill the voids and reestablish fixity. Resinous foams have a clear advantage in such work, however, because of their light weight and greater bond strength, as long as the expansive pressures can be tolerated by the existing ties or temporary support provided. Many historic structures are clad with molded terra-cotta. Such decorative units often consist of cellular sections, many of which were typically unfilled during original construction. Figure 13.5 shows resinous foam being injected into such voids to integrate the cladding and underlying substrate as part of a seismic strengthening.

FIGURE 13.5 Ornamental cellular cladding contains significant voids.

Grouting is an essential element for adding new reinforcing steel to an existing masonry structure. The reinforcing is typically placed into holes drilled from either the top or the outside corners of the structure, or occasionally from the space around large openings. Although most such work is performed in connection with strengthening old structures, it is sometimes performed in modern structures as well, primarily those constructed with hollow cement masonry blocks. These units are made with cells that, when laid up, form continuous vertical passages, which are typically spaced every 8 in. (203 mm).

It is common to place reinforcing steel in some, but not all the cells. Where reinforcing has been placed, the cell should be completely filled with a concrete grout poured in from the top during construction. Additional steel is sometimes required in the strengthening of such structures, and in some instances required reinforcing is not installed during original construction. In either case, new reinforcing steel can be placed in holes drilled from the top and grout injected, usually from the bottom of the cell until it exits the top. Because of the volume of grout required to effectively mobilize the new reinforcing, such work is done with cement grouts very similar to those that would have been used in the original construction. These typically contain small gravel and can often be placed with a standard small-line concrete pump.

13.1.4.1 INJECTION INTO MASONRY

With injection into masonry, in order to minimize the risk of displacing any of the masonry elements, grout pressures are typically quite low, seldom more than 5 or 10 psi (0.3 or 0.7 bars). As in all grouting, a pressure gauge should be provided as near to the grout nipple as possible, as illustrated in Figure 13.3. Because of the relatively low injection pressure and its rather quick decay once pumping ends, it is usually not necessary to provide shutoff valves at the nipples. Rather, they can be plugged with corks or other suitable stoppers as soon as the connection is broken.

13.1.4.2 DEALING WITH LEAKS

Because masonry is inherently porous, it is virtually impossible to grout without at least a few leaks, and even when best practices are used and precautions are taken, many will probably occur. Although cementitious grouts can be washed off prior to their hardening, most resinous products cannot and will leave permanent stains anywhere they have leaked or been spilled.

For this reason, it is wise to include repointing or sealing of the joints as part of the work when grouting in masonry where finished faces must be preserved. This should always be performed ahead of the grout injection, so as to minimize any leaks. Such work must be done meticulously and with the greatest care to completely fill all cracks and defects. Time spent in this effort will usually be more than offset by that not required for sealing leaks during injection. And, more important, leaks that do not occur will not deface the surface.

In its natural state masonry is quite porous, so grouters must be vigilant and constantly on the lookout for leaks during injection. These can become even more problematic in older structures that have experienced earthquakes or other disturbances causing loosening or outright cracking of the units. Such was the case with the very old unreinforced masonry cathedral appearing in the photo in Figure 13.3. Notice the row of wedges in the arch bed joint, which had opened up to result in a gap of more than 1/4 in. (6 mm) in thickness. The wedges are backing up dry oakum, which had been pressed into the fault prior to the start of injection, in a manner similar to that described in Section 12.6.5.5 for rock grouting.

Although complete repointing of the masonry joints is preferable, at the very minimum faults that are likely to leak during injection should be precaulked. In this process it is a good idea to leave occasional "windows" to allow the grout movement to be observed. This permits its progress to be followed during injection and can serve to ensure that all of the faults are being properly filled. Tails of oakum can be left hanging from the inspection windows for convenient caulking once grout appears.

13.2 FILLING CRACKS AND JOINTS

Grouting methods are widely used to fill cracks and joints both in repair of existing structures and construction of new ones. The purpose of such work is to block water flow, to provide a structural filler to resist compressive forces, or to bond the two sides of a fault together to create a monolithic section. The methods and materials used will vary with the individual applications and must be carefully chosen to satisfy the purpose for which the injection is being made.

13.2.1 Injection Materials

Both cementitious and resinous grouts are used for crack and joint grouting. In the construction of massive concrete sections such as dams, the construction joints are often grouted to compensate for the inherent drying shrinkage of the concrete. This is usually not done until the concrete is well aged so as to have already undergone the most significant amount of contraction. Where precast elements have been erected, the joints often require filling. These may be essentially tight joints that have to be adhered together, or they may be open by an inch (25 mm) or more, needing only to be filled. And then, of course, there are random cracks that occur in virtually all concrete.

13.2.1.1 EPOXY

Epoxy resins are the most frequently used crack repair grouts. This is, no doubt, due to their high bond strength, which restores the monolithic integrity. Probably more significant, however, is the massive promotion by special interests with a stake in the material. The result is that it is widely promoted for the filling of all cracks and joints in concrete, regardless of their cause, and, to a lesser degree, those in masonry.

The concrete repair industry is bombarded with continual promotions urging the injection of "high-strength epoxy" to reintegrate all cracked concrete back to its monolithic state. In addition to welding cracks, a further feature often mentioned is prevention of deteriorating media from penetrating the section, where corrosion of reinforcing steel might be initiated. Although epoxy does provide some outstanding features, it is not the magic cure-all some interests would like us to believe, and its use should be carefully considered.

Epoxy is a fairly rigid material, as discussed in Section 7.4.1, and is capable of attaining exceptionally high bond strength. Moreover, it can be formulated to contain 100 percent solids, which can be essentially free of shrinkage. Thus, cracks that have been so repaired will, as a rule, be tightly welded together. This is desirable in those instances in which such welding is required and will not cause further distress to the member. These instances, however, represent only a small fraction of the total crack repairs performed. Most cracks are the result of shrinkage and/or normal movements in the concrete section or structure. In many cases, these should not be adhered tightly because this can cause further distress.

Flexibility is important in most crack repair, and thus filling with a less rigid material is often preferable. As previously discussed, the stiffness of cured epoxy is ever changing. It is controlled by temperature and will vary widely with temperature changes. Even if a "flexible" epoxy were to be injected into a particular crack, it would become more rigid as the temperature drops. Conversely, it will soften at elevated temperatures. The temperature of the repair environment must thus be considered when epoxies are contemplated for use.

13.2.1.2 POLYESTER RESIN

Rigid polyester resins act much like epoxy, with the exception that they exhibit high shrinkage upon curing. They are advantageous in that they are generally easier to work with and far more forgiving of imprecise proportioning and mixing than the epoxies. Excessive shrinkage usually precludes their use, however. It will typically result in either a loss of bond to one or both sides of the opening, or a cohesive failure within the filler itself. For this reason, polyester resins are not widely used to fill structural openings.

13.2.1.3 CEMENTITIOUS GROUT

Both common portland cement and ultrafine products have been used to fill cracks and joints, especially in dams and other water-retaining structures. Most cementitious compositions shrink and thus do not develop the very high bond strength of the resinous materials, so they are employed primarily for filling applications. For openings of less than about 1/8 in. (3 mm) wide, neat cement grouts are generally used. As the gap widens, sand can be added. Both the grain size and the proportion of the sand can be increased with widening gaps. For openings greater than about 1/2 in. (13 mm), standard masonry mortar is used. Because bleed cannot be tolerated in this type of application, only stable grouts should be used, and in order to minimize shrinkage, water-reducing admixtures are highly recommended.

13.2.1.4 URETHANES

As discussed in Section 7.3.2, polyurethane grouts come in numerous types and performance capabilities. They are the first choice of many grouters for remediating leaks through concrete. Although cured urethanes are most often in the form of flexible solids or foams, such resins that develop into rigid forms are also readily available. Their greatest use is in water control grouting, which is discussed in Chapter 17.

13.2.2 Crack Repair

A huge amount of crack repair by injection is done daily around the world. The principles common to all grouting apply equally to injection into cracked concrete and masonry, but, for whatever reason, most conventional grouters have not historically performed this type of work. In fact, much of that which is accomplished is performed by specialist concrete repair contractors that do few, if any, other types of grouting. Nonetheless, crack injection remains one of the most widely used procedures in the repair of concrete.

13.2.2.1 EPOXY CRACK INJECTION

As discussed earlier, even though it may not be the best material in all cases, epoxy is heavily promoted for crack repair and there are many contractors who perform such work exclusively. Before even considering such an application, however, it is wise to first ponder the reason the cracks exist. All concrete expands as temperature increases and contracts as it falls. In addition, all new concrete will shrink upon drying. Significant shrinkage will usually occur for a year or more. Thus, proper design and construction of structures should allow for this expansion and contraction of the hardened mass.

Accordingly, movement joints are usually provided to allow for the orderly movement of a structure. Often, however, sufficient joints are not provided, or those that are do not function properly, thus causing cracks. If the cracks are welded together without first providing for continued movement such as that which first caused them, new cracks are likely to develop. Likewise, if a movement joint is filled with a material that binds the two sides firmly together, a new crack will likely develop beside it, as illustrated in Figure 13.6. Therefore, only cracks that are not subject to significant future movement should be firmly fixed in place by epoxy injection.

Once it has been determined that a crack is indeed dormant, or alternate movement joints have been provided, it is advisable to further determine whether there is a valid structural need to reintegrate the section into a solid mass. If not, it is probably better not to perform the injection. If, however, it is determined that a

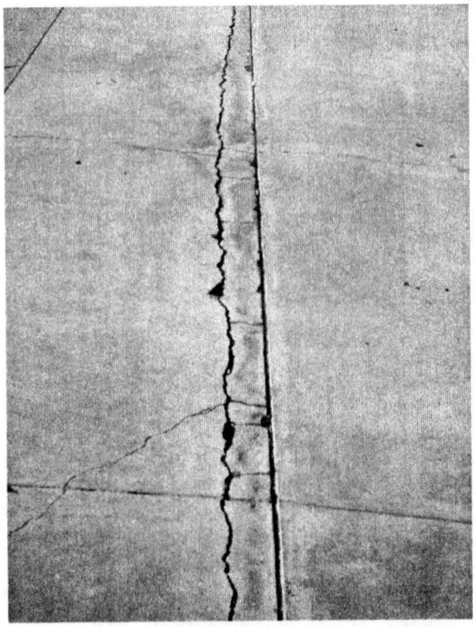

FIGURE 13.6 This new crack appeared following welding of joint together with epoxy grout.

monolithic section should be reestablished, it is then important to verify that it will be possible to attain a proper bond. Should the crack be contaminated with dirt or other debris, it may not be possible to inject it properly. Obviously, any impacted debris must be removed before the resin can be injected. Furthermore, the crack surfaces must be clean if the epoxy is to bond to them.

Removing debris from cracks is not easy, and they are often impossible to clean adequately, especially in concrete slabs on grade. If cleaning is to be attempted, vacuuming is the recommended method. Blowing with compressed air should not be allowed. Although this will displace the debris near the surface and the crack may appear to be open, in most instances the debris that exists at depth has simply been impelled further into the gap and packed more tightly.

Flushing with water, an acid solution, or solvent is often proposed. This approach has not historically proven to be very effective, however, and there are downsides to using any of these liquids. Even though many epoxies will bond to a wet surface, they generally do not develop a bond as great as they would under dry conditions. Furthermore, the exothermic heat that develops can cause boiling of any latent fluid in the crack. Acid should not be allowed in reinforced concrete, as it can cause corrosion of the steel. The pH of concrete is normally very high, which provides an invisible protective membrane around the reinforcing. Should the pH fall, corrosion is likely.

If the cracks are flushed with solvent, it is imperative the solvent used is compatible with the particular resin formulation that will be employed. Not all solvents are suitable to be in contact with every epoxy formulation. The only condition in which solvent flushing is appropriate is where oils or grease have penetrated into cracks, such as foundations under machinery and the like. Flushing in such cases is probably advantageous, but it is important to be aware that regardless of the effort, the ability to obtain a good bond of the injected resin is doubtful at best.

Prior to injection, exposed surfaces of cracks must be sealed and some sort of injection ports provided. To ensure penetration for the full thickness of the element being repaired, the ports should be placed at a distance apart that is slightly greater than the thickness. It is desirable to place ports on both sides of the element. Injection is nearly always made on one side only, but the progress of the grout into the defect can be readily monitored through the ports on the opposite side during injection.

Cracks as narrow as about 5 mils (0.13 mm) in thickness can be readily permeated by the available low-viscosity epoxy injection resins. It is actually possible to permeate smaller gaps, but this requires extraordinary effort. As the thickness of a gap increases to more than about 30 mils (0.8 mm), the port spacing can be expanded. Because of the great difference in physical properties between epoxy and concrete, this material should generally not be used when the defects are thicker than about 0.2 in. (5 mm).

Once the crack is sealed and the ports provided, the injection can proceed. As is the case in virtually all grouting, it should be started at one end and continue in a continuous manner to the opposite extreme. It is usually preferable to start at the end that is at the lowest elevation and work upward. This may be accomplished with a single injector, or several adjacent ports can be filled simultaneously, as illustrated in Figure 13.7. Regardless of the number of injectors used, with exception of the first one to be connected to a given crack, no other ports should be connected until the advancing resin has appeared at their location.

Where multiple lines are used, injection should always start with a single line at the lowest port. Once grout appears at the next higher port, a second line can be attached. Again, with appearance of the resin at the next port, attachment of a third line, and so on, until the maximum number of injection lines has been

FIGURE 13.7 Simultaneous injection into multiple ports.

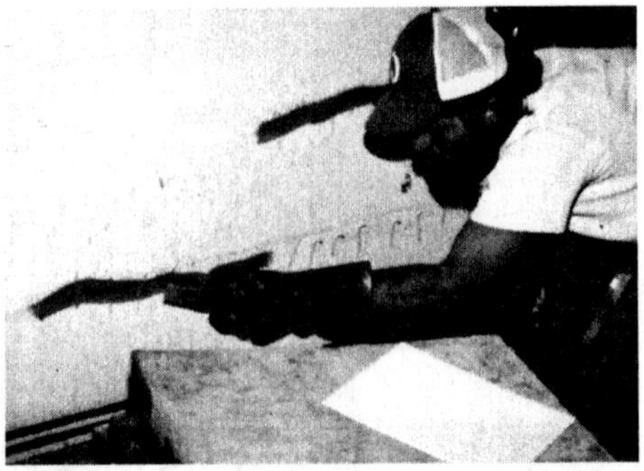

FIGURE 13.8 Masking tape is removed before sealer hardens, leaving an open crack port.

connected. Then, the lowest injector is leap-frogged ahead as resin appears at the next higher port. This procedure has a substantial advantage, in that sustained pressure is kept on a considerable portion of the defect, which will make up for any material that is absorbed into the substrate over time. It also facilitates more rapid injection, which, of course, has obvious benefits.

There are two very different methods for providing injection ports, and each requires a particular approach to the injection itself. The *skip seal* technique involves placing short pieces of 1/4 in. (6 mm) wide masking tape across the crack at each location where a port is desired. Once this has been accomplished, a sealing material, most often a nonsag epoxy or polyester resin, is spread with a wide spatula over the crack. The spatula is held tightly to the surface so that the sealer is forced into the outer portion of the gap and the profile of the surface is not changed, as illustrated in Figure 13.8. Before sealing material hardens, the masking tape is removed so as to provide gaps in the seal.

With the skip seal method of porting, the injection device must be physically held against the open port, which is, no doubt, the reason for its being dubbed the *fist-to-the-wall* method. The injection device is equipped with some sort of a seal on the outlet, most often a simple rubber O-ring, but the user must nonetheless hold the device in precise alignment and with considerable thrust (Figure 13.9) to prevent leakage around the injector. This is obviously a strenuous effort for the operators, especially with ports that are very slow takers, requiring continued injection for a considerable amount of time. Once the injection is completed, the port must be closed. This is typically accomplished by rubbing a bar of paraffin wax over it so as to provide a wax plug immediately in from the surface.

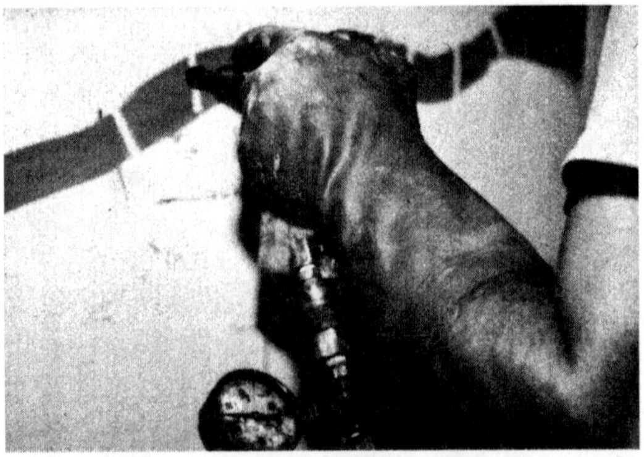

FIGURE 13.9 Injection head must be held in perfect alignment and firmly against port.

The skip seal method has some considerable advantages. It is by far the fastest and most economical method to seal a crack and provide ports. Because the resulting seal is smooth and perfectly flat to the receiving surface, there is essentially no cleanup or other restoration required following injection.

There is, however, one substantial disadvantage, in that the injection device must be physically held against the surface for the entire injection interval. Where fairly wide cracks exist, this is not a great problem, as the resin flows freely and does not take very long to reach the next port. But with slow takers, the injection may be required to continue on the same port for an extended amount of time, often 20 minutes or more. Under such conditions the technicians tend to tire and, not infrequently, to move to the next port without completely filling the first one. In addition, injection can be made in only a single port at any one time.

Another shortcoming of this method is the inability to positively shut off a port upon completion of injection. Although a wax seal will stop the resin from running out by gravity, it is not sufficient to withstand any pressure such as can result from continued injection at the next port. For these reasons, this method should generally not be used, except in very special cases where the sections are fairly thin and the cracks reasonably wide so as to accept the material fairly rapidly.

The preferred method involves installation of some sort of physical port to which a firm connection can be made. Several different types of ports, manufactured especially for epoxy injection, are available. They all work in a similar manner. The ports are secured over the crack at the desired locations, as shown in Figure 13.10. Port installation is often done with the use of a hot glue gun. Once in place, sealing material is applied to the remaining crack intervals, as shown in Figure 13.11. During this operation the sealer is also built up around the ports to provide additional fixity and ensure against any leakage. This procedure obviously requires more time, material, and cost. In addition, because the sealing material is built up on the face, it must be removed in situations where surface restoration is required.

FIGURE 13.10 Preformed plastic ports applied over cracks.

The substantial advantages of this method far outweigh its limitations, however, especially in respect to the completeness of filling and the

FIGURE 13.11 Filling in sealer material between preformed ports.

FIGURE 13.12 Injector is held in place by friction with this port system.

overall quality of the work. The attachable ports permit use of multiple injection lines, as illustrated in Figure 13.7, which can be maintained for as long as required without exhausting the workers. Further, a positive stopper is installed upon completion of the injection. This is usually in the form of a rubber or cork friction stopper, or in some cases, one that is mechanically connected. One method involves the crimping of a steel sleeve installed over a flexible tubular section of the port.

Connection to the ports can be achieved by friction, through use of a tapered nozzle pushed into a somewhat flexible port, as shown in Figure 13.12, or mechanically, with some sort of connector. Friction attachment is considerably more efficient than mechanical attachment. This would probably be the case with any grout, but is especially so in dealing with epoxy. Because these materials are allergenic, heavy rubber gloves must be worn by the workers so as to avoid direct skin contact. Moreover, the materials tend to be sticky, so they are likely to build up on the gloves. The use of makeup tools, or the manipulation required to make mechanical connections, is thus severely hindered.

Actual injection can be performed by mixing the epoxy in small batches and injecting them one at a time, or continuously with proportioning pump in-line mixing equipment. Batched material is forced into the cracks by either a pump or a pressure vessel. Because epoxy starts to produce exothermic heat as soon as the two parts are combined, the mixture can become excessively hot, causing premature set, when under pressure in a standard pressure pot. The pot shown in Figure 13.13 has been fitted with a refrigeration coil on the interior. With the use of a thermostat, the temperature can be regulated as desired. It is best kept on the low side, at about 60°F (16°C), when the mixture is in the pressure vessel so as to increase the set time.

The penetrability of most epoxies increases with higher temperature, so it is allowed to become warmer once the mixture is in the injection line. If the work is to be performed in a cold environment, the injection line can be coiled

FIGURE 13.13 Thermostatically controlled refrigerated pressure pot used for injection.

immediately ahead of the outlet and a heat lamp shined on it. This is not the fastest or most efficient injection method, but it is particularly well suited to jobs in which very slow injection is required in order to properly fill very fine fissures.

For most work, a proportioning pump in-head mixing system will result in the most efficient operation. In this system the proper proportion of each component is pumped to the injection head, where the components are mixed immediately prior to entering the port. To ensure proper batching, positive displacement pumps must be used. Piston pumps are well suited to this service. Rotary gear pumps are much less expensive and are thus frequently used in the lower-cost equipment. These are not recommended, however, as they tend to wear unevenly and thus easily get off-ratio with use. As discussed earlier, epoxy must be precisely proportioned if the cured resin is to develop its maximum potential strength and durability.

Many pumping systems for epoxy injection are commercially available. Some are very large and heavy, which is a serious disadvantageous for field use. Easily handled, more compact units are recommended for most work, such as that shown in Figure 13.14. Transparent material reservoirs, like those shown in Figure 13.14 should be mandatory on all such pumping systems, as they allow visual observation of the relative levels of the different components. Failure of the levels to fall proportionately provides immediate evidence of improper proportioning, allowing prompt corrective action to be taken. As improperly proportioned material will not develop its best properties and may even not harden, it is essential that it not be impelled into the cracks. Thus, such a simple, essentially no-cost requirement is more than justified.

Two combination port-injection systems have been developed in Japan, and both are available in the United States. They allow long, continuous injection times and thus use long dwell-time epoxy formulations. The Balloon Injection Concrete System (BICS) system involves combination port injectors that are installed during crack sealing (Figure 13.15). These are fitted with a short expandable hose and a one-way valve through which the epoxy is injected, expanding the hose as a balloon. A generally uniform pressure of about 45 psi (3 bars) is maintained on the resin. If all the resin in the injector

FIGURE 13.14 Transparent reservoirs on proportioning pump injection units.

FIGURE 13.15 BICS injectors mounted over cracks as part of concrete bridge repair.

is depleted, more can be added through the valve.

The Micro-Capsule system involves a glue-on port to which is mated a small spring-forced caulking cartridge. The spring is sized so as to impart a generally steady pressure of about 45 psi (3 bars). The plunger is mechanically restrained prior to the capsule being mated to the port. Once secured in place, it is released so as to provide continuous pressure until it has completely emptied. Both systems have a distinct advantage in that they maintain pressure continuously at each port location. The principal limitation is the relatively high cost of the hardware. In addition, there is a waste of excess epoxy that has been placed in the injectors but is not required by the cracks.

There is some contention as to the optimal pressure for injection. Some grouters recommend high pressures exceeding 100 psi or more (6.9 bars), believing that this makes the work go faster. However, many engineers express concern that high pressure will cause a crack to open farther and cause more damage. Yet neither of these positions may be valid. Unless the crack is filled with water, which is virtually never the case, the pressure is always zero at the leading edge of the resin. My own evaluation has established the optimal pressure to be within a range of 40 to 60 psi (2.8 to 4 bars) at the port. This is in line with extensive research in Japan, where a range of 45 to 50 psi (3 to 3.3 bars) has been established as optimal. Permeation of epoxy requires time, and unreasonably high pressures do not greatly speed it up.

Although it is easy to seal the cracks where they are exposed on a surface, it can be difficult when they are behind obstructions, such as door and window frames, where they are not accessible. This need not be a problem, however, as sealing can be performed by drilling a hole through the element and inserting a piece of snug-fitting, flexible backer rod such as that commonly used in caulked joints, as illustrated in Figure 13.16.

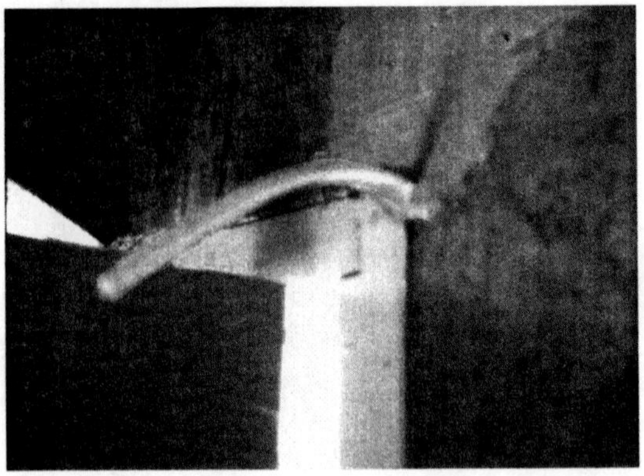

FIGURE 13.16 Foam backer rod in a drilled hole across the crack will block against epoxy leakage.

13.2.2.2 CONSTRUCTION JOINTS IN MASS CONCRETE STRUCTURES

All new concrete shrinks as it dries, and this shrinkage will cause opening of the joints. Such behavior is not always acceptable, most notably in concrete arch dams, wherein the entire dam section must behave in compression for the design to function. In addition, in water-retaining structures, open joints will result in unacceptable leakage.

Neat cementitious suspension grouts are used almost exclusively for filling such joints. The grout is often introduced through piping networks that have been installed during construction to allow its optimal placement. The open-ended piping must terminate in the joint and is usually placed so as to introduce grout in the interior, allowing it to flow and vent to the exterior faces. Care must be exercised in placing the piping system, and especially in concreting around it, to not allow concrete to enter or block the open ends that access the joints. Although much work has been performed whereby the grout was not contained on the exterior, but allowed to run out of the near surface portion of the joints, this is not a desirable practice. The surface joints can easily be sealed in a manner similar to that used for epoxy injection, which will result in both complete filling and a much cleaner operation.

13.2.2.3 JOINTS BETWEEN PRECAST ELEMENTS

Precast concrete elements are usually placed with gaps at the joints. These openings are then later grouted so as to form a monolithic section. They may be small, about 1/4 in. (6 mm) in width, or can be open as much as several inches. Such gaps always extend to the surface of the elements, so some type of sealing or formwork must be provided to contain the grout. Cementitious grouts are used almost exclusively for this work. They may be in the form of neat cement suspensions, or, for larger openings, sanded plastic-consistency mortars. As in all grouting, injection should be started in the lowest portion and proceed in an orderly manner to the opposite extremity.

13.3 CONTACT GROUTING

Contact grouting involves the filling of generally small voids behind or under a rigid lining or conduit. The grout is typically injected through holes that penetrate the lining. These may be formed by casings inserted prior to placement of the concrete, or may be drilled thereafter. Voids typically filled by contact grouting are relatively thin, and cementitious suspension or slurry grouts are used. Because the quantities are not great, small injection lines, seldom larger than 1 in. (25 mm), are most often used. As in the grouting of all voids, at least one injection hole is required as is a means of venting air, water, and diluted grout. Venting is usually provided by additional grout holes, but these are not required as long as the void can vent into the surrounding formation. In isolated voids where venting cannot occur, vacuum grouting, as discussed in Section 2.6, can be used.

13.4 POSTTENSION TENDON GROUTING

Concrete exhibits high compressive strength but does not provide significant tensile strength. This is the purpose of the reinforcing steel that is commonly used. For the reinforcing to be mobilized, it must stretch slightly, resulting in the concrete's yielding or sagging a little. Such sagging can be prevented by stretching the steel reinforcement before it is loaded—a process known as *prestressing;* the steel so treated is referred to as a *tendon.* Prestressed tendons are most often high-strength stranded wire cables, but solid bars can also be used. The steel can be stretched in a casting bed before the concrete is placed around it (referred to as *pretensioned* steel) or after the concrete has been placed (referred to as *posttensioned* steel).

Prior to the stressing of posttensioned tendons, they must not be bonded to the concrete. This is typically prevented by encapsulation within a snugly fitting encasement or by installing the tendons into *ducts* consisting of pipe or a similar tube. Most very long and high-capacity tendons are so installed. To bond these tendons to the surrounding structure and to protect them from corrosion, the ducts are normally grouted following the stressing operation. Figure 13.17 shows such an operation in progress, where the tendons of a large bridge section are being grouted. The single greatest enemy of posttensioning tendons is corrosion. For the grout to provide acceptable corrosion protection, it must be dense and must completely fill the space.

Grouting has historically been treated as a stepchild of the prestressing industry, where installation is typically performed by ironworkers who have neither a knowledge of nor a liking for cementitious mixtures. To a large extent, unstable suspensions consisting of only cement and water have been used. The result has been extensive accumulation of bleed water, initially,

FIGURE 13.17 Each of the tendon ducts will be grouted from the end anchorage shown around the periphery of this bridge section.

and then air at the upper locations of tendon ducts. A further problem unique to posttension grouting is *wick* bleed. A standard for wire tendons is the seven-wire strand. It consists of a central *king wire,* around which six perimeter wires are wound. Water tends to wick up around the king wire by capillary action, thus contributing to the bleed.

The result of much unsatisfactory past performance of tendon grouting has been the formation of an active industry for repair of such structures. It was not until 2001 that a guide specification for posttension grouting finally arrived. The bulk of the publication, *Post-Tensioning Institute Committee on Grouting Specifications* (2001), deals with grout materials and mixtures and emphasizes the problems created by bleed of unstable grouts. In the Specifications, four classes of grout are enumerated, A through D, the particulars of which are shown in Table 13.1.

Posttensioned tendons can be placed in virtually any direction or configuration. In horizontal members such as beams, they are usually draped so as to provide the optimal restraint. Where beams are continuous over several supports, the tendon ducts can have multiple drapes, which results in several high and low areas. All low areas should be provided with injection ports and valves, and all high areas should contain vent ports with valves or other positive means of capping. As in all grouting of voids, injection should first be made in the low-

TABLE 13.1 Material and Testing Requirements for Posttension Duct Grouting

Class	Exposure	Maximum W:C Ratio	Constituents	Required Testing
A	Nonaggressive	0.45	Cement	Setting time Grout strength Pumpability and fluidity Wick-induced bleed
B	Aggressive	0.45	Cement Fly ash Slag Silica fume High-range water reducer Calcium nitrite Other admixtures per mfg. recommendations	Setting time Grout strength Pumpability and fluidity Wick-induced bleed Permeability Volume change Bleed Corrosion Wet density
C	Any	0.45	Prepackaged formulations	
D	Any	To be determined by design engineer		

est area so the grout can travel in an orderly manner to the higher vents.

13.4.1 Rock and Soil Anchors

Another wide area of use for posttensioning is in the construction of rock and soil anchors. Here the tendons are embedded into drilled holes that penetrate to a competent layer of geomaterial, where they are anchored. This anchorage is typically provided by grout, which is injected into the annular space between the tendon and the drill hole wall. In this usage, it is imperative that the grout be strong and durable. And, of course, it must completely fill the anchorage zone space and have intimate contact with the wall of the hole if good restraint is to be realized. In this regard, the capacity of anchors in soil can be substantially increased by pressurizing the grout so that the soil is displaced, resulting in a final section larger than the original hole.

High-capacity rock and ground anchors can be very bulky and composed of a large number of *bundled* tendons, as illustrated in Figure 13.18. A stay-in-place tube for the grout injection is usually attached to and placed with the tendon bundle. This tube can be seen exiting in the center of the individual bundles in Figure 13.18. For posttensioned anchors, the steel strand in the anchorage zone, which is always at depth, must be firmly embedded. The portion between the anchorage and the surface must be allowed to stretch, however, so it is typically encased within a sheath to prevent bonding and provide corrosion protection. Positive corrosion protection is imperative, so the sheathed bundle of tendons often requires the grout injection lines to be attached on the exterior.

In simple installations where the tendons are reasonably short, the grout line can be simply pushed down beside the tendons once they are in place. Where the tendons extend to great depths, however, it should be attached to the bundle to ensure accurate placement. For very deep installations, such as that shown in Figure 13.19, where a concrete dam is being anchored to the underlying rock, multiple grout lines may be provided to allow the point of injection to ascend in successively higher intervals during the course of filling. In such instances, it is cheap

FIGURE 13.18 High-capacity, multistrand tendons with the grout injection line in the center.

FIGURE 13.19 Deep multistrand anchors such as this often will have injection tubes terminating at different levels.

insurance to supply an extra injection tube to the bottom for use in case the main tube becomes damaged or plugged.

To increase the anchorage restraint, individual tendons are often alternately spread apart with hardware and bundled tightly together. When put down a drill hole, they will always be fitted with *centralizer* hardware to ensure their position in the middle of the hole. Although this hardware is essential for proper positioning of the tendons, it does create space restrictions, which provide an easy opportunity for entrapment of air and water from the grout as it is placed and begins to harden. This is a particular problem with some spreader hardware that is flanged, and thus forms a ready trap. Selection of hardware with a maximum amount of open area, and void of isolated pockets that will encourage the entrapment of air or water, is thus highly recommended.

For these installations, durable, high-strength grouts are essential, as are stability and absence of bleed prior to hardening. Relatively thick cementitious suspensions and slurries are most frequently used. Although neat cement mixtures are the most common types, finely sanded grouts may provide superior performance. The inclusion of up to about 40 percent fly ash will improve both penetrability and resistance to pressure filtration. The amount of mix water should be minimized as practicable, and the inclusion of water-reducing admixtures is essential, as well as viscosity modifiers to control bleed. Because of the strength and durability requirements, bentonite should never be used to provide stability, as in other grouts.

13.4.2 Repairing Defective Tendon Grouting

Because posttensioned ducts are completely enshrouded, the sufficiency of the tendon grouting is difficult to determine. For most tendons in common structures, and especially those that are positioned in essentially horizontal configurations, an insufficiency will probably not be noted until corrosion has become so great that the required tendon restraint becomes compromised. However, tendons that are composed of a single steel bar, inclined downward from horizontal and/or in large ducts, can often be remediated through injection so as to fill the ungrouted area.

In instances such as when less grout has been used than required, a plug has formed, or equipment failure has occurred during grouting, the insufficiency of the work will be obvious. In cases where less than acceptable filling is suspected, either invasive or nondestructive exploration will sometimes expose less than complete filling. Where this condition is found, remedial grout injection will be required. The actual injection is fairly simple; finding and gaining access to the voided areas, however, is not.

The location of an unfilled duct can be approximated through an analysis of the volume of grout actually placed versus total volume of the duct, if these factors are known. The grout will nearly always have settled into the lower portions of the duct, and in ducts that are steeply inclined or vertical, it will typically be in the bottom. In situations where the quantity of grout injected and the duct volume are not known, ultrasonic testing can sometimes give a good idea of the locations of void areas.

If a small hole can be drilled, either through the stressing plate on the higher end of the tendon or at an angle so as to enter the duct obliquely, a small-diameter steel rod can be inserted until it encounters the grout. Cold rolled steel bars of 1/8 in. (3 mm) diameter placed through 3/16 in. (5 mm) drill holes work well for this. In situations where a large amount of bleed has resulted in a considerable length of ungrouted tendon, it is likely that a thin layer of soft and spongy material laitance can be felt on top of the grout. This will be easily discerned with probing of the bar, and should be expected.

Once the suspected location is found, the concrete cover must be *carefully* removed so as to expose a short section of the duct. *Carefully* is emphasized, as it is essential that the tendons not be nicked or otherwise damaged. Extreme care is required, because the tendons can be very near or directly against the inside wall of the duct and some ducts have very thin and somewhat flimsy walls. Once the duct has been encountered, the exposed portion should be removed to reveal either grout or a void.

Once a void is found, it can be delineated by running a fiber-optic camera to each extreme of the ungrouted area. Modern fiber optics allow a camera to be threaded into spaces only a fraction of an inch across. The fiber line can bend so as to go around fairly sharp corners, including standard pipe fittings. In one case, this technology was able to penetrate 3/4 in. piping that contained four 90° elbows. The camera provides a picture of what lies ahead as it is inserted so that appropriate adjustments can be made.

Once the end locations are known, further penetrations can be made to expose both extremities of the ungrouted area. The duct is then penetrated to expose the tendons and facilitate connections. A good idea is to run a wire or small steel rod between these two extremities to confirm that they are continuously open. Grout injection and vent nipples can then be installed and the duct wall restored. The removed concrete should then be replaced via dry packing or another acceptable method. Once the patches have cured, grout injection can be made in a normal manner, as illustrated in Figure 13.20, always working from the lowest area upward, with venting at the farthest and higher end. Obviously, a good-quality stable grout should be used for such remedial work.

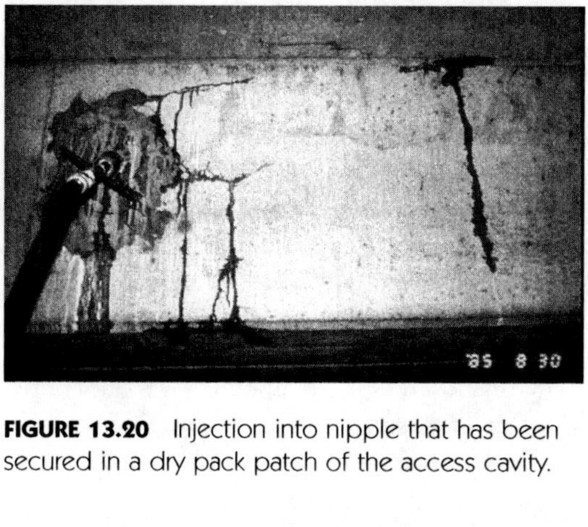

FIGURE 13.20 Injection into nipple that has been secured in a dry pack patch of the access cavity.

13.5 GROUTING BEDPLATES

Grouting is often performed under bedplates and machinery bases to ensure uniform and firm support. In many instances, these bases must be set and maintained at precision tolerances. Accordingly, the grout material must be strong and durable, as well as stable, free of bleed, and essentially nonshrinking. Both cementitious and resinous materials are used. Because of the limiting properties of resinous grouts, as discussed in Section 7.4, such grouts should not be used where the thickness will be very great, say more than about 1/2 in. (13 mm).

Many preblended proprietary products, both cementitious and resinous, are marketed and widely available for this type of work. For large applications, an acceptable cementitious mix can be designed and proportioned either on-site or in a ready-mix truck. However, because of the importance of avoiding settlement of the solids and shrinkage of the grout, a refined mix design is required and preinjection testing should be performed to fully qualify the mixture. Moreover, because of the extreme performance requirements of this work, large-scale preinjection testing is advisable, even where preblended mixtures are to be used. The test method presented in Section 8.6.1 and illustrated in Figure 8.9 has been used in a large number of applications, and it is recommended for any previously untried grout mixture proposed for use.

The object to be grouted should be accurately set to the required tolerance and firmly secured in place. The amount of restraint necessary will depend on the grout pressure to be used. Typical base plate grouts do not require much pressure to be simply forced from one side to the other; however, the quality of the work will be improved if a positive pressure can be used and maintained for a short period of time. This will typically be within a range of 10 to 25 psi (0.7 to 1.7 bars). Maintaining a continuous pressure on the grout requires the void space to be completely enclosed so as to withstand the contemplated pressure level, as can be seen in Figure 13.21.

A large steel plate being established in an industrial plant, on which precision equipment will be assembled, provides a good example of precision setting and firm anchorage in place. Each of the bolts seen protruding from the plate in Figure 13.22 is being used to secure the plate during the grouting. The taller threaded rods are anchor hold-downs, whereas the shorter bolts are spreaders to maintain the exact position off the foundation slab. The hold-down rods have been threaded into couplings that are attached to anchor bolts deeply embedded in the foundation.

FIGURE 13.21 Void space must be completely enclosed to enable the maintenance of a sustained pressure.

FIGURE 13.22 Steel plate is firmly held in a precise position during grouting.

An 8,000 psi (55.2 MPa) high-strength cementitious grout was used to fill the approximate 2 in. (50 mm) space between the foundation and the 2 in. (51 mm) thick plate. Starting at one end, it was injected progressively into three holes along the center longitudinal axis of the plate. Although not easily visible in the photo, pieces of 2 × 4 lumber have been placed continuously in a 2 in. (51 mm) wide recess around the plate. The wood wedges distributed around the perimeter are to be used to wedge the 2 × 4s tightly against the plate once good-quality grout has arrived and the gap is no longer needed for venting. Securing these bulkheads in place will contain the grout and allow for higher pressures to be used as the work progresses.

In an unusual application, a spring-retained base plate for a space simulator (shown in Figure 13.23) was secured in place with a 16 in. (40 cm) thick layer of cementitious grout. The various environmental conditions that were to be applied to the interior of the sphere were considered so severe that the 4 in. (100 mm) thick plate was expected to warp and lift slightly off the massive concrete foundation. Accordingly, a grid of closely spaced anchors was fabricated into an enclosure for the grout. The anchors were all restrained by powerful springs so that

FIGURE 13.23 Floor of the space simulator is retained by spring-loaded devices anchored in the grout.

the bottom plate of the simulator could deform in waves, but would always return to its proper position on the foundation. The bottom plate assembly, including all of the anchors, was fabricated upside down (shown in Figure 13.24).

The spring-loaded anchors were to intermesh with reinforcing steel that was protruding from the concrete foundation. Thus, the forces were transferred from the anchors, through the grout, to the foundation. Both the strength and completeness of filling were thus critical to the proper performance of the completed assembly. The large quantity of grout required was supplied in ready-mix trucks. It contained equal parts, by weight, of cement and concrete sand, as well as water-reducing admixtures and methylcellulose viscosity modifier to provide a stable grout with a water : cement ratio of about 3.8 and a slump on the order of 6 in. (150 mm). It attained the required compressive strength, exceeding 7,000 psi (48.3 MPa) in 28 days.

FIGURE 13.24 The anchors were installed to the inverted base prior to grouting.

The entire steel structure was erected prior to the grout being placed, which provided ample restraint to the plate during grouting. The grout was injected through a 1-1/2 in. (38 mm) pipe so as to be deposited in the exact center of the plate. Additional radial pipes were installed around the periphery of the assembly. These terminated only 8 ft (2.4 m) in from the edge and were supplied as spares in the event that the entire area was not completed from the central point of injection. None of these supplemental pipes were required, and they were allowed to fill with the advancing grout, being plugged with tapered wooden plugs once grout appeared.

As previously mentioned, resinous grouts are best reserved for special applications where minimal quantities are required. Such was the case where the steel base assembly of a new installation did not bear properly on the concrete foundation. A gap, which varied in thickness from zero to about 3/8 in. (9 mm), existed throughout the area of the 8 ft (2.4 m) wide assembly (illustrated in Figure 13.25). The gap was sealed around the periphery with thixotropic-consistency epoxy mortar. Short pieces of 1/4 in. (6 mm) plastic tubing were sealed in at regular intervals on one side to serve as injection ports. Short breaks in the sealer were left at intervals of about 2 ft (0.6 m) on the opposite side to act as vents. Epoxy resin was injected from the ports on one side until it returned from the opposite side vents, which were then stopped with standard corks.

FIGURE 13.25 The narrow space between the foundation and the steel base is filled with epoxy resin.

In another case, heavy steel bearing plates failed to connect properly during erection. A gap, varying from zero to about 1 in. (25 mm), remained after the bolts had been taken up tightly. This would compromise the ability of the assembly to withstand the high compressive loads that were anticipated in operation. A heavy steel bar was welded continuously on both plates so as to fully contain the void area. A threaded hole to receive standard 1/2 in. (13 mm) threaded pipe as provided in each corner.

A high-strength silica-filled epoxy was then injected from one corner until it returned from the remaining three ports as illustrated in Figure 13.26. As the epoxy grout flowed freely from these vent ports, they were plugged with standard pipe plugs. Cured epoxy resin is particularly susceptible to creep under continuous compressive stress. By filling the epoxy with fine silica sand, the resin proportion of the total grout mass was reduced, thus lessening potential creep. Furthermore, the substantial steel bar

FIGURE 13.26 Silica-filled epoxy is restrained by the heavy bar connecting the steel plates.

that was placed around the plates to cover the gap and permanently welded in place fully contained the resin so that it could not be displaced.

13.6 ABANDONING CONDUITS AND TANKS

Grout filling is often used in the abandonment of underground pipelines, tanks, and similar substructures. In these operations, the strength of the grout is usually of little importance, but stability and resistance to shrinkage are critical because of the primary objective of providing 100 percent complete filling with no remaining void or trapped content. Most tanks and some pipelines have been used for the transport or storage of volatile compounds, which are normally lighter than the fill grout. Remaining remnants of these will tend to float to the highest areas as the grout displaces them during injection, and they must be provided a means of escape.

Large amounts of grout are often required, so ready-mixed concrete or mortar is generally used. To minimize shrinkage, the water content must be held to the absolute minimum, requiring the inclusion of water-reducing admixtures. Bleed is usually not allowed, so a viscosity-modifying admixture should also be used. As strength is typically not an issue, minimal cement is required; however, to attain flowability and pumpability, a minimum equivalent of about five bags of cement or other finely divided material per cubic yard is required. Fly ash is actually more efficient than cement in improving flowability, so it can be substituted for much of the cement. Many concrete batch plants can supply stock mix designs of low density cellular *flowable fill*, as described in Section 6.3.6. These usually provide the necessary attributes of stability and shrinkage resistance and are ideal for many applications.

Most underground tanks are fitted with both fill and vent lines. These are typically on the opposite ends of a tank. Although the fill line can be a short distance from the end, the vent is usually immediately adjacent to it. It is thus best to inject into the fill line. This will drive any remaining content to the vent, and it will be pushed up that line as the tank is filled. Prior to grout injection, the vent must thus be intercepted at the ground surface, and arrangements must be made for capturing and handling any liquids that may be discharged. To minimize such discharge, it is important to empty the tank to the greatest extent possible before grout injection.

Pipelines must, obviously, be accessed at sufficient intervals to allow for injection and venting. Unless sufficient manholes, cleanouts, or other accesses exist, this will require excavation and penetration. With modern concrete pumps and flowable fill mixtures, it is possible to fill such lines for a distance of 500 ft (150 m) or more, depending on the size and condition of the line. Under ideal conditions, this distance can be extended to more than 1,000 ft (300 m).

Abandoned pipelines can present many challenges, however. Although it is fairly unusual for significant amounts of liquid spoils to gather, many abandoned pipes contain dirt, debris, and/or thick sludge. This material tends to build up ahead of the advancing grout front, and if it is present in large enough quantities, it can completely plug the line during the injection. It is thus a good idea to inspect the line for such material before beginning injection. The simplest method is to simply shine a strong light down the line and visually observe the condition. This is easy to do on a sunny day. An ordinary mirror can be positioned to pick up the sun's rays and direct them down the line. If the pipe access is too deep to directly pick up the sun's rays, a mirror on the ground surface can be used to direct the light to another at the pipe level. If the pipe is reasonably straight, any debris or other condition that might impede the filling will be observed.

Should the pipe have dips that cannot be seen, it is possible that they contain either excessive sludge or perhaps a liquid. Unless this material occupies a fairly small amount of the total length, resources should be maintained onsite to capture and handle any that flows from the vent during filling. Where conditions are questionable, it may be advantageous to run a TV camera down the line. Although this involves considerable expense, it will provide accurate information about the conditions and disclose any debris that may be present.

In situations where considerable debris is found, the line should probably be cleaned ahead of injection, or the distance between access points shortened. Consider the following example of the effects of debris on an advancing grout. High pressure was required during abandonment of a 300 ft (90 m) long section of 6 in. (150 mm) drain line. Upon filling, about 10 gal (38 L) of liquid sludge was first to arrive. This was followed by a several-inch-long plug of relatively dry debris and an essentially no-slump extrusion of grout mixed with debris. The original grout in this case was a high-slump, flowable fill mixture.

13.7 FILLING ANNULAR SPACE/BACKFILLING

Flowable fill materials are often used to fill large annular spaces and other isolated subsurface voids. A common requirement is encasement of pipes or other conduits that penetrate tunnels, or perhaps a larger pipe or casing, which is known as *backfill* grouting. The same material and pumping procedures used for abandoning pipes and conduits are usually applicable here. One exception, however, can be the weight of the fill grout. This can become critical in preventing flotation of the inner line. Lightweight flowable fill is therefore an ideal material for this requirement. It can be either mixed on-site or delivered in ready-mix trucks. Because a large portion of the total grout is air, it is often most economical to mix on-site as needed. The subject of backfill grouting is discussed further in Chapter 16.

13.8 FILLING FABRIC FORMS

Fabric forms, which are essentially closed bags, to contain grout are commonly used in a variety of applications. They are especially advantageous when working under water, as the bags contain the grout travel and retard any washout of the cement. Such forms can be fabricated into any shape or required geometry. Originally, they were developed to contain unstable fluid suspension grouts, prior to the ability to compound and pump the low-mobility and cohesive stable grouts now readily available. In addition to containing the grout, they have a further advantage in that they will permit excess water to permeate and drain, while containing the solids of the grout. This allows the placement of generally

FIGURE 13.27 Grout-filled fabric forms are widely used for scour protection.

overwatered grout for applications that require the good strength and durability of low-water cement mixes. The trend now, however, is to use ready-mixed cement sand mortars, most often with a slurry consistency.

By far the most frequent use of fabric forms is in lining earthen surfaces to prevent erosion or scour, as shown in Figure 13.27. They are especially useful in this application, as they can be rapidly spread into place by unskilled workers, as shown in Figure 13.28, and can be applied to large areas quickly. Where required, divers can easily place them under water. Once the forms

FIGURE 13.28 Prior to filling, fabric forms can be easily positioned by unskilled workers.

are in place, grout is injected through inlet portals that are provided in the most accessible location, usually on the top. The fabric can be formed to provide an essentially flat surface or one that is highly textured, as shown in Figure 13.27, which absorbs currents or wave action and thus adds considerably to the protective value.

Another common use of these forms is in repair of piling, where they have the advantage of being relatively light in weight and thus easily handled by divers, as shown in Figure 13.29. For this use they are generally supplied with a zipper so that they can simply be zipped to closure, once wrapped around the pile. In these applications, it is helpful to fabricate a rigid steel hoop by which the forms can be suspended, as illustrated in Figure 13.29. This suspension is mandatory in installations that do not encapsulate all the way to the mud line but, rather, are limited to the tidal zone, which is quite common. The grout can be injected from special ports fabricated into the bottom or, more typically, injected down pipes from the top, as shown in Figure 13.30.

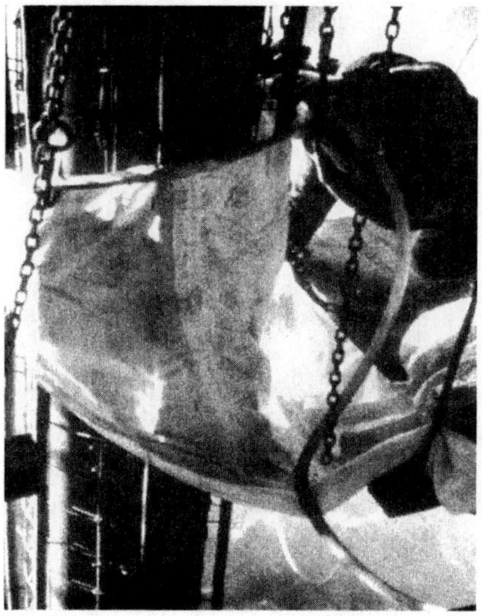

FIGURE 13.29 A zippered fabric form being placed around a pile by divers.

FIGURE 13.30 The form is suspended in place from a special steel hangar.

13.9 PREPLACED AGGREGATE CONCRETE

Preplaced aggregate concrete is cast by first depositing large aggregate into the forms and then pumping a suitable grout into it to fill the pore space. Although grout is most often composed of cement and fine sand with appropriate admixtures, in earlier times epoxy resins were sometimes used in order to achieve rapid setting. With the ready availability of rapid-setting cements, as discussed in Section 5.2.5, the use of resins, which are more difficult to deal with and much more costly, is no longer justified.

Preplaced aggregate concrete has many advantages, which can result in higher-quality construction and a cost savings in some situations. Because the aggregate is placed first, the forms are completely filled. Each piece of aggregate is in direct contact with its neighbors so that one cubic yard of concrete contains a full cubic yard of aggregate. Shrinkage of the composite concrete can thus not occur, resulting in greater bond strength where the new concrete is being added to old.

Larger-size aggregate can be used, as it is no longer necessary for it to be suspended in a plastic mixture. This results in the large aggregate making up a greater proportion of the total composite, which requires less mortar and thus less cement paste. Less paste translates to lower shrinkage and improved durability, as well as greater economy. Further, the aggregate can be more easily placed into areas of limited access and tightly packed around penetrations and beneath overhead obstructions. The procedure is ideal for applications such as radioactive shielding, which requires very high density aggregate. This would separate out of conventional concrete, but can be readily used when placed prior to the mortar.

Properly injected, the grout will completely fill all pore areas, regardless of restrictions, making the procedure ideal for casting concrete in difficult positions and configurations. The procedure is particularly useful for casting underwater concrete. In this situation, as much time as necessary can be used to properly place the aggregate, which is nonperishable. The time required to place the mixed grout, which is typically less than one-half the total volume of the mass, will be considerably less. And the grout will displace the water as it is injected, so little washout will occur.

Notwithstanding the many advantages, there are some applications in which use of preplaced aggregate concrete should be questioned. In underwater placements where the water is strongly polluted with sulfates or other contaminants, it is better not to use this procedure. Where contaminants are present, but only moderately, its use might be practicable, provided that the aggregate is deposited and grouted over a short period of time or even in the same day. Where contamination is suspected, laboratory tests to verify applicability of the process are strongly recommended.

Although the high content of coarse aggregate and its particle-to-particle contact avoids shrinkage of the mass, the injected grout will require excessive water, which results in shrinkage and increased permeability. Therefore, it is not very satisfactory for use in harsh environments, such as pulp mills and chemical processing plants. This has been a serious limitation historically, but it is not as great now that high-quality, stable, yet highly penetrable, grouts are available. Be aware, though, that the quality of the grout is of great importance when the work is to be done in an area of extreme exposure.

13.9.1 Coarse Aggregate

Coarse aggregate is usually graded so that essentially all will be retained on a 3/8 in. (9.5 mm) sieve, with the maximum size usually 1-1/2 in. (38 mm). The maximum size limitation to 1-1/2 in. (38 mm) is not due to technical limitations, but rather because that is the largest-size concrete aggregate commonly available. If the section is thick enough and the reinforcing steel spaced sufficiently far apart, there is no technical reason that larger aggregate cannot be included. Larger aggregate is, in fact, advantageous, as it reduces the amount of cementing mortar required. In one instance, in casting a massive thrust block, cobbles up to about 8 in. (200 mm) were included in the aggregate.

Because the aggregate is not in the concrete, where it would affect mobility and workability, and the injected grout completely fills underparticle space, shape and gradation become less important. The important criterion, aside from quality, is the pore volume. It is clearly desirable to have a minimum amount of pore space, so as to lessen the quantity of grout required. In practice, void contents within a range of 40 to 50 percent are typical. The lower percentage, of course, is preferable; however, the gradation must take into account the cost of manufacture and supply.

To ensure good bonding, it is essential that the aggregate be thoroughly washed so as to be free of any dust or other surface contaminants. Unlike conventional concrete, in which the action of mixing will clean the aggregate surfaces and promote good bond of the mortar, dirty aggregate can adversely affect the strength and overall quality of preplaced aggregate concrete. Much of the aggregate now commonly available for concrete is washed and of adequate cleanliness. It is imperative to carefully monitor this, however, and provide supplemental washing at the job site as required. In large applications, cleaning is best accomplished on a trammel-type wash screen. On smaller projects, the aggregate can simply be washed with a strong blast of water as it is being removed from the stockpile for placement. A small pressure washer is especially useful for such cleaning.

13.9.2 Formwork

Formwork must be of sufficient strength to withstand the injection pressures, and must be essentially watertight so as not to allow excessive leakage of the very penetrable grout. For large formed areas prefabricated panels are generally used. These must be free of holes or other places where the mortar might leak. Although it is common practice in conventional formwork to provide tie holes somewhat larger than the ties, this should not be done with preplaced aggregate. Joints between the individual form sections should be provided with a gaskets or other appropriate seals, as should the contacts with other construction.

On small and/or difficult-to-access sections, the forms are often built sequentially a few feet at a time, along with placement of the aggregate. To minimize leakage, tongue-and-groove lumber is often used, as are foam gaskets between sections or at contacts with already completed construction. Where sufficient space allows, standard caulking guns can be used to fill any holes or cracks, but they must be filled by hand application in more restricted areas. A mixture of portland cement and a light oil, such as form-release oil, has been successfully used for such filling where rapid setting is not required. Where a fast set is needed, casting plaster works well.

In some applications, such as placement of new preplaced agregate concrete (PAC) against cracked or particularly porous concrete or masonry, it is beneficial to maintain a small positive pressure head on the grout once all of the pore space has been filled. This will encourage intrusion of grout into the substrate defects, thus increasing the bond. Where this is done, however, the forms must be able to withstand the higher sustained pressure, which will require much more robust construction.

Such a tactic was used in the repair of some beams where excessive honeycomb had occurred during construction. The faulty concrete was removed as practicable, but there was concern that further concealed voiding might exist behind the reinforcing steel. Special steel hoops, as shown in Figure 13.31, were anchored into the top portions of the beam. These were designed and

FIGURE 13.31 Special steel saddles were fabricated to support formwork.

spaced so as to withstand a positive force of 15 psi (1 bar) on the formwork. To fill any remaining voids in the substrate, a positive head of 12 psi (0.8 bar) was maintained for five minutes at the conclusion of the grout injection.

13.9.3 Injection Ports and Pipes

For vertical sections less than about 1 ft (0.3 m) thick, injection ports can be placed in wooden forms by installing pipe nipples into drilled holes. A standard 3/4 in. nipple screwed into a 1 in. (25 mm) hole will provide ample fixity. Where used, such nipples should be placed in a grid of about 2 ft (0.6 m) vertically by 4 ft (1.2 m) horizontally. For sections with vertical thickness up to about 3 ft (0.9 m), nipples need only be placed in the bottom. A nipple must be provided at each extreme end, as well as in any corners that are more than about 2 ft (0.6 m) horizontally from another nipple. Obviously, all nipples that will be below the final grout level must be equipped with valves, as illustrated in Figure 13.31.

For thicker sections, horizontal pipes can be used. These should be inserted into very snug holes drilled through the form and should reach to within about 6 in. (150 mm) of the back of the cavity to be filled. Where this distance exceeds about 5 ft (1.5 m), additional sets of pipe should be provided so that one set will be available terminating at each 5 ft (1.5 m) of section thickness.

Vents will be required for all isolated areas, such as under blockouts or embedments or overhead roof areas. These may be provided by holes drilled through the forms or small-diameter copper or plastic tubing leading out from the potential trap area. These vents must be suitably plugged once good-quality grout runs from them. Plugging is readily accomplished with corks, tapered wooden plugs, or other appropriate stoppers. Where flexible tubing is used, it can simply be crimped or bent back and tied, if required.

For open areas, vertical injection pipes are embedded in the aggregate as it is deposited. These are typically of standard Schedule 40 pipe with a diameter within a range of 3/4 to 2 in. (25 to 50 mm). They are successively pulled as grout is injected and rises in the aggregate mass. Depending on the volume and configuration of the section and the gradation of the aggregate, the pipes are spaced a distance varying from about 4 ft (1.2 m) to 10 ft (3 m) each way. The spacing is closer for thin, narrow, or irregularly shaped placements, as well as for those that require very high quality. In massive placements, especially where large-size aggregate is used, the greater spacing is applicable. For sections more than about 3 ft (0.9 m) thick, a row of pipes should be provided within about 6 in. (150 mm) of each face.

Initially, the injection pipes should extend all the way to within about 6 in. (150 mm) of the placement bottom. Standard Schedule 40 pipe and couplings are most often used for depths to about 60 ft (18 m). For greater depths, flush wall tubing should be used. This can be in the form of a standard Diamond Core Drill Manufacturers Association (DCDMA) casing or custom-made sections from thick walled pipe or tubing. Very deep pipes can require considerable withdrawal force, so standard pipe with oversized couplings must be avoided. Moreover, the couplings must be of sufficient strength to withstand the withdrawal. For very deep placements, separate sets of shorter pipes might be provided for injection into the upper zones if any of the deeper pipes break or are otherwise lost during withdrawal.

13.9.4 Placing the Aggregate

Aggregate should be placed to the densest possible state, practicable to minimize the amount of mortar required. Where the space to be filled is open and there is ample space between the reinforcing, large aggregate can simply be dumped from the top until the section is filled. Where this is done, common sense must prevail, however, and the aggregate should not be dropped from such a height that breakage will occur. Where much reinforcing exists however, a tremie pipe should be used to prevent displacement of the reinforcing due to the impact of the aggregate.

Conversely, where space is restricted or reinforcing congested, it may be necessary to hand place and compact the aggregate. For the overhead beam repairs illustrated in Figure 13.31, galvanized metal lath was progressively wired to the reinforcing as the aggregate was hand placed. Although compaction is usually not required, vibration of the forms and/or rodding with a probe can often increase densification. As an example, Figure 13.32 shows a worker using a shovel handle to pack the aggregate tightly under a large bearing plate.

Maintaining cleanliness of the aggregate is crucial. The placing methods must not cause breakage or the development of dust or other contaminants. Where it is dropped onto soft surfaces, especially on top of underwater sediments, a geotextile filter cloth should first be placed so as to prevent contamination from fines. Once placed, the aggregate should never be flushed with water from the top. This will pick up any remaining dust or other fines and carry it to the bottom of the placement, where it will be deposited on the aggregate. Such contamination is virtually impossible to remove and will result in very poor quality in that zone.

FIGURE 13.32 Packing of large aggregate under the plate with a shovel handle.

The aggregate should be moist at the time of the grout injection. If it has been allowed to dry since placement, rewetting will be required. Likewise, if the work is being performed in a very warm environment, it will be advantageous or even necessary to cool the aggregate prior to grout injection. This is best handled by wetting it with cool water. The required water should be pumped through the injection ports or pipes so that it slowly rises from the bottom of the placement.

13.9.5 The Grout Mixture

Obviously, the grout must develop good strength and durability, be highly penetrable, and yet free of excessive shrinkage or bleed. With the present admixture technology, such a grout can be readily compounded. It will typically contain cement, fine sand, fly ash or a natural pozzolan, and admixtures to lower the water requirement, possibly retard the setting time, improve penetrability, and control both shrinkage and bleed potential. Although any standard cement can be employed, Types I and II are most commonly used. The sand should be sound, well graded, and, for most work, finer than a No. 16 sieve.

The cementitious material will normally be a combination of portland cement and Class F pozzolan. Because the pozzolan greatly improves the penetrability of the grout as well as retarding the set, its proportion is usually on the order of 25 to 35%, by weight, of the cement. The grout will typically contain 1 to 1-1/2 parts of sand, by weight, of the cementitious material. Greater proportions of sand can be used when the minimum size of the large aggregate increases. The pozzolan will contribute considerable strength to the grout, but not at early ages. This must be recognized, and in instances where high early strength is required, the proportion of pozzolan can be reduced or eliminated completely. Although job site batching is most common, high-quality preblended grout materials are available and should be considered, especially for small and very high quality applications.

In much earlier times, preplaced aggregate concrete was a proprietary process. To promote the uniqueness of the procedure and to derive continuing profits once the patents expired, the content of some of the mix constituents was generally kept secret and they were promoted by proprietary terms. The pozzolan was often referred to as *Alfesil* and treated as if it were a very special and expensive compound. We now know it was simply common fly ash such as is now used.

Various admixtures including aluminum powder, used to produce a gaseous expansion, were combined into a single product that was referred to by the proprietary term *Intrusion Aid.* Although mention of such expansive agents continue to appear in the literature and they are still frequently specified, we now know that there are serious side effects to their use, including an increase of bleed. With our current technology and the ability to compound high-strength stable grouts of good durability, there is no rationale for their continued use.

With time, the generic description of Intrusion Aid became *fluidifier.* It typically contains a water reducer, a set retarder, and other admixtures to reduce settlement of the solids in the grout and improve pumpability. These terms are still used in specifications and other technical material. Although many admixture distributors routinely handle all of the needed ingredients, few market a special combination as a fluidifier. It is thus prudent to abandon the use of such verbiage and recognize that the rheology of grout for preplaced aggregate concrete is no more or less special than that of any cementitious grout. As with all grouts with special requirements, unless the proposed mix is of proven quality, laboratory tests should be carried out to confirm that it will satisfactorily fulfill the intended needs.

What is now the ASTM C 939 funnel test, as described in Section 8.1, was developed long ago, explicitly for controlling flowability of grout used for preplaced aggregate concrete. Experience has shown that grout with a flow time between 20 and 24 seconds is ideal for such use. Longer flow times will result in grout of higher strength, but it will not penetrate the aggregate nearly as efficiently. In combination with larger minimum-size aggregates, flow times of as much as 26 seconds have been found satisfactory, however.

13.9.6 Injecting the Grout

Fundamental to quality performance, the grout must rise in a slow, orderly manner to uniformly fill all of the pores in the large aggregate. Injection will always start at the lowest point of the placement. It can proceed via either of two progression mechanisms, horizontal layer or advancing slope. The horizontal method is generally used for sections of considerable height, as illustrated in Figure 13.33. With this method, grout is injected over the entire area in discrete horizontal layers of a few feet in thickness. If the injection is through ports or horizontal pipes penetrating the forms, all of the ports at any given level will be filled until grout exits those at the next higher level. This will be continued in a uniform manner, starting at one end and working toward the other.

FIGURE 13.33 Horizontal placement is usually used for sections of considerable height.

Where the injection is through vertical pipes, it will likewise always start at the deepest location and continue in an orderly progression from end to end. At the conclusion of injection into a particular pipe, that pipe will be raised to within about 1 ft (0.3 m) of the top surface of the grout. It is essential for the tip to be embedded in previously placed grout, however. The setting time of the grout must be sufficient that it does not stiffen excessively during the turnaround time between injections. This is handled by delaying the set of the grout by the inclusion of retarding admixtures, cooling the mix, and/or the use of multiple grout pumps. Where multiple pumps are employed, however, it is essential that the injection points be selected so as to promote an orderly advance of the grout.

In most conditions, and with the use of a properly retarded grout mix, the pipes should stay open for several hours between injections. There will be instances, however, when sufficient time has elapsed or environmental conditions prevail so as to cause the grout to stiffen in the pipes prior to their next planned injection. Although there are no good solutions for this condition, when it occurs the pipes can be lifted, so that the bottoms are above the grout level, and then opened up by inserting a metal rod for the full depth. When this is done, the distance the pipes have been lifted should be noted.

Immediately prior to the next injection cycle, the pipes should be driven back to their original level. Although this sounds easy, in practice it is not, and penetration is often blocked by a large piece of displaced aggregate under the tip. In such instances, the pipes should be driven as far as possible. To minimize such problems, the pipes should not be raised more than necessary to clear the previous grout level. The first solution that many workers conceive to resolve such problems is injection of water down the pipe. This should not be allowed under any condition. It will cause washout of the cementing materials in the area of the pipe tip, as well as separation of the sand, resulting in sand pockets or streaks.

The advancing slope technique is generally used where the placement involves a large horizontal area relative to height, as illustrated in Figure 13.31. With this method, the injection is made, again, starting at the lowest point. It is continued at that initial point until grout reaches the top of the placement, as confirmed visually or by its exiting from vent holes. At this point the grout will have developed a slope of about 1 ft (0.3 m) vertically to 5–10 ft (1.5 to 3 m) horizontally. Where the injection is under water, the slope will be somewhat gentler.

Injection is continued so as to move the advancing slope laterally until it reaches the opposite end. Once this occurs, the next higher line of holes is injected in a similar manner. This process continues until the section has been completely grouted. Where injection is through the forms, all ports must be equipped with valves. Once the grout has stiffened sufficiently that it will not flow, the valves are removed and washed out.

It is essential to follow the grout travel and know the location of the advancing boundary at all times. This can be indicated by seeps in the formwork, by observation of small vent/inspection holes through the form, monitoring of sounding wells embedded in the large aggregate, or observation of grout vent/return lines. Where the mass being injected is above water and enclosed in formwork, small holes, typically 1/2 in. (13 mm) in diameter, can be drilled through the forms. The advancing grout front can then be observed at it reaches a given hole. Once good-quality grout exits, the holes are plugged with corks or other appropriate stoppers.

Sounding wells are simply pieces of slotted tubing placed vertically in the large aggregate so as to allow penetration of the grout and thus reveal its location. The tubes are typically 2 in. (50 mm) in diameter and contain holes or slots of sufficient size to allow the grout ready access. When working in the dry, a simple inductance probe can be lowered to instantly determine the grout level as desired. When the injection is submerged, a device that sinks in water but is buoyant on the denser grout can be used. For confirmation of the complete filling of isolated "roof" pockets, a grout return line should be run to a convenient observation area that is higher than the top of the pocket, as shown in Figure 13.34. This is typically a piece of 1/2 or 3/4 in. (13 or 19 mm) tubing, which starts at the highest elevation within the pocket. Once the grout has completely filled the defect, it can be observed exiting the line, which is then plugged.

Grout is typically injected at a rate of 1 to 4 ft^3 (0.03 to 0.12 m^3) per minute. The actual pumping rate used depends on the porosity of the large aggregate, the size and configuration of the section being injected, and, of course, the grout rheology. Obviously, the injection system must be sized according to the pumping rate, but delivery lines usually fall within a range of 1 to 2 in. (25 to 50 mm) diameter so as to provide a grout velocity between 2 and 4 ft (0.6 and 1.2 m) per second.

Although economy dictates the use of the highest practicable pumping rate, if the rate is excessive, the grout can tend to skip over pore areas, leaving ungrouted aggregate. In extreme cases, the aggregate can be essentially displaced or uplifted, resulting in both excessively rich and deficient grout zones. Many have attempted to pump through a manifold so as to inject into several holes at the same time. As with other types of grouting, this is not feasible because slight variations in the back pressure virtually always result in uneven filling.

It is, however, sometimes advantages to use a "Y" fitting that divides the main injection line into two separate branches, each equipped with a valve. Where this is done, injection is not made simultaneously from both branches, but rather from only one at any given time. Yet this approach can save time because the inactive line can be moved and connected to the next port, requiring only the opening of a valve when injection is to begin there.

No vibration or other consolidation is required during the injection. In fact, in a number

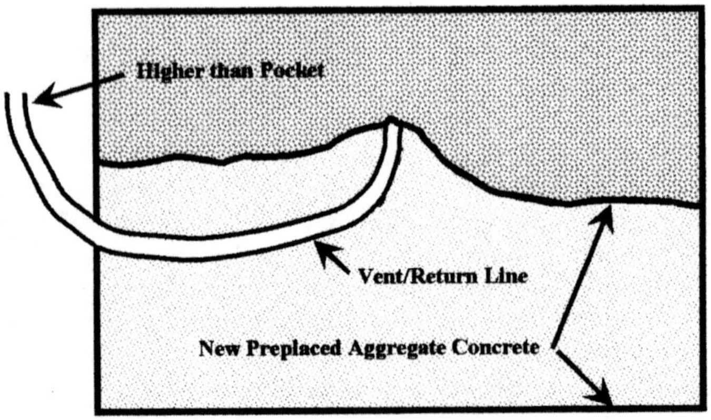

FIGURE 13.34 Vent/return lines should be installed in all overhead isolated pockets.

of trials it was found that the use of conventional high-frequency external form vibrators resulted in the development of many bug holes on the finished surface due to the generation of air bubbles at the grout-form interface. Low-frequency tapping with a hammer, however, can improve the grout wetting and coverage at the base of the injected mass or at the joints on the ends.

13.9.7 Quality Control

The quality of preplaced aggregate concrete depends on proper placement of the aggregate and the complete filling of its pore space with a quality grout of proper consistency. As previously mentioned, the penetrability of a grout is evaluated by the ASTM C 939 Flow Cone test. For evaluation of the grout's strength, standard cylinders or cubes can be made. It is important to keep in mind, however, that the strength of the concrete is dependent on the quality of the aggregate, as well as its complete embedment within the injected grout, so the strength of the grout alone will not necessarily indicate that of the concrete.

For confirmation of the strength of the composite concrete, standard 6 × 12 in. (152 × 305 mm) cores should be prepared in accordance with ASTM C 943. A special split-cylinder mold is placed vertically on an end plate and filled with the same large aggregate as used in the production work. The mold is tapped as it is filled so as to ensure good compaction of the aggregate. Once the mold is completely filled, a top plate is positioned and the end plates are tightly clamped together. Both plates contain threaded holes to receive pipe fittings.

The cylinders are maintained in a vertical position, and an injection port fitting is secured onto the lower end, as illustrated in Figure 13.35. Grout is pumped into the mold until it exits the vent hole in the top plate. The specimen is left undisturbed until the grout has hardened, after which it is demolded. It can then be tested in the same manner as ordinary concrete specimens. Although this test was developed for use in the laboratory, it can be satisfactorily performed in the field and is the only method to provide valid information as to the strength of the composite concrete, other than taking cores of the hardened product.

FIGURE 13.35 Setup for conducting ASTM C 943 test specimens in the field.

13.10 AN UNUSUAL STRUCTURAL APPLICATION

Several grouting technologies were used in the abandonment and removal of a deep concrete pit that contained highly radioactive materials. The pit was in a building in California that was part of a facility that produced nuclear medicines. It had a horizontal cross section of about 4 × 5 ft (1.2 × 1.5 m), extended more than 30 ft (9 m) from the top of the slab-on-grade floor, and was filled with water. This was used to shield radioactive emissions from elements that were stored in the bottom of the pit. An accident involving the rupture of a small canister of radioactive material occurred. Immediately following the accident, all contaminated elements, including the ruptured canister and its contents, were dumped into the pit and the building was sealed for a number of years.

The owner assembled a group of engineers and contractors to devise a plan to safely remove the pit and its contents, which were to be hauled to a disposal site in Nevada. Initially, exposure levels near the pit were so high that workers could be near it for only a short time period. A fundamental requirement in the planning was to increase the level of shielding overlying the debris in the bottom. This was accomplished through a grout injection program performed in three phases.

The first operation was to gently place a layer of highly stable cement suspension grout in the bottom of the pit so as to contain the hot contents and prevent their being stirred up by subsequent work. The grout had to be resistant to washout, yet mobile enough to readily spread out and encapsulate the debris. To accomplish this, a sanded grout stabilized with a high dosage of methylcellulose, and of low water content, was used. In addition to the admixtures, it contained equal parts of cement, a natural pozzolan, and fine sand. It was pumped through a 3/4 in. pipe that was suspended about 1 ft (0.3 m) above the pit bottom, and a volume of grout that would provide about an 8 in. (203 mm) continuous layer was deposited.

After the initial grout had hardened, a steel assemblage was installed directly on top. The assemblage provided for compartmentalizing the next 4 ft (1.2 m) of the pit depth. It also facilitated the placement of grout under high sustained pressure so as to create an intimate bond with the existing pit walls and provided a means to attach lifting hardware to the resulting mass. To divide the space into smaller vertical compartments, 2 in. (50 mm) thick steel plates were placed horizontally at distances of 1 and 4 ft (0.3 and 1.2 m) above the base, as shown in Figure 13.36.

FIGURE 13.36 The steel assembly with sealing tubes prior to installation.

Outward-facing open channels were installed around the periphery of the plates and fitted with continuous flexible rubber tubes connected to piping that would eventually extend to the surface. Once the assemblage was in place, the tubes were inflated with pumped grout so as to form permanent seals, which effectively isolated the compartments. A stable cement-bentonite grout was used for this operation, wherein the objective was to provide a positive seal but maintain some flexibility so that it would not be broken by further disturbance.

Once this was accomplished, a high-density grout was injected into the 3 ft (0.9 m) thick upper chamber. Because shielding effectiveness is directly proportional to the density of the material, the use of the highest-density grout that could be pumped was desirable. After experimenting with several mixtures, one that provided a density of about 215 lb per ft^3 (3444 kg/m^3) and yet was pumpable was adopted. It contained equal parts, by weight, of cement, fly ash, and a mixture of 60 percent minus 16 mesh crushed barite sand and 40 percent 1/16 in. (1.5 mm) lead shot. The water requirement was minimized through use of a water-reducing admixture, and to keep the heavy aggregates in suspension, a heavy dose of methylcellulose was included.

This grout, which had a consistency similar to a mason's mortar, was injected to completely fill the space. The two heavy horizontal steel plates, which contained this placement, were designed to withstand a pressure of 150 psi (10 bars), with the only limiting factor being the strength of the grout-filled seals. This allowed a sustained pressure of 100 psi (6.9 bars) to be held on the grout once the cavity had been completely filled, ensuring intimate contact with the walls of the existing pit. This was necessary, as the design called for full load transfer of the underlying structure to the grouted-in-place steel assemblage. As this was critical to the success of the work, a full-scale mock-up including the lower portion of the pit was constructed prior to the actual production work.

That grout was allowed to cure for 14 days, after which the lower compartment of the assemblage was injected with a thick neat cement suspension grout. This grout also contained water-reducing and viscosity-modifying admixtures so as to be completely free of bleed. It was subjected to an extended continuous pressure of 45 psi (3 bars) for one hour following complete filling of the cavity. Piping for the various injections was built into the assemblage and extended to the surface, as shown in Figure 13.37, during its installation. Filling of the compartments was done by injection into the bottom of one corner and continued until good-quality grout exited the vent pipes in the top of the other three corners.

FIGURE 13.37 Injection lines for the various grouting operations were extended to the surface.

Following this isolation grouting, the cohesionless sandy soil surrounding the pit was solidified with chemical solution grout and an excavation was made down to the level of the top plate. The upper portion of the pit was then demolished, and excavation continued to the bottom of the pit. To reduce the radioactive intensity, lead sheets were wrapped around the remaining portions of the pit as they were exposed.

FIGURE 13.38 The resulting mass on a transport truck for disposal.

Lifting hardware was then attached to the top steel plate, and the remaining portion of the pit was lifted to the surface.

Here, heavy steel beams extending several feet beyond each end of the mass were placed immediately against each side. Formwork was constructed around the entire mass, providing for an 18 in. (0.45 m) annular space on the sides and 12 in. (0.3 m) coverage over the top. This space was filled with 1/2 in. (13 mm) diameter spherical lead shot. Vertical injection pipes for grout injection were installed as the shot was placed. A neat cement grout was then injected to form a very high density preplaced lead aggregate concrete.

Once this had cured, the mass was lifted by means of the steel beam extensions and placed on a thick precast slab of the same lead aggregate composition. At this point the radioactive levels had been reduced sufficiently that the assembly was considered to pose no further hazard. It was moved onto a waiting transport truck, as shown in Figure 13.38, for haulage to the disposal site. And thus ended a three-month effort that had included several very different grouting applications.

14

CHAPTER

Groutjacking

14.1 **SLABJACKING**
- **14.1.1** Injection Holes
- **14.1.2** Grout Mixtures
  - **14.1.2.1** Low-Mobility Mortarlike Grouts
  - **14.1.2.2** Slurry-Consistency Grouts
  - **14.1.2.3** Polymer Foams
- **14.1.3** Jacking Mechanics
- **14.1.4** Grout Injection
- **14.1.5** Monitoring and Control
- **14.1.6** Range of Use

14.2 **GROUTJACKING STRUCTURES**
- **14.2.1** Lifting Mode
- **14.2.2** Lifting Mechanisms
  - **14.2.2.1** Lifting with Compaction Grout Bulbs
  - **14.2.2.2** Lifting with Horizontal Hydrofracture
  - **14.2.2.3** Lifting from a Created Void
- **14.2.3** Groutjacking Tolerances

GROUTJACKING IS AN all-encompassing term used to describe the parting of a soil or rock mass, or the displacement of a structure, as a result of injection pressure during grouting. The displacement may be accidental, or it can be deliberate, usually where raising a settled improvement is required. Where intentional, it is usually performed in a controlled manner, and with the proper technique it is possible to raise settled improvements to fairly close tolerances. In most instances, the jacking results in an upward heaving of the ground surface, although it can also occur in any configuration and accidental lateral displacement is not uncommon, especially when injection is near a retaining wall or a downslope. Where grouting is performed above a tunnel or other underground space, the displacement can be into that space.

The earliest reported use of intended groutjacking was in the early 1930s, when J. W. Poulter developed the "Mudjack" machine. Poulter's interest was in stabilizing the pumping joints of concrete highway pavement. This was accomplished with his early machine, wherein a "clayey loam soil" mixed with water to form a thin slurry was injected under the pavement slabs. With experience, it was found that the slabs could actually be raised. The process was dubbed *mudjacking,* and that term remains in common use in some areas.

The mudjack had limitations in both the type and consistency of the grout it could handle, and early work often had to be redone because of poor durability of the grout. Accordingly, studies to develop more durable grout materials were carried out by many organizations and government agencies. Throughout the 1940s such studies considered a variety of grout ingredients, including cement, pozzolan, hydrated lime, fly ash, limestone dust, asphalt bitumen, and calcium chloride. Virtually all of the work involved the addition of such ingredients to the basic clayey soils to compound "pourable" mixtures of "creamy" consistencies that could be handled by the mudjack, which was then the only commercially available machine for such work.

Most of the early research was performed by organizations involved with highways and other concrete paving, with the goal of finding a suitable material to repair and prevent further *pumping* of paving joints. Because even the best of the soil-based slurries was not very durable over time, a search for alternate materials ensued. This led to the use of hot asphalt bitumen. which tended to be more durable, as it provided a seal against water penetration. The bitumen was heated to about 450°F (232°C) and injected at pressures varying from about 25 to 60 psi (1.6 to 4 bars). Although it was widely used from the mid 1940s to the latter 1950s, the substantial disadvantages and hazardous nature of the work—due to the high temperature of the material under pressure—as well as the relatively high cost, led to its demise.

In the early 1950s it was found that care had to be exercised in the filling of subsurface voids, as the surface would heave if pumping continued once they were filled. Moreover, it was found that jacking would also result during compaction grouting if injection was continued sufficiently long at a given location. This work was being accomplished with a cement-sand grout similar to a mason's mortar. The mortar was of significantly lower mobility than the slurry materials historically used, which limited the distance of their travel from the point of injection. Because the slurry would cover a wide area prior to the initiation of lift, jacking often occurred at unwanted locations. With the lower-mobility grouts, however, somewhat precise placement could be made, as the grout remained near the point of injection.

A landmark project was undertaken in 1957, wherein a large concrete box culvert had been undermined by flooding during construction, resulting in settlement and underlying voids. My former firm was retained to fill the voids with grout and informed the owner's engineers that there was a good chance that the structure could also be groutjacked back to its proper elevation during the work. The engineers were skeptical.

We requested a survey party to be present during the void grouting just in case the structure should tend to rise, but the request was bluntly denied as it was believed that this certainly could not occur and that monitoring was not warranted. While the voids were being filled (Figure 14.1), the inspector noticed a separation of the form board marks on the culvert from those on an adjacent wall. The culvert was rising. He immediately called his superiors, and in short order not only had a survey party arrived, but also a large crowd of officials, including both the chief engineer and the head of inspection. The culvert was jacked back to proper position to the surprise and satisfaction of all.

This experience led to many future applications involving the precise raising of settled

FIGURE 14.1 Once the voids were filled, the culvert was raised to grade.

structures as well as slabs and pavement. The necessity of grout deposition in the most optimal locations is fundamental to reliable groutjacking. This is readily obtained through use of stiff mortarlike grout. Not all of the holes will be used for lifting, so once the proper elevation has been reached, a more penetrable grout is generally used to fill any remaining space. Some grouters continue to perform this work with slurry-consistency grouts, especially for jacking slabs. However, these grouts do not allow anywhere near the degree of reliability and control as their stiff mortarlike cousins do, and thus their use is generally not recommended.

14.1 SLABJACKING

Highway work continues to represent a large proportion of slabjacking now performed, although concrete slabs of all types, from small residential sidewalks to the largest industrial floors, are routinely leveled. In 1977, the Committee on Grouting of the Geotechnical Engineering Division of the American Society of Civil Engineers published a report, *Slabjacking—State-of-the-Art.* Many examples of use were reported, including the leveling of a flood-settled supermarket floor, a tennis court, a warehouse floor, and a quarry-tile-surfaced plaza area at a university medical center.

Although the degree of accuracy achieved in lifting varies both geographically and between different contractors, it has long been established that working to typical new construction tolerances is readily achievable. An interesting example was reported in *Western Construction Magazine* (1969). A number of settled areas of pavement distributed over a 12.1 mi (19.5 km) section of Interstate Highway 15 through Cajon Pass, California, were raised. The individual settled areas varied from 20 to 300 ft (6 to 90 m) in length and were depressed by as much as 5 in. (127 mm). These depressions were raised to an accuracy of 0.005 ft (1.5 mm), which was well within new construction tolerance requirements.

Slabjacking should not be confused with *subsealing,* although the two terms are sometimes used interchangeably. The former refers to the purposeful raising of a pavement surface, usually to a stated tolerance, whereas the latter simply involves injection to fill any void and stabilize yielding slabs, usually at joints. Although a minimal amount of lifting is sometimes called for in subsealing, it is generally not very great or closely controlled. The grout used in subsealing is virtually always of a slurry or thick fluid consistency.

14.1.1 Injection Holes

Injection hole layout is dictated by the thickness, strength, condition, and configuration of the slab. Holes are normally established on a grid, with spacing varying from about 5 to 10 ft (1.5 to 3 m) each way. Badly cracked or very flexible slabs require a closer spacing than thick, rigid ones. For slabs with a thickness of 4 in. (100 mm) or less, a spacing of 5 to 6 ft (1.5 to 1.8 m) is used. The individual rows of holes are typically staggered, as illustrated in Figure 14.2.

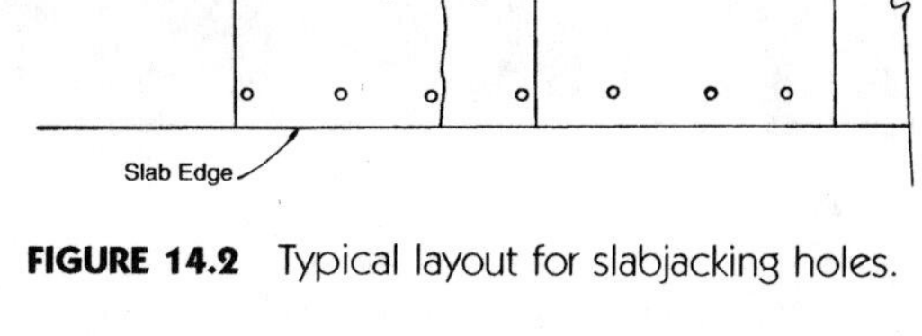

FIGURE 14.2 Typical layout for slabjacking holes.

The rows nearest the slab edges and working joints are typically set in about 1 to 2 ft (0.3 to 0.6 m) from them. The spacing may be varied for individual holes that border a crack or joint. Here, the hole is offset so as to be immediately adjacent to the defect in order to take advantage of the weakened plane already existing.

The holes are usually within a range of 1.5 to 2.5 in. (38 to 64 mm) in diameter. They are most often drilled with pneumatic rotary-percussive rock drills, which is by far the fastest and least expensive method, although diamond core drills are sometimes used, especially in cases where restoration of the slab surface is important. For example, where terrazzo, quarry tile, or similar finishes must be penetrated, as illustrated at the left in Figure 14.3, the cores are labeled as to their original location and orientation. Upon completion of injection, they are replaced in proper position, and the surrounding space is filled with a resin or mortar colored to blend with the surroundings, as illustrated at the right in Figure 14.3, where the hole can barely be seen in the center tile.

14.1.2 Grout Mixtures

As mentioned earlier, a variety of grout mixtures continue to be used for slabjacking. Whereas mortarlike mixtures may be favored because of their ability to stay in the area of injection and allow closer control of the work, the usefulness of the cementitious slurries, for applications which require little if any lifting is recognized. Foaming polymer grouts have also been used and have become heavily promoted in recent times.

FIGURE 14.3 Decorative surfaces can be penetrated with a core drill (left) and the core replaced to blend with the surroundings (right).

14.1.2.1 LOW-MOBILITY MORTARLIKE GROUTS

Grouts very much like those used for compaction grouting (described in Section 11.1.1.2), are entirely satisfactory for slabjacking, except that gravel-size aggregates should not be included unless a significant void exists and/or a large lift is to be made. The aggregate should thus fall within the envelope provided in Figure 2.7. Cement is included, usually in a proportion of about 10 to 12 percent of the aggregate by volume. Alternately, standard masonry sand can be used, although this will result in a somewhat harsher and less penetrable grout. The penetrability of a mixture, required to start lift, can be greatly improved, however, by the addition of fly ash in an amount equal to one to two times that of the cement.

A frequent requirement when work involves highway pavement is rapid return to traffic, and concern is often expressed that this may affect the strength of the grout. Based on experience with hundreds of thousands of square feet of pavement successfully jacked, this is not a valid concern. One must remember that these low-mobility grouts require considerable pressure to inject. Once in place, the grout is fully contained, such that it is not easily displaced. Furthermore, the moisture is usually absorbed by both the grade and underside of the slab, so that the grout becomes immobile.

Slabs are typically raised to their proper levels using only a portion of the grout holes that have been drilled, and because the grout remains near its point of injection, the underslab void is not completely filled once it has been raised. The remaining holes are thus used to fill any remaining space with a more mobile grout. This is normally the same mixture used for lifting, except with a greater amount of water. It is usually pumped into one hole until it penetrates to and returns out of several nearby holes.

14.1.2.2 SLURRY-CONSISTENCY GROUTS

Slurry-consistency grouts are compounded with a wide variety of both basic constituents and proportions thereof. The only common denominator is the use of portland cement, although its proportion may vary from 5 to 50 percent. Other ingredients include both fly ash and natural pozzolan, hydrated lime, rock dust, and fine sand.

The use of these grouts for slabjacking seems to be dominated by state highway departments, and there has apparently not been a great deal of collaboration between them. This is not surprising, considering the answers to a questionnaire sent to all 52 highway departments in the United States as part of the *State-of-the-Art* report by the Committee on Grouting (1977). With the exception of four adjacent states in the southeast, no two departments indicated use of the same grout constituents or mixtures, although virtually all were of slurry consistency. This may be the result of some departments' difficulty in changing old practices. Once their benefits are discovered, it is likely that the lower-mobility grouts will be used more widely. A possible exception will be the continued use of mixtures containing high proportions of fly ash in geographic areas where such material is plentiful at little or no cost.

14.1.2.3 POLYMER FOAMS

The use of foaming urethane for slabjacking has been variously promoted since the early 1980s. The original process was protected by U.S. patents, and the lone promoter states that the procedure continues to be protected by those and updated patents, so that only one source is available for this application. In practice, the components of a very rapidly reacting plural-component formulation are pumped through separate hoses to a mixing head, from which the material is impelled directly into small drill holes. The components are usually heated, and special proportioning pump equipment is required. The resin is injected through closely spaced small-diameter holes in discrete increments. With each increment, a small amount of lift is achieved. The process is repeated until the surface has reached the desired grade.

The procedure offers distinct advantages: very small holes, on the order of 1/2 in. (13 mm) in diameter, are required, and the amount of raw material that must be handled and pumped is only a small portion of the final in-place volume of the expanded foam. And because foaming to final volume is almost instantaneous, the work proceeds quite rapidly. This not only alloys fast completion, but also results in a relatively clean operation. Because the added material is of low density, minimal weight is added to the underlying geomaterials, which can be important where settlement has already occurred.

Countering these advantages, however, are several limitations, the greatest of which is probable lack of complete filling of the cavity. Because the injected material reacts and reaches its maximum volume in increments shortly following injection, significant voids remain after treatment. This was found to be the case when a bridge approach slab near the Denver airport that had been so treated continued to settle and was subsequently removed. The foam grout was observed to have developed into pedestals at the drill holes, leaving a large portion of the total area untreated.

Yet there are places where the procedure can be very advantageous and should be considered. An example is the lobby of a large hotel, where settlement of the slab-on-grade floor occurred near the registration desk. The area had to remain in service virtually full time. Cross slits

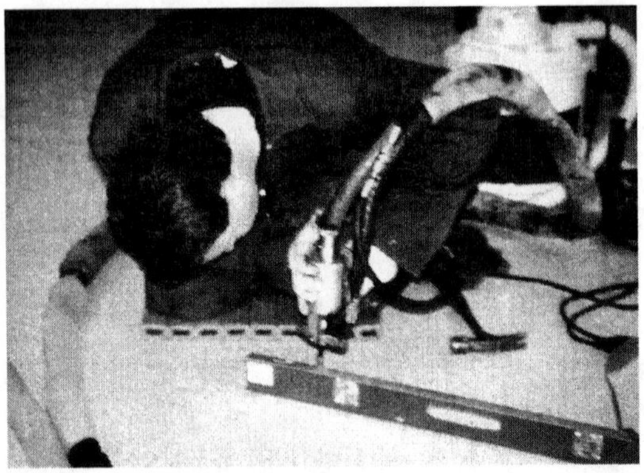

FIGURE 14.4 Injecting polymer foam grout through a small hole under the carpet.

about 4 in. (100 mm) long were cut in the carpet covering at each injection location. These were folded back, and 5/8 in. (16 mm) diameter holes were drilled through the slab. Injection was then made as illustrated in Figure 14.4. The carpet was folded back and glued down so that little sign of the work remained. In all, an area of about 600 sq ft^2 (55 m^2) was treated in a period of less than seven hours, the work was done in the early morning when minimal lobby activity occurred. Because the loads were light, complete filling was not as important as avoiding interference with the facility's normal operations.

14.1.3 Jacking Mechanics

Returning a slab to grade without causing new cracking requires consideration of the original settlement mode. The forces placed on the underside and within the slab should be as gentle as possible. This dictates uniform lifting in an essentially reversed order of the original settlement, usually starting in the area of greatest settlement. Unbalanced forces transferred into the slab during lifting will be proportional to the contact area of the advancing grout along the slab bottom. This suggests that wide hole spacing and highly mobile grout will be advantageous.

Unfortunately, this is not feasible in most applications. Greater forces are required at the area of greatest settlement, to both initiate the lift as well as maintain it. If a highly mobile grout is used so as to affect a large contact area, the lift will tend to occur at the outer areas where restraint is not as great. Obviously, this precludes an orderly return to proper grade and is, in fact, the most frequent reason for failure to accomplish the jacking objectives.

The old rule stating that "the shortest distance between two points is a straight line" is applicable to the performance of slabjacking. As a slab settles, it tends to lengthen. This may be indicated by an opening of the surrounding joints or the appearance of new cracks within the slab. As an example, Figure 14.5 shows an idealized sag in a slab. Cracks have occurred at points A and D on the ends of the slab and are wide open as the top of the slab is in tension at these points. One or more compensating cracks will probably have occurred in the lower portions of the dip, shown as points B and C in Figure 14.5. At these points the top of the slab is in compression, so the cracks will appear to be very small, and may not even be visible to the unaided eye.

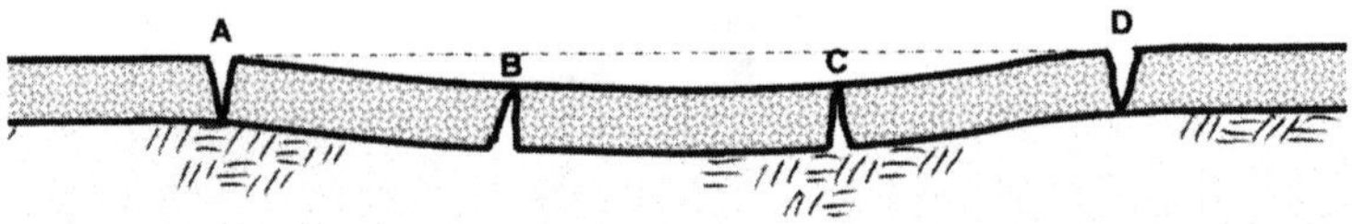

FIGURE 14.5 Typical cracking pattern in an idealized dip.

Slabjacking, if properly performed, should not cause new cracks in the concrete; however, cracks such as those indicated at points B and C will likely open as the sag is removed. Should they be filled with mud or other debris, such as results from the pounding of traffic, they will be opened wide. Conversely, as the sag is raised, the boundary cracks at points A and D will tend to close. For this reason, it is wise to clean any existing debris out of these cracks prior to the start of injection. This is readily accomplished with a blast of compressed air.

If debris remains in the cracks and joints, or if the work is done on a hot day when the concrete is thermally extended, it is likely that the existing length is insufficient for the straightened slab. In this case, the slab will bind as it is being raised. Once such binding begins, the injection pressure will increase. If the pump is not stopped immediately upon such a sign, ever higher pressure will build until the slab heaves violently, usually to an excessively high elevation.

It is thus important to continuously monitor pressure and to stop pumping if slab binding is indicated. Should this occur, a relief space must be created by saw cutting full depth, chipping, or some other means. If binding occurs on a particularly hot day, it may suffice simply to stop the operation and resume in the early morning hours when the temperature is lower. It often happens that cracks for the full width of a slab occur concurrently with an explosive rise. Failure to recognize this mechanism has resulted in many excessively raised slabs, which are usually excessively cracked, with no means of pushing them back down to proper grade.

Usually, there are voids on the undersides of slabs that have settled. Unfortunately, these are seldom under the point of maximum settlement. During lifting, the grout acts like hydraulic fluid in a jack. The magnitude of lifting force is in direct proportion to the area of the grout on the underside of the slab. Once upward movement occurs, a continuous void area will be developed; however, this does not occur until initiation of lift. If no void exists around a starting hole, one must be created. This can be formed by jetting with either compressed air or water under pressure, as illustrated in Figure 14.6. The use of air is much preferred, as injected water will create quite a messy operation.

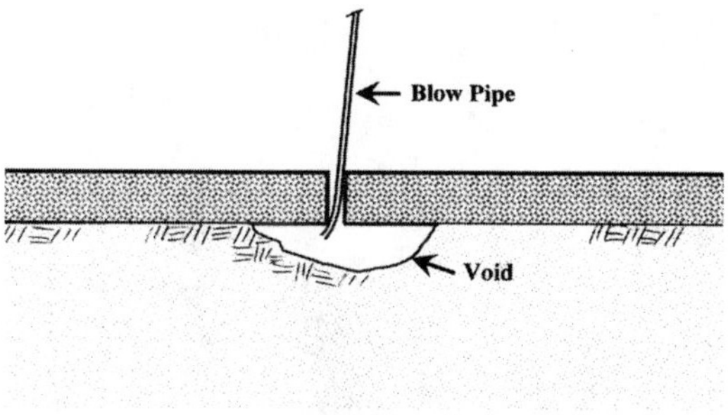

FIGURE 14.6 A void is created under pavement to start lift.

14.1.4 Grout Injection

Once movement has been initiated, it is important to not lift more than about 1/4 in. (6 mm) at any one point at one time, if cracking is to be avoided. This requires constant moving of the injection header from one hole to another as the work progresses. A good procedure that vastly improves production is bifurcation of the main hose into two separate branches, each with a valve, as illustrated in Figure 14.7. Grout is not pumped through both headers simultaneously, as it will always go to only one hole as a result of variation of resistance between the different locations. However, when injection is to be switched from one hole to another, the valves can simply be changed simultaneously. The first header can then be moved to the next hole in line, so that it will be ready to switch on when required. This procedure greatly reduces the amount of pump down time for moving the

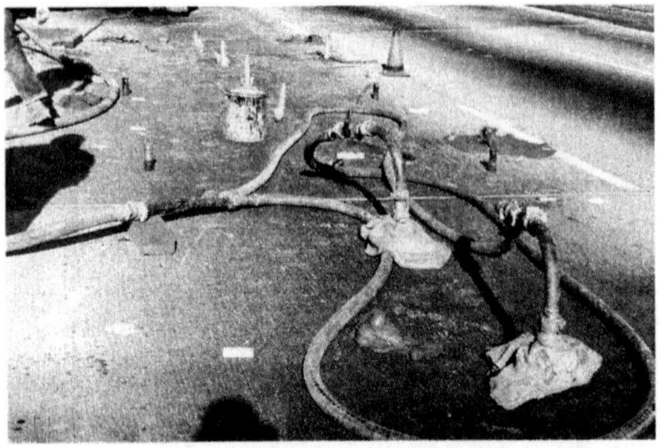

FIGURE 14.7 A "Y" fitting from a main delivery line divides into two branches.

headers and allows the pump to keep running almost continuously.

There are several different methods of connection to a hole. When thin slurry grouts are used, caulking a nipple into the hole or the use of a packer is generally required. Where the grout is more plastic (stiffer), self-sealing handheld injectors or pipe nipples that are secured into the hole with wooden wedges will suffice. A preferable header setup is shown in Figure 14.8. This includes an appropriate gauge and a valve for control, as discussed. The injector is secured into the hole with a few wooden wedges. The wet

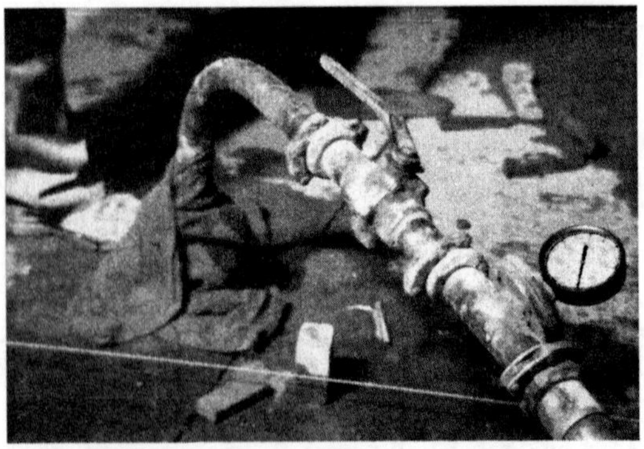

FIGURE 14.8 Typical header setup. Wet burlap bag around connection protects against splatter.

burlap bag wrapped around the connection is to block any spray of grout that may leak between the wedges and to limit the amount of grout spray in the unlikely event that the connector should come loose.

Securing with wedges is fast and easy. Although this may not resist very high pressures, it is more than adequate for those used with most slabjacking, with the exception of the momentary high-pressure spike often experienced at the start of a hole. Using a two-line system, a worker is always at the header to open the valve. By simply pushing down on the valve as it is opened, the penetrating nipple is wedged tighter in the hole so that it will restrain that higher pressure level. If the nipple does not stay secured, the worker will feel it and can pull it free so that no spray of grout occurs. And, of course, even if it does, the wet burlap will prevent it from going far.

A handheld injector requires a worker to hold it in the hole full-time while pumping. It contains a flexible rubber tube, which is inserted into the hole and sealed by expansion resulting from the grout pressure during injection. It must be continually thrust into the hole by the attending worker. To reduce worker fatigue, long handles are usually fixed to these injectors so that bending over is not required of the worker. Although this type of device does facilitate the frequent moves essential to orderly lifting, a pressure gauge is generally not used and the only knowledge of the pressure being developed is gained by the "feel" of the worker, who is constantly leaning on the injector. Leak-free injection and detection of excessive pressure are thus totally in the hands of that worker. And, unfortunately, a hand holding the injector is not available for other monitoring of the work, which is essential if close tolerances are to be achieved.

Depending on the roof area of the void adjacent to the first hole injected, a pressure spike of up to several hundred psi may be experienced as pumping starts. This will quickly relax, however, as soon as lift occurs. Once started, the typical pressure required to maintain orderly raising varies between 30 and 120 psi (2 and 8 bars). The lower levels prevail with thin slabs in which holes are spaced relatively closely, whereas higher levels should be expected on thicker and stiffer pavements with greater hole spacing. The pressures are also lower when working with thinner, more mobile grouts than with the thicker mixtures.

As in all grouting, a sudden drop of pressure and/or long, sustained injection at little or no pressure is indicative of the grout's flowing to a void or some other escape route. Although subslab voids are not uncommon with settled pavement, once they are filled, the area of lift must move to different locations, so careful observation is mandatory. Of course, a pressure drop may not be due to voids being filled, but rather to leakage at the edge of the slab or into a pipe or other substructure. For this reason, pressure gauges should be required at the grout header for all work.

A pumping rate of up to about 2 ft^3 (56 L) per minute can be used in the early lifting and/or when filling voids. A much lower injection rate is in order as the slab approaches its final grade, however. The final pumping rate will determine the accuracy of the lift. Where high accuracy and very low tolerances are required, the final injection rate should not exceed about 1/2 ft^3 (14 L) per minute. Obviously, for optimal performance, the grout pump must be capable of the full output range that may be used. Moreover, easy and uniform adjustment of that output while continuously pumping must be provided.

14.1.5 Monitoring and Control

To achieve an orderly lift and attain the desired elevation tolerance, the relative position of every area of the rising slab must be known at all times. This requires the use of monitoring methods that are simple and yet will provide immediate information. Many monitoring procedures applicable to compaction grouting, as described in Section 11.1.1.4, are equally relevant here. However, in slabjacking, the best single tool to observe changes in elevation is a series of simple string lines, spaced about 4 or 5 ft (1.2 or 1.5 m) apart and tightly stretched over the area to be raised. These are set a short distance above the final required grade, and lifting is continued until a gauge block of the proper height just comes in contact with a line, as illustrated in Figure 14.9.

A system used successfully for many years employs gauge blocks that are 3/4 in. (19 mm) thick, 1-1/2 in. (38 mm) wide, and 3 in.(76 mm) long. The line is set 3/4 in. (19 mm) above the intended final grade. The approximate relative position of the slab at any particular time can be ascertained by placing the blocks either on edge or on end. Where the starting sag exceeds 3 in. (76 mm), multiple blocks can be used. In practice, several blocks are distributed at intervals of about 3 or 4 ft (0.9 or 1.2 m) under each line.

Occasionally, the final profile of the slab will be in a vertical curve, so that using blocks of equal height is not suitable. String lines can still be used, although gauge blocks of varying height will be required and their specific positions must be predetermined and maintained throughout the

FIGURE 14.9 Gauge block under string line is used to establish the surface elevation.

lifting operation. In this situation, the exact distance from a straight line to the final slab elevation is calculated for each interval of length along the line. Blocks of the proper height are then procured and placed in the respective positions. It is imperative they remain in their exact positions throughout injection, which can be readily ensured by initially marking each position.

For frequent work that involves lifting to other than straight lines, rather than cutting special blocks, a set of precut blocks of different height can be used. These are typically graduated in height intervals of about 0.005 ft (1.5 mm). Spacing along the line is then calculated, based on the height increment variation of the blocks. Handheld calculators can be prgrammed so that calculations are easily made in the field as required. Grouters using such differential blocks usually have sets that range from about 3/4 in. (19 mm) to 6 in. (152 mm) in height.

The lifting resistance of a slab is reduced at locations of cracks and joints. Early detection of any movement at these points is thus imperative. Such impending movement can often be felt before actually starting. Furthermore, sounds emanating from the area of movement can often be heard. Although it is not possible to have workers with their fingers on every bit of joint, it is good practice for the personnel to be aware of potential movements and to check areas where movement is indicated or anticipated by placing a finger over the joint.

It is generally desirable to correct vertical offsets at joints during the work, and thus a constant knowledge of the movement patterns is requisite. Such knowledge can be gained by placing simple devices, such as pairs of wooden blocks or wedges, adjacent to each other, as shown in Figure 14.10. As a reference of the initial position, a line should be placed on the higher block before injection starts. Further marks can be made as the lift progresses, and it is useful to indicate the time on each line as it is drawn. When this making is done on several sets of blocks along a joint, a good assessment of the upward movement can be made at any time.

FIGURE 14.10 Marked blocks can be used to observe differential lift at joints.

As previously discussed, open cracks or joints usually exist at the ends, or around the periphery, of a depression. These should be cleaned out prior to injection so they can close as the depressed area rises. It is a good idea to observe changes in the width of such discontinuities as the work progresses. The easiest way is by placing masking tape across the joint before starting injection. A joint's closing will be indicated by a buckle of the tape. Once movement is confirmed, another piece of tape can be applied to indicate any further narrowing of the gap. Again, it is useful to mark the time when each indicator is placed so as to allow better assessment of the overall operation. This procedure is illustrated in Figure 14.11, in which a buckle of the near piece of tape indicates closure of the crack. Another piece has been placed nearby, which is still taut, indicating no further movement. In the unlikely event that the crack should open wider,

FIGURE 14.11 Masking tape will indicate changes in crack or joint width.

the tape will either be broken or loose adhesion on one side.

When working with more mobile mixtures, it is useful to follow the movement of the grout prior to the initiation of lift. This can be accomplished by visually observing the underslab conditions with the aid of a flashlight or by directing sunlight into the hole by way of a small mirror, as illustrated in Figure 14.12. Placing thin wooden sticks that bear on the subgrade and protrude from the holes can also be useful. These are best placed so that they all lean in the same direction as they exit the hole. If grout comes in contact with such a device, it will usually be displaced, changing the lean orientation.

Unique to the work performed on some state highways is a *Benkelman beam* to monitor vertical deflection at joints under moving loads, both before and after injection. This is essentially a tubular beam that is grounded independently of the pavement. One or more precision extensometers are mounted over the area being evaluated. Although the concept is good from the standpoint of the information it can supply, its use is not very practical within a grouting environment where hoses are dragged about and personnel are in constant motion.

FIGURE 14.12 Under-slab conditions can be observed by lighting with a mirror tilted into the sun.

14.1.6 Range of Use

Much of the literature available on slabjacking, and indeed most of the larger applications, are in connection with highways and similar pavements. There is, however, a volume, probably as large or even a greater, used on other types of work. The process is extensively employed to correct settlement problems in virtually every type of structure, as well as exterior slabs on grade. It sees frequent application in emergency response to settlement problems, often following floods. Because it is substantially less costly than replacement of slabs, it is even used around single-family residences to raise settled drives, walkways, and patios, and there is a substantial number of contractors in large cities who perform residential work exclusively.

FIGURE 14.13 Shelving and contents were raised with the floor.

FIGURE 14.14 Slabjacking can easily be accomplished around and under obstructions.

Not only is it possible to raise slabs, but often the equipment and fixtures that are mounted or supported by them as well. Figure 14.13 shows an example in which shelving and contents are being raised simultaneously with the floor on which they bear. This work was performed in the evidence storage locker of a large courthouse complex, and removing the contents presented a major problem for the administrator. As can be seen, the fixtures were simply left in place, protected only by a plastic cover, and raised with the floor. In all, about 1200 ft^2 (110 m^2) of floor, which had settled up to 3 in. (75 mm), was releveled in two shifts of work performed at night when the facility was otherwise unoccupied. The mixer and grout pump remained outside the building, requiring 250 ft (75 m) of hose to access the work area.

Not only can equipment on a floor be lifted simultaneously, but it is also practical to perform slabjacking around and under equipment and other obstructions, as illustrated in Figure 14.14. Here, settled slabs in a tallow rendering plant are being raised. The facility is often washed down with water, which was ponding in the settled areas and creating a safety hazard. About 1,500 ft^2 (138 m^2) of settled floor area was leveled in a two-day period while the plant remained in full operation.

Contrary to popular notion and the unnecessarily messy operation of some contractors, as with all grouting, slabjacking need not be an untidy operation. Many projects have been performed in occupied space or around delicate equipment that could not tolerate dust or splattered material. Figure 14.15 shows work in progress in a room filled with racks of highly sensitive electronic equipment. A broken water main had caused settlement of up to 5 in. (127 mm) and left some substantial voids under the floor. Because of the extreme sensitivity of the equipment, which remained in full operation during the work, as part of the extra

FIGURE 14.15 Slabjacking need not be messy as illustrated in this electronic equipment room.

care taken throughout the operation, secondary protection was provided by panel enclosures, as can be observed. Also visible is the clean floor, which remained shiny throughout the operation.

14.2 GROUTJACKING STRUCTURES

A wide variety of types of structures can be raised by groutjacking. The procedure is usually the most economical method for precise leveling of settled works, and it can usually be executed with fewer disturbances than other remedial methods. It is important to first establish the reason for the settlement, however. If it was the result of faulty underlying soil that is subject to additional consolidation, more settlement will probably occur. If groutjacking is performed without first attending to the deficient soil, the future settlement will actually be exacerbated by the additional weight of the grout that is deposited during the lifting effort.

Most groutjacking is thus performed as part of a larger effort, the principal objective of which is to densify the faulty underlying soil. In the United States, this is most frequently accomplished by compaction grouting, with the lifting performed as a part of that program. There are, however, cases in which settlement has resulted from unusual events such as earthquakes or floods and the underlying foundation soils are otherwise adequate. In such instances, only the lifting operation need be performed.

Because groutjacking will always result in an addition of weight to the underlying soils, which can be considerable, it is essential that the soil's ability to support the increased load is adequate. There have been more than a few instances in which the most efficient method to level a structure was groutjacking, but because of a question relative to the soil's ability to withstand the additional load, other means were used. This obviously is not a problem where the soil has been improved as part of the remedial work, but can be a limitation where marginal materials are left untreated.

Because virtually all structures are quite stiff and the unit bearing loads generally high, lifting from immediately under the foundation is not often feasible unless special preparations are made for containment of the grout. This is not much of a limitation, however, as it is usually not desirable to raise the surface improvements only. Most structures have utilities, drains, and other piping embedded in the underlying foundation soils. If the structure were jacked up from the subgrade only, the utility connections would be broken or badly distressed.

Because of the rigidity of most structures and the impracticability of lifting at numerous places simultaneously, the total lift must be achieved in many small increments, much as employed in slabjacking. Here, again, it has been found advantageous to lift the upper soils along with the structure, as this tends to extend the area of lift from any one injection point. This operation is best accomplished with low-mobility grouts, although some work has been reported in which flowable grouts have been used.

In earlier times, it was quite common to perform lifting operations with highly mobile grouts, but the results were somewhat hit-or-miss. Although a few projects were successful and highly promoted, most were failures due to difficulty in containing or controlling the grout. These latter cases were seldom mentioned, and unless one was close to the grouting industry, the occurrence of such failures would likely not be known. In more recent times these grouts have been used with a fracture mechanism, as discussed in Section 2.3, which allows somewhat better control. The stiffer grouts do provide a clear advantage, however, as they are far more resistant to uncontrolled travel and subsequent unrestrained leakage and are overall significantly more controllable.

14.2.1 Lifting Mode

Because comparatively high grout pressures are required to lift heavy structures, procedures must be used that facilitate confinement of the grout. It is for this reason that the lowest-mobility grout that is pumpable is usually the best choice. Likewise, injection at depth is usually better than near the surface, as the confinement obviously is greater. Another advantage of deep injection is that the utilities and other substructure will undergo minimal disturbance and will be raised with the structure. All the prevailing conditions must be balanced, however, because the load to be displaced likewise increases, due to the weight of the overlying soil or rock, as the depth of injection increases.

Ideally, a structure would be raised from many lifting points uniformly distributed throughout its footprint. This is not practicable with grouting, however. Manifolding a single hose from the pump to several branches has often been attempted, but seldom found to be satisfactory. The reason is the varying loads that almost always exist at the different locations, such that all of the grout flows to the single point of least resistance. Although it is not practical to inject from a large number of locations, multiple pumps can be used, as well as separately valved multiple branches, as discussed in Section 14.1.4 and illustrated in Figure 14.7. Virtually all structural groutjacking is performed in multiple small increments, thoughtfully sequenced throughout the structure.

14.2.2 Lifting Mechanisms

Obviously, the grout pressure required, and thus the degree of confinement needed, will be proportional to the horizontal roof area the grout is able to affect at any given time. The exact procedures to be used will depend on the particulars of the structure and, especially, the foundation unit loads, the nature of the underlying geomaterial, and the type, quantity, and depth of existing substructure. One of three general lifting mechanisms will usually be involved, however.

14.2.2.1 LIFTING WITH COMPACTION GROUT BULBS

Where compaction grouting is used to stabilize a faulty soil under a settled structure, the total uplift force increases with expansion of the horizontal area of the injected grout mass, as discussed in Section 9.4.1, and illustrated in Figure 9.6. The soil stress condition changes as densification occurs, resulting from the applied horizontal forces of the growing grout mass. Upon injection of sufficient grout, these factors will combine to result in jacking of the overlying soil and, obviously, any structure on or substructure within it.

The stress state at any given location is defined by the resistance to static forces on that soil. At any given point, there will be stresses from many different directions. These are interrelated, and if the stresses on all possible planes through a point are analyzed, one that is higher and another that is lower than any of the others will be identified. The direction of these two extremes, which are known as *principal* stresses, are always perpendicular. The plane of the minor principal stress, that is the least resistance, is typically a vertical plane. This results in the preferential direction of any fracture being a vertical plane, which, of course, will not be useful for jacking in a vertical attitude.

Compaction grouting, however, applies high horizontal forces to the soil, and at some point, as these increase, the plane of minor principal stress will become horizontal. Any hydrofracture that is to occur will then be preferentially in a horizontal orientation. With the greater upward force resulting from the growing horizontal cross section of the grout, this fracture will propagate and sustain an upward jacking force. This mechanism will, of course, repeat with each stage of

FIGURE 14.16 Railroad tunnel and adjacent retaining wall was lifted several inches during compaction grouting of the underlying faulty soil.

grout injected; thus, the total amount of uplift required can be achieved in many small increments so as to not cause excessive stress in the overlying structure.

Figure 14.16 shows a cut-and-cover railway tunnel and a high adjacent retaining wall that had suffered from excessive settlement on one side. To correct the soil deficiencies, compaction grouting was performed to depths of about 90 ft (27 m) from the bottom up, in 1 ft (0.3 m) stages. Because the soil confining pressure is high at depth, very high injection pressures, and thus an excessive amount of grout, were required to lift from the lower stages. Once the injection ascended to a depth of about 40 ft (12 m), however, the structures were lifted in small increments to a total amount of several inches.

In a somewhat different example, a swimming pool was settling because of consolidation of an improperly compacted layer of underlying fill at a depth of about 10 ft (3 m). Because of the shallow depth, compaction grouting was performed from the top down in 2 ft (0.6 m) stages to densify the culprit soil. Although longer stages are normally used in compaction grouting from the top down and would have been adequate for the required densification, the need to jack one side of the pool shell nearly 2-1/2 in. (63 mm) dictated the use of a greater number of short stages.

14.2.2.2 LIFTING WITH HORIZONTAL HYDROFRACTURE

It is sometimes possible to create a horizontal hydrofracture by injection into a short portion of an open grout hole, as illustrated in Figure 14.17. The fracture is initiated by the pressure of a small amount of highly mobile grout, followed by low-mobility grout like that used in compaction grouting. It is important to restrict the quantity of the mobile grout, as an excessive amount can cause other uncontrollable fractures to develop. Normally, only about 2 to 3 gal (7.5 to 11 L) of thin grout is required to initiate the fracture. This typically consists of neat cement and water, which are mixed by hand in a bucket and poured down the casing immediately prior to injection of the thicker grout.

When grouting into previously untreated soil, the hole should be tightly cased to within a few inches of the bottom so as to expose only that small increment to the grout pressure. With only a few inches of hole open, a horizontal rather than vertical fracture will occur. Even

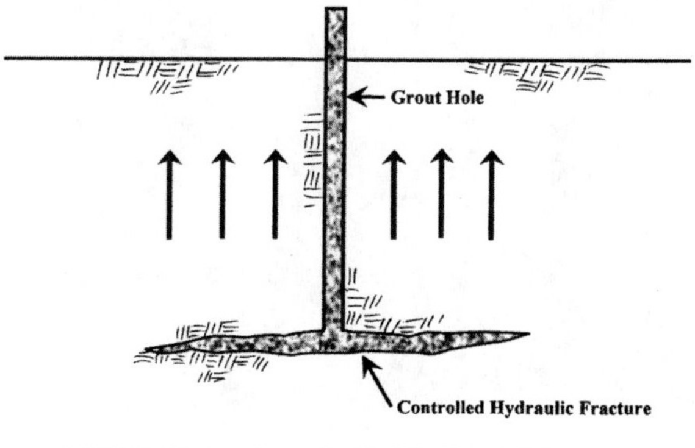

FIGURE 14.17 A controlled horizontal fracture can initiate jacking at depth.

though the normally weaker vertical plane will exist, it will probably not be affected, because of the very short portion of the hole that feels the grout pressure.

Where compaction grouting has been previously done but the required lift not accomplished, a horizontal fracture is somewhat easier to achieve. The groutjacking holes should be drilled at the exact locations of the previous compaction grouting holes so as to penetrate the previous grout. When pressure is applied, a horizontal fracture will tend to form at the interface of two previously grouted stages. This was the method used to raise pile-supported columns of a parking structure to within 1/16 in. (1.5 mm) after it settled, while still under construction, because of the drilled shaft piles failing to penetrate to competent soil. Compaction grouting was used to densify the deficient area but did not lift the columns to the specified grade. The holes were redrilled and a few gallons of slurry were dumped in, followed by immediate connection and injection of stiff mortarlike grout (Figure 14.18).

14.2.2.3 LIFTING FROM A CREATED VOID

A void from which to lift can be physically created by angular drilling, tunneling, or another excavation method. This has the advantage of delineating both the configuration and the area of grout deposition for the lift and facilitates prediction of the results. In the simplest application, holes are drilled either horizontally or at a shallow incline under the structure or element to be lifted. A long, heavy double fireplace in a residence had settled excessively, pulling the adjoining portions of the building down as well. To selectively raise it back to grade, two sloped holes were drilled under the fireplace, as illustrated in Figure 14.19. One was cased to a point slightly beyond the near edge of the footing, and the other extended nearly to the far edge, a distance of about 11 ft (3.3 m). Grout was then injected, and the fireplace was easily jacked back to its proper position. By having the casings terminate at each end, the element could be adjusted to absolute level.

In a similar but much larger application, 12 in. (0.3 m) diameter horizontal holes were drilled under a huge metal furnace, shown in Figure 14.20. It was settling differentially on one side, which was causing distress at the connections of numerous overhead utilities. The owner was adamant about not removing the unit from active production, so repairs were carried out with the furnace in full service and filled with a

FIGURE 14.18 Pile-supported column is raised to grade along with the surrounding soil.

FIGURE 14.19 Inclined holes are used to raise the concentrated load of a large double fireplace.

FIGURE 14.20 Horizontal holes were bored under the raft foundation of the metal furnace to facilitate grout injection.

partially molten metal. The combined weight was said to be slightly more than 500 tons (450 metric tons), requiring an exceptionally great concentrated lift. Two horizontal borings were placed from a pit existing on one side of the furnace. To reach under the foundation, the holes penetrated a retaining wall directly adjacent to the concrete raft supporting the furnace, the upper portions of which are shown at the bottom right of the Figure 14.20.

Soil around the borings was hollowed out on the back side of the wall to provide a key for concrete plugs, which were then cast so as to isolate the injected grout. Pipe sections of 2 in. (50 mm) were cast into the plugs to allow connection of the grout headers. One pipe terminated a short distance past the plug, and another extended to the far end under the opposite side of the unit. Two grout plants were employed to simultaneously inject the four pipes. A "Y" fitting, illustrated in Figure 14.7, was attached to each 2 in. (50 mm) diameter pump line. This fed two 1-1/2 in (38 mm) branch lines, each of which was connected to one of the borings and fitted with a valve at the point of injection.

Once the boring cavities were filled with low-mobility grout, the lift was started. The valves were cycled so as to inject the entire length of both borings in a generally uniform and simultaneous manner. As the proper elevation was approached in the first location, the points of injection were adjusted by throttling the valves so as to bring all of the area to proper level. The owner had been warned in advance that the operation was unprecedented and that there was a small chance that proper leveling might not be achieved. He was willing to take the gamble, however, and, fortunately, once lifting commenced, it went very smoothly and was completed in less than an hour's time.

A uniform settlement of about 2 in. (50 mm) occurred in the case of another industrial furnace (Figure 14.21), shortly after completion of construction. Although the settlement was fairly uniform, it resulted in misalignment with some connected equipment and therefore required correction. The unit was founded on a very rigid, heavily reinforced 4 ft (1.2 m) thick continuous concrete footing. Access was restricted, however, and was possible only on two adjacent sides. There was insufficient space to drill inclined holes or to excavate a satisfactory pit to allow horizontal drilling.

Accordingly, two small access pits were excavated on one side of the foundation to facilitate

FIGURE 14.21 Poor access required the excavation of small tunnels from which to inject the grout under the foundation.

tunneling under it. Small tunnels, about 2.5 ft (0.7 m) wide by about 3 ft (0.9 m) high and immediately under the concrete, were then excavated, extending from the access pits to the opposite side. The tunnels and access pits where then backfilled with pumped concrete to within a few inches of the foundation. Prior to the concreting, a pair of 2 in. (50 mm) pipes was placed in each tunnel immediately under the foundation so as to terminate near each corner. Once the initial concrete fill had hardened, the remaining space between it and the underside of the foundation was plugged with concrete and the remaining portions of the pits filled with compacted soil.

A low-mobility grout, similar to that used for compaction grouting, was then pumped from two separate grout rigs. As in the preceding case, a "Y" fitting was supplied so that each branch could be independently controlled. Injection was made so as to lift the foundation in a uniform manner. More than a week was required for preparation, but once grouting started, the furnace was raised to the desired elevation within only a few hours pumping.

In another interesting application, earthquake-induced settlement of a bridge column founded on a large raft foundation was corrected. Access was substantially restricted, in that the existing grade sloped up to a busy highway on one side and down to the shoulder of another on an adjacent face. Considerable excavation was required, including some shoring, to expose the top of the footing, as shown in Figure 14.22. Eight equally spaced vertical injection holes were drilled through the foundation and advanced about 1 ft (0.3 m) into the underlying soil. Circular voids immediately under the foundation were then eroded through the use of high-pressure water impelled from a special tubular lance with right-angle nozzles. The lance was rotated by hand as water was pumped at a rate of about 12 gal/min (0.76 L/sec) and 8,000 psi (552 bars) pressure. Round cavities on the order of 3 ft (0.9 m) in diameter were eroded.

FIGURE 14.22 Cavities were eroded with high-pressure water to facilitate lifting of this bridge column.

Stiff mortarlike grout was then injected. Four separate grout pumps were used, and each was connected to two holes that were individually valved. All valves were left open at the start of lifting. The plan was to close one on each pump once the jacking was initiated. The pumps nearly reached their maximum pressure capability at about 1,600 psi (110 bars) to start the lift, but once movement occurred, it continued at a fairly constant pressure of about 700 psi (48 bars). Because the rise was orderly and uniform, all of the valves were left open. This may be one of very few cases in which simultaneous injection was successful through more than one hose at a time. It was, no doubt, strongly influenced by the generally equal loading at the grout hole locations and the very stiff foundation mat.

14.2.3 Groutjacking Tolerances

As with slabjacking, groutjacking of structures can be performed to very close tolerances. But the unfortunate fact is that a considerable amount of the equipment in regular use cannot be slowed and/or pump uniformly enough to work very accurately. Furthermore, many contractors who claim expertise, especially very large contractors, have neither made the effort to learn how to perform the work precisely nor provided sufficient training for their working crews to do the work effectively. Thus, although many structures have been raised to very tight tolerances and it is possible lift and/or level to within about 1/16 in. (1.5 mm) of specified grade, such performance should not be expected of all grouting contractors, even if they claim to do it routinely. The best assurance of receiving quality performance is to contractually provide a stiff monetary penalty for failure to achieve the specified elevations.

Precise leveling is fairly easy in situations where the foundation is fairly rigid, of a generally regular shape, and can be accessed so as to place grout holes and thus inject grout in the particular areas desired. It becomes much more difficult in situations where poor access or other restrictions prevent optimal grout placement, as in the case of the furnaces illustrated in Figures 14.20 and 14.21. And it is sometimes almost impossible to correctly raise a large structure that has been previously repaired.

Reinforced concrete bends and deforms over an extended time period. The deformed shape tends to develop "set," whereby returning to its original state over a short time period is difficult. Furthermore, during deformation, cracks develop. If these have been filled either naturally with debris or intentionally in repair, they will not be able to close. New stress paths can be created in the structural system, and new cracks and faults may develop. Settlement stress patterns developed over long periods of time often cannot be reversed quickly, so small adjustments over an extended time may be required. Any fill in the defects must be removed or alternate breaks provided to enable the structure to return to its original state. Significant settlement nearly always results in major redistribution of stresses within the structural system. A fully competent structural engineer should thus be involved when lifting of such structures is contemplated.

As in new construction, specifying unreasonably tight tolerances in groutjacking existing structures can be counterproductive and virtually always increases the cost for the work. In many cases, reestablishing the original elevation for all portions of a structure may not be in the best interests of the owner. Where a structure covers a large footprint but has settled only on one end, it is often best not to bring that end all the way to its original elevation, but rather to lift as required to feather the variation out so that it is not noticeable. Volumes have been published in regard to acceptable deflections in structures, and these generally work out to a variation off level of about 1/8 in. (3 mm) in 1 ft (0.3 m), which is generally excessively conservative.

Perhaps the best example involves a tract of small residential buildings built on generally unstable bay deposits in the San Francisco Bay area, in the late 1940s or early 1950s. The existing soils were so poor the builders realized that deep foundations would be required if unacceptable settlement was to be prevented, but the cost would preclude development. The solution was to construct small, nearly square homes on a rigid mat foundation. A jacking pit was constructed under each corner of the mat to enable easy adjustment when unacceptable differential settlement occurred. The developer provided a warranty to return and level any structure that settled. The economics of the project were based on making adjustments once the differential exceeded about 1/8 in. (3 mm) per ft (0.3 m). To everyone's surprise, no calls were received until the settlement approached about twice that level. Because the foundation was so rigid, distortion of the doors and windows did not occur, and absent that event, the deflection was not readily noticeable.

A common tolerance that is easy to achieve is lifting to within 1 in. (25 mm) of the specified grade. Reasonably easy for competent grouters is lifting to within 1/4 in. (6 mm) of the required elevation. Significantly greater accuracy is sometimes required, however, and has been achieved on many projects. A survey on the day following jacking of the parking structure column shown in Figure 14.18 established that it was 1/8 in. (3 mm) lower than specified. The contractor was required to regrout so as to raise it to within 1/16 in. (1.5 mm) of assigned grade, which was accomplished. Operating requirements for machinery like that shown in Figure 14.23 may call for it to be absolutely level. All portions were, in fact, brought to within less than 1/16 in. (1.5 mm) of absolute level, which was said by the owner to be as precise as the original construction.

In many situations it is best to simply jack the settled elements sufficiently to look and feel good to the casual observer. For example, the main reason for groutjacking the settled addition to the masonry building shown at the left of Figure 14.24 was the visual impact of the opening where it joined the main structure. The rather small settlement had resulted from minor settlement of the new foundation and was of little structural significance. It had, however, been tied into the foundation of the original structure, so it was not able to settle uniformly at the juncture. The far side of the structure was thus raised only an amount required to close the joint, as illustrated at the right in Figure 14.24.

Regardless of the contemplated tolerances, unless absolutely necessary for the anticipated serviceability, it is often best to leave flexibility in the leveling requirements until the work is actually performed and the various constraints to jacking are observed. This is especially important in dealing with structures that cover large areas, have very flexible foundation systems, or have been previously repaired cosmetically, with former cracks filled or otherwise repaired.

FIGURE 14.24 Settlement of addition resulted in objectionable opening of joint (left), and it was closed when settled area was groutjacked back to grade (right).

FIGURE 14.23 Operation of this machine required leveling to a very close tolerance.

15

CHAPTER

Grouting in Pipes and Conduits

15.1 ACCESS RESTRICTIONS

15.2 SAFETY ISSUES

15.3 SMALL INACCESSABLE CONDUITS

15.4 WORKER-ACCESSIBLE CONDUITS

15.4.1 Water Leakage Control

15.4.2 Strengthening Surrounding Soil

15.4.3 Stabilization and Groutjacking

15.4.4 Jacking to Reestablish Roundness

PIPES AND CONDUITS are almost always embedded in soil, so the methods of grouting around them are no different from those used in other soil grouting applications. Likewise, the reasons for grouting within a conduit are not much different from those for grouting in more accessible places:

- To stop leakage into or out of a line
- To strengthen or stabilize the surrounding ground
- To groutjack deformed sections back to proper cross section
- To raise settled portions to the proper alignment

There is one substantial difference in working in such confined space, however, and that is the logistics and other aspects of accessing and working inside these tubular environments. A great deal of grouting is performed from the ground surface around pipes and conduits. Working from the surface is unquestionably the simplest and usually most economical procedure, but there are instances in which free access to the overlying ground surface is not available for one reason or another, so the work must be accomplished from within. Furthermore, in correction of water leakage into or out of a line, small discrete injections with pinpoint accuracy can be made from within the conduit, whereas much larger grout quantities would be required in working with less accurately placed grout holes from the surface.

When contemplating work in pipes, consideration must be given to the type of conduit and its mechanical behavior with the ground. There are two very different types of conduit, rigid and flexible. Most concrete pipe is designed to be rigid, whereas corrugated metal pipe (CMP) and many composite pipes are flexible. Where the diameter is small, say less than about 1 ft (0.3 m), this is not very significant, but it can become an important issue in the case of large diameters. Rigid grouts should generally not be injected adjacent to flexible conduits, as this can create distress in the otherwise resilient composite soil-conduit system.

15.1 ACCESS RESTRICTIONS

Obviously, working in the interior of a conduit requires openings for access. These may be in the form of existing inlets or manholes, or excavation and penetration may be required. Personnel must traverse long distances to get to the work site, along with equipment and materials, which can be difficult and time-consuming in the usually cramped, often wet conditions. And, of course, the grout must be pumped for the full distance. Sewers and storm water drains typically have manholes spaced every several hundred feet, but other lines often have few if any normal entries.

Grout can usually be pumped up to several hundred feet, so existing entries will suffice in many cases. Where sufficient access does not exist, however, excavations and openings will be required. Sometimes, especially in large conduits, it is possible to move workers and equipment to the area of work even though the distance might be excessive for grout pumping. In such cases, a small-diameter drill hole may be bored from the ground surface to enter the line near to the work area. This can provide access for the grout lines and necessary utilities.

If the work site is very distant from open access, light will be required, as will a communications system. Cramped conditions, as well as special procedures required for handling the exhaust of internal combustion engines, generally prohibit such engines use in enclosed spaces. This means that sufficient electric current and/or compressed air will be required at the work site. Of greatest importance, is the issue of worker safety, as such environments often lack sufficient fresh air and/or may contain high levels of poisonous or explosive gas, in addition to providing less-than-ideal access.

15.2 SAFETY ISSUES

The interiors of pipes and conduits are considered confined spaces, and they can offer hostile atmospheres in which to work. Many people die each year from ill-conceived entry into such environments, including some that are well trained and should know better. Testing the atmosphere to ensure a constant supply of fresh air is common sense, as well as being required by most safety regulations. It is imperative to understand any applicable safety codes and to obtain the required permits for entry into confined spaces. In the United States, such regulations are in effect under the Occupational Safety and Health Administration (OSHA). These include not only the furnishing of personal safety equipment, but special training in its use.

Any space in which the oxygen level is less than 19.5 percent or greater than 23.5 percent is considered hazardous. Oxygen levels of less than 17 percent in the air we breathe will cause slow deterioration of vision. As the amount declines to less than 14 to 16 percent, serious physiological deterioration will occur and likely impede the ability of a worker to exit the area. Conversely, flammable and explosive conditions will occur when the oxygen level rises significantly above 21 percent. Air quality detection instruments should therefore always be maintained and frequently used in such hazardous areas. In addition, a non-entry attendant should always be on duty outside at the point of entry. Among this person's duties is to maintain regular communication with the workers on the inside to ensure that they are continuously alert. In particularly hazardous exposures, the maintenance of escape apparatus and even the employment of standby rescue personnel may be sensible and are often compulsory.

In some instances, simply opening a manhole or another access device on each end of the area to be occupied will be sufficient to maintain a natural draft of fresh air. More often than not, however, forced ventilation will be required. This will make the already restricted access way even more cramped, as illustrated in Figure 15.1. And, of course, when air is forced into one opening, another must be maintained for the exhaust. A good practice is to contact the local fire-rescue department that serves the project area.

FIGURE 15.1 Oftentimes required ventilation ducts make restricted access even more cramped.

Its personnel are usually more than happy to meet on the job site before entry is made to review the plans and provide applicable advice. Even more important, should an accident occur, they will be thoroughly familiar with the site and the operation if they have already visited the location. The importance of safety and conformance with all applicable safety regulations cannot be overemphasized when work is to be performed in confined spaces.

15.3 SMALL INACCESSIBLE CONDUITS

Pipes with diameters less than about 18 in. (0.45 m) are inaccessible to all but the very smallest workers, so work within them must be done remotely. In earlier times, this was extremely difficult, if not impossible, but with the advanced development of closed-circuit TV and robotic tooling, it is now practical to perform many operations in small-diameter lines. The greatest amount of such work is in connection with the maintenance of sewers, where TV inspection and remotely controlled grout injection are routine. Strengthening of soil is not often required in the case of small pipelines and is difficult to perform, especially in situations where water infiltration also exists. This is because most of the stronger and more rigid grouts appropriate for strengthening are not well suited to use with the robotic equipment and procedures. Where infiltration is occurring, very fast setting grout is required; however, it must set into a soft gel so that the robotic packer is not grouted in place.

Water or other solutions, either entering or leaking from the lines, are normally of some concern. If wastewater in the line is leaking out, it can contaminate the surrounding soil and ultimately affect the quality of the groundwater. Yet any liquid that infiltrates will add to the volume of flow that must be handled and most often treated, both of which can have a high cost. Interestingly, this type of grouting is usually performed by specialist sewer maintenance contractors, who often do this exclusively and do not perform other types of grouting. This is, no doubt, due to the distinct safety requirements and very specialized equipment required for work in small lines.

The basic operation and much of the equipment used work on the same principle, whether in cleaning a line, inspecting it by TV, or injecting leaking areas with a remotely controlled packer. Many operations can be performed from only a single point of access, whereas others require winches at separate openings to pull the various tools back and forth. In small-line sewers and drains, unless the line is known to be fairly clean, the first operation is usually clearing of any debris. This is commonly performed by jetting the line with special high-pressure nozzles that are pushed through the pipe on the end of a hose. Special nozzles are fitted on the back of some cleaning heads so as to provide thrust for penetration into the conduit. Such jetting equipment is typically part of a self-contained truck-mounted unit with full power operation for inserting and withdrawing the cleaning jets. It usually has a storage tank of 1000 to 2000 gal (3785 to 7570 L) capacity to supply the required jetting water.

Where roots, bonded solid matter, or other obstructions to the jet occur, *rodding* or *bucket cleaning* will be necessary. Rodding consists of pushing rotating cutters into the pipe on a flexible shaft, commonly referred to as a rod. There are many types of tools and cutters that can be mounted on the head of the rod. The rods may be sectional or continuous, but the latter are more typical of modern sewer rodding units, which are usually truck mounted, with necessary remote power to propel the tools. Bucket cleaning involves the dragging of a tubular bucket apparatus back and forth through the line by way of cables extending from powerful winches at both the upstream and downstream manholes. The bucket typically has a hinged door on each end so that debris can enter it on the end from which it is being pulled, but is retained on the trailing end. Bucket cleaning is the most aggressive method and is usually reserved as a last resort when other methods have not proved satisfactory.

A particular problem in sewers and storm water lines is penetration of roots through joints and other defects. Roots are naturally attracted by the constantly moist conditions and can literally fill the inside of a pipe. Considerable effort is required to completely remove such roots. Because they will return in short order, they are usually treated with a herbicide once they are cut back. Additionally, leakage sometimes occurs at breaks and offset pipe joints in the lines. Although such features present difficulties, they can usually be opened and restored through internal grouting.

Once the line is clear, an initial TV inspection may be performed to determine the general degree of leakage, if any. Modern video equipment uses color TV and high-resolution cameras that provide an accurate picture of the actual conditions. Any liquid leaking into the line will be plainly visible on the screen, as illustrated in Figure 15.2, but leakage out of the line will be detectable only with the use of a pressure test. If excessive leakage is already known to exist, all of the joints may be visually inspected, air-pressure tested, and grouted in a single operation. This is usually performed with the use of TV inspection and chemical grouting rigs built specifically for sewer grouting and typically contained in a single mobile van.

FIGURE 15.2 Infiltration will be seen plainly on the TV screen.

In operation, a double packer is inserted into the line. Ahead of the packer is a coupled TV camera, all as illustrated in Figure 15.3. The TV will transmit a picture of in-line conditions to a screen at the control center in the van (Figure 15.2). In addition, a permanent continuous tape of the entire inspection is usually made for showing to a group or for future reference. Once the in-line equipment is inserted into the pipe,

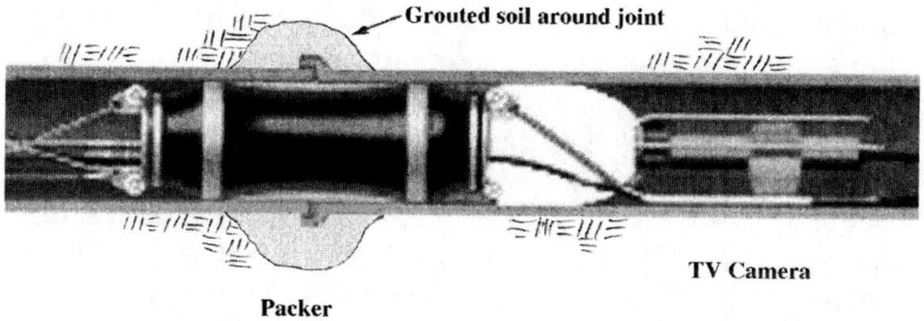

FIGURE 15.3 All TV inspection and grouting equipment for sewer maintenance is usually contained in a single van.

the entire operation is performed from within the van and usually requires only two workers to sustain.

Modern TV injection packer assemblies are sophisticated units that generally have multiple hoses to support the various functions. Because very rapid setting grouts are typically used, they are usually mixed in the packer chamber, and thus separate chemical supply hoses are employed. To these are added air hoses, to inflate the two packer sealing elements and possibly the interior portion of the packer sleeve between the elements, an instrumentation line for pressure testing, and, of course, the power and control cables for the TV. All of these are in addition to the pulling cables often required to propel the assembly through the pipeline. Because at least some limited flow will continue in most pipes during the rehabilitation process, the packer must have a hollow center to facilitate such flow. The various cables and hoses must be reeled in and out simultaneously, which is typically accomplished by special reels in the van.

The packer is used for both pressure testing and to isolate the joint or other defect area for subsequent grout injection. Because open and offset joints commonly occur, the packer must be capable of sealing over as well as passing offsets and gaps of 1 in. (25 mm) or more. For proper seating, the pipeline must be exceptionally clean. Even a small amount of sand or other sediment or debris on the invert can prevent satisfactory sealing. Once in place, both ends of the packer are tightly sealed by inflation with compressed air. To optimize sealing, a high air pressure is desirable; however, an excessive pressure can damage the pipe. Inflation pressure is thus usually limited to about 25 psi (1.7 bars).

Pressure testing is usually performed with compressed air and must be carried out at a pressure at least 5 psi (0.3 bar) greater than that of any surrounding groundwater. Once the pressure stabilizes, it should be held on the defect for a minimum of 15 seconds. If it drops more than about 2 psi (0.14 bar) during this time period, the joint will be deemed faulty and require grout injection.

Where grouting is called for, it is usually accomplished immediately, while the defect area is still isolated by the packer. To keep the work moving, the grout should set rapidly so as to free up the packer for moving to the next defect. In addition, the grout must be highly penetrable and remain so until immediately prior to gelling, as discussed in Section 7.1. The reacted gel should be sufficiently weak that any excess left in the packer chamber and adhering to the interior wall of the pipe can be readily removed once the injection is completed. About the only grout that fulfills these requirements is polyacrylamide solution grout, which was thoroughly discussed in Section 7.3.3.1; this is by far the most frequently used grout for such work.

As discussed earlier, the gel time of this grout can be very closely controlled through the proportioning of the two components. This requires a proportioning pumping system that displaces correct portions of the two parts precisely.

Because penetrability of the different joints and other defects will be ever variable, the ability to pump at a variety of rates is also required. These rates usually fall within a range of about 1/4 to 10 gal (0.94 to 37.9 L) per minute. The required grout pressure will depend on the depth of the line and the pressure of any groundwater, but will seldom exceed about 50 psi (3.4 bars). The pumps typically used are capable of working to pressure levels on the order of 100 psi (6.9 bars), which is far in excess of what is required for most work.

On completion of injection, the sealing pressure of the packer elements is reduced to a point that the packer can be pulled back and forth over the grouted area so as to remove excessive grout that may be adhered to the pipe wall. The air pressure test is then repeated, using the same pressure criteria used before the injection. Test pressure is usually based on the depth of the line but seldom exceeds 5 or 6 psi (0.34 or 0.39 bars). Because the same isolated chamber is used for both air testing and grout injection, there is always a risk that grout has entered and contaminated the test system. Therefore, upon completion of grouting and before the packer is moved to the next location, the air testing system should be purged of any grout and confirmed to be free of obstruction. Pressure gauges should have returned to zero, and proper operation of the monitoring system must be confirmed. Where there is any question of the test system's accuracy, the packer should be moved to a section of pipe free of defects and sealed in place, and the test pressure applied and successfully held.

15.4 WORKER-ACCESSIBLE CONDUITS

Workers can enter and perform in large-sized conduits, but the efficiency with which they operate is directly related to the inside diameter of the line. If they lie straight on a wheeled dolly, workers can fit into pipes as small as about 18 in. (0.45 m) in diameter. They will not be able to turn around, nor will they have much mobility, however, and they will be able to perform only those operations that can be accomplished with their arms stretched out above their heads. Before the development of the modern TV and remotely controlled tooling, a painful amount of such work was necessary, but this is no longer the case. Repairs to small lines are thus usually carried out without entry.

As the size of the line increases to more than about 24 in. (0.6 m), all but the largest workers will be able enter on hands and knees and work somewhat effectively. As the diameter approaches 54 in. (1.4 m), it will be possible for most workers, although hunched over, to walk through and maneuver inside the space. And, of course, as the size increases, the restrictions to orderly movement become less profound. Aside from affecting the workers themselves, conduit size will also influence the type of equipment that can be used. Size has a direct effect on the drilling of grout holes through the lining wall and into the surrounding soil, as both the size of the drilling equipment and, in particular, the length of the drill rod joints will be dictated by the internal diameter of the line, as illustrated in Figure 15.4, in which a handheld drill is being used.

Regardless of the conduit diameter, the space available for storage of tools and materials will always be restricted, and it will often be impossible for anything to pass an active work site. Therefore, the use of multiple crews is often prohibited, extending the time period required for even the simplest operations. A further constraint is dealing with water and perhaps sludge, which often exists on the invert of such lines. This both hampers movement and makes it difficult to find tools that may have dropped, as well as presenting safety concerns with the possible use of electrically powered tools.

FIGURE 15.4 Space limitations usually require compact drills and short drill rod sections.

15.4.1 Water Leakage Control

For leakage correction, it is preferable to grout from inside a conduit if it is sufficiently accessible, as this facilitates drilling grout holes to intercept the leakage paths with good precision. In addition, it allows visual observation of the variations of leakage and, it is hoped, its reduction as grout is injected. Such an operation is not always a wonderful experience for the workers, however, who must perform not only in cramped conditions, but with the presence of excessive water as well. Furthermore, some of the most efficient materials for water control are the water-reactive polyurethane grouts, discussed in Section 7.3.2, but their use can be difficult if not impossible where the space is very wet or humid.

Many of these grouts will gel or foam as soon as the base resin comes in contact with moisture. Although this is an advantage once the grout is injected, it can be a huge problem when the water is literally showering down upon the workers and virtually everything else, as illustrated in Figure 15.5. Because the quantities of material conventionally used in water control grouting are small, they are usually proportioned and mixed near the injection site. This obviously can be a problem with water-reactive grouts when working in the very moist environment of a leaking pipeline. The grout must be mixed and pumped from a distance, or dispensed from sealed containers into closed inlet ports of proportioning pumps located near the leak, if premature reaction is to be avoided.

In dealing with badly leaking joints such as illustrated in Figure 15.5, it is often advantageous to force the leakage to a concentrated potion of the joint in order to minimize the number of holes that must be drilled. This can be accomplished by packing the defect with dry oakum that has been previously saturated with a rapid-setting, single-component, water-reactive urethane.

FIGURE 15.5 Wide areas of leakage can sometimes be channeled to a concentrated area by caulking.

The saturated joint seal must be compressed into the joint quickly, as the resin will begin to expand and foam almost immediately upon coming in contact with the water. Because the resin will continue to expand as long as available space allows, it must be tightly packed into the void so as to provide a very tight plug. With this procedure, it is often possible to channel the leakage to more concentrated areas that are easier to access for the final grout injection.

15.4.2 Strengthening Surrounding Soil

In the construction of large-diameter pipelines it is difficult to properly compact the backfill in the lower quadrants due to interference from the overlying pipe. The result can be serious distress, resulting in deformation of the cross section and the occurrence of longitudinal cracks near the springline. When this occurs, the most practical solution is densification of the deficient soil by compaction grouting. Although this is usually more economical to perform through holes drilled from the surface, where it is prohibited by access limitations, it can be done from within the conduit, as illustrated in Figure 15.6.

FIGURE 15.6 Improvement of poorly compacted backfill in the lower quadrants of large conduits is a frequent requirement.

Permeation grouting is sometimes considered where the soils are granular or sand has been used for backfill; however, one must use caution in such a selection. Pipe is typically laid on a granular bedding material. This material may be sand but is more often gravel, especially where the bottom of the trench was below groundwater level during the original construction. Thus, a large amount of highly penetrable but quite competent material may exist under the pipe. Should permeation grout find its way into such material, very large quantities could be pumped needlessly. This is another example of an important axiom: *It is not the quantity of grout pumped, but the effectiveness that matters.*

Injection is usually accomplished through one or more lines of grout holes, drilled in a radial configuration, along the flawed length of the line. Because the section of defective soil is in the bottom quadrants of the pipe, it is relatively easy to drill. Usually, only a small volume of soil is deficient, so the grout holes are seldom required to be drilled to more than about 3 ft. (0.9 m) in depth and they are typically injected in a single stage.

15.4.3 Stabilization and Groutjacking

Pipelines sometimes settle because of their being founded on faulty soil, changes in the groundwater level, or the placement of additional fill or other weight on top. Although such settlement is sometimes of little consequence, for sewers and drains that must flow by gravity, serious problems can result. Groutjacking, either alone or, more typically, in combination with compaction grouting, can often be used to correct such problems. The lifting of conduits is, of course, performed from below, and concentrated forces are usually directed against the underside of the invert. Because the top of the line will be restrained by backfill and the weight of any over-

FIGURE 15.7 Vertical shoring is often required when lifting large conduits.

lying structures, it will strongly resist raising. The result will be a tendency for the round cross section to be squashed, whereby, the vertical axis shortens. To control this tendency, vertical shoring should be installed prior to injection, as shown in Figure 15.7.

Figure 15.7 illustrates compaction grouting inside an 80 ft (24 m) long section of large drainage line that had settled nearly 1 ft (0.3 m) from its original grade. The settled area was found to lie over a mass of very low density peat as much as 12 ft (3.6 m) deep, over which the pipe alignment traversed. This was underlain by a very dense sand deposit. The compaction grouting was accomplished in four stages, working from the top down. Lift did not occur until the last stage, however. To avoid excessive overstressing of the pipe or shoring, only about 2 in (50 mm) of lift was allowed at any one location at any time The holes thus had to be reestablished several times to permit the many increments of injection required to achieve the proper alignment.

Normally, to minimize subsequently required drilling, the holes would have been washed out upon completion of injection. In this case, however, because of ponding and a muddy mess that would have been created in the settled portion of the line, this was not done and the holes were redrilled all the way from the surface as required. Relatively high pressures of up to 600 psi (41 bar) were required for lifting the conduit and about 14 ft (4.2 m) of overlying soil. Lifting was performed in as many as six separate increments and the line was satisfactorily raised to within about 1/2 in (13 mm) of its proper gradient.

15.4.4 Jacking to Reestablish Roundness

Although excessive deformation of rigid pipes usually results in structural failure where significant longitudinal cracks develop, flexible conduits do not show such obvious signs of distress, but can fail completely if excessive deformations occur. It is thus important to make sure that such deformation is not allowed to develop with these lines. Although it is not easy, sectional deformations can usually be at least partially corrected by groutjacking. Where such distress has occurred, poorly compacted supporting soils usually exist, and these must be improved as part of the remedial program.

A serious problem resulted when a deep fill was placed over an existing 84 in (2.1 m) diameter corrugated metal drainage culvert. The pipe had both settled and become seriously deformed, raising concerns that total failure might occur. By the time the deficiency was discovered, a major highway had been built over the surface and opened to traffic. To avoid traffic interference, and because the fill contained large rocks that would complicate drilling, it was decided to execute the work completely from within the culvert.

Compaction grouting was accomplished by drilling ten longitudinal rows of radial holes around the periphery of the pipe, as shown in

Figure 15.8. The holes were placed at a spacing of 10 ft (3 m) for the length of the conduit. To arrest the settlement and provide a good foundation, the four rows of grout holes below the invert level were extended all the way to bedrock, as illustrated. These varied in depth from about 8 to 15 ft (2.4 to 4.5 m). The higher holes were drilled a distance of 15 ft (4.5 m) out from the wall of the pipe. They were drilled and grouted in stages of 5 ft (1.5 m), starting at the pipe wall. Because of the restricted space, drilling was performed with handheld, pneumatically powered drills, as illustrated in Figure 15.4.

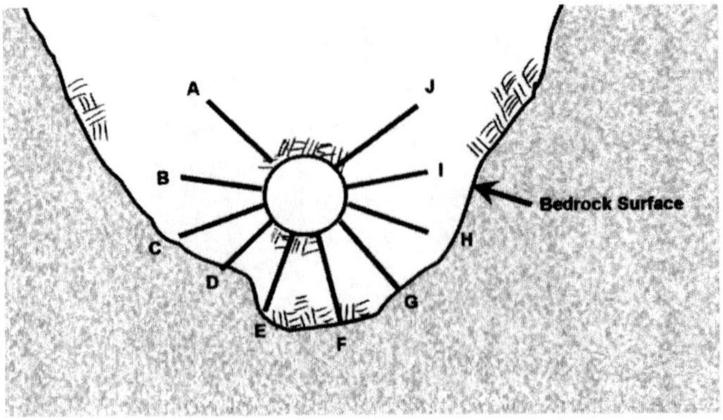

FIGURE 15.8 Ten rows of grout holes were drilled.

In hole rows E and F directly under the pipe invert, a conventional compaction grout mix that contained about 12 percent portland cement was used to provide positive support for the pipe invert. As the CMP-soil structure is designed to be flexible, no cement was used in the higher holes. To maintain pumpability and provide for the fines normally represented by cement, an increased proportion of silt-sized material was included in the aggregate.

In addition to differential settlement of the invert, the culvert suffered from a generally horizontal flattening type of deformation, in which the transverse axis was nearly a foot longer than the vertical axis. Although the pipe was on a sufficient gradient to prevent ponding of water, the hope was to raise the invert portion that had undergone the greatest settlement to reduce stresses within the shell. An even more important objective was to reverse at least some of the cross-sectional deformation by groutjacking once the adjacent soil was densified.

These separate requirements presented a huge challenge in sequencing the order of injection. To prevent escalation of the settlement as a result of the added grout weight, densification of the underlying soils would normally be performed first. However, the desired jacking of the invert that would normally be performed during injection of these holes would tend to increase the cross-sectional deformation. The procedure adopted was to first grout one stage in hole rows C and H on either side of the conduit, followed by completion of rows E and F in the invert. It was hoped that sufficient improvement of the soils on either side of the conduit would occur, so that those underlying the invert could be densified and at least a little vertical jacking accomplished without additional deformation.

It was encouraging that the length of the horizontal axis was reduced by about 3 in (75 mm) during initial injection. Unfortunately, however, it was accompanied by a slight increase in settlement. The first two stages of the bottom two holes, E and F, were then injected. This had a positive effect on the settlement, and the invert was raised by about 1 in. (25 mm). However, a horizontal deformation of similar magnitude occurred, thus increasing the length of the horizontal axis. The final stages of the bottom holes were then injected, followed by careful grouting of those on the sides.

Inside the shell, spring-loaded, calibrated-gauge struts were used to monitor the roundness throughout the injection period. The original plan was to employ eight of these, one placed vertically, one horizontally, and the others equally spaced between. This was found to create virtual fences in the culvert, making movement of the personnel nearly impossible. The intermediate struts were then eliminated, leaving only those that were horizontal and vertical. These were found to be sufficient to monitor profile changes during injection, and yet able to be passed, although with effort, by the workers. Three sets of these were used longitudinally, with one placed near the point of injection and the others about 5 ft (1.5 m) on either side. Continuous monitoring of the elevation of the overlying ground surface was also performed, but no changes in elevation were noted at any time during the work.

FIGURE 15.9 The overhead holes were the last grouted.

The horizontal deformation was almost completely eliminated with this work, without any measurable change in the invert profile. The first stages of the highest holes, A and J, were then injected (Figure 15.9), but they accepted little grout even with a high pumping pressure of about 500 psi (34.5 bars). This high pressure had little effect on the cross-sectional dimensions, which demonstrated the effectiveness of the overall work. This being the case, the last two stages of the upper holes were eliminated. The grouting effort had effectively relieved distress and returned the round configuration of the tube, but the settled portions of the invert gradient were raised only minimally.

16

CHAPTER

Grouting of Underground Structures

Chapter contributed by Raymond W. Henn, Lyman Henn, Inc.

16.1 GEOTECHNICAL AND STRUCTURAL GROUTING
- **16.1.1** Geotechnical Grouting
- **16.1.2** Structural Grouting

16.2 Excavation Methods
- **16.2.1** Underground Excavations in Soil (Soft Ground)
- **16.2.2** Underground Excavations in Rock

16.3 LINING SYSTEMS

16.4 CONCLUSIONS

THIS CHAPTER GIVES a brief description of various grouting methods commonly used during construction of engineered underground structures. An overview is also given of underground excavation methods, lining systems, and typical operational requirements of the completed underground facility. There is an emphasis on how the interaction of these methods, systems, and requirements can affect the design and execution of underground grouting programs.

Various grouting methods and terminologies specific to underground excavation methods and lining systems are reviewed. However, no attempt is made to revisit subjects such as grout rheology, materials, mix designs, mixing and pumping equipment, or grouting methods in soil or rock, because these subjects are covered elsewhere in this book and are equally applicable to grouting in underground structure applications. There is, however, some grouting equipment specially designed to be used in underground construction. This equipment is basically the same as that used in grouting from the surface, except that it is designed to be utilized in the smaller spaces available underground.

Depending on the site-specific geology, groundwater conditions, lining systems, and operational parameters of the completed underground facility, grouting requirements during construction can represent a sizeable portion of a project's overall costs and schedule. This important point is often overlooked or underestimated by engineers and contractors alike. Failing to consider or underestimating grouting costs and the time required to perform such work can lead to project cost and schedule overruns and cause disputes and claims between the owner, engineer, and contractor.

Engineered underground structures can be grouped in four basic categories:

1. Portals
2. Shafts
3. Tunnels
4. Chambers

Underground civil engineering, as well as mine development and mining operations, utilize these four categories of structures. Other less common underground structures such as raises, cross passages, adits, rooms, caverns, drifts, winzes, and stopes can be included in one of these four basic categories for grouting purposes.

Portal and tunnel structures located at shallow depths, approximately 9.3 m (30 ft) or less, can be constructed using the "cut-and-cover" method. Portal structures can also be constructed using the "open cut" method. Tunnel and chamber structures located at greater depths are excavated with the use of excavation methods performed from underground, utilizing a portal or shaft for access and muck (excavated materials) removal. In these cases the tunnel or chamber excavation is said to be "mined." The terms *cut-and-cover tunnel* and *mined tunnel* are common in the technical literature. Shaft excavation is commonly referred to as "shaft sinking." A typical civil underground or mining project generally requires at least two, and can often require three or all four, of these categories of underground structures.

Engineered underground structures can be located completely in soil, completely in rock, or in both soil and rock. A common sequence of construction is to start the excavation in soil and effect a transition into rock. An example of this common transition is a shaft started in soil from the ground surface and possibly encountering a weathered rock zone, followed by sound rock with depth. Another example is a tunnel started at a portal in soil then making a transition to a mixed face condition, having both soil and rock present at the same time, and making a further transition to a complete rock face condition. There are, however, times when a tunnel excavation can make a transition from soil to rock or rock to soil several times in a single tunnel drive.

An example of a geologic setting in which this may occur is a tunnel alignment that traverses first through a rock ridge, then into an alluvial valley, and then back into a second rock ridge. These points are made to alert the reader that an individual underground construction project, or even an individual structure category of that project, such as a shaft, portal, or tunnel, can require geotechnical grouting of both soil and rock or of a geologic material that can be classified somewhere between soil and rock.

Although a project may be located completely in soil, the soil can range from clay to coarse clean sand or gravel. This variation in soil types will require different grouting methods and materials at different locations and times during the excavation process. Therefore, the on-site grouting equipment and the experience of the project's grouting personnel must often include the capability to work equally effectively in a wide range of soil and rock, as well as weathered rock.

16.1 GEOTECHNICAL AND STRUCTURAL GROUTING

Both geotechnical and structural grouting can be required during construction of an individual underground facility. Geotechnical grouting, the grouting of geologic materials surrounding the underground structure, can be performed during various stages of the underground excavation and lining construction. Examples of these stages are grouting performed before excavation begins, called *pre-excavation* grouting; grouting performed during excavation; and *postexcavation* grouting, which is performed after excavation and/or lining installation is complete. Geotechnical grouting may be necessary during only one stage of the underground excavation and lining construction. However, it can be required during two or all three stages of construction. The types of geotechnical grouting commonly utilized in conjunction with underground construction include:

- Jet grouting
- Compaction grouting
- Compensation grouting
- Cementious or chemical permeation grouting

Geotechnical grouting can be performed from the ground surface, underground from within the construction area, or both. It is always preferable to grout from the ground surface whenever possible, because any grouting done from within the underground construction area will almost always conflict with the ongoing excavation and lining processes. However, sometimes conditions such as depth of the underground excavation, as well as interferences and obstructions such as buildings, roadways, utilities, and other infrastructure at the surface or below ground, can make drilling and grouting from the surface impractical. In these cases the geotechnical grouting must be performed from underground.

Structural grouting used in association with underground facilities construction includes grouting methods called *backfilling* and *contact grouting*. Both of these structural grouting methods involve the filling of voids behind the initial support or final lining systems. Backfilling consists of placing one of a variety of materials, such as a neat cement or sanded grout, conventional concrete, cellular concrete (foam grout), or a flowable fill into the annulus space behind a final liner. Contact grouting usually entails injecting a neat cement or sanded grout mix into a relatively small void. The quantity of contact grout required for a project is usually small as compared with the quantity of backfill material required.

16.1.1 Geotechnical Grouting

Geotechnical grouting is used to modify the ground surrounding the underground structure being constructed. Geotechnical grouting is performed in association with excavation and lining of engineered underground structures to:

- Reduce the permeability of surrounding soil
- Reduce the conductivity of surrounding rock
- Reduce or eliminate groundwater inflows into the completed facility
- Reduce or eliminate water, other liquids, and gas outflows from the completed facility
- Improve the ground's strength and structure
- Increase efficiency of the excavation and lining construction
- Improve safety of the workers and construction equipment
- Help limit surface settlement
- Protect other existing underground structures, utilities, and surface structures
- Develop structure/ground interaction

As stated earlier, geotechnical grouting can be performed at various stages or times during construction of an underground structure. These stages of grouting are called:

- Pre-excavation grouting
- Grouting during excavation
- Postexcavation grouting
- Post-lining installation grouting

Examples of the geotechnical grouting methods utilized for pre-excavation grouting and grouting performed during excavations of underground facilities include:

- Jet grouting—used in soil
- Compaction grouting—used in soil
- Compensation grouting—used in soil and rock
- Permeation grouting—used in soil and rock
  - Cementitious grouts
  - Chemical grouts

Figure 16.1 shows pre-excavation jet grouting being performed from the ground surface for a tunnel constructed in soft ground (soil) condi-

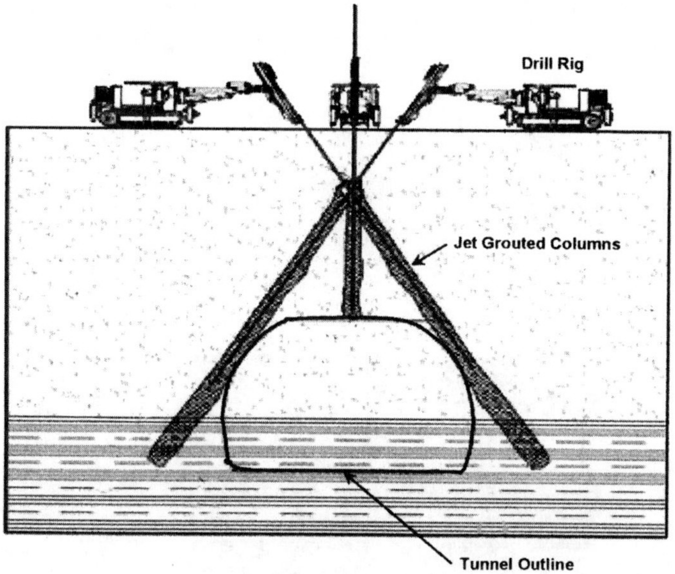

FIGURE 16.1 Jet grouting used from the surface for soft ground tunneling.

Geotechnical grouting methods commonly utilized for postexcavation grouting and post-lining installation grouting of underground facilities include:

- Compaction grouting—used in soil
- Compensation grouting—used in soil and rock
- Permeation grouting—used in soil and rock
- Cementitious grouts
- Chemical grouts

Figure 16.4 shows the post-lining installation of a radial grout curtain installed through a cast-in-place concrete tunnel lining for permeation grouting of the surrounding rock.

tions. Pre-excavation permeation grouting being performed from a tunnel boring machine (TBM) is illustrated in Figure 16.2, and Figure 16.3 shows compaction grouting being performed during excavation of a soft ground (soil) tunnel.

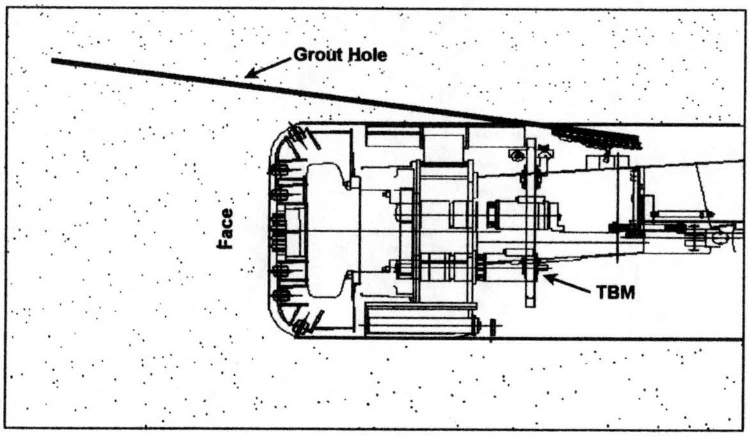

FIGURE 16.2 General equipment layout of probe/grout hole drill.

16.1.2 Structural Grouting

Structural grouting is the filling of the annulus void between the final lining system and the initial support system, or the excavated ground surface. It can also refer to filling voids between the initial support system and the excavated ground surface. The terms *backfilling* and *contact grouting* are used to describe these void-filling processes. Much more detailed and complete definitions of these two structural grouting methods are given by Henn (2003). Simple definitions of backfilling and contact grouting are offered here:

- Backfilling is the filing of a void(s) or annulus created during the excavation and lining installation processes (Figure 16.5). The backfill material can be neat cement or sanded grout, conventional concrete, cellular concrete (foam grout), or flowable fill. The size and shape of the void, and therefore the approximate volume of backfill material required, are usually known before backfilling begins. The normal voids requiring backfill are relatively large, ranging from approximately 2 to more than 24 in. (50 to more than 610 mm). The lateral extent of the void is usually fairly constant but can vary.

- Contact grouting is the filling of a void(s) created during the excavation and lining installation processes. Contact grouting can also be required to fill voids that remain after completion of backfilling or placement of cast-in-place concrete lining. The contact grout material is usually a neat cement or sanded grout. The size, shape, and lateral extent of the void, and therefore the volume of contact grout required, are not usually known before the work begins. The normal depth of such voids is relatively small, ranging from approximately 1/16 to 6 in. (1.5 to 152 mm).

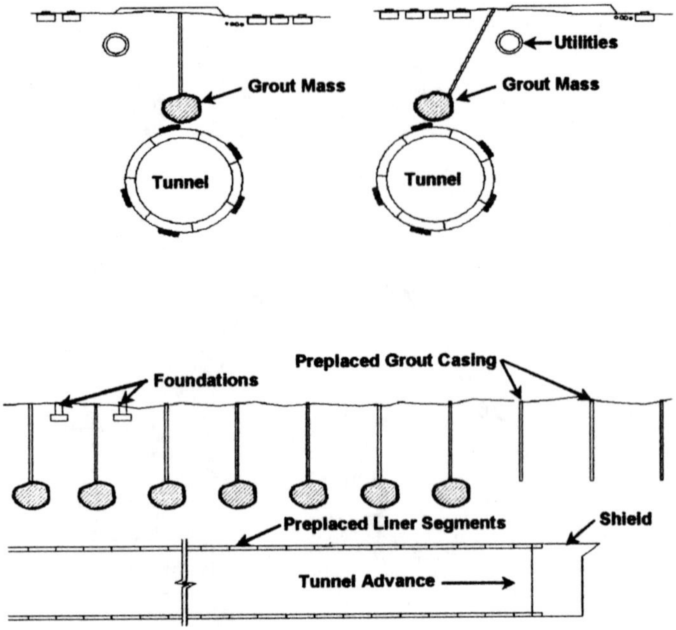

FIGURE 16.3 Compaction grouting—section showing holes drilled vertically and angled (top) and longitudinal view (bottom).

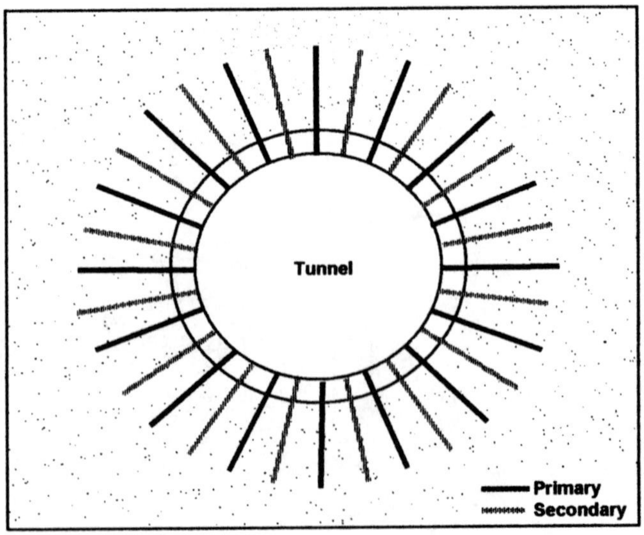

FIGURE 16.4 A typical section of tunnel grout curtain.

Examples of backfilling operations include filling the annulus between the surrounding ground or initial support system and:

- A precast concrete segmental liner system
- A piping system or penstock installed in a tunnel
- A riser pipe installed in a shaft

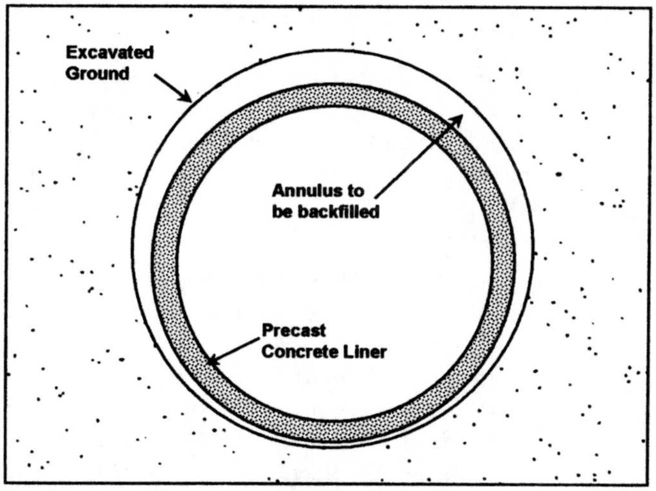

FIGURE 16.5 Example of backfill grouting behind a nonexpandable precast concrete segmental tunnel liner.

Examples of contact grouting include filling voids between the excavated ground or initial support system and lining, such as:

- Steel liner plates used to line shafts and tunnels
- Previously placed backfill or cast-in-place concrete tunnel and chamber liners

These voids are usually in the crown (roof) of a tunnel or chamber, located between the ten o'clock and two o'clock positions.

Contact grouting can also be used to fill voids between the initial support and the excavated ground, such as:

- Behind steel ribs with horizontal wood lagging installed as initial tunnel support
- Behind ring steel with vertical wood lagging installed as initial shaft support

16.2 EXCAVATION METHODS

The method(s) used to excavate the various categories of engineered underground structures can have an important impact on the types and scope of the geotechnical and structural grouting required. The excavation method or methods selected are based on project specifics such as:

- Geology
- Groundwater conditions
- Expected ground behavior during and after excavation
- Cross section and shape of excavation
- Length of a tunnel
- Depth of a shaft
- Height and width of a portal or chamber
- Site access
- Geographic location of the project
- Local area labor and construction practices
- The general economic and bid environment at the time of the bid

16.2.1 Underground Excavations in Soil (Soft Ground)

Soils can range from clay-size particles to boulders, and every particle size and combination of sizes in between. The excavation can be located entirely above or below the water table, or some parts of the excavation can be located above and others parts below. All of these factors will affect the selection of the excavation method and the lining system, as well as the timing, types, and quantities of grouting required.

Typical soil excavation methods used for mined underground structures include:

- Hand mining—portals, shafts, tunnels, and chambers
- Machine mining—shafts
  - Augering
  - Drilled pilot hole and reaming
  - Crane with clam bucket

- Track or wheeled loader
- Hydraulic excavator
- Roadheader
- Machine mining—tunnels
  - Digger shield
  - Roadheader—shielded or unshielded
  - Earth pressure balance tunnel boring machine
  - Slurry tunnel boring machine
  - Microtunneling
- Machine mining—portals and chambers
  - Track or wheeled loader
  - Hydraulic excavator
  - Roadheader
  - Dozer with ripper(s)

In noncohesive soils and excavations below the water table, the ground is usually modified before excavation begins with the use of methods such as dewatering, ground freezing, or any of a variety of geotechnical grouting methods. However, as an alternative to ground modification, the effects of soil behavior and groundwater can also be controlled by using specialized tunneling methods that use compressed air, earth pressure balance (EPB) tunnel boring machines, or a slurry shield (SS) tunnel boring machine. Many projects, however, require a combination of ground modification methods and specialized tunneling methods. The ground modification and excavation methods employed, as well as the lining system used, are all interrelated and likewise have a great impact on the project's geotechnical and structural grouting requirements.

For a hypothetical example of typical relationships, consider a soft ground tunnel located below the water table with relatively shallow cover (the vertical distance from the top of the tunnel excavation to the ground surface). The tunnel is excavated using an earth pressure balance tunnel boring machine (EPBM) in conjunction with a bolted, gasketed, nonexpandable precast concrete segmental, one-pass lining system. Depending on the specific soil and groundwater conditions, the project while using specialized tunneling and lining methods (EPBM and a bolted, gasketed segmental system), may also require performance of the following geotechnical and structural grouting methods:

Possible Geotechnical Grouting Methods

- Preexcavation permeation, jet, compaction, and compensation grouting around or under existing utilities, roadways, surface structures, and other existing underground structures
- Compaction grouting during excavation to mitigate "real-time" surface settlement
- Compensation grouting before and/or during excavation to mitigate surface settlement
- Postexcavation and liner installation compaction or compensation grouting to restore existing utilities, roadways, surface structures, and other existing underground structures that settled as a result of the tunneling process.

Required Structural Grouting

- Backfilling the annulus between the excavated ground and the outside face of the nonexpandable precast concrete segment rings. The backfilling would be performed immediately after erection of the segment ring was completed. Backfill injection is started as the segment ring is leaving the tailshield of the EPBM.

16.2.2 Underground Excavations in Rock

Rock, like soil, can have a wide range of engineering properties and structures that affect selection of the excavation method and lining system, as well as the timing, types, and quantities of grouting required. As a general rule, rock excavated using the drill-and-blast method will require more postexcavation grouting than that in which machine excavation was used. Typical rock excavation methods used for underground structures include:

- Drill and blast—portals, shafts, tunnels, and chambers
- Machine mining methods—shafts
  - Raise boring
  - Blind hole drilling
  - Roadheader
  - Continuous miner
  - Hydraulic impact hammer
  - Pilot hole with reaming
  - Down-the-hole hammer drilling
- Machine mining methods—tunnels
  - Tunnel boring machine
  - Roadheader
  - Continuous miner
  - Hydraulic impact hammer
- Machine mining methods—portals and chambers
  - Dozer with ripper(s)
  - Roadheader
  - Hydraulic impact hammer
- Other excavation methods—portals, shafts, tunnels, and chambers
  - Hydraulic rock splitters
  - Expansion agents in drill holes

Geotechnical grouting can be performed before rock excavation begins, during excavation, or after excavation or liner installation is complete. Most geotechnical grouting of rock is done for groundwater control. However, grouting can also be used to stabilize and strengthen weak, fractured, or blocky rock.

The anticipated geology and groundwater conditions, coupled with the excavation method and the lining system, are all interrelated. These relationships will have a large impact on the geotechnical and structural grouting requirements. For a hypothetical example of these relationships, consider a hard rock tunnel located several hundreds of feet below the water table. In such a geologic setting, fractures, joints, bedding planes, faults, and other geologic features or structures encountered by the excavation may be expected to produce high volumes of groundwater at high pressures. If, in this example, the tunnel was excavated with a TBM and lined with a cast-in-place concrete system, any of several geotechnical and structural grouting methods would most likely be employed.

Probe hole drilling in front of the TBM cutterhead would be used to alert the contractor and engineers of the presence of groundwater and of potential pressurized inflow rates before the TBM excavation actually encounters the water. The probe hole(s) can also be used to check for gas in gassy underground conditions. If groundwater is encountered in amounts greater than the predetermined and specified quantities allowed, the probe hole(s) and additional grout injection holes, as required, would be used to permeation grout ahead of the excavation, ("pre-excavation grouting"). This grouting is almost always performed using a portland-cement–based suspension grout. If geologic conditions warrant, grouts made with ultrafine cement rather than portland cement can be used.

After this initial round of pre-excavation grouting is completed, a second round of probe hole(s) drilled in front of the TBM cutterhead is performed to check for remaining groundwater ahead of the excavation. If groundwater is again encountered in amounts greater than the predetermined and specified quantities allowed, a second phase of pre-excavation permeation grouting is performed. This probe hole drilling/pre-excavation grouting is continued until an acceptable reduction in groundwater flow from the probe holes is achieved. The same method of probe hole drilling and pre-excavation permeation grouting can be used for tunnels excavated with other methods, such as the drill-and-blast and roadhead methods. Heuer (1995) provides a method of predicting groundwater flows into tunnels.

Additional geotechnical permeation grouting can be performed from within the tunnel, at or behind the excavation face, to address specific areas of groundwater inflows. These inflows are usually associated with geologic features or structures such as joints, bedding planes, faults, and shear zones. This type of grouting is often referred to as *feature* grouting.

Geotechnical grouting can also be performed after excavation and lining installation are complete. This postexcavation/lining grouting is usually referred to as *consolidation* grouting or as a *radial grout curtain.* In the preceding example, contact grouting would most likely be required after completion of the cast-in-place concrete liner placement.

16.3 LINING SYSTEMS

Lining systems used to construct underground facilities are classified as *initial* or *temporary support systems* and *final lining systems.* When an underground excavation is initially supported and a final liner is installed later, the lining method is called a *two-pass* lining system. When only one lining is installed, which serves as both the initial support and the final lining, it is called a *single-pass* lining system. Two examples of a lining system that can be used as a single-pass system are a bolted, gasketed precast concrete segmental system and a liner plate system.

These two lining systems are commonly used to line shafts and tunnels. They can also be used as initial support where a final liner was installed within the initially supported excavation. In this application, a two-pass lining system is employed. Some engineered underground structures constructed in sound hard rock are left unlined. This practice is referred to as leaving the excavation *bald.* Leaving a hard rock underground opening (structure) unlined is most common in facilities that have little or no human entry. Examples of facilities that can remain unlined include water tunnels and underground storage chambers for water, gas, and petroleum products. These unlined facilities, however, often still require grouting.

The lining system(s) selected and the timing of its installation will have a direct effect on the timing, types, and quantities of geotechnical and structural grouting required. Table 16.1 provides examples of common initial support systems and typical structural grouting requirements. Table 16.2 provides examples of common final lining systems and typical structural grouting requirements. In both Tables 16.1 and 16.2, under the column titled "Structural Grouting Required," note the comment regarding contact grouting: "May require contact grouting. . . ." The word *may* is used because even though there will almost always be a void for each lining type listed, a question may arise as to whether the void needs to be filled with contact grout or can be left open. A decision on whether to perform contact grouting should be made for each particular project, based on the project-specific design and the operational requirements of the completed facility. Contact grouting underground can be a very costly and time-consuming operation; therefore, it should be required only with good reason. Henn (1996) provides a more detailed discussion of contact grouting methods.

16.4 CONCLUSIONS

The construction of an engineered underground structure will almost always involve one or more geotechnical and/or structural grouting methods. Depending on the site-specific geology, groundwater conditions, and operational

TABLE 16.1 Initial Support Systems

Type of Initial Support	Geology	Structural Grouting Required
SHAFTS		
Soldier piles and horizontal lagging	Soil	May require contact grouting behind lagging
Soldier piles with steel plates	Soil	May require contact grouting behind plate, usually near surface
Ring steel with vertical lagging	Soil and rock	May require contact grouting behind lagging
Liner plates	Soil and rock	May require contact grouting behind plates
Rock bolts	Rock	None
Shotcrete with wire mesh	Soil and rock	None
Shotcrete with steel or poly fibers	Soil and rock	None
TUNNELS		
Steel ribs with horizontal lagging	Soil and rock	May require contact grouting behind lagging
Rock bolts	Rock	None
Shotcrete with wire mesh	Soil and rock	None
Shotcrete with lattice girders	Soil and rock	None
Spiling	Soil and rock	None
Expandable precast concrete segments	Soil and rock	May require contact grouting behind segments
Nonexpandable bolted, gasketed concrete segments	Soil and rock	Will require backfilling between the outside face of segment and the excavated ground
Cable bolts	Rock	None
PORTALS		
Shotcrete with wire mesh	Soil and rock	None
Shotcrete with steel or poly fibers	Soil and rock	None
Soil nailing	Soil	None
Soldier piles with horizontal lagging	Soil	May require contact grouting behind lagging
Soldier piles with steel plates	Soil	May require contact grouting behind plates, usually near surface
Mechanically stabilized earth (MSE) walls	Soil	None
Rock bolts	Rock	None
CHAMBERS		
Rock bolts	Rock	None
Shotcrete with wire mesh	Soil and rock	None
Shotcrete with steel or poly fibers	Soil and rock	None
Steel sets with lagging	Soil and rock	May require contact grouting behind lagging
Shotcrete with lattice girders	Soil and rock	None
Cable bolts	Rock	None

TABLE 16.2 Common Final Lining Systems

Type of Final Lining	Types of Structures	Structural Grouting Required
Cast-in-place concrete	Portals, shafts, tunnels, chambers	May require contact grouting
Nonexpandable concrete segments	Shafts, tunnels	Backfilling required
Liner plate (cast iron or steel)	Shafts, tunnels	May require contact grouting
Precast concrete pipe (RCP)	Shafts, tunnels	Backfilling required, may require contact grouting
Welded steel pipe	Shafts, tunnels	Backfilling required, may require contact grouting
Prestressed concrete cylinder pipe	Shafts, tunnels	Backfilling required, may require contact grouting
Polymer concrete pipe	Shafts, tunnels	Backfilling required, may require contact grouting
Corrugated metal pipe	Shafts	Backfilling required
Other corrugated metal shapes	Portals, tunnels, chambers	Backfilling required, may require contact grouting
Fiberglass pipe	Shafts, tunnels	Backfilling required, may require contact grouting
Jacked pipe and casings made of various materials	Tunnels	May require contact grouting
Shotcrete with wire mesh	Portals, shafts, tunnels, chambers	None
Shotcrete with steel or poly fibers	Portals, shafts, tunnels, chambers	None
Shotcrete with lattice girder	Tunnels, chambers	None

requirements of the completed structure, the grouting program can be a large part of the project's overall cost and schedule. For this reason the project's owner or the owner's engineer should engage the services of a team of underground construction experts. At a minimum, the team should include an experienced underground/tunnel construction engineer, an engineering geologist knowledgeable of underground construction, and a grouting professional well versed in the latest underground construction and grouting methods and materials. The team should be organized and in place early in the planning and preliminary engineering stages of any underground construction project. This will allow the project to receive the maximum benefit from their collective expertise.

A more in-depth and detailed discussion of grouting of underground structures, covering underground grouting methods and equipment, is presented in *Practical Guide to Grouting of Underground Structures* (Henn, 1996) and *AUA Guidelines for Backfilling and Contact Grouting of Tunnels and Shafts* (Henn, 2003).

CHAPTER 17

Leakage Control in Structures

17.1 THE LEAKAGE PATH
- **17.1.1** Leakage Routes
- **17.1.2** Locating the Source
  - **17.1.2.1** Dye Tests
  - **17.1.2.2** Dealing with Ponded Water

17.2 GROUT HOLES/INJECTION PORTS
- **17.2.1** Injection Holes in Common Structures
- **17.2.2** Injectors in Permeable Construction
- **17.2.3** Leaks in Massive Structures
- **17.2.4** Drilling Into Large Water Flows

17.3 THE GROUT MATERIAL
- **17.3.1** Polyurethane Formulations
- **17.3.2** Acrylic Resins

17.4 GROUT INJECTION AND MONITORING

17.5 EXAMPLES OF USE

GROUTING IS the most frequent method of repair to control leakage in structures. Below-grade construction is always subject to groundwater entering the interior. Although not nearly as well recognized, repair of leakage from intentionally constructed features such as swimming pools, ponds, and other waterscapes on above-grade levels is also a huge problem, which is frequently repaired by grouting. Although simple leaks through cracks and joints in underground structures are easily located, this is not the case with much structural leakage, and finding the source often requires more effort and time than actual control.

Unlike the leaking of water through geomaterials, leakage in structures is nearly always through small and discrete passages, most often in the form of cracks or joints. Relatively small amounts of grout are thus usually required, so the unit cost of the material is of much less importance. Accordingly, such water control grouting is typically accomplished with specialized chemical grout systems available from a number of sources. Because generally small quantities are required, the injection holes and connecting hardware are usually much smaller than their counterparts used in geotechnical grouting, as are the equipment and injection rates employed.

17.1 THE LEAKAGE PATH

Water showing up as a leak in some area of a structure has followed a given path from its source. The area where the leak shows up is known as the *negative* end, or in the case of a concrete or masonry elements, the negative side. Although this location is virtually always visible and known, it is of relatively little importance in correcting the leak. Rather, it is knowledge of the origin location, or *positive* end of the leak, that is fundamental to effective correction. And, of course, knowledge of the actual route is important, especially where it might be long or circuitous. To be effective and permanent, the grout seal should be placed into the leakage route starting at the inlet or positive end. In fact, all that is really needed in many instances is a good seal at the inlet. Although the negative location is always known and usually easy to access, if the plug is put there, the permanence of the fix will be in question, assuming that it is even possible to effect stoppage with such an effort.

In simple situations such as a leaking pipe joint, the origin is fairly obvious; it is the outside of the joint, and the leakage path is a short straight run through the joint. While a large flow of water may be entering the pipe and it can justly be called a serious leak, its repair will be quite straightforward, because of the grouter's knowledge of, and easy access to, the origin. Conversely, many leaks have very long and sometimes roundabout paths from their sources. This makes finding such a location significantly more difficult and sometimes impossible. If a leak is plugged at the negative end only, however, it is likely that the water will be redirected somewhere else along its route and reappear at a different location. Most grouters dealing with water control work have experienced the frustration of "chasing leaks," drying up a leak at one location only to have it reappear nearby, usually within a short time.

17.1.1 Leakage Routes

There are four basic routing modes of water leakage in structures. Perhaps the most common and certainly the easiest to remediate are cracks and joints through the concrete that provide a direct path for the water to travel. The next group of defects, which are significantly more difficult to deal with, are similar except for being supplemented with one or more waterproof membranes on the back side. In this situation, the source is a defect in the membrane, but this can be a considerable distance from the crack or joint through which the water is appearing. The water first enters a break in the membrane and then runs in a path between the membrane and the concrete until it encounters a weak spot or opening in the concrete, through which it appears. An effective repair requires mending the membrane, but the location of the break in it is not known and can be anywhere within a radius of 10 ft (3 m) or more in any direction from the visible leak. The actual path may be an essentially straight line, or it can be a long circuitous route.

There are also very small fissures and narrow cracks, often not readily visible without aid of a magnifier, which although very small, can nonetheless conduct significant amounts of water, given enough time. These are sometimes in the form of porous concrete, and thus no discriminate paths exist; rather, a general saturation of the negative surface occurs. Concrete is essentially impermeable if it is cast with a water : cement ratio of less than about 0.5 to 0.6 and properly cured. It just happens that this is about the water content required to place the concrete, so it need not be exceeded, yet it often is when rigid inspection is absent during construction. As a practical matter, most concrete in publicly owned or other significant structures has been mixed and placed so as to be watertight. In these structures, leakage is almost always through cracks or joints, which are identifiable.

These major structures, however, are nearly always protected by a high-quality external membrane, making the leakage source ever more difficult if not impossible to find. To exacerbate the problem, the exact seepage area is frequently not very obvious, as the leaking fluid can wet a large surface area but be so slow that the exact place of entrance is not easily found. Conversely, structures that have been cast with overwatered concrete often exhibit leakage that can literally saturate a large surface area. When this occurs, the source defects are usually very large in area, so discrete treatment is nearly impossible.

17.1.2 Locating the Source

The source of the seepage must be determined to allow proper treatment. Where there are multiple sources and/or leaks, an effort should be made to match each leak to its source. In many situations, the area of the leakage source is not known at first, but is found when drilling for the grout injection. This can, however, result in holes that do not intercept the seepage zone, and considerable experimentation with holes at different trajectories may be required. The effectiveness of grout injected at the source or well into the flow path is superior, so it is worth the extra effort required to intercept the path.

Perhaps the most difficult sources to find are those occurring where there are multiple layers of concrete in combination with waterproofing membranes. This is commonly the case in structurally supported plaza decks. In one instance, leakage appeared on the underside of a structural slab that was overlain with a continuous built-up waterproofing membrane, over which 4×8 ft (1.2×2.4 m) sheets of composition protection board was laid. Plastic sheeting was placed over the board, followed by a sand-absorbent layer, and finally the finish concrete slab. Water leaking from the surface saturated the sand layer and traveled over each of the various intermediate layers until it found a penetration, so that the leak occurring on the underside was a long way from the original site of entry. In such cases it is common to attempt to build up a continuous membrane of grout outside the waterproofing layer, and in this instance this was done in the horizon of sand.

Leakage through concrete dams and similar water-retaining structures often appears to be

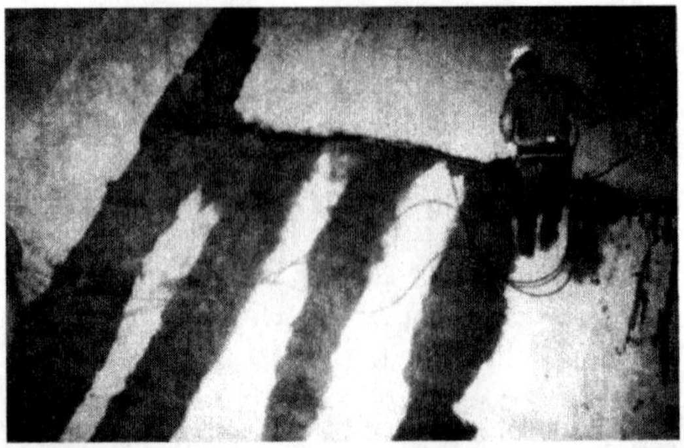

FIGURE 17.1 Apparent leaks in horizontal joints can be deceiving.

through both horizontal and vertical joints, as illustrated in Figure 17.1. The horizontal joints are usually quite tight, however, because of the weight of the overlying construction, and any opening allowing water movement is often only near the face as a result of drying shrinkage of the surface concrete. In such cases, the main source of seepage is often found to be the vertical joints, even though the exposed surfaces of the horizontal joints are also filled with water. Therefore, it is usually a good idea to initially treat the vertical joints, extending in the horizontals only a short distance on either side. Often, this will control all of the leakage so that separate treatment of the horizontal joints is not necessary.

17.1.2.1 DYE TESTS

Dye tests are often performed, in which the water is colored at its suspected source so that it can be readily identified if and when it exits the leakage site. Different colors of dye can be used in the case of multiple suspected sources. Many dyes are available that will color water at relatively low concentrations of less than about two parts per million. Red, green, yellow, and violet dyes are especially effective at these low concentrations. Although a blue dye will not be easily discerned by itself, where multiple possible sources may exit a common leak, it can be useful in combination with yellow to identify the different the sources. When yellow dye is placed in one source and blue in another, the exit color will be green if the two sources are actually combining, as blue and yellow together result in a green color. When the source is groundwater from outside the structure, the colored water can be placed in borings drilled at regular intervals adjacent to the outside of the structure.

17.1.2.2 DEALING WITH PONDED WATER

Particularly difficult to diagnose leaks, which often confound grouters, originate in pools, ponds, or water courses within structures. These features are typically constructed with multiple layers of waterproofing membrane to ensure against leakage. When leakage does occur, however, the path is often long and circuitous, as it penetrates the spaces between the various layers. Unless obvious sources are observed in a pool, it is not uncommon for grouters to flounder as they attempt to treat the leakage from the negative side only. Interestingly, however, this type of leak can be amongst the easiest to correct for those with the knowledge and ability, as the leak source—namely the pool—is known and accessible.

Virtually all open bodies of water are subject to evaporation, so that the surface level will go down even though no leakage has occurred. The amount of water lost to evaporation varies with the environmental temperature and humidity, so the amount of makeup water required to maintain a given level will be ever variable. It does, however, follow established trends. It is thus good practice to meter and record the amount of makeup water so as to establish normal trends. When the quantities appear excessive, a simple *bucket test* can be conducted to determine just how much, if any, of the water loss is due to leakage. This is accomplished by partially filling a bucket with water and placing it in the larger body of water so that the fluid

FIGURE 17.2 Bucket test to determine the water lost to evaporation.

levels are identical, as illustrated in Figure 17.2. The amount of water lost to evaporation will be the same both inside and outside the bucket. Leakage from the pool will be confirmed if its surface level goes down faster than that within the bucket. It is necessary to place a weight in the bottom of the bucket to hold it down in the water.

When leakage is identified, it is then necessary to make a detailed survey of the underwater surfaces. Unless the pool is of minimal depth, not more than about 1 ft (0.3 m), divers are required. There is always suction at a leakage source, and if it is substantial, it can often be felt by drawing a hand over it. Where it is small, it can frequently be identified by simply holding a piece of lightweight plastic such as Saran wrap over the suspect area. The plastic will be drawn tight against the surface anywhere water is being lost. Where the seep is very slight, so that little vacuum is created, dye testing is required.

Although the use of dyes to color large volumes of water at possible sources, combined with observation at the leaking areas, is fairly common, not so common but of even greater value is deposition of discrete small amounts of dye into the water adjacent to the suspected leaking area. This is accomplished by placing single drops of dye at regular intervals of about 6 in (152 mm) each way over the suspect area. The individual drops are observed as they are placed. Most will dilute into the water, shading it evenly in all directions; however, even a very slow leak can provide sufficient vacuum to draw in the colored water, identifying its location.

This is tedious work, and it requires divers with the patience and discipline to systematically cover the affected areas. The dyed water is dispensed one drop at a time, usually from a large dropper or syringe. The diver must, of course, resurface from time to time to resupply his color source. In a particular project, about 2-1/2 days were required to identify the leakage source in this manner. Once found, however, it was plugged with a very small amount of grout in a period of less than two hours. Previous attempts to cure the leakage from the negative surface, even though requiring several days and consuming a large quantity of grout, had failed to correct the problem.

17.2 GROUT HOLES/ INJECTION PORTS

As previously mentioned, because the spaces that must be filled with grout are discrete, relatively small quantities are usually required. Therefore, the pumping rate can be much slower, which allows the injection holes to be small as well. This is important in water control grouting, as very high injection pressures are commonly used so as to penetrate into the fine fissures that are often present. The smaller cross-sectional area of the injection port results in a lower total back-pressure force, which makes it easier to secure the needed injection hardware.

Because injection is preferably made at or near the leakage source and the surface is wet, surface-set ports, which are widely employed in structural epoxy injection, are not satisfactory for most water control work. Here, injection holes should penetrate to near the back side of the leaking section, which is fairly easy to accomplish in most structures where the member thickness is not great. In massive structures, however, long holes are often needed. The length of hole will affect its size, as both cutting removal and accuracy of alignment are affected by the diameter. Larger holes must thus be used where long lengths are required. Very large water flows are often encountered when working in massive structures. These may also require larger holes in order to inject the grout at a rate sufficient to overwhelm the flow.

17.2.1 Injection Holes in Common Structures

For holes less than about 1 ft (0.3 m) long, a 3/8 in. (9.5 mm) diameter hole is sometimes used, although 1/2 in. (13 mm) diameter is more common. A diameter of 5/8 in. (16 mm) is adequate for depths up to about 2 ft (0.6 m), beyond which a 3/4 in. (19 mm) or larger hole will be required. It is not practical to couple extension rods for these small-sized holes, so the depth of drilling is usually limited to the length of the drill that can be obtained. Although not available over the counter, it may be possible to special order single-length drills of 1 in. (25 mm) diameter, up to 4 ft (1.2 m) in length.

Becauase of the potential for electrical shock, the wisdom of using electrically powered tools in such close proximity to water may be questioned, but the fact is that most contractors use electric impact rotary drills for this work. Lightweight pneumatic or hydraulic powered drills are readily available, however, are usually more powerful and faster and certainly safer to use from the standpoint of possible electric shock. They are therefore the only type of drill recommended for such use.

FIGURE 17.3 Typical staggered holes angled to intersect the defect.

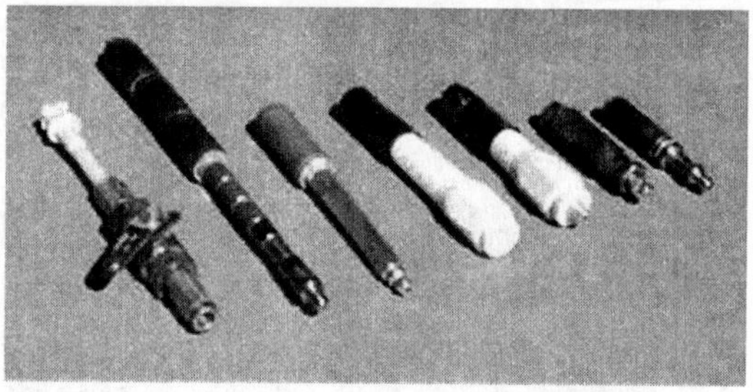

FIGURE 17.4 A variety of small packers are available for injection ports.

Holes should be oriented so as to intersect the defect at an angle. For simple cracks and joints they are typically staggered from side to side, as illustrated in Figure 17.3, with a spacing of about 1 ft (0.3 m). The spacing will vary, however, depending on the width of the defect. Where it is wide open, say more than a few millimeters, spacing can be spread out to 2 ft (0.6 m) or more, whereas it might be tightened to only about 3 in. (75 mm) where the crevices are very tight. As soon as grout is placed, the leakage flow should stop. Should leakage remain between injection points, additional holes, usually split spaced, can be readily provided. Be aware, however, that most well-experienced contractors prefer to err on the conservative side and establish an initial spacing based on the closest intervals they believe might be needed.

As with all grouting, a nipple or other device is needed to make a connection to the hole. Because the holes frequently have water draining out and the surfaces are usually wet, it can be difficult to cement in a nipple. Small packers are thus commonly used. Most suppliers of water-control grouts also distribute a variety of such devices. These vary from one-time tapered hammer-in ports limited to 3/8 in. (9.5 mm) or 1/2 in. (13 mm) holes, to reusable mechanical packers supplied in sizes from 1/2 in. (13 mm) through 3/4 in. (19 mm). As illustrated in Figure 17.4, a variety of lengths are available and there is a wide choice of effective packer element lengths to withstand various injection pressures.

Because of the high pressures often used, these injectors are usually supplied with a grease-type fitting for connection. These are most often the common zerk or button-type lubrication fittings. Although widely used, and certainly adequate for much work, the lubrication fittings do present one substantial disadvantage. Because they are essentially one-way valves, neither water nor diluted grout can run out during injection of the neighboring ports. This limits the information retrieved during injection and can compromise its effectiveness. Although not commonly employed, valves can be fitted onto the injectors and are recommended where it may be beneficial to follow the return flow.

17.2.2 Injectors in Permeable Construction

Seepage through porous concrete or masonry in the absence of cracks or other obvious defects is one of the most difficult conditions to remediate. This situation will require virtually saturating the

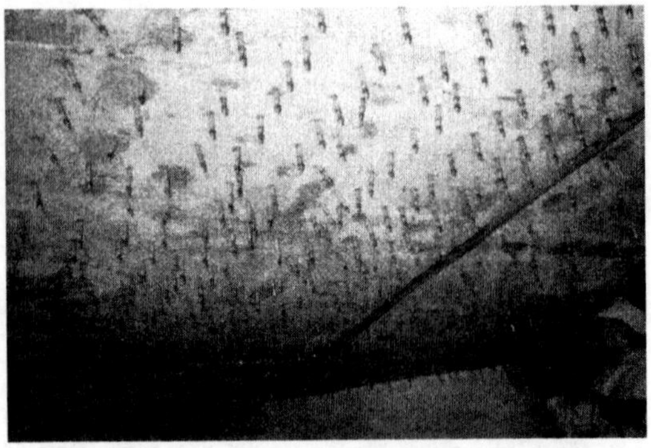

FIGURE 17.5 Porous areas will require a grid of closely spaced ports.

mass with the grout, and a continuous grid of usually closely spaced holes, as shown in Figure 17.5, will probably be required. Holes in such a grid are most often oriented normally to the surface, but individual holes may be inclined into areas of particularly high leakage or where there is indication of a possible fine crack. Generally, the holes should extend completely through the section so that in addition to saturating the mass, the grout will tend to form a continuous membrane on the back side.

There is always a temptation to effect a continuous grout membrane with a minimal number of widely spaced holes, similar to the practice used in forming impermeable curtains in soil. This is not a good idea, however, as it will require substantially larger quantities of grout at a high cost and virtually no benefit. In addition, as the spacing expands, the risk of having areas of missed contact with the substrate increases. Although grouting in soil seldom requires a 100 percent stoppage of all leakage, absolute stoppage is a frequent requirement for work in structures. It is thus far more effective to make many discrete injections at close spacing. This will not only form a suitable membrane, but will saturate and, to some degree, stabilize the porous mass.

17.2.3 Leaks in Massive Structures

Leakage into massive structures often takes long circuitous routes, and it can be nearly impossible to ascertain the flow path on the interior. Thus, the same basic rules apply, and holes are typically placed to cross existing joints on the interior, which are usually either horizontal or vertical. Where leakage from specific features such as construction joints is suspected, the appropriate angle should be established that will most likely intersect it. When only the general location is known, holes should be placed at a variety of angles, as shown in Figure 17.6, in an effort to better define it. Although percussion drilling is the most frequently used method, in situations where the location of joints and similar features are unknown, core drilling that allows visual examination of the features should be considered.

17.2.4 Drilling into Large Water Flows

When it is necessary to drill to depths greater than about 3 ft (0.9 m) and/or space does not allow the use of long bit lengths, coupling drill steel is required. The smallest size in which coupling steel is commonly available is 7/8 in (22 mm),

FIGURE 17.6 Holes at different angles to intersect a distant leaking joint.

and the smallest bit is for a 15/16 (24 mm) diameter hole. In order to readily expel the drill cuttings, however, prudence dictates that a minimum hole size for any but the shallowest holes be at least 1 in (25 mm).

Leakage into structures can be great, as illustrated in Figure 17.7, making it difficult for the working crews to function. In such situations it is beneficial to initially drill one or more holes to the lowest elevation of the leak so as to provide a preferential flow path and minimize the adverse effects of the widespread overhead flow. Because of the difficulty of installing nipples and packers in flowing water, it is best to secure oversized nipples into starter holes at the beginning of drilling before encountering the large flows. The drill string can then be contained within the nipple, in the same manner as used in grouting rock. Where considerable flow is expected, a full flow opening ball valve should be installed on the nipple before continuing the drilling. With this preparation, the valve can be quickly closed as soon as the drill string has been removed, shutting off the flow.

Where a particularly sizeable flow is exiting a large or wide defect, it is advantageous to cover the defect with a gasketed steel plate fitted with nipples and valves to allow the water to pass freely. These must be sufficiently large to allow easy exit of the entire stream of water without

FIGURE 17.7 Severe leaks diminish worker efficiency.

FIGURE 17.8 Open valves on a steel plate can relieve water pressure to allow fixing to the surface.

exerting excessive pressure against the plate during its installation. Once the plate is secured in place, the valves can be closed, as shown in Figure 17.8.

17.3 THE GROUT MATERIAL

As discussed in Section 7.3, water control in structures is most often achieved with one of a wide variety of grouts especially formulated for this use. These grouts are nearly all based on one of two different chemical families, polyurethane or acrylic polymer. These are typically much more expensive than the grouts commonly used in geomaterials, but because the quantity required is relatively small, material cost becomes less important while effectiveness is critical.

The grout to be used should be chosen on the basis of the needed end result, the size and nature of the defect, and the pressure of the leaking water. By far the most frequently used grouts are urethanes, but practice varies widely as to the particular type or formulation. These materials can effectively fill thick and wide open voids with a tough foam or gel. Once contacted by water, most will foam, and the resulting compound is easily identified. Spillage and overflow of the hydrophilic formulations will bond tenaciously onto concrete surfaces and can be very messy to deal with, requiring abrasive or water blasting to remove.

The acrylic compounds are generally much more penetrable and thus advantageous to use in tight formations. For larger defects, they can be filled and thickened with ultrafine cement, talc, or another similar finely divided powder, but this is not their best use and a urethane would likely be more appropriate. Most of the acrylics are clear and appear similar to water. They are thus not easily recognized when mixed with water or flowing out of defects, so they do not create much of a mess if they are spilled on, or leak from, a surface.

Epoxy resins are not widely employed for water-control grouting, but have limited use, primarily in situations where improvement of structural integrity is also needed. These materials are usually injected as low-viscosity fluids, but can be acquired in virtually any consistency or easily thickened in the field by intermixing finely divided powder such as talc. An epoxy develops a tenacious bond on any surface it contacts, and spills are not only unsightly but extremely difficult to clean up. Experienced contractors will make use of a number of different formulations from each group of materials to provide the best performance, considering the individual project requirements.

17.3.1 Polyurethane Formulations

Hydrophilic water-reacted resins are the most commonly used of the polyurethane formulations. Their final properties are dependent on the chemistry of the particular formulation, the amount of water with which it has been mixed, and the temperature of the grout in place. With small proportions of water, they can form a strong, expansive foam. As the proportion of water increases, the foam will become weaker, followed by the development of a tough gel. This will become progressively weaker until it reaches an emulsion form. Because hydrophilic polyurethane grouts are attracted to water, they will develop a good bond to wet surfaces. These particular formulations are heavily marketed and often promoted as easy to use and fail-safe to apply. Although this can be true, these grouts are often not the best choice, and even when they are, overly simplistic application will probably result in less than good performance.

The cured state that occurs will vary according to the different water contents, and the particulars will vary according to the formulation. The formulations should therefore be mixed according to the individual manufacturer's instructions and with a controlled proportion of water. This is typically accomplished with stream mixing at the hole collar by way of a positive-displacement, proportioning pump system. Because relatively inexpensive pumps are readily available that will provide high pressure to inject these grouts as a single solution, much work is so accomplished. This is not recommended, however, because of the usually rapid reaction and the relatively poor performance of the materials when mixed with other than the optimal water content. If the resin is injected into a wet crack with no additional water, it will usually foam and cure. The final properties, however, are often inferior, usually due to an insufficient proportion of water or inadequate mixing.

Hydrophilic urethanes can be formulated to achieve a wide variety of mixed consistencies, ranging from a thin waterlike substance to thick goop, resembling toffee. Although most are water activated, two-component chemical cure formulations are also available. These provide the distinct advantage of a controllable set or gel time. Because this time is always short, their practical use is limited to injection with two-component proportioning pump equipment.

Most distributors market a number of different formulations, each with its own properties; however, a simple basic formulation is the most widely used. Although it may be the easiest to apply, it is often not the most effective, especially when injected without the advantage of proportioning pumps that can closely control the proportion of water. These materials will react with a wide range of different water contents, but to achieve their best properties require accurate proportioning and through mixing. Many hydrophilic urethanes continue to take in available water after their initial reaction. This obviously weakens the compound, so they are not always the best materials for applications that are not completely restrained.

Hydrophobic urethanes require only a small amount of water to initiate reaction. Once reaction is initiated, they will not take in additional water and will usually develop into a rigid foam that is not subject to shrinkage upon drying. Because these grouts repel water, they will not develop much, if any, bond to a wet surface. They can be formulated to provide a high expansion rate, up to about 1000 percent if unrestrained, which has an important economic effect on their use. They require only minimal water to react. In humid environments, the resin can start to react in an open container before injection, making its effective use questionable.

17.3.2 Acrylic Resins

Like the urethanes, acrylic-base grouts come in many different formulations. They all have two or more components that must be properly proportioned and mixed during injection. As a class, these grouts are much more penetrable than the urethanes and are thus particularly useful for treating very fine defects. The two or more components they contain are premixed in separate stock solutions. These are then combined in varying proportions to obtain the desired properties and setting time. Special mixing and proportioning equipment is required for their application. Many of these grouts are extremely sensitive to temperature. The gel time can be up to 30 minutes or more at cold temperatures, but almost instantaneous at the high end of injection temperatures.

Acrylic grouts all cure into a gel. The strength of the gel can vary from soft to very tough, depending on the concentration of the stock solutions and the final proportioning. The gel has the advantage of tending to expand if in contact with water, so it maintains tight contact to wet surfaces. To emphasize this property, these materials are often referred to as *hydrogels.* Unfortunately, they also shrink if they are allowed to dry out. The shrinkage will stop and expansion resume if they are rewetted, but there is typically a time delay between the reapplication of moisture and full expansion.

The reaction of the acrylic grouts is ideal for placement. They show very little thickening until immediately before reaction, as discussed in Section 7.3.3.1. Upon reaction, a soft gel first develops, which firms up over the next several minutes. This behavior facilitates working with very fast gel times, as even if gel were to occur in the delivery system, it could be pumped out within the first few minutes. Extremely alert and experienced workers are required, however, if these advantages are to be realized. Because acrylic-base grouts are generally attracted to water, they will develop a good bond to wet surfaces. They do require greater ability, as well as special equipment, to apply properly, so they are not nearly as widely used as the urethanes. They will provide superior performance under many conditions, however, especially where high penetrability is needed.

17.4 GROUT INJECTION AND MONITORING

An important function of a grouter is to minimize the amount of grout and injection that will be required. Channeling leakage to a relatively small area where possible, greatly facilitates injection of the surrounding area by removing much of the water flow. Where there is leakage through joints or other large apertures, a relief hole can be drilled near the base. The parts of the joint that are relatively free of flow can then be caulked or patched with a cement mortar. Such treatment alone may suffice to prevent further leakage, but at the very least, it will minimize the loss of grout during the subsequent injection.

Grouting must be sequenced to fit the water movement and leakage patterns. Although there are no hard rules, it is generally the case that where seepage is minimal, injection starts at the lower elevations and progresses in an orderly manner upward. If large flows are occurring, injection should start in the least active areas and progress toward the area(s) of greatest flow. Obviously, changes in the leakage distribution must be continuously observed and evaluated as the work progresses. Most work in connection with active leaks benefits from fast gel times, especially when the water is moving rapidly. As a general rule, the faster the flow, the better it will be for the grout to react immediately upon or during the completion of pumping.

During injection it is fairly common to experience a change in the color of the leakage water, followed by the appearance of dilute grout and, finally, good-quality grout issuing from the leaks. It is useful to identify these changing properties. This is not much of a problem with urethane grouts, as they usually exude in frothy off-white foam that is clearly identifiable, as illustrated in Figure 17.9. Such is not the case with acrylic formulations, however, many of which appear as clear water. Some manufacturers include a reversible dye in their products, causing them to be noticeable for a short period of time, after which it fades out. Dye can easily be mixed into those grouts that are not readily identified and differentiated from water.

Both the penetrability and reaction time of the grout are affected by temperature. For example, with a 20° drop in temperature, from say 70° to 50°F, (21° to 10°C), the viscosity of most urethane grouts will roughly double. The set time will also increase measurably. It is thus important not to allow the grout temperature to fall to a very low level. In cold environments, the grout should be stored in a heated enclosure so that the temperature does not descend to less than about 70°F (21°C). Electric band heaters on the containers can also be used for warming.

FIGURE 17.9 Return of polyurethane grouts is plainly visible.

The temperature at which the grout will be required to react must also be considered. Prevailing temperatures in the defects will be about the same as that of any seepage water. Where the leakage flow is not too great, hot water can be pumped through the defects immediately prior to grout injection. Obviously, however, this will not work where large, cold flows are occurring. In these cases, a two-component chemical-cured grout must be used. Both urethane- and acrylic-base grouts are available, but because the viscosity of the urethanes is especially sensitive to low temperatures, the use of an acrylic should be considered.

Many grouts, especially those in the urethane family, can be prepared and injected as a single solution with simple pumps. For most work, this is not recommended, however, as it generally prevents the use of optimal setting times and does not allow changing the proportions during injection. Such adjustment is useful to allow variation of the reaction time in response to changing flow conditions. Furthermore, with some urethanes, a single pump will not facilitate the inclusion of enough water to obtain the best properties of the reacted grout. Adjustable proportioning pumps are thus strongly recommended for most water-control work.

17.5 EXAMPLES OF USE

The breadth of the different water-control grouting requirements is wide. Many applications are straightforward and can be accomplished with only a few regularly placed holes and simple equipment and injection procedures. Others, however, require many holes at different trajectories, and enlightened progression and sequencing of the injection requiring highly skilled workers as well as sophisticated equipment. Although many applications are completed pretty much as preplanned, others present ever-changing conditions that require constant variation of the procedures. Some examples of common, as well as more specialized applications, follow, demonstrating the wide variability of water-control grouting in structures.

A waterstop in the deck of a parking structure had failed, resulting in seepage dripping onto the parked automobiles. Even more significant, it was allowing salt-laden water to permeate into the concrete, which could cause serious deterioration. A row of 1/2 in. (13 mm) holes was established on each side of the joint. They were inclined so as to intersect the joint at about mid-thickness, as illustrated in Figure 17.10. Water-reactive hydrophilic polyurethane grout was injected with a proportioning pump system. It contained the correct proportion of water to form a tough, flexible gel. Injection was started on one end and progressed sequentially to the other. Repair of the 80 ft (24 m) long joint was completed by a two-man crew, which is common in applications for which chemical solution grouts are ideal.

FIGURE 17.10 Angled holes intersect the interior of joint.

Water was leaking around a pipe penetrating the wall of an underground utility room. A hydrophobic urethane was injected from the back side, allowing the leaking water to initiate a reaction and be carried into the joint. Only a few hours were required to correct this otherwise troublesome leak. This was a very small installation, but without the specialized water control grouts and the ability to inject them, this application could have become a major operation.

Excessive leakage of construction joints in an underground metro tunnel was causing a safety problem by raining onto the trains and overtaxing the sump pump system. Hydrophilic polyurethane grout was injected from ports placed so as to intersect the joint in the outside one-third of the section, as illustrated in Figure 17.11. To contain the grout in the nominal 3/4 in. (19 mm) wide joint and minimize leakage onto the interior surfaces, it was first caulked

FIGURE 17.11 Open joint was caulked to minimize leakage during injection.

with dry oakum. The oakum packing was terminated a few inches from the base of the joint to allow free draining of the water and diluted grout until the end of injection, when final closure was made.

The mortar of a brick manhole was severely deteriorated, allowing leakage of water into the lower portions. The objectives of the proposed grouting were not only to stop the infiltration, but also to penetrate the poor-quality mortar to stabilize it and reintegrate the masonry section. An acrylic grout combined with ultrafine cement was mixed into a thin slurry and injected from a ring of holes near the surface, as shown in Figure 17.12. The 1-1/2 in. (38 mm) holes extended completely through the masonry wall, which restrained a very low permeability clayey soil. By filling with cement, the reacted grout would possess significant strength, but the fluid portion would have the ability to penetrate into the porous construction.

Water was leaking through the joints of a very old concrete gravity dam. The water froze in the winter, causing disruption of the near surface concrete, primarily in vertical joints, as can be observed in Figure 17.13. Grout holes of 1 in. (25 mm) diameter were drilled at an angle so as to intersect the joint at a depth of between 5 and 6 ft (1.5 and 1.8 m) from the dam face. A hydrophilic urethane grout was injected, starting at the top of the dam and working downward so as to push the water downward. Immediately ahead of the injection, wedges and oakum packing material were forced into the wider joint areas to minimize the leakage of grout onto the surface.

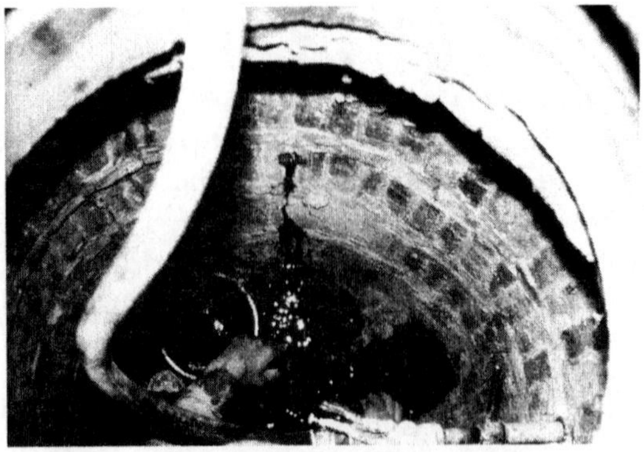

FIGURE 17.12 A cement-filled acrylic grout integrated poorly bonded brick.

FIGURE 17.13 Holes drilled so as to intersect leaking joint several feet from the surface.

FIGURE 17.14 Excessive seepage forced the closure of museum.

In 1984, leaking water forced the cancellation of tours and closure of the 130-year-old Miller Brewery museum in Milwaukee, Wisconsin. The museum was constructed in the 1950s in underground brick caves, originally used to store beer and abandoned since 1906 when refrigeration first became available. The arched caves are constructed of massive limestone blocks, with an inner lining of common clay brick, as shown in Figure 17.14. The walls vary in thickness up to as much as 44 in. (1.1 m) and were overlain by as much as 60 ft (18 m) of earth overburden.

Leakage had been a recurring problem through the years, and many previous attempts to solve the problem had failed. Among others, these included extensive programs to inject both bentonite slurry and sodium silicate-base grout. These efforts ivolved grout holes drilled from the surface over and around the structures. The walls were finally sealed with a acrylic-base grout, working from the interior. This grout was selected for its high penetrability and nearly translucent appearance, which would thus avoid staining the existing masonry.

Prior to injection, the existing brick joints were repointed to minimize grout leakage. Holes were drilled on an 18 in. (0.45 mm) grid over the entire area. The spacing was slightly adjusted as required to keep the holes in the masonry joints. This made for easier drilling, but more important, they would be hardly noticeable upon patching with mortar at completion of the work. The holes were 1/2 in. (13 mm) in diameter and penetrated through the brick masonry well into the underlying rock.

Grout was pumped at a slow 1 gal (3.8 L) per minute so as to promote good penetration. A proportioning pump system was used to impel the correct quantities of the solutions to the grout header, where they were stream mixed, as shown at the left in Figure 17.15. The proportions were controlled so as to result in a gel time of about one minute, just enough time for the grout to permeate sufficiently. Injection continued in each hole until the grout appeared on the interior surface of the brick or in adjacent grout holes. The grout pressure was not to exceed that of the overburden. This was not a difficult restriction, however, because of the slow pumping rate.

The slow and meticulous injection called for considerable patience on the part of the grouting crew, as did the absolute requirement not to deface the existing walls or polished slate floor. The entire floor area was protected by heavy plastic tarps, and sensitive care was taken in the proportioning and pumping of the grout, as shown

FIGURE 17.15 Solutions were stream mixed at the grout port (left), and mixing and pumping equipment sat on the tarp-protected floor below (right).

at the right in Figure 17.15. The effort resulted in complete success, in that the leaks were essentially stopped for the first time in many years, allowing the museum to reopen. In all, some 8,000 gallons (30,200 L) of grout were injected into more than 2,000 holes. Upon completion, the holes were patched with mortar compounded to blend with the original construction.

A 75-year-old concrete arch dam in North Carolina presented several grouting challenges. Leakage water was flowing into an interior gallery at the base of the dam at rates as high as 950 gal per minute (3600 L/minute). It was thought that the considerable leakage was through horizontal construction joints containing poor-quality concrete, resulting from several months-long gaps during the original construction and a number of other construction-related defects. The exposed surfaces of high water: cement ratio concrete were subject to severe freeze-thaw damage during winter pauses in the work. The original construction records provided insight into a number of problems and various attempts to alleviate them, including experimentation with admixtures to improve workability (a novel idea in 1927) and the placing of dry cement in the joints. Although very little was actually known about the sources of the leakage and the condition of the joints, it was believed that the running water had likely further deteriorated the already weakened joints.

Such movement of water in horizontal planes can result in uplift forces within the joints, adversely affecting the gross stability of the dam. Because relatively little was actually known about the interior condition, a combination exploration-remedial program, which included grouting of any encountered defects, was established. The objectives were to fill any existing voids, permeate any porous concrete to improve its durability, bond the individual blocks together, and, of course, to control the leakage. Because none of the established water-control grouts would provide for all of these needs, especially the requirement of high compressive and bond strength, a moisture-tolerant epoxy resin was selected.

Primary holes were established so as to intersect the joints on 4 ft (1.2 m) centers. These were staggered in two rows intended to penetrate the target joint a short distance. Because one of the primary aims was to achieve a better understanding of the interior concrete condition, all drilling was by diamond coring. The 36 mm (1.4 in.) diameter cores were first examined in the field, then placed in core boxes and retained for further reference. The 46 mm (1.8 in.) holes were drilled from the base gallery within the dam, using an independently powered hydraulic coring rig, as shown in Figure 17.16. They varied in length from about 10 to 80 ft (3 to 24 m). Because of cramped conditions in the gallery, short lengths of drill rod were required, the longest being 5 ft (1.5 m).

Packers were installed so as to seal about 4 ft (1.2 m) in from the face of the gallery wall. A tube was provided that extended the grout line to within a few inches of the first defect, as revealed by the retrieved cores. Cement grout was then injected to fill the annular space around the tube to just short of the main defect. This provided greater fixity to resist the injection pressure, as well as to prevent the epoxy grout from

FIGURE 17.16 Core drilling in dam gallery.

entering any unintended minor defects, and allowed the use of low-cost cement grout for filling. Epoxy grout was then injected into the primary defects.

It was first thought that the fissures were generally tight, so a highly penetrable epoxy formulation was contemplated. Upon retrieval of the cores, however, some wide open joints and voids were found to exist. Where these occurred, a finely divided talc powder was intermixed with the resin to add body, reduce the propensity to flow freely in the larger faults, and lower the material cost. The epoxy was mixed in small batches of about 5 gal (19 L) and injected from the gallery, using a vertical air piston pump.

Defects encountered in the initial work were somewhat more extensive than first anticipated. Although the horizontal joints were defective, as expected, one vertical joint also required extensive work. The conditions encountered required drilling and treatment of many secondary holes, and in some instances, even tertiary holes were required. The drilling of these revealed many sealed defects resulting from the primary grouting. The average fissure was found to have a width of 1 to 6 mm (0.04 to 0.25 in.), but filled defects as thick as 10 mm (0.39 in.) were also found. Upon completion, the leakage was reduced to an easily managed volume of about 35 gal per minute (132 L/min.).

A very different type of application was made in the remediation of leaks in a large marine exhibit. The 22 ft (6.6 m) deep, 1.5 million gal (5677 m^3) lagoon shown in Figure 17.17 contained a number of bottlenose dolphins for public exhibit. Upon filling the lagoon and delivery of the dolphins that were to be its occupants, water was being lost at a rate of about 3,000 gal (11,000 L) per hour, in spite of a thick epoxy lining which appeared to be intact and free of defects. To make matters worst, there was significant seepage into an underground viewing area and even around the viewing windows. Such leakage was unacceptable, but with the large mammals already in the pond, it could not be drained.

FIGURE 17.17 Dolphin pond was losing 3000 gal of salt water per hour due to leakage.

To identify the leakage sources, a detailed inspection was made by divers, who joined the dolphins in their pool. It included the deposition of individual drops of dye, systematically, in a tight grid immediately above the lining. The dye identified several potential leaks, all of them so small that they could not be observed with the unaided eye. These were marked upon discovery for later corrective treatment.

Holes of 1/2 in. (13 mm) diameter were drilled in each defective area with a small pneumatic rock drill. These varied in both depth and trajectory so as to intersect the various faulty areas, but were within a range of 4 to 12 in. (101 to 304 mm) in length. Where the defects were so small that intersection at depth was not feasible, the holes were placed vertically within the fault area. Individual packers equipped with zerk inlet fittings were placed in the holes.

A single-component water-activated hydrophobic polyurethane chemical-solution grout, formulated for nearly instantaneous reaction, was then injected. This particular grout was selected because the contractor had had a great deal of positive experience with it on other underwater installations and, in particular, because

FIGURE 17.18 Underwater injection made without disturbing the dolphins.

it would foam immediately upon coming in contact with water, so that any leakage would turn to low-density foam and float to the surface where it could be easily removed. The high-pressure piston-type grout pump remained above water, adjacent to the pond, and fed the grout through 3/8 in. (9.5 mm) hose to a diver, who made the connection to the packer ports, as illustrated in Figure 17.18.

The injection rate of the grout was about 1-1/2 gal (5.7 L) per minute, with pressure ranging from a low of 300 psi (20 bars) to nearly 3,000 psi (200 bars), depending on the particulars of the individual defects. Injection usually continued until the grout was observed returning out of minute defects in the lining. Its return was visually obvious by its whitish-yellow color. It also foamed as soon as it came in contact with the water, with individual pieces swiftly floating to the surface. A worker armed with a swimming pool skimmer fished them out as they surfaced. In addition, a floating containment ring was kept on the surface to trap any deleterious material that might surface.

The work was painstakingly tedious and required the workers to have much patience in methodically covering the entire underwater surface of the pool. In all, about 80 gal (300 L) of grout was placed over a 46-day period. Such repairs require workers who are not only highly skilled in the technology, but qualified divers as well. In addition, special equipment is needed to drill and inject under water. And, of great importance to the owner, no grout or other contaminant that might sicken the dolphins or mar the aesthetics of the decorative facility could be allowed to escape. Although the corrective work was detailed and tedious, it was 100 percent successful in finding and correcting the leakage in a neat and orderly manner.

CHAPTER 18

Grouting in Extreme Environments

18.1 TEMPERATURE EFFECTS ON GROUT
- **18.1.1** Cementitious Grouts
- **18.1.2** Noncementitious Grouts

18.2 EXTREME CHEMICAL EXPOSURES

18.3 PROVISIONS FOR TEMPERATURE EXTREMES
- **18.3.1** Hot Environments
- **18.3.2** Cold Environments
  - **18.3.2.1** Protective Shelters
  - **18.3.2.2** Heat Sources
  - **18.3.2.3** The Grout Temperature
  - **18.3.2.4** Water Supply
  - **18.3.2.5** Delivery Line Protection

EXTREME ENVIRONMENTS present many challenges to the grouter. Virtually all grout materials are affected by temperature, as is the productivity of the working personnel. Operations in confined spaces present special hazards as to quality of the air the workers breathe, as well as the risk of explosion. At very high elevations, the air is much thinner, so the performance of both humans and machines is affected. All of these factors can result in the requirement for special equipment and/or procedures, which should be considered in planning for the project.

18.1 TEMPERATURE EFFECTS ON GROUT

Virtually all grout mixtures will react, set, or cure more rapidly as the temperature increases. For many, there will be a practical limitation beyond which an immediate or flash set will occur. Conversely, reactions are much slower as the temperature decreases. Although the level at which a grout will fail to set or cure varies, the reactions of virtually all grouts are subject to a minimum temperature. Modification of mixture proportions or inclusion of modifying ingredients for some grout formulations, especially those in the chemical solution and resinous categories, can improve their properties at temperature extremes. All will require special consideration, however.

Freezing weather always complicates the work. Grout obviously cannot be injected into frozen ground or defects that are filled with ice. Beyond that, the materials, and especially water, usually require heating. Once mixed, the grout must be protected so that it does not freeze in the delivery system. Special "winter grade" fuel or additives may be required for the proper performance of the power equipment. Heated shelters will likely be needed for the mixing and pumping equipment, material storage, and warming for the personnel. Where extremely cold weather is anticipated, the necessary provisions should be made in advance to facilitate immediate implementation as needed. It is too late to reasonably respond once grout has frozen in the system or water becomes unavailable. Many projects have undergone needless delay and have even been shut down for an extended time while the contractor's staff worked at thawing things out after a freeze. Preemptive action is always superior to merely responding after the damage has been done.

18.1.1 Cementitious Grouts

As in Section 5.1.3 and illustrated in Figure 5.2, the temperature of the grout has a profound effect on setting time. Hot environments can have especially dramatic effects on all cementitious grouts. The hydration reaction will be significantly more potent, and thus set, curing, and drying time lessened. Although the reaction behavior varies among different cement types, and even among different cements of a given type, the set time of all will be reduced, and very significantly, as the temperature rises much above about 100°F (38°C). Type III cement, which is extensively used in grouting because of its finer grind, is especially affected because of its greater surface area available for hydration activity, whereas most blended cements are less affected because of their content of pozzolanic materials, which are not nearly as reactive.

The maximum allowable temperature for cementitious grouts is typically placed at a level between 80° and 100°F (27° and 38°C). Except in very unusual situations, exceeding this temperature level should not be allowed, as setting time can become uncontrollably rapid and strength and durability will be negatively affected. As a practical matter, higher-temperature grouts can be workable in situations where they are being used rapidly, so that mixed grout is used within no more than about 30 minutes of mixing. Where the grout will not be quickly used, however, the lower temperature limits should prevail.

Whereas hot temperatures result in rapid setting and generally complicate injection, cool temperatures slow the reaction, so they do not interfere with the injection; however, cementitious compositions will not set as their temperature approaches freezing. From a practical standpoint, the minimum grout temperature should at no time descend to less than about 40°F (4°C). Because the hydration reaction is significantly slowed by low temperatures, the size of batches and their age become less important.

18.1.2 Noncementitious Grouts

Like cement, nearly all other grouts set and/or harden as the result of a chemical reaction. This reaction is nearly always exothermic; that is, heat is produced as a result of the reaction. The time required for reaction to start, as well as the rate of its development, varies among different compounds and their particular formulations, as well as the temperature of the mixed material. Reaction will always accelerate in hot environments, especially in the case of epoxies and some other resinous grouts. In addition, as many of these materials are strongly exothermic, they can generate a great amount of heat during reaction, so temperature control of the mixed grout is imperative. Interestingly, the penetrability of most resin grouts improves markedly as their temperature increases. Conversely, virtually all will thicken and become significantly more difficult to mix and pump at lower temperatures.

18.2 EXTREME CHEMICAL EXPOSURES

In addition to issues of personal safety when working in hazardous air or confined spaces, there are other factors to consider. Extremes in the chemistry of a formation into which grout is injected, can have a significant influence on setting time and can, in fact, cause the grout to have either a very rapid flash set or no set at all. For example, a cement grout was injected under the bedplates of some equipment in a food processing plant. After three days, the grout still had not set, even though the prevailing temperature was quite warm. Investigation disclosed remnants of liquid sugar under the machine. Sugar is a powerful retarder of portland cement and can delay the set in quantities as low as 0.5 percent, by weight, of the cement.

Setting of portland cement is also affected by its exposure to many salts, especially those of copper and lead. In addition, severe retardation can be expected with exposure to even small levels of sucrose, sodium phosphate, sodium borate, and sodium arsenate. Other salts, such as calcium chloride, will have an accelerating effect. It is thus good practice to test the compatibility of portland cement with the local water and the formation that is to be injected when preparing for work in processing or industrial plants or other sites where extreme chemical exposure might exist.

The set and hardening properties of a chemical-solution grout can also be affected by the chemistry of both the mix water and the formation into which it is placed. For example, sodium silicate will gel rapidly even absent a reactant, when pumped into soil that is strongly acidic. Exposure to salt and/or salt water can have a significant effect on the gel time and curing of most acrylic-base grouts. The gel time will be greatly reduced with exposure to sodium chloride, calcium chloride, most sulfates, and phosphates. Again, appropriate tests should be performed to ensure compatibility of the proposed grout when planning work in areas of extreme chemical exposure.

18.3 PROVISIONS FOR TEMPERATURE EXTREMES

Very hot and very cold temperatures are the most frequent extreme environments encountered by grouters. Provisions for working in such environments are well established, and routine procedures are commonly used, when working with concrete. Much of that technology applies to grouting as well, especially when dealing with cementitious grouts. Accurate control of the set or gel time of grout is far more important than it is for the set time of normal concrete, however. Consequently, even greater effort to control the grout temperature is required.

18.3.1 Hot Environments

The set time of cementitious grouts in hot environments can usually be controlled by simple strategies such as shading the grout mixing area, insulation on the delivery line (as illustrated in Figure 18.1), storing the cement away from direct exposure to sunlight, cooling any aggregates by water spraying, using set-retarding admixtures, and using light-colored injection lines. More stringent efforts will be required at very high temperature levels, however. These may include using chilled water and/or substituting shaved ice for part of the mix water and covering the injection lines with a reflective insulation, especially where they are very long. With

FIGURE 18.1 Insulation of the grout delivery line to prevent excessive heating.

exceptionally high temperatures, it may be necessary to treat the grout with admixed liquid nitrogen, which is a very potent heat reducer. Obviously, the size of the grout batch should be limited so that it is promptly consumed and not allowed to age excessively. Although not very practical for most work, the cooling of an enclosed batch plant is also an option in extreme conditions.

The set time of chemical-solution grouts can often be controlled through adjustment of the proportions of the various ingredients, or through inclusion of modifying compounds. Large solution tanks are best painted silver or white so as to reflect heat and limit its absorption. Smaller material containers should be stored out of the sun and in cooled enclosures when the temperature is extremely high or the reactions particularly sensitive. For this approach, refrigerated trucks and large van trailers are usually readily available for rental in larger metropolitan areas and make excellent storage shelters. Tap water is usually reasonably cool, but if it is stored in surface tanks or otherwise warmed, it can be cooled by in-line refrigeration.

18.3.2 Cold Environments

Cold temperatures, and especially sustained periods of temperatures below freezing, are perhaps the most frequent and troubling extremes the grouter encounters. Not only do low temperatures retard or completely stop the required chemical reactions of the grout, but if sufficiently low, can freeze water as well as grout within a delivery system. Heating the basic materials and/or the mix water is thus advisable and often required. The freezing of both mix water and drilling fluids is a real problem that must be mitigated. To keep both machinery and workers operating efficiently, major tasks, including drilling, grout mixing, and pumping, are best performed in heated protective shelters. In addition, the grout delivery system must be insulated or otherwise protected from freezing.

18.3.2.1 PROTECTIVE SHELTERS

Because good-quality protective shelters involve considerable expense, there is a strong incentive to attempt working without shelter or to minimize the scope of the shelters. This is especially likely as winter approaches and neither the extent nor the timing of cold extremes is fully understood. The natural tendency of many contractors is to put off supplying proper protection as long as possible. It is not unusual to see the erection of makeshift shelters, such as shown in Figure 18.2, as the weather deteriorates. In that instance, haphazard measures were taken to provide protection for the grout plant operator and a portion of the mixer, but most of the equipment was left unprotected. Consequently, freeze damage occurred, and the worker shown is actually repairing the pumping mechanism as a result of freezing in the piston drive waterbox. More often than not, such minimal protection proves to be insufficient, resulting in excessive delays while proper winterization is performed upon the arrival of really cold conditions.

Without appropriate protection, work can be brought to a sudden halt with the occurrence of sustained freezing temperatures. Water ex-

FIGURE 18.2 Makeshift shelter protects the operator but not the equipment.

pands with great force when frozen, so the resulting ice formation within confined spaces of the equipment or the injection system can cause serious damage. Delays of several days can be required to establish a proper working environment, thaw out and repair damaged equipment, clear the water and material lines of frozen matter, erect proper heated enclosures, and insulate pertinent supply and grout lines. On important work, such interruptions cannot be tolerated, so prudence dictates the provision of proper protection well in advance of any expected inclement weather.

Shelters should be provided for both mixing and pumping, as well as for drilling and the injection header. There are two types that are commonly employed, standard prefabricated shelters, as illustrated in Figure 18.3, and those built by the contractor for the specific project needs, as shown in Figure 18.4. Prefabricated shelters typically consist of regularly spaced standard frame bents with a fabric cover and are widely available from several manufacturers in widths of up to 100 ft (30 m) or more and virtually any length. They are particularly well suited for enclosing large open spaces. Job-built enclosures are best used for small and irregularly shaped shelters.

Two important elements in the design and/or selection of a shelter are the volume of

FIGURE 18.3 Standard modular protective shelter.

FIGURE 18.4 Shelter built around grout plants and material supply.

enclosed space and infiltration of light from the exterior. The volume of space will dictate the size of the heaters that will be required, and the opacity of the structure must provide sufficient light to enable the safe prosecution of work on the interior. Generally, it is advantageous to minimize the amount of enclosed space to only that actually needed, so as to limit the heating requirement. Although the common design of standard shelters usually provides considerable height, and thus space to heat, job-built structures can be assembled to meet only the actual space requirements. A good example is shown in Figure 18.4, in which a custom shelter has been erected around the grout plant.

To allow light to penetrate, a clear or translucent material is commonly used for cladding. The shelter shown in Figure 18.4 is covered with standard 10 mil thick clear polyethylene sheeting, which is highly translucent, whereas that illustrated in Figure 18.3 is of a translucent white canvas on the higher portions. Canvas usually does not transmit as much light as polyethylene sheeting, but is durable in all weather conditions and will last for a long time. Conversely, the clear polyethylene will degrade when exposed to the ultraviolet rays of sunlight. It is thus typically used only where service will not be required for more than about a year's time. Although thick insulating blanketing will minimize heat loss, it is seldom practical for short-term requirements. It is relatively heavy, which not only requires a stronger framework for support, but also greater effort to install. In addition, such cladding is totally opaque and thus does not allow light to penetrate.

For protection during drilling, especially where large drill rigs will be employed, custom enclosures are often used. These need only enclose the drill mast and operating personnel as long as sufficient space is provided for the required drill rod extensions. Such an enclosure is shown in Figure 18.5. As can be observed, it tapers from the top of the drill mast to about an 8 ft (2.4 m) height at the end of the available drill rod, so as to minimize heating requirements. This allows for easy supply of drill rod, as well as facilitating rapid erection, and, further, minimizes the exposed area that must withstand wind forces. The translucent canvas cover permits ample sunlight to penetrate and light the work area, as shown in Figure 18.6.

18.3.2.2 HEAT SOURCES

A variety of temporary heat sources are available. The simplest is an open-top drum in which

FIGURE 18.5 Special enclosure attached to drill mast to protect the drilling operation.

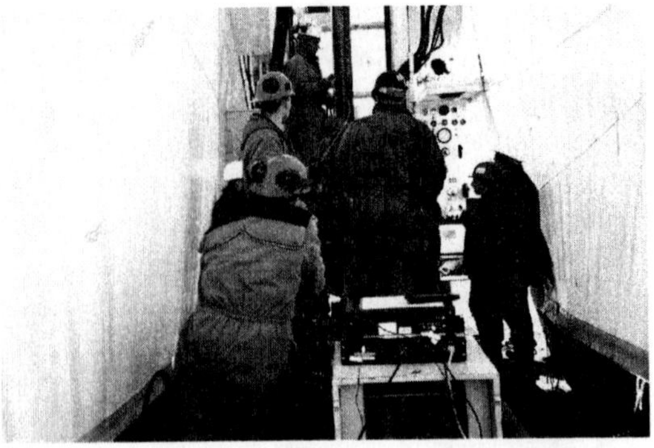

FIGURE 18.6 Translucent cladding allows the sunlight to enter.

a fire is maintained, sometimes referred to as a *hobo heater*. Either wood or coal can be used as fuel. Although such heaters can provide some warming for the workers, they are not very effective at very low temperatures and not safe to use inside enclosures. The next simplest source is the *salamander*, also known as an *orchard heater* or *smudge pot*. These heaters are fueled with oil, commonly kerosene or diesel fuel. They produce a relatively large amount of pollution and are thus not legal for use in many areas. For heating small areas, or individual containers of grout material or small tanks, either electric resistance or propane-gas-powered equipment is available. In general, electrically powered heaters are not nearly as efficient as propane-powered units. Furthermore, the substantial amounts of heat commonly needed require a large amount of electricity, complicating both its supply and distribution.

Most heating thus depends on propane gas as the source. Small portable heaters, such as shown in Figure 18.7, are widely available with heat output capacities varying from about 150,000 Btu/hour to 350 Btu/hour. Although heat is produced by burning gas, a small electrical supply is also required to power a contained blower, which directs the hot air in a given direction. The propane can be supplied from small

FIGURE 18.7 Highly portable propane-fueled heaters and tanks.

portable tanks, as illustrated, or from large tanks through appropriate hose lines. Because propane burners are very compact and yet produce a good volume of heat, they are the most frequently employed. Several units may be used in a single shelter, preferably directed at the locations where workers are most often present. Much lower output radiant heaters, which can be mounted on portable propane tanks, are also available. Although these can be useful to warm individual workers, they do not put out much heat and are thus not widely used.

For the heating of large enclosures, high-capacity skid- or trailer-mounted heaters are available and are commonly rented or leased for the duration of a project. They are available in a variety of capacities, and their output is typically within a range of about 250,000 BTU/hour to more than 4,500,000 BTU/hour. Such heaters are usually not vented, so they must remain outside the enclosure. A temporary duct is thus required to direct the heated air into the enclosed space. These heaters are most commonly fueled by propane gas, but oil-fired units are also available.

18.3.2.3 THE GROUT TEMPERATURE

Control of the grout temperature is important for two reasons: first, it is temperature that controls the set or gel time, and, of course, freezing in the delivery system must be prevented. In addition to mixing within heated shelters, the easiest way to manage the grout temperature is through control of the constituent materials. Heating of any mix water is probably easiest to accomplish. Furthermore, bags and other containers of the various ingredients can be stored in heated enclosures. Large tanks or bulk containers should be well insulated.

Bulk aggregates should be covered to prevent saturation by rainwater or snow. Clean sands and larger aggregate can be heated by way of heated pipes or culverts run through the bottom of piles. Heat for these tubes is typically supplied by portable propane heaters, such as those shown in Figure 18.7. Because heat always rises, it is important to place the tubes in the bottom of a pile and provide a continuous airtight cover on the top. Although sands will readily drain so as to allow the heat to permeate, silts and silty aggregates such as those used in compaction grouting are usually so fine that moisture is not easily expelled. These materials must usually be agitated, as heat is applied, to be effectively treated. This is usually best accomplished with circulation through a portable drum kiln, as illustrated in Figure 18.8.

FIGURE 18.8 Portable drum kiln used to heat and dry fine-grained aggregate.

Electrical band-type heaters can be used on drums or smaller containers of fluid or resinous injection material. These are commonly available to fit on 55 gal drums or 5 gal pails, but can be obtained by special order or adapted to virtually any size container. Such heaters can also be installed on grout mixing or supply reservoirs and are often used in this way even in mild conditions to provide the greater penetrability of most resins at higher temperatures.

18.3.2.4 WATER SUPPLY

A dependable water supply is fundamental to most drilling and grouting operations. This will require insulation of all aboveground pipes and hoses, as well as those within the freeze zone of the surface soils. Even with this precaution, the temperature of the water can be quite low, negatively affecting the set time of the grout in which it is used. Heating of mix water is thus sometimes required, and almost always advantageous, when working in low-temperature environments. Although water must simply be held at a temperature greater than 32°F (0°C) to prevent freezing, higher temperatures are desirable for water to be used as a constituent of most grouts.

Two very different modes of water heating are used: in-line, in which the water is heated as it runs through the heating element; and storage, in which the heating elements are submerged within the water in a storage tank. In-line heaters are available in a wide variety of sizes and methods of heat production. They can be powered by electricity, oil, propane, or natural gas. Propane-fired heaters are by far the most efficient and practical for the majority of grouting applications. Immersion heaters are generally electrically powered. They typically require a large power supply, and their rate of temperature increase is usually quite low. For these reasons they are not frequently used in grouting.

In-line heaters can be used for direct heating of the water as it passes through the line on a one-time basis, or in combination with a storage tank, whereby the water circulates from the tank, through the heater, and back to the tank. One-time passage, of course, limits the amount of temperature rise that can occur, whereas continuous circulation involves many passes of the water through the heating element. Heaters operating in circulating systems are usually controlled by a thermostat in the inlet piping or the storage tank, so the temperature of the water can be maintained at a rather constant level. Swimming pool heaters, which are designed specifically to handle large flows of water, are ideal for this kind of application. They are normally supplied with self-controlling adjustable thermostats and are best maintained in well-insulated enclosures.

All components of water systems in cold environments must, of course, be insulated. Because of the continuous dwell time of water in storage tanks, they should be especially well protected. Heavy insulating blankets normally used for curing concrete in cold exposures perform well in such applications, as illustrated in Figure 18.9. Note the large propane tank in the lower left corner used to fuel the heaters. Water heaters are typically used in continuous service and maintained in controlled environments. Under these conditions, freezing of water in the heater

FIGURE 18.9 Water tank insulated with heavy concrete curing blankets.

is precluded. In grouting, however, the heaters are sometimes not in warmed enclosures and breaks in the operation often result in their being turned off for extended periods of time. Any water that remains in the elements can thus freeze, causing serious damage. For this reason, drainage valves that enable easy removal of any latent water should be provided.

18.3.2.5 DELIVERY LINE PROTECTION

Grout delivery lines often extend out of, or between, protective shelters. When this occurs, they must be insulated or otherwise protected to avoid freezing of the grout therein. Grouters are sometimes of the opinion that grout will not freeze in the line because of its movement and common exothermic reaction once mixed. This is not accurate, however, and many jobs have been severely delayed because of such freezing. In one instance, freezing within the lines of two different crews occurred within a few hours in the especially cold early morning hours, resulting in all work coming to a sudden halt for more than four days. The time required to properly insulate the lines was minor as compared with that required to "drill" the frozen grout out of the sections of rigid delivery line, as shown in Figure 18.10.

FIGURE 18.10 Frozen grout in rigid delivery pipes had to be drilled out.

The line can be insulated in a variety of ways, including wrapping in building insulation or concrete curing blankets, or bedding and covering with straw or any other suitable material. These measures will severely limit moving or mobility of the line, however. The most effective method, and the only one that facilitates easy movement, is encapsulation within standard pipe insulation. This material is usually supplied with a lengthwise split to facilitate application. It can be used with either rigid pipe or flexible hose. Securing with duct tape or another suitable means is required. Where movement will be minimal, it can be secured only occasionally. Near the grout header, where considerable flexibility is typically required, much more frequent securing or even continuous wrapping will be required, however.

CHAPTER 19

The Use of Explosives in Grouting

19.1 INTRODUCTION
19.2 PRINCIPLES OF EXPLOSIVES USE
19.3 FORMATION OF EXPANDED HEAD PILES AND GROUND ANCHORS
19.4 CLEARING OBSTRUCTIONS TO GROUT FLOW
19.5 ACCESSING SUBSURFACE FISSURES AND VOIDS
19.6 IN-PLACE PERFORATION OF INJECTION CASING
19.7 IN CONCLUSION

19.1 INTRODUCTION

The use of explosives to facilitate grouting is really the story of one man and the work of his firm, Argus GeoLogistics, Inc., of Warren, Michigan. This company has advantageously married the two technologies for more than 50 years. As a child, Bob Gronowicz could virtually always be found on the job sites of his father's masonry contracting business. These included a large number of unusual and difficult repair projects, and the young Gronowicz grew up not only with a wealth of construction knowledge, but with the mind-set that no job was impossible and that otherwise impossible jobs were simply a series of ordinary ones.

His introduction to explosives occurred during World War II while serving in the United States Army paratroops in Europe. He became somewhat enthralled by the shapes of holes resulting from hand grenade explosions and began to make mental note of the relationship of how the grenade was thrown to the results of the following explosion. As his curiosity expanded, he experimented by dropping grenades into pipes and tubes that he found in the war-damaged debris. He discovered that explosive fractures could be made predictable by scoring or otherwise weakening the tubes prior to explosion, in a manner not unlike the way a mason cuts a brick. He had done this many times, scoring the brick at the location of the desired cut and then hitting it with a hammer.

By trial and error, Bob learned that various materials responded differently to the explosive charge. Aluminum tended to completely disintegrate into small pieces, cast iron would break in a manner similar to glass, and mild steel would fracture and expand into a "bird cage"-like form. He was able to control the shape of the resulting bird cage by varying the amount and location of the explosive powder, which he

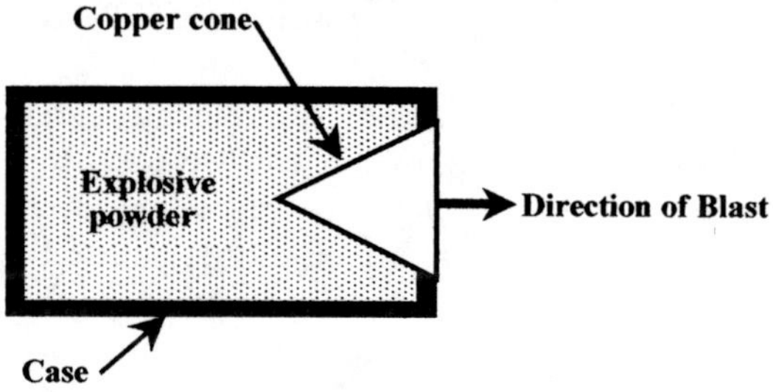

FIGURE 19.1 Hollow cone in the explosive determines the direction of the charge.

had removed from the grenades. Perhaps of greatest interest, however, was the discovery that the explosive force would be directed into a concentrated jet when the powder was placed in the bottom of a champagne bottle. The effect of such a shaped charge (illustrated in Figure 19.1), now widely recognized in the explosives industry, is commonly known as the *Monroe effect.*

He returned to civilian life and his father's business, and while working on a job to reline boilers on locomotives, he noticed a publication from the Portland Cement Association, *Concrete for Railroads,* in the engineer's office. Always curious, he was enthralled by an article that described the "Mudjack" stabilization of a mile-long stretch of railroad track. The line had been constructed over very soft soil deposits that were within the limits of a former swamp.

The Argus firm had been working on construction of a large paper warehouse that was served by a rail siding. Inside the building was a 1,000 ft (300 m) long siding, depressed so that the floors of the rail cars and the building were at the same elevation, thus facilitating the loading of huge rolls of paper. Once the cars were loaded, they were moved to an outside storage siding. The center portion of the inside track well had settled, greatly complicating the movement of the loaded cars, which was then done by manual labor with the use of special pry bars. Soil test borings disclosed that the track was built on an old streambed.

Young Bob, having carefully studied the Mudjack article, noting especially the number of workers and the equipment shown in the accompanying photographs, made some quick mental calculations and told the engineer that his firm could raise the track back to level in a couple of days. This sounded good to the engineer, and he promptly authorized the work to be done. But Argus had not done such work previously, nor did it possess the required equipment. So a Mudjack machine was rented, and the work was started.

Holes were drilled under the track, and casings similar to those shown in the Portland Cement Association brochure were installed. The Mudjack had limited pressure output, however, and there was difficulty in getting the grout intrusion started. So Bob, using his explosives experience, placed one-quarter stick of 40 percent dynamite into the holes to loosen the soil ahead of injection. The soil did not loosen; instead, a horizontal fracture developed at depth, which not only accepted the grout, but resulted in nearly effortless lifting—almost too effortless, as was discovered after ten days of pumping. The workers had actually raised the tracks too high, but the engineer was delighted. He found that they could block the wheels of the cars once the locomotive pushed them in. Then, when they were ready to go outside, they would roll on their own when the blocks were removed.

The plant engineer told his colleagues the wonders of these grouters, and this led to ever more work. And the use of explosives to facilitate grout placement became a routine procedure. As experience was gained, it was determined that the "mud" being used, which was composed primarily of topsoil, was not very durable. Argus thus starting using a cementitious grout similar to masonry mortar; however, the Mudjack machine was not able to pump it.

This turn of events led to a search for better equipment, which included experimentation with several different types of pumps, pressure vessels, and so on. Eventually, the company developed its own equipment to mix and pump appropriate cementitious grouts. And to learn how to better control the action of its explosives, it built a small laboratory for research, resulting in the ability to compound explosive materials into shaped charges capable of accomplishing virtually any particular objective.

As word got out about the extraordinary accomplishments of the firm, ever-more challenging projects came its way and, as always, unique solutions, very often involving explosives, were developed. Bob Gronowicz and Argus may be the only grouters in the world to routinely employ explosives in their work. To protect their truly proprietary methodology, they did not until now go out of their way to share this experience with the grouting fraternity at large.

Realizing, after more than 50 years of continuous work, that his future activities will be restricted, Bob has provided the many case histories that follow, as well as all of the photographs included. His varied grouting innovations, many of which involve explosives, are of significance and have facilitated valid solutions to countless real-world problems. It is hoped that with the knowledge presented here, other grouters can build upon this positive experience.

19.2 PRINCIPLES OF EXPLOSIVES USE

For most grouting work, only a very small amount of explosive material is used. It is, however, formed into a shape that will direct all energy to a preplanned target area. That shape virtually always includes an inverted cone, which results in the previously discussed Monroe effect. Although the basic conical shape is always present, the volume of explosive powder and the exact shape of the cone are custom-formed to provide the optimal results for each individual application.

Depending on the job requirements, the explosives are housed in either of two basic configurations. Most commonly, they will be placed in a casing, over a preformed cone, as illustrated in Figure 19.1. The case can be metal or die cast, of either glass or plastic. A copper directional cone is typically created from a copper sheet, which is usually about 1/8 in (3 mm) thick. It is formed to the desired shape and the joint is soldered. The explosive is placed over the directional cone, as shown in Figure 19.1. When used in areas where the temperature is above about 140°F (60°C), such as around industrial furnaces or smelters, the case is encapsulated inside an insulated body.

When the explosive is detonated and the cone is instantly compressed, squeezed, and disintegrated in a forward-moving jet, the greatest intensity of the blast will always be at a distance from the base of the cone equal to its height, as illustrated in Figure 19.2. Thus, by adjusting the geometry of the cone, the area of maximum force or disruption can be predictably established. One or more charges can be combined on an explosive *tool* so as to provide the desired results. In grouting, it is fairly common to have several charges positioned so as to fracture the ground radially away from the grout hole.

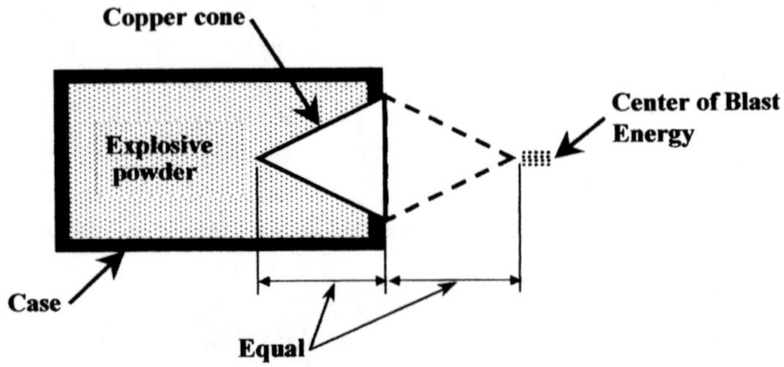

FIGURE 19.2 Center of blast intensity will be proportional to the cone depth.

FIGURE 19.3 Several individual charges making up strings will result in linear splitting or fracturing.

Where it is desired to produce linear fractures, several separate charges can be assembled in single lines, as shown in Figure 19.3, to produce a linear tool that will fracture or weaken a zone in the desired plane. Where multiple charges are used, the tool is often referred to as a *string*.

Alternately, the explosive can be placed in holes drilled into a solid steel bar so as to form a *gun*, as shown in Figure 19.4. In this configuration, the holes are first filled with explosive, into which the cone-shaped liner is pressed. This design is especially beneficial where vertical cuts or an elongated bulb are required.

The evolution of a Monroe-effect charge is now well understood within the explosives industry. Within about 6 microseconds of the detonation, the charge reaches full speed and pressure as it approaches the apex of the cone. At this point, the front is moving at about 25,000 ft (7,507 m) per second and has developed pressures on the order of 4 to 5 million psi (275,862 to 344,827 bars). The cone first collapses inward and then disintegrates as the superheated explosion gases propel it forward at a speed on the order of 8000 ft (2402 m) per second. This all occurs within about 10 microseconds of detonation. The metal particles are forced toward the axis of the cone as it collapses and disintegrates, moving forward in a compact jet with a diameter of only about 1/8 in. (3 mm), as illustrated in Figure 19.5. At this point, the velocity of the front is on the order of 21,000 ft (6,300 m) per second.

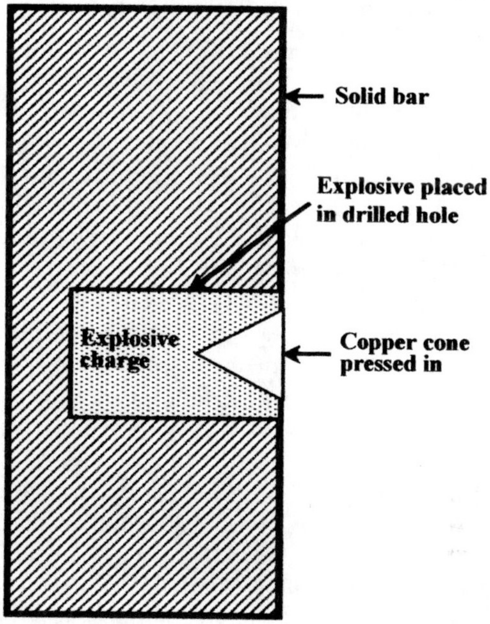

FIGURE 19.4 A gun can be made by pressing a preformed cone into a hole filled with explosive.

The jet has a temperature approaching 1900°F (1000°C) and, with its high velocity, can penetrate metal and, of course, soil or rock. Although the explosive jet is only about 1/8 in.

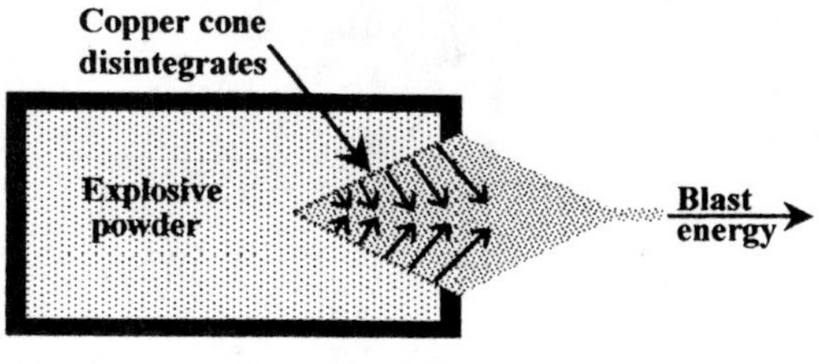

FIGURE 19.5 Energy resulting from disintegrated cone is propelled into a concentrated jet.

(3 mm) in diameter, it can create a hole through steel of 1/2 in (13 mm) or more. As the explosive jet penetrates into rock or soil, it forms a hole through compaction, in a manner similar to a nail being driven into wood. Just as one or more nails can cause a split when driven into wood, geomaterials are likewise fractured, and part of the technology Argus has developed involves the overlapping action of several trajectories to form a predictable fracture.

As mentioned earlier, it is common practice to have more than one charge in a single explosive tool. For example, four separate charges would be used in a *tool* intended to blow four holes through a casing or within a given hole, as illustrated in Figure 19.6. Each charge is, of course, placed in the proper direction for its intended target. The individual charges can all be detonated simultaneously or sequentially in any desired order. The amount of time delay between detonation of various charges can be as desired, but is usually on the order of only a few milliseconds.

The preferred explosive is RDX, also known as Cyclonite or Hexogen. This is normally supplied as a white crystalline solid or powder. It is commonly mixed with waxes or polymers such as silicone to provide a puttylike material that can be easily molded or formed. Although it is highly stable in storage, it is considered one of the most powerful and brisant of the known explosives. As an aid to grouting, very small quantities are used, 5 to 15 g being sufficient for most applications. It is typically detonated by means of an electrically controlled blasting cap, which is also packaged in the explosive tool.

The use of shaped charges for perforation of oil well casing and fracturing of the surrounding rock has become well established in the petroleum industry. For such applications, many standard explosive tools are commercially available, and, in fact, the American Petroleum Institute (1994) has produced a publication, *Standard Procedure for Evaluation of Well Perforators*, API RP-43, that outlines many of the different explosive perforating tools commonly used. These tools, however, are usually not applicable to grouting, as they are designed for use in oil wells, which are considerably larger in diameter and extend to much greater depths than typical grout holes.

Using similar technology, Argus custom designs and fabricates explosive tools specifically for the individual grouting requirement. For use in individual holes, these are always smaller and more compact than their commercially produced cousins. Because they are custom designed and fabricated for each application, the strength and direction of the charge can be precisely targeted. Often, the explosives are encased within a pipe or drill casing that has been scored or otherwise weakened so as to fracture and expand in a preplanned manner.

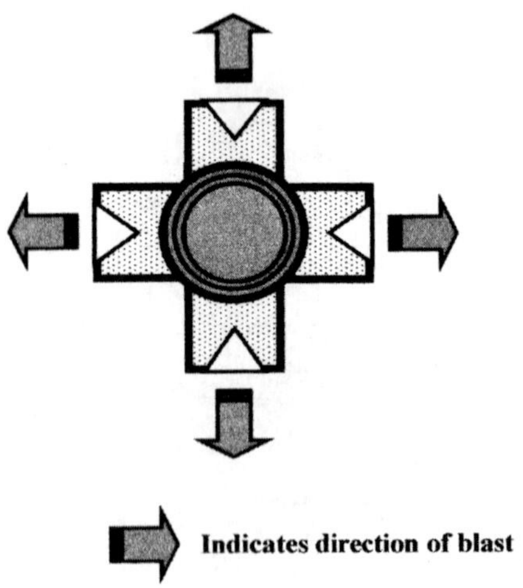

FIGURE 19.6 A typical four charge explosive tool.

19.3 FORMATION OF EXPANDED HEAD PILES AND GROUND ANCHORS

Controlled expansion of the tips of micropile casings, combined with grout injection, can provide a substantial increase in the piles bearing capacity. Likewise, similar expansion of the anchorage end of tieback anchors will increase pullout resistance. Argus has developed several basic configurations of expanded casings, all of which are formed explosively during or immediately preceding grout injection. In most instances the expansion is enhanced by preweakening or notching of the steel tube. Figure 19.7 shows a simple expanded flare shape, which is obtained by preweakening a single line at the base of the tube prior to detonation. A very different type of expansion is illustrated in Figure 19.8, in which both four- and eight-leaf anchor versions are shown.

An appropriate explosive charge is installed at the previously weakened tip of the shell. The tube is then filled with grout, and with the grout under pressure, the charge is detonated. Grout injection continues until a precalculated amount of grout has been placed. Figure 19.9 shows a pile that has been extricated by excavation so as to reveal the tip condition. Therein, the pile casing has been expanded into a four-leaf configuration. As can be seen, the spaces between the leaves are filled with grout.

FIGURE 19.7 Weakening by scoring at a single location prior to the blast results in a belled casing.

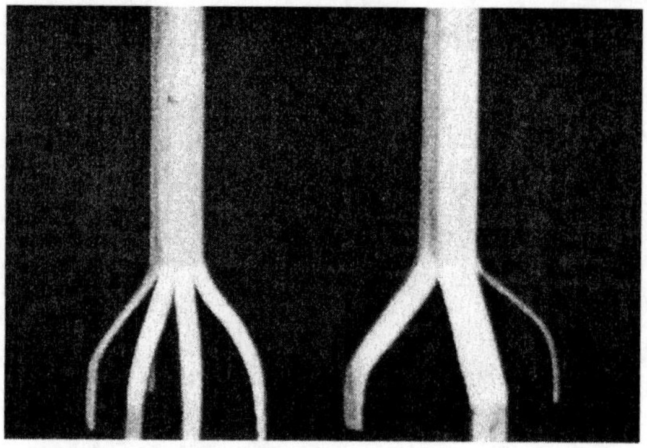

FIGURE 19.8 Typical results from weakening casing at several locations.

Through the years, Argus has completed several hundred expanded-tip micropile jobs. Depths have varied from 20 ft (6 m) to as much as 60 ft (18 m). Although soil conditions dictate the pile depth, the capacity is strongly influenced by the pipe size and tip condition. Much previous work involved standard Schedule 80 steel pipe of either 3 or 4 in. (76 or 102 mm) size. The dimensions of such pipe and the typical supporting capacity of the resulting piles are shown in Table 19.1.

The indicated capacities are the minimums expected from the respective pipe sizes. In many

FIGURE 19.9 Four-leaf expanded pile base extricated from the soil by excavation.

TABLE 19.1 Composition and Capacity of Standard Schedule 80 Steel Pipe Micropiles

	O.D.		I.D.		WALL THICKNESS		WEIGHT/FT.		PILE CAPACITY	
Pipe Size	**in.**	**mm**	**in.**	**mm**	**in.**	**mm**	**lb**	**kg**	**tons**	**kg**
3	3.50	89	2.90	73.7	0.30	7.6	10.25	4.65	12	10,900
4	4.50	114	3.82	97.0	0.34	8.6	14.98	6.79	14	12,727

cases, greater capacities are obtained. In one instance, the design capacity for 32 ft (9.6 m) deep ordinary micropiles constructed of 4 in. (102 mm) Schedule 80 pipe was 16 kips (7272 kg). A load test resulted in plastic deformation at only 8 kips (3636 kg), and settlement of a 0.5 in. (13 mm) when loaded to 12 kips (5455 kg), only 75 percent of the design value.

Argus then installed piles similar in every respect, except that the tips were expanded explosively with simultaneous grout injection. On the basis of data that had been gathered during prior work, the tip was expanded to an anticipated diameter of 15 in (38 cm). Subsequent load tests proved the piles capable of supporting the 16 kip (7272 kg) design load, with no settlement being noted. In fact, 100 percent of the elastic shortening of 0.08 in. (2 mm) was recovered upon release of the test load.

Typical of the explosively expanded micropile installations Argus has made were those required for an addition to the Fisher Mansion in Grosse Point, Michigan. Twenty-seven 4.5 in (114 mm) diameter piles were driven to a depth of 34 ft (10.2 m). The pile shell was standard 4 in. (102 mm) Schedule 80 steel pipe. It was driven with a pneumatically powered Vulcan Model 100 vibratory pile driver. An expendable aluminum guide point/plug was placed at the tip of an explosive tool for protection of the charge and to keep the casing clear of spoil during driving, as shown in Figure 19.10. Once in place to full depth, the pipes were filled with cement-sand grout designed to obtain an unconfined compressive strength of 5000 psi (34.5 MPa). With the pipe filled and under pressure, the explosives were detonated to create an expanded grout tip and withstand a 16 kip (7272 kg) load.

Many and varied projects involving expanded tieback anchors have been completed by Argus. The provision of a flare tip greatly increases the capacity of anchors, which allows fewer and/or shorter anchors to provide the necessary restraint. The expanded tip not only provides greater bearing area, but also densifies the surrounding soil. This results in cost savings much greater than the cost of the explosive and allows the work to be completed in less time.

A typical example involved a retaining wall that was tipping outward and near failure at the entrance of a historic cemetery in the downtown core area of Detroit. Argus proposed 44 ft (13.2 m) long grouted tieback anchors. They were drilled at a 30 degree inclination, as shown in Figure 19.11, and were positioned so as to not come in contact with any of the buried remains. To ensure the latter, a seismic survey was performed prior to laying out the exact anchor

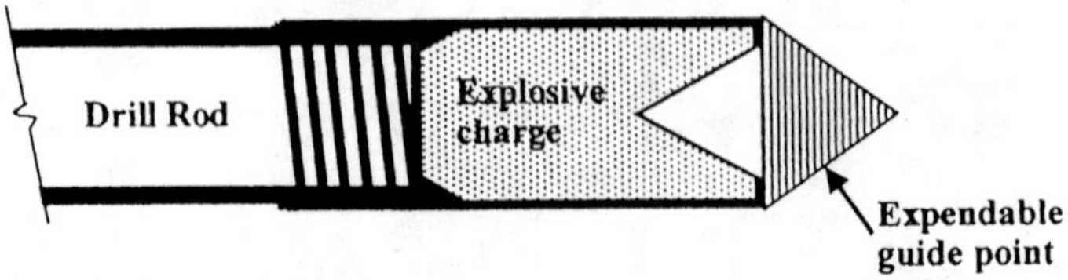

FIGURE 19.10 Expendable aluminum guide plug typically used when inserting casing.

FIGURE 19.11 Augering of tieback holes.

locations. The importance of this survey was emphasized when it was found that several coffins were not actually placed as indicated by the positions of accompanying grave stones.

To place the tiebacks, 4 in. (100 mm) diameter holes were first drilled to full depth with a continuous flight auger. A standard 3 in. (76 mm) diameter Schedule 40 steel pipe was then inserted full depth into the hole. The tip of the pipe had been preweakened, and an explosive tool containing four equally spaced explosive charges was installed immediately behind an expendable cast aluminum guide point, similar to that illustrated in Figure 19.10. Once inserted, the pipe was filled with grout and with the grout under pressure, the charge was detonated. Injection continued until 6.5 ft^3 (0.18 m^3) of grout was placed. This was based on injection of a grout mass with a diameter of about 19 in. (48 cm).

Upon completion of the grout injection, a 10 ft (3 m) long section of 1 in. (25 mm) high-strength threaded steel bar was inserted. Once the grout had hardened, steel plate washers and nuts were installed to restrain any future movement of the wall. New decorative precast panels were then installed to hide both the original wall and the tieback hardware.

Another project developed as a result of water leakage into a below-grade exhibit hall that was part of a large hotel/convention center. The 14 ft (4.2 m) high retaining walls surrounding the hall were constructed of concrete block masonry. For whatever reason, the cells of the blocks had not been filled during construction as required. For this reason and because of other deficiencies, including the omission of structural pilasters, a brick veneer had been added to some portions of the interior walls. A large swimming pool in near proximity to the walls was found to be the leakage source, although this was not initially understood.

A waterproofing contractor engaged to correct the leakage used an expanding urethane foam grout. During initial injection, the walls began to move inward as a result of the foam pressure. This caused a cessation of injection and reevaluation of the wall's structural capacity. They were determined to be unstable, and a repair plan was formulated that included the installation of 60 tieback anchors, designed to retain a tensile force of 10 kips each. Much of the area adjacent to the wall contained mechanical equipment, severely restricting access for the work. Because of the access restrictions and tight working area, it was desired to obtain the required restraint with the smallest possible anchors, which led to the adoption of explosive expanded anchors.

These were constructed by first augering 4 in. (102 mm) diameter holes to a depth of 40 ft (12 m), as illustrated in Figure 19.12. A standard 3 in. (76 mm) Schedule 40 steel pipe fitted with an explosive tool was then inserted. The tip of the pipe had been preweakened so as to provide an eight-leaf tip configuration upon detonation. The explosive tool consisted of four shaped charges, each containing 3 g of RDX explosive, which were fired simultaneously once the pipe was filled with a cement-fly ash grout. With pressure of about 50 psi (3.4 bars) held on the grout, the explosive was detonated. Because the density of the soil was quite low, 20 ft^3 (0.56 m^3) of grout was injected. This effectively

FIGURE 19.12 An especially fabricated drill rig was used for drilling anchor holes.

flared the casing and forced grout out into the surrounding soil to develop the necessary pull-out restraint.

A 10 ft (0.9 m) long piece of 1 in. (25 mm) diameter Dywidag Threadbar was then pushed into the fresh grout. Once it gained sufficient strength, a system of vertical steel whalers (Figure 19.13) was placed. The anchors were then stressed by applying a torque of 500 ft lb (69 kg m) to the nuts.

FIGURE 19.13 Tie-back rods provide restraint for whaler system.

19.4 CLEARING OBSTRUCTIONS TO GROUT FLOW

One of the first theatres in the United States to have air-conditioning was in Detroit, Michigan. The early air-conditioning system included two wells that were about 300 ft (90 m) deep. One was placed in the basement of the building, and the other was located outside and immediately adjacent. Cool water from the inside well would be pumped up for cooling, and once warmed, would be circulated down the other well. As time progressed, the theater was converted to a furniture store, and the wells were long forgotten until one day when the store owner noticed a foul odor emanating from the basement.

A well service firm was contacted to correct the condition. It misdiagnosed the problem, however, which was highly toxic hydrogen sulfide gas in the wells. The company's workers welded a steel cap to the 8 in. (203 mm) casing of the interior well and pumped air into it, utilizing a 600 cfm (282 L/sec) compressor. Sadly, one of its employees had remained in the basement and was overcome by the gas, which led to his death.

Argus was retained by the fire department to seal the wells. It accomplished this by placing a shaped explosive tool in the bottom of the outside well. It was then filled with rapid-setting calcium aluminate cement grout. With pressure on the grout, the explosive charge was detonated so as to blow out the well screen and create an open passage to the aquifers feeding the wells in the surrounding rock. Grout was injected until it appeared at the other well in the basement. Before its appearance, however, about 3,000,000 gal (11,355,000 L) of contaminated water flowed into the basement from the inside well. Obviously, this had to be pumped and properly disposed of, which in itself was a major undertaking. Although the wells had originally been equipped with proper screens and filters, the extended inactivity resulted in severe fouling. It is doubtful that the grout flow could have been initiated were it not for the explosive tool.

19.5 ACCESSING SUBSURFACE FISSURES AND VOIDS

In cases where concealed underground fissures or other voids exist, interception with drilled grout holes is problematic at best. In instances where a drill hole does not contact the defect but is close to it, explosive energy can often create a passage for the grout. Such was the case when deep-seated voids were found during foundation construction for a new high-rise building in downtown Detroit.

The new 32-story building was founded on bell-bottomed cast-in-place concrete piles, the tips of which were to be founded in "hardpan" at a depth of 105 ft (31.5 m) below the ground surface. While the belling tool was being disconnected from the drilling rig upon completion of the last pile, it accidentally fell to the bottom of the hole. Retrieval activities disclosed that the tool had not only fallen to the 105 ft (31.5 m) drilled depth, but had actually punched beyond the tip of the shaft, and into a previously undetected low-density peat deposit. Further investigation revealed the existence of similar randomly spaced deposits under several of the other piles.

Grouting was selected as the best remedial method. Accordingly, a series of grout holes was drilled into the suspect area. Although they extended to a depth of 115 ft (35 m), many of the holes failed to intercept the low-density deposits. They were nonetheless injected in an effort to "break through" to any nearby faulty zones. Such breakthrough with grout pressure alone was not possible even at a pressure of 4000 psi (276 bars). Argus thus used shaped explosive charges to "open up" the hardpan and create fractures to any nearby low-density deposits. The explosives were packaged into *perforating charges*, each consisting of 5 g of RDX explosive, directed so as to create a hole about 3/4 in. (19 mm) in diameter. Four charges were formed into each tool to create holes oriented at 90 degrees, as shown in Figure 19.6.

The hardpan was successfully fractured, allowing grout injection into otherwise isolated faults. With this procedure, some 6000 ft^3 (170 m^3) of grout was injected in an around-the-clock effort that continued uninterrupted for 23 days. The grout consisted of cement, fly ash, and silica fume slurry. It was mixed in a paddle mixer, to which the three components were fed directly from bulk material tankers with feed augers. Injection was by a modified Mayco ball-valve concrete pump with injection pressure on the order of 900 psi (62 bars).

To monitor grout flow and detect any interconnection between holes, an ordinary condom was placed on the top of each injection casing. Pressure transferred from the casing being grouted to the tip of another not yet injected, would inflate the condom, which was readily visible, as illustrated in Figure 19.14.

FIGURE 19.14 Common condom (left) will inflate if grout or other pressure enters casing (right).

Many pockets of pressurized methane gas underlie the surface of eastern Michigan adjacent to Lake Huron. This presents considerable risk for those drilling exploratory holes or other excavations at depth. Such was the case for a crew drilling test borings for a replacement of the Military Street bridge in the heart of the commercial district of Port Huron, Michigan. Upon penetration of a clay layer at a depth of 60 ft (18 m), bubbling and blowing of the drill circulation fluid occurred, indicating that pressurized gas had been encountered. The crew immediately reported to the local fire department and started to withdraw the drill string. During withdrawal, a flame ignited, but self-extinguished.

Upon arrival, the fire department ordered the crew not to move the drill unit and to leave the area. While the firefighters stood by, four city blocks within the downtown shopping area were evacuated. The evacuation continued for the next two days while a remedial plan was formulated. It provided for Argus workers clothed in Nomex fire resistant suits to perforate the casing at depth and to grout both the casing and any voids that might exist. With firefighters standing by at full alert (Figure 19.15), a one-way perforator tool was lowered to the bottom of the 4 in (100 mm) casing the drillers had left in place. The hole was then filled with a cement-bentonite grout, and the perforator was detonated as injection continued. A grout quantity equal to 1.25 times the calculated volume of the drill casing was injected. The area was then declared safe, and the drillers were allowed to return and retrieve their rig.

FIGURE 19.15 Grout plug is cautiously placed as firefighters stand by with hoses ready for action.

In another instance, a drilling crew working inside a Chrysler automobile plant in Detroit was not so fortunate when it encountered pressurized methane while redrilling a previously abandoned test boring. The initial hole, which was adjacent to a main supporting column, had been advanced to a depth of about 75 ft (23 m), when methane gas was encountered. As the workers were abandoning the hole, an overhead bridge crane passed by, sparking an explosion and flash fire. Although minor damage resulted from the explosion, the fire was self-extinguishing. The crew completed the hole abandonment by gravity filling with a cement-water mixture.

Using a hollow-stem continuous flight auger, replacement drillers were cautiously advancing the hole when they detected a minor amount of methane at a depth of about 30 ft (9 m). They shut down the drill rig for about two hours to allow the gas to dissipate. As they started the engine of the rig to continue the work, an explosion and large fireball engulfed the rig, as shown in Figure 19.16. Because it had successfully handled similar problems for the owner, Argus was immediately called and its crews arrived on scene within 40 minutes, while the fire was still burning.

As the fire was brought under control, with the firefighters standing at the ready, the remnants of the drill rig were pulled out of the way, allowing an inspection of the adjacent column, Figure 19.17. For an immediate although temporary seal, the Argus crew made connection to the hollow stem of the auger, which remained in the hole, and pumped about 200 gal (756 L) of

FIGURE 19.16 Explosion-engulfed drill rig on fire.

a bentonite gel mixture so as to fill the hole and any adjacent voids. The bentonite grout was mixed in batches with the following constituents:

Ground gelling bentonite	60 lb (27 kg)
Bentonite pellets	60 lb (27 kg)
Water	50 gal (189 L)

This resulted in a very stiff but flexible gel, which would provide a good seal. Initially, the bentonite pellets acted only as aggregate filler, but they would continue to absorb moisture and thus swell with time, which would result in an ever-tightening seal. The mixture would, however, remain soft, allowing removal of the auger, but nonetheless seal against further venting of the gas.

FIGURE 19.17 Adjacent column can be seen as the fire is extinguished.

With the immediate hazard under control, a permanent fix was formulated. That plan provided for drilling an inclined hole from outside the structure so as to intersect the abandoned test hole at a depth of about 50 ft (15 m), as detailed in Figure 19.18. This hole was intended to vent any existing gas out of the building while a more permanent seal was placed in the original test hole. The auger was then removed by screwing it out of the bentonite-filled interior hole, with the bentonite remaining to prevent gas from escaping into the building.

The 4 in. (102 mm) vent hole was inclined at 50 degrees off horizontal so as to intercept the abandoned hole at a depth of about 50 ft (15 m). Once it was completed, a slightly inclined hole was advanced adjacent to the original test hole, as shown in Figure 19.18. It intersected the original hole at a depth of 43 ft (12.9 m) and was continued to a final depth of about 75 ft (22.5 m). An explosive tool containing four directional charges, equally spaced around the circumference, was then lowered to the bottom.

A cement-fly ash slurry grout was injected, and the explosive tool was detonated once the casing was filled with grout, which was held under a sustained pumping pressure of 150 psi (10.3 bars). The action of the explosive in effect created a weakened horizontal plane that extended for several feet around the original drill hole. This facilitated the grout filling of any nearby pockets that might have contained trapped gas. Injection continued until a total of 30 ft^3 (0.84 m^3) of grout was placed. The outside vent hole was monitored for methane during the entire operation, although none was detected.

19.6 IN-PLACE PERFORATION OF INJECTION CASING

A very different type of project involved construction of foundations for several permanent beacon "light cells" about three miles off-shore in both Lake Erie and Lake St. Clair. Steel sheet pile cells, 35 ft (10.5 m) in diameter, which extended through about 70 ft (21 m) of water, were to be filled with concrete. Frequent high winds and bad weather were expected, and these could cause interruptions in the work. Preplaced aggregate concrete, with the large aggregate placed first, was determined to be the best method of concreting, as interruptions in the aggregate filling were acceptable.

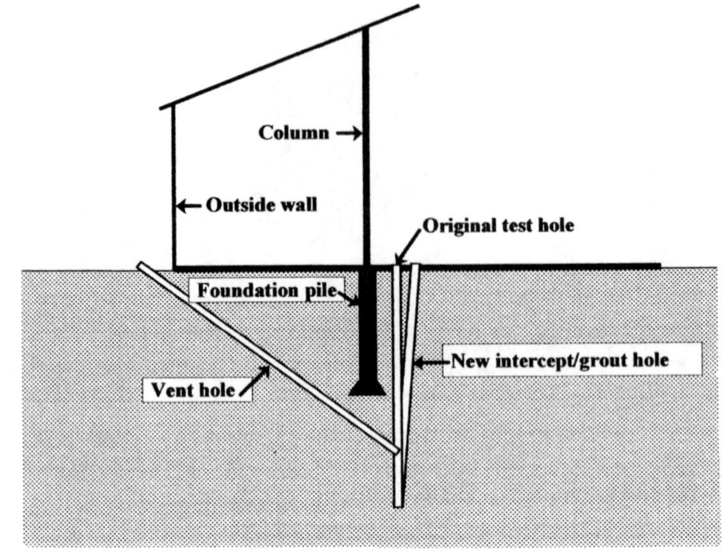

FIGURE 19.18 Cross section illustrating location of the grout and vent holes.

This method offered further advantages, in that the size of the aggregate is not limited, so larger rock could be used, resulting in a greater volume of aggregate and less cement. With the lesser amount of cement, the heat of hydration of the large concrete mass would be reduced, significantly improving its hardened quality. Although pauses in the aggregate filling were tolerable, any interruptions during grout injection could be a problem, especially if the grout were to harden in the pipes so that they became unusable.

It was originally contemplated to work around the clock for the several days that would be required to grout a cell. This plan, however, presented safety problems related to working in the dark, especially in the event of the inclement weather that was common in the area. It would also result in higher costs. The solution was to work only during daylight hours, but to facilitate this approach, a system not adversely affected by interruption was devised.

Standard 3 in. (76 mm) Schedule 40 steel pipes extending vertically to the bottom of the cell were secured in place on a grid of about 5 ft (1.5 m), as shown in Figure 19.19. A single slotted *sounding* pipe was installed in the center of

FIGURE 19.19 Injection pipes were encased in the aggregate fill.

the cell to monitor the height of the rising grout as it was injected. The entire cell was then filled with plus 2, minus 4 in., limestone aggregate, utilizing a barge-mounted crane and clam bucket.

Grout was then mixed, utilizing four 200 gal (756 L) capacity double-tank vertical paddle mixers. These fed directly into the hopper of a single Mayco ball-valve concrete pump. The grout was pumped into a common manifold that supplied four separate injection lines, as shown in Figure 19.20. In this manner, four different casings were simultaneously grouted and lines were moved between the different grout pipes as required to fill the cell uniformly. At the end of each shift, a sponge was pumped into each injection pipe to clear it to the approximate elevation that had been grouted.

FIGURE 19.20 Grout pump fed common manifold that led to four separate injection lines.

At the beginning of work the following day, an explosive tool containing four shaped charges, consisting of 3 to 4 g of Cyclonite explosive were lowered to the bottom of the injection pipes. These were shaped and detonated so as to cut the pipes at the top elevation of the previously placed grout, allowing injection to continue.

The grout mix consisted of:

Cement	1 bag, 94 lbs (42.7 kg)
"Torpedo" sand	3 ft^3 (84.9 L)
Intrusion Aid admixture	1 lb (0.45 kg)
Water	7 gal (26.5 L)

The Intrusion Aid admixture acts as a water reducer/plasticizer, as well as a gas-producing expanding agent, to minimize the formation of air voids under the relatively large-size aggregate. The mix design provided a grout with slump of about 6 in (152 mm) using the standard ASTM C 143 test.

Grout materials were transported to the work site by a barge, which was moored adjacent to the work. Aggregate was normally supplied in bulk, and the cement was in standard 94 lb (42.7 L) bags. The sand was proportioned by volume into batch-size quantities, which were fed to the various mixers by a conveyor belt that rotated, as required, to each mixer location.

The ability to cut the injection pipes at any desired point allowed the work to progress in stages over a period of several days. This scheduling removed much risk from the concreting operation and greatly enhanced safety, as well as increasing the efficiency and lowering the cost of the operation. In all, six separate cells were constructed. Grout injection for each cell required about seven days, with the crew working an average of about 12 hours a day. The ability to interrupt the work as desired was possible only through use of the shaped explosive charges to cut the injection pipes as desired.

19.7 IN CONCLUSION

As demonstrated, there can be a substantial advantage in using explosives to facilitate grouting. One must be aware, however, that explosives improperly handled can result in catastrophic accidents. Special licenses and permits, as well as approved assembly and storage facilities, are required in most jurisdictions. A good reference on working with explosives and appropriate safety practices is the American Petroleum Institute's (1967) *Recommended Practices for Oilfield Explosives Safety, RP-67.* Although it is not directed specifically to grouting, this work is nonetheless germane to the work and provides much good, commonsense advice. The term *explosive* can carry negative connotations; explosives themselves are objectionable to many people and are often banned from use. For this reason Bob Gronowicz refers to his explosives work as *energy conversion.*

CHAPTER 20

Emergency Response Grouting

20.1 BEFORE INJECTION STARTS

20.2 TYPICAL EMERGENCY SITUATIONS
20.2.1 Sinkholes
20.2.2 Subsurface Water Flow/Piping
20.2.3 Subsurface Soil Loss
20.2.4 Structural Failures
20.2.5 Extinguishing Underground Fires

BECAUSE GROUTING is widely used to stop or control the flow of water, arrest or correct settlement damage, and stabilize faulty soil, it is often called upon in emergencies to solve such problems as they occur. Although grouting methods may provide the best solutions available, they are not cure-alls and discretion dictates that they not be employed as magical solutions. As in all grouting, proper performance requires both knowledge of the in situ conditions and thoughtful planning for optimal solutions. Unfortunately, emergency response seldom allows sufficient time for valid assessment or appropriate planning. Rather, a frantic urge to immediately "do something" usually prevails. This can be counterproductive, however, and injection without consideration of the particular requirements can not only be ineffective but can also create a greater problem and extenuated distress.

20.1 BEFORE INJECTION STARTS

When emergency grouting is called for, grouters will be well served to always remember one admonition: "Keep cool—don't panic!" Regardless of the seriousness of the circumstances, thoughtless injection can actually aggravate a situation. Inappropriate grouting can result in accelerated and/or increased damage, permanent harm to the formation, and wasted resources. Whereas decisions must often be made on the fly, one should thoughtfully consider all available alternatives before the commencement of any work. There are instances in which no injection at all may be the most prudent decision. Emergency work constitutes an area of grouting in which past experience is particularly important. The most experienced people available should thus be involved in the rapid decisions that must be made.

It is important to always keep in mind that whatever is injected in haste must be dealt with once the emergency is over. Examples abound. Upon appearance of a sinkhole in a major intersection, ready-mixed concrete was pumped into the throat. A 10 ft (3 m) length of rigid 3 in. (76 mm) pipe was attached to the end of a boom pump hose. It was lowered so that the hanging vertical pipe was centered over the hole, and pumping was initiated. The boom was lowered slowly, allowing the pipe to penetrate into the loose soil debris a few feet. When about 16 yd^3 (12 m^3) of concrete was pumped, the concrete surfaced in the crater. Injection was then stopped, and all parties rejoiced that the problem was resolved.

Further investigation the following day disclosed that the sinkhole was centered over and caused by a bad joint in an existing 24 in. (0.6 m) diameter storm sewer. The concrete had filled the remaining pipe for a distance of about 36 ft (10.8 m) in each direction. Removal and replacement of that pipe caused significantly more interference to the movement of traffic than the sinkhole had and, of course, involved a high cost as well. In this case, no injection would have been a far wiser and certainly much less costly action.

In another case, a high-rise structure still under construction was suffering excessive differential settlement. The settlement resulted from consolidation of fine-grained soil underlying the very heavily reinforced raft foundation, by a depth greater than about 35 ft (10.5 m). In panic, an around-the-clock compaction grouting operation with multiple crews was initiated in an effort to arrest the settlement. Unfortunately, the injection was made from the underside of the foundation to a depth of about 30 ft (9 m), *above* the underlying culprit soils. A total of 20,500 ft^3 (574 m^3), representing a weight of about 1500 tons, of grout was injected. This amounted to an equivalent surcharge load on the order of 2 ft (0.6 m) of soil.

As would be expected, the rate of settlement increased, in this case by a factor of 3, causing significantly increased damage to the structure. Not only was the grouting effort a total waste, but the much greater damage to the structure carried a very high cost to repair. Injection of large quantities of grout into the ground, absent a good profile of the existing soil conditions, is risky at best and should not be done. In this instance, no grouting at all was appropriate, and had it not been attempted, the huge cost of the final remedial work would have been greatly lessened.

In another instance, a very expensive hillside residence was threatened by a landslide, the head of which had extended to within about 20 ft (6 m) of its foundation. The slide had already consumed a lower structure and had been increasing in size for the past few days. Because groundwater was thought to be a major contributor to the problem, the home owner's engineer, in haste, called for injection of a chemical solution grout cutoff to block groundwater movement above the residence. This work was in progress when the rim of the slide grew so as to consume the building. Although it is unlikely that the partially completed grouting accelerated the growth of the slide, if it had been successfully completed, it would have restrained the downslope drainage, actually increasing the driving forces at the upper slide boundary. This would have certainly resulted in deterioration of the situation. Again, no grouting would have been a far wiser move, as was apparent once the emergency ended and more thoughtful consideration prevailed.

In 1992, flooding of the ancient 50 mi (80 km) long freight tunnel system in downtown Chicago represented a huge disaster. The flood was caused by a previously driven pile in the Chicago River penetrating one of the tunnels. The existence of the penetration was known, and in fact the penetrating pile and growing mound of mud around its base had been videotaped, but no corrective action had been taken as the system had been previously abandoned. The flooding of many basements and disrupted utility service resulted in a virtual shutdown of the downtown area. In the chaos that followed, a number of unsuccessful attempts were made to stop the flow. Among them was the deposition of a large amount of concrete into the tunnel. When all else failed, experienced mine troubleshooting divers were brought in. They successfully stopped the flow by constructing a bulkhead on either side of the break. This activity was greatly hindered, however, because of the irregular surface of concrete remaining on the tunnel floor. Again, work that was done in haste proved to have a negative result and significantly added to both the cost and the time required to remedy the problem.

In another case, piping through a water-retaining earth embankment resulted in leakage of about 1500 gal (5700 L) of water per minute. This was a serious situation, and many of the owner's senior engineers were on-site. No single source for the leakage could be identified, but divers did discover suction along a considerable length of a joint in the concrete lining. A hole cored through the lining identified an under-slab void of only about an inch in depth. Accordingly, a ready-mixed grout, without any large aggregate, was pumped to fill the void. It was rapidly washed through the embankment, however. A second hole was drilled, which encountered a void of several feet in depth, indicating the existence of a single large passage. This would require a much less mobile grout material and very rapid pumping in order to overwhelm the flow. A call was made to the ready-mix plant to hold the incoming trucks until three were available to be dispatched together with a mix containing large aggregate.

During this lull in pumping, the leakage increased and became more turbid, causing even greater unease. A senior supervising engineer, unaware of the planned strategy, called the ready-mix plant demanding that any available trucks be immediately dispatched with the original mobile mix. Two trucks were sent but were never used, because of the impropriety of the mix. The action made in panic actually delayed effective control of the flow by about a half hour. Fortunately, the remedy, although delayed, was successful. A complete failure could very well have occurred, however, as a result of the delay in obtaining the proper mix in sufficient quantity.

In another case, a large quantity of bagged bentonite was deposited into a whirlpool behind a water-retaining embankment. This measure was successful in slowing the flow so that more permanent repairs could be made. Yet it greatly increased the time required and the cost of the work. Bentonite mud can cause hydraulic fracturing of the soil if sufficiently pressurized. Subsequent evaluation indicated compaction grouting to be the best solution for remediation of loose horizons within the embankment, and was so performed. To reduce the risk of hydrofracture due to the bentonite, a very slow pumping rate was adopted, however, greatly increasing both the time required for the remedial work and its cost. Although the bentonite was successful in controlling the leakage, it did complicate the final fix.

20.2 TYPICAL EMERGENCY SITUATIONS

There are a wide variety of emergency situations that can benefit from grouting. The basics of material selection and choice of injection methodology should generally be no different than for any other rationally designed grouting application. Some deviation may be in order early on, however, because of inability to immediately obtain the optimal equipment and material. Any injection that is made should be clearly thought out, with consideration given to dealing with the injected material once the emergency is over.

20.2.1 Sinkholes

Sinkholes occur when friable soil, usually in combination with water, escapes into some sort of subsurface void. The void may be man-made—such as a defective pipeline or other substructure, mine, or tunnel—or natural, as in the case of cavernous voids in soluble rock. Although much rarer, voids can also be formed in soil because of erosion by passing water. Sinkholes can develop suddenly, and many continue to increase in volume with time. They usually originate as a result of eroded soil flowing into an underlying void space. Because water movement through soil is a common ingredient to the formation of a sinkhole, it is crucial to avoid surface ponding and to provide good drainage in any development over karstic terrain.

Sinkholes have been known to swallow entire buildings and other structures, and when they occur in built-up areas, can destroy utilities and surface improvements as well. The sinkhole shown in Figure 20.1 developed as a result of tunnel construction at considerable depth below the bottom of the hole. In the process, it consumed not only the street but several structures and all the utilities, including the large sewer shown in the center. A question always arises with this type of failure: Did it result from an uncontrolled soil run into the lower conduit, or was it a consequence of leakage from the overlying line?

Naturally occurring sinkholes are common in geographical areas underlain by karstic rock, especially where the overlying soil is granular and saturated. In these areas, sinkholes are a natural component of the geology, appearing as natural surface depressions. Development is best avoided in areas known for prolific sinkhole activity. In special cases where economics dictate building, mitigation measures should be taken ahead of other construction. These usually involve limited grouting to densify and/or solidify the overlying soils near the top of the rock.

FIGURE 20.1 Sinkhole resulting from loss of ground during underlying tunneling.

Grouters are often called soon after the development of a sinkhole, especially when it is in a developed area or near a structure. The immediate response is often to pump something into the crater and fill the void into which soil is escaping, even though its nature, volume, or configuration is unknown. Such action is not recommended, however, as it will likely fail to improve the situation, can result in even greater damage, and, of course, will have a considerable cost. Although some form of grouting will usually be a component of remediation, it is unwise to make any injection until, at the very least, the nature of the void that initiated the sink is understood.

If it is found that the culprit void was manmade, the condition that caused it must first be identified and corrected. If it has resulted from the natural dissolution of underlying rock, the depth to the throat opening through which the soil is escaping and some idea of the size and configuration of the gap should be established. This investigation is normally performed by geotechnical engineers and geologists and usually includes boring and probing, both within the limits of the sink as well as in adjacent areas. A variety of seismic exploratory methods, such as ground penetrating radar (GPR) and electrical resistivity tomography, are also commonly used. And, of course, a review of the area's topography and, particularly, the records of any prior sinkhole activity must be included.

Often, the first thought of grouters, and unfortunately, engineers and geologists who should know better, is to completely fill any voids found under or in the area of a sinkhole. This option is particularly favored by contractors being paid according to the quantity of grout pumped. Such thinking is faulty, however. Solution channels in rock can extend to huge caverns that will take immense amounts of grout, even though filling them is usually not necessary to correct the problem. Rather, only the throat opening need be closed or capped to prevent further subsidence. Once this is accomplished, the remaining hole can be properly filled with imported soil. Knowledgeable and honest grouters know this and will not promote unnecessary injection; unfortunately, however there are opportunists in the industry who will keep on pumping as long as the owner continues to pay. In a recent example, an unscrupulous contractor was found to be drilling holes through good-quality solid rock in excess of 70 ft (21 m) deep in order to find a void to fill.

To more clearly understand the nature of karstic terrain, envision caves that open to the surface. These are natural solution features of rock and no different from those that exist under the ground surface. They can be very large, deep, and interconnected to a wide network of similar spaces. Often, the entrance openings are small and can continue as relatively small passages for a considerable distance, but then lead to mammoth cavities. To fill such huge cavities will serve no useful purpose, but the entrance throats must be sufficiently sealed so as to preclude loss of overlying soil. These openings can be no more than a few inches in width and seldom exceed about 3 ft (0.9 m). The sealing of such openings can be accomplished either by filling the entrance area or by forming a cap in the soil overlying the opening. This may be accomplished through the densification of compaction grouting or solidification by permeation grouting. And, of course, any repair must include provision of good drainage so that surface ponding does not occur.

20.2.2 Subsurface Water Flow/Piping

The movement of water through soil and rock is a major problem in much subsurface construction. Although it is normally controlled in a preplanned manner, sudden breakthroughs and/or excessive quantities can create an emergency situation. A large inflow suddenly occurred during the advance of a hard rock tunnel. Only minor leakage had been previously experienced, but the sudden inflow from a highly fractured zone of rock forced a stoppage of work. A combination of cement and polyurethane grout was injected by a specialty contractor, reducing the flow to a manageable level and allowing work to resume.

In a less dramatic but even more costly example, driven concrete piles penetrated an impermeable layer of soil underlying a large excavation during construction of a federal courthouse. The base of the excavation was about 25 ft (7.5 m) below and immediately adjacent to the Sacramento River. The inflow of water was controlled satisfactorily, until the foundation piles were driven. Once they penetrated an underlying impermeable zone, unacceptable amounts of water surfaced in an artesian flow.

Following several attempts of injecting simple grouts, simultaneous injection of sodium silicate solution and cement grout into casings installed immediately adjacent to the piles was performed. The grout was stream mixed at the collar of each casing, as shown in Figure 20.2. This led to immediate stoppage of the intruding

FIGURE 20.2 Stream mixing of grout at the hole collar stopped water flow.

water. Two casings that extended to a depth of 35 ft (10.5 m) were placed on opposite sides of the individual piles. They were driven as near as practicable to the piles with an air-track-mounted hammer. A temporary steel plug was inserted into the base of each casing to keep it clear of soil during driving and was knocked out upon reaching the final depth. Grout was injected at a pressure based on 2 psi (0.14 bar) per foot of overburden depth. A total of about 100 gal (380 L) of grout was required for each pile.

Water moving through soil can cause erosion and relocation of the individual grains or units of soil. This is a major factor in the development of sinkholes, as previously discussed, and can occur in water-retaining embankments such as dams and levees. In the latter, the development of a water course is commonly referred to as *piping*. It typically starts as simple permeation of the water in a preferential path through the soil. As it continues, the finer particles are eroded and transported in a downstream direction. Progressively larger particles are affected, until an actual voided channel is formed. As the cross section of the void is enlarged through continued erosion, an ever-increasing amount of water is transmitted, causing accelerated soil loss. Unless the process is halted in a timely manner, a point will be reached at which total failure of the embankment occurs.

Such leakage is most often discovered by observation of excessive seepage flowing from a drainage system or downstream of an embankment. Where leakage flow is substantial and/or at a reasonably shallow depth, a whirlpool will often develop in the water over the source. Such a sighting clearly identifies the source location, which greatly facilitates direct injection to stop the water movement. However, lack of such a visible source typically indicates a leak that is originating from many smaller sources or one that is submerged in very deep water. This can also be the case where the upstream face of the embankment is paved and the source water is entering from a considerable length of distributed joints and/or cracks.

In such instances, identification of the leakage source is much more difficult, and direct injection is not always feasible. The best approach in such situations is to inject grout holes upstream of the leakage exit in an effort to intercept it. This usually involves placing one or more lines of grout holes through the embankment at a fairly close spacing of about 10 ft (3 m), as shown in Figure 20.3. Split spacing at closer intervals may be required. Sometimes divers can locate the underwater source, but this can be a dangerous assignment, as they can be drawn into the leak if it is large. In the case of seepage through a lining, divers can usually discern suction in the particular joints or cracks that are involved.

In such situations, it is often possible to core drill a hole through the lining to intercept the main water passage, although several holes may be required prior to reaching the target void. The void will be readily known once it has been encountered, however, as a strong suction on the core barrel will occur, and it can, in fact, be extremely difficult to remove the core barrel from the hole. In such cases, it may be best to simply disconnect the bit from the drill and allow it to be consumed. Diver safety must be carefully con-

FIGURE 20.3 A line of grout holes upstream of the leak exit will often intercept the water course.

sidered in such operations, as a failure of the lining may result in an outrush of water, sucking the diver in. It is thus wise to consider providing a movable steel grillage or other positive restraint to protect against such an occurrence.

A rough estimation of the length of the leakage route can often be made through differential temperature evaluation. Should the water temperature noticeably rise between the pool and leak outlet, a rather long and/or torturous route is indicated, whereas no temperature change suggests a rather direct route. This can be important in the selection of a remedial strategy. A highly cohesive low-mobility grout containing large aggregate is in order where direct injection into a substantial leak channel can be made. Conversely, where the leakage is through only relatively small apertures or porous ground, a typical compaction grout mix that will force the soil into a denser condition will likely be more appropriate.

A nearly universal procedure that has been used for many years in dealing with piping leaks is to build a "chimney" at the outlet. This is typically made of sandbags, as illustrated in Figure 20.4, and the main objective is to reduce the leakage head. The amount of any reduction is minimal at best, and the benefit is questionable. Such containment can be useful, however, in that it somewhat contains solid matter that has been carried by the water. This can aid in the collection and identification of such solids, which is especially useful should they contain constituents of any injected grout.

FIGURE 20.4 A sandbag "chimney" will reduce the leakage head and contain escaped grout.

When large volumes of water are leaking, it is important to be able to measure the initial amount of flow as well as any changes in the flow. This is best accomplished through installation and monitoring of a weir, as illustrated in Figure 20.5. Although a rectangular weir opening can be used, a standard 90 degree V-shaped opening will increase the accuracy of measurement at low flow rates, which, it is hoped, result as grouting continues. V-shaped weirs cut from standard plywood sheets can handle flows up to about 3,000 gal/min (190 L/sec). These are usually secured in place within a dam built of sandbags. Plastic sheeting placed over the interior sandbag face and extending well into the bottom is recommended in order to reduce leakage through the dam which will adversely affect the accuracy of measurement.

There are two widely divergent schools of thought on how best to stop massive flows of water. The more traditional approach is to use conventional pourable grouts, often in combination with

FIGURE 20.5 A standard V-shape weir to measure the flow rate.

hot bitumen. A more practical approach, however, is to use ready-mixed grout containing viscosity-modifying antiwashout admixtures. Because of continuing erosion, time is critical in the management of most leakage problems. Conventional grouting equipment usually requires considerable time to mobilize and has somewhat limited pumping rates, whereas ready-mixed materials and very high output concrete pumps are commonly available on short notice. In addition, they are far more economical, and with modern admixture technology, very cohesive limited-mobility mixes can be compounded.

Rapid placement of the grout is imperative where massive flows of water exist. Although the more traditional approach depends on at least some of the injected grout remaining within the leakage channel so that it builds up with time and eventually stops all flow, with a sufficiently high pumping rate the leakage can be literally overwhelmed. Modern concrete pumps can output in excess of 200 yd^3 (153 m^3) per hour. They are readily available with multisegment articulated placing booms that can extend to nearly 200 ft (60 m). This provides enormous flexibility and allows quick positioning of the delivery line without the need to lay out hose or assemble rigid line. For example, in a situation occurring in 2001, some 37 yd^3 (28 m^3) of grout were injected in less than 10 minutes time (Figure 20.6) to successfully stop a subsurface water flow of about 1500 gal/min (95 L/sec).

Because these very standard resources are now so readily available, it is practical to use multiple pumps, subject only to the availability of sufficient space to maneuver the ready-mix delivery trucks. For very rapid placements, it is necessary to be able to position two trucks simultaneously at each pump hopper. Each truck requires only about three minutes to completely empty, so very fast positioning and exiting must occur. In situations where there is insufficient space at the main pump location, the grout can be supplied by other pumps located at greater distances where more maneuvering room is available.

FIGURE 20.6 Ready mixed grout was rapidly pumped to stop leakage.

20.2.3 Subsurface Soil Loss

Large amounts of soil can be eroded by sudden bursts of water, as from a broken pipeline. Such sudden exposure to water can also cause hydroconsolidation of the soil. Many granular soils will densify and decrease in volume upon saturation. This most often results in surface settlement, but can also develop as a void under a structure, another hard surface, or a more competent soil layer. The most common cause of these developments is sudden flooding due to broken pipes or conduits or inadequate surface drainage. Such subsurface voids can cause settlement or complete failure of overlying structures and other improvements and thus warrant prompt remedial attention.

The breakage of a large water main under the sidewalk immediately adjacent to the main entrance of the Pantages Theater in Hollywood, California (Figure 20.7, left), threatened its closure. The escaping water had created a void that extended from the underside of the sidewalk to

FIGURE 20.7 A broken water main under the walk (left) caused a cavernous void (right).

a depth of about 6 ft (1.8 m). It was roughly 5 ft (1.5 m) wide and extended for more than 30 ft (9 m) along the building's foundation, as illustrated in Figure 20.7 (right), with a total volume of about 32 yd^3 (24 m^3). The ornamental sidewalk, which contains the famous Hollywood Stars, had, amazingly, not settled or shown any other distress. Immediate reestablishment of its support was considered crucial, however.

The owner did not want any sign of the problem or its repair to be visible from the surface, as that would negatively affect business. Furthermore, to allow for future penetration and repairs, filling with any material that would not be easily excavated was prohibited. The solution was to inject a very low strength grout consisting of a natural pozzolan, silty sand, and just enough water to form a stiff mortarlike consistency. All work was accomplished in the early morning hours before the theater opened. For grout injection, a hole was drilled from an adjacent tree well to intersect the void, so that no sign of the work was visible from the surface when the theater opened in the evening.

A similar emergency with a very different cause occurred when the float valve controlling supply to a large elevated water storage tank malfunctioned in the early morning hours. Several thousand gallons of water overflowed, resulting in hydroconsolidation of the underlying soils and completely undermining the foundation for one of the four supporting columns. Minor settlement of the surrounding ground surface was also evident, and there was concern about possible soil disturbance around and under the remaining legs. To exasperate the problem, heavy rain was forecast to occur momentarily.

Repair involved immediate shoring of the tower leg, with steel beams spanning the eroded pit. A tent was then erected and surface dikes placed to prevent intrusion of any rainwater, as shown in Figure 20.8. Concurrent with this work, the void underlying the foundation was filled with a typical compaction grout mix, followed by drilling and injection of a series of compaction grouting holes under and around it. Once this had been completed and the soil returned to a competent state, compaction grouting was performed in eight holes uniformly spaced around the foundations of the remaining three legs, in order to detect and correct any deficiency.

20.2.4 Structural Failures

Grouting is often called upon to alleviate problems resulting from structural deficiencies. Such

FIGURE 20.8 Emergency shoring and a tent erected to preclude flooding while grouting.

deficiencies are often caused by the failure of underground conduits and structures. The 1996 blowout of a pressure effluent line at a paper mill on the Powell River in British Columbia, Canada, required immediate attention. The top of the 48 in (1.2 m) diameter line, which normally carries about 2 million gal (7570 m^3) of effluent per hour, was about 22 ft (6.6 m) below the ground surface. Effluent was gushing to a height of 10 ft (3 m) above the ground surface, with an estimated leakage of about 500 gal (1890 L) per minute. The leakage was above a "Y" structure in the conduit immediately adjacent to the river and could not be shut down without idling the mill. Divers were not available, and although TV inspection was attempted, high turbidity prevented receipt of a clear picture. The native soil was known to contain cobbles and boulders up to about the size of a desk. It was thought that much of the finer-grained soil had been carried away with the effluent, leaving significant voids between the rocks and around the conduit.

The emergency measures adopted involved injection of a water-activated hydrophilic urethane grout to fill the large voids. The particular formulation selected would expand at a low pressure to form closed-cell foam. This facilitated complete filling without applying undue pressure to the already distressed conduit. The grout was placed through eighteen 3/4 in. (19 mm) pipes that were worked into the soil with a handheld pneumatically powered drill. The line could not be shut down for more than a two-hour period, so this work was performed during several such breaks. An inflatable packer was fabricated with the intent of bridging any openings or offsets in the conduit joints. It was not able to be inflated tightly, however, because of suspected excessive offsets at the joints, but did restrain the grout somewhat, as evidenced by the urethane foam adhering to its surface upon removal.

The urethane injection worked generally as planned and was successful in decreasing the leakage to about 20 gal (76 L) per minute. To gain further reduction, a much more penetrable acrylamide grout was injected into an additional 15 holes. This resulted in complete stoppage of flow to the surface and, presumably, complete blockage of the subsurface breaks, the details of which remained unknown. The emergency grouting was successful in that it was accomplished without loss of plant operation and provided a satisfactory repair, allowing the plant to remain in full production for the next eight months until a regularly scheduled maintenance shutdown, during which the line was replaced. Upon exposure, a 4 in. (102 mm) offset was found on one branch of the Y, as well as a 1 in. (25 mm) offset on the other. Both had been completely sealed by the grout.

In an unusual project, the failure of an 8 ft (2.4 m) diameter vertical polypropylene tank threatened shutdown of a major building. The tank was buried under the basement and served as a wastewater collection sump. It had been placed, in two sections, into an oversized drilled hole, with a horizontal joint at mid-height. The top was covered with a heavy steel plate, and the annular space filled with pea gravel. While the tank was in service, failure of the joint had allowed some gravel to enter it, resulting in damage to the pumps. The problem had to be corrected quickly, as it was essential to the building's occupancy.

The pumps and top were removed and the tank drained. It was then temporarily filled with sand to restrain any possible deformation from the grouting that was to follow. An expansive two-component, chemically cured, hydrophobic urethane foam was injected through 14 injection pipes placed into the pea gravel, around the periphery of the tank. The 1/2 inch (13 mm) pipes were specially fabricated and were fitted with regularly spaced 1/8 inch (3 mm) holes through the wall and a bottom plug. Grout was injected at a rate that would ensure the exit of an equal volume through each of the holes simultaneously. To prevent any upward displacement of the tank, vertical steel shores were installed, as shown in Figure 20.9. The work was successful in that the pea gravel was immobilized and the tank rendered free of further leakage.

The concrete slab-on-grade floor of an upscale restaurant lobby in Los Angeles suddenly dropped several inches during the busy lunch period. The lobby area, which was filled with people, was immediately adjacent to the retaining wall of a several-level underground parking garage. The underlying backfill had apparently settled over a period of several years, leaving a void under the slab, which was apparently bound in place until it finally failed. This, of course, required the restaurant to be closed. It was of particular concern to the building owner, who was subject to a substantial penalty should the area not be restored to full use within a period of 24 hours.

The carpeting was removed, decorative walls and fixtures were protected with plastic sheeting, and conventional slabjacking was carefully prosecuted, as illustrated in Figure 20.10, to raise the floor to proper grade. The work was executed with care, to make sure that no grout splatter or other damage to the area occurred. Note the wet burlap sacks that were placed around the injection pipes to contain any escaped grout. All work, including replacement of the carpet, was completed within 20 hours of the initial failure, and the facility opened as scheduled the next morning.

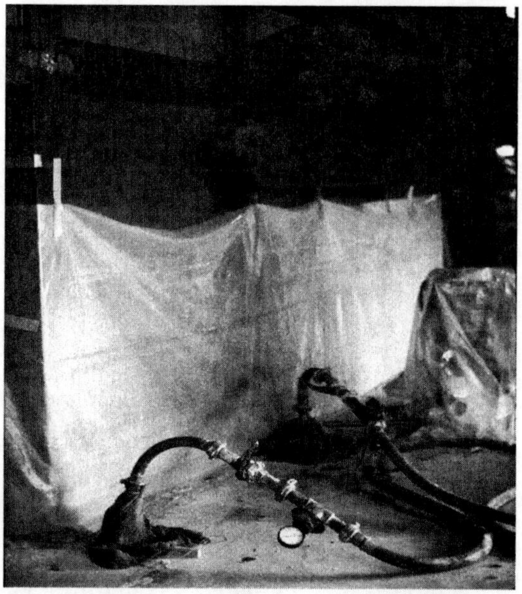

FIGURE 20.10 Emergency slabjacking in progress. Note protection on decorative wall.

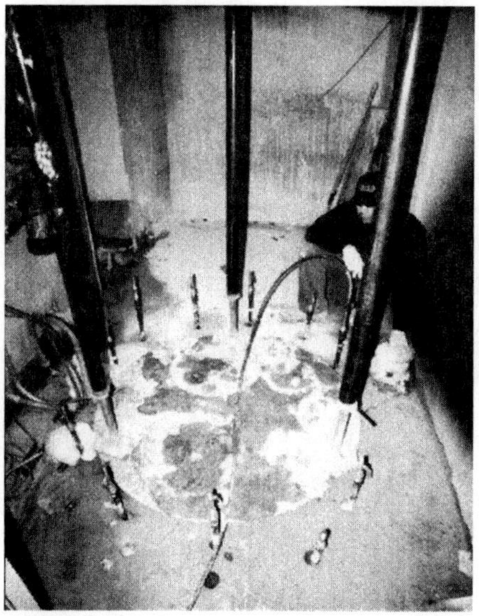

FIGURE 20.9 Grout injection around a sand-filled and shored open-top tank.

20.2.5 Extinguishing Underground Fires

Underground fires sometimes occur in sanitary landfills, old mine works, and the like, through spontaneous combustion. Although plain water is often injected to extinguish such burning, it flows freely, is absorbed by the soil, and leaves little residual to completely cut off oxygen to the burning matter. A cementitious or clay grout, however, will tend to coat any surfaces with which it has come in contact, leaving residue that can be built up over time to completely smother the fire.

A particularly difficult situation occurred under a Norfolk Southern rail site near Washburn, Tennessee. The area had been used historically to dispose of spent ash and clinker from the old steam locomotives. In those earlier times however, considerable amounts of carbon, which could support further combustion, remained in the burned-out clinker. Through the years the fill, which had reached a depth of 60 ft (18 m) in some areas, had become covered with vegetation and was long forgotten.

In 1995 remnants of the coal started to burn, as evidenced by smoke seeping from the ground. Within a month, overlying trees were dying and the amount of noxious smoke had increased to unacceptable limits. By this time the ground surface had become so hot that worker access was limited. The solution was injection of a cement suspension grout through closely spaced small-diameter casings, as shown in Figure 20.11. The fire appeared to be extinguished after ten days of continuous injection. It reignited after about two weeks, however, requiring another four days of injection. Subsequent observation proved the grouting to be completely successful in permanently extinguishing the fire.

FIGURE 20.11 Smoke rising from the ground as the grout is injected.

PART II

Grouting Design and Control of Grouting

CHAPTER 21

Understanding Geology

21.1 GEOLOGIC AGE

21.2 TYPES OF ROCK
- **21.2.1** Igneous Rock
- **21.2.2** Sedimentary Rock
- **21.2.3** Metamorphic Rock

21.3 GEOLOGIC STRUCTURE
- **21.3.1** Primary Features
  - **21.3.1.1** Texture
  - **21.3.1.2** Hardness
- **21.3.2** Secondary Features
  - **21.3.2.1** Folds
  - **21.3.2.2** Joints
  - **21.3.2.3** Faults
- **21.3.3** Secondary Mineralization
  - **21.3.3.1** Solutioning
  - **21.3.3.2** Weathering
  - **21.3.3.3** Infilling

21.4 THE ROCK CONFIGURATION/GEOMETRY
- **21.4.1** Bedding
- **21.4.2** Direction and Attitude of the Strata
  - **21.4.2.1** Strike
  - **21.4.2.2** Dip
- **21.4.3** Outcroppings and Overhangs

21.5 STRESS STATE

21.6 DERIVATION OF SOIL
- **21.6.1** Transport Mechanisms
- **21.6.2** Depositional Mechanisms
  - **21.6.2.1** Aeolian Deposits
  - **21.6.2.2** Water-Deposited Soils

21.7 CLASSIFICATION OF SOIL
- **21.7.1** Unified Soil Classification System
- **21.7.2** Fine-Grained Soil Consistency
  - **21.7.2.1** Atterberg Limits
- **21.7.3** Field Identification and Description

21.8 GEOTECHNICAL CONSIDERATIONS

THE SCIENCE OF GEOLOGY involves study of the historic formation and alteration of the earth, and especially of the geomaterials (soil and rock) of which it is composed. These are subject to continuous change through natural processes that principally occur over very long periods of time. There are two major interests in geology, petroleum and mineral discovery and classification, and engineering geology. It is engineering geology that pertains to an understanding of the near surface deposits that may be grouted, and the geologists who are so involved are generally known as *engineering geologists.*

Tectonic activity deep within the earth's crust causes continual movement of the near-surface geomaterials. Such activity dictates the formation of various types of rock as well as the nature of the discontinuities and other internal defects that are always present. With time, the surface rock may deteriorate to form soil. This process is usually strongly influenced by weather activity, which tends to degrade and transport the surface rock particles. The structure of the soil is the result of the particular rock of which it is composed, as well as its depositional and formational history. These are strongly influenced by the effects of water and wind as well as by chemical alteration.

Although engineers generally concentrate on the present structure and the mechanical properties of soil, geologists focus more on the historic origin and subsequent modification of a particular formation. All soil started as rock. Through time, the rock was weathered and broken down into increasingly smaller pieces. It may also have been subject to chemical alteration. An understanding of these processes is within the purview of geology, so the capabilities of a geologist are not limited to dealing only with rock, but include consideration of soil as well.

Unfortunately, in some arenas, there is dissention as to the role of the geologist in grouting investigation, design, and production. Some large organizations assign such work to engineers only, and others, to geologists. And many in each group are of the opinion that this work should be done only by their own. In nearly 50 years of investigation of foundation failures of both initial, and remedial construction, a common element that is nearly always present points to a failure to recognize geological influences as a cause of the deficiency. It is clear that engineers and geologists must work together. They very much compliment each other in both their fundamental knowledge and in the manner in which they interpret geotechnical conditions, and they are equally important to the overall work.

21.1 GEOLOGIC AGE

As mentioned earlier, geological processes occur over very long periods, and millions of years are but a scant moment in geologic time. Geologic history has been divided into 14 primary periods, the oldest of which is about 4500 million years. Each primary period is further divided into *epochs,* which are generally in line with the appearance of different species of life. Our interest, from the standpoint of grouting, will involve only recent geological time periods of less than about 300 million years, and will obviously be limited to the mechanics of the formation with which we are dealing. Understanding the details of that formation can be enhanced by knowledge of the life forms, weather patterns, and the like, that were present at its emergence. The periods of geological age are listed in Table 21.1.

In mapping and sectioning a given site, the geologist separates the different materials into geologic *units* according to their makeup and age. Each different unit is given a label, which typically consists of two or more letters. The first letter usually indicates the period of origin of the particular material in that unit. For example,

TABLE 21.1 Geological Ages

Period	Epoch	Age (millions of years)
Quaternary	Holocene	
	Pleistocene	1.6
Tertiary	Pliocene	5
	Miocene	24
	Oligocene	38
	Eocene	55
	Paleocene	66
Cretaceous	Late	138
	Early	
Jurassic	Late	
	Middle	205
	Early	
Triassic	Late	
	Middle	240
	Early	
Permian	Late	290
	Early	

very recently formed units begin with "Q," indicating that it is from the Quaternary period, or up to 1.6 million years old.

21.2 TYPES OF ROCK

All rock is classified in one of three broad categories: *igneous, sedimentary,* and *metamorphic.* The origin of each differs from the rest, as does their structural makeup, and thus the type and extent of joints and other defects. Consequently, the investigative effort required to establish the general rock condition, and especially any unusual defects that may be present, will vary according to the rock type. Similarly, the grout injection parameters will also vary. A basic understanding of the origin of a given formation is thus fundamental to optimal grout treatment.

21.2.1 Igneous Rock

Igneous rock forms when *magma,* which is molten rock, is pushed up to or toward the earth's surface, where it cools and crystallizes. When thrown out as by a volcano, it is commonly referred to as *lava.* Volcanoes are essentially vents through which rock and associated gasses and ash erupt. Rock that issues from such eruptions can be either in a molten form or in the form of rock fragments within a molten flow. Such rock can have extreme variation in its physical properties not only due to its initial chemical composition, but also resulting from variations in its flow and subsequent cooling. Moreover, molten rock can contain gases that expand upon release, forming blowholes or bubbles in the cooling formation. In addition to rock, volcanoes spew ash and cinders, which may or may not solidify, as well as a frothed glass known as pumice.

Not all magma reaches the surface, and it can be emplaced into the earth's crust as *intrusive* rock. The path of such intrusive flow will be the course of least resistance for the particular magma. Magma varies widely in degree of fluidity or viscosity, as well as in its exact mineral makeup. As most of the gases will vent to the top of the volcano, intrusive deposits are usually quite dense, but they will contain some fractures as a result of cooling. Depending on the manner of its formation, volcanic rock will develop a variety of textures and other properties. That which originates at great depth typically consists of crystals or grains and is commonly referred to as *porphyritic.* This rock is often massive, with no discernable bedding, and its joints are widely spaced. Shallow intrusive deposits, however, can develop many closely spaced cooling cracks or fractures, which are commonly perpendicular to the flow axis.

Rock that results from lava flow can, in contrast, exhibit a wide variety of densities and can even be quite porous and permeable, often sufficiently so to float on water. Volcanic ejecta can flow as either very thin or quite viscous liquid, can contain solid rock fragments, or can advance in an avalanche-like mass. It can even develop into long, hollow lava tubes. The ejecta will always contain considerable gas and produce ash, cinders, and other uncemented particulate material. Most areas of volcanic origin have experienced multiple eruptions over very long time periods. Consequently, other particulate materials and sediments are often intertwined within the pyroclastic deposits.

The top of a lava flow will typically cool much faster than the lower parts, which results in the formation of a crust. As the mass continues to move, the crust will often break into angular and jagged floating fragments composed of mostly *vesicular* material. These fragments can form great islandlike deposits, with portions often projecting above the flowing molten rock. Upon coming to rest and cooling, these projections can be widely distributed or may be in fairly close proximity. The resulting rock is known as *flow breccia.* Spaces between the individual pieces can become significant channels for water, given sufficient time to develop.

Magma rising in a crack or fissure will form a *dike,* which assumes a relatively thin, plate-like long and narrow shape. Dikes typically occur in joints that cut across the bedding or foliation planes of the parent rock. A similar inclusion, which forms between layers of sedimentary rock in the shape of a sheet, is known as a *sill.* Sills are thin in comparison to their areal extent and are parallel, or nearly so, to the enclosing bedding or foliation of the original rock. Multiple sills can occur and are often connected by dikes, which are the feeding channels for the advancing magma. Although dikes are often in a flat configuration, they can be tilted at high angles. Dikes and sills usually consist of very hard dense rock and are typically much more resistant to weathering and erosion than the surrounding parent rock.

Granite and diorite are the most common intrusive igneous rocks. Extrusive volcanic rocks include rhyolite, andesite, basalt, and volcanic tuff and breccia. With the exception of tuff and breccia, igneous rock tends to be hard and strong. It is not easily eroded, and the walls of its joints are usually hard and durable. Existing joints and other defects are typically open, and infilling is not common unless it is from some external source. At depth these rocks tend to be massive with widely distributed fractures, but near the surface they can be highly fractured and overstressed as a result of secondary processes, which are discussed shortly.

21.2.2 Sedimentary Rock

As the name implies, sedimentary rock results from the settlement of remnants of older rock through water. Surface rock weathers and disintegrates in the presence of air and water. Most of the rock is broken down to a small size that is picked up in creeks and moved downstream to larger rivers and, often, on to the ocean. The larger pieces are rolled along the bottom of the streams, where their edges are rounded, while the finer fraction is carried in suspension. Some minerals are dissolved completely, however, and transported in solution. All of these remnants settle in the water to form distinct lenses known as *strata.* The contacts of the individual strata are known as *bedding.* Although the minerals of most sedimentary rock are moved and deposited by water, they can also be transported by wind or ice.

As the depth of the deposits increases, some of the water is squeezed out of the deeper sediments by the weight of the overlying minerals and they become bound by cementation into rock. Mixed sands and gravels become conglomerates, clean sands become sandstone, silts and clays form shale, and the calcareous content precipitates to limestone. Sedimentary rock often contains animal fossils or remnants of plant life common to the time of its deposition. These allow identification of the relative age of the formation.

Although shales are generally of low permeability, sandstone and conglomerates can be highly permeable and limestone can contain very large voids and solution cavities. And, of course, all of these rock formations are subject to faulting and folding, which result in joints and fractures that will facilitate the free movement of water. As a general rule, however, shales are less brittle than sandstone and thus not as likely to be highly divided. Some sandstone is weakly cemented, which can result in a propensity for internal erosion. Furthermore, where the cementation is weak, the walls of defects can slough so as to become filled with the loose remnants, which are also highly prone to internal erosion.

21.2.3 Metamorphic Rock

Metamorphic rock is of secondary origin, derived from other preexisting rock by mineralogical, chemical, and/or structural changes, essentially in the solid state. These changes occur in response to marked changes of temperature, pressure, shearing stress, and chemical environment, generally at depth in the earth's crust. Rocks of this type are thus considered as constituting a separate descriptive group. Formed by varying degrees of temperature and pressure, recrystallization of existing minerals to form new minerals, and changes over long periods of time, metamorphic rock is typically very dense and hard. This causes it to fracture easily when deformed or uplifted to the surface, which usually results in clean, open joints that are easily grouted.

Because it is dense and hard, metamorphic rock tends to be quite resistant to drilling, resulting in slow progress and rapid wear of the drilling tools. Drilling progress is typically much slower in metamorphic rock than in penetrating the original rocks from which it is derived. Metamorphic rock tends to be visibly crystalline and usually has well-defined cleavage or foliation, so that breaks tend to exhibit a planar structure caused by a flattening of the grains of the parent rock, resulting in flat-sided slabs. Common metamorphic rocks include gneiss, quartzite, schist, slate, talc, marble, and serpentinite.

21.3 GEOLOGIC STRUCTURE

The geological structure is indicative of the tectonic and formational/depositional origin of the rock. Through observation of this structure, along with the depositional history, a geologist can attain a good idea of the subsurface particulars of a given site. Such knowledge is obviously requisite to optimal planning of a grouting program.

21.3.1 Primary Features

Although certain features are common to a given rock type, there are several criteria that are common to all rock and are thus used to classify individual specimens. Of these the formational/depositional history is obviously paramount. Other characteristics that are significant for all rock types include the texture and hardness.

21.3.1.1 TEXTURE

The texture of rock is defined by the size and arrangement of the particles of which it is composed. For most rock, the particulars of the individual particles are discernable by visual observation. The texture is defined by grain size, as follows:

Very coarse	Greater than 2.0 mm
Coarse	0.6 to 2.0 mm
Medium	0.2 to 0.6 mm
Fine	0.06 to 0.2 mm

When the grains are all of about the same size, the texture is further defined as *well sorted*, whereas *poorly sorted* indicates that the grain size is variable. For very fine grained rock in which the individual grains are to small to be visible, the texture is described by the feel and luster.

21.3.1.2 HARDNESS

The hardness of a specimen is determined by scratching it with a knife and/or striking a handheld specimen with the blunt face of a geologist's hammer. Standard definitions are as follows:

Very soft	Peels with a knife, crumbles under firm hammer blows
Soft	Scratches with a knife, breaks easily with a firm hammer blow
Medium hard	Resists a knife scratch, breaks with firm hammer blows
Hard	Requires repeated hammer blows to break
Very hard	Resists breakage with hammer blows

21.3.2 Secondary Features

The earth is not static, but is rather in an endless state of transformation. This results in continual movement and development within the near surface layers. Warping, tilting, uplift, and depression of the surface rock occurs continuously. Such movement usually develops very slowly over the long time periods of geologic history, so that the changes are not noticeable. Movements can occur suddenly, however, as in earthquakes and landslides. Secondary features are those aspects of the rock fabric that have resulted from this continuous development. These include folds, joints, and faults, which can be open or infilled with other materials. Both the strength and permeability of a particular rock formation are directly related to the character of these secondary geologic features. An understanding of such features is thus fundamental to the design and prosecution of an optimal grouting program.

21.3.2.1 FOLDS

Folds are undulations of the rock in the earth's crust, which have formed over very long periods of time as a result of tectonic activity. Individual folds can be in virtually any attitude or configuration, and they may be compact or extend for several miles. The most extreme forms of folds are *synclines* and *anticlines.* A syncline is a fold that is upwardly concave, such as in a valley, as shown at the left in Figure 21.1. Downwardly concave bedding, such as that appearing at the right in Figure 21.1, wherein the beds or layers dip down in opposing directions, are known as anticlines. Folds can occur over very short distances, resulting in sharp inclinations, or they can be very gradual with only slight tilting. In extreme cases they can be overturned, as illustrated in Figure 21.2.

Folds can result in considerable fracturing of the rock in a parallel orientation to their axis. Such parallel fractures also encourage the development of ample defects in a perpendicular direction. This is one of the formative criteria for the routing of streambeds in which erosion of the rock is greater in such areas of structural weakness. Consequently, it is generally not a good idea to found a dam or tunnel in a parallel orientation along the axis of such features.

FIGURE 21.1 Upward-treading strata form a syncline (left) while downward-treading strata form an anticline (right).

21.3.2.2 JOINTS

Joints are fractures or partings in rock, across which there has been little or no displacement. They often appear in *sets,* in which groups of several will trend in a given direction and a nearly parallel configuration, when viewed in plan. Two

FIGURE 21.2 Synclines and anticlines can be so severe as to overturn.

or more joint sets sometimes occur, usually at approximate right angles to each other. Such groupings are known as *joint systems.* Most joints are closed so that little or no gap exists between the two joint planes. In others, however, the joint planes are distinctly separated so as to form definite fissures. In some cases these fissures can become infilled with fine-grained material, usually deposited from the movement of water throughout the joint system.

Joints can be initiated by tension, compression, shear, or torsion, although those caused by tensile stress are the most common. This is, no doubt, due to the fact that rock is weakest when under tension. Particularly prominent are such joints along the crests of anticlines where high tensile stresses result from the folding of the formation. However, they can also result from compressional forces that accompany folding, especially in highly stratified sedimentary deposits. Near-surface shale deposits are often closely jointed and of high permeability. These are commonly quite weak, however, and not able to withstand the pressure of grouting. Surface leaking is a particular problem in such cases. Thus, the grouting may best be performed from a gallery after the formation has been sufficiently restrained by prior construction, or the upper shale excavated and removed, following grouting of the lower deposits. The provision of a positive cutoff such as a slurry wall may be a better method where such geologic conditions occur.

Joints are common in igneous rock as a result of tensile stresses generated by contraction during cooling of the molten rock. Both the pattern and extent of such cracking vary according to the size and shape of the igneous mass, the rate of cooling, and the manner of deposition. Intrusive masses tend to fracture into large blocks or prisms, whereas the more finely grained masses such as thin sills tend to develop closely spaced joints. An intriguing and somewhat unusual structural development results in the formation of long, hexagonal, near vertical columns of rock. Although not often encountered, these can be especially challenging to grout.

In granite and other coarse-grained igneous rock that is highly stressed, joints can be caused by large compressional forces. These are commonly at more or less right angles to each other, so as to divide the formation into roughly cubical blocks. Granite is often massive, with the joints widely spaced, especially at depth. It is particularly susceptible to development of *exfoliation* or *slabbing* joints, however, which are the result of stress relief, generally parallel to the exposed rock face. These develop due to load reduction, either through natural processes such as erosion or by man-made excavation. This is a particular problem in the case of steep-faced canyons, which is readily recognized by the existence of open joints parallel to the face, as illustrated in Figure 21.3.

Where such jointing occurs, the spacing generally increases with depth until a competent *bottom* is reached. Because the surface rock layers, which can be either thin or quite massive, tend to be displaced outward by even small forces, grout injection in such rock requires extreme care and a generally low grout pressure is in order. Furthermore, continual monitoring of

FIGURE 21.3 Open joints parallel to a steep face are indicative of overstress foliation of the rock.

the pressure behavior is needed, and injection should be terminated with any sudden pressure loss, which indicates the likely occurrence of groutjacking. In extreme cases, rock bolting and/or anchoring may be beneficial.

21.3.2.3 FAULTS

Faults are joints in which one side of the fracture surface has been displaced relative to the other so as to form an offset. The amount of offset can range from only a few millimeters to several tens of feet and can be in any direction. This movement, referred to as *slip*, may have occurred gradually, over a long time interval, or suddenly, as in an earthquake. When the direction of slip is other than horizontal, the ground surface is displaced so that *tilting* occurs. This, of course, also changes the surface elevations and contours and can drastically change the drainage patterns.

Faults can occur as discrete fractures, but more often involve a number of adjacent breaks so as to form a *fault zone*. Some faults have been subjected to so much stress and slip that zones of ground, crushed, or broken rock several feet thick have been produced. This debris is known as *fault breccia* unless it has been very finely ground, in which case it is referred to as *gouge*. Penetration of such zones during drilling can be problematic, as they are likely to cave and wedge-shaped pieces of hard rock can tend to lock the drill bit in place.

Faults can be open, but they are more often tightly closed. Because they occur in rock that is frequently highly compressed, the faces of the fracture are often slick and they can contain scratches or grooves in the direction of the movement. This condition is referred to as *slickensides*, a primary tag that defines a fissure as a fault. Individual faults sometimes contain thin seams of clay gouge, and faults that were once open may be filled with a variety of minerals transported and deposited by water over time.

Existing faults are most recognizable in sedimentary rocks as offsets of the bedding planes and are easily recognized. Faults occur in all rock types, however. Many extend through the ground surface layer where near vertical surface differentials known as *scarps* (Figure 21.4) are easily discerned. The discontinuities commonly associated with faults can often be observed in road cuts or in test trenches. The age of the movements that created the fault can usually be determined by careful examination of the offsets and any fossils or plant remnants that exist in the various strata.

FIGURE 21.4 Near vertical offset at surface of fault is known as a scarp.

21.3.3 Secondary Mineralization

The forces of nature are constantly acting on all surface rock and soil. Groundwater is constantly moving within the soil and underlying rock. Although the results of this action are often not perceptible in the rather short time periods of observation, significant change occurs over geological time. Just as the structure of rock changes with time, so do the mineral composition and the physical properties. These factors can have a dramatic influence on the need for grouting as well as the groutability of a given formation.

21.3.3.1 SOLUTIONING

Many rock species are soluble. Limestone, dolomite, some gypsum, and calcareous sandstone dissolve quite slowly, so that significant losses are not obvious over short time periods. Over long intervals of geologic time, however, considerable volumes of these rocks can dissolve and be lost. Such dissolution is capable of creating cavernous voids and, indeed, this is the mechanism by which caves are formed. Small passages in the soluble rock become enlarged through continued dissolution until they develop into cavernous voids. At the rock surface, the *throat* of the void enlarges until it eventually becomes so great that the overlying soil is no longer capable of arching over it. At this point the soil collapses, creating a sinkhole. Herein is the genesis of the sinkholes that are so common in certain geographic areas, most often overlying limestone, and that grouters are often called upon to remediate.

Other rock is subject to more rapid dissolution and can loose considerable mass over relatively short periods of time. Anhydrite, some gypsum, and salt, for example, are highly soluble. These minerals are often contained in sedimentary deposits in which the joints are quite tight and thus resistant to the injection of grout. With only small amounts of exposure to water, however, they can be dissolved over fairly short time periods. Even very fine fissures that are apparently too small to initially accept grout can, with time, become enlarged through dissolution to the point that they transmit major quantities of water. Dissolution of gypsiferous rock and soil has led to piping distress and, in some cases, complete failure of a number of water-retaining embankments.

21.3.3.2 WEATHERING

Rock, as well as soil, is constantly being attacked by the forces of nature. Physical factors such as wetting, drying, freezing and thawing, temperature changes, and erosion by both water and wind cause disintegration of that which is exposed or near the surface. Chemical processes break the cementation and otherwise alter the properties of rock. Together, these mechanisms can literally change the structure and character of rock. Fissures can be enlarged or welded shut, cementation can be added or dissolved, and many rock species can be decomposed to friable soils.

Water that has penetrated the joints and intricacies of rock can dissolve or alter the minerals. If the water freezes, it expands with great force, which causes the rock to fracture, and when such action takes place in a closely jointed rock, complete disintegration occurs. The chemical and physical actions of weathering usually progress simultaneously, so that the deteriorating effect is synergistic. Together, they break the rock down into small pieces, in essence preparing it for the subsequent erosion that will allow it to be moved to new locations. And as this occurs, new surfaces will be left for the process to recur. The degree of weathering has a considerable impact on the need for grouting and, indeed, on the particular grout that might be applicable to inject, if any. Standard terminology has been adopted to describe the degree of weathering, as follows:

- *Unweathered.* No physical or chemical alteration of the rock visible, with only slight staining of the surfaces; few, if any, fractures
- *Medium weathered.* Partial alteration of minerals; staining extends into rock; slight softening of the surface, with some cracks and fractures
- *Highly weathered.* Chemical alteration and discoloration throughout the rock; soft surface and many fractures; rock texture only partly preserved

When rain, hail, and/or wind beat on the weaker portions of the rock surface, disintegration occurs. Much of the disintegrated material is picked up by either wind or water and is transported to a new location. Wind can carry finer particles and is responsible for the formation of dune sand. However, water may have leached minerals out of the underlying joints, which would then be open and require grout injection. The history and nature of a formation can often be determined by remnant minerals, such as iron stains, that may be left behind after leaching.

21.3.3.3 INFILLING

All of this activity can carry the eroded material into existing joints and fractures, both on the surface and at greater depths. Near surface joints and fractures can be infilled with fine sand, silt, or clay, and these can contain organics. Deeper defects are filled primarily with minerals carried by water, which is usually in the form of clay. Grouting is, of course, not possible in such filled defects. Although sometimes attempted, removal of the deleterious material from the crevices is extremely difficult and often impossible.

21.4 THE ROCK CONFIGURATION/GEOMETRY

The constant folding and warping on the outer shell of the earth creates a virtually infinite variety of orientations of the bedding and jointing features. Satisfactory filling of these defects with grout requires not only that they be located, but also that their inclination or attitude is known. With this knowledge, grout holes can be oriented so as to optimally intersect the individual features. Standard methods for measuring and definition of the attitude of geomaterials are well established.

21.4.1 Bedding

As previously mentioned, sedimentary rock is formed with layers of materials that are deposited one at a time. The individual layers typically consist of materials of quite different grain size and, in some cases, different origin. Solids settle in water preferentially; coarse grains settle rapidly, but the settlement rate becomes lower with a progressive reduction in particle size. The very fine fraction, which is clay, is held in suspension for extended time periods. Furthermore, the velocity of the water or wind that deposits such material varies, so the deposition rate is not uniform. Therefore, the individual layers of sedimentary deposits, which are known as *strata,* are commonly of different thickness. When the thickness is less than 4 in. (100 mm), the deposit is said to be *thinly bedded,* and deposits of greater thickness are said to be *massive.*

Similarly, the grain size, shape, and texture are ever-variable. When composed of clay and other fine-grained material, the strata are usually quite thin and form shale. Sandstone, which is composed of the medium-sized grains, can be either massive or thinly deposited. In either case, the different strata will have a distinct interface, known as the *bedding plane,* and the composite feature is referred to simply as *bedding.*

Although sedimentary rock is formed in a generally level configuration, tectonic activity will result in its folding, as previously discussed, so bedding planes can exist in virtually any orientation. Depending on their origin and history, sandstone and shale can be either soft and weak, or strong and brittle. Likewise, the bond of their bedding planes can vary from very strong to extremely weak. It is not unusual to find interbedded deposits of shale and sandstone. Because the bedding planes were subjected to the weight of the overlying materials when formed, they are usually quite tight and resistant to grout injection, notwithstanding the fact that they usually represent a weakness in the formation.

21.4.2 Direction and Attitude of the Strata

To clearly understand the slope or attitude of a formation and the positioning of its joints, both the inclination of the stratum and the direction in which it is heading must be considered. These values are defined as the *strike* and *dip,* respectively, and together define the attitude of a stratum relative to a level plane. Although strictly speaking, earth's surface is curved and usually not level, for purposes of delineating the rock structure, it is considered to be a level plane.

21.4.2.1 STRIKE

The strike of a formation is defined as the direction or trend of a bedding or fault plane as it intersects the horizontal. It is always expressed as a compass bearing and is descriptive of the *direction* or *azimuth* of the intersecting bedding planes. It will always be at an angle of about 90 degrees from the direction of the maximum slope of the bedding or a fault.

21.4.2.2 DIP

Dip is the maximum inclination of the bedding planes of a formation from a level plane or *strike line*. It is given as an *angle*, in degrees. Obviously, if the strata are horizontal and level, they will have zero dip, and the dip cannot exceed an angle of 90 degrees.

21.4.3 Outcroppings and Overhangs

The surface of bedrock is not always uniformly folded or left in smooth planes that are easy to characterize. Furthermore, igneous and metamorphic rocks are usually not formed in layers. Consequently, many rock surfaces are neither flat nor uniformly oriented. And as the rock disintegrates to form soil, some areas are invariably more resistant than others, so that they remain intact and project, or even overhang the predominant slope. These are referred to as *outcroppings*, as illustrated in Figure 21.5, or, as the case may be, *overhangs*. Any surface water will tend to run in the low areas as the wear progresses, cutting them even deeper by erosion. These actions can have a significant influence on the structure and character of the rock and any residual soil that is created.

FIGURE 21.5 Outcroppings of a rock form a very irregular slope.

21.5 STRESS STATE

Understanding the stress history of rock is important in grouting, because near surface rock that is overstressed can easily be jacked and displaced during injection. And because granular rock is typically created at great depth, when found near the surface it is often overstressed as a result of the much greater overburden pressure it withstood in its earlier life. Thus, great care must be taken during grouting of such rock, and especially any defects that approach a parallel treading to the ground surface.

On the other hand, rock that remains at depth is always subject to the weight of the overlying formation. When tunnels or other underground works are excavated, they tend to provide stress relief for the surrounding rock. Should there be faults or joints that run parallel to the direction of the excavation, entire blocks of the rock can be displaced. As stress in the near rock is relieved, yielding of that more distant can occur. It is therefore crucial to have a good understanding of the rock structure when making such excavations, and especially when injecting grout in near proximity to them.

In one instance, a large-diameter tunnel was being excavated through a moderately jointed rock. Grouting was used to preserve some overlying water bodies, and the amount of seepage into the excavation was severely restricted. Without regard for the stress relief potential of the excavation, or the frequency or orientation of the jointing, pumping pressures were set on the basis of the overburden mass. During grouting in one particular section, a massive block of rock was suddenly displaced into the excavation. Simultaneously, a significant increase in water infiltration occurred, presumably due to a loosening of the joint structure that had previously been restrained by the displaced block. This mishap was directly attributable to a lack of appreciation of the influence of the joint structure and the effects of its stress relief, which had resulted in use of an excessively high grout pressure.

21.6 DERIVATION OF SOIL

The forces that wear on rock, breaking it down so as to form soil, have been mentioned earlier. The particulars of the soil that is produced are dependent on the original makeup of the rock and the particular mode of deposition. Soils that are formed in place through disintegration of the rock lying immediately under them are known as *residual* soils. Such deposits can be readily identified, as they typically graduate from topsoil on the surface to increasingly coarser texture with depth until they finally reach the level of the deteriorating rock. By far, however, the greatest amount of soil has been deposited remotely from the original rock of which it is derived, as the result of some mode of transport.

In addition to the disintegration of rock into granular-size debris, much larger pieces can be produced by exfoliation and *rock fall.* Talus, which is made up of irregularly shaped fragments of rock, is the result of pieces falling from receding cliffs, due to various forms of weathering. The individual chunks can be either soft or brittle, depending on the properties of the parent rock. Over time, talus will weather to a progressively finer state, with that which is weaker and/or softer deteriorating most rapidly.

21.6.1 Transport Mechanisms

Soil constituents can be moved by any of several basic mechanisms. Finer-grained sand, silt, and clay particles can be picked up by wind and carried long distances. Water is by far the most active transporter of soil, however, and it can move virtually all manner of broken-down rock, up to and including cobbles and even large boulders. Streams start at the higher elevations of a terrain. Initially, they are small and carry only the finer rock fragments. The individual streams join to form larger channels as they descend and are capable of carrying ever-larger pieces of debris. Finer particles are often carried in the water, and their original angular shape may be somewhat preserved.

As erosion eats away at their support, larger chunks fall into the streams. These larger fragments, along with gravel and the larger sand fraction, are impelled along the bottom. In the process, they are further disintegrated by the grinding action of their movement. This causes the corners to become rounded, and surfaces may become somewhat polished. The channels branch into succeedingly larger streams until they are capable of carrying very large amounts of solids. Although some of the solids may drop out to form new riverbeds or valleys, much will continue the journey to eventually be deposited in the ocean.

In some instances, the loose debris on steep slopes becomes so saturated that it will be liquefied and move downslope in a mass, known as a *mudflow.* Similar material not saturated, but propelled downslope by gravity or sheet flow, forms a *colluvial* deposit. Chunks of rock that ravel off steep faces can gather at the base of the face or slope and are known as *talus.* Glaciers can pick up and carry large amounts of solids, which fall out as the ice melts, to form *glacial till* or *moraine*. Large masses of rock and/or soil covering a wide area can creep slowly downslope. And, of course, there are landslides as well, where large masses of geomaterial fall suddenly. Many of these mechanisms work together, making it difficult to understand clearly all the travel a material has undergone.

21.6.2 Depositional Mechanisms

Regardless of the method of transport geomaterials may experience, they reach their final location by one of a number of depositional mechanisms. The exact method indicates much about the structure and performance of the resulting soil, including its groutability. Soils made from wind-carried materials are known as *aeolian* deposits. These can be uniformly graded or of a well-sorted structure, depending on the exact mechanics of their makeup. Those resulting from water transport are often sorted and composed of layers of different gradation and physical makeup. When movement is downslope by gravity, however, the different sized grains usually remain intermixed.

21.6.2.1 AEOLIAN DEPOSITS

Soils carried by wind are deposited in one of two general ways, either as sands deposited into dunes, or finer particles settling as dust. Sand dunes are formed when wind-blown sand encounters an obstacle, such as rocks or vegetation, protruding from the ground surface. Initially, the sand piles up against the obstruction, and, once started, the dune will grow ever larger as long as the wind continues and a sufficient supply of sand is available. Dunes are of various sizes and shapes, depending on the strength and direction of the winds.

Once formed, sand dunes are not stationary, but ever-moving as long as sufficient wind continues to blow. Sand from the windward face of the dune is constantly picked up and deposited on the back side, where it drops and comes to rest at the angle of repose for the particular sand. The slope of the windward face is much gentler, usually no more than 10 or 15 degrees from horizontal, and varies according to the power of the wind. Because of the differences in wind speed, the sizes of the grains picked up and released vary, so that they are deposited on the back side in layers according to their size. This manner of deposition and the additional differences of wind direction result in variation of both the direction and inclination of these micro-layers, which will form a complex labyrinth of inclined and cross bedded strata on the interior of the dune.

Smaller particles can be carried by wind in the form of dust. These are typically in the size range of silt and clay and are known as *loess*. Loess can be deposited in layers varying from nearly zero in thickness to more than 100 ft (30 m). The layers have the unique property of being able to stand in virtually vertical bluffs even though the constituents are not cemented and are completely friable. The particle size of loess is so small and the structure so tight, that injection of even the most penetrable grouts is not practicable.

21.6.2.2 WATER-DEPOSITED SOILS

A reduction in the velocity of moving water can result in settlement of at least some of the solids it contains. Even a slight reduction can cause some suspended material to settle, as in a stream bed. And, of course, if a rapidly moving stream of water on a considerable gradient suddenly slows down or comes to rest, as in a wide, flat valley or lake, virtually all of the suspended matter will be released. The larger, denser particles will be the first to separate, and these will be followed by those of progressively smaller sizes. The clay mineral that can remain in suspension for a long time will be the last to separate. These are referred to as *alluvial* deposits, and are all subject to stratification because of the different settlement rates of the variously sized constituents. The deposits may have been built up from many episodes, or they may be the result of a single event. The number of events that led to the formation of a particular deposit can be determined by evaluating the bedding and, in particular, the number of clay layers.

There are different forms of masses derived from alluvial deposition. The most common are defined as *alluvial fans, plains, deltas,* and *terrace deposits.* Alluvial fans occur when the relatively rapid movement of a stream is drastically slowed, as by entering a flat valley where the water is able to spread out rapidly over a wide, flat area. As this occurs, settlement of the solids accelerates because of the greatly reduced velocity of the water. Such deposits normally radiate from the confluence of the stream, covering a large area. Alluvial plains are composed of two or more fans from different streams that have joined and overlapped so as to form a continuous expanse. Deltas are formed in much the same manner, except that the deposition is into the ocean, a lake, or other body of water. This results in a dilution of the solids content of the original stream, which can increase the rate of sedimentation, but the mechanism is the same.

Terrace deposits are formed by changes in the flow of streams that have deposited alluvial material over a long period of time. Most streams leave at least some portion of the solids they carry along the way, especially upon encountering relatively flat terrain or a change of course. Over time, fairly wide expanses of such sediments can be built up. With a change in the flow, as by reduction or an increase in the velocity that causes a portion of the stream bed to be eroded to a lower flow level, these terrace deposits will be left behind. Terrace deposits are subject to the same sorting of grain size as other alluvial deposits, except that the clay lenses may be absent. This is because they have often been made up of sediments from moving water, and the very fine clay particles usually do not settle out of water that is moving. Moreover, because the flow behavior of a stream is ever-variable, its sediments usually have greater variation than is found in fans, plains, and deltas.

Soils that slide directly from higher elevations, such as colluvial and mudflow deposits, are generally unconsolidated and often require improvement. Over periods of geological history these can form deep natural fills. They start, however, as loose material on steep slopes. Very often bedrock crops out, and the broken-down material, which likely includes organics from vegetation, will fill in around such outcroppings. If any of the bedrock overhangs or contains cavernous voids, as shown in Figure 21.6, very loose conditions or even outright voiding can occur in and beneath them, as the new soil particles are unable to fill tightly around such impediments as they are deposited. Such conditions are very difficult to determine by normal geotechnical test borings, as they are sporadically located. These deposits are often improved by compaction grouting, but their features can result in very unpredictable and sometimes substantial grout takes.

Soil materials can also be transported in the ice that forms glaciers. When the ice melts, the

FIGURE 21.6 Rock overhangs can result in poor quality filling and even subsurface voids.

solids, which can be substantial, will settle. These will form either of two compositions, *glacial till* or *moraine.* Till is unique in that it is heterogeneous and completely free of stratification. The individual components, whether large boulders, fine sands, or even clay, are deposited immediately upon dropping out of the melting ice. Also distinctive to till is the shape of the gravel and larger-sized pieces. These may be angular and in the same basic condition as when broken and encapsulated in the advancing ice. Larger pieces that may have been dragged along on the underside of the mass can be somewhat rounded and nearly always contain distinctive surface striations.

Moraines are formed much like till, but they develop in a very haphazard manner. This is due to their further transportation by water after dropping out of the melting ice, so that there is no consistency in either composition or uniformity. The surfaces of moraines are often littered with pits and mounds and completely void of any uniformity or order. Sometimes they appear primarily as rock debris without the finer fractions. These may have come from glaciers that originally contained little or no fine material, or where the deposits were so thin that the fines washed out.

21.7 CLASSIFICATION OF SOIL

Soils are classified primarily according to their grain size and texture. Sand and gravels are considered to be *coarse-grained* material, and silts and clays are classified as *fine-grained.* Whereas the individual grains of sands and silts are of chunk-like geometry, clays are noteworthy in their flat flake-like configuration. Granular soils will have a continuous pore structure that is normally filled with either water or air. The individual grains are generally in contact with one another, so the structure will not change with variation of the water content and the pore system will allow permeation or movement of water. The soil can therefore be intruded with thin grout, solidifying the individual grains through *permeation grouting.*

Clay, however, is composed of microscopic clay platelets that form a lamellar structure. Because of their shape and small size, any water that encompasses these platelets becomes part of the structure. Thus, some clay soils can exhibit a large change in volume with changes of the moisture content. Clay platelets are said to be *colloidal,* that is, they are so small that they will not settle very rapidly in water and can, in fact, require many months to so do. Clays are thus *cohesive* soils, whereas those of a granular structure are considered *noncohesive.*

In reality, most soils are mixed and contain a combination of different sized grains and soil types. The level of cohesiveness of mixed soils therefore varies and is expressed as the degree of *plasticity,* which is a function of their water content. Soils such as clean sands are noncohesive and are further described as *nonplastic,* whereas those that contain clay are considered to be *plastic.* The volume and condition of the pore structure of soil thus varies greatly, as does the degree of plasticity for those soils that contain clay. Granular noncohesive soils can be *permeation* grouted, but plastic soils usually do not accept grout, and thus improvement of these soils is limited to *fracture* grouting, or chemical change.

21.7.1 Unified Soil Classification System

In order to define the varieties of soil textures, the *Unified Soil Classification System* was developed. It was first devised by Arthur Casagrande at Harvard University and was formally adopted by both the U.S. Bureau of Reclamation and the Army Corps of Engineers in 1952. The system is now used almost universally. It divides soil into two broad groups, *coarse-grained* and *fine-*

TABLE 21.2 Size and Nomenclature for Coarse-Grained Soils

	Classification	Size
Rock	Boulders	12 in. (0.3 m) and larger
	Large Cobbles	6 in. (150 mm) to 12 in. (0.3 m)
Gravel	Small Cobbles	3 in. (75 mm) to 6 in. (150 mm)
	Coarse	3 in. (75 mm) to 3/4 in. (19 mm)
	Fine	3/4 in. (19 mm) to No. 4 sieve (4.75 mm)
Sand	Coarse	No. 4 sieve (4.75 mm) to No. 10 sieve (2.0 mm)
	Medium	No. 10 sieve (2.0 mm) to No. 40 sieve (425 microns)
	Fine	No. 40 sieve (425 microns) to No. 200 sieve (75 microns)

grained. The coarse-grained fraction is further divided as shown in Table 21.2.

All material finer than a No. 200 (75 micron) sieve is considered as *fines.* Although the size of these materials is not further defined in the Unified System, older systems have placed the upper limits of clay at somewhere between 2 and 5 microns, with everything larger being considered silt. In reality, only the clay that is colloidal adds significant plasticity, which is generally undesirable. This is usually in the fraction less than 3 microns in size, although such measurements are difficult to make in ordinary laboratories and there is a considerable difference of opinion as to the exact size. The influence of the clay content of a soil is thus usually defined by the plasticity and is determined by the *plastic limit,* which is discussed shortly.

The Unified Soil Classification System defines the grain size groupings and uniformity by letter designations: *G* for gravel, *S* for sand, *M* for silt, and *C* for clay. To further describe the uniformity of the various components, the additional designations *W* and *P* are used, which mean *well sorted* and *poorly sorted,* respectively. To further define the condition of clays, the letters *H* and *O* are used to designate *high plasticity* and *organic content,* respectively. Although not a complete presentation, the more salient delineations of the Unified Soil Classification System are provided in Table 21.3.

21.7.2 Fine-Grained Soil Consistency

Fine-grained soils and especially clay cannot be sieved, so their properties are established by their consistency at different moisture contents. In the case of clay, the consistency will very greatly as a function of the amount of moisture it holds. When very wet, it will appear as a viscous fluid. Should it be very dry, it will have properties of an unyielding solid. In between these moisture extremes, it will be variably moldable. To define the moisture contents where these different consistencies are reached, simple tests known as the *Atterberg limits tests* have been established.

21.7.2.1 ATTERBERG LIMITS

Four arbitrary *states* in which a plastic soil may exist are established by the standard Atterberg limits tests: *liquid, plastic, semisolid,* and *solid.* The tests determine the amount of water, as a percentage of the weight of the particular soil, that is needed to reach each state. The values of these water contents are known as the *liquid limit, plastic limit,* and *shrinkage limit.* The tests involve only that portion of the soil that passes

TABLE 21.3 Unified Soil Classification System

Symbol	Typical Names	Field Identification Procedures
GW	Well-graded gravels, gravel-sand mixtures, little or no fines	Wide range of grain sizes and substantial amounts of all intermediate particle sizes
GP	Poorly graded gravels, gravel-sand mixtures, little or no fines	Predominantly one size or a range of sizes with some intermediate sizes missing
GM	Silty gravels, gravel-sand-silt mixtures	Nonplastic fines or fines with low plasticity
GC	Clayey gravels, gravel-sand-clay mixtures	Plastic fines (for identification procedures, see CL below)
SW	Well-graded sands, gravelly sands, little or no fines	Wide range of grain size and substantial amounts of all intermediate particle sizes.
SP	Poorly graded sands, gravelly sands, little or no fines	Predominantly one size or a range of sizes with some intermediate sizes missing
SM	Silty sands, sand-silt mixtures	Nonplastic fines or fines with low plasticity
SC	Clayey sands, sand-clay mixtures	Plastic fines

		IDENTIFICATION PROCEDURES ON FRACTION SMALLER THAN NO. 40 SIEVE SIZE		
Symbol		**Dry Strength (Crushing Characteristics)**	**Dilatancy (Reaction to Shaking)**	**Toughness (Consistency Near PL)**
ML	Inorganic silts and very fine sands, rock flour, silty or clayey fine sands, or clayey silts with slight plasticity	None to slight	Quick to slow	None
CL	Inorganic clays of low to medium plasticity, gravelly clays, sandy clays, silty clays, lean clays	Medium to high	None to very slow	Medium
OL	Organic silts and organic silty clays of low plasticity	Slight to medium	Slow	Slight
MH	Inorganic silts, micacious or diatomaceous fine sandy or silty soils, elastic silts	Slight to medium	Slow to none	Slight to medium
CH	Inorganic clays of high plasticity, fat clays	High to very high	None	None
OH	Organic clays of medium to high plasticity, organic silts	Medium to high	None to very slow	Slight to medium
Pt	Peat and other highly organic soils	Readily identified by color, odor, spongy feel, and, frequently, by fibrous texture		

a No. 40 sieve (425 micron), so it is illustrative of the behavior of the finer portion of the soil only.

When saturated, clay behaves like a liquid. As it dries out, however, a point will be reached where it begins to exhibit resistance to shear. This point is defined as the liquid limit, at which the soil begins to act as a moldable plastic. Upon further drying, it will reach a condition where it is no longer plastic but becomes somewhat brittle. This is described as the plastic limit. And with continued drying, the soil will be marginally moldable and will crack as it is worked, at which point it is in the semisolid state. When the moisture has been reduced so that the soil will no longer shrink, it is said to be at the shrinkage limit.

To establish the liquid limit, soil specimens at various moisture contents are placed in the brass cup of a special testing device. A V-shaped groove is placed on the top of the specimen and it is repeatedly dropped, so as to apply equal impacts, until the specimen liquefies, as indicated by the closing of the groove. The liquid limit is that level of moisture at which complete closure of the groove occurs with a prescribed number of impacts. The plastic limit is reached when 1/8 in. (3 mm) thick threads of the soil cannot be rolled out on a glass plate without cracking.

Values derived from these tests can be used to calculate index properties of a particular soil. The *plasticity index* (PI) is established by subtracting the plastic limit from the liquid limit. This property is often used in the specification for aggregate for compaction grouting in which a strict limitation of the clay content is required. The *shrinkage index* (SL) is determined by subtracting the plastic limit from the shrinkage limit.

21.7.3 Field Identification and Description

The materials and conditions found in all test borings should be continuously logged. Field logs should include the Unified Soil Classification System symbols for the various materials encountered, as well as data on the field consistency and strength. In addition, the number of blows of the standard sampling hammer per foot of specimen should be noted. This is known as the Standard Penetration Test (SPT), and the number of blows per foot are defined as the *N* value. The moisture content of the sample should also be provided, as well as the depth to groundwater, if encountered. Because the samples of fine-grained materials are subject to wide variations due to both the condition of the constituent material and moisture content, standard procedures have been developed for both their identification and notation. Table 21.4 provides guidance for such field identification and notation.

TABLE 21.4 Field Identification of Fine-Grained Soils

Consistency	Identification Procedure	SPT—Blows/ft
Very soft	Loses shape under its own weight	< 2
Soft	Easily penetrated several inches by thumb	2–4
Firm	Penetrated several inches by thumb with moderate effort	4–8
Stiff	Readily indented by thumb, but penetrated only with great effort	8–15
Very stiff	Readily indented by thumbnail	15–30
Hard	Indented with difficulty by thumbnail	> 30

21.8 GEOTECHNICAL CONSIDERATIONS

Detailed discussion of soil and rock properties is beyond the scope of this book, but it is important for the geotechnical grouter to have a good understanding of the principles by which these materials are understood. Typically, a geologist will always be involved when the grouting is into rock. More often than not, grouting in soil is under the purview of the geotechnical engineer. Yet because many soil problems are the result of geological processes, the involvement of a geologist can be of great benefit. Finally, engineers and geologists need to appreciate the knowledge that each discipline brings to a given problem. Their background and abilities are complimentary, not competitive. Any competition imagined is purely perceived and nonproductive. Each has an important influence on the optimal success of most geotechnical grouting projects.

CHAPTER 22

Investigations for Grouting

22.1 PRELIMINARY CONSIDERATIONS
- **22.1.1** What Are the Objectives?
- **22.1.2** Information Needed
- **22.1.3** Data Review
  - **22.1.3.1** Prior Investigation Reports
  - **22.1.3.2** Geologic Maps and Public Records
  - **22.1.3.3** Original Plans and Construction Records
  - **22.1.3.4** Historical Aerial Photos
- **22.1.4** Site Access/Restrictions
  - **22.1.4.1** Access Limitations
  - **22.1.4.2** Staging Area
  - **22.1.4.3** Coordination with Other Activities
  - **22.1.4.4** Hours of Work
  - **22.1.4.5** Environmental Requirements
- **22.1.5** Existing Improvements
  - **22.1.5.1** Protection Requirements
  - **22.1.5.2** Substructure

22.2 SOIL GROUTING
- **22.2.1** Soil Conditions
  - **22.2.1.1** Test Borings
  - **22.2.1.2** Exploratory Excavations
  - **22.2.1.3** Sampling
  - **22.2.1.4** Laboratory Tests
  - **22.2.1.5** In Situ Evaluation
- **22.2.2** Settlement Repair
  - **22.2.2.1** Existing Elevations
  - **22.2.2.2** Structural Damage
  - **22.2.2.3** Jacking/Leveling Requirements
  - **22.2.2.4** Utilities

22.3 ROCK GROUTING

22.4 GROUTING IN STRUCTURES

22.1 PRELIMINARY CONSIDERATIONS

As mentioned throughout this book, grouting should be planned just like any other construction operation. Its planning is unique, however, in that the grout will be injected into spaces not visible and of unknown configuration and/or structure. In the case of faults, voids, or pore spaces, neither the extent nor the volume can be seen or easily understood. These conditions must be investigated sufficiently to gain a good understanding of the existing, as well as the achievable final properties. Invasive inspection, such as through borings or test excavations, is thus normally required. Although there are noninvasive methods, they are primarily limited to developing supplemental information and usually require invasive methods for calibration of the particular system.

22.1.1 What Are the Objectives?

Clear objectives should be established prior to the start of any grouting program. Why is the grouting being done, and what properties are expected of the improved substrate? Is the improvement to be permanent or temporary, such as the solidification of cohesionless soil to enable tunneling or other construction? If grouting will be performed to reduce water seepage, how much reduction is required and under what conditions? How will the reduction be measured? If strengthening a formation is the goal, what final strength is required? How is it to be evaluated and confirmed? The precise purpose and objectives of the proposed grouting should be thoroughly considered and expressly stated before commencement of any work.

Under emergency conditions, such as in reducing piping through an embankment or settlement around an active sinkhole, the time available for consideration and the availability of methods and materials are often limited. The objectives and associated risks should nonetheless be considered. Blindly pumping grout can have adverse results, and the risks and ramifications should be recognized and considered before the work begins.

22.1.2 Information Needed

Grouting design requires a good knowledge of the existing conditions, especially those relating to void content and structure. The needs will vary, of course, with the type of grouting and its purpose. In rock, the treading of the joints must be known, as well as their spacing and the aperture widths. The ease with which water moves through the formation is also pertinent. Flow paths have likely developed along joint intersections and increased in size as a result of long-term erosion and polishing. Although not easy to acquire, knowledge of these features is fundamental to effective filling with grout. Flow patterns can change with weather or, perhaps, as a function of the pool elevation of an adjacent impoundment. An understanding of such changes is important and can be gained through assessing changes in water levels in monitoring wells, as illustrated in Figure 22.1.

FIGURE 22.1 Changes in groundwater levels should be monitored and noted.

Soil presents perhaps the greatest challenge for grouting design because of its nearly infinite range of properties and the availability of multiple grouting processes and improvement mechanisms. A rough idea of permeability is required for virtually all grouting technologies and is, of course, fundamental to the design of permeation grouting. Density, moisture content, depth to water, particle shape and size distribution, and consolidation potential are all important properties to understand. The presence of rock, cobbles, or boulders, and the existence of organics or foreign materials must also be understood.

Grouting in soil is often performed in connection with settled structures or in developed areas in which man-made structure exist both above and below the ground surface. Aboveground improvements can affect access to and progress of the work, and substructure can interfere with hole layout and may be subject to damage if encountered during drilling. Underground pipes may have existing breaks, or breaks can be caused by injection forces, so that they become filled with grout. Surface improvements can be displaced, unacceptably jacked, or otherwise damaged; thus, their location and condition must be known. In short, a complete knowledge of all existing improvements is necessary if they are to be preserved during grouting.

Unlike most construction planning, the planning of grouting programs should not end upon starting the work. A great deal of information can be obtained during both drilling and injection. In a sense, each grout hole can act as an exploratory hole, complementing the information gained in the initial planning. New information will often suggest beneficial changes to the work as originally conceived, which may result in a better finished product or a cost savings. In cases of remedial grouting, it is possible to determine the cause of the problem solely by close observation of the drilling and grout injection behavior. Thus, grouting contract provisions should always allow for, and facilitate, potential changes in the work as dictated by the information gained as the work progresses.

22.1.3 Data Review

For most sites, at least some data describing the topography and other aspects are available, and in developed areas, there will be a considerable database. When grouting is proposed in connection with a geotechnical failure or distressed structure, prior investigations have likely been made. In addition, if grouting is to be done in connection with a significant structure, many individuals will have knowledge that can contribute to an understanding of the conditions. Fragments of information are often received, which in themselves are of little value, but when combined with other data, become significant. Thus, all information collected should be retained.

22.1.3.1 PRIOR INVESTIGATION REPORTS

Grouting is often accomplished in connection with or around existing structures or on major projects that have been in the planning for many years. In many cases, previous subsurface investigations have been made, either for the subject project or for others nearby. Existing reports often contain significant information that is useful to the development of the current work. For example, grouting is often performed in connection with tunnel construction under urban areas. Prior geotechnical reports are almost certain to have been produced for the planning of significant structures nearby. Furthermore, excavations will have been made during construction of those structures or for underground utilities. Any such available material should be examined in the early stages of project planning.

22.1.3.2 GEOLOGIC MAPS AND PUBLIC RECORDS

Geologic maps are available for many parts of the world and virtually all of the United States. The amount of useful information will vary with the individual area, but at minimum will include elevation contours, details of earthquake faults, and seismic history, as well as a general layout of the surface features. For developed areas and many watersheds, significantly more information is usually available, including rock types, depths, the strike and dip of bedded rock, and the location of known faults. Maps and data for areas in the United States are available from the United States Geological Service. For near surface features, the U. S. Department of Agriculture maintains maps showing the predominant soil type and, often, its depth, permeability, and other features.

In addition, the individual states' Office of Mines and Geology or Office of State Geologist have a variety of maps and other data available. In many cases, these include detailed subsurface mapping. More important, the local staff can be a great source of information for other available resources. Most states maintain records of water well drilling, which provide the depth to bedrock, general type of rock, and, of course, information as to the volume of inflowing water. For major wells, detailed drilling logs are commonly available. In urban areas, local building and/or engineering departments usually have an extensive database of local conditions.

A very comprehensive guide to available geological information in the United States is the *National Directory of Geoscience Data Repositories,* available from the American Geological Institute (AGI). Here, listings of information from 120 different sources throughout the United States are provided by state. Full contact information for the various sources is provided.

22.1.3.3 ORIGINAL PLANS AND CONSTRUCTION RECORDS

Grouting work is sometimes performed near to, or within, a structure. If the original construction plans and other pertinent material are available, they can often provide insight into existing conditions as well as any distress that is occurring. Be aware however, that things are not always built in conformance with the plans. And even the *as-built* or *record drawings* are often incorrect. Although they are intended to show the exact location and routing of hidden utilities and underground lines, one must be extremely careful and dubious of the information shown. The intended practice is to prepare these records as the work is done, so as to have accurate information upon completion. In reality, however, they are often not prepared until all construction has been completed, usually as a condition for the responsible party receiving final payment.

If the as-built drawings are incorrect, much damage can resulted from reliance on such information. In one case the as-builts for a major courthouse indicated that a main sprinkler feed had been placed as shown on the original plans. The location of an 8 in. (203 mm) cast iron pipe was clearly indicated, with an accompanying note that it was a minimum of 3 ft (0.9 m) below the floor slab subgrade. Grouting was being performed to correct settlement of the structure and to allow its continued operation; only one courtroom was made available to the contractor at one time. Noisy work was accomplished during the nighttime hours in order to limit any disturbance.

In the early morning hours, while workers were drilling through the floor slab, the main was hit and broken, resulting in rapid flooding of the area. The pipe was found 16 ft (4.8 m) to the east of its designated location. Furthermore, it was not of cast iron, but of more fragile asbestos cement "transite" and the top was at the subgrade elevation, not 3 ft (0.9 m) below, as shown. The contractor had marked the location of all utilities shown on the as-built plans on the floor surface. Because of the incorrect drawings, he was relieved of responsibility for the rather considerable cost of repairing the pipe and the resulting flood damage.

The emergency slabjacking mentioned in Section 20.2.4 and illustrated in Figure 20.10 provides another example of faulty record drawings. The restaurant was on the ground level of a huge landmark building only a few years old. This was a centerpiece structure of the Century City development in Los Angeles. At the time of the failure, the same contractor was employed on another landmark building directly across the street. In addition, the owner had a large in-house engineering and construction staff that was also on-site. Thus, the very people who supervised and inspected the problem structure were readily available when the slab dropped. They had the record drawings out when the grouting contractor arrived on-site, but no underground utilities were shown in the work area.

Only about 10 holes were required through the slab, and these were drilled with a pneumatically powered rock drill. Upon penetration of the very first hole, the drill literally dropped, indicating a void of considerable depth. A steel rod was inserted and "felt" a hard surface about 4 ft (1.2 m) below. Inspection with a flashlight disclosed a piece of sheet metal on the underside of the slab and a concrete slab below. And, of special concern, air under low pressure could clearly be felt exiting the hole. All progress in performing this "emergency" work came to a standstill while the actual condition was evaluated.

After several hours, the foreman who had supervised placement of the slab arrived and immediately solved the problem. "That's where that culvert was . . . we poured the walls and bottom slab and then put corrugated metal across the top for a form." "What is a g__ d___ culvert doing here?" demanded a senior engineer. It seems that placing a 60 in. (1.5 m) diameter reinforced concrete pipe warm air return, shown to be about 40 ft (12 m) away on the record drawings, would have interfered with the construction schedule, so the routing was changed, as was the mode of construction. Had that foreman not been available, or had he not remembered what had happened a few years before, the repairs would not have been completed in time to save the owner a huge "loss of use" charge.

Ergo: Do not rely on drawings even when clearly marked "as built."

22.1.3.4 HISTORICAL AERIAL PHOTOS

Aerial photographs have been taken at various times over most built-up areas and are commonly commercially available. They usually start as very high resolution photos covering relatively large areas, but any portion can be enlarged to show the smallest details. Much can be learned about the history of a site by comparing photos taken at different times.

An interesting example involved severe settlement of a structure. It was part of a new development on a large parcel of land that had laid vacant for many years. Available photos showed a pit being excavated about 50 years before, supplying borrow for a new road about 5 miles (8 km) away. The photo even showed dump trucks carrying the soil while the pit was active. The next photo, five years later, showed the area being used for farming, with that activity surrounding the pit. After a further five-year period, the pit had been filled in and the farmer's crop covered its former location. Apparently, the farmer had simply pushed loose fill into the pit to improve his field. And after 40 years, when a developer built two upscale homes on the site, its former existence was not discernable.

22.1.4 Site Access/Restrictions

Restrictions as to working time and access often prevail, especially in built-up areas. These conditions will affect the manner in which the work is undertaken, the time required, and its cost. They should thus be identified early in the investigation stage, to allow appropriate consideration during project planning.

22.1.4.1 ACCESS LIMITATIONS

When grouting was to be applied to a new dam, truck traffic on the only access road winding up a canyon to the site was limited to daylight hours. Because the grouting was continuous around the clock, sufficient material had to be delivered the previous day. And, of course, sufficient area had to be available for its storage. In another instance, an access road led directly to the grouting area, but truck traffic was not allowed because of a weight restriction over some buried conduits. Grouting is often performed in portions of structures that otherwise remain in use during the work. Access to the work areas is often restricted in order to avoid excessive disturbance to the regular occupants.

22.1.4.2 STAGING AREA

Contractors require a sufficient area to mobilize and store equipment and materials, as well as space for the mixing and pumping equipment. These requirements vary widely according to the size and type of the project but must always be considered, as they affect both schedule and cost and can dictate the particular equipment and methods to be used. Ideally, ample space will be available at a work site, but this is not always possible. For example, a particular roadway tunnel remained in service during rehabilitation. Only one lane could be blocked at any given time, requiring the grout plant to be located on a truck that was easily moved. Material that was stored off-site required shuttling to the work area in small quantities.

Another project involved massive quantities of chemical solution grout, requiring large storage tanks at the grout plant, as well as sufficient maneuvering room for tank trucks to make deliveries. The only available space large enough was about two blocks away, requiring a temporary pipeline for the grout and, of course, precluding the use of short-dwell-time grouts. Similarly, the grout plant is often located in out-of-the-way places when work is performed in and around structures, so as to be out of sight of the occupants and users.

22.1.4.3 COORDINATION WITH OTHER ACTIVITIES

Very often, grouting is only one of many activities occurring on a site, which must be coordinated so that conflicts are avoided. A serious problem arises when trucks or wheeled equipment cross grout hoses during the work. This will cause a pressure spike, which can initiate jacking or other damage in the formation. If a wheel crosses on or near a coupling, the connection may be broken, causing interruption of the work. Special routing and/or protection for hoses is often required.

22.1.4.4 HOURS OF WORK

In urban areas, and especially near occupied structures, the hours of work for grouting are often restricted. They may be limited to nighttime or weekends when the normal activities of the area are absent. Noise restrictions can also limit hours of work, and these are often in direct conflict with the requirements for working near occupied structures, as they may prohibit noise in other than daylight hours. Work may also be constrained during special events or on local holidays. All such restrictions should be considered during project planning, particularly in development of the schedule and determining the time required.

22.1.4.5 ENVIRONMENTAL REQUIREMENTS

Environmental requirements always control the discharge of excess water, but can go far beyond in their effect on grouting. In one instance, ultrafine cement grout was proposed for the repair of very porous concrete in an old dam. The controlling authorities would not allow it, however, for fear of adversely affecting downstream fish. The work was thus accomplished with epoxy resin, which increased the cost several times. A subject that seems to arise ever more often is the high alkalinity of portland cement. Although reports of injury or damage caused by this material seem to be lacking, environmental officials often show concern, requiring discharge to have a neutral pH, which can involve treatment and significant cost.

22.1.5 Existing Improvements

When grouting is performed in, under, or around structures, they must usually be protected from damage or defacement. Often certain rooms or areas are not accessible for the work, so that inclined or perhaps horizontal holes from pits or shafts are required. A substantial advantage of grouting as a remedial method is the operational flexibility available. Holes can be placed in different trajectories from available locations. Both the type of grouting to be performed and the details of its progress are influenced by the nature of any special requirements.

22.1.5.1 PROTECTION REQUIREMENTS

Grouting is said by many to be inherently messy, which is unfortunate in that work is often accomplished by other means to avoid the perceived mess. The work of grouting need not be any messier than other common construction

FIGURE 22.2 Protection of finished surfaces during grouting.

operations, and many grouting contractors routinely work in and around structures without the undesirable disorder or defacement. Simple measures, such as placing a wet burlap bag around an injection point, as illustrated in Figure 20.10, will prevent splatter with grout if a connection breaks or leak develops. Simply picking up spilled grout before it is dragged around by hoses will go a long way in keeping a site clean. Drill circulation flush should be captured and contained, as described in Section 10.2.2.1 and illustrated in Figure 10.31.

Surfaces to be preserved should be protected when extensive work is to be accomplished nearby. This is often as simple as covering them with plastic sheeting. Where damage is likely to occur, such as on the lower portions of walls, supplemental protection of plywood or collapsed pasteboard boxes, shown in Figure 22.2, can be used. Couplings of hoses can scratch floor coverings, but this is easily prevented by simply wrapping them in rags or similar soft material. Grouting can be performed without undue mess, and such requirements should be clearly mandated during initial planning.

22.1.5.2 SUBSTRUCTURE

A great concern in grouting in urban areas is damage to underground pipes and other substructure. The most frequent problem is penetration during drilling, but displacement of pipes and filling them with grout are other risks. To prevent damage, the location of all underground substructure must be known before the work begins. This should be clearly shown on the plans, including size and approximate depth, if known. As previously discussed, substructure is often placed other than as shown on the original project plans, and especially on the as-built drawings. Independent location of all substruce, well in advance of drilling and injection, is thus strongly recommended. Pipe-locating instruments that can be used from the surface to accurately determine the layout of buried lines are readily available, and professional pipe-locating firms are usually active in urban areas.

Even where the location of a sewer or drain line is known, it is prudent to run water through the system during grouting. This should be done from a location upstream of the injection to a downstream observation point such as a clean out or manhole. Where downstream access is not readily available, excavation to the top of the pipe and selective opening of a hole is advisable. Continuous water flow, combined with occasional observation of the downstream opening, can provide early warning should any grout enter the system. A change in the color of the flow can be observed before any extensive filling with grout, allowing relatively easy clearing by immediate flushing with water.

Contractors should be held responsible for damage to underground lines when their accurate location has been provided. It is unfair, however, to assign responsibility for damage to lines not known to exist or for which the location is unknown. Specifications often call for the contractor to assume full responsibility for all substructure, whether known to exist or not. This is unreasonable and usually results in large contingency amounts being added to the cost of the work. Therefore, *the location of underground structure should be determined before work begins and should be clearly provided in subsequent drawings.*

22.2 SOIL GROUTING

Investigations for contemplated grouting in soil must provide an accurate portrayal of the existing properties. Because grout will add weight to the treated formation, it is imperative to ensure continuity of a competent layer, the depth of which must be identified. In addition, if the work is in a built-up area, the condition of adjacent structures should be documented prior to grouting. The documentation of elevations and any existing cracking or other distress is important.

22.2.1 Soil Conditions

The properties of the soil are fundamental to determining its amenability to improvement by grouting, as well as to selection of the best method to be used. The particulars of the soil also dictate the grouting parameters, such as injection rate, pressure, and so forth. Sufficient exploration and testing to establish the existing conditions are thus required. Grout indiscriminately injected into soil can be damaging, and, as a rule, injection should not be made without first ascertaining the soil's properties and propensity for improvement by grouting.

22.2.1.1 TEST BORINGS

Sufficient exploratory borings must be made to allow an accurate depiction of subsurface conditions. The number and layout of the borings will depend on the consistency of the area soils and the nature of the proposed injection. Where grouting is performed in connection with differential structural settlement, the relative underlying conditions of different areas must be evaluated, so a greater number of probes are often required.

All borings must be of sufficient depth to encounter competent supporting material. Because surface elevations are subject to change, reference elevations should be provided for all borings. Depth to water and standing water levels should be noted, as well as significant changes in the drilling penetration rate. Any instances of encountering organic matter or rock and other unusual events should also be noted. The number of blows ("N" values) for all samples should be recorded.

A rational grouting program can be neither planned nor carried out without a good knowledge of the existing conditions. Excuses for not providing adequate information, such as lack of access for drilling equipment, or terminating a hole because of caving or hitting a rock, are not acceptable. Disastrous events, as well as large cost overruns, have resulted on projects due to inaccurate or insufficient advance information.

22.2.1.2 EXPLORATORY EXCAVATIONS

Test pits or trenches can be made in situations where holes are not readily drilled or where direct observation of the soil structure is important. Hand-excavated pits can be readily dug to a depth of about 6 ft (2 m), although this is seldom sufficient for most grouting. Pits and trenches as deep as about 20 ft (6 m) can be easily made with mechanical excavators, as can much deeper excavations, albeit with some difficulty and extensive efforts required for temporary support. For greater depths, tunnels can be excavated, although this is usually limited to large-scale and important applications. In addition to obtaining soil samples, photographs should be taken of all exposed excavation walls.

22.2.1.3 SAMPLING

Samples representative of the soil conditions encountered by the borings must be taken. The frequency will be dictated by the variation of the soil, the objectives of the proposed grouting, and the accuracy of estimated cost required by the client. At minimum, at least one sample, and preferably more, should be taken of each soil type or condition encountered. Permeation grouts tend to flow to and through more permeable strata, often completely missing those of lower permeability unless special efforts are made. Moreover, grout penetration can be retarded, and even completely stopped, by layers of clay or other very low permeability soil. Even very thin seams of clay can block movement of the most penetrable grouts. Furthermore, when grout is injected below the water table, groundwater must be able to escape at a rate at least equal to that of the grout placement and the driving pressure.

It is not unusual for faulty, fine, or silty sands to exist over deposits of clean and/or coarse-grained materials, which, although quite adequate and not needing improvement, are significantly more permeable. There have been many instances in which grout intended for permeation of the upper layer actually traveled to the underlying more permeable material. Thus, there must be sufficient sampling to disclose these conditions. A good rule of thumb for sampling frequency is, "If in doubt, take more." Although sampling does take time, it is a small portion of the overall boring cost, so ample specimens are best acquired while the drill rig is already on the hole. Information not obtained because of a lack of samples can ultimately prove costly. When permeation grouting is contemplated, an understanding of even the microstratigraphy of stratified soil is imperative.

22.2.1.4 LABORATORY TESTS

Although soil can be visually classified in the field, laboratory tests are required to obtain a good understanding of its properties. Of great importance for most grouting is the particle size distribution and resulting permeability. Consolidation properties are also important when compaction grouting is considered. Where unusual soil environments or extreme groundwater conditions are suspected, pH and the content of sulfates and bicarbonates should also be determined. All remaining soil from the tested samples should be retained, at least until the grouting is completed, so that they are available should questions about their properties arise.

Laboratory tests should include unit weight, relative density, moisture content, grain size analysis, consolidation, and Atterberg limits in the case of fine-grained soil. Where the proportion of minus 200 mesh fraction is significant, hydrometer tests should be performed to further classify those fines. Of great importance to grouters is a soil's permeability; laboratory-determined permeability may not be indicative of the actual soil, however. The reason, no doubt, is the virtual impossibility of obtaining truly undisturbed samples, the inability to remold laboratory specimens to accurately reflect the soil structure, and the boundary effects of relatively small specimens. Reliance on laboratory-determined permeability is thus best avoided. Where it is important to ascertain permeability, full-scale field pumping tests should be conducted.

22.2.1.5 IN SITU EVALUATION

Nothing can substitute for real soil samples; however, various in situ tests can provide supplemental information. Of greatest value is the cone penetrometer test (CPT) for use in soil relatively free of rock. CPT testing is fast and relatively inexpensive where large amounts of comparative data are desired. When it is used, at least a few borings should be driven to allow correlation of data. CPT probes can be made from lightweight rigs, in some cases mounted in standard pickup trucks, but because these rely on screw in anchors for reaction, setup time is significant. Large CPT trucks, which generally weigh 25 to 30 tons (Figure 22.3) can be set up and operate rapidly, making them particularly useful where many probes are needed. CPTs provide an excellent means of comparing the before- and after-grouting conditions of soil densified by compaction grouting.

The science of ground-penetrating radar (GPR) is advancing rapidly. It involves electromagnetic pulses that are generated on the ground surface. They reflect from the various interfaces at depth to a retrieval instrument also on the surface. Equipment is available that can

FIGURE 22.3 Large CPT rigs can setup and probe rapidly.

traverse the surface, radiating and receiving repetitive pulses, so as to obtain a continuous record of subsurface interfaces along the route. These instruments, basically measure variations in the electromagnetic velocity of the materials encountered. The procedure is especially useful for identifying clear interfaces, such as the presence of boulders, buried pipelines, or subsurface structure, within a soil mass. Where the velocities of soil strata are clearly different, identification of boundaries may be possible; however, this condition seldom occurs.

Seismic exploration methods can also discern differences in the properties of a formation. That most commonly used, evaluates the velocity of stress waves traveling through a formation, and is referred to as *seismic tomography.* Simply stated, sound travels faster through a hard medium than through a soft one. Thus, a dense or solid mass usually produces a higher velocity than low-density soil. By analysis and interpretation of the velocity, and attenuation of a large number of sound waves driven at different inclinations through a formation, a profile of the relative density or stiffness of the various components can be constructed.

In application, the pulsing between two bore holes is evaluated. A transmitter in one of the holes initiates a pulse, which is picked up by a receiver in the adjacent hole. Typically, the transmitter is moved vertically in the borehole so as to direct pulses at a variety of angles. The receiver usually remains at one location for a given set of data. By changing the receiver depth, multiple sets of data are obtained with which a seismic profile can be developed. Either shear waves, which are perpendicular to the borehole, or compressional waves, which are parallel, can be used. Compressional waves are most frequently employed, however, as their analysis is simpler. Good coupling of the pulse to the formation is crucial to the receipt of valid data. This is usually accomplished by cementing 1 1/2 or 2 in. (38 or 51 mm) PVC pipe into oversized drilled holes. Nonshrink grout must be used, as even minor shrinkage can prevent valid wave propagation. Cement-bentonite grouts are commonly used.

Although stress wave velocities are usually determined with direct paths between two bore holes (cross-borehole survey) or by analysis of reflective rays, some work has also been accomplished by directing waves from transmitters and receivers, both placed on the ground surface. The effective depth of this *surface procedure* is a function of the distance between the transmitter and the receiver. As the accuracy of the resulting plots, known as *tomograms,* is dependent on the number of individual ray paths and resulting nodal intersections, downhole evaluation is preferred.

Analysis of ray path data requires extensive processing, which is practicable only with specialized computer software. Thus, such work is commonly performed by firms possessing highly specialized knowledge and equipment. For this reason, plus the requirement for very large numbers of different ray paths to provide accurate determinations, the procedure is somewhat expensive, and it is usually reserved for especially large or important applications. Where the budget allows for acquisition of sufficient data, however, it has been found to accurately portray existing conditions and provide valuable information.

22.2.2 Settlement Repair

Repair of settlement damage is the most frequent use of compaction grouting, which usually includes groutjacking to reverse some or all of the subsidence. The design of such work requires the same information needed for other types of soil grouting. In addition, both the mode and the amount of settlement, as well as the scope of the resulting structural distress, are needed, because these factors affect hole layout and sequence of injection.

22.2.2.1 EXISTING ELEVATIONS

Where grouting is considered for remedying structural settlement, a level survey should be made, with sufficient entries to enable a good understanding of the settlement patterns and differentials. The relative elevations of floors are best expressed by contours, as illustrated in Figure 22.4. This can be readily accomplished through the use of a manometer system (water level). In such an operation, a reservoir of water or other fluid is placed at a generally central location and appropriate elevation. One end of a small flexible tube, usually 3/8 or 1/2 in. (9.5 or 13 mm) in diameter, is connected to the reservoir, with the other end attached to a rigid rod provided with a ruler, as illustrated in Figure 22.5. Fitting a hardened sharpened point on the bottom, as shown in the inset, greatly reduces the time required for a survey, as it can be poked through carpet, or between sections of floor covering, negating the need for correction calculations.

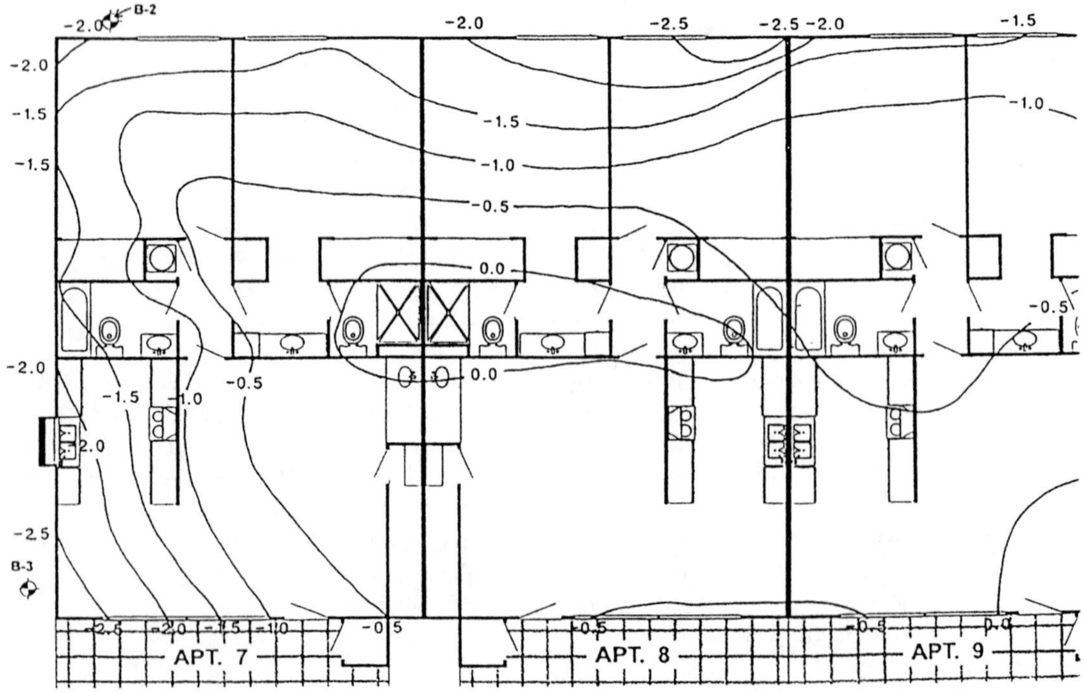

FIGURE 22.4 Elevation contours on plan facilitate groutjacking.

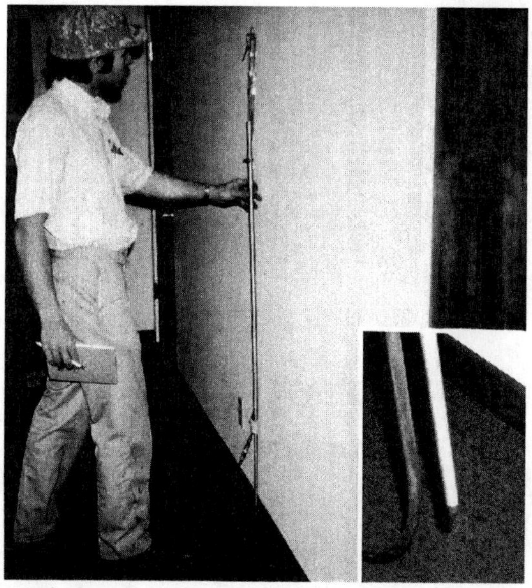

FIGURE 22.5 Manometer system facilitates contour mapping of floor levels.

22.2.2.2 STRUCTURAL DAMAGE

In assessing structural damage, information relative to cracking or other signs of structural distress must be provided. Significant cracking should be mapped, with approximate widths indicated. The locations of previous cracks that have been filled, as well as other repairs, should be provided if known, as their presence will have a dramatic effect on any groutjacking or releveling. Differential settlement usually causes significant load path and stress distribution changes within a structure. For example, racking at joints will cause a rigid frame to behave as a hinged frame. Where structural distress is observed, an evaluation should be made by a qualified structural engineer.

22.2.2.3 JACKING/LEVELING REQUIREMENTS

The requirements, including acceptable tolerances, for releveling structures should be determined. These can be based on either aesthetic or serviceability criteria. Settlement of off-level floors is seldom noticed. Rather, it is distress around the openings, sticky doors, water running the wrong way on drainboards, and the like, that initiates concern. Although many structures can be releveled with precision, as discussed in Chapter 14, this is not always possible where previous cracks have been filled or other repairs have been made. Any required jacking should thus be based on reasonable needs.

22.2.2.4 UTILITIES

As discussed earlier, grouting can usually be performed in near proximity to underground utilities without undue risk of damage, as long as their location is known. Where settlement has occurred, however, existing sewers and drains may have shifted so as to have little, or even reverse, drainage. Groutjacking at depth will lift these lines as well as surface improvements, whereas jacking of only the structure, will tend to pull it away from the buried lines, often causing breaks. This not only causes damage, but also fails to correct any displaced drainage of the lines that are buried. The condition and flow direction of subsurface drains should thus be determined and stated.

22.3 ROCK GROUTING

Requirements for grouting in rock are considerably simpler than those for grouting in soil, but are not always easier to document, especially in complex formations. The dip and strike of sedimentary rock should be provided. In all cases, the predominant distribution and configuration of joints are required. Cavities and other defects in soluble rock should be mapped as practicable. Much of this information will be disclosed during the grouting, but only if the holes penetrate the defects. A good knowledge of the angular treading of the defects is thus required for the appropriate layout of the grout holes.

22.4 GROUTING IN STRUCTURES

As with grouting in rock, grouting in structures always involves the filling of cracks or voids. Structures, however, have finite element boundaries that are usually not very great in dimension. Exploration is thus easier. Although the completeness of advanced knowledge is not nearly as important to proper performance, it may be required for cost estimating purposes. Cracking is usually visible and easily mapped. Invasive investigation by either core drilling or chipping will be required to evaluate internal defects. Ultrasonic testing can be effective for locating void areas. This testing is conducted similarly to cross hole tomography, as discussed in Section 22.2.1.5, except that the pulse sender and receiver are placed in direct contact with the element. Ground-probing radar can also be employed in a similar manner.

CHAPTER 23

Design of Grouting in Soil

Chapter contributed by Michael J. Byle, Gannett Fleming, Inc.

23.1 INTRODUCTION

23.2 GROUT FUNCTIONS IN SOIL

23.3 MASS IMPROVEMENT
- **23.3.1** Uniformity of Mass Improvement
- **23.3.2** Permeation
- **23.3.3** Jet Grouting
- **23.3.4** Compaction Grouting

23.4 MODULUS/STRENGTH IMPROVEMENT
- **23.4.1** Volume of Treatment
- **23.4.2** Settlement Mitigation
  - **23.4.2.1** Compaction Grouting for Settlement Control
  - **23.4.2.2** Permeation Grouting for Settlement Control
  - **23.4.2.3** Tunneling Compensation

23.5 PERMEABILITY REDUCTION
- **23.5.1** Permeation Grouting
- **23.5.2** Jet Grouting

23.6 STRUCTURAL ELEMENTS
- **23.6.1** Grouting for Underpinning
  - **23.6.1.1** Permeation Grouting Underpinning
  - **23.6.1.2** Limited-Mobility Grouting Underpinning
  - **23.6.1.3** Jet Grouting Underpinning
- **23.6.2** Excavation Support

23.7 KARST MITIGATION
- **23.7.1** Structural Columns
- **23.7.2** Structural Rafts
- **23.7.3** Grout Filler (Void Filling)
- **23.7.4** Grout Sealing
- **23.7.5** Cap Grouting

23.8 DESIGN SUMMARY

23.1 INTRODUCTION

Grouting is a versatile tool with many applications. These include controlling or reducing settlements, increasing bearing capacity, reducing or preventing water flow, underpinning of foundations, earth support and shoring, increasing liquefaction resistance to earthquake forces, stabilizing subgrades, creating struts or other structural elements, stabilizing karst, and others. There is no limit to the ways in which grouting can be used. The intent of this chapter is to highlight commonalities in grouting applications and outline appropriate approaches and methods for their design.

Grouting design is not like structure design, in which dimensions and material properties can be accurately constructed and directly measured to yield a high degree of certainty. Design of grouting is inexorably linked to constructibility.

The design must consider the limitations of the grouting process and accommodate these limitations. In general, perfect uniformity cannot reliably be achieved; therefore, designs should not be based on 100 percent uniformity. The uniformity of grout injections will always be subject to field performance issues and variability in the subsurface conditions. Although some of these factors can be predicted, allowances must be made to manage the remaining uncertainty.

There are few laboratory tests that can accurately measure the strength of grouted soil. Permeation testing in the laboratory is difficult because of inability to obtain undisturbed soil samples. Once secondary structure is destroyed and the sample remolded, laboratory testing will almost always achieve more uniform grout permeation than is possible in the field. Anisotropy (nonuniformity) is a significant factor, because fluid grouts will travel along the path of least resistance in the soil, leaving some areas ungrouted. Even limited-mobility grouts will react to strength anisotropy, yielding nonuniform displacement as the injected grout preferentially forces its way into weaker zones in the soil. In situ anisotropy is not easily tested in the laboratory, and appropriate adjustments must be made to soil parameters and the expected performance. These factors must be explored and evaluated to determine the preferred approach to obtain the required result.

23.2 GROUT FUNCTIONS IN SOIL

Grouting may accomplish a number of different functions in the ground. The most common reasons for grouting are to control settlement, reduce permeability, and increase bearing capacity. To achieve these objectives, the grout may increase the soil strength and modulus, reinforce the mass, or form structural elements that directly carry load, bypassing the surrounding soil. The analytical and design approach for each of these functions is different.

Permeation grouting functions as soil improvement primarily by adding adhesive/cohesive material to the void spaces within a soil mass. In this manner, permeation grouting may be used for general soil improvement, to create zones within the soil mass that are structurally stronger and/or have reduced permeability. The grout may form a substitute stronger structure connected through the void spaces within the soil. Alternately, the grout may bond to existing soil grains to form a new composite material akin to preplaced aggregate concrete. Furthermore, in special circumstances the grout may alter the soil chemistry to increase its strength and reduce its settlement potential. Permeability reductions are achieved by the grout filling or sealing pathways within the soil to inhibit or restrict fluid flow.

Compaction grouting is a form of limited-mobility grouting that improves the soil by displacing and compacting the soil to increase stiffness and shear strength and reduce settlement potential. The density increase caused by displacement due to compaction grouting will also increase the stiffness of the soil, resulting in a higher elastic modulus and reduced elastic settlement under load. Other types of limited-mobility grouting are used to form grout inclusions in the soil that, depending on the grout composition, are generally much stiffer than the surrounding soil and attract load to reduce stress in a stratum, to act as structural foundation elements, or to serve as reinforcement of the soil.

Jet grouting replaces a portion of the soil, creating a soilcrete mass. The properties of the soilcrete product of jet grouting are directly related to the original soil properties, inasmuch as the soilcrete is composed of grout blended with a portion of the existing soil. The soilcrete may be created in columns that act either structurally or in a composite fashion to reinforce a stratum. Jet grouting columns can be overlapped to create monolithic masses, walls, or buried horizontal slabs.

In general, the design of the grouting is evaluated on the basis of the function desired and the intended means of achieving it. The primary approaches are mass improvement, soil reinforcement, and formation of isolated elements. Assessing the approaches is key to good grouting design and to achieving an economical and effective grouting solution.

23.3 MASS IMPROVEMENT

Soil improvement involves modification of a site's soils to improve their performance in meeting a design objective. This can be achieved by uniformly modifying the soil or by creating discrete elements that act in concert with the soil mass. Where discrete elements are used, they must be small and closely spaced so that there is no effective load concentration. In the design of mass improvement, what matters is the composite behavior of the grouted soil volume, and discrete elements may be used only to provide a composite benefit that acts generally without regard to the location of individual elements. Obviously, the discrete element approach is not an effective choice for controlling permeability, but for strength- and settlement-related applications it can be effective and economical.

23.3.1 Uniformity of Mass Improvement

Creating a relatively homogeneous mass of improved soil is a common and fairly reliable practice where soil conditions permit. This approach is typically obtained with the use of permeation, compaction, or jet grouting. Where a reasonable degree of uniformity is achieved, the design can be based on the average improved soil properties. Achieving uniformity requires a well-defined knowledge of the structure and characteristics of the soil formation to be grouted. The ability to inject grout of predictable quantity and quality will be affected by the following physical factors of the subsoil into which it is injected:

- Homogeneity of the soil
- Presence of secondary structure providing preferential paths for grout
- Cohesive and frictional strength of the soil/rock
- Compressibility of the soil
- Tendency of the soil to build up and dissipate pore pressures
- Presence and proximity of subterranean structures or openings
- Isotropy/anisotropy of soil properties
- Existing stress field in the soil
- The presence of free surfaces such as slopes or excavations

All of these factors must be evaluated to assess the viability of grouting and the physical limitations of individual grout injections so as to design a grouting program that achieves continuity in the soil mass.

23.3.2 Permeation

A practical limit to permeation grouting can be estimated, based on time of injection and applied head for Newtonian chemical grouts, by the equation proposed by Maag (1938), as follows:

$$t = \frac{\mu n}{3kHr}\,(R^3 = r^3)$$

Where μ = viscosity of grout in centipoises
n = porosity of soil
k = soil permeability
H = head of grout
r = radius of injection hole
R = radius of injected grout at time t

For Bingham fluid grouts, the limiting radius is based on the pore diameter, Bingham yield stress, and head acting on the grout, as described by Bruce (1994), as follows:

$$R_L = \frac{\delta_w g H d}{4\tau_s} + r$$

Where R_L = limiting radius
δ_w = density of water
g = acceleration due to gravity
d = effective diameter of average pore
τ_s = Bingham yield stress

Injection locations are established to maintain a sufficient overlap of the grout injections to ensure reasonable continuity. Usually a 10 to 25 percent overlap is reasonable, though the spacing may be varied appropriately depending on the tolerance for ungrouted anomalies in the particular application. The spacing should be set to ensure that the design will perform properly, considering the largest anomaly that can remain ungrouted.

23.3.3 Jet Grouting

Creating a uniform mass in using jet grouting may be accomplished by continuous overlap of jet-grouted soilcrete columns. Typically, the column diameter is determined by the soil conditions and the energy of the jet grouting system used. Estimates of achievable column dimensions may be obtained from knowledgeable contractors or from prior experience. Typical conventional single-jet grouting may achieve column dimensions of 8 to 52 in. (0.2 to 1.3 m), depending on the soil conditions, rate of withdrawal, and jet pressures. Achievable column diameters in cohesive soils are typically about one-half the size obtainable in noncohesive granular soils. Special jetting systems utilizing multiple jets and/or special nozzles may achieve much different results. Uniformity within columns may also vary because of the grout-encapsulating clay lumps or by the shadowing effect of obstructions within the soil mass, such as pipes, foundations, or other foreign bodies. Actual column dimensions should be established in a test column for design verification prior to production grouting. Columns should be overlapped at 15 to 25 percent of the column diameter to ensure reasonable interlocking of the columns to create a uniform mass. Reduced overlaps may be considered where the application can tolerate ungrouted inclusions.

23.3.4 Compaction Grouting

The distribution of improvement surrounding a single compaction grout injection decreases radially from the point of injection, as illustrated in Figure 23.1 (top). To achieve a uniform im-

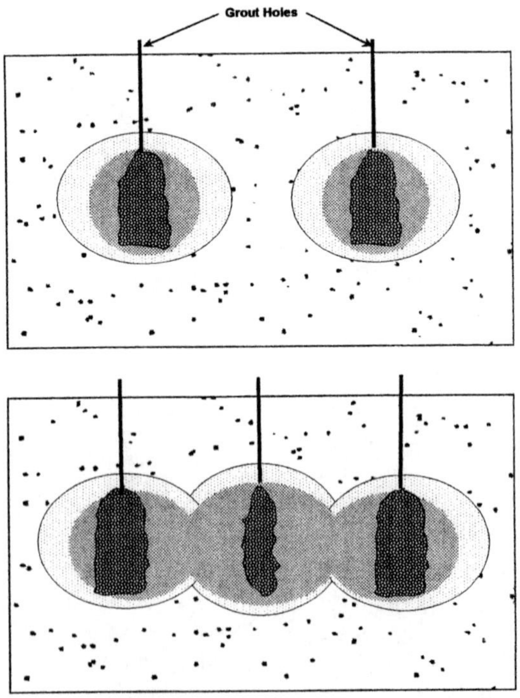

FIGURE 23.1 Distribution of compaction around compaction grouting injections (top) for isolated injections and (bottom) for close split-spaced injections.

provement, multiple injections must be spaced sufficiently close that the force applied by successive grout injections reacts against the already improved mass resulting from previous injections. For this reason, compaction grouting patterns must be designed so that the holes are injected from the perimeter toward the center of an area to be improved, and the holes must be split spaced, as shown at the bottom of Figure 23.1. The spacing of grout injections has been reported to range from 8–12 ft (2.5 to 3.7 m) (Al Alusi, 1997), 8 ft (2.5 m) (Bandimere, 1997), and 6 ft (1.8 m Stilley (1982). In general, the selected spacing is determined from site-specific conditions. The sensitivity of the site to movement will determine the largest injection that can be safely made. A rule of thumb is to space the injections three to five times the diameter of the injected mass.

23.4 MODULUS/STRENGTH IMPROVEMENT

Although modulus and strength are not always directly related, the improvement in one is often associated with improvement of the other. Increasing the soil modulus may be required for reducing settlements, supporting pavements, or altering seismic response. Increasing strength may be required for load bearing or slope stability improvements. The first step in designing grouting programs is thus to establish the amount of improvement required and achievable, and then to define the required limits of improvement to achieve the desired performance.

The amount of improvement that can be achieved depends on the grouting method selected and the properties of the soil. Although jet grouting typically achieves unconfined compressive strengths up to 300 psi (2.1 MPa) in silts and clays, it can achieve unconfined strengths exceeding 2000 psi (13.8 MPa) in coarse granular soil. Elastic moduli comparable to those of stiff clay or approaching those of concrete can be obtained, depending on the subsoils present.

Likewise, permeation with ultrafine cements, portland cements, acrylates, urethanes, and other systems can achieve a broad range of strengths in various media. Chemical solution grouts generally increase the cohesion and modulus of the soil, but may reduce the angle of internal friction. Moreover, many of these grouted soils will experience creep under sustained high loadings, having long-term strengths of one-quarter to one-half of the unconfined compressive strength. For these methods, bench tests and field test grouting are necessary to determine the actual achievable strength, using different methods and appropriate creep-adjusted strengths. The advice of an experienced, competent grouting contractor, familiar with the local conditions expected, may be the best source for preliminary design data.

Compaction grouting improvement is subject to fewer variables. Compaction grouting may increase the internal friction angle of soils, based on increasing the relative density. The improvement may be estimated from Myerhof (1956) equations, as follows:

$$\phi = 25 + 25D_r \text{ for soils with} > 5\% \text{ fines}$$

$$\phi = 30 + 25D_r \text{ for soils with} < 5\% \text{ fines}$$

Where ϕ = angle of internal friction in degrees
D_r = relative density

The initial relative density can be estimated from the standard penetration test (SPT) and a target relative density established to obtain the required internal friction.

Soil modulus can be estimated from empirical relationships or measured from in situ testing. Empirical correlations between modulus and relative density may be used to assess the required improvement needed to obtain the desired performance.

23.4.1 Volume of Treatment

For any ground improvement method aimed at increasing the strength or modulus of a soil, the area to be treated must be defined to achieve the project objectives. The volume to be treated is estimated, based on elastic stress distribution theory, to provide adequate support for the area in question. Such support often requires treatment beyond the limits of the footings or structures to be supported.

Depth of treatment must include consideration of the effect of the increased soil unit weight, including the weight of the grout, on deeper strata. Improvement of deeper layers may be necessary to support the added weight of the overlying mass of grouted soil. Where a softer layer will remain below the grouted volume, the settlement of the multilayer system must be computed for each layer in the system. Differential settlement between the center and the edges of the loaded and grouted area must be considered. Computation of the settlement for a circular loaded area is as follows:

$$S = C_d \alpha \sigma B \frac{(1 - \mu^2)}{E_u}$$

Where S = estimated settlement
C_d = a shape factor equal to 0.64 for circular area
σ = uniform loading stress
B = diameter of the loaded area
μ = Poisson's ratio
E_u = Young's modulus of the upper soil layer
α = correction factor for the relative stiffness and thickness of the layers, as shown in Figure 23.2.

Where grouting will increase the ground stiffness so that it is much greater than that of the surrounding soil, the grouted soil may act in a structural manner, carrying the load to a deeper layer. In this case, both settlement and bearing in the deeper strata under the full load distributed over the grouted area should be checked. The grouted area should be appropriately sized to avoid excessive stress on the deeper strata.

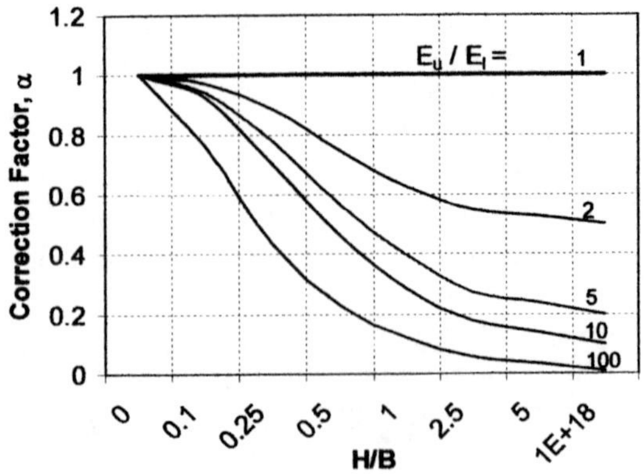

FIGURE 23.2 Settlement correction based on the data from Burmister (1965) for E_l layer of infinite depth and $\mu = 0.4$.

23.4.3 Settlement Mitigation

23.4.3.1 COMPACTION GROUTING FOR SETTLEMENT CONTROL

Compaction grouting can reduce or mitigate the settlement potential of soil in a number of ways. The most direct is by displacing and compressing the soil an amount equal to or greater than the expected settlement, thus, in effect, precompressing the soil. The increase in density caused by displacement due to compaction grouting will also increase the stiffness of the soil, resulting in a higher elastic modulus and reduced elastic settlement under load. The grout itself forms inclusions in the soil that, depending on the grout composition, are generally much stiffer than the surrounding soil and will attract load, thereby reducing stress and associated settlement in the surrounding soil.

Generally, only the densification is considered in compaction grouting design. This is largely due to the ease of verification and the ability to analyze and predict performance with relatively simple and widely accepted methods. A simple approach to evaluating the improvement required to control settlement is described in the following paragraphs.

COMPACTION DISPLACEMENT. For compaction grouting applications, the design must provide the amount of density increase required and the volume of material to be densified. The simplest approach to determining the required density increase is to compare the initial relative density to that required to achieve the desired performance. The required volume change is estimated from the unit weights associated with the initial density and the desired relative density, by the following equation after Byle (2003):

$$V_d = \frac{\gamma_f - \gamma_o}{\gamma_f} * 100$$

Where V_d = percent displacement volume required
γ_f = final desired total unit weight
γ_o = initial in situ total unit weight

For all practical purposes, the volume change desired will be equal to the volume of limited-mobility grout injected.

Stresses and strains are radially distributed about the point of grout injection (Byle, 1991). The stress distribution is generally assumed to follow a spherical or cylindrical distribution in homogenous soils, depending on how the grout is injected. This assumption is generally valid, provided that the grout formulation is sufficiently immobile to remain in a tight spherical or columnar mass. However, there is little evidence in the way of field measurements to assess actual stress distributions. The assumption of radial distribution of stresses and strains is generally not valid where soil has sufficient secondary structure to cause irregular grout shapes or for grouting that produces other grout shapes such as wings or lenses.

All initial strains induced by applied stress are elastic until the stresses exceed the yield stress for the substrate material. Elastic compression is reversible, and hence may be temporary if the induced lateral stresses can be released. This will be an important factor where small volumes of soil are compacted and at the perimeter of the grouted area. The elastic range is important for computation of increased shear strength using common Mohr-Coulomb failure envelopes.

Plastic strains are not reversible, and result in a permanent change in the soil structure. Plastic strains may also change the Mohr-Coulomb failure envelope for some soils, which should be carefully considered when calculating the grouted soil strength. In the case of sensitive, cemented, loess, or other strain-softening soils, displacement grouting can result in significant strength reductions by breaking down the soil structure and reducing its cohesive strength. This strength reduction can in many cases be overcome, however, by the increases in density and frictional strength.

Materials having anisotropic stress-strain properties will strain in an anisotropic manner under uniform stress. The direction and orientation of the anisotropy will be directly reflected in the strain achieved by an injection at a uniform pressure. Movement can always occur toward a free surface (such as the ground surface, a slope, or a vertical cut) without a material strain where confinement is insufficient, where strains exceed the tensile strength of the soil, or where loadings exceed the limits of arching.

In saturated and near-saturated low-permeability soils, consolidation occurs under the residual stresses induced by the grout injection. However, the generation of excess pore pressures in these conditions may seriously limit the amount of grout that can be controllably injected. Densification due to consolidation can be estimated by conventional consolidation theory, based on the solids partial pressure (p_gN) only (not the injection pressure). The residual stress will decrease as consolidation occurs. In some cases the permeability of the injected grout may be significantly higher than that of the substrate soils, and the grout will act as a drain after injection.

INJECTION VOLUME. The total volume of grout required can be estimated by multiplying the total volume to be treated by the percent displacement volume calculated earlier. There will be some difference between the volume of grout injected, as measured at the pump, and the actual volume of displacement in the ground. This is due to volume reduction because of the compression of air voids in the grout under pressure and the expulsion of water and/or paste through system leakage, pump bypass, inaccurate flow measurement methods, and loss to the ground at the point of injection. This volume differential is specific to the equipment, procedures, and mixtures used. Proper control of limited-mobility displacement grouting generally includes some fluid loss (Byle, 2000). The amount of this loss is typically small and not significant relative to the design.

23.4.3.2 PERMEATION GROUTING FOR SETTLEMENT CONTROL

A laboratory testing program will be needed to determine the exact strength behavior of the grout-soil interaction. A number of different grouts and grouting techniques can be used for permeation grouting and can lead to a wide variety of grouted soil properties. The methods are discussed elsewhere in this book. Ideally, this behavior will be validated by test grouting in the field, before any production grouting, unless it is well understood from prior experience in similar conditions.

PERMEATION EFFECT ON SETTLEMENT. Permeation will change the modulus and strength of the soil. For elastic settlement the magnitude is reduced in proportion to the increase in stiffness. The stiffness increase may be measured in the laboratory, but should be verified in the field. Where an entire soft stratum is grouted, the reduction in settlement can be estimated from the following equation:

$$\Delta S = S_i * E_i / E_g$$

Where S_i = the initially calculated settlement for the proposed loading
E_i = the initial elastic modulus
E_g = the elastic modulus after grouting

Where the grouting is performed on only a portion of a deep stratum or where deeper layers of lower-modulus materials remain after grouting, settlement analyses should be based on an appropriate layered stiffness model.

The grout mass resulting from most chemical grouts will be subject to creep, under stress over extended periods of time. This creep must be accounted for when using these grouts to control settlement. An appropriate reduction in the modulus should be made to allow for creep over the life of the project.

Where permeation or fracture grouting is used to alter soil chemistry or otherwise affect the consolidation properties of a soil, the beneficial effect must be assessed on the basis of the revised consolidation properties obtained from bench scale testing. Settlement analysis may be preformed using conventional methods, provided that the grouted area encompasses the entire zone of stress influence beneath the loading.

23.4.3.3 TUNNELING COMPENSATION

Grouting is commonly used to compensate for lost ground in tunneling to prevent settlement at the ground surface. This may be accomplished with the use of hydraulic fracture grouting or limited-mobility grouting (compaction grouting). In this application, the grout is injected above the crown of the underground opening to displace the lower soils into the opening, to compensate for lost ground outside the tunneling shield or supports. This is typically accomplished as a method specification inasmuch as the volume of grout required is specifically related to the tunneling practices, groundwater, and soil conditions.

23.5 PERMEABILITY REDUCTION

Grouting is commonly used to reduce the permeability of soils. Seepage cutoffs may be achieved by creating lines or panels of overlapping jet grout columns, or barriers of permeation-grouted soils. Limited-mobility grouting, including compaction grouting, will have little influence on the permeability of a soil and is rarely if ever used for this purpose.

23.5.1 Permeation Grouting

Permeation grouting reduces permeability by filling and sealing pathways and pore spaces within the soil. The primary design issue involves determining the reliable permeability value that can be obtained and the required grouting program to ensure continuity of the grouted soil. These estimates are based on groutability, as well as field and laboratory estimates of postgrouting permeability. Conventional Darcian flow equations may be used to assess the area requirements for grouting. The spacing and distribution of injections are based on the physical limits of grouting, as discussed in Section 23.3.2.

23.5.2 Jet Grouting

The permeability of the soilcrete created by jet grouting may provide a suitable groundwater cutoff in granular soils. Jet grouting is not commonly used for permeability reduction alone, but often provides this function as an adjunct to temporary construction support, subgrade stabilization, or contaminant encapsulation. Deep mixing technologies are rapidly replacing jet grouting for cutoff construction, except where horizontal cutoffs are required at depth. Jet grouting has significant flexibility in creating horizontal and vertical cutoffs through relatively small holes, but does require large equipment, thus limiting its use inside and near structures.

23.6 STRUCTURAL ELEMENTS

The primary use of jet grouting and noncompaction applications of limited-mobility grouting is to create structural elements in the ground. Structural shapes may be created from single and overlapping soilcrete columns by injecting limited-mobility grouts or by controlled injection of permeation grouts to stabilize an area. The primary geostructural applications of grouting include temporary and permanent excavation support, foundation underpinning, and augmentation of structure foundations.

23.6.1 Grouting for Underpinning

Where suitable soil conditions exist, grouting can be used to provide temporary or permanent support for structure foundations. Structural underpinning can be created by permeation of the soil beneath the foundation or by creating a foundation element by replacement (jet grouting) or displacement (limited-mobility grouting). The strength of a permeation or jet-grouted underpinning is based on the grouted material's strength, as described in Section 23.4. Formation of limited-mobility grout columns is controlled by the properties of the grout and the ability to displace the surrounding soil.

23.6.1.1 PERMEATION GROUTING UNDERPINNING

Permeation grouting introduces grout into the pores of the soil without any essential change in the original soil volume. Underpinning with permeation grouting can be performed without affecting the existing condition of a structure. The soil beneath the foundation can be grouted to improve the bearing capacity. Grout is injected at a relatively low pressure to avoid disturbance in the original soil structure. The degree of improvement by permeation grouting is driven by the properties of the existing soil, specifically, the geometry of the pores. Low-viscosity fluid grout that can permeate the soil matrix must be used. Permeation with sodium silicate grout is appropriate for temporary support and where excavation of the grouted soil may be required in the future. Some of these and other chemical-solution grouts may be used for permanent support where subsurface conditions are compatible with the long-term stability of the grout. Portland cement or ultrafine cements may be lower-cost options to chemicals. An example of the use of permeation grouting to improve bearing is shown in Figure 23.3 (left). Permeation grouting may be completed from beside, or by drilling through, the structure.

23.6.1.2 LIMITED-MOBILITY GROUTING UNDERPINNING

The difficulty in the use of limited-mobility grouting for underpinning lies in determining the effective cross-sectional area for a computation of stress, because the grout mass diameter is a function of the resistance of the surrounding soil or rock to displacement, except in very uniform conditions. Without detailed study, one cannot be entirely sure what the actual injection dimensions will be prior to injection. The injection size will vary according to such factors as confinement, anisotropy, and other nonhomogeneity factors (Byle 2003).

There are two basic approaches to handling this uncertainty. The first, and most conservative, is to design the strength on the basis of the diameter of the injection hole. There is a high degree of certainty that this diameter will be achieved, and any additional injection quantity serves to increase the factor of safety. This may be a reasonable approach for critical structures where high-strength grout can be used at reasonable spacings. The second method relies on a statistical approach and assumes that an average injection diameter will be achieved and any deviations in the grout diameter will be randomly distributed. This is a common approach

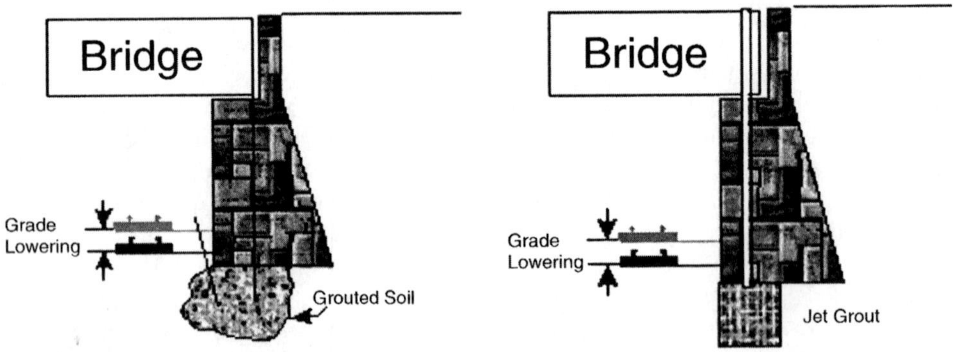

FIGURE 23.3 Grouting to underpin a bridge abutment by permeation or jet grouting.

and is reasonable for grouting in nonlayered deposits such as random fills, colluvium, or broken rock and where there are a large number of columns so that a reduced capacity in any one column will not be critical. In this case, a minimum injection per stage must be specified and stages must be small enough that a reasonable ensurance of continuity is obtained.

For layered deposits, a third, more complex, approach may be used where the hole diameter scheme is not feasible. This involves a detailed evaluation of the subsurface layering to estimate the probable behavior of grout injection in each layer. Additional controls, such as reduced stage lengths and/or reduced pumping rates, may be used in specific layers to alter the grout behavior in critical zones and produce the required dimensions. Consideration should be given to using top-down grouting, in which each stage is drilled through and injected below the previous stage, where layers of substantially different stiffness would permit grout to migrate away from the target layer to softer, less stiff layers above or below. Layered soils will require detailed grouting design and specifications that are closely monitored in the field to validate the design assumptions. Test grouting and excavation of control columns may be required.

For grouted columns, the capacity is controlled by end bearing and skin friction, as in to other types of piles. End bearing may be enhanced by injection of a larger bulb of grout at the column base. For most granular soils, the skin friction may be conservatively taken as the shear strength of the improved soil using normal at-rest earth pressure. The grout injection will induce higher residual stresses that will increase the frictional resistance above the at-rest earth pressure; however, the residual stresses decrease over time to a fraction of their original value (Schmertmann and Henry, 1992). Where columns are spaced sufficiently close to produce a uniformly stressed volume of soil, the corrected residual stresses may be used in the computation of column frictional capacity. For single columns, the at-rest earth pressure should be used.

23.6.1.3 JET GROUTING UNDERPINNING

Jet grouting may be used as a method to extend a foundation deeper to improve bearing capacity. In this case, the soil beneath the foundation is jet grouted to form a structural element to carry foundation loading to a deeper stratum and improve its bearing capacity. The underpinning is accomplished by repeated injections to create a solid mass of overlapping columns. An example of jet grout underpinning is shown in Figure 23.3 (right). It can be completed by drilling immediately adjacent to, or through, a foundation to provide more complete support. The design must consider the temporary fluid condition of the soilcrete to ensure that the foundation remains stable during treatment. Jet grouting beneath loaded foundations must be sequenced to provide continuous support and allow ample curing time for columns to gain adequate strength before adjacent grouting. Jet grouting may also be used to increase the size and weight of gravity-retaining structures. In this case, the soil behind an abutment or retaining wall is grouted to form a soilcrete mass. The soilcrete mass acts as an extension of the abutment to improve its stability. Structures modified by jet grouting should be instrumented and monitored continuously during the period of jet grouting.

23.6.2 Excavation Support

Grouting may be used as an excavation support, to augment conventional shoring, or to replace it completely. Reinforced grout columns may be used to form a cantilevered or braced wall to provide temporary or permanent support for excavations. Permeation grouting may be used to strengthen a soil to provide stable unsupported cuts. Conventional methods of analysis may be used to assess local, global, and internal stability of the grouted support system.

23.7 KARST MITIGATION

Mitigation of subsidence and sinkholes in karst constitutes a large portion of grouting applications. There are a number of different approaches that involve filling openings, spanning openings, and sealing off erosional pathways. Often, more than one approach to karst mitigation can successfully eliminate the existing hazards.

The performance mechanism of the grouting must be considered. The grout may act as a structural inclusion, bonding agent, filler, or sealer in the soil or rock. Limited-mobility grouting may also act to displace soft soils and/or densify them in the process. Several structural support approaches are described in the following paragraphs.

23.7.1 Structural Columns

Columns may be created with the use of jet, deep mixing, or limited-mobility displacement (LMD) grouting. These columns may act as structural supports that either directly support a foundation or act in arching to support the soil over solution features in the rock. LMD grouting in this application may induce lateral residual stresses that enhance arching (Schmertmann and Henry, 1992), as well as filling voids in the upper rock.

23.7.2 Structural Rafts

Another structural support mechanism is the creation of a mat or raft of grouted material that is capable of distributing surface loadings and spanning deep openings. For subsidence mitigation, these are best located at or near the top of rock. However, where local bearing capacity is an issue, they may be created near the bearing elevation. Rafts may be created by permeation grouting, whereby granular soil particles are cemented together by a fluid chemical or hydraulic cement that has low enough viscosity to uniformly permeate the soil without displacing it. This is generally not a practical option in fine-grained cohesive soils, however.

In most soils, effective rafts can be created by jet grouting and/or deep mixing. Soilcrete columns created by jet grouting, or by soil mixing in an overlapping pattern, can form contiguous masses of soilcrete, which can span voids and serve as a barrier to piping. Jet grouting has the advantage of being able to create a variable-thickness mat at depth, whereas deep mixing must extend to the ground surface.

A raft may also be constructed within the rock to support foundations bearing on the rock, which may be riddled with solution openings (Figure 23.4). In this application, the limited-mobility grouting technique is used to inject high-strength grout into voids and solution features within the rock. The grout displaces loose or soft material and creates structural linkages between rock masses and across voids, tying pinnacles together and supporting rock floaters. The design of these rafts requires injections to be made at sufficiently close spacing so that any intervening voids can be safely spanned.

23.7.3 Grout Filler (Void Filling)

Grout may be used as a filler material merely to fill voids. The filling of openings reduces the potential for the collapse of domes, impedes subsurface erosion, and reduces the permeability of formations. This is not considered a structural fix, because the grout need only occupy space

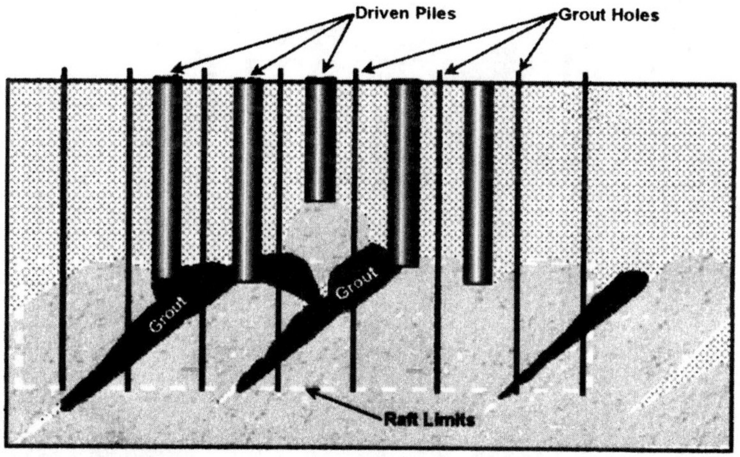

FIGURE 23.4 LMD grouting to form a rock raft for support of driven piles.

grout appears to migrate excessively through interconnected openings, it can be stiffened or gravel added to help choke off smaller channels. For very large openings, it is usually appropriate to place standard concrete by pumping, or under gravity, through a large-diameter pipe into the void.

Hot melts such as bitumen have also been used for filling and sealing voids. Hot melts are less commonly used than other grouting techniques because of safety considerations related to pumping high-temperature materials and the special equipment and expertise required.

and have sufficient erosion resistance to remain in place. Neat cement grouts may provide some strength, but may not displace soft materials and cannot provide a consistently reliable support in common karst conditions. Furthermore, common fluid suspension grouts are so penetrable that they can easily run into and become lost in the relatively large voids common in karst.

The type of grout selected for this application should be matched to the size of the opening and the permeability of the formation. For economy, grouting should be designed to keep the grout within the area required. Fluid grouts may have low enough viscosity that they will flow away through interconnected voids during and after placement. This can sometimes be limited by controlling the gel or set time of the grout. Sodium silicate or other chemical admixtures can be added to portland cement-based grouts to control set time and prevent excessive grout travel. Sand or fly ash is commonly added to filler-type grouts to reduce the required quantity of cementing agent and, hence, the cost of the grout. For large openings, a limited-mobility displacement grout is appropriate. Where

23.7.4 Grout Sealing

Where grouting is primarily intended to reduce permeability, a number of different methods may be required. In general, large voids must be filled with a filler grout prior to introduction of more expensive and mobile permeation grouts. At dam sites, groundwater can be flowing through connected voids at high enough velocity to prevent sealing with hydraulic cement-based grouts. On some occasions initial closure can be established by pouring gravel through a large pipe into the voids. This can be followed by limited-mobility grouting or a sanded portland cement grout with sodium silicate and/or antiwashout additives. The final sealing can be done with conventional permeation grouting. Depending on the permeability requirements, the final permeation grouting may be done with the use of common portland cement, ultrafine cement, or chemical-based grouts. Complete sealing of karst is possible, although it is challenging and will likely be very expensive.

23.7.5 Cap Grouting

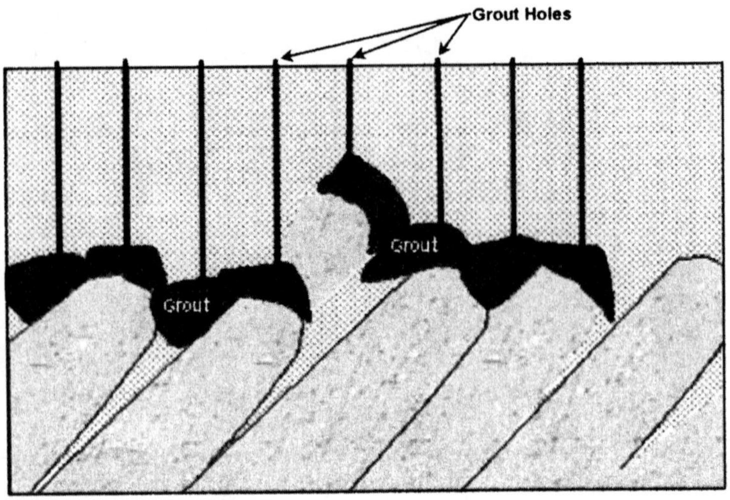

FIGURE 23.5 Cap grouting.

A hybrid method for treating karst is to inject grout at the top of the rock in an attempt to fill voids and seal openings at the rock surface (Figure 23.5). When properly executed, this method can be effective. *Cap grouting,* as it is called, is most commonly done with ready-mixed concrete or mortar, but both sanded port-land cement grout and limited-mobility cement grout are sometimes used. This can be a relatively inexpensive method, as it does not require drilling into rock and only a limited zone at the top of the rock is grouted. However, it will not address the issue of "floaters" (unsupported rock blocks) and must be done at fairly close injection spacings to obtain reasonable certainty of closure.

23.8 DESIGN SUMMARY

The desired result of a grouting project is frequently achieved by a number of different grouting methods and materials. Grouting contractors often tout those for which they have a capability as being the best solution for all conditions. This is seldom the case, and designers and owners must be careful not to be taken in by the hype of a sales pitch or slick ad. The choice of a grouting method should be based on an understanding of the mechanics with which it operates and how the injected grout will interact with the existing soil or underlying rock. For karst mitigation, cap grouting, grouted columns, or raft solutions may each provide benefit to a particular site. Which is the most appropriate will depend on a number of factors, including contractor availability, material costs, drilling equipment available, settlement tolerances, acceptable level of risk, size of project, and so forth.

Where conditions are amenable to a number of different grouting techniques, the most significant factors for choice will be the locally available resources, potential contractor's familiarity with applicable techniques, and knowledge of the site conditions. Grouting is not a panacea, and it can be a Pandora's box. There is no substitute for a thorough subsurface investigation. If one does not have a good knowledge of the subsurface, followed by appropriate consideration of the function and behavior of the grouting during design, the unexpected will probably happen; furthermore, contractors cannot possibly be expected to give a realistic price for the work. If the nature of existing conditions and the intended function of the grouting is not clearly understood, the work may not be effective and can actually cause damage. In grouting design, one must always be wary and suspicious of potential unintended consequences.

24 CHAPTER

Design of Grouting in Rock

24.1 THE EXISTING GEOLOGY
- **24.1.1** Rock Type and Structure
- **24.1.2** Fracture Orientation
- **24.1.3** Fracture Aperture, Spacing, and Surface Condition
- **24.1.4** Hydrology and Groundwater Movement

24.2 STRENGTHENING APPLICATIONS
- **24.2.1** Consolidation Grouting
  - **24.2.1.1** Fan Arrays of Holes
- **24.2.2** Surface Treatment

24.3 SEEPAGE CONTROL
- **24.3.1** Dam Foundation Curtains
  - **24.3.1.1** Curtain Depth
  - **24.3.1.2** Hole Layout
  - **24.3.1.3** Grouting Galleries

THERE ARE MANY fundamental differences between the design of grouting in soil and that of grouting in rock. Although additional weight of the grouted zone is of fundamental importance in soil, it is seldom a consideration in rock, as quality, and thus the supporting capability of rock, tend to increase with depth. And the aperture opening and volume in rock nearly always become less with depth. Moreover, the quantity of grout injected into a typical unit of rock is generally less than that injected into an equivalent unit of soil, and, of course, rock is a generally stronger formation capable of supporting higher loads. Because rock grouting always involves the filling of existing voids, the only type of grouting done is permeation. But here lies the critical issue: identifying the location and the details of the particular rock defects to be improved.

In earlier times, little real analysis or design was afforded grouting programs; rather, one or more lines of holes would be placed. They would typically be spaced 20 ft (6 m) or more apart, with the space being continually halved in a split spacing fashion until the takes lowered, even with a thin grout. Sometimes the hole spacing would become ridiculously close, so that with the normal variations in drilling, the bases of two adjacent holes might even overlap. Injection would start with very thin grout. If it penetrated freely—that is, with little or no pressure—it would be incrementally thickened. Huge quantities of grout were sometimes placed in single holes, without logic, except that of the contractor who was being paid by the bag.

This practice was more than wasteful in regard to both required construction time and costs. Holes often missed the defects; sometimes they were drilled much deeper than required, again based on some rule of thumb that often failed to take the geologic situation into consideration. The grouts commonly used were unstable, being subject to sedimentation and bleed, which often resulted in only partial filling and low durability. Properly performed, grouting into rock should provide complete filling, with a stable grout, not subject to excessive bleed or shrinkage, placed in holes that are positioned to optimally penetrate the existing defects. Furthermore, the injection pressures should be based on the rock quality and defect configuration, and the highest safe pressure should be used. This will, of course, vary, and is dependent on the parameters of the particular rock, which will vary between different sites and can vary from place to place on the same project.

24.1 THE EXISTING GEOLOGY

Although there is a similarity of behavioral patterns within each rock type, the actual extent of defects and their properties vary infinitely. To be successful in a grouting program, one must have a good sense of the type, orientation, and condition of the existing formation defects. This usually requires extensive geologic exploration, including appropriate mapping and test boring. An investigation must provide information on the orientation, frequency, and condition of the various joints and defects in order to allow the design of a rational grout hole layout.

Holes must intersect defects if those defects are to be properly treated. This will often require inclination of the holes. Concentrated areas of voids or other geological anomalies may require different treatment than the surrounding terrain. The importance of a thorough evaluation by well-qualified geologists cannot be overstated, as quality of design is wholly dependent on a good knowledge of the subsurface. In fact, rational design *cannot* occur without first establishing the in-place conditions through comprehensive investigation. Thus, the participation of a qualified engineering geologist on the design team is essential for optimal performance.

24.1.1 Rock Type and Structure

As discussed in Sections 21.2 and 21.3, the composition and behavior of the three primary types of rock, igneous, sedimentary, and metamorphic, differ widely. Similarly, the nature of defects will vary in both extent and condition within rock of any one type and/or age. Although variation is to be expected between different types, it can also be extensive within a type. With knowledge of the rock's age and depositional history, a geologist can provide a good idea of the primary existing features. Of fundamental importance in planning a grouting program are aperture width, orientation, and extent of weathering.

Weathering can change rock surfaces and cause cementing of the joints tightly, or disintegration may result from exposure to the atmosphere or dissolution by passing water. Eroded matter can be transported into the defects, becoming impacted so as to resist grout intrusion. Defect surfaces are affected by both rock type and weathering, so that they can be either quite smooth or very rough, both of which affects the ease with which the grout will move over them. Should the rock surface be dry, it will tend to pull moisture from the grout, inhibiting its movement. Of course, the rock type determines the likely nature of defects and their orientation as well as interconnection.

24.1.2 Fracture Orientation

The orientation(s) of rock defects must be known to enable the proper design of the hole layout and orientation to intersect them optimally. Although it is much easier to drill vertical holes, as well as control their trajectory, they will not be useful if they fail to intersect the primary defects. Inclined holes will be required for many applications. The proper inclination can vary, depending on the complexity of the joint systems. Where many different defect types and/or orientations occur, a variety of hole inclinations may be required.

The orientation of the defects also dictates the level of pressure that can be safely used during injection. As discussed in Chapter 9, the size of a grout mass influences the propensity to jack or displace the formation, and, as illustrated in Figure 9.7, voids with large horizontal areas will be subject to jacking at much lower pressures than those that are essentially vertical. Determination of the allowable grout pressure is dependent on a knowledge of both the rock properties and defect geometry and the orientation.

24.1.3 Fracture Aperture, Spacing, and Surface Condition

The ability of gout to travel once it has been introduced into a void is dependent on the aperture width, spacing, and surface condition. The degree of defect interconnection also has a major influence on the distance of grout travel. These factors must be considered to determine the optimal hole spacing, grout mix, and volume limitations, if any. Grout will obviously travel more freely in large voids than in small ones. And it will travel more freely over a smooth, regular surface than one that is undulating and rough. The resistance to travel varies every bit as much as the variability of defects, so no positive rules are applicable to all situations. Table 24.1 provides guidance for grout travel under various conditions. These data are based on a cementitious suspension grout, with equal parts of water and cement, being impelled under a pressure of 100 psi (6.9 bars).

The degree of interconnection, if any, is also a major contributor to grout travel. Large, interconnected defects will generally take grout easily and sometimes allow it to run freely, far beyond the needed location. Volume limitations should be established where this is likely to occur, but the validity of such limits is wholly dependent on a good understanding of the defect size. In the early phases of important work, sufficient core holes should be placed after initial grouting to ensure that the limitation quantity is reasonable. This may be assessed during drilling of adjacent grout holes, but additional confirmation holes may also be required.

Although excessive grout travel can be a problem, well-connected defect networks are significantly easier to fill than distributed smaller voids. Correction of these major features also provides the greatest ratio of benefit to grouting effort, whether the purpose of the injection is

TABLE 24.1 Approximate Grout Travel Distance in ft (m) in Different-Size Openings of w:c 1 Grout at 100 psi (6.9 bars) Pressure

	CRACK WIDTH IN. (MM)							
CRACK CONDITION	**1/16**	**(1.58)**	**1/8**	**(3.17)**	**1/4**	**(6.35)**	**1/2**	**(12.7)**
	ft	**m**	**ft**	**m**	**ft**	**m**	**ft**	**m**
Smooth surface, wet	9	2.7	15	4.5	30	9	>100	30
Rough surface, dry	5	1.5	8	2.4	20	6	>80	24

strengthening or seepage control. The size and volume of the defects also dictate the appropriate grout mix, as sedimentation, bleeding, and shrinkage have a far greater effect on large masses than on those that are small. Thicker mixes are appropriate for large defects and are likely needed in order to limit travel and minimize bleed and shrinkage.

24.1.4 Hydrology and Groundwater Movement

Where seepage is a reason for the grouting, an understanding of the aerial hydraulic regime and relative conductivity of different parts of the site is important. This will usually require pump and pressure tests in drilled holes during the final phase of the geological assessment. Once hydraulic conductivity values have been assigned, flow nets can be constructed using established modeling techniques for flow through porous media, which is discussed by Cedergren (1989). With the wide availability of user-friendly software packages, seepage issues are usually best handled with the use of finite element or finite difference models. This subject is discussed further in Chapter 26.

24.2 STRENGTHENING APPLICATIONS

The degree of benefit realized from grouting in strengthening applications drops off sharply as the width of the unfilled fractures decreases. For example, although the numbers will vary with different defect conditions, filling the widest 50 percent may provide 70 to 80 percent of the achievable improvement. Generally, the filling of very small defects typically contributes little benefit. Thus, the thoroughness of an injection does not ordinarily have to be nearly as great as in seepage control grouting. Because grout travels much more readily in wider defects, fewer grout holes will be needed and they can generally be spaced a greater distance apart.

Hole spacing should thus consider the defect widths. Where the widths are great, say more than about 1/2 in (13 mm), the grout may travel as much as, or more than, 100 ft (30 m) from the point of injection. Where the objective is to fill only these major defects, the primary holes can reasonably be spaced at an even greater distance. Where the improvement is required in a narrower section, however, hole spacing will be based on width of the improved zone and volume limitations will be in order. Thick grouts and generally low pressure should also be used where travel must be limited in large defects.

24.2.1 Consolidation Grouting

The filling of major fractures for strengthening a large zone of rock is commonly referred to as *consolidation grouting*. The term *blanket grouting* is sometimes used interchangeably, although it is more applicable to the filling of defects in relatively shallow sections and usually where seepage reduction is also required. The largest use of consolidation grouting is in dam foundations. Here, treatment ordinarily extends to a depth of 20 or 30 ft (6 or 9 m) from the rock surface. Usually performed separately from the dam curtain, it typically covers a large area and sometimes the entire footprint of the dam. In the curtain area, it will increase confinement for the subsequent grouting, but its primary value is to strengthen the mass to reduce consolidation of the foundation under the weight of the new dam.

Concrete dams are rigid, so that they must be uniformly supported, whereas embankment dams are capable of some flexibility. Earth embankment dams have another potential weakness, however: internal erosion, which can occur much more freely if space is available for the eroded material to be deposited within a short distance. Consolidation grouting can thus be important under all dams, and especially under the core of earth embankments built over badly fractured or porous rock. Although consolidation grouting in connection with water-retaining structures may be primarily for strengthening, seepage control can also be an objective.

Grout holes are generally laid out in a square pattern, with a final spacing on the order of 10 ft (3 m). They are injected in a split-spacing sequence with either two or three layers of hierarchy. The space may be further split in areas of large take. Because the extent of rock defects typically reduces as depth increases, consolidation grouting holes should generally be injected in stages from the top down. If the depth of such treatment is less than about 10 ft (3 m), injection in a single stage might be adequate.

Where a curtain will also be built, the orientation of the hole rows should be the same and spacing should be designed so that the curtain holes split spaces between the consolidation holes. This work has historically been accomplished with vertical holes and a uniform spacing, but the same rules apply here as in all rock grouting. The orientation of the holes must be designed to obtain maximum connection to the particular defects to be filled, and the spacing should consider the conductivity of the fracture system. Furthermore, additional holes or an alternate orientation may be called for to treat special situations. As with all rock grouting, this requires a good geological assessment.

24.2.1.1 FAN ARRAYS OF HOLES

Work performed from within the interior of a tunnel or other isolated location may require arrays of holes to be fanned out from a concentrated point. Although this is often done, it must be appreciated that such an arrangement is far from optimal. Obviously, the spacing between holes will vary with the distance from the drilling location. Accordingly, it will often be feasible to inject only the most distant portions of alternate holes. Such layouts should be thoughtfully designed so as to provide acceptable coverage in all areas. It should be noted that the long lengths and often considerable depths of tunnels can preclude a sufficiently thorough definition of all the rock defects ahead of construction. Accordingly, grout injection may be used for exploration ahead of the heading as well as for remedial treatment.

24.2.2 Surface Treatment

Where significant fractures extend to the rock surfaces, it must receive some form of pretreatment. In earlier times this would often consist of *slush grouting*, that is, simply distributing common grout over the surface so that it would penetrate and fill any cracks or other defects by gravity. Selective gravity filling of significant surface defects is good practice, but more economical and effective methods are now available for covering the entire surface.

Wet mix shotcrete utilizing transit-mixed concrete, as discussed in Section 12.4.3, is both fast and convenient. With modern boom pumps, the distribution nozzle can be supported, moved, and positioned by the boom, so it need only be guided by a single worker. Concrete boom pumps are commonly available with a reach on the order of 200 ft (60 m), allowing coverage of large areas from a single setup location. Because distribution of such concrete is both fast and economical, the application of thick overlays can be justified. Moreover, the thickness can be varied so as to result in a relatively smooth surface that facilitates both drilling and injection.

So-called dental or regularizing concrete, also discussed in Section 12.4.3, has traditionally been used where the rock surface was particularly rough. This operation can now be combined with the application of pumped shotcrete so the surface can be regularized in a single operation. There are advantages in ironing out rough foundation surfaces that go far beyond the benefits for grouting. Indentations tend to collect debris, which must be cleaned out before construction can be continued. In the case of concrete dams, the joint surface must be squeaky clean, and the effort required will be substantially reduced with a more regular starting profile.

Earth embankments are constructed in layers of reasonably uniform thickness, each of which must be compacted. This is easily accomplished when they are placed in continuous large open areas, but getting started on a rough surface with isolated depressed areas is challenging. Hand compaction or very small machines are usually required, which are both costly and time-consuming. Moreover, the uniformity and quality of the resulting embankment usually suffer.

Although provision of a considerable regularizing layer does involve substantial expense, it can be largely offset by cost savings in subsequent construction and the work quality can be notably improved. Such a layer can be depressed into a prepared trench and/or the thickness increased in the area of the grout curtain, improving restraint for the grout and eliminating need of a separate grout cap, as discussed in Section 12.4.4. This can permit a higher pumping pressure at the shallower depths and thus allow a faster injection rate. Moreover, jacking of a very rough rock surface is difficult to monitor, and substantial movement is common before it is detected. A rigid layer over the surface will crack, even with small movements, so jacking damage can be substantially eliminated. In short, more attention should be paid to surface treatment and any economical analysis must include the value of *all* the subsequent benefits. This is equally applicable to both strengthening and seepage-control grouting.

24.3 SEEPAGE CONTROL

Aside from its use in dam foundations, much grouting to reduce seepage is performed in connection with the construction of tunnels and other underground structures. In such work, some leakage can be handled by collection and pumping, so working to a high reduction standard is not usually needed. The cost of a unit of seepage reduction increases significantly as the standard rises, but a considerable degree of reduction can be achieved with a relatively small effort. It is thus important to balance the cost against the benefit for such work as the major leakage is brought under control.

24.3.1 Dam Foundation Curtains

Dam foundation curtains represent by far the greatest amount of grouting for seepage control and generally require a much higher standard of performance than is common for other applications. With current technology, it is possible to reduce seepage in virtually any rock mass to very low levels, although cost increases with performance standards. The required standard must be rationally established early in the design phase. Many factors, including the value of the water, must be considered in making this determination. The subject has been extensively discussed by Houlsby (1990); much of his thinking was previously presented in Section 12.2.3.3. More recently, Wilson and Dreese (2003) have proposed the seepage reduction that can be obtained through four different levels of investigation, planning, and grout injection. They state that level 4 (the highest) procedures can lower permeability to about 1×10^{-8} m per second (0.1 Lugeon), which is negligible.

24.3.1.1 CURTAIN DEPTH

Traditionally, curtain depth has been established by some rule of thumb, usually based on the height of the dam plus some constant. One such formulation, widely used by the U.S. Bureau of Reclamation, is one-third the height plus a value between 25 and 75. The exact constant value selected is a matter of judgment of the designer. Such design criterion will lengthen the flow path of any seepage, which is important from the standpoint of safety of the dam structure, but do not realistically address the seepage cutoff. This factor depends on the nature and depth of defects in the foundation rock. On the basis of actual grouting logs, it is obvious that much injection has occurred in holes that were unnecessarily deep. And then, if the success, or more important, the lack of success, of many grouting efforts to reduce seepage is examined it is clear that much work has been of insufficient depth.

Rational design must be based on the actual conditions of the underlying foundation rock and the established depth of treatment required. To be reasonably effective, the curtain must be of sufficient depth to tie into a zone of acceptably low permeability rock. Depending on the geological makeup of the particular site, this can be either greater or less than depths determined by any rule of thumb. In addition, more than one depth may be appropriate for different sectors of a given site. And there is always a chance that isolated geologic features existing at a depth greater than the main curtain will require special treatment.

24.3.1.2 HOLE LAYOUT

Perhaps the most important single step in the design process is choosing the number of rows of holes, their spacing and orientation, and the closure criteria. All sites differ, but in general terms most contain some defects significantly more open than others. The first grout injected preferentially intrudes into and along the larger defects, and the distance of travel can be great. Meanwhile, smaller fissures have perhaps been intruded slightly and "slicked over" as the grout was filling those areas of least resistance. Depending on the amount of time that has passed while the low-resistance grout was placed, the limited grout within and covering the smaller defects will likely have taken up sufficiently that further intrusion does not occur.

Thus, although the largest voids have been treated, often for distances far greater than necessary, many of the smaller defects remain open. During injection of the next-higher order holes, the process is repeated, except that now the preferential route of least resistance is the larger defects left open after the previous filling. And, again, smaller fissures can remain unfilled. The effective travel distance of the grout in these smaller features, however, becomes progressively shorter as the aperture openings decrease; thus, the finer defects remaining open near the primary holes cannot be reached, as they are too distant. This is the reason that split spacing often continues until the holes are so close that they may actually cross over each other at depth because of the normal inaccuracies of drilling. In this regard, specification of rotary drilling which is generally more accurate, might be in order for holes deeper than 150 or 200 ft (45 or 60 m).

It is for this reason that the efficacy of single-row curtains is so unpredictable. Although a few have performed very well, many have proven to be less than effective. With the normal inaccuracy of drilling, lower portions of holes can be several feet from the planned location. By increasing the number of holes, as with multiple rows, there is less opportunity for excessive spacing at depth. And, of course, the greater the number of holes, the more likely they are to wander off course in opposing directions so as to somewhat compensate for each other.

Three-row curtains are thus preferable and are adequate to allow seepage reduction to a high standard. Normally, the outer rows will be injected first. This will likely fill all the major defects so as to confine grout travel when the center closure row is grouted. Because only relatively small defects remain, the horizontal defect area affected by a given hole will be reduced sufficiently that higher injection pressures can be safely used without the risk of jacking the formation. This arrangement is optimal for most curtains, the one exception being in the case of rapidly moving subsurface water. In this instance, injection should start with the row of holes downstream of the flow. The center row should be next treated, with closure reserved for the upstream row. In such cases, it may also be prudent to split space to a closer spacing in the downstream row.

Hole spacing typically starts at about 40 ft (12 m) for the primary holes, with a final spacing of 10 ft (3 m) for the tertiary, and down to 5 ft (1.5 m) for the quinary if used. Yet this *common* practice fails to consider adequately the nature of the defects. In badly fractured formations, it is preferable to limit the grout travel to avoid large placements beyond the needed zone of improvement, while at the same time filling all of the major defects from the primary holes. Therefore, injection of the outside rows should consider excessive travel, and appropriate volume limitations might be placed. Assuming uniformly fractured rock, travel distance will generally be equal in all directions, so the holes should be spaced no farther apart than about 60 percent of the anticipated travel.

The distance of grout travel has been subject to the speculation of many. Ewert (1985) suggests that it can vary from as little as a few centimeters (> 1 in.) to as much as 10 m (33 ft). However, in one instance grout showed up in a well more than 300 ft (90 m) from the point of injection. Furthermore, quality grout has been run through 1200 ft (360 m) of 3/8 in. (9.5 mm) tubing at pressure of less than 180 psi (12 bars), in well controlled tests. Certainly, it is reasonable to expect travel of similar distances in any major geologic defects that may prevail. It is likely that much of the grout historically injected has run to areas far beyond the intended zone of benefit.

The drilling of holes is expensive and does require time, so the amount of drilling should be held to the least reasonable extent. However, what is reasonable must be carefully evaluated. Grout injected outside the planned boundaries also entails an expense of both cost and time. Limitation of the quantity pumped into more closely spaced initial primary holes, can often offset the costs of the additional drilling. But, more importantly, it can greatly increase the effectiveness of the work. Again, rational hole spacing must be based on the nature of the defects as revealed by a thorough geologic evaluation of the site.

24.3.1.3 GROUTING GALLERIES

Grouting from galleries within dam structures is discussed in Section 12.4.5. There are both substantial advantages as well as disadvantages to their inclusion and use. Although injection can be delayed until sufficient dam structure has been completed to provide maximum confinement for the grout, allowing higher injection pressures, the cost of the grouting will be greater. However, because it is always possible to perform further grouting, less thorough initial treatment may be justified. Consideration of a grouting gallery should include the type of underlying geology, the standard of seepage reduction required, and the likelihood of any future grouting requirement. Where drainage galleries are planned, they can be combined with those used for grouting.

CHAPTER 25

Quality Control, Records, and Verification

25.1 QUALITY CONTROL
- 25.1.1 Drilling Observation
- 25.1.2 Grout Material
- 25.1.3 Injection Behavior
- 25.1.4 Site Surveillance

25.2 INJECTION RECORDS
- 25.2.1 Manual Records
- 25.2.2 Automated Records
- 25.2.3 Be Prepared for Changes

25.3 VERIFICATION OF GROUTING EFFECTIVENESS

MUCH HAS BEEN WRITTEN and said about the verification of grouting, suggesting it should be accomplished separately from the actual work. This is incorrect, as the best verification is through analysis of the injection records that result from appropriate quality control efforts during injection. These subjects are interrelated and cannot properly be dealt with separately.

25.1 QUALITY CONTROL

The effectiveness of a grouting effort is often obvious, as can be observed visually when leakage is stopped or a settled structure is grout jacked to grade. The results of much grouting, however, are not so obvious. For example, where cohesionless soils have been solidified by permeation grouting to facilitate tunneling, soil zones satisfactorily treated will normally not be discovered until they are reached by the tunnel heading. Likewise, the results of soil strengthening at depth cannot be directly observed.

The completeness and quality of a grouting program begin with appropriate preinjection planning and a clear knowledge of the formation to be improved. Such as effort continues with careful and accurate monitoring of the work at all times and modification as dictated by the observed grout behavior during injection. If these matters have been conscientiously attended to, a high-quality finished product is virtually ensured. If, however, as is all too often the case, injection begins without a good knowledge of the formation and there is little or no monitoring of the work, less-than-quality performance is likely.

In earlier times, monitoring of grouting was often performed with methods that lacked accuracy, and data receipt was often subject to unavoidable delay. With modern computer technology, the earlier limitations have been largely eliminated. There is now no excuse for poor monitoring or lack of timely adjustment of the work as required to obtain a predictable quality product. It must be realized however, that subsurface conditions vary, and practicality and economics usually limit our understanding of the beginning soil. The details of unknown conditions will be made known, however, by grout behavior during injection.

25.1.1 Drilling Observation

Correct layout and drilling of the grout holes is fundamental to quality performance. This may seem easy, but keeping track of the numerous holes and the individual grout stages can be a daunting task. And, of course, the holes must be drilled in the proper sequence. As discussed in Section 10.2.3, attaining proper trajectory of the drilling is extremely difficult, and holes are seldom drilled to the accuracy we would like. The high cost of achieving great accuracy, however, causes us to accept the fact that not all holes will be precisely aligned.

Hole location should always be within about 1 in. (25 mm) of the planned layout. Locations are typically identified by a stake or a paint mark on the surface. To distinguish between holes of various orders, different colors are useful; say, red for primary, yellow for secondary, and so on. Each hole should be numbered for identification. It is useful to provide a suffix to each number to identify the hierarchy—for instance, "P" for primary "S" for secondary, and so forth. It is very easy to lose track of the hole identification, especially in soil, so it is usually worthwhile to provide offset reference points.

The drill rig must be firmly positioned and the mast set to the exact specified alignment prior to the start of production. From a practical standpoint, we are limited in controlling accuracy once drilling is under way, so care in the drill setup is important. To ensure that the holes are advanced to the proper depth, it is a good idea to set aside the exact number of drill rod sections required.

Much can be learned during drilling that affects the work quality. Although sampling is usually not done in grout holes, observation of the drill penetration rate can be informative, and, of course, any drop of the drill head, which is indicative of a void, should be noted, as should any loss of the drill circulation flush. The nature of the cuttings returned in the flush should be noted, as well as the presence of any organics or foreign material. Experienced drillers can tell a good deal about the conditions encountered by the "feel" of the drill. They should be encouraged to make note of any unusual feelings. Core drilling can be used for holes in rock or structures, wherein obtaining additional information is important. Because such holes are much slower and more expensive to drill, they should be used only where more data are required, however.

All data should be noted on a suitable record. This is best accomplished on a drilling log form prepared especially for the work. At minimum, the hole number, date and time, depth, drilling rate, and particulars of the return flush should be included. As with all field records, the forms should facilitate the required notation by providing appropriate columns or spaces needing only checkmarks or numbers to be inserted. For most work, recording is done by the drillers; however, on important projects, especially in the early stages, a full-time inspector may be used.

25.1.2 Grout Material

Rigid material requirements are common in construction, and many specifiers tend to be equally rigid in specifying and controlling materials in grouting. This is not always required, however. Although strength is the most important parameter for most cementitious mixtures, it is relatively unimportant in much grouting, in which resistance to shrinkage and to bleed are primary. Reason must thus be exercised in testing and control of the grout.

Testing methods for grouts are described in Chapter 8 and are not repeated here. Suffice it to say that it must be certain that the grout meets the requirements for the particular work. Tallying the materials delivered to the project against the quantity of grout used will keep contractors honest and ensure the strength of the grout mix. Where bagged cement is used, empty bags can be counted at the end of a given hole or perhaps daily. The density of cementitious suspensions can be checked with the mud balance described in Section 8.2, or with a flow-through density meter when computer monitoring is used.

An exception to the norm occurs in structural grouting, where strength is usually important. Here, standard specimens can be made in the case of cementitious grouts, followed by testing in the laboratory. Where resinous materials are used, requiring them to be delivered in the manufacturer's original containers is recommended. Again, a count of empty containers will confirm the amount of material used and, in the case of multicomponent systems, the use of the correct proportion of each. This is especially important where proportioning pump systems are being used.

In compaction grouting where a very stiff mixture is required, slump testing is often called for excessively. With experience, personnel will be able to visually evaluate the adequacy of the grout. The reason for low slump is to prevent hydrofracture of the formation, and this will be disclosed by the pressure behavior. Careful observation of the pressure behavior is by far the most important factor, and it is counterproductive to become unduly concerned with slump as long as it appears to be about right, visually, and no hydrofracture is indicated.

25.1.3 Injection Behavior

Much has been said earlier relative to the importance of continuous monitoring of grout behavior during injection. As the cost of the work is directly linked to the rate of injection, it is advantageous to all involved to use a pumping rate as high as practicable. A rate too high will result in incomplete filling, hydraulic fracturing, or jacking of the formation, however, so the injection parameters must be continually observed. Ideally, pressure at each stage should increase slowly and steadily until injection is completed. If the rate of pressure rise is too fast, or if it increases abruptly, the pumping rate should be immediately reduced and adjusted to maintain a slow, uniform increase.

Conversely, if the pressure level remains static or slowly deteriorates during injection, it is likely that the pumping rate can be increased. Of course, if a sudden pressure drop occurs, hydraulic fracturing, jacking, or leakage of the grout is indicated, and immediate reduction of the injection rate or complete cessation of injection is in order. Although occasional rapid pressure reductions will occur on many grouting jobs, their occurrence should be infrequent, and immediate corrective actions the norm. When very sensitive work is being performed, such as injection into a water-retaining embankment, even minor hydraulic fracturing cannot be tolerated, so low pumping rates are in order. On such sensitive applications, there should also be ample instrumentation and monitoring of the grouting site. Of special importance is the monitoring of pore water pressure, which is directly related to the injection rate and resulting pressure of the grout.

Figure 25.1 is an actual record of injection behavior of a continuously computer-monitored bottom up, compaction grouting operation. The pressure spike occurring at the beginning of the record indicates the elevated pressure required to start the grout flow. This is normal, resulting from thixotropy (cohesion) of the grout, as discussed in Section 4.2.1. The nearly complete pressure losses indicated result from pulling the casing to the next higher stage, which is also

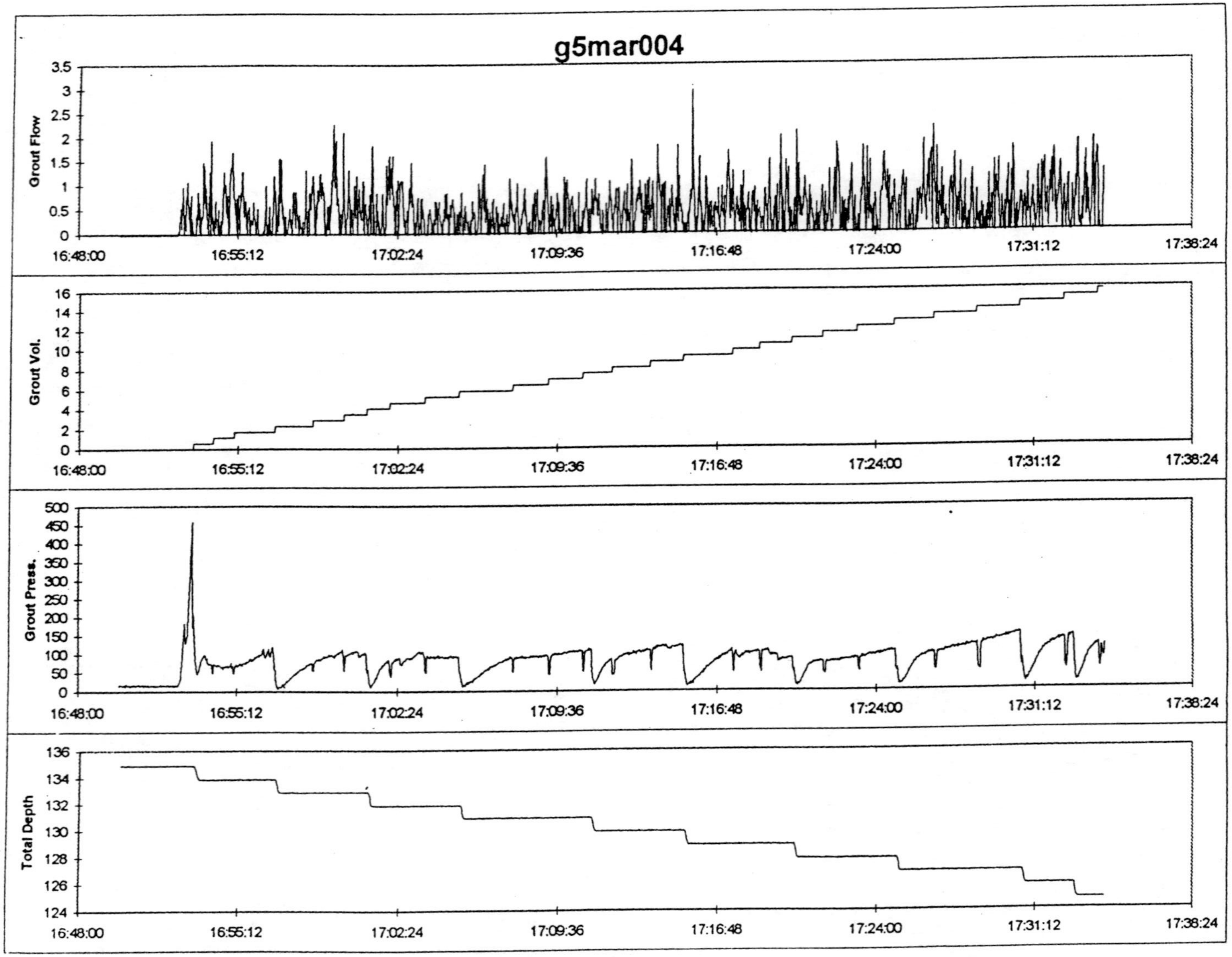

FIGURE 25.1 Computer-generated record of grout parameters.

indicated by the rise in depth, shown on the bottom curve. The smaller dips of pressure during individual grout stages indicate pump strokes. Note that these line up with the increases in grout quantity shown on the Grout Volume curve. Such real-time records, showing all the injection parameters, provide a good knowledge of the grout deposition and potential effectiveness. Hydraulic fracturing, loss of control, or other undesirable events would be indicated by a sudden pressure change, assuming that the pumping rate was uniform.

In permeation grouting, it is usually prudent to place a limitation on the grout to be placed at any one stage. When unreasonably large quantities of grout are taken in a number of holes, the spacing is likely too great. For economic reasons, the greatest hole spacing that provides the intended result should be used. Firm establishment of hole spacing is thus often delayed until initial injection takes place. Even on the same site, more than one hole spacing may be found optimal.

Although pressure behavior at a constant pumping rate discloses invaluable information relative to the formation, knowing the pressure without knowing the injection rate, or variations therein, is of absolutely no value. The widespread practice of reporting only the highest injection pressure reached on a given grout hole, absent the injection rates or pressure variation, is of no practical use. It is unfortunate that many professionals who should know better have blindly accepted such data.

Experienced grouters, and especially grout pump operators, know that a given pressure can be reached by simply increasing the pumping rate momentarily. Thus, high pressures are often inappropriately created and reported, usually to the economic benefit of the contractor. Relative pressures *and* injection rates are very instructive and should be routinely monitored and recorded in real time. This can be effectively done either manually or with automated recorders or computer processing.

25.1.4 Site Surveillance

Continual observation of the ground surface and of overlying structures and substructure is imperative. Virtually all grouting imparts elevated pressures to the formation. This results in a risk of possible hydraulic fracturing, upheaval, and/or lateral spreading. Such deformations can affect surface improvements or substructure, resulting in damage. It is thus essential that both the ground surface and improvements be continuously observed during injection. Likewise, underground structure, and especially sewer and drainage piping, should be monitored to ensure that they have not been displaced or filled with grout. Most damaging movements are immediately preceded by a significant loss of pressure, as discussed earlier. It is thus important to be particularly observant of the work area upon sudden pressure loss.

Many methods are available to monitor the elevation of the surface and structures. To be effective, however, the selected method must be capable of surveying very large areas and providing immediate notice of any movements. Thus, although conventional surveying can be useful, it usually is not satisfactory for primary surveillance. Multistation manometers, whereby any number of tubes can be connected to a single reservoir, with the terminal end fixed to the ground surface or on a structure, have been successfully used on thousands of projects. The water level is marked on or adjacent to the tube prior to injection. Any vertical movement that occurs will cause a change in the water level, which is readily observed. To enhance visibility, the water can be colored, most easily with a few drops of food coloring in the reservoir.

Manometers are extremely flexible in use, as any number of lines can be employed. Figure 25.2 shows tube terminals with a number of different mountings. One simply attached to a post driven into the ground is shown on the far left. To its right, a tube is fixed to a wall with duct tape. Note the ruler taped beside it to facilitate measuring any differences in level. Next on the right are tubes from

FIGURE 25.2 Variety of mountings for monometer terminals.

several different reservoirs, and on the far right is a tube fixed to a stand at an angle on the ground to facilitate reading from a standing position.

Although they do not provide immediate indication of movement, conventional survey instruments are useful, especially in scanning large areas. Arrays of prisms can be preset throughout an area, or on one or more structures, to facilitate reading by a standard instrument. These are usually total stations and can be automated to make regular sweeps, sending the readings by modem to a central receiver in a convenient location.

Rotating laser instruments in combination with multiple targets can provide an indication of vertical movement. Any number of targets can be monitored with a single instrument. Although these devices can be useful, there are several limitations. First, accuracy varies greatly between different instruments, and in the case of those of lower cost, is usually not satisfactory. Second, the targets must be mounted on stands sufficiently stiff to prevent them from being disturbed by wind. The targets typically give an audible signal upon movement, after which they are manually reset to the new level. And here is the greatest limitation: Even the highest-quality instruments provide neither a usable measure of movement nor a cumulative measure of total movement.

Simple, inexpensive devices can also be very effective. String lines tightly stretched across the injection area make excellent elevation references. Small pieces of masking tape fixed over cracks will either deform or break upon any change of crack width. For measuring the width, telltales with rulers or crack monitors, as illustrated in Figure 25.3, can be secured over the crack. Close visual observation of the surface over injections will often signal pending movement, and audible sounds often accompany movements in structures or rigid pavements. Thus, nothing can replace the eyes and ears of experienced and competent grouting personnel.

25.2 INJECTION RECORDS

Complete records of all grouting operations are imperative to achieve optimal results. Their greatest value is during the actual work, in providing a timely appraisal of injection behavior,

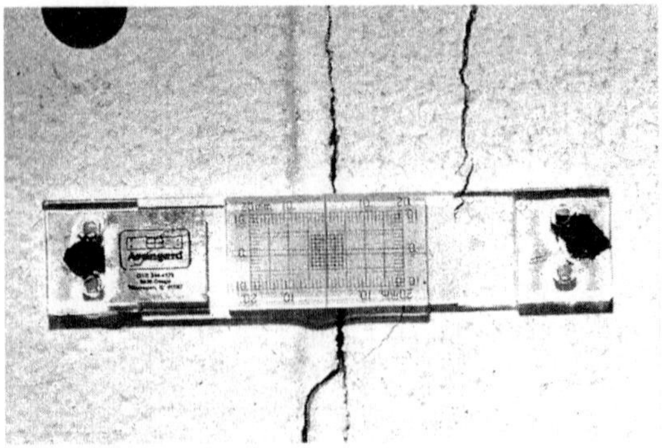

FIGURE 25.3 Simple monitors will disclose crack movements.

which gives an idea of the actual conditions existing in the formation. This permits quick evaluation by the grouting personnel on-site, allowing effective response. The records can also be of value in providing pertinent information for any future work. Moreover, in light of the unfortunate litigation so common in today's world, a good record of all operations is important.

At the very minimum, records should indicate grout pressure, volume injected, and the time at regular intervals. Obviously, the date, time, hole number, and grout stage depths should always be provided. It is also convenient to have grout flow rate, which can be back calculated when time and volume are known. With suspension grouts, continuous monitoring of the grout density will provide the water:cement ratio for quality assurance.

25.2.1 Manual Records

If records are being generated manually, entries should made at all pressure movements so that the pressure behavior can be plotted against time. Appropriate forms should be used in which a minimum amount of field entry need be made, such as the example in Figure 25.4. Simplicity, such as requiring only a checkmark or numbers, will greatly improve the output of field technicians. Regular summaries showing the highlights of all holes and their interrelation should also be prepared. Many grouting professionals tend to analyze in great detail the activity of individual grout holes, but sometimes fail to observe the trends of overall performance. It is important to occasionally stand back and view the overall grouting operation. Understanding how the individual holes behave and interact often clarifies the scatter in individual hole data.

25.2.2 Automated Records

On large or important projects, either continuous chart recorders, or automated data acquisition systems with direct connection to a computer, should be used. Computer software programs are now available, and others are being developed, especially for the monitoring of grout injection. Moreover, some grouting specialty contractors have their own proprietary programs. A substantial advantage of computer monitoring is the ability to immediately output the data in a suitable form for closer observation or further analysis. Data formats that are computer generated can usually be adjusted in an exaggerated scale to aid interpretation of specific events.

Even where data acquisition is manual, it is advantageous to enter it into a computer database, such as Microsoft Excel, to facilitate graphical output. Figure 25.5 illustrates bar graphs from one of the first holes grouted on an actual project. The grout take and pressure are shown, side by side, as a function of depth. Note that the takes were quite low at depths greater than about 28 ft, even at relatively high pressure. Following similar behavior on several distributed primary holes, it was decided to reduce the hole depth for the remaining work.

It is useful to post such graphs according to their relative position on a wall, as illustrated in Figure 25.6. Trends such as high grout take being limited to a fairly small zone, as in the illustration, then become easy to recognize. Records posted for each hole or hole stage allow even lower-level technicians to understand the grout deposition. The ability to quickly comprehend the overall status facilitates the determination of how best to proceed. For example, the posted data shown in Figure 25.6 suggests that the depth of future holes need not be as great as long as they extend through the zone of large take.

25.2.3 Be Prepared for Changes

Aside from the fact that geology and geotechnical engineering are not perfect sciences, from a practical standpoint, it is never possible to have a complete understanding of the subsurface conditions into which grout will be injected. As

FIELD DATA – DRILLING AND GROUTING

PREPARED BY ______________________ JOB ______________________ JOB NO. __________

DRILLING HOLE NUMBER __________ STAGE __________ DATE DRILLED __________

GROUTING MATERIAL __________ HOLE NO. __________ SEQUENCE __________ DATE GROUTED __________

Depth	Material	REMARKS	Quantity	Rate	Press.	Time	REMARKS

FIGURE 25.4 Example of form for manual records of grouting data.

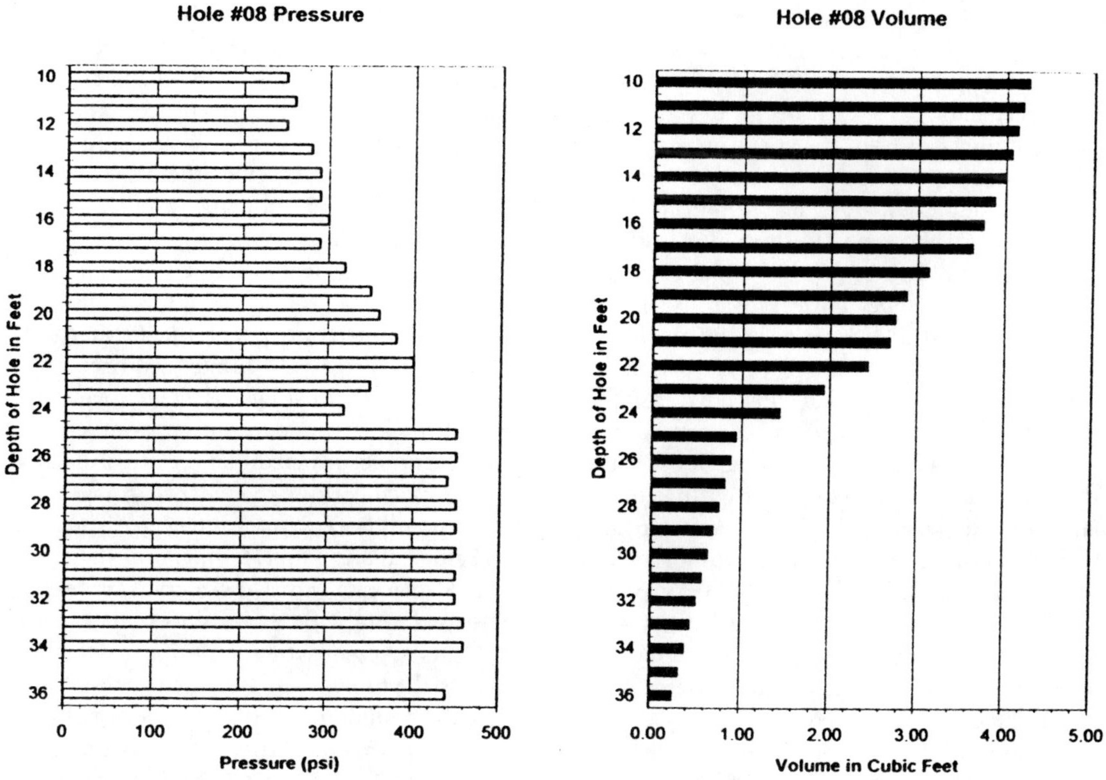

FIGURE 25.5 Graphical presentation of manually derived data entered into an Excel spreadsheet.

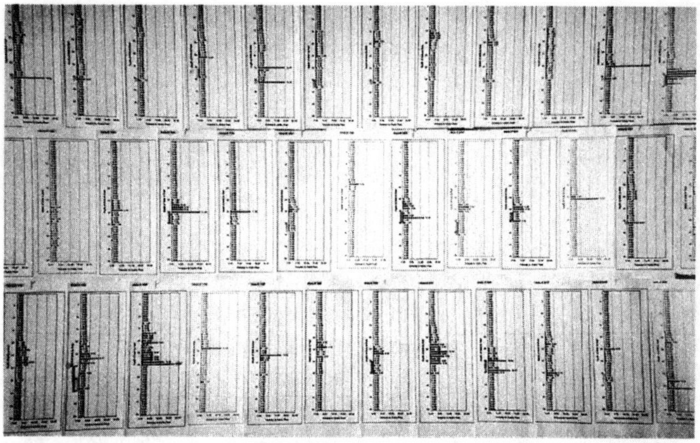

FIGURE 25.6 Graphical records assembled in proper order clearly show injection trends.

mentioned throughout this chapter, the flexibility of a grouting program is thus mandatory if optimal performance is to be realized. It is not unusual to find large grout takes in only a portion of the zone to be treated, as in the illustration.

In a memorable project, remedial grouting was performed to control settlement of some railroad tracks. Many years before, a small tunnel had been excavated about 30 ft (9 m) under the tracks, in which a sewer pipe was placed. The soil was very sandy, and although the records were sketchy, it was believed that the tunnel, which was only about 25 ft (7.5 m) long, was solidly shored with continuous sets of timber planks. It was said to have been backfilled with "jetted" sand. Payment for the cementitious suspension grout was based on the number of bags of cement used, the estimated quantity being 350 bags.

Penetration of about 3 in. (75 mm) of timber confirmed the expected shoring. A sudden drop of the drill steel by nearly 1 ft (0.3 m) upon penetrating the timber, however, was a surprise. The owner's inspector was advised of the likely overrun, and a filled grout was suggested as more appropriate. The inspector doggedly insisted that the contract estimate was correct, with a cryptic comment to the effect that "loose sand is what you felt." Four thousand bags of cement later, the voids were filled. The contractor's profit was obviously grand, but, more important, a valuable lesson was learned. Subsurface conditions are not always what we expect. Don't ever believe that what you *think* you know (as did the arrogant inspector) is really so. Surprises do occur in the grouting business.

Another interesting example involved the failure of a four-year-old sewer, which was constructed quite deeply in an area of generally poor soils and high groundwater. Following the development of a sinkhole, internal inspection disclosed some minor distortions of the invert alignment, suggesting questionable stability of the pipe bedding of some 12 in. (0.3 m) of 1 in. (25 mm) gravel. The repair provided for injection of cementitious grout into the bedding gravel to solidify it into rigid concrete. After a few days of uneventful injection, a large pressure drop occurred. The first thought was that grout was entering the pipe, but inspection found that not to be the case. Careful evaluation determined that no other substructure existed in near proximity; injection proceeded, but with very close observation of the pressure behavior. This suggested the filling of a small open pipe, for about 4 ft^3 (113 L) of grout, after which normal behavior returned. Subsequent discussions with the inspector of the original work disclosed that the contractor had occasionally embedded slotted pipe in the bedding as part of dewatering during construction. There were no records of this detail, and had it not been for the memory of the inspector, the otherwise harmless anomaly would have remained a mystery.

A similar incident was not so harmless. In the early 1960s, a grouting firm was retained to do remedial grouting under the main courthouse in Los Angeles. A tunnel had been excavated under the existing basement to provide access from a new underground parking structure. The tunnel passed under the heavily loaded slab-on-grade floor of an evidence storage area, by a depth of only about 4 ft (1.2 m). The floor had settled about 3 in. (75 mm) over the alignment. Slabjacking with a low-mobility grout was to be used to raise the floor to its proper grade, followed by compaction grouting of the soil down to the tunnel roof. Because this was an active public building, all work was done at night. The owner's maintenance staff was concerned about possible damage to the facility, so the owner's in-house plumber was present during the work.

One night the grouting foreman observed pressure behavior indicative of grout entering an open pipe. He halted injection and informed the plumber. On his manual record form was a comment to the effect, "Stopped grout—told the plumber I think I'm in a pipe," with the time noted. About ten minutes later a comment appears, "Plumber says everything OK," with the time noted. Injection was resumed, but a further note, after about ten more minutes, stated, "Told plumber know I'm in something—stopped grouting—hole abandoned." More notes followed, addressing further word from the plumber that all lines were checked and clear; there was no reason to stop grouting.

Some months later, with the first rain of the season, the grout was indeed found to be "in something"—a 10 in. (254 mm) cast iron pipe, serving as collector for the entire roof drainage system and crossing the tunnel alignment, was filled with grout. During repair, it was discovered that the probable cause of the original settlement had been leakage from that pipe as a result of disturbance during tunneling. Fortunately for the contractor, the vigilance of an extremely competent foreman and his notation on the records provided relief from responsibility for what turned out to be a very expensive event. As stated throughout this book, continuous monitoring and recording of pressure behavior should be mandatory all for grouting work.

In another instance, differential settlement of several inches of a large commercial building had occurred. The building, which was only about a year old, was built on one edge of a deep canyon fill. The geotechnical investigation did not reveal obviously poor soil, but called for compaction grouting as the fill was not up to specified density. Primary grout holes were spaced about 20 ft (6 m) apart along the outside of the foundation that had experienced the greatest damage. Drilling indicated nothing noteworthy, and only minor amounts of grout, with very uniform pressure buildup, were injected in the first several holes. But then a surprise: At a depth of about 40 ft (12 m), one of the holes exhibited very irregular pressure behavior and took a large quantity of grout. This was followed by similar behavior at a depth of about 35 ft (10.5 m) in another hole.

Reevaluation of the site grading data suggested a pioneering haul road might have treaded up the canyon under the building. On the basis of this assumption, grout holes were established over the suspected haul road location. Indeed, large quantities of grout were taken in only one or two stages, at various depths, in those holes. Plotting indicated a uniform gradient, typical of a haul road. Apparently, loose slopewash and fill existed on the downhill side of the haul road. With this knowledge, the grouting program was reevaluated and further work limited to the area of the suspected haul road. The cost of the work was reduced to less than half of that expected, resulting in a very happy owner. Although the contractor had the job cut short, he more than made up for the lost revenue by referrals and future work from the owner and his geotechnical engineer.

25.3 VERIFICATION OF GROUTING EFFECTIVENESS

Much has been written and discussed about verification of the effectiveness of grouting, and in fact the American Society of Civil Engineers, GeoInstitute Committee on Grouting, had an entire session on the subject in 1995, the proceedings of which are available (Byle and Borden, 1995). This attention and the frequent talk of verification by grouters as part of their promotion, notwithstanding, the sad fact is that relatively little effort has been exerted by many to truly understand the quality (or lack thereof) of their work, especially in the United States.

From a practical and economic standpoint, the most meaningful verification of the results of virtually all grouting occurs as the grout is being injected, by continuous observation and control, especially in regard to the rate of injection and pressure behavior. No amount of postinjection investigation or testing can duplicate the benefit of careful monitoring and control during injection. There are, however, a variety of postgrouting tests that can, to varying degrees, confirm the completeness and general quality of the work.

If the leaking areas can be seen in water control work, reduction of visible leakage is certainly acceptable verification. If the site of leakage is not visible, however, such as in a dam curtain, reduction of flow in the drainage system or downstream boreholes will have to be observed.

Where permeation grouting has been applied to strengthen soil, test excavations or borings can be made. The completeness of solidification will be observable, and specimens can be procured for laboratory evaluation. It must be remembered that the fundamental strength under a continuous load of chemically-grouted soil is only a small portion of that indicated by most standard laboratory compression tests, as discussed in Section 8.6.2. Such masses should thus be subjected to long-term creep tests.

For compaction grouting, in which grout remains in individual globules, density testing of the improved soil between the injection points can be performed. Although, in the early days, both split spoon specimens for laboratory testing and standard penetration tests were used, these are time-consuming and are now seldom employed. Cone penetration testing following grouting has proven reliable in determination of postgrouting density. Its one limitation is the required access for relatively large equipment, so it may not be feasible inside or under structures.

Where penetration tests are used, probes or tests are made midway between injection points, which theoretically would be the zone of least improvement. It is thus imperative to preserve the exact location of the injections, which often requires special effort. If the injection is through a paved surface, the hole locations can easily be observed. If the holes are drilled through a soil surface, however, they can become obliterated quite readily. Markers, such as long bolts embedded in the grout, or surveyor's flags can be used. Normal operation of equipment and the dragging of hoses over the surface can easily destroy hole markers, so they must be tough and resilient.

Geophysical methods, such as described in Section 22.2.1.5, have been promoted and successfully used on some projects for verification. Ground-penetrating radar and crosshole tomography are the most commonly used. Because of the considerable expertise required, these methods are expensive and time-consuming, so they are best reserved for special situations.

CHAPTER 26

Numerical Analysis of Grouting

Chapter contributed by Dawn Shuttle, University of British Columbia, and Mike Jefferies, Golder Associates, University of Manchester Institute of Science and Technology (UMIST).

26.1 ABOUT NUMERICAL ANALYSIS
- **26.1.1** Why Analyze?
- **26.1.2** Why Numerical Analysis?

26.2 FRACTURED ROCK GROUTING
- **26.2.1** The Nature of Fractured Rock
- **26.2.2** Discrete Fractured Network Models
- **26.2.3** Input Data for DFN Models
- **26.2.4** Simulation of Grouting with DFN Models
- **26.2.5** Performance Evaluation

26.3 COMPACTION GROUTING
- **26.3.1** Overview
- **26.3.2** Finite Element Software for Compaction Grouting
- **26.3.3** Expected Compaction
- **26.3.4** Analysis of Ground Response to Grouting

26.1 ABOUT NUMERICAL ANALYSIS

26.1.1 Why Analyze?

Numerical analysis is unusual, at least in current grouting practice. Design, as indicated in Chapters 23 and 24, depends a lot on precedent, with judicious adjustments for the conditions of a particular project. But estimation of quantities and costs remains problematic. The direction of grouting depends on comparing the actual ground response with what was expected and then making prudent adjustments. All in all, the engineering of grouting is strongly biased toward precedent and past experience. Yet this reliance on experience is no different than that used in other ground improvement processes, virtually all of which are empirically based.

The situation with grouting has begun to change in recent times, however, and several groups around the world have begun to take a more engineered approach. "More engineered" means more quantified, less subjective, less dependent on experience (or, more often, opinion, if one is honest), and something that can be explained to other engineers without the need for them to have 20 years' prior grouting experience. This chapter is about current developments toward an engineered approach, which depends on analysis.

Of course, a reasonable reaction may be to ask just how this approach differs from the design approach and the issues raised in Chapters 23 and 24. The answer is that probable geological aspects are going to be quantified and not just identified. The grout behavior during injection will be quantified and the geology modeled. This process is often called analysis, although a fairer term is *simulation*, as the models allow "what-if" evaluations during design.

The question then arises, Why analyze? The answer is the ability to use the aforementioned what-if simulations. It allows a clear expectation of what can be achieved for any proposed grouting method/protocol, and it permits the engineer to provide a least-cost or best-performing solution (as appropriate). This is proper engineering design. Analysis also provides a method of looking at what has been achieved, although it will not replace proof tests.

This chapter considers two situations in which analysis can yield substantial insight: fractured rock grouting and compaction grouting. As discussed in Chapters 23 and 24, the required results from grouting will have been established, such as the required hydraulic impedance of a curtain in fractured rock or the required final ground density with compaction grouting. The issue then becomes achieving additional insight in how grouting can be performed to get to these required end results.

26.1.2 Why Numerical Analysis?

The reason grouting has depended so heavily on experience and precedent is that there is no simple design methodology for grouting, such as the equivalent of bearing capacity analysis for soils or beam strength theory for structures. Even grouting of uniform sand with chemical grout is a complex situation, with the potential for grout dilution by mixing and/or the grout front not moving uniformly from the injection point. When fractured rock is considered, the information needed is far beyond that which can be captured with hand calculations. There are also situations in which the grout strengthens the ground. In ground strengthening there is a subtle interplay between volumes injected, soil movement, and confining stress that defy hand calculations.

This difficulty with the limits of hand calculations is overcome by the ready availability of inexpensive personal computers that can run complex (and usually finite-element or finite-difference based) software. These numerical approaches form the subject of this chapter.

Space does not allow complete explanation of the software in detail or even a full exposition of the theory. However, it provides links to relevant websites, and both freeware and demonstration software are available for downloading. This chapter is intended to show what can be done, why it should be done, and to provide links to get started.

26.2 FRACTURED ROCK GROUTING

26.2.1 The Nature of Fractured Rock

Chapter 24 introduced the importance of understanding the extent and properties of defects (i.e., fractures in rock). This is not simple, however, as fractures in rock form complicated networks with ranges of properties. They often have fractal-like characteristics, in which the closer one looks, the more detail one sees. The key issue is to what extent these features control liquid flow. When the fractures are sufficiently pervasive and interconnected, an equivalent hydraulic conductivity can sometimes be defined, so the rock can be modeled as a soil by using what is called an equivalent porous media (EPM) model. However, more commonly, at any given scale there are a limited number of discrete fractures, and these control flow. Grouting then becomes an issue of intersecting and sealing the key fractures, which is as much a geometric as a hydrogeologic problem, leading to the following questions:

- What grout hole pattern will most efficiently intersect the fractures that matter?
- What are the expected spatial pattern and hydraulic properties of the fractures that will not be sealed in the grouting program?
- What is the maximum grout particle size needed to seal the key fractures?
- Where will injected grout spread, and what grout volumes, pumping rates, and pressures will be required?

The situation is more complex than even these questions suggest. Fracture aperture (which determines grout volumes) varies over the surface of the fracture. Fracture infillings and roughness both affect the spread of grout. Fracture size, orientation, transmissivity, aperture, and interconnection all influence the effectiveness of grouting. All of these parameters are intercorrelated and thus difficult to characterize. Fractures are never uniform, and the large number of fractures involved in typical grouting problems makes this simulation an inherently stochastic problem. Normal site characterization describes only a small percentage of the actual fractures.

Engineering analysis for modeling 3D patterns of conductive fractures is called discrete fracture network (DFN) analysis. DFN analysis provides a series of tools to derive the spatial pattern and properties of fractures throughout the rock mass to be grouted (Dershowitz et al., 2002). This analysis requires integration of geological, hydrological, and engineering information. Chapter 24 introduced the importance of understanding the extent and properties of defects (i.e., fractures in the rock).

The first concept encountered when looking at fractures is that no single value of a parameter describes any particular aspect. For example, consider the area of a feature in its plane. Measuring the fracture area in a volume of rock shows a distribution of values, so that one can talk about an average area and perhaps a standard deviation of area, but there is no such thing as a single fracture area. Likewise, fracture aperture, orientation, and transmissivity are ever variable, constituting a range, not a single value. It is thus common to represent these ranges using probability distributions, which are seldom the actual distributions. Recognition and quantification of the distribution of fracture characteristics, with probability distributions, distinguishes a numerical approach from the more geological descriptive view put forward in Chapter 24. And this is accomplished with a discrete fracture network (DFN) approach.

26.2.2 Discrete Fractured Network Models

The discrete fracture network (DFN) modeling approach originated in the mid-1970s at the Massachusetts Institute of Technology (MIT) (Baecher and Einstein, 1975). It has since been applied to model rock masses for a wide variety of applications, including water supply, fractured oil reservoirs, contaminant transport, rock tunnel and slope stability, and grouting. Development of the DFN approach is documented in the literature, the key contributions including those of Dershowitz and Einstein (1979), Long (1984), Swaby and Rawnsley (1996), and Ivanova (1998). Fractures are generally represented as polygonal objects in three-dimensional space, as shown in Figure 26.1, with the aperture treated as a property of the object (i.e., fracture). More recent DFN modeling represents nonplanar fractures using tessellated surfaces. The geometric, mechanical, and hydraulic properties of fractures are evaluated using custom DFN procedures developed over the past 25 years. Generally, DFN models are based on a combination of deterministic fractures (i.e., fractures that are known completely), conditioned fractures (i.e., fractures for which some properties such as location are known, but other properties must be defined on the basis of population statistics), and purely stochastic fractures. In this way, the discrete fracture network directly models the geometry and hydraulic properties of discrete fractures in foundation rocks within the degree of confidence of the geological knowledge of the site.

The DFN method is also able to simulate flow of water through fracture networks and flow of grout into discrete fractures. Flow and transport in fracture networks is solved by discretizing the fractures as triangular finite elements, so that the heads and fluxes can be calculated by using interconnected two-dimensional elements in a three-dimensional model. The flow property of each finite element represents the combination of the transmissivity due to the natural in situ aperture and the injected grout volume. Figure 26.1 illustrates how grouting is simulated.

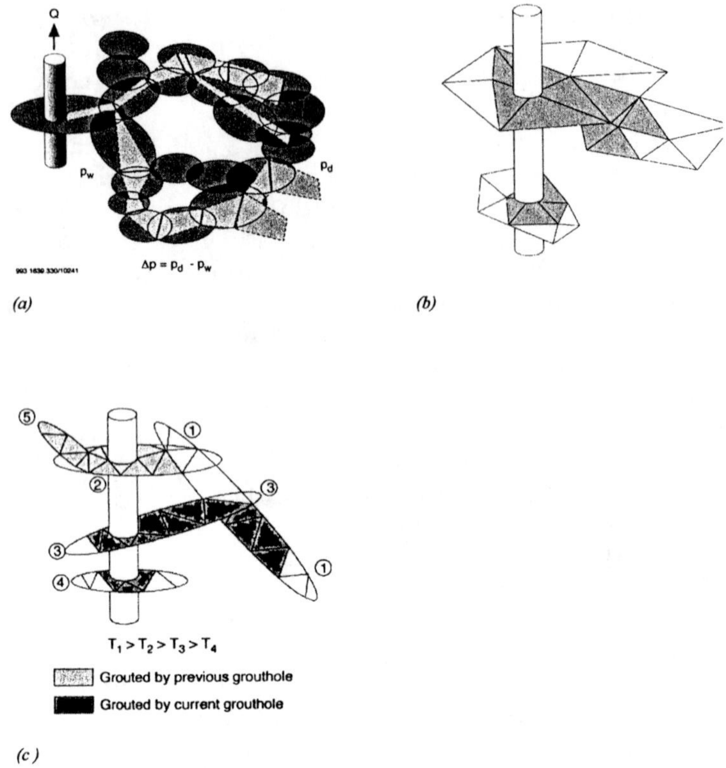

FIGURE 26.1 Simulation of grouting in DFN models: (a) Network of fractures from grout stage; (b) grouting simulation; and (c) grout pattern for a borehole section.

The DFN approach is ideally suited for quantitative analysis of foundation grouting. The geology is captured with all its associated uncertainties; the grouting is simulated with grout penetration based on the realized geology and grouting parameters (hole spacing, orientation, grout grain size and rheology, injection pressure); and the groundwater flow through this grouted network can then be simulated to assess the effectiveness of the grouting.

DFN models are created and simulated numerically; a leading package for this work is FracMan. FracMan is based on original work at MIT and now includes a complete suite of modeling tools, data analysis packages, and fluid flow simulators. Fractures are represented by planar ellipses (or finite sided approximations) with the areas, aspect ratios, apertures, and so forth, chosen from the specified probability distributions. This means that there is no single model itself with a particular feature in a single place. Rather, *realizations* of the specified model are created, with calculations carried out on each realization. The results of all these calculations are then expressed as a probability of a particular outcome. Thus, where a particular feature is known to be in a particular place (e.g., a fault crossing a borehole), then that can be specified to constrain the model. Complete information on FracMan can be obtained from the website *http://www.fracman.com.*

26.2.3 Input Data for DFN Models

FracMan puts numbers on what a geologist sees, and the approach to generating the data for a DFN model will be familiar to any geologist. DFN models can be implemented at any level of site characterization and can be successively refined as additional data become available. The initial modeling can be based on purely generic information and geologic context. As structural geology, hydraulic tests, and grout pretest data are collected, the DFN model can be updated. Because this is inherently a stochastic approach, the increased data translate directly to decreased uncertainty in model predictions.

Data analysis for the DFN approach, which is summarized in Dershowitz (1995), requires two basic types of data:

1. Structural geology, describing the geometry of discrete fractures, faults, dikes, karsts, and discrete stratigraphic units
2. Hydraulic data, such as pump tests, well tests, and grout tests

The primary sources for structural geologic data are lineament maps and borehole logs. Borehole imaging logs are particularly useful. These can be analyzed to derive fracture spatial pattern (geostructural and geostratigraphic), fracture intensity, fracture size (radius) distribution, and fracture aperture distribution. The primary sources for hydraulic data are falling head tests, grouting pretests, and hydraulic head measurements. These hydraulic tests can be analyzed to derive the fracture transmissivity distribution, the storativity, and fracture transmissivity-aperture correlations. In the absence of site-specific data, generic correlations can be used. For example, Dershowitz et al. (2002) have developed generic correlations between fracture size, aperture (storage, mechanical, and flow), transmissivity, and storativity. Hakami (1995) has developed fracture roughness profiles, which can also be used to provide generic correlations between mechanical and hydraulic fracture apertures.

26.2.4 Simulation of Grouting with DFN Models

Grouting is simulated by including the grout holes in the model in the same place, orientation, and sequence as proposed for the construction work, and then simulating the injection of grout into fractures intersecting or connected to the grout holes. In this approach, the effectiveness of any particular grouting program can be addressed directly, inasmuch as discrete fractures not directly or indirectly connected to these holes are not grouted, but provide potential flow pathways through the foundation and constructed curtain.

The two approaches for simulation of grout distribution to fracture planes are rheological modeling (Gustafson and Åsa, 2000) and empirical modeling (Shuttle et al., 2000; Shuttle and Glynn, 2003). Grout penetration depends on a complex combination of grout grain size and rheology, grouting pressure, and fracture transmissivity, roughness, and connectivity. In the rheological approach, grout is modeled as a non-Newtonian fluid on a fracture-by-fracture basis within the model.

These simulations are complex and time-consuming. From a practical perspective, a limited number of rheological simulations are carried out to provide an empirical basis for simulating the distribution of grout into connected fractures. Such an empirical relationship can also be developed based on the basis of site-specific or generic grout injection data, as was done by Shuttle et al. (2000). In this approach, grout is distributed to the fracture surface area according to an empirical algorithm derived from field observations. For each fracture connected to a grout hole, the possible grout take volume is estimated on the basis of the fracture transmissivity and aperture, the connected fracture volume, and the grout grain size and properties. This grout take is then distributed to the fractures connected to the grout hole according to a heuristic algorithm, ordered according to fracture transmissivity and distance from the grout hole being simulated, as illustrated on Figure 26.1(a).

Following injection into the fracture elements immediately adjacent to the borehole, the remaining grout is injected into the most transmissive fracture attached to the borehole, one element at a time. Grout is injected into successive elements until (1) the fracture is completely filled, (2) a more transmissive fracture is intersected, which is preferentially filled, or (3) grout penetration reaches the limit for the chosen injection pressure and grout rheology. At the end of each injection stage, the transmissivity of each grouted element is set to a value representing the in situ grouted fracture transmissivity (typically, many orders of magnitude smaller than the median transmissivity of the ungrouted fracture network).

Figure 26.1(b) illustrates the resulting three-dimensional grout pattern from a single borehole. Grouting is carried out in sequence, using the same sequence of grout sections as used or proposed for the field. As a result, if grout from a previous borehole section has filled elements connected to the current borehole section, those elements already have a low transmissivity and will not be further grouted. By modeling the sequential filling of fractures with grout, this approach replicates fractures that are not reached by grout, even for very close borehole spacing. As a result, the effective postgrouting permeability calculated with the use of the grouted DFN model has the potential to be more realistic than the estimate obtained by tests of a few confirmation boreholes, providing a more representative picture of the entire grouted mass.

The detailed grouting model described here is inherently stochastic, inasmuch as the majority of fractures in the DFN model are generated from population statistics, and not measured directly. As a result, the actual performance of each individual grout hole is not predicted; rather, the model shows the expected behavior of the overall curtain. Details predicted include the percentage of stages that will take little or no grout, the percentage of stages that will take large amounts of grout, and the mean and standard deviation of grout take per hole. Discrete fractures not directly or indirectly connected to these grout holes are not grouted, but provide potential flow pathways through the foundation and constructed curtain.

Grout penetration is based on fracture aperture, fracture transmissivity, and grout cohesion/viscosity. Hole stages with a hydraulic conductivity of less than the target value (generally of the order of 10^{-7} m/s) are usually treated as ungrouted, with this limit depending on the cement modeled in the simulation. Fractures are discretely placed into triangular finite elements, with grout injected to every finite element directly intersecting the borehole, most transmissive element first, as illustrated in Figure 26.1(a). The volume of grout taken by each element is based on the element volume, which is calculated as the product of the element aperture and the element area.

26.2.5 Performance Evaluation

DFN grouting simulation naturally estimates the geometry and properties of fractures that will remain open following grouting, and this leads to an estimation of grouted rock mass performance. DFN grout simulation provides a probabilistic estimate of grouting reliability, in that each simulated DFN represents one possible realization of the rock mass. The percentage of DFN realizations that perform within desired levels is then an estimate of the reliability of a proposed grouting program. This is particularly valuable when grout curtains are designed for contaminant containment or prevention of piping.

DFN models are especially insightful in evaluating the role of ungrouted fractures. Typically, once grouting has been completed, the quantity of grout injected into each stage and the results of hydraulic conductivity tests in confirmation holes are known. This does not measure the effectiveness of the grout curtain, however, and Figure 26.2 illustrates some of the potential issues. High-transmissivity grout sections often require a thicker grout that preferentially enters the largest fractures in a grout hole, but leaves smaller-aperture fractures ungrouted. Furthermore, grout can completely miss fractures that form a portion of the conductive fracture pathways not connected to the grout holes. Grout can block fractures in confirmation boreholes without adequately sealing the neighboring defects, giving a false indication of adequate grouting. Entire fracture sets can be missed by both grout holes and confirmation boreholes because of relative orientations or periodic spacing. All these issues can be missed by conventional grouting evaluation, which relies on "effective" reduction in conductivity of a limited number of hydraulic tests on confirmation boreholes.

All the aforementioned concerns are readily addressed by the DFN approach, however, as it naturally captures all uncertainties and issues. For example, grout curtain effectiveness can be calculated as the equivalent permeability of the fractures remaining open after grouting. A unit gradient is applied across the grouted model, and flow is calculated by the finite element method to derive the equivalent hydraulic conductivity of the grouted mass. An example of the results of such flow simulations is illustrated in Figure 26.3, which shows the distribution of hydraulic head in an ungrouted portion of the fracture network. The nonuniform head

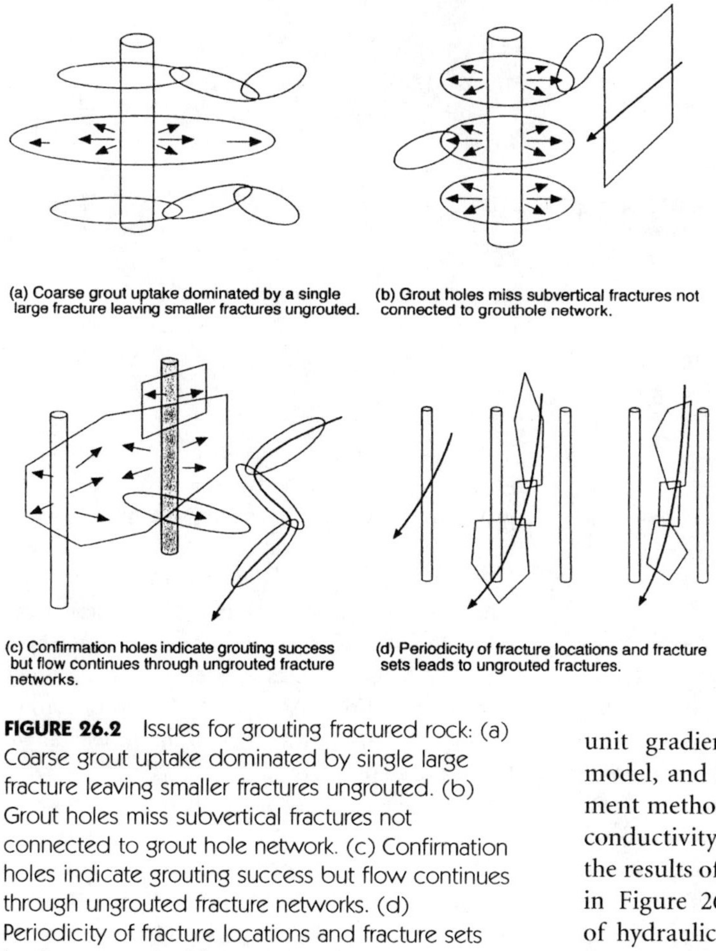

FIGURE 26.2 Issues for grouting fractured rock: (a) Coarse grout uptake dominated by single large fracture leaving smaller fractures ungrouted. (b) Grout holes miss subvertical fractures not connected to grout hole network. (c) Confirmation holes indicate grouting success but flow continues through ungrouted fracture networks. (d) Periodicity of fracture locations and fracture sets leads to ungrouted fractures.

distribution in the grouted rock mass is evident and obviously raises interesting questions about evaluating grout curtain effectiveness with the use of relatively few piezometers across a grout curtain. There is a clear potential for being misled, which can be avoided by performing DFN modeling as the work progresses.

Subsequent hydraulic modeling of retained head across the simulated curtain measures the effectiveness of grouting with the proposed grout design and protocol. The stochastic approach allows calculation of grout curtain reliability and the probability that significant conductive pathways through the curtain will remain following grouting. Reliability is determined by generating multiple realizations of the stochastic DFN model and calculating the frequency of occurrence of significant ungrouted pathways.

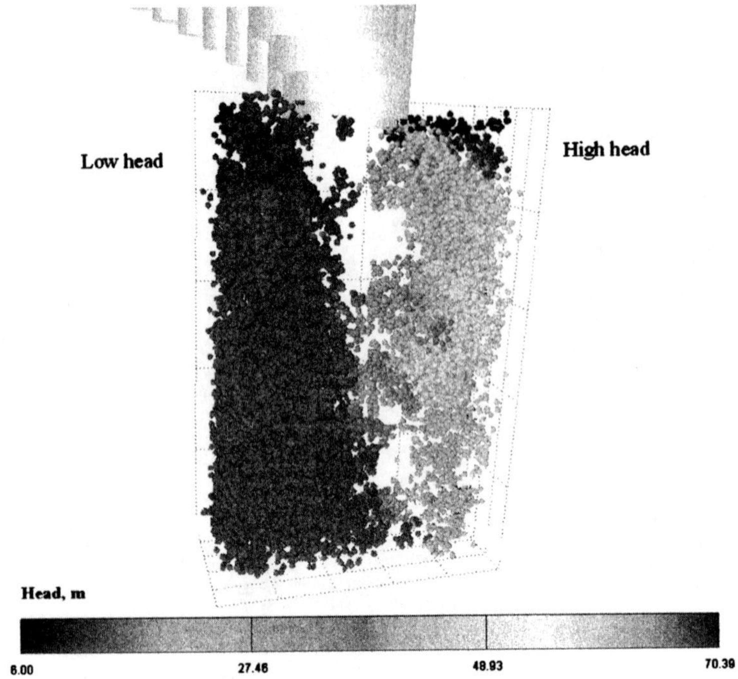

FIGURE 26.3 Computed distribution of head in ungrouted fractures from DFN model of the grout curtain and surrounding rock.

26.3 COMPACTION GROUTING

26.3.1 Overview

When compaction grouting is contemplated, interest often centers on determining the density change that can be achieved. During or after the work, it is usual to want to know what density was actually achieved. Analysis of compaction grouting can help with both tasks and can also indicate to what extent postgrouting penetration resistance changes are caused by increased ground strength as opposed to increased horizontal confining stress.

Analysis of compaction grouting is a challenging task because the soil adjacent to the grout bulb experiences shear strains of several hundred, or even thousands, percent. Moreover, because soil behavior depends on density and compaction grouting changes this density, a good soil stress-stain (constitutive) model that captures density effects is essential for representative calculations. As good soil models cannot be treated analytically, large-strain finite element techniques are required.

The few soil constitutive models that account for the effect of changing density during shear, as needed for reasonable analysis of compaction grouting, are all based on the state parameter (ψ) as the basic index of soil behavior.

The state parameter is simply the difference between the current void ratio of the soil and the critical state void ratio at the same mean stress. Unlike relative density, soil behavior in terms of the state parameter is unaffected by fines content or stress level. The state parameter was suggested nearly 40 years go by the late Peter Wroth (of critical state soil mechanics) and shown to be useful by Been and Jefferies (1985, 1986). Simply put, a very dilatant and dense soil is one with $\psi \approx -0.3$, the negative sign arising because the soils void ratio is less that that for the critical state. In contrast, a very loose and easily statically liquefiable soil would have $\psi > +0.1$. Soils can exist in a whole range of states between these practical limits, and ψ indicates just where in this range of behavior a soil lies, regardless of soil type. The state parameter is a bit like relative density in a conceptual sense, except that it actually works as a unifying index, unlike relative density.

Although compaction grouting is usually carried out with a grid of holes, possibly grouted in a primary-secondary sequence, analysis to date has not looked at the interplay between hole arrangement and sequence. Instead, the two published analyses (Shuttle and Jefferies, 2000; Kovacevic et al., 2000) both considered the ground response to a single hole—in effect, what is achieved with primary grouting. This restriction allowed the convenient adoption of radial symmetry, so sophisticated constitutive models could be implemented numerically in an easy and computationally efficient manner. Rapid parametric studies on-site also became possible, allowing "what-if" simulations of ground response to grout injection based on the measured grout pressure-take behavior. The results presented are based on a sufficiently slow injection rate to provide a drained response in the target soil. The effect of faster injection rates inducing a partially drained response has been modeled, but not yet thoroughly dealt with.

26.3.2 Finite Element Software for Compaction Grouting

Finite element analysis of compaction grouting is numerically challenging, but need not prevent routine use of the approach. There are standard commercial packages that can be adapted to compaction grouting, given a little diligence with the constitutive model. The Flac program marketed by Itasca is a popular package with consulting engineers that can be used for detailed analysis. The adoption of Flac requires adding a constitutive model, using Flac's add-in capability. Details of a suitable constitutive model, NorSand, have been published (Jefferies and Shuttle, 2002), with code fragments that can be downloaded from the Geotechnique website via the Internet. But perhaps more useful, the code used to produce examples provided shortly, can be downloaded from the website of the Geotechnical Research Group at the University of British Columbia (UBC).

The downloadable code from UBC comprises an executable file that takes the simulation parameters as an input file and writes output files of void ratios, stresses, and so forth, that can be opened with a spreadsheet for plotting and engineering assessment. Details on the procedures and file types are in the file *comapctgrout.zip* that can be downloaded from the UBC website, *http://www.civil.ubc.ca/research/geotech/index.html*. This zip file also contains the executable code together with the downloadable code, as the file *comapctgrout.zip*. A description of what these inputs are, how they are measured, and what can be computed is given in the following paragraphs.

In terms of geometry, the program idealizes compaction as taking place in a uniform soil having an effective horizontal stress prior to grouting, with the soil at a chosen initial state parameter ψ_0. The grout hole diameter must be specified. In principle, it is the effect of one grout stage that is computed, with the grout injection given in units of ft^3/ft or L/m, depending on whether Imperial or SI units are chosen.

The soil model used in NorSand (Jefferies, 1993, 1997; Jefferies and Shuttle, 2002) is a simple constitutive model based on the critical state and requiring seven parameters. These parameters are the soil properties, which are shown in Table 26.1, together with the ranges commonly encountered. Most of these parameters are familiar within modern geotechnical engineering, and there is further discussion of them in the downloadable files.

Key to a reasonable simulation is correct assessment of the initial conditions. The in situ horizontal geostatic stress has always been difficult, and its determination remains controversial. Horizontal geostatic stress is almost never at the widely quoted $K_0 = 1 - \sin(\phi)$. As a rule of thumb, $K_0 = 0.7$ seems to fit measurements in loose sands, with overconsolidation, preloading, vibration, and/or densification leading to higher values. The initial state parameter is readily determined directly from CPT data, which, it is hoped, form part of the precompaction site investigation. Been et al. (1997) give a method for estimating ψ from CPT data, with Shuttle and Jefferies (1998) providing a formal framework; ψ is related to the logarithm of the stress-normalized tip resistance.

26.3.3 Expected Compaction

One purpose in modeling compaction grouting is to determine realistically achievable densification targets and how those targets may be a function of the grouting intensity (quantity of grout per unit length of grout hole). Figure 26.4 illustrates simulations using the software that can be downloaded from UBC for a silty sand. The figure shows void ratio versus distance from the axis of the grout hole at various stages during the grouting, so that the evolution of density change can be seen.

Results for three initial conditions—very loose ($\psi = +0.05$), loose ($\psi = +0.00$) and medium dense ($\psi = +-0.10$)—are shown in Figure 26.4. Because there are three different densities, there are three different initial void ratios. The first two sets of simulations show the effect of grouting in loose soils, such soils being expected to be improved by compaction grouting. There is quite a concentration of the com-

TABLE 26.1 NorSand Parameters and Typical Values for Sands

Parameter	Typical Range	Remarks
CRITICAL STATE LOCUS (CSL)		
Γ	0.9–1.4	"Altitude" of CSL, defined at 1 kPa
γ	0.01–0.07	Slope of CSL, defined on base e
PLASTICITY		
M_{tc}	1.2–1.4	Critical friction ratio, triaxial compression as reference condition
H	50–500	Plastic hardening modulus, often f(ψ)
χ_{tc}	2.5–4.5	Relates peak dilatancy to ψ; often taken as 3.5 Triaxial compression as reference condition
ELASTICITY		
G	40–200 MPa	Elastic shear modulus, usually measured geophysically
ν	0.1–0.3	Poisson's ratio, commonly 0.2 adopted

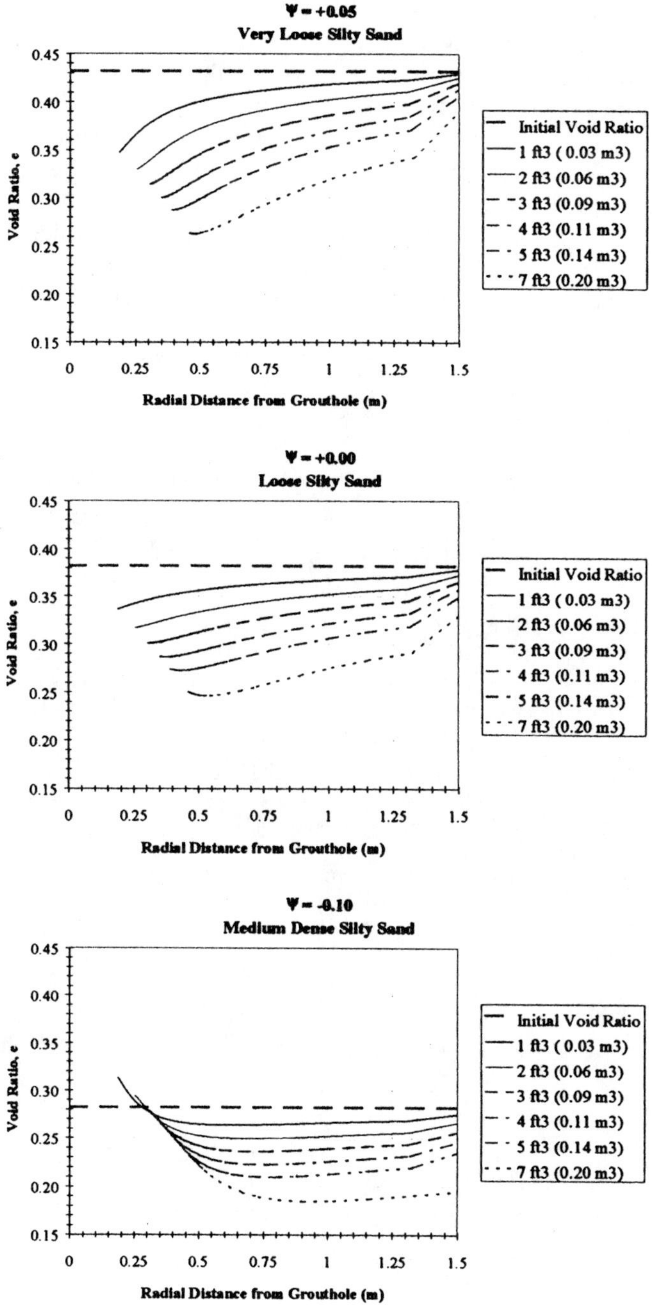

FIGURE 26.4 Example of void ratio change induced by compaction grouting.

paction effect close to the grout hole at low injected volumes, but once the quantity of injected grout rises to about 4 ft^3/ft (113 L/0.3 m), the compaction effect becomes pronounced and extends at least 3 ft (0.9 m) from the grout hole. Notice that the void ratio change induced by this compaction grouting is an average of about 0.1 void ratio units, with the very loose silty sand at 5 ft^3/ft (142 L/0.3 m).

The third set of simulations was undertaken to show that compaction grouting can densify soils that are initially more dense than loose. The initial condition was moderately dense initial state $\psi = -0.1$, which is actually quite dilatant, with a peak friction angle typically about 38 degrees. The initial stages of grouting do indeed loosen the soil near the grout hole, as the soil dilates upon shearing. However, the increasing means stress overcomes the initial dilation and compaction ensues once the injected grout exceeds 3 ft^3/ft (85 L/0.3 m). Of course, only modest densification is achieved in this initially rather dense soil. The purpose of this simulation is to illustrate that dense soils are not harmed by compaction grouting.

An interesting aspect of this last analysis is that it indicates why compaction grouting is effective. That effectiveness is sometimes questioned because of the idea that grouting simply compresses the soil, so only minor density improvement can be obtained. This is erroneous, however, as the effects of shear must be considered. Soil is sheared by expansion of the grout bulb, not just compressed. When sheared this way, soil responds to increased stress at the slope of the critical state locus, not to constrained compression stiffness. Roughly, compression during shear is three to five times more effective at inducing density change than during a simple increase in mean stress, which is why compaction grouting works.

26.3.4 Analysis of Ground Response to Grouting

Finite element analysis can be used to infer ground conditions from compaction grouting, provided that the injection pressure and grout flow rate are measured with electronic transducers. An example of such data is shown in Figure 26.5. Ground conditions are inferred by guessing the soil state and initial stress conditions, computing the expected pressure versus grout injected relationship, and then comparing the computed relationship with that measured. Guesses are iterated until a close match between simulation and measured ground response to grout is obtained, with these parameters then set so as to directly indicate the ground condition.

Analysis of electronically measured grouting parameters is an alternative and complementary approach to waiting until grouting is completed and then doing further penetration tests. There is the added advantage that this iterative modeling approach estimates the achieved stress state as well as the achieved density. Figure 26.5 indicates an example of the achieved match.

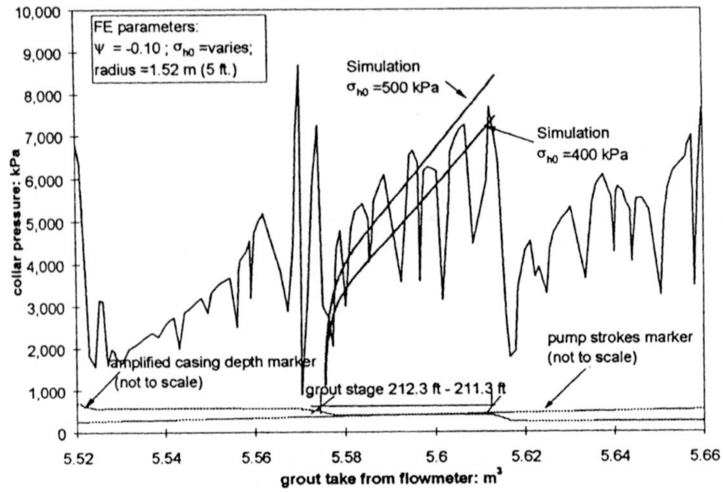

FIGURE 26.5 Examples of the match between computed and measured grout injection.

CHAPTER 27

Specifications and Contracts

27.1 SPECIFICATION REQUIREMENTS
- **27.1.1** Prescriptive Specifications
- **27.1.2** Performance Specifications

27.2 CONTRACTING ARRANGEMENTS
- **27.2.1** Contractor Prequalification
  - **27.2.1.1** Prequalification Submittals
- **27.2.2** Negotiation with Chosen Contractor
- **27.2.3** Payment Basis
  - **27.2.3.1** Time and Material
  - **27.2.3.2** Cost Plus Fixed Fee
  - **27.2.3.3** Lump Sum
  - **27.2.3.4** Unit Price

27.3 FULL-SCALE TRIALS

ALTHOUGH GROUTING is very much a technology, it is also a specialized area of activity for most owners and their engineers, most of whom do not practice it extensively. There are available, however, many geotechnical engineering firms that have professionals with considerable grouting experience, as well as several individual consultants. Yet perhaps those in the best position to maintain expertise in the technology are the contractors who regularly perform the work. Many have the specialized knowledge, equipment, and experienced personnel to do this work well. Specifications should not be so limited or stringent that well-qualified contractors will be precluded from using their innovations and special expertise.

Unfortunately, however, many grouting contractors completely ignore the sound engineering fundamentals or the established grouting technology, and some continue to shroud their operations in mystery. We now have high-tech "magic" purveyors who use the technical jargon espoused by well-educated professionals to promote their activities, often in professional organizations and technical publications, as well as through persuasive advertising. This latter group often sponsors seminars, where they provide supposedly nonproprietary handouts and guide specifications. Unfortunately, their field performance very often falls far short of the effectiveness of their marketing activities.

One must thus use extreme care in accepting advice from grouting contractors or, as many refer to themselves, geotechnical specialty contractors. One should especially resist using the "guide specifications" distributed by such operators for tentative projects. Such documents virtually always favor the supplier and very often provide clever language that will eventually be used as the basis for a claim or extra payment. So pervasive and sometimes biased are contractor-supplied specifications that their origins can often be identified upon viewing. Unfortunately, such guide specifications are often technically erroneous, and their use can result in poor project performance, regardless of which contractor is awarded the work. For example, a recent specification for compaction grouting, prepared by a well-known geotechnical engineering firm, stated that "additives such as . . . bentonite . . . can be included in the grout at the contractor's option." As has been well established, and discussed in Chapter 2, inclusion of bentonite in compaction grout will cause the grout to behave as a fluid in the ground, presenting a serious risk of hydraulic fracturing of the soil and loss of control of the grout deposition.

27.1 SPECIFICATION REQUIREMENTS

Specifications must be well thought out and directed to the requirements of the actual project. They can be either prescriptive or performance-based, but must clearly define the intent of the work and the limits of the zone to be grouted. The extent and detail required depend on the method of contracting to be used, which is discussed shortly.

27.1.1 Prescriptive Specifications

Prescriptive specifications dictate exactly what the contractor is to do and how to do it. This is the surest way to achieve an acceptable end product when the contractor cannot be preselected. Prescriptive specifications do, however, discourage contractor development and restrict the use of any special talents or abilities a contractor may have. Thus, although the end product will be acceptable, superior performance will not likely occur. Where competitive bidding is used, the best contractors may decline participation, as they will not be allowed to exploit their unusual knowledge and talent. Prescriptive specifications must be specific and technically correct, which requires the writer to have considerable expertise in the required grouting procedures. The work in which they are used will require full-time observation by inspectors fully knowledgeable of the methods to be employed. Prescriptive specifications are best used where open bidding is required or control over the selection of a contractor is limited.

Because the final product depends on strict control of the contractor's operation, prescriptive specifications must be detailed, thorough, and technically correct in every respect. Specific areas that should be addressed include the following:

- A tentative layout of the grout injection holes. Provisions for changing either the number of holes, or their spacing or location, must be provided, however.
- Straightness or verticality requirements for the holes, and the range of hole diameters that are acceptable.
- Grout mixer requirements. High-shear, colloidal mixers should be required for cementitious suspension grouts.
- Material requirements. These will vary with the type of grout used. For example, bentonite or other clay should be specifically prohibited from use with compaction grout. Either the specific grout mix or the generic classes of acceptable solution grouts for water-control grouting should be stated.
- Acceptable range of pumping rates, and provision that it will be uniform at all required rates.
- Range of grout pressure required and method of pressure selection for individual holes and hole stages.
- Requirements for monitoring and recording the injection parameters during grouting. Consider the mandated use of a record format, such as shown in Figure 25.4, or a similar layout specially developed for the particular project or grouting method being considered. Note that continuous monitoring of pressure behavior should be mandatory in most grouting,

which should be recorded along with the time of each record entry (with the exception of some water-control applications). Continuous, real-time computer monitoring is preferable; however, many contractors lack either the knowledge or the ability for its application, and it is thus usually reserved for large or important projects.

- Requirements for workers and supervisors of the grouting crew. Some contractors, especially large companies, send only a supervisor and/or perhaps one key worker, hiring locally to form the crew. This results in less than good performance, because of the special skills and interworking of crew members required in grouting. This practice should be prohibited by specification.

The efforts needed for administration and inspection of work performed under prescriptive requirements are much greater than those for work based on performance specifications. Specification requirements are often contested, field relations can be difficult, and requests for change orders and extra payment are common. Upon completion of a project, disputes and contractor claims often arise, which virtually always increase the cost to the owner.

Every provision of a specification merits thoughtful consideration. The exhibit on pages 546–559 presents guidelines pertinent to the various provisions. Although they apply directly to compaction grouting, much of the discussion is also applicable to all grouting and demonstrates the requirements and reasons which must be considered.

27.1.2 Performance Specifications

Performance specifications require only that a stated improvement of the formation result, and leave the means and methods to the contractor. This encourages contractors to use their greatest ability to achieve the required benefit. All responsibility is assigned to the contractor, which makes administration much easier and greatly reduces the required inspection effort. This approach does require a clear statement of the required improvement and must designate the means to be used for verification. Herein lies the greatest limitation, in that effective tests may not exist. Furthermore, any confirming tests will be short-term, whereas the work is often expected to perform long-term. Moreover, the beginning condition must be accurately stated, which generally means a comprehensive investigation up-front.

Performance specifications encourage innovative contractors to both show and benefit from their special abilities and, in the case of a truly competent contractor, will usually result in superior performance. The designer and specifier will also be greatly relieved of responsibility for any problems encountered during the work. When especially competent contractors are used, the work will typically be completed faster and more economically. Moreover, because the means and methods are chosen by the contractor, the risk of extra work claims is greatly reduced.

Performance specifications work best for work that allows easy determination of satisfactory completion. Water control, where the leakage is obviously visible, is a good example; the seepage is either stopped or not. Groutjacking of structures is another, wherein the tolerances achieved are easily measurable. Grouting in geomaterials, however, can be difficult and expensive to verify. In rock, core drilling is required, which is both time-consuming and costly, and soils can present even greater problems. Permeation grouting for strengthening usually does not require 100 percent filling of pore space. Determination of the required filling may be difficult, and evaluating voids remaining after injection nearly impossible. The use of performance specifications is thus best left to applications where results are readily determinable, or where a well-qualified contractor of proven ability and high principal has been employed.

GUIDE SPECIFICATIONS FOR COMPACTION GROUTING

Using the Bottom-Up System

1.0 Scope of Work

Compaction grouting shall be performed as generally depicted on the plans, as outlined in these specifications, and as directed by the engineer. Changes may be called for in both the exact layout and number of grout holes as conditions encountered during the work are evaluated. The purpose of the work is to densify compressible soil within the specified work area.

Define scope of project. If groutjacking of settled improvements is to be included, so state and provide final elevations and required tolerances. The work can be done on a prescription or performance basis. Prescriptive specifications should provide the pumping rate and the maximum pumping pressure required. Performance requirements usually require the soil to be improved to a given density, usually based on either final STP or CPT resistance values.

2.0 Access and Site Conditions

The contractor shall visit the site and independently verify any access- or work-related restrictions. The type and condition of the soil to be improved is delineated in the geotechnical report of __________ __________ dated _____.

Provide any restrictions to working time or area of work. Compaction grouting can be performed with little interference with a facility's normal operation, but if required should be clearly defined. Refer to plans if applicable. All available geotechnical data should be provided to perspective contractors with the plans and specifications. Because compaction grouting adds considerable weight to the treated soil, it is imperative to extend the grouting to a competent soil or formational material. The anticipated depths should be stated, and provisions for additional depth to reach such a competent layer should be made.

3.0 Treatment Area and Depth

The soils shall be grouted at the locations and to the depths outlined in the plans and specifications. The grout hole locations and depths are based on the best available geotechnical and structural exploration data; however, not all grouting locations have been explored in detail. The precise limits of the area to be grouted and the exact location and required depth of the individual grout holes are not positively known but will be revealed as the work progresses. The engineer at his discretion may move the location of the proposed holes, change the proposed depths to be grouted, add new holes or eliminate some of the proposed holes shown on the plans.

Generally, a plan showing the anticipated grout hole locations will be provided. Perspective contractors should also be furnished all available geotechnical information. Any information relative to the existence of buried rocks, boulders, abandoned foundations, or other substructure should be included.

4.0 Subsurface Pipes and Utilities

The location of existing underground utilities is indicated on the plans. The contractor shall protect these lines and repair at no cost to the owner any that are damaged if located within 1 ft of the location designated on the plans. The contractor will not be responsible for damage to utilities that are not shown on the plans, or that are located more than 1 ft distant from the location indicated on the plans.

Contractors should not be held responsible for substructure for which the location is unknown. Although specifications sometimes assign all risk for such damage to the contractor, this will come at a high price and is seldom in the best interests of the owner. The use of a pipe locating specialist service should be considered where extensive substructure exists.

The contractor shall access existing sewer or drain lines both upstream and downstream of the area to be grouted. Clear water shall be run into the upstream access and observed downstream so that any leakage of grout into the line will be observed. Grout injection shall immediately stop upon the appearance of leakage into the line, and it shall be thoroughly flushed out and cleared of any grout.

The flow in wastewater pipes can usually be observed by accessing downstream manholes, cleanouts, or excavated inspection holes.

5.0 Materials

5.1 Grout Mixture

The grout mixture shall consist of cement, water, and aggregate. The slump shall not exceed 1.5 in. when measured using the ASTM C 143 slump test cone.

Do not specify slump determination as per the C 143 test, as it requires rodding of the grout as the mold is filled. Such rodding will result in holes in the somewhat cohesive compaction grout mix.

A minimum of _____% cement, by weight, of the aggregate shall be used.

Most compaction grout includes 6% to 12% cement; however, cement is not required subject to provision of enough fines in the aggregate to ensure the required pumpability.

The grout shall obtain a minimum unconfined compressive strength of _____ psi when tested in accordance with the requirements of ASTM C 39.

The strength of the grout is of relatively little importance, as the purpose of the injection is to compress and densify the adjacent soils. A common requirement for compressive strength is 100 psi. Specification of excessively high compressive strengths provides little benefit and can add appreciable cost. In many cases no strength requirements for the grout may be preferable.

Additives to improve pumpability, such as concrete pumping aids, clay, and similar materials, shall not be used.

Regardless of slump, such additives can cause the grout to behave in the ground like a fluid, resulting in hydraulic fracturing and loss of control.

5.2 Cement

Cement shall be Type I or Type II, per the requirements of ASTM C 150. Other types of cement may be used, subject to the engineer's approval.

It is important to keep in mind the purpose of the grout, which is to expand in a homogeneous mass so as to displace and compact the adjacent soils. High strength is seldom required, so the properties of the cement, if used, are of relatively little importance.

The cement can be delivered to the job site in bags or by bulk delivery. Bulk-delivered cement shall be stored in appropriate silos or other containers specially designed for cement storage. Cement storage facilities shall be subject to the engineers' approval.

5.3 Aggregates

The grout aggregate shall consist of naturally occurring round-grained materials conforming to the grain size distribution shown in the envelope on Figure 6.15.

Proper grout aggregate gradation is fundamental to satisfactory application of compaction grouting. A copy of Figure 6.15 is attached. Use of crushed or angular aggregates does not adversely affect the properties of the hardened grout, but can result in pumpability problems. Such use should not be prohibited, but grout pumpability should be the responsibility of the contractor and will not be a problem with natural round-grained materials.

The inclusion of clay-size particles in the aggregate will not be allowed. The portion of the aggregate passing a No. 40 sieve shall not possess a plasticity index (PI) greater than 15 when tested in accordance with ASTM 424.

Clay in the grout will likely result in the loss of injection control and hydraulic fracturing of the soil. This is the number one cause of compaction grouting failures and must not be allowed to occur.

5.4 Water

Water shall be free of excessive amounts of salts or other impurities that adversely affect the set or hydration of the cement in the grout mixture.

Again, it is good to keep in mind that the strength of the grout is usually of relatively little importance.

6.0 Drilling and Grouting Equipment

6.1 Drilling Equipment

Drilling equipment shall be capable of advancing the grout holes to the required depth through the existing soil and rock materials, as well as any natural or man-made obstructions. The term *drilling* herein used

Some contractors prefer to drive a casing, utilizing a disposable plug on the bottom end that is knocked out prior to grout injection.

shall include driving methods for casing placement. Either rotary or rotary percussive drilling equipment can be used. Circulation flush shall be accomplished with water or an air-water foam. The use of other drilling fluids, or mud, will not be allowed.

The remnants of many drilling fluids, and especially mud, can result in hydraulic fracturing of the soil and loss of control of the grout injection when pressurized by the grout.

6.2 Grout Casing

The casing shall be a minimum of 2 in. (50 mm) internal diameter and possess sufficient strength to withstand the drilling/driving, grout injection, and withdrawal forces required for the work. It shall have flush joints on both the interior and exterior surfaces. The joints may be threaded or welded. The maximum casing joint length shall be 5 ft. (1.5 m).

Larger casing will have no adverse effect on the work, but is obviously heavier and more difficult to handle.

Longer casing lengths preclude efficient reading of the grout pressure gauges that are placed on the top of the casing. Longer lengths are also difficult to handle and present a safety risk.

Any casing or holes lost because of broken or bent casing, insufficient casing strength, or the contractor's inability to pull the casing, shall be abandoned, and new replacement holes shall be drilled and grouted at the contractor's expense.

The casing is subject to extraordinary forces during grout injection and withdrawal. Common drill casing is often not of sufficient strength to withstand these forces, especially in rocky soils or in near proximity to a hard boundary.

6.3 Casing Withdrawal System

The casing withdrawal system shall be capable of withdrawing the casing in the required stages under both normal and extraordinary conditions. Any hole lost because of the contractor's inability to pull the casing will be replaced with a new hole at the contractor's expense.

Casings occasionally bend, break, or become stuck, especially in deep holes and/or rocky conditions. Contractors are more likely to use adequate equipment and casing systems if they must bear the cost of such problems.

6.4 Grout Batcher/Mixer

The grout mixing system shall be capable of precisely proportioning the mix constituents and blending them into a homogeneous grout of uniform consistency. It shall be capable of continuously batching and mixing the grout in sufficient quantity, without interruption due to inadequate batching or mechanical limitations. Automatic volumetric proportioning and mixing systems complying with the requirements of

Continuous production of a grout of uniform consistency (slump) requires both quality equipment and a competent operator.

There are many manufacturers that supply equipment complying with these standards. Many grouting contractors use homemade or, in some cases obsolete, batching units that are not in compliance. The specifier is cautioned to review both the standard and the equipment proposed for use to ensure compliance.

ASTM C 885 and/or ACI 304 will be acceptable. Where used, the automatic metering systems shall be calibrated in the presence of the engineer prior to the beginning of grout injection and at any additional times required by the engineer. Properly calibrated volumetric containers and scales required for proper calibration shall be provided by the contractor. Batch-type mixers may be used, subject to provision of an accurate means of proportioning the individual grout constituents.

Some contractors prefer to use batch-type mixers, often in combination with a preblended grout material furnished in bags.

6.5 Grout Pump

The pump shall be of the piston type and have a grout pumping piston no greater than 4 in. in diameter. It shall provide positive displacement of the grout and be capable of pumping at variable rates of 0.5 to 3.0 ft^3 (14 to 85 L) per minute at continuous pressures of up to 1000 psi (690 bars).

Larger-size cylinders result in inadequate velocity of the grout through the system and allow for excessive leakage at the seals because of their greater circumference. In some applications, greater pressure capability and/or a greater or lesser pumping rate will be required. Specify accordingly. Keep in mind that line friction will result in a pressure loss of 1 to 4 psi per foot of delivery line.

The pump shall be in first-class operating condition and capable of complete filling of the grout cylinders on each stroke. It shall be equipped with a force-feed or other mechanism as required to ensure such complete filling. Short stroking of the pump will not be allowed, and any grout pumped during such malfunctions will be at the contractor's expense.

A remote off-on control for the pump shall be provided at the grout injection point. An accurate system to measure the quantity of grout pumped during any time interval shall be provided. The grout volume measuring system shall be calibrated by the contractor each day of pumping at the beginning of work and at any other time that short stroking or a change of the grout pumping rate is suspected.

Small-line concrete pumps are most often used for compaction grouting. These pumps are usually equipped with very long material cylinders. Although quite adequate for plastic concrete, complete filling with the relatively sticky grout mix is often difficult.

6.6 Grout Delivery Line

The grout delivery line shall consist of high-pressure flexible hose, or a combination of hose and rigid pipeline, and be watertight under pressure. It shall be a maximum of 2 in. in diameter. All components of the delivery line, including coupling clamps, shall be in good condition and capable of handling the pumping pressures to be used with a minimum safety factor of 2. A pressure test of the delivery line to confirm its conformance with these specifications, shall be performed as required by the engineer.

Watertightness of the delivery line is essential. Even a slow drip from a coupling can cause excessive water loss from the grout, resulting in downstream plugs.

Most compaction grouting is done with a 2 in. (50 mm) delivery line, and 1-1/2 in. (38 mm) lines have been successfully used. Larger lines result in insufficient grout velocity, which delays the arrival of fresh grout to the header and can contribute to line blockages.

6.7 Pressure Gauges

Pressure gauges shall be installed at the grout pump and at the grout header (top of hole casing). All gauges shall be accurate and in good working order. They shall be protected from grout intrusion by suitable gauge protectors. Gauges shall have a minimum dial diameter of 3 in., and the maximum pressure range shall be not greater than 150% of the anticipated maximum grout pressure. A sufficient number of spare pressure gauges shall be maintained on the job site, and any gauge of questionable accuracy shall be promptly replaced.

A 3 in. (75 mm) dial is about the smallest that can be readily read from a reasonable distance.

Gauge readings are not accurate in the upper and lower 25% of the dial range.

7.0 Data Acquisition and Reporting

7.1 Required Monitoring Information

The contractor shall maintain and provide the engineer a continuous log delineating the drilling and grout injection parameters for each hole. Minimum data provided on the log shall include:

- Date(s) and time the hole was drilled.
- The drilled depth
- Note of any obstructions encountered or unusual events occurring during drilling
- Depth drilled and elevation of both the top and bottom of the hole
- Date(s) and times grouting of the hole was started and ended

- Injection pressure at the pump and at the hole collar in not more than three-minute intervals from the time grouting starts
- The injected volume at each three-minute interval
- Grout termination criteria
- Location and amount of any uplift or displacement of any structure or the ground surface during grouting
- Notation of any other observations relating to the grouting

The name of the person preparing the logs and the current date shall occur on all pages of the log, and the pages shall be numbered chronologically from the start of work. The current time shall be noted at each log entry.

Pressure and rate of injection are directly related. The pressure will increase with an increase of the injection rate and will lower as the injection rate is reduced. Both values are of equal importance, and one without the other is of limited usefulness. Knowing the time when events occur allows calculation of the pumping rates, which are fundamental to understanding the effectiveness of the work. Further, knowledge of the times of various events allows analysis of the project execution and production, should this be required following completion of the work.

7.2 Daily Report

The contractor shall provide a daily report delineating the activities for each day that his people are on the job site. The log shall provide, at minimum, the following:

- Date
- Weather
- Holes drilled
- Holes grouted
- Materials received and time of delivery
- Names of any visitors and time on job
- Details of any accidents, injuries, or other unusual events, including the occurrence time

Copies of receipts or delivery tickets for all materials delivered to the job site shall be provided to the engineer.

Knowing the quantities of material delivered to the job site allows a check of the payment quantities billed.

7.3 Movement Monitoring System

The contractor shall provide instrumentation to detect any movement of the ground surface or any structure within a radius of 30 ft (9 m) from any hole being grouted. The monitoring devices shall be capable of detecting movements of 1/32 inch (0.75 mm) or more in any direction.

The instrumentation may include but is not limited to stringlines, telltales, manometers, lasers, and optical devices. Backup surveyor's instruments shall be provided in such quantity as to allow evaluation of the movement of all structures without the necessity of moving the instrument location during injection. The original position of all structures shall be established prior to grout injection by the placement of suitable markings or targets thereon.

The contractor shall dedicate experienced and fully qualified personnel to monitor the instrumentation so as to prevent any damage to the site or structures. Any damage that does occur due to insufficient instrumentation or monitoring effort shall be repaired or replaced by the contractor at no cost to the owner.

This is often lacking and a frequent cause of damage during grouting.

8.0 Communication System

The contractor shall provide a communication system that allows immediate voice contact between the grout hole, pump operator, and monitoring personnel. Any damage that occurs as a result of lack of immediate communication shall be repaired or replaced by the contractor at no cost to the owner.

Lack of communication is another frequent cause of damage during grouting.

9.0 Sequence of Work

The sequence of drilling and grouting of the individual holes will be at the direction of the engineer. In general, alternate primary holes will be grouted first, followed by the intermediate secondary holes.

10.0 Drilling

10.1 Establishing Grout Holes

The grout holes can be effected either by drilling or driving of a temporarily plugged casing. The method used shall be capable of penetrating rocks and any other geological obstructions. Regardless of the method used, the casing shall be in tight, intimate contact with the surrounding soil of the resulting hole so that it is firmly held in place and resistant to ejection from the grout pressure, and/or leakage of grout around the perimeter.

The exact method of effecting the holes is not significant in most compaction grouting work. On some projects, however, specification of the exact drilling method is in order. For example, hydrocollapsible and other dry soils require moisture in order for compaction to occur. The requirement of a drilling method using water for the circulation flush can provide the necessary moisture.

10.2 Hole Location
The exact location, depth, and spacing of the grout holes is subject to change as directed by the engineer during the performance of the grouting program.

10.3 Control of Drilling Circulation Flush
Water or other circulation media shall be captured and directed to an approved disposal location. Ponding of uncontrolled drilling flush or wastewater in the work area will not be permitted. Settling tanks or other devices to separate drill cuttings and other solids from the water shall be used when required.

Ponding of water and mud in the work area is a big problem with some contractors. Excessive water ponding on the surface can penetrate and weaken the upper soils so as to compromise their ability to restrain the grout forces. It also adversely affects orderliness of the work.

10.4 Water Injection
The injection of water into the drilled holes in addition to any remnant water from the drill circulation may be required or prohibited by the engineer, depending on the exact conditions encountered.

Dry soils may require additional water for adequate compaction.

Contractors typically inject water to loosen stuck drill steel or casing, and this must be reasonably controlled. The injection of excess water can weaken large volumes of soil at depth, resulting in unacceptable surface settlement.

10.5 Drilling Log
The contractor shall prepare a log of each grout hole, which delineates the nature of the geomaterial penetrated. The log shall include the depths of hard or soft soil zones, any existing voids, reduction or loss of circulation flush, encountering of rocks, boulders, and any other significant conditions. Copies of the log shall be provided to the engineer upon completion of drilling of each hole.

It is good practice to consider every hole an exploratory hole. Although the acquiring of detailed information is not practical during drilling of grout holes, a cursory description of the materials and conditions encountered can add valuable information to the known data.

11.0 Grout Injection
11.1 Depth Confirmation
The actual depth of the open grout hole shall be confirmed by measuring with a properly weighted measuring tape immediately prior to connection of the grout delivery line. The measured depth shall be noted. If it is less than the planned depth, the hole shall be redrilled to proper depth prior to grouting.

Occasionally, driller's depth records are incorrect. In addition, caving or squeezing in of the holes can occur prior to grout injection.

11.2 Sequence

The sequence in which the holes are drilled and grouted is subject to the approval of, and may be modified by, the engineer. Grout injection shall not be initiated into any hole within 12 ft (3.6 m) of a hole previously grouted within the preceding 12 hours. In general, holes located near a downslope or retaining wall shall be injected prior to those holes located more distant therefrom.

The sequence in which the holes are injected has a profound influence on the success of the work and the avoidance of further damage. Generally, perimeter holes should be grouted prior to those on the interior. Zones that provide lesser lateral restraint, such as those near downslopes or retaining walls, should always be grouted first.

11.3 Grout Staging

Holes shall be injected in ascending stages, starting at the bottom and working upward. No stage shall be injected until the immediately underlying stage has been completed. Individual stage lengths shall be a minimum of 1 ft (0.3 m) and a maximum of 4 ft (1.2 m) in height.

Most contractors pull the casing in 1 ft (0.3 m) increments, which is valid for many jobs. Upward heave of overlying structures is a common refusal criterion. Whereas the heave resulting from one grout stage is insignificant, the cumulative heave of many stages can be considerable. Where this may be undesirable, specification of larger grout stages is recommended. The improvement in density resulting from grouting is not diminished with the use of stage lengths up to about 4 ft (1.2 m). The stage length should consider the contractor's casing system, however. Stages of 1, 1.5, or 3 ft (0.45 or 0.9 m) are appropriate where 3 ft (0.9 m) lengths of casing are being used, but 1 or 2.5 ft (0.3 or 0.75 m) would be more reasonable where 5 ft (1.5 m) joints are employed.

11.4 Access Requirements

Access to the mixing, pumping, and injection locations shall be provided to the engineer or his designated representatives at all times.

11.5 Injection Rate

The grout injection rate shall be as directed by the engineer. It is expected to be within a range of 0.5 and 2.0 ft^3 (14 and 56 L) per minute, and at an average of 1.5 ft^3 per minute for the entire work. An adjustment will be made in the contract price if the average injection rate varies more than 10% up or down from the expected 1.5 ft^3 per minute.

Contractors usually want to pump at the fastest rate possible in order to minimize the cost of the work. However, an excessively high pumping rate can promote early surface heave, negate effective densification, and lead to hydraulic fracturing. Fine-grain soils require a slower pumping rate than coarser grain deposits.

11.6 Grout Refusal Criteria

Grout injection into any stage of any grout hole shall be discontinued as directed by the engineer. General refusal criteria can be any of the following:

- Pumping at a pressure of 1000 psi (69 bars) or more for a period of three minutes.

Some grout holes develop high pressure at the start of injection but take grout freely at significantly lower pressure after a few minutes of pumping.

- Sustained pumping at a pressure of ________ psi (________ bars) or greater.

Most soils are effectively compacted at sustained pressure within a range of 350 psi to 500 psi (24 to 34 bars). Pressure limitation criteria should generally not be set until the work has started and injection behavior in the actual job site soils has been observed.

- Unwanted displacement of the ground surface or an adjacent structure of 1/32 in. (0.75 mm) occurs except where intentional lifting of a structure is intended.
- A grout volume of ______ ft^3 has been injected.

In reasonably uniform soil where the grout volume required to obtain the required density can be rationally determined, specification of an appropriate volume limitation for the injection is advisable. In situations where the existing soil density is variable or where actual voids may exist (such as nested boulders), widely varying grout quantities will be injected and a volume refusal criterion is usually not in order. In such cases the use of a pressure criterion is usually more prudent.

11.7 Improperly Grouted Holes

Any grout hole that is lost or damaged, does not reach the design depth, or is not continuously grouted as a result of equipment deficiencies or mechanical failure, inadequacy of the grout mix, improper drilling, mixing, or injection procedures, shall be backfilled and replaced by another properly installed hole at no cost to the owner.

11.8 Groutjacking

The existing ______ structure(s) shall be raised to the grades shown on the plans, further delineated in these specifications, or as directed by the engineer. All lifting shall be performed uniformly around the structure and in small increments so as to minimize further damage to the structure.

Jacking of settled surface improvements is often performed during the compaction grouting work. Not all contractors are suitably skilled at such lifting, however. Where it is to be accomplished, appropriate specification provisions are requisite, as is selection of a contractor of proven ability. It is often not necessary to return a structure to an absolute level position, but rather to simply adjust the differential elevations so that they are not readily noticeable.

11.9 Hole Completion
Completed grout holes shall be filled to the ground surface with grout placed under a minimum pressure of 5 psi (0.34 bars).

12.0 Site Maintenance and Restoration

12.1 Housekeeping
The contractor shall keep the site clean and tidy at all times. Site improvements shall be protected from damage or becoming soiled through suitable temporary covering. Spilled grout shall be promptly picked up and moved to an appropriate waste storage area. Hoses, delivery lines, and other items that are not in immediate use shall be neatly stored in a manner that will not impede the ongoing work. All trash, used cement bags, etc., shall be collected and neatly stored for disposal. As soon as a reasonable quantity of such waste material has gathered, it shall be promptly removed from the site. Water and waste grout shall be promptly collected and disposed of. Water shall not be allowed to pond in the work area.

12.2 Site Restoration
Upon completion of the work, all waste shall be removed from the site and the site restored to as near its original condition as possible. Any remnants of drilling fluid or grout that have splattered on improvements shall be completely removed. Where suitable removal is not practicable, the affected areas shall be recoated or replaced to the satisfaction of the engineer.

13.0 Submittals

Prior to the start of work the contractor shall submit to the engineer the following information. No work shall be commenced prior to the engineer's approval of the various submittals.

13.1 Grouting Plan

- Description of all grouting equipment proposed for use, including, but not limited to, mixers, grout pumps, delivery lines and appurtenances. Information shall include the make, model, year manufactured, and general condition of each item. If items are to be rented, written verification should be submitted verifying that the unit will be available from the renter. Inspection of the listed equipment may be made prior to award of a contract.
- A description of all equipment and instruments to be used for surveys and monitoring of the ground surface and adjacent structures during the work.
- Names and qualification statements for all personnel who will be employed in the drilling, grouting, and monitoring operations. These include all drillers, mixer and pump operators, grout header technicians, and all monitoring personnel. All listed personnel must have a minimum of three years' continuous experience in similar grouting work.
- A description of the drilling methods and equipment to be used.
- A description of the casing and casing withdrawal system to be used, including a review of prior experience in the use of such casing in similar soils and to similar depths as in the contemplated project.
- Grout material sources, including grain size distribution, PI or hydrometer test results of the aggregate to be used, and the proposed mix design.
- Methods and equipment for calibration of the proportioning of the grout constituents.
- Methods and equipment for calibration of the grout quantity pumped and pumping rate.
- Statement of proposed sequence of operations.

- Description of grout injection operations.
- Method of structural lifting and description of proposed groutjacking operations.
- Copies of proposed drilling and grouting record forms.

13.2 Monitoring Procedures

- General plan and description of proposed monitoring operations.
- Description and specifications of monitoring equipment and instruments.
- Proposed methods to obtain and record monitoring data.
- Copies of proposed monitoring data forms.

27.2 CONTRACTING ARRANGEMENTS

The use of open competitive bidding for most grouting work should be avoided if possible. It serves as an invitation to lesser-qualified and often shady operators, of which unfortunately, there are many. Although public agencies are usually limited to open bidding, it is possible even for them to require meaningful prequalification of contractors for specialized work such as grouting, and this approach has been successful on many public projects.

27.2.1 Contractor Prequalification

At a minimum, bidding should be limited to appropriately prequalified contractors. Prequalification should not be given without a thorough appraisal of the individual applicants, which will include contacting the owners and/or engineers of *several* prior similar projects. *Several* is emphasized, as references given by some in the past have been found to be less than honest. In addition, referrals may be friends or relatives, and even shoddy operators have occasional projects that turn out well. Continuity of a contractor's activity is also important. One who has not preformed similar work for an extended period of time is not likely to have working crews organized for such an operation.

27.2.1.1 PREQUALIFICATION SUBMITTALS

The use of a prequalification submittal form is strongly recommended. It is also a good idea to visit the contractor's headquarters and perhaps other active job sites. The accuracy of statements made in the submittal can thus be confirmed, as well as the condition and orderliness of the contractor's equipment and facilities. The importance of due diligence in validating a contractor's qualifications cannot be overemphasized. The form with which this begins should include, at minimum:

1. Company name
2. Years in business

3. Years performing the particular type of grouting
4. Bank and surety
5. Insurance information
6. List of five projects similar to the contemplated project in size, scope, and intent. Information provided should include the location; dates work started and ended; owner's name, address, and phone; owner's representative or engineer, including the individual's name, address, and phone number; description of the work; and contract amount.
7. List of key personnel who will be assigned to the contemplated job, including superintendent, foremen, drillers, grout pump operator, mixer operators, header technicians, and monitoring technicians, including backup personnel. Provide for each: name, position, length of employment; former positions and dates held; prior experience in similar types of grouting; special education, training, or certification, if any.
8. List of equipment to be used, including backup units. Make, model, capacity, age, condition, and location where unit can be inspected, and whether owned or rented. If rented, provide name of renter and a written commitment that the equipment will be available for the project following award of a contract.
9. Scope and details of any company training programs.
10. Scope and details of any *routine* Quality Control programs.
11. Scope and details of any company testing or research and development facilities and details thereof.
12. Particulars of company safety program, including a listing of each accident, employee injury, or incident of property damages within the last five years.
13. Details of any lawsuits filed against the company or its principals within the last five years.

In assessing the submittals, all entries should be independently verified; less than honest responses are often given. It is distressing so much dishonesty exists, but it does and must be recognized. For example, a particular contractor responded that he was going to use a certain make and model of batching/mixing plant on a project. He further stated that it would be rented from Hertz Equipment Rental in Denver, Colorado, and he had a commitment that the equipment would be available should he receive the project. A simple phone call to the designated renter revealed that it did not own any such plants, but advised calling an 800 number that connected to its national equipment inventory. A call to that facility disclosed that the firm did not own any such equipment, nationwide.

In another case, a contractor was awarded a project based on his proposal to use continuous computer monitoring that involved a "proprietary software program specifically for compaction grouting," which only he possessed. A visit to the project revealed that, indeed, a computer was set up on a table in the injection area. Unfortunately, the only data that was being recorded was the hole number, which was entered manually, and pressure readings transmitted from a pressure transducer at the grout collar, operating only sporadically. There was no mention of injection rate, and, of course, this rendered the pressure values useless. The claim of use and the setup on the site did indeed look impressive. Unfortunately, it was simply a ruse meant to imply special proficiency, which the contractor did not possess. The proprietary software was useless, and the entire operation appeared to be a sham.

The fact is that there are some brilliant marketing and proposal writers in the grouting industry. Unfortunately, they are not always so brilliant in carrying out the work, and sometimes their presentations are downright dishonest. It is not unusual, once on the project, to find that the submission of often frivolous claims appears to take precedence over performance of good work. One must be very cautious in accepting information from grouting contractors, and independent confirmation of the appropriateness and accuracy is essential. Fortunately, there are some extremely good firms available that are both honest and knowledgeable. These are good sources of information and a pleasure to work with once on a job. We should seek them out and fully appreciate their being on our projects.

27.2.2 Negotiation with Chosen Contractor

Where possible, rather than employ competitive bidding, it is strongly recommended that a contract be negotiated with thoroughly qualified and honest contractors. Often these are regional firms that do not advertise but, rather, depend on recommendations and repeat work. Locating them may require considerable effort, but it will be well spent, in that a better-quality finished product will be delivered in a timely manner and a lower ultimate cost will likely result.

In grouting, existing conditions are almost never sufficiently understood ahead of the actual work. Thus, some unknowns will remain after the work starts. The extent of the unknown factors will be directly related to the thoroughness of the investigation, and the thoroughness of the investigation directly affects its cost. When contractor selection and mode of payment allow flexibility of the work including unit quantities, less precision is required in the estimate and thus there is a lower cost for the investigation.

Owners are nearly always looking for the lowest price and the use of a low bidder may be tempting. It is the final cost that matters, however, and quality work with honest dealings, free of conflict both before and after job completion, is nearly always least costly. It is hard to understand why, although sufficient money for proper investigation, design, and construction is seldom available, generous funds appear for the litigation that often follows. Dollars are much better expended on quality performance than on dickering with a less-than-good contractor.

27.2.3 Payment Basis

Because of the many unknowns common to grouting work, choosing an equitable payment method is not easy and, in many cases, there is no really good way. Grouting involves filling of voids, of which the size, distribution, or volume is seldom well understood ahead of actual injection. In addition to the unknown volume, conditions encountered once work commences may call for changing the number of grout holes, as well as the number and lengths of stages.

27.2.3.1 TIME AND MATERIAL

Where a single quality contractor is used, payment on the bsais of time and material is probably most equitable to all parties. Many contractors regularly work on this basis and will furnish rate sheets outlining crew makeup and equipment furnished, at either an hourly or daily rate. Materials are usually charged at cost or with a markup varying from 10 to 15 percent to cover the cost of handling. Although a guaranteed final cost for the owner is not provided, reliable contractors will provide an estimate of the total cost range. This is advantageous to the owner, in that he or she pays only for the actual work accomplished, free of contingent amounts often included in firm price bids.

27.2.3.2 COST PLUS FIXED FEE

Payment on the basis of cost plus a fixed fee is usually reserved for larger projects; the owner is billed for the actual costs, to which are added a fixed fee. The fee may be a lump sum for the entire work or a percentage to be applied to the costs. This arrangement is particularly useful when the quantities of material or the extent of the work are difficult to ascertain ahead of actual progress and/or design continues during the work. It is commonly used for large projects where all expenses and accounting are handled through the job site office.

27.2.3.3 LUMP SUM

A lump sum price for completing the work can provide the owner with a firm figure, but can also present challenges for the contractor in making estimates. Established contractors understand risk and will always include contingency amounts for any conditions not clearly understood. Thus, the initial price will be higher than the likely reasonable cost to complete the work. Consequently, unless the work involves few unknowns, the total cost to the owner can be greater than that incurred if more equitable means are used. In addition, and especially if the contractor's costs should exceed those expected, claims are likely, which means that a more thorough investigation and planning effort should be made.

27.2.3.4 UNIT PRICE

By far the greatest amount of grouting is done on a unit price basis. This method is, however, the most subject to abuse, particularly in grouting where unknowns commonly prevail. Contractors typically evaluate estimated unit quantities very carefully. Their bids will be unbalanced to reflect any quantities they expect to vary appreciably. Items expected to increase will be bid at a higher than normal price. The excess will be deducted from those that are expected to underrun. Through clever unbalancing, these items can appear to reflect the lowest cost, but the cost will actually be higher for the completed job.

Grout injection nearly always involves unknown material quantities. The amount injected can be easily manipulated through subtle variations of the mix or pumping rate, however. It is fairly easy for a contractor's workers to either increase or decrease the amount of grout injected in a given time period. Consequently, good practices may be abandoned when there is a financial incentive to pump either a high or low amount of grout. Payment on the basis of grout quantity is thus not recommended. A better approach is to pay for material furnished, with pumping *time* a separate item.

The reasonableness of the contract quantity estimate is another issue. Adjustments of unit price are typically made if an item varies unreasonably. What is reasonable is contentious at best in grouting, in which variation of 25 percent or more is not unusual. The courts have often considered a 10 percent change to justify a price adjustment in construction, and, unfortunately, grouting is treated the same as any other construction. Another potential problem is that substantial variation in the grout quantity may be used to justify a "differing site conditions" claim. Again, paying for injection time by the hour will greatly lessen the incentive for manipulation of quantities or claims.

A further problem, especially where grout takes are expected to be low, is that a contractor may set unreasonably high prices for drilling while underpricing the injection items. When this is done, it provides a strong incentive to minimize the grout quantity by early hole refusal, which is fairly easy to do by throttling the pump momentarily or manipulating the mix. Where the drilling price is unreasonably high, the mixing and pumping must be very closely observed.

Finally, the item most often abused is *mobilization.* Grouting requires special equipment and/or setup and is often performed in remote areas. A mobilization item is thus often warranted. The bid amount should be reasonable, however, and should reflect the actual cost involved. All contractors load the mobilization price to the greatest amount they expect to get by with, as it is usually part of the initial payment. For example, a contractor just completing a project will incur little expense in mobilizing for further work at the same site or another nearby. Yet, unless somehow restricted, the contractor may bid a much higher price.

Some public agencies have tried to control the problem by assigning a fixed amount to mobilization. This is unfair, however, to the contractor who must travel a long distance or otherwise incur costs greater than the assigned amount. Conversely, it is an unfair advantage to nearby contractors with substantially lower costs. The best solution is to require bidders to substantiate their mobilization price by an anticipated or actual cost breakdown. Payment for mobilization can also be conditioned upon submittal of the actual cost, with documentation provided at the time of billing.

Contractors of lesser quality and those unsure of themselves want the largest number of bid items possible, as this gives them the greatest opportunity for manipulation. For example, drilling can be based on a single unit of cost per foot or can include many different units, such as lump sum setup, casing or header nipples by the foot or pound, drilling in overburden, drilling in rock, and so on. Quality contractors understand their costs and are not usually planning a claim during the bid period, so they are content with a lesser number of items. Drilling should normally be bid by the foot (meter), including setup and all incidentals, grouting injection preferably by the hour, grout material by the actual units used, and mobilization by lump sum subject to confirmation of actual costs. Other significant items may be separately priced, but incidentals should be included in the primary items.

27.3 FULL-SCALE TRIALS

Unless the contemplated work is quite routine, full-scale trials should be considered, especially where the formation profile and/or the existing conditions are not well understood. Such trials are often performed as part of the design and supervised by the engineer. Successful completion of a trial, with testing and/or exposure of the grouted formation, will not only confirm that the intended objectives can be met, but can also allow a more precise understanding of the work to be done. Where more than one grouting method or material is being considered, all should be evaluated by trial injections.

Trials can also facilitate more accurate estimates of both materials and cost and, equally important, allow appraisal of the contractor's field operations. If a different contractor is awarded the production work, in the case of important or unusual requirements, that contractor should successfully perform a similar trial. A word of caution is in order, however: Unless trials show the need for changes in the anticipated methodology, allow no changes in the contractor's personnel, equipment, or methodology for the full-scale work. There have been instances in which grouting contractors have performed brilliantly on test applications, but, once qualified, have used less costly procedures, resulting in inferior performance of the production work.

CHAPTER 28

The Games Contractors Play

28.1 MARKETING EFFORTS
28.1.1 Direct Advertising
28.1.2 Press and Publications
28.1.3 Seminar Sponsorship
28.1.4 Academic Research
28.1.5 Trademarks and Patents
28.2 PROPOSAL TRICKERY
28.3 ONCE ON THE JOB
28.4 DEFENDING BAD WORK
28.5 THE GOOD PLAYERS

GROUTING CONTRACTORS have a long history of operating under a veil of secrecy and have generally not shared experiences with each other. There may have been some justification for this in earlier times, but such an attitude is certainly not reasonable now. Grouting technology is well established, and although individual firms no doubt have operational methods they prefer to keep confidential, most knowledge in the field of grouting is common and easily obtained. Yet multiple attempts to form an organization to represent the grouting industry in the United States have failed, primarily because the more prominent contractors have withheld their support. Although basic mechanisms are the same, there is considerable variability in the operations of different contractors.

Unfortunately, many engineers and grouting inspectors lack sufficient knowledge of the technology to feel confident in discussion, allowing themselves to be intimidated by fast-talking "experts." Unlike the situation in Europe and Asia, where most grouting is performed by large, well-established organizations, the greatest amount of this work is performed in the United States by relatively small local or regional firms. Many of these are extremely competent and honest operations, whose jobs result from good work and the recommendations of satisfied clients. These truly good grouters seldom advertise and are typically unknown outside their home area. Not only do they provide the best work, but they also tend to be the most innovative and make the greatest contributions to advancement of the technology.

Then there are the opportunists, the wheeler-dealers who may or may not understand the technology, but succeed by manipulation and less than honest dealings. They are masters at juggling bid unit prices, presenting unwarranted claims, and generally taking advantage where possible. Fortunately, this group is not large, but the chicanery of its members badly tarnishes the reputation of good grouting professionals.

There are also the large *geotechnical contractors*, who tend to support elaborate advertising and promotional programs, including sponsorship of educational seminars and university research. Unfortunately, such educational events are bound to be biased in favor of the sponsors, and even independent researchers tend to work with the information and materials they have been provided by their sponsors. The performance of this group of operators is often less than good, which also gives a bad name to grouting. If an engineer or owner *perceives* that he or she is dealing with the very best, yet receives poor performance, that person is not likely to recommend grouting in the future.

28.1 MARKETING EFFORTS

The names of a few large operators tend to be seen frequently and have become the equivalent of household names in grouting, primarily as a result of brilliant marketing. Although the marketing ability of a firm may be exceptional, it does not signify competency to carry out the projects that result. It thus behooves us all to be cautious in our first impression of products and firms and to question our real knowledge of their quality and competence and how it was acquired. We need only ask: If a firm really is the performance leader, why does it spend large amounts of money to tell us so? Word of exemplary performance will always spread fast, negating the need for expensive advertising.

28.1.1 Direct Advertising

Advertising is part of life, and we are subjected to it nearly nonstop every day. Ideally, advertising allows an organization to get its name out and to inform the public about its goods or services. Name familiarity is said to be fundamental to the success of firms producing goods and services used by the public. Traditionally, this has not been the case in engineering and construction, in which the primary source of work has been either the competitive bidding process or repeat business and recommendations by satisfied clients. Only in recent times have we become recipients of expensive advertising and promotional programs. Because extensive advertising is costly, we must be wary of accepting the story conveyed, as the advertiser has paid dearly for it.

Perhaps the best illustration can be drawn from another industry, the manufacture of organs. I once heard the president of a large organ manufacturing firm and the vice president of research and development for another engaged in a lively debate. The two firms were the largest in the industry, alternating from year to year in respect to sales volume. By any measure, they were both well known and successful.

One of the firms maintained an expensive advertising budget, but did virtually no research and made little effort to improve its product. The other spent virtually nothing on advertising, but was the acknowledged leader in new developments. The first insisted that development activities were unimportant as long as potential customers *perceived* a firm to be a leader, which resulted from advertising. The other was of the opinion that word spread by happy customers, who had actually experienced the finest sound firsthand, was more convincing. Neither of the foes was able to change the other's thinking. Obviously, however, great satisfaction personally experienced is superior to *paid* advertising to influence our thinking.

28.1.2 Press and Publications

Most of us receive more than a few industry publications, which often contain interesting articles and other information. The articles are made to appear as if they had been written independently by the editorial staff or alleged author. Although this is sometimes the case, many such articles actually originate from paid advertising agencies, so that what appears independent is not. Furthermore, performance that appears in the literature to be outstandingly successful may in reality have been less than acceptable.

The administrative vice president of my former contracting firm was also the engineering editor of a well-known construction periodical. Although he occasionally went out on a project to prepare an article, many that were published originated in advertising agencies. As he sorted his mail, he discarded many of the large brown envelopes from ad agencies without even opening them. Yet occasionally one would be opened and the article read and included in the magazine as if written by the editorial staff, often free of any editorial changes. Those were the ones from ". . . a really good writer and you would never know it was a commercial".

In a more personal example, while going through mail, I ran across an interesting article about corrective work performed on a sea wall. It caught my eye, as the methods used were inappropriate and shouldn't have worked at all, let alone as wonderfully as was presented. While reading the article, I received a call from an attorney. The conversation started to the effect "Mr. Warner, I represent a client that just had a sea wall repaired by epoxy grouting and it isn't working . . ." to which I replied "Do you mean the ______ project at ______ and done by ______." The attorney was amazed that another grouter could be so informed of work gone badly, and I was relieved to know that what should not have worked, didn't. The article was obviously written by a special interest not competent enough to know the work was destined to fail. Sadly, however, less informed readers had probably been impressed enough to use the featured contractor and suffer a similar fate.

One would think we could rely on peer-reviewed papers presented by professionals in technical journals. Unfortunately, even here we must be wary. A case in point is *Guide Specifications for Chemical Grouts* by the Committee on Grouting (1968), which appeared in the *Journal of the Soil Mechanics and Foundations Division of the American Society of Civil Engineers.* It is basically an exact copy of "General Specifications for Field Use of AM-9 Chemical Grout" (Karol, 1963), distributed by the American Cyanamid Company, which at the time was holder of the patent rights to acrylamide grout, described in Section 7.3.3.1. The only difference was replacement of "AM-9" with "chemical grout" wherever it appeared in the text.

This supposedly authentic and independent guide virtually precluded the use of any chemical grout other than AM-9. Fifteen pages, including 20 references, which I and four others submitted (Discussion, 1969) challenged the guide's provision for only one grout material, and questioned the applicability of the publication. Yet many project specifications were issued in the ensuing years, which copied the guide verbatim. Unfortunately you cannot unring a bell.

28.1.3 Seminar Sponsorship

Commercial interests often sponsor educational seminars, usually limited to invitees. These are portrayed as being noncommercial and sometimes even include university professors to give a keynote address or provide some of the instruction. Printed handouts often include guide specifications for various types of grouting. Even though a nominal charge might be made to give the impression of independence, the material presented is usually biased and the guide specifications are sometimes technically wrong. One must thus be aware of the source of any guide specifications and wary in their use. Although worded to appear innocuous, these often contain language that clearly benefits the originator.

28.1.4 Academic Research

A clever marketing ploy to gain credibility is to sponsor academic research, wherein the sponsor contributes material and perhaps the use of equipment. Along with these benefits, of course, comes considerable advice and council. The result is often biased reporting and misinformed readers. Particularly unfortunate was a paper recommending the inclusion of bentonite in compaction grout (Borden and Groome, 1984). The work was sponsored by a contractor that even supplied a pump and other equipment. It involved pumping various mixtures through a tightly coiled hose to determine the required pressure, without regard for the behavior of the grout in the ground. This was done, in spite of numerous prior publications stressing the importance of not allowing clay to be included in compaction grout.

Many publications have discussed ultrafine cement, using as examples MC-500, MC-300, and MC-100, all from a single source, which just happened to be controlled by a large grouting contractor. There is nothing wrong with reporting on products; however, in this case, MC-100 and MC-300 never existed outside the laboratory. By including them in "research" implies that a range of different materials have been comparatively evaluated when, in fact, only one is available. It is not a coincidence that the controlling owner had close connections to the researchers and, of course, provided his own materials. Unfortunately, such work ignores other cements of different derivation and misleads readers, who assume that the products investigated are representative of available materials.

The same can be said of numerous papers on sodium silicate grout and grouted soils, in which only the sodium silicate is identified. We know, of course, as presented in Section 7.2.2, that the properties of grouted soils are greatly affected by the reactant material. By neglecting to identify it or simply stating that it is "proprietary" does nothing to advance the science, and the only intent is to advance the sponsor.

28.1.5 Trademarks and Patents

Many terms have been trademarked or otherwise registered to give the impression that they represent something special, available only from the registering source. Most of the time, however, they simply refer to commonly used grouting terms. Examples include *Soilfrac,* in reference to fracture grouting; *Pressgrout,* for grouted micropiles; *Powergrout,* for compaction grouting; and *Soilcrete,* to describe the results of jet grouting.

Patents and trademarks are frequently referred to in grouting contractors' literature, suggesting something special that only they can do. Examination of a patent, however, often discloses little if anything special, and sometimes the patent is not even applicable to the reference. A copy of any patent in the United States can be obtained from the U.S. Patent Office for a small fee. Most recent patents are available on the Patent Office website on the Internet and are downloadable. All patents are placed in classifications and subclasses. Unfortunately, those pertaining to grouting appear under many different classifications; however, every patent includes the classification in which it is listed, at the top of the first page. The classification shown for a given patent can be used to find all others in that classification. Also at the top of each patent is a list of similar patents that have been researched.

Experience has shown that the patents pertaining to grouting are often limited in scope and seldom protect the range of activity one might expect from the holder's reference. Thus, before giving credence to proprietary rights resulting from a patent, it is a good idea to obtain a copy of the patent and see exactly what it claims.

28.2 PROPOSAL TRICKERY

Grouters often attempt to gain a competitive edge during the bid or proposal period. Section 27.2.3.4 discusses the practice of juggling of unit prices, which is very prevalent. Other areas of abuse include substitution of specified materials or equipment and proposal of a grouting method or method of ground treatment different from that specified.

Where the contractor is to be chosen on the basis of a technical proposal, one must evaluate the proposal very carefully and independently confirm all important content. Some firms are brilliant at drafting proposals suggesting that they are superior, but field performance is often disappointing. A trick of one well-known firm is to state that the job is so important that independent outside expertise should be retained. And because of this company's fine reputation, it has received commitments from well-known independent experts who would otherwise be unavailable. Letters from the experts agreeing to consult are included in the proposal. Such consultants can hardly be considered independent. If a project is of such importance that outside consultation is justified, the consultants should be selected and paid for by the owner or the owner's engineer.

Elaborate presentation of proposed computer monitoring is sometimes provided and obviously designed to impress. The true intent is not always noble, however. I once overheard a major firm manager exclaim to colleagues as they exited a meeting where their faulty performance was confirmed by their own computer output, "this will never happen to us again." Although the firm boasts of its inhouse computer ability, in this particular case it was subcontracted. Details of the required output were mandated by contract and continually observed and analyzed by the owner's staff.

A few years later, while reviewing proposals for another project, I was particularly interested to see an impressive presentation on the same firm's proprietary "fourth generation software." It included impressive looking charts of the purported output, but on close perusal, nothing special was found and the most important parameters of pressure and pumping rate in real time were not provided.

28.3 ONCE ON THE JOB

Upon receipt of a contract, the attitude of many professional-appearing firms can change greatly. Required materials or equipment are often questioned and substitutions requested. Personnel who were named in the proposal suddenly become unavailable, and the highly experienced staff that was promoted seems to vanish. And then there are the jugglers, who can be extremely clever in changing the project's scope or attributes to their own advantage.

Proposals were received from three invited contractors for some important remedial grouting. The job involved the settlement of a large foundation due to a sinkhole. The situation was an embarrassment for the geotechnical engineer who missed some underlying solution cavities in his investigation, but, more important, it threatened timely completion of the major project. Time was critical, and work was required to continue around the clock without a break until the job was completed. Several types of grouting were included, requiring multiple crews. The job would be a major undertaking for any firm, and the contract was awarded to the second-lowest bidder who was believed to be the most qualified.

On the day the job was to begin, the owner became alarmed to discover that the contractor had not mobilized, but, rather, had submitted a request for 23 changes in the specifications. When the contractor was told that his behavior was not very professional, he replied, "I'm always professional until I get the job; then I'm just a damned contractor." Although there are many very fine grouting contractors, there are many others bringing disrepute to all. And some of the worst actors present themselves very professionally until the work starts.

Arbitration was used to resolve a dispute over payment for excessive grout used on a project. The work involved solidification of sand under the spread footings of an existing building with sodium silicate chemical grout. Usable space was being developed in the basement, requiring undermining of the footings by only a few feet. The contractor stated in his bid that the plans were unclear to him, but he believed the work could be completed with no more than a stated amount of grout, with any additional being charged at a given price per gallon. The price quoted for the additional grout was more than three times the going rate. The justification for the high cost was a *royalty* for use of the *proprietary* mixture, which was not identified. Onsite, grout was simply "jetted" with open-ended pipes, in such a way that it was not placed where required but, rather, to greater depths. In fact, more grout was placed below the specified zone on the first day of work than was to be used for the entire job.

The grouter filed a claim against the general contractor who refused to pay for the overage, and back charged the grouter for the resulting delays. Upon excavation, many ungrouted areas were found. But causing greater distress was the emanation of an ammonia odor from the grout, which became so strong that work had to be shut down and decontamination crews retained. Under oath in a deposition, the contractor, after continuous resistance, finally identified the grout reactant as formamide and sodium bicarbonate. These were common reactants and certainly not proprietary, although in earlier times they were subject to a patent of the Diamond Alkali Company. There was no justification for the high charge, and the reactant's being proprietary was a sham, as is all too often the case.

On another significant project, grouting was awarded on the basis of a contractor's technical proposal, which included glowing descriptions of the highly qualified personnel to be used. Real-time computer monitoring was required, as was a blend of several ingredients for the grout aggregate. Because consistency was important, routine particle size distribution tests were required. Instead of staffing with brilliant qualified personnel as proposed, many locals were hired, including a young high school dropout assigned to monitor the computer system. She had no training, had not previously heard of grouting, and could not understand *all of those squiggly lines* on the screen. To keep the work moving, the owner had to separately monitor the work.

In another instance on the same project, a radio transmission was overheard between the contractor's superintendent and the young project engineer on one of his first jobs.

Superintendent: Take the new hire to the lab. She'll be the testing tech.

Engineer (haltingly): Don't you think we need someone with a little experience?

Superintendent: Do as I say!

Unfortunately, contractors are not always as they appear in their proposals and promotional material. Specifications and contracts must thus be specific and enforced on-site. The use of penalties for straying from the proposal and the contract will improve performance, but employment of truly good contractors, of proven principles, is far better.

28.4 DEFENDING BAD WORK

Perhaps it should be expected that contractors may defend their transgressions by producing experts. The sad fact is, however, rather than having extensive knowledge of grouting, some *experts* appear to simply be masters of deceit. It is well recognized in litigation that experts can be retained to support virtually any position. This is true in grouting as well. In one instance, the defendant contractor retained a university professor to develop a complicated analysis to absolve him of responsibility for seriously damaging a structure. The obviously incompetent crew had displaced a downslope, splitting the overlying structure and offsetting a wall by several inches. Fortunately, the owners had a true expert who was not intimidated by the professor. In this instance the contractor paid for the damage, but in many other cases he likely escaped liability as a result of the professor's hype.

28.5 THE GOOD PLAYERS

The foregoing text may leave the impression that all grouters are bad news. This is far from correct, however, and many highly principled firms producing excellent work are available. Unlike their not-so-wonderful counterparts, good contractors tend not to advertise and are thus not readily found. In between these extremes are the average operators, neither outstanding nor poor. The contrast between the good and the bad, however, is great.

This is amply illustrated by two projects performed in 2002. In June there was an emergency involving the California Aqueduct, in which a piping leak of about 1500 gal/min (5670 L/min) had developed. The leak was stopped with pumped ready-mixed grout, and the water was lowered for further repairs. A line of cone penetrometer test (CPT) holes disclosed widely variable conditions of the earth embankment, and defects in the underlying rock were suspected because of the geologic setting. In late afternoon, a call was made to a grouter known for good performance. With an exploratory program under way and incomplete knowledge of the conditions or work to be done, he was told that crews were needed immediately to work around the clock. They should be prepared to do compaction grouting to a depth of about 60 ft (18 m) and permeation grouting of the underlying rock to more than 100 ft (30 m); whatever equipment might be needed should be mobilized at once.

Even with such short notice, a convoy of trucks and equipment arrived the next morning: two large service trucks, each pulling a tilt trailer with a drill rig; two smaller service trucks, one pulling a compaction grouting mixer and the other pulling a pump. Additional equipment, including a large compressor and permeation grouting plants, arrived over the next several hours, as well as fully experienced personnel who worked together like a finely tuned machine. Within 20 minutes of their arrival, the first hole was being drilled. Grout injection started a few hours later, and work continued for the next several days without a break. The service trucks (Figure 28.1), were each equipped with a large compressor and tool boxes filled with the numerous small items that might be needed. The work site was kept clean, as was the equipment illustrated in Figure 28.2.

On a compaction grouting project in September, the contractor had been retained much earlier with ample time for planning, but was onsite "mobilizing" for the entire week prior to the start of the work. The equipment was delivered on hired transport trucks. It was well used and not in very good mechanical condition, which was apparently recognized and provided for by the

FIGURE 28.1 All necessary tools and accessories were in tool boxes on the service truck.

FIGURE 28.2 Equipment and worksite were kept clean and orderly.

presence of a full-time mechanic. Frequent line blockages plagued one of the pumps. After a few days, the project manager was notified that the wear plate was excessively worn. He retorted that the pump had just been rebuilt in the shop and was in first-class shape. After a few more days of problems and as many as 19 blockages in a single shift, the mechanic replaced the wear plate. It was not only excessively worn, but completely broken in two (Figure 28.3). Before the job ended, that "rebuilt" pump required many repairs including complete replacement of the agitator.

The contractor chose to use a standard NX casing, which proved inadequate, due to breakage during withdrawal. This occurred within the first week of work and continued to the end, three months later. The arrogant contractor refused to change his ways and obtain stronger casing. Not only did it break, but its being in 10 ft (3 m) lengths complicated pulling during the bottom-up injection. The precarious position required of workers to disconnect it, as shown in Figure 28.4, resulted in the header often falling, breaking the gauges. The situation was made worse by workers hired locally and lacking grouting experience. The mechanic was kept more than busy from the beginning to the end of the project. And the work quality suffered because the ground penetrated by broken casing was not grouted in the upstage sequence. This required much reworking, without a good knowledge of the remaining casing depth. The incompetent performance proved to be expensive for the owner and extremely stressful for the entire engineering team. Completion was delayed and the cost considerably increased.

There are many fine grouting contractors. Seeking them out and employing them will result in high-quality work, performed neatly, and virtually always at a lower final cost.

FIGURE 28.3 Excessively worn wear plate was broken in two.

FIGURE 28.4 Long casing joints made breaking precarious and falling header was often damaged.

PART III

Drilling and Grouting Equipment

CHAPTER 29

Understanding Power Transmission

29.1 POWER TRANSMISSION FUNDAMENTALS
- **29.1.1** Rate of Movement
- **29.1.2** Force
- **29.1.3** Torque
- **29.1.4** Power

29.2 MECHANICAL SYSTEMS
- **29.2.1** In-Line Coupled
  - **29.2.1.1** Rigid Couplings
  - **29.2.1.2** Flexible Couplings
- **29.2.2** Gearboxes
  - **29.2.2.1** Speed Reducers
  - **29.2.2.2** Worm Gear Drives
- **29.2.3** Roller Chain and Sprockets
- **29.2.4** Belt Drives
- **29.2.5** Variable Speed Drives

29.3 HYDRAULIC SYSTEMS
- **29.3.1** System Fundamentals
  - **29.3.1.1** Flow and Pressure
  - **29.3.1.2** Hydraulic Fluid
  - **29.3.1.3** Hydraulic Circuitry
- **29.3.2** Valves and Fittings
- **29.3.3** Hydraulic Pumps
  - **29.3.3.1** Gear Pumps
  - **29.3.3.2** Vane Pumps
  - **29.3.3.3** Piston Pumps
- **29.3.4** Rotary Motors
- **29.3.5** Hydraulic Rams and Jacks

29.4 PNEUMATIC SYSTEMS

MANY CONTRACTORS modify otherwise conventional equipment, and a few build custom units from scratch. Although this can be advantageous, there have been many instances in which failure to follow basic mechanical fundamentals has resulted in less than good performance. Effective planning and modification require a basic understanding of power transmission, as well as knowledge of the available components.

29.1 POWER TRANSMISSION FUNDAMENTALS

The principles of design for power transmission systems are based on the load, speed and rate of operation. An understanding of these fundamentals is therefore requisite to rationally design or alter a particular drive system. Relatively simple formulas are used to establish the necessary elements. Once these are determined, the particular method of power transmission and the various elements needed can be rationally selected.

29.1.1 Rate of Movement

Velocity and *acceleration* are used to describe the rate of movement, which is fundamental to determining power requirements and selecting the drive system components. Velocity is the rate of linear motion of a mass and is equal to the distance of travel divided by the time required to traverse that distance. It is expressed as follows:

$$V = S/t$$

Where V = velocity
S = distance
t = time

The velocity or *speed* of a drive element, such as a belt or chain, is fundamental to selection of the drive components. The speed of a shaft or wheel is typically expressed as *revolutions per minute* (RPM), and this is a function of linear velocity. RPM is determined by dividing the linear velocity by the circumference of the shaft or wheel.

Acceleration is the rate of increase of velocity in a given time interval. It is determined by dividing the velocity increase by the amount of time required, and is expressed as follows:

$$a = (VF - VI)/t$$

Where a = acceleration
VF = final acceleration
VI = initial acceleration
t = time

29.1.2 Force

Force is a pressure or load applied to a mass and can be thought of as pushing or pulling on that mass. It has both a magnitude and a direction, and when applied will cause either deformation or displacement. Acceleration produced by a given force will be directly proportional to its magnitude. Greater force will promote higher acceleration and, conversely, as the mass to which it is applied grows, the acceleration slows. This relationship is expressed as follows:

$$F = Ma$$

Where F = force
M = mass
a = acceleration

29.1.3 Torque

Torque is twisting force that causes rotation of a shaft or wheel. It is independent of time and equivalent to force in linear movement. It is expressed in ft lbs (m kg) and measured by the magnitude of pull, as a function of distance from the center of a shaft, as follows:

$$T = FD$$

Where T = torque
F = force
D = distance

A practical example is the force needed to break a joint of drill steel. Greater force is required with a short wrench (short distance) than with a longer one. With a given wrench, however, the required force is the same regardless of how fast it is applied.

29.1.4 Power

Power defines the amount of work accomplished in a given amount of time. It is expressed in horsepower (hp) and is the product of torque and speed. The value of horsepower originated with Boulton and Watt to define the power of their newly invented steam engine. They determined that an average horse could work continually at a rate of 22,000 ft lbs per minute. They increased this by 50 percent to define mechanical power as follows:

$$1 \text{ hp} = 33{,}000 \text{ ft lbs/min}$$

The horsepower required to perform a given amount of work in a given time is expressed as follows:

$$hp = \frac{\text{ft lbs/min}}{33{,}000}$$

A shaft can be turned with a wrench or other lever to find the horsepower or torque requirements. By determining the pull force required for one minute, as by a spring-loaded scale, and multiplying by the length of the wrench or lever used, the ft lbs/min can be determined. Inserting this into the preceding formula will enable calculation of the required horsepower. Always keep in mind that horsepower is directly related to time. If the RPM of a given shaft is doubled without any change in torque, the required horsepower would also double.

29.2 MECHANICAL SYSTEMS

Generally, mechanical power transmission is the most efficient. Because the driven component must be directly linked to the driver, however, there is little flexibility as to location or orientation from the power source. Furthermore, the entire system must be properly balanced and must remain so if excessive wear and damage to the power source is to be avoided. With exception of belt drives, shock loads will be transmitted to the power source with all direct mechanical systems.

29.2.1 In-Line Coupled

Mechanical couplings are used to connect axial shafts that are transmitting torque and turning about the same centerline. As the driven shaft is directly coupled to the driver, these are the most efficient systems, transmitting virtually 100 percent of the available power. In the case of rigid couplings, the two shafts must be maintained absolutely concentric to prevent vibration and damage. Because this is extremely difficult and angular distortion and axial movements often occur, a variety of flexible couplings have been developed and are widely used. These are able to absorb varying amounts of shock, so such loads are dampened before reaching the power source.

Several types of both rigid and flexible couplings are available, and optimal selection can be a daunting task. Criteria that must be considered include horsepower, torque, speed (RPM), environmental conditions, maintenance requirements, and the nature of both the power source and the driven equipment. The power and speed capabilities of different couplings vary, as do the amount of space required and the ease of installation and replacement. Perhaps most important is their acceptance of any variation in alignment that may occur between the two shafts.

29.2.1.1 RIGID COUPLINGS

For light-duty applications, simple one-piece *sleeve-type couplings* (Figure 29.1, a) are widely used because of their simplicity and low cost. They are available for the same or different size shafts and are provided with locking setscrews. Keyways can be provided, but they are not usually required for the light-duty installations in which these couplings are used. They can be difficult to replace as sliding down the shaft on one

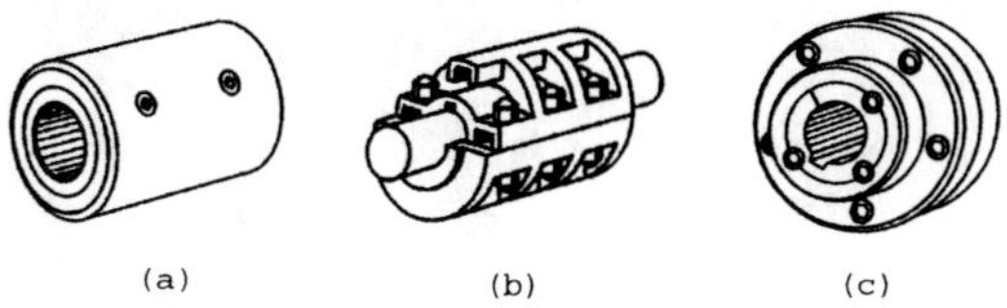

FIGURE 29.1 Rigid couplings: (a) sleeve type, (b) split rib, and (c) flanged.

side is required, and one of the shafts must be displaced.

Longitudinally *split ribbed couplings* (Figure 29.1, b) form a sleeve when bolted together. They come in a wide variety of sizes but are usually limited to use with identically sized shafts. Keyways are provided and are easy to install and replace because of the bolted split construction. These couplings are used in all kinds of applications, including heavy-duty types.

For heavy-duty applications, *flange-style couplings* (Figure 29.1, c) are most commonly used. They consist of two flanges that are bored and keyed to fit the shafts, which can be of the same or different diameters. The flanges are individually slipped over the respective shafts and bolted together. This requires an offset of at least one of the shafts for installation.

29.2.1.2 FLEXIBLE COUPLINGS

Flexible couplings can flex while rotating to accommodate misalignment between the connected shafts. Many different types are available, the three most common of which are illustrated in Figure 29.2. All are available for a variety of shaft sizes and can accommodate different size shafts. Two fundamentally different types are offered, metallic and elastomeric. Metallic couplings provide flexibility through mechanical flexing, wherein loose-fitting parts are allowed to roll or slide against one another, and usually require lubrication. Elastomeric units are provided with nonmetallic elastomeric filler between the metal parts, which allows flexibility of the joint.

(a) (b) (c)

FIGURE 29.2 Flexible couplings: (a) chain style, (b) Morflex, and (c) jaw type.

Elastomeric materials degrade when heated, limiting their use at elevated temperatures

Chain-style couplings consist of hardened hubs fitted with integral sprockets that make up back-to-back, as shown in Figure 29.2, a). A double connected roller chain is wrapped around the sprockets and connected with a pin. Movement within the chain and between the sprocket and chain allow up to about two degrees misalignment. There is some slack with rotation reversal, so they are best used for continuous duty with minimal starts and stops or reversal. The chain must be lubricated and is typically provided with protective covers that rotate with the coupling. Chain couplings are easy to service and replace in the field.

Morflex-type couplings (Figure 29.2, b) consist of two halves, each attached to a hub that is typically keyed and can accommodate any size shaft. Each side of the coupling contains chambers filled with a flexible neoprene "biscuit," through which bolts are placed to connect the two pieces. The bolt holes are oversize through the metal, but the biscuits, which are completely contained and always under compression, fit tightly. This allows flexing of the two sides, and they can withstand up to about 5 degrees of angular misalignment. They are very rugged and can be used at temperatures ranging from 0° to 200°F (−18° to 93°C). Although slightly more expensive than other couplings, they are very durable, as the elastomer is completely contained, and they have given excellent long-term service.

Jaw-type flexible couplings are composed of two hubs fitted with interlocking jaws. A gap is provided between the jaws to receive a star-shaped insert, which can be either bronze or a synthetic elastomer. The flexibility of the insert and any space between it and the jaws allows angular distortion up to about 3 degrees. Inserts are available in a variety of materials, including polyurethane, neoprene, and oil-impregnated bronze. Depending on the insert material, they can operate at temperatures between 55° and 250°F (13° and 121°C). Their low torsional stiffness provides good dampening for shock loads.

29.2.2 Gearboxes

Gear systems are used to change either the speed or the direction of rotation. This is accomplished by the meshing of a variety of gears, usually within a closed *gearbox.* The box is required, as meshing gears must be constantly lubricated to prevent excessive wear. The shaft penetrations must include seals, and a means of checking and adjusting the lubricant level is required. Even with the best of lubrication, gear systems lose power as a result of frictional forces. The value of the loss is expressed by the efficiency, which is the ratio of output power to input power expressed as a percentage.

For gearing of parallel shafts, either spur gears, wherein teeth are parallel to the shaft, as shown in Figure 29.3, a, or helical gears, in which the teeth are at an angle (Figure 29.3, b), are used. Spur gears are the most economical and can be used with ratios of 1:1 to 5:1. They cannot carry a load as great as the other styles, however. Helical gears are very efficient and can be used with greater ratios of 1:1 to 10:1. Where a change of shaft direction is required, beveled gears, as illustrated in Figure 29.3, c, are used. These are available in several designs, including straight and helical bevel, which is similar to the helical style shown in Figure 29.3, b. For changing the shaft direction 90 degrees, worm gears, which are discussed shortly, can also be used.

FIGURE 29.3 Standard gear types: (a) spur, (b) helical, and (c) beveled.

Gearboxes must be firmly mounted to ensure smooth operation. The power source is typically connected with a flexible coupling. The output shaft, however, is often fitted with a chain sprocket or pulley. In such installations it is important to consider the *overhung load* capability of the particular unit. This is the side or radial force applied at a right angle to the shaft, which imparts a high load on the shaft bearings. If this load is excessive, early failure of both the bearing and the oil seal can occur. The overhung load capacity of gearboxes is usually provided in the performance data of the manufacturer and is typically based on the overhung load being applied at the middle of the shaft extension.

29.2.2.1 SPEED REDUCERS

The amount of speed reduction is a function of the number of teeth on mating gears. For example, a gear with 10 teeth would have to rotate ten times to turn a meshing gear with 100 teeth, a full rotation. In the simplest form, speed-reduction gearboxes consist of one set of simple gears mounted on respective input and output shafts. For high levels of reduction, one or more *jackshafts,* each with large and small gears, can be included between the main shafts. Such simplicity is not always practical, however, and much more complicated arrangements involving multiple gears and shafts are common.

29.2.2.2 WORM GEAR DRIVES

A continuous helical gear on a shaft is known as a *worm,* and it mates with a *worm gear,* as illustrated in Figure 29.4. Worm gears provide especially great speed reduction, which minimizes the number of gears and shafts required. For example, a worm typically has the equivalent of only three teeth, which are known as *starts.* Worm gears run exceptionally smoothly and quietly and are capable of withstanding shock loading or momentary overloads. A single worm

FIGURE 29.4 Worm gears provide large speed reduction.

TABLE 29.1 Standard Sizes of Roller Chain

Size Designation	PITCH in.	PITCH mm	Standard Sprockets Number of Teeth
25	1/4	6.4	9 to 112
35	3/8	9.5	9 to 112
40	1/2	12.7	9 to 112
41	1/2	12.7	9 to 112
50	5/8	15.9	9 to 112
60	3/4	19.2	9 to 84
80	1	25.4	9 to 80
100	1-1/4	31.8	10 to 80
120	1-1/2	38.1	11 to 60
140	1-3/4	44.5	11 to 60
160	2	50.8	11 to 60
200	2-1/2	64.5	—
240	3	76.2	—

drive has the input and output shafts at a right angle, enabling easy change in shaft configuration. Worm gear drives are unique in that they continue to generate the tooth form just as in the original cutting, rather than wear it away as in other gear configurations.

29.2.3 Roller Chain and Sprockets

Chain drives provide positive rotation as well as the ability to somewhat dampen shock loads. They are available in many sizes, which are defined by *pitch,* which is the distance between the individual roller pins. Standard chain is available with pitches ranging from 1/4 in (6 mm) to 3 in. (76 mm), as indicated in Table 29.1, and can consist of a single *strand* or up to four side-by-side strands. The basic chain is supplied in long lengths with fixed pins; however, a pin can be removed to break a link and obtain any desired length. Removable links, which are available in either half or full links, are then used to reconnect the chain into a continuous loop.

Sprockets are available in many sizes and with a variety of hubs so as to fit most any size shaft. As with gears, speed reduction is determined by the difference in number of teeth between the drive and the driven sprocket. The most frequently used sprockets for common shaft sizes are stocked with a finished bore and a keyway. A much wider selection of sizes is available, however, to be used with separate stock hubs.

For good performance, roller chain requires lubrication. Both the horsepower transmitted and the maximum RPM of the sprockets depend on the lubrication method. For low speeds and minimal loads, manual or drip lubrication can be used. Where higher performance is necessary, either steady-stream or submerged bath lubrication is necessary. The exact requirements depend on the maximum feet per minute of travel and vary according to both chain and sprocket size. Major manufacturers provide catalogs with charts indicating appropriate selection parameters.

29.2.4 Belt Drives

Belt drives are the simplest and least costly types available. Whereas we are all familiar with the standard V belt used on our cars, belts are also available with different widths and cross sections as well as special designs. In industrial applications, as many as ten matched belts may be used side by side. *Timing belts* contain gearlike teeth on the interior, matched to grooves in the *sheaves* or pulleys, providing positive power transmission without the need for lubrication. Belt drives are particularly suited for high speeds and are ideally run within a range of 1000 to 5000 ft/min (300 to 1500 m/min). They can accept more shock than any other type of mechanical drive. Although simple belts are available at most any hardware or auto parts store, manufacturers' catalogs should be consulted when designing or modifying equipment.

Belts stretch with use, especially at high temperatures, requiring the ability for adjustment of the distance between centers. The distance between centers should generally not be less than the diameter of the larger sheave. As with gears and sprockets, different size sheaves can be used so as to vary the speed of the driven shaft up to a ratio of about 8:1. The use of very small diameter sheaves should be avoided, as they significantly lessen efficiency and will reduce the belt's longevity. The contact ark on the smaller sheave should be at least 120 degrees. Belts should be kept clean and properly tensioned. When operating, the tight side should be in a straight line, with a very slight dip of the slack side, and tension should be regularly checked and adjusted as required. Normally, belts should not be used when the environmental temperature exceeds 160°F (71°C).

29.2.5 Variable Speed Drives

The ability to adjust drive speed is a frequent requirement for grouting equipment. Variation can be achieved through either special adjustable-speed enclosed boxes or with adjustable pitch pulleys. The mechanics of the enclosed boxes are usually proprietary and will vary among manu-

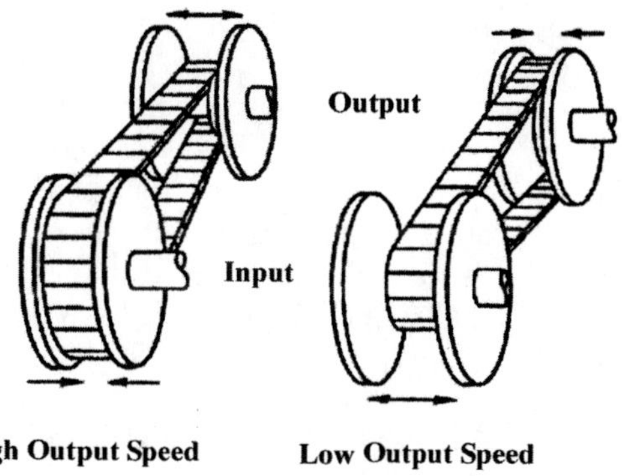

FIGURE 29.5 Speed variation with variable pitch pulleys.

facturers. The faces of adjustable pulleys (Figure 29.5) are not fixed but can move, relative to one another. There are two types, one in which the distance between shafts is uniform and the widths of both pulleys are adjustable, the other requiring only one adjustable pulley but variation of the shaft separation distance.

In the first case, the movable face of one of the pulleys is spring loaded so that it is constantly forced into the belt. The face separation of the other pulley is mechanically controlled, typically with a hand crank. As the mechanically controlled face is moved inward, the diameter of the belt contact increases. To compensate, the width of the other pulley must increase, resulting in a smaller effective diameter. The speed ratio with this arrangement can be varied from zero to about 9:1.

Single adjustable pulleys used with moving bases work similarly. They are spring loaded, so the faces keep a positive pressure on the belt. The belt is sized so that the faces will be at their maximum distance apart when the shafts are most distant. This results in the minimum effective diameter for the belt to traverse the adjustable pulley. As the shaft spacing is decreased by way of a mechanically adjustable mount, the movable face is forced in by the spring pressure so as to increase the diameter of belt travel. With this arrangement, any size pulley can be used on the fixed end and ratios of up to about 5:1 can be achieved. The adjustable pulley can be positioned on either the input or output shaft. This setup is quite simple and very effective, and it has been used on many proportioning pump systems built for plural-component grouts.

29.3 HYDRAULIC SYSTEMS

Power is transferred in hydraulic systems through the pressure of the hydraulic fluid, which is specialized oil. The pressure is generated by a hydraulic pump and typically directed through hose lines to either a cylinder or jack, or a rotary motor. Hydraulic cylinders and motors are compact, require little space, and can be placed in any configuration, providing great flexibility of use. Efficiency is greatest when the distribution lines are short and all of the elements are assembled into a fixed system. It is possible, however, to pump the oil long distances to power remotely located apparatus.

29.3.1 System Fundamentals

Typical hydraulic systems consist of a reservoir or tank for the oil, a pump to pressurize it, and the apparatus that is to be powered, usually a motor or ram. In between are the fluid delivery and return lines and an assortment of filters, valves, and other controls. The oil temperature rises as it passes through the various elements and apparatus that are powered. Power transmission efficiency lowers as the oil's temperature rises, which can cause a dramatic reduction in performance. For this reason, many hydraulic systems include a heat exchanger to remove excessive heat from the oil.

Hydraulic systems can be straightforward, such as a pump operating a single ram or jack. Conversely, they can be complex where the operation of many different rams, motors, and such is required. Many machines are fitted with multiple hydraulic pumps, often piggybacked on one another. Although they often utilize the same oil reservoir, they can form separate systems. A variety of valves, flow splitters, and bypass mechanisms are required where the speed of the various elements must be controlled, and the propensity for the oil to heat increases with the complexity of the system.

Hydraulic oil must be kept scrupulously clean to prevent excessive wear of the components. High-efficiency filters are thus an essential part of any but the simplest of systems. For this reason, and because air must not develop within the system, hoses for portable hydraulically powered tools are fitted with one-way valves so as to always contain the oil. This makes the already stiff hose heavy to handle, and great care must be taken to ensure that the couplings are clean when joining lines. Poor performance of hydraulically powered apparatus is nearly always the result of insufficient oil pressure or flow, or of excessive heat or debris in the system. These deficiencies will also cause unnecessary wear of the components.

The operating speed of motors and other tools is directly controlled by the oil flow rate. To decrease speed, the flow rate must be lowered, typically through bypassing to the reservoir. The result is less efficiency and reduced power delivered to the tool. In addition, bypassing fluid is a significant heat generator, so the remaining flow is less efficient. Variable output hydraulic pumps are available and used on some equipment, but because of their greater cost, splitting or bypassing fluid remains the most common method for speed control.

29.3.1.1 FLOW AND PRESSURE

Accomplished work is a product of the hydraulic oil pressure times the flow rate. Therefore, a given amount of work can be accomplished with either a low pressure system combined with a high flow rate, or one with high pressure but less flow. Because lower pressures are easier to develop and control, and promote less wear on the components, most unitized hydraulic systems operate on this basis. These *high-volume low-pressure* (HVLP) systems operate at between 1500 and 3000 psi (103 and 207 bars), with flow rates as great as 50 gal/min (3.2 L/sec) or more.

In the case of portable apparatus, it is necessary to string out hose lines. They must remain filled with oil in order to prevent contamination, and their weight is directly related to size. Further, as the line lengthens and/or fittings are added, the generated heat increases. To minimize weight, smaller lines are preferable and are acceptable, provided higher pressure is used. Thus, systems and tools working on (HPLV) flow have been developed. In these systems, pressures of 5000 to 6000 psi (345 to 414 bars) and flow rates less than 15 gal/min (0.95 L/sec) are typical. Such systems are preferred for remote applications requiring hose makeup in the field, as the significant reduction in required flow allows the use of smaller hoses, which are lighter and much easier to handle.

Several manufacturers of small portable hydraulic tools, such as saws, drills, and impact wrenches, have formed the Hydraulic Tool Manufacturers Association to provide guidelines for hydraulic parameters and standardized fittings. These tools are described as *low-pressure hydraulic tools* (LPHT), which operate at 1000 to 2000 psi (69 to 138 bars) and are divided into three categories according to flow:

Type I—5 gal/min (19 L/min)
Type II—8 gal/min (30 L/min)
Type III—12 gal/min (45 L/ min)

Hydraulic jacking systems, on the other hand most commonly operate and are rated at 10,000 psi (690 bars).

A basic part of hydraulic system design is to match the pressure and flow values of all of the system components. This is fairly easy with unitized systems, by using the original manufacturers' parts or their equal for any repair. Problems develop, however, when jury-rigging with parts and components of different capabilities. For example, LPHT hose used for a jacking application could easily burst because of the higher pressure. Many problems with hydraulic equipment have been caused by failure to recognize and provide the proper flow and pressure parameters.

On a particular project, the contractor had several hydraulically powered machines. He had removed the original engines from each, substituting a single large hydraulic pump to power all the units. Hoses were run from this pump to the different machines, apparently without consideration for the individual requirements. Especially adversely affected was an auger-type mixer, in which the auger was turning excessively fast. It was so fast that grout was not properly mixed, resulting in continual problems in the pump hopper and downstream. In a similar case, there was repeated breakage of the coupling between the drive motor and the main shaft of a mixer. Apparently, the pressure, and perhaps the flow, of the central hydraulic pump was excessive, overstressing the machine components.

In another instance, a portable hydraulic pump manufactured to hydraulic tool manufacturers association (HTMA) Type I requirements, providing for 2000 psi (138 bars) pressure was used for powering withdrawal jacks for upstage grouting. The jacking force was insufficient to pull the casing, which should have been evident, as at 2000 psi (138 bars), the jacks were working at only 20 percent of their rated capacity. In a similar example, the hydraulic system of a small concrete pump used for compaction grouting was tapped to power the withdrawal jacks. Again the capacity was insufficient, as the system pressure was 2,500 psi (172 bars), far short of the required 10,000 psi (690 bars) for the rated capacity.

Volumetric batching plants for concrete are often used in grouting. Here, the volume output is a small portion of that for which the machines were built, so the speed is often reduced by bypassing fluid. Unfortunately, this heats the fluid, which lowers its efficiency. Where such equipment is going to be used only at the slower output rates, the hydraulic pump should be replaced with one of equal pressure but lower flow rate.

Understanding both the pressure and flow requirements is fundamental to building, modifying, or repairing hydraulic equipment. Their importance cannot be overemphasized, as proper operation is wholly dependent on them. An excellent reference for anyone involved with hydraulic power transmission is the *Fluid Power Society Lightning Reference Handbook* (2001).

29.3.1.2 HYDRAULIC FLUID

Fundamental to proper operation of a hydraulic system is the use of quality hydraulic fluid. It must be of the correct viscosity, remain stable at the operating pressures and temperatures, provide good lubricity, be resistant to compressibility (low bulk modulus), and, obviously, be compatible with the seals and other system components. Virtually all oils manufactured for hydraulics are of high grade and contain stabilizers and antiwear additives. They do, however, tend to thicken at low temperatures and thin as the temperature increases. The standard test temperature at which viscosity is determined is 100°F (38°C), and this is an ideal operating temperature for such a system.

Hydraulic systems have two mortal enemies, both involving the oil. These are heat and dirt in the oil. Heat is generated in all hydraulic systems as a result of friction between the oil and the system components. A particularly large contributor is bypassing fluid in order to slow down the flow, and thus the speed of the driven apparatus. Moreover, friction within the walls of the hoses, especially when the oil penetrates couplings, can be substantial. Good practice dictates that the line velocity not exceed about 10 ft (3 m) per second. The oil temperature should never exceed 140°F (60°C), and operation between 100° and 120°F (38° and 49°C) is preferable. High temperatures thin the oil excessively, greatly compromising its ability to transfer energy. It also reduces lubricity, which will cause greater wear of the components and break down the integrity of elastomer seals and hoses. To control heat, many hydraulic systems have heat exchangers, and these should always remain connected and in good condition.

Any particulate matter in the oil will greatly increase wear of the pump and other system components. For this reason, one or more high-quality filters should be installed at both the suction and pressure sides of the pump. It is imperative that these filters be regularly inspected and replaced as indicated. Excessive contaminants can cause pump cavitation if they are allowed to build up on suction screens, inhibiting flow, so these should be cleaned periodically. And if appropriate oil cleanliness is not maintained, a mechanic's nightmare will occur, with intermittently sticking valve spools and rapid wear of the components.

Oil cleanliness is especially important with remote systems in which the hose lines are made up and broken in the field. Here, it is extremely important to meticulously clean the couplings as they are made up. Hose ends should be connected once the hose is rolled up. System filters should also be more frequently checked and renewed. The use of multiple filters should be considered. Early in my experience with hydraulic power transmission, short life of the components was a serious problem. Through controlling the oil temperature with a heat exchanger and providing multiple filters and frequent servicing, the problem virtually disappeared. The importance of keeping the oil clean cannot be overstressed.

29.3.1.3 HYDRAULIC CIRCUITRY

Hydraulic power transmission is now well developed. Although simple applications are common, modern equipment can have very involved control systems. Two distinctly different types of circuitry are used on concrete and grout pumps: *open loop* and *closed loop*. In closed-loop systems, a pulsating pump passes pressurized fluid from one drive piston to the other continually. Leakage within the system depletes the fluid volume, so a separate *charge pump* is required to replenish it. Excess replenishment oil goes over a low pressure relief valve and through a heat exchanger, which is the only cooling. Closed-loop systems require smaller reservoirs and are capable of higher operating speeds, but they do not switch the direction of flow very precisely, especially at the lower end of their operating range, which is common in grouting.

Oil in open-loop systems flows from the pump to directional valves, which send it to the motors or cylinders, from which it returns to the reservoir. Multispool valves can be used so that several mechanisms are switched simultaneously, providing precise control. Cooling of the oil is easily accomplished as it returns to the reservoir. The nonuniform pumping illustrated in Figure 9.8 resulted from a closed-loop system. The problem disappeared when it was replaced with a pump using open-loop circuitry.

29.3.2 Valves and Fittings

The routing of hydraulic fluid is controlled by a variety of valves. In the simplest form, these simply open or close, but more often they involve metering or bypassing fluid to adjust or change the operating speed of the driven apparatus. The operation of a simple valve routes the fluid to a single circuit, whereas multiple-spool valves can divide the flow and send it to multiple locations. The performance of a hydraulic system is directly linked to the circuitry that is controlled by the valves. They are a major source of heat generation in the fluid, and so must be thoughtfully selected. Replacement valves in existing systems should be identical to the originals, if possible, and if not possible, they should be of the same type and capacity. For new or modified systems, valves should be kept as simple as possible.

29.3.3 Hydraulic Pumps

The type of hydraulic pump affects both the cost and the performance of a system. Most systems use positive displacement pumps, which provide either fixed or variable displacement output. There are several different types, but those used in grouting equipment are usually gear, vane, or piston pumps. Service requirements, and especially the necessary operating pressure, will dictate the type of pump to be used.

29.3.3.1 GEAR PUMPS

Gear pumps are the least costly, but they are also of low efficiency unless operating at near their highest rated speed. They consist of two meshing gears, one of which is driven by the power source, as illustrated at the left in Figure 29.6 The gears are snugly enclosed within a housing and side plates, sometimes called *wear* or *pressure plates*. A vacuum is created as the gear teeth unmesh at the pump inlet. This attracts fluid, which is carried between the teeth to the outlet, where it is forced out. The running clearance between the gears and the housing results in *slip*, an almost constant loss of fluid, causing poor performance at low operating speeds. The flow rate of gear pumps varies from about 0.2 to 150 gal/min (0.76 to 567 L/min) with pressure capability up to about 2500 psi (172 bars). Gear pumps are very compact, requiring minimal space, and are easy to install.

29.3.3.2 VANE PUMPS

A number of sliding vanes are mounted in slots of a rotating *rotor* that is mounted off-center in a housing or *ring*, as shown at the right in Figure 29.6. The vanes are thrust against the ring by centrifugal force or some form of positive pressure such as springs, and slide in and out of

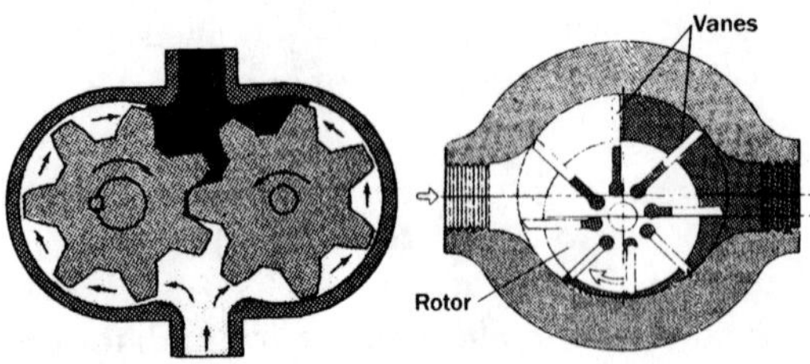

FIGURE 29.6 Common hydraulic pumps; gear on the left, vane on right.

the rotor slots. The volume of the chambers between the vanes reduces upon rotation because of eccentricity of the shaft, pressurizing the fluid. Vane pumps are very efficient and remain so for long periods of time, as the vanes move outward in their slots to counter wear and are always in intimate contact with the ring. They are available with output pressure to about 4000 psi (276 bars) and volume output from 0.5 to 250 gal/min (1.9 to 945 L/min). They are also very compact and, being very efficient, are probably the most frequently used type of pump.

29.3.3.3 PISTON PUMPS

The pistons of *axial* piston pumps reciprocate in a parallel direction to the drive shaft and piston block. Most are multipiston pumps and use check valves to direct the fluid from the suction to discharge. As the cylinder block is turned, *shoes* on the pistons bear against the angled *swash plate* (Figure 29.7, left), which converts the rotary motion to linear. The piston stroke is dependent on the swash plate angle. It can be adjusted in *variable displacement* or *pitch* pumps to change the fluid output. With the plate perpendicular to the pistons, output is zero, but increases as the angle becomes greater. *Bent axis axial piston* pumps operate similarly, except that the swash plate is at a fixed angle, which is equal to the bend in the axis, and is thus not adjustable.

The pistons of *radial* pumps are arranged radially around a piston block and move perpendicular to the driving shaft. The pistons reciprocate in cylinders of a rotating block that is eccentrically encased within a larger reaction ring and case (Figure 29.7, right). As the block turns, the piston strokes, taking in fluid while moving outward. The fluid is pressurized as the length of open bore decreases. Either cylindrical or ball pistons are used, and the pumps are described as check valve or pintle valve, depending on their porting arrangement. Pump displacement can be varied by movement of the reaction ring for which there are a variety of controllers.

Commonly available piston pumps can develop pressure to about 7500 psi (517 bars), with flow output up to 100 gal/min (379 L/min). These are, without question, the most versatile

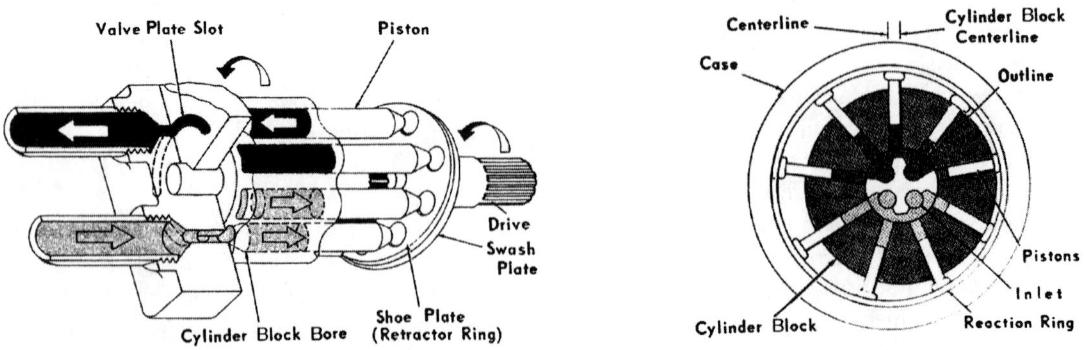

FIGURE 29.7 Piston hydraulic pumps; variable displacement axial on left, radial piston on right.

and highest-performing hydraulic pumps available. They are also the most complicated and costly. When used in closed-loop systems, they also require a *charge pump* to make up for any fluid lost to slippage.

29.3.4 Rotary Motors

Rotary motors have pressure faces coupled to the output shaft and porting timed so as to provide continuous rotation. They may be gear, vane, or piston types and appear similar to hydraulic pumps, but the basic mechanics are different. Pumps rely on volumetric rather than mechanical efficiency, whereas motors require mechanical efficiency in converting the fluid face pressure to rotary motion and torque. Hydraulic efficiency is also required, however, to keep the shaft turning at a constant speed regardless of the load. Higher fluid displacement for a motor of a given size will provide the highest efficiency. It is possible to use a pump as a motor by simply reversing the fluid connections, but although torque will be developed, the efficiency will be very poor.

The efficiency and output power of fluid motors is greatly affected by the cleanliness of the oil. Even minor contaminants in the oil cause wear of the moving parts, increasing internal leakage or slip. Increased internal leakage causes increased heat, further decreasing efficiency. With time, the rate of wear increases, accelerating the efficiency degradation. Gear-type motors generally wear until they no longer provide sufficient torque to continue operating. Vane motors behave similarly, but can have sudden catastrophic failure if the vanes are damaged, such as by bouncing on a rough ring or becoming wedged between wear plates. Piston motors are highly resistant to wear and generally provide longtime service. As emphasized earlier, maintaining quality fluid, free of contamination and/or excess heat, will go a long way toward ensuring the long life of all hydraulic motors.

29.3.5 Hydraulic Rams and Jacks

Hydraulic cylinders are available in an almost endless range of sizes and capacities for use as either rams or jacks. Their working pressure varies from less than 1000 psi to 5000 psi (69 to 345 bars) maximum for those made for general use. As previously mentioned, cylinders for jacking applications are typically rated at 10,000 psi (690 bars). Cylinders are either single or double acting, whereby they are powered in both directions. Single-acting cylinders can be powered to either push or pull and require some sort of force for retraction. Those made for jacking applications generally contain a spring return and are available with capacities up to 1000 tons.

29.4 PNEUMATIC SYSTEMS

Pneumatic power is considered a form of fluid power, and it exhibits many similarities to hydraulic oil power transmission. Compressed air has a substantial advantage over oil for use in grouting, in that the air is basically weightless and it need not be maintained in the system. Furthermore, it is easy to blow any debris out air hoses before connection, so contamination need not be a problem. Both cylinder rams and rotary motors can be powered pneumatically. An air motor slows down upon a load increase, but the output torque simultaneously increases until stall occurs, and a stall condition can be maintained without damage. The output speed can be controlled by throttling the air inlet and the direction of rotation changed by simply switching input and outlet ports.

Both radial piston and vane-type motors are available. Vane-type motors are the most widely used, however, as they are substantially less costly and very compact in size. The power of vane motors is highly dependent on speed, and maximum performance is typically at the fastest rated speed, usually within a range of 1500 to 3000

RPM. Performance typically deteriorates rapidly at speeds lower than about 1000 RPM. Both four- and eight-vane motors are widely available, and they perform about the same, with one exception. If a four-vane motor stops directly on dead center, it may not start when air pressure is reapplied. Should this occur, the problem is usually solved by simply turning the shaft slightly, as with a pipe wrench.

Piston motors have exceptional starting and low-speed performance and will provide high torque at virtually any speed. Their cost is many times that of vane motors, however, and they are bulky and much heavier for equivalent output. Because performance is not affected by speed, they are ideal for powering drills and grout pumps where operation at many speeds is required. I have purchased many old air track drills for the sole purpose of recycling the piston air motors that power each track. Piston motors are extremely durable and provide long, trouble-free operation.

There are two requirements for working with compressed air, whether powering rams motors or pneumatic tools: cleanliness of the air and sufficient lubrication. As mentioned earlier, simply blowing the hoses out before making a connection will ensure clean air. In-line oilers are advisable at the compressor for lubrication. Moreover, it is a good idea to pour about a teaspoon of lightweight oil into the motor or tool prior to connection. Experience has shown that this step can greatly decrease downtime for pneumatic equipment.

CHAPTER 30

Drilling Equipment and Accessories

30.1 DRILL RIGS
- **30.1.1** Handheld Drills
- **30.1.2** Mini Drill Rigs
- **30.1.3** Self-Contained Hydraulic Drill Rigs
- **30.1.4** Pneumatically Powered Percussive Drills
  - **30.1.4.1** Wagon Drills
  - **30.1.4.2** Air Track Drills
  - **30.1.4.3** Down-the-Hole Hammers

30.2 DRILL HOLE CASING
- **30.2.1** DCDMA Drill Casing
- **30.2.2** Special Casing Systems
- **30.2.3** Standard Steel Pipe
- **30.2.4** Carbon Steel Tubing

30.3 DRILL ROD
- **30.3.1** Rotary Drill Rod
  - **30.3.1.1** DCDMA W Series Rod
  - **30.3.1.2** DCDMA WJ Series Taper Joint Rod
  - **30.3.1.3** Wireline Drill Rod
  - **30.3.1.4** Custom Lightweight Rod
- **30.3.2** API Drill Rod
- **30.3.3** Percussion Drill Rod
  - **30.3.3.1** Heavy Percussion Rod
  - **30.3.3.2** Hex Sinker Drill Steel

30.4 DRILL BITS
- **30.4.1** Rotary Drilling Bits
  - **30.4.1.1** Drag/Fishtail Bits
  - **30.4.1.2** Roller Cone Bits
  - **30.4.1.3** Diamond Core Bits
  - **30.4.1.4** Continuous Flight Augers
- **30.4.2** Percussive Drill Bits
  - **30.4.2.1** Equipment-Mounted Drill Bits
  - **30.4.2.2** Bits for Sinker Drills

30.5 SLEEVE PORT PIPES (TUBE-À-MANCHETTE)

30.6 SPECIAL DRIVE NEEDLES

30.7 CIRCULATION (MUD) PUMPS

MANY DIFFERENT TYPES of drilling equipment and accessories are commercially available, although little has been designed specifically for grouting. Fundamental to the drilling of grout holes are drill rigs, casings, rods, and bits. In addition, mud pumps to pressurize the circulation flush, mud tanks, and the appurtenant hoses and fittings are required.

30.1 DRILL RIGS

All drill rigs use some type of mast or track on which the powered drill head traverses to force the drill string into the formation, and withdraw it to receive additional drill rod extensions. The mast is usually adjustable so as to assume any inclination required for the desired hole trajectory.

Rotary and rotary percussion drill rigs are similar in this respect; however, the equipment used with percussion drills must be heavier and more robust than that used only for rotary drilling. Thus, although most percussion rigs can be fitted for either drilling method, such is not the case for many rotary rigs. Combination rotary/rotary percussion drill heads are becoming commonly available and are supplied as standard equipment by many manufacturers, allowing the use of either method with the same basic machine.

Most drill rigs utilize a *top drive* head; that is, the drive motor or hammer is attached directly to the top end of the drill rod. There are many different types of standard drill rod, and many specialist contractors have developed their own designs, usually for the purpose of minimizing the weight of the *drill string*. The length of the individual *joints* of rod, is usually dictated by the height of the drill rig *mast*. Where there is sufficient overhead clearance, masts using 10 ft or even 20 ft long joints are common. Even longer joints are sometimes used, but generally in combination with very large drill rigs and larger diameter holes than are commonly used in grouting. Where overhead space is restricted, shorter masts and drill rods must obviously be employed.

Some rigs do not have masts, but, rather, short power feeds provided with a chuck. The drill rod telescopes through the chuck, which grips it, as shown in Figure 30.1. The required rotation and down thrust are transmitted through the chuck. The stroke length of such drill feeds is usually only a few feet. Although such machines were common in earlier times, they are not now widely used, except in areas of limited access or overhead clearance.

The primary factors by which drill rigs are classified are maximum *torque* and *thrust* on the drill string. For many types of work, the pullback or withdrawal force is also important, as well as the maximum spindle speed. Although there are no standards for classifying drill rigs, they may conveniently be placed in broad categories, as indicated in Table 30.1, based on their torque and thrust capabilities. These values must be considered approximations only, as there is significant variation among different manufacturers.

FIGURE 30.1 Power feeds with hollow chucks are used in confined areas.

30.1.1 Handheld Drills

Handheld drills are often remotely powered by compressed air or hydraulic power. Before the present advanced development of the directional drilling industry, pneumatically powered *bore* motors were widely used for horizontal drilling of small utility lines. Several manufacturers built these motors, which were basically adaptations from simple air-powered portable drills. Early horizontal boring was done substantially in conformance with present rotary drilling methods, and water was used almost exclusively as the circulation flush. Because the operators would be standing while pushing horizontally in the di-

TABLE 30.1 Torque and Thrust Categories of Drill Rigs

Category	Gross Weight Pounds (Kilograms)	Torque Foot-Pounds (Kilogram-Meters)	Thrust Pounds (Kilograms)
Handheld	<150 (<68)	60–300 (8.3–41.5)	—
Mini	<5,000 (<2270)	140–1,000 (19.4–138.3)	100–2,000 (45–909)
Small	5,000–12,000 (2,270–5,450)	2,000–5,000 (276–692)	2,000–3,000 (909–1364)
Medium	12,000–20,000 (5,454–9,090)	6,000–10,000 (830–1,380)	3,000–6,000 (1,364–2,727)
Large	20,000–35,000 (9,090–15,900)	10,000–15,000 (1,380–2,070)	6,000–12,000 (2,727–5,454)
Huge	>35,000 (>15,900)	>15,000 (>2,075)	>12,000 (>5,440)

rection of drilling, there was no reason to keep the drill head short, and the water swivel was typically located between the drill motor and the drill rod.

In the early 1950s, such a boring motor was used to drill vertical grout holes. Although it worked well for horizontal holes, several limitations had to be overcome to adapt optimally to vertical drilling. The height of the drill above the base working level was important, as the working range was practically limited to the reach of the drillers, as illustrated in Figure 30.2. The water swivel was thus moved to the back (top) of the motor, which required machining a hollow shaft through the motor and the provision of adequate seals to prevent leakage. This resulted in a 6 in. (150 mm) increase of drill rod length that could be used, greatly increasing efficiency.

In addition, heavier thrust bearings were required to withstand the higher forces of vertical drilling. Once these modifications were made, the use of handheld bore motors to produce both vertical and inclined holes became common. Through the 1970s and 1980s there were three separate manufacturers of *wet head* pneumatic drills in the United States, Thor, Ingersoll Rand, and Chicago Pneumatic. Although these items are no longer manufactured, they regularly appear in the used equipment market and

FIGURE 30.2 Drill rod used with hand-held motors is limited by the reach of the drillers.

should be of interest to grouters who work in confined areas.

The most important considerations of handheld tools for vertical drilling are:

- Weight
- Sufficient torque
- Rotation speed
- Reversibility to facilitate makeup of the drill string

Although these tools were traditionally fabricated with a cast steel case, to minimize weight, both Thor and Ingersoll Rand provided optional aluminum bodies. This reduced the weight from more than 50 lb (23 kg) to as little as 35 lb (16 kg), depending on the model. For a drill of a given weight, the torque and rotation speed are directly related. As speed increases, torque becomes less. A minimum torque of about 60 ft lb (8.3 kg m) is required for efficient operation, and many drills have greater output. A minimum operating speed of about 150 RPM is required, and 300 RPM is desirable.

30.1.2 Mini Drill Rigs

As previously discussed, a very large proportion of grouting is performed in confined areas, often around, inside, or under structures. The compactness and maneuverability of the equipment is thus an important consideration. Historically, many specialty contractors either modified commercially available equipment or totally built their own rigs. Figure 30.3 shows one of these, a light, independently powered drill rig working inside a building.

FIGURE 30.3 Light-weight independently powered drill rig inside a building.

To minimize weight and increase compactness, the power source can be remote, using either pneumatic or hydraulic power transmission. Hydraulic power is the more efficient, but it is not nearly as easy to use as compressed air. Hydraulic systems work at significantly higher pressures than pneumatic systems. They require much stronger and heavier hose including one or more layers of metal jacket, making them heavy and stiff. In addition, unlike weightless air, hydraulic oil adds weight to the hose, requiring greater effort to handle and store.

Unfortunately, emptying hydraulic oil from a hose when not in use is not a viable option, as it must be kept absolutely clean and free of debris. Air hoses can be blown free of any debris prior to connection to a drill, whereas the interior of a hydraulic hose will always be coated with oil, attracting dirt that cannot be easily removed. Hydraulic hoses are thus fitted with valved couplings that close upon being broken. This effectively seals in the oil, keeping it from becoming contaminated, but adds considerable weight. Hydraulic-powered percussive drills are thus usually self-contained. In those instances when a separate hydraulic power source is used, it is typically located in near proximity to the drill so as to minimize the length of required hose. Although the power transmission efficiency of hydraulic oil is significantly greater than that of compressed air, the latter is much easier to deal with on a construction site. Both are thus widely used in drilling and grouting.

The small rig illustrated in Figure 30.3 is pneumatically powered. Although it is heavier and more powerful than the handheld drill units discussed earlier, a similar motor is used. It is raised with a pneumatically powered winch attached to the base. Thrust is limited to the weight of the bore head and drill string, which is adequate for much soil. Where greater thrust is required, weights can be hung on the drill head carriage. If greater thrust is required for only a short time, such as in penetrating a rock, the operator can hang weights on the drill carriage or pull down on it. Although the use of lightweight "mini" rigs is usually limited to areas of restricted access, there are places where they can be advantageous, even though such an area is accessible to larger equipment. The penetration rate is not as fast as that of a larger machine, but the total time required to complete a hole can be less, because of much faster moving between holes and setup.

Figure 30.4 shows another custom-built mini drill rig. A very large piston-type pneumatic motor powers the drill head, and three separate pneumatic winches are mounted on the deck. One is used to raise the drill head, another to provide down thrust, and the third to move the skid-mounted rig by securing the winch line to any fixed object in the direction of the desired move and winching it into place. This rig is much heavier than those shown previously and can exert about 1100 lb (500 kg) of thrust without the provision of additional weight. Where greater thrust is required, weights can be placed on the deck near the mast. On one project, this rig was used to install 40 ft (12 m) deep pressure-injected piles, which were drilled with an 8 in. (203 mm) continuous flight auger.

The frequent requirement for drilling in confined locations has not been unnoticed by drilling equipment manufacturers. Several have introduced very compact and maneuverable machines, specifically designed for work in confined areas. A variety of remotely powered hydraulic drills are available from a number of manufacturers. A particularly narrow machine is the Atlas Copco Mustang A-32 CNS, which consists of a tracked drilling machine and a separate wheeled hydraulic power source. The hydraulic power unit can be towed behind the drill, as illustrated in Figure 30.5, and the units, which are connected by hoses, can be 50 ft (15 m) or more apart. The tracked drill is only 29.5 in. (750 mm) wide, so it can get through narrow doorways and into very tight locations. For such a tiny machine, it will develop an amazingly high torque of 3070 ft lbs (424 kg/m).

FIGURE 30.4 Custom-built rig is driven by large piston air motor and air-powered winches.

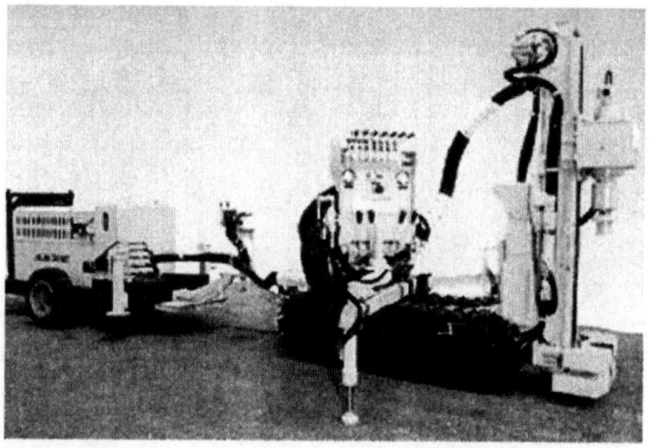

FIGURE 30.5 Compact drill rig tows its independent power source, which can be remotely located.

30.1.3 Self-Contained Hydraulic Drill Rigs

Self-contained, fully hydraulic drills are particularly efficient and are thus widely used to produce grout holes in rock. Their operation is generally limited to surfaces not more than about 30 degrees off level, which inhibits their use on very steep sites, such as dam abutments. To enable use on steep terrain, beveled platforms mounted on wheels or rails, illustrated in Figures 12.15 and 12.16, are sometimes used.

These rigs are available in a wide variety of sizes and capacities. On the lower end of the scale are machines with gross weights on the order of 10,000 lb (4545 kg), such as the Davey Model DK 50 and the Casagrande Model M3E, illustrated in Figure 30.6. For narrow access, the tracks of the Davey drill will retract to only 30 in. (762 mm), although its normal working width is 40 in. (1016 mm). The M3E Casagrande rig is also compact, with a width of only 34 in. (864 mm), and yet provides a hefty maximum torque of 6000 ft lbs (824 kg/m) with 6700 lb (3045 kg) thrust. These rigs are among the smallest of their type, yet they are more than adequate to drill most grout holes in either soil or rock.

Many rig manufacturers will individualize their standard units to the customer's particular needs, such as providing custom mast height and type of drill head. The compact rig shown in Figure 30.6 is fitted with a short mast to enable working inside a building with limited headroom. Some manufacturers will supply a basic drill model on different mountings. For instance, the Atlas Copco Mustang Model A-32 drill can be acquired as a self-contained track mount, a skid mount, or the separately powered unit shown in Figure 30.5. Large and very powerful self-contained rigs, such as the track-mounted Casagrande C 8 shown in Figure 30.7, are available where large or deep holes are required. These rigs are not commonly used for run-of-the-mill grout holes, although they are often employed for operations such as jet grouting, installing micro piles, and drilling for tieback anchors.

FIGURE 30.7 Large powerful self-contained hydraulic drill rig for large or deep holes.

FIGURE 30.6 Small self-contained hydraulic drill rig inside a building.

30.1.4 Pneumatically Powered Percussive Drills

The simplest form of rotary percussive drill is the common handheld pneumatic rock drill, often referred to as a jackhammer but more properly defined as a *sinker drill.* Such drills are widely used for grouting in concrete and masonry structures, and sometimes used for shallow grout holes in geotechnical applications. These tools are available in weight classes of 15 lb to 85 lb (6.8 to 39 kg). The heavier tools will always have a "T" handle, but the smaller units can have a "D" handle, which facilitates drilling of horizontal or overhead holes. Heavier sinker drills can be mounted on a "jackleg," which is simply a telescoping tube to which the drill is attached. It is extended through the force of a compressed air cylinder, so as to take all or most of the weight and apply thrust. Another form of mounted sinker drill is the "stopper," which is basically a rotary percussive hammer integral to a jackleg. The jackleg of these units is in line with, and directly opposite, the drill bit so that its extension will provide thrust for drilling. Drills of this type are used in small tunnels or other situations where the necessary reaction for the leg exists.

Rotary percussive drills are also available on powered mounts, referred to as *drifters.* Drifters are basically tracks that contain a movable mount for the drill. They are available in a variety of lengths up to 20 ft (6 m) or more. A continuous screw feed, which is powered by a separate air motor, typically propels the drill mounting, although the feed of some larger drifters is by chain and powered sprocket. One or more drifters can be mounted on a support vehicle to form a *drill jumbo* for work in tunnels or in other situations where repetitive setups are required. Some manufacturers make specialty rigs that contain one or more drifters.

30.1.4.1 WAGON DRILLS

Wagon drills are simply masts mounted on wheeled carriages, originally designed to allow the use of longer drill steel and the mounting of heavier hammers than could be manually handled. They are typically steered and pushed into place manually. The mast is usually adjustable, all the way from subhorizontal to vertical in the traveling position, and up to about 25 degrees either way transversely. Movement of the drill carriage is accomplished by a powered chain drive. Although not widely used or even commercially available now, wagon drills can be very advantageous for drilling holes on fairly level surfaces. They are especially useful for holes that are more than a few feet deep, as the length of the drill steel used is limited only by the height of the mast.

30.1.4.2 AIR TRACK DRILLS

The successor of the wagon drill was a similar adjustable mast, mounted on a self-propelled tracked carriage, commonly referred to as an *air-track drill,* as illustrated in Figure 30.8. For drilling deep grout holes, heavier air hammers are advantageous and are available on these machines. They have a low center of gravity and are inherently stable, even on steep or uneven surfaces, so tipping is not a great problem. By operating the crawler tracks in opposing directions, they can turn "on a dime" and are thus highly maneuverable. Their articulated masts are fully powered, which enables great flexibility in setup. Track drills are available with considerable reach off the undercarriage, enabling hole locations up

FIGURE 30.8 Air track drills are very maneuverable and provide considerable reach beyond tracks.

to 15 ft (4.5 m) or more away from the accessible setup location.

Because of their low center of gravity and relatively light weight, air track drills can be readily winched up the steepest of slopes, as illustrated in Figure 30.9. In this regard, they are often fitted with one or more pneumatically powered "tuger" winches, the cables from which can be attached to any anchor or "dead man" along or on top of a slope, thus negating the need for a separate winch carrier.

Because of their greater efficiency, fully self-contained hydraulic drills are now used to do much of the work traditionally performed by air track drills. The latter remain popular, however, because of their agility and ability to operate on slopes and other difficult-to-navigate terrain. In fact, these units can be literally hung from above and can drill in the most inaccessible places, as shown in Figure 30.10. Track drills are available in a variety of sizes and mast arrangements. The largest are capable of drilling to about 200 ft (60 m) with a conventional top hammer; however, as with any percussive drill, efficiency drops with depth and the required length of drill rod.

FIGURE 30.10 Relatively light weight and low center of gravity allow air tracks to be literally hung.

FIGURE 30.9 Air track held on steep abutment slope with a winch.

As with rotary drilling, the cuttings are expelled from the hole by circulation. Track drills are widely used for drilling blast holes, in which the cuttings are blown from the hole with compressed air. This is generally not good practice for grout holes, however, as the cuttings tend to be blown into the cracks and defects, blocking grout penetration. The problem increases when drilling in saturated conditions, as the cuttings tend to combine with moisture to form a slurry, which is hard to expel from the hole and will effectively seal over any defects it coats.

Furthermore, if the circulation becomes blocked, for example, by a thick paste of wet drill cuttings, horizontal defects may be displaced by the high-pressure air. Consequently, water is most frequently used for the circulation flush. In cases where it does not provide good cutting removal, or the fluid pressure at the bottom of the hole must be minimized, air-water foam can be used. Because the drill rig, rod, and accessories must withstand the ruthless pounding of the hammer, they must be much heavier and more robust than those used for rotary drilling.

30.1.4.3 DOWN-THE-HOLE HAMMERS

As mentioned earlier, the drill string absorbs considerable energy as it descends to the bit. Thus, the efficiency of percussive drilling suffers as the depth of the hole increases. To overcome this limitation, for holes with depths greater than about 100 ft (30 m), hammers are available, which are lowered to the bottom of the hole on the end of a drill string and connected directly to the bit. Rotation is supplied by the conventional rotation of the drill string, but percussive action occurs at the hole bottom. Less thrust is required on the drill string, as these hammers tend to pound their way downward in the hole. As they impart less bending stress to the drill string, straighter holes are produced with these hammers than with top hammer drills.

The smallest holes produced with down-the-hole percussive hammers are on the order of 3 in. (75 mm), which are about the largest holes commonly used in grouting. Furthermore, there is a risk of exhaust air entering the defects and causing damage to the rock. This risk is much less with the newer hammers, however, which operate on water pressure rather than air pressure. These water turbine hammers were tested during construction of the Olivenhain Dam in Southern California, and no adverse effect was found. Although still not widely employed in common grouting, they do see use for deeper and larger holes, especially those for grouted anchors and piles. Down-the-hole hammers are commonly available in sizes up to about 24 in. (610 mm) in diameter. Larger sizes can also be obtained, and some manufacturers build single assemblies that contain multiple hammers for the drilling of very large holes.

30.2 DRILL HOLE CASING

Commonly available drill casing was designed for drilling requirements only. The rigors of grouting, especially in upstage grouting using hydraulic withdrawal jacks, impose much greater forces, often exceeding those the casing can withstand. Many specialty contractors have thus developed their own proprietary casing systems. These are of many different origins, extending from ordinary steel pipe to high-precision carbon steel tubing with advanced threaded coupling systems.

30.2.1 DCDMA Drill Casing

A very long time ago (1929) the Diamond Core Drill Manufacturers Association (DCDMA) organized and set standards for the dimensioning and nomenclature of drill rod, casing, and appurtenances used for diamond drilling. Although there have been some modifications, these standards continue to the present and are recognized throughout the world. The system nomenclature provides for the drill rods, bits, core barrels, and so forth, of a given size designation, to fit inside casing of the same designation. Size is defined by the first letter in the designation, as shown in Table 30.2.

Following the size designator is a second letter signifying the group of matching tools. It is usually "W," which is the standard nearly always used. For casing the designation "X" indicates higher strength for holes of greater depth; however "W" group tools are all compatible with those casings. Occasionally a third letter is included, indicating the specific nature of a component, such as a special thread design.

TABLE 30.2 Letter Designations and Size of DCDMA Standard Downhole Systems

Letter Designation	Inches	Millimeters
R	1.0	25.4
E	1.5	38.1
A	2.0	50.8
B	2.5	63.5
N	3.0	76.1
K	3.5	88.9
H	4.0	101.6
P	5.0	127.0
S	6.0	152.4
U	7.0	177.8
Z	8.0	203.2

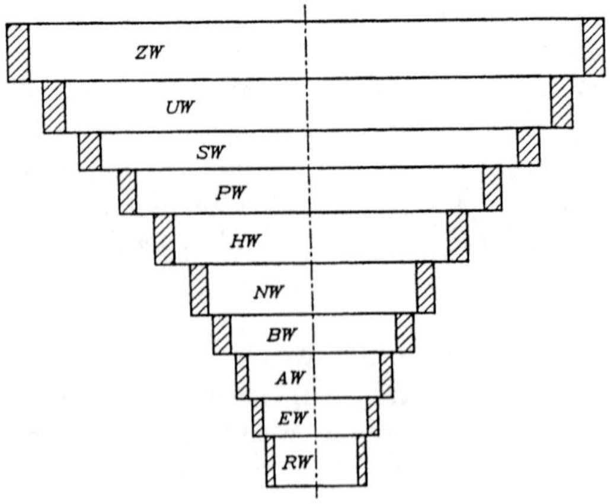

FIGURE 30.11 DCDMA casing nests so any size will telescope into next larger size.

These size designations and standards continue to be widely used in exploratory boring, and because of the availability of the systems, are used in the grouting field as well. The casing is provided with square-threaded flush joints and is commonly available from many suppliers in sizes A through P and in lengths of 5 and 10 ft (1.5 and 3 m). Other sizes or lengths can be obtained on special order. The casing is designed to "nest"; that is, each casing size will telescope into the next larger size, with a small amount of additional clearance, as shown in Figure 30.11. This allows use of the next smaller size casing through a casing string, where hole size is to be reduced with depth.

Dimensional data on the most commonly used diameters of drill casing, are shown in Table 30.3.

30.2.2 Special Casing Systems

Although the standard casing is readily available and easy to use, it was neither designed nor intended to withstand the rigors inherent in some grouting. Extraction of the casings from deep compaction grouting holes, in particular, imparts extraordinary force to the casing, and coupling failure is not uncommon. For this reason, some contractors have developed proprietary casing and, in some instances, entire drilling systems as well. An example of robust casing widely used in compaction grouting is shown in Figure 30.12. It consists of a 1/4 in. (6 mm) thick wall tube, with double-taper threaded joints that can be made up or broken quickly.

30.2.3 Standard Steel Pipe

Some grouters prefer to use standard steel pipe for casing, as it is far less costly than a proper flush joint tube. Although this can be satisfactory for shallow holes (less than about 20 ft (6 m)), and is sometimes used for depths up to 60 ft (18 m) or more, the larger diameter of the couplings requires an oversized hole, resulting in lack of good casing contact with the soil. This can allow grout to rise outside the casing, rather

TABLE 30.3 DCDMA Nomenclature and Nominal Dimensions for Drill Casing

	EW		AW		BW		NW		HW	
Size	**in.**	**mm**	**in.**	**mm**	**in.**	**mm**	**in.**	**mm**	**in.**	**mm**
O.D.	1.812	46.0	2.250	57.1	2.875	73.0	3.500	88.9	4.500	114.3
I.D.	1.500	38.1	1.906	48.4	2.375	60.3	3.000	72.6	4.000	101.6

than being deposited as intended, and can also increase the required withdrawal force.

Constant use, and especially driving standard coupled pipe, tends to cause breakage in the root of the threads, which is the weakest area. Pipe threads, by design, are on a slight taper, so the joint tightness against leakage will increase as the male thread is tightened into the female. Although this is advantageous in piping, it results in a separation of the pipe ends even when the threads are turned to the maximum degree of tightness, as illustrated in Figure 30.13 (bottom). To minimize this problem, electrical conduit couplings can be used rather than the standard pipe couplings. Unlike pipe joints with their tapered threads, joints of electrical conduit have parallel threads, which allow the pipe sections to be turned to the point that the ends are in contact with one another, as shown in Figure 30.13 (top), so the pipe itself takes much of the working force, rather than the threads alone.

FIGURE 30.12 Heavy flush-walled robust casing used for compaction grouting.

Steel pipe is manufactured to a variety of nominal sizes, starting at 1/8 in. (3 mm) and expanding to more than 8 in. (203 mm), depending on the particular grade. It should be noted that the nominal size is indicative of neither the inside nor the outside diameter. The actual inside diameter of most pipe can be more or less than the nominal size, depending on the grade. The outside diameter will always be substantially greater than the nominal size.

The unit weight of pipe is proportional to the wall thickness and is defined by the pipe *Schedule Number.* The outside diameter is always the same for any given pipe size in order to allow threading with standard tools. Thus, the

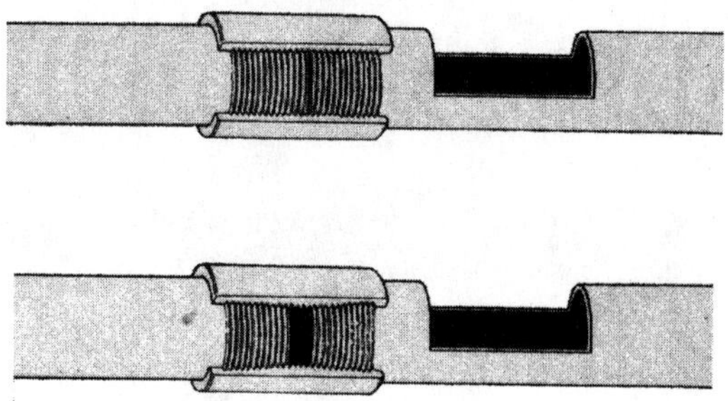

FIGURE 30.13 Standard pipe threads are tapered (bottom); electrical conduit has parallel threads (top).

TABLE 30.4 Dimensions and Weight per foot of Standard Schedule 40 Steel Pipe

Nominal Size/Inch	OUTSIDE DIAMETER		WALL THICKNESS		INSIDE DIAMETER		WEIGHT/FT (M)	
	in.	(mm)	in.	(mm)	in.	(mm)	lb	(kg)
1/8	0.405	(10.287)	0.068	(1.727)	0.269	(6.833)	0.24	(0.357)
1/4	0.540	(13.716)	0.088	(2.235)	0.364	(9.246)	0.42	(0.625)
3/8	0.675	(17.145)	0.091	(2.311)	0.493	(12.533)	0.57	(0.848)
1/2	0.840	(21.336)	0.109	(2.796)	0.622	(15.799)	0.85	(1.265)
3/4	1.050	(26.670)	0.113	(2.870)	0.824	(20.930)	1.13	(1.681)
1	1.315	(33.401)	0.133	(3.378)	1.049	(26.645)	1.68	(2.500)
1-1/4	1.660	(42.164)	0.140	(3.556)	1.380	(35.052)	2.27	(3.378)
1-1/2	1.900	(48.260)	0.145	(3.683)	1.610	(40.894)	2.72	(4.047)
2	2.375	(60.325)	0.154	(3.912)	2.067	(52.502)	3.65	(5.431)
2-1/2	2.875	(73.025)	0.203	(5.156)	2.469	(62.713)	5.79	(8.616)
3	3.500	(88.900)	0.216	(5.486)	3.068	(77.927)	7.58	(11.279)
3-1/2	4.000	(101.600)	0.226	(5.740)	3.548	(90.119)	9.11	(13.556)
4	4.500	(114.300)	0.237	(6.020)	4.026	(102.26)	10.79	(16.056)
5	5.563	(141.300)	0.258	(6.553)	5.047	(128.19)	14.62	(21.755)
6	6.625	(168.275)	0.280	(7.112)	6.065	(154.05)	18.97	(28.227)

inside diameter is reduced as the thickness of the wall increases, which is defined by an increasing Schedule Number. The most common pipe schedules, in order of increasing attributes, are Schedules 40, 80, and 160. Other common designators for these weight classes are as follows:

Schedule 40—Standard Weight, abbreviated as *STD*
Schedule 80—Extra Heavy, abbreviated as *XS* or *EH*
Schedule 160—Double Extra Heavy, abbreviated as *XXS* or *DBL. EH*

Table 30.4 provides the weights and dimensions of Schedule 40 common steel pipe.

The wall of Schedule 80 pipe is thicker than that of Schedule 40 pipe, as shown in Table 30.5.

Standard pipe is also available in different grades of steel and overall quality. In manufacture, pipe and tubing are either *seamless* or contain a linear seam for their entire length. *Welded* pipe is made by forcing a strip of steel that has been heated to the welding temperature, through a die or welding roll, which forms it to a tubular shape. The edges are forced together by the forming rolls, resulting in a fusion weld. This is usually accomplished by continuous electric resistance welding.

As its name implies, seamless pipe does not have a seam; it is made by piercing a solid steel round to produce a rough, heavy-walled tube. The rough tube then goes through a rolling and finishing process that brings it to final form. The piercing action requires very high quality steel that has a uniform and fine-grain structure, as

TABLE 30.5 Dimensions and Weight of Standard Schedule 80 Steel Pipe

Nominal Size/Inch	OUTSIDE DIAMETER in.	OUTSIDE DIAMETER (mm)	WALL THICKNESS in.	WALL THICKNESS (mm)	INSIDE DIAMETER in.	INSIDE DIAMETER (mm)	WEIGHT/FT (M) lb	WEIGHT/FT (M) (kg)
1/8	0.405	(10.287)	0.095	(2.413)	0.215	(5.145)	0.31	(0.461)
1/4	0.540	(13.716)	0.119	(3.023)	0.302	(7.671)	0.54	(0.804)
3/8	0.675	(17.145)	0.126	(3.200)	0.423	(10.744)	0.74	(1.101)
1/2	0.840	(21.226)	0.147	(3.734)	0.546	(13.868)	1.09	(1.623)
3/4	1.050	(26.670)	0.154	(3.912)	0.742	(18.847)	1.47	(2.187)
1	1.315	(33.401)	0.179	(4.574)	0.957	(24.308)	2.17	(3.229)
1-1/4	1.660	(42.164)	0.191	(4.851)	1.278	(32.461)	3.00	(4.464)
1-1/2	1.900	(48.260)	0.200	(5.080)	1.500	(38.100)	3.63	(5.401)
2	2.375	(60.325)	0.218	(5.537)	1.939	(49.251)	5.02	(7.470)
2-1/2	2.875	(73.025)	0.276	(6.782)	2.323	(59.004)	7.66	(11.398)
3	3.500	(88.900)	0.300	(7.620)	2.900	(73.660)	10.25	(15.252)
3-1/2	4.000	(101.600)	0.318	(8.077)	3.364	(85.446)	12.51	(18.615)
4	4.500	(114.300)	0.337	(8.560)	3.826	(97.180)	14.98	(22.290)
5	5.563	(141.300)	0.375	(9.525)	4.813	(122.25)	20.78	(30.921)
6	6.625	(168.275)	0.432	(10.77)	5.761	(146.33)	28.57	(42.512)

well as good ductility. Steel in seamless tube products is thus usually of much higher quality, than that of the more common welded seam pipe.

Common steel pipe, which is the least expensive and most widely available, is typically of welded construction. It is normally manufactured in conformance with ASTM Standard A 120, which provides for pressure only, and is not intended for close coiling, bending, or high-temperature service. Schedule 40 is by far the most common pipe and the only one stocked by many dealers. Pipe in conformance with A 120 can be either seamless or welded, but because welded-seam pipe is less costly to manufacture, it is more common. It is readily available in nominal sizes of 1/8 in. (3.18 mm) to 4 in. (101.6 mm), in both Schedules 40 and 80.

Although not widely stocked, ASTM A 53, Grade B pipe can be either welded or seamless. It is suitable for welding, bending, and general fabrication and requires tensile, hydrostatic, flattening, coiling, and bending tests. Limited chemical analysis of the steel is also required.

A premium-grade pipe will meet the requirements of ASTM A 106. It must be seamless and conform to more stringent requirements of the steel chemistry. This pipe is used for high pressure and temperature service applications and is subject to tensile, bending, coiling, and flattening tests and more complete chemical analysis. It is available in sizes from 1/8 in. (3.18 mm) through 1-1/2 in. (38.1 mm) in Schedules 40 and 80. It is also available in Schedule 160 in sizes of 1 in. (25.4 mm) through 1-1/2 in. (38.1 mm).

A standard for pipe used in the gas and oil production industry is API 5L. Pipe conforming to this standard can be either welded or seamless. It requires hydrostatic, tensile, flattening, and bending tests during manufacture. It also calls for stringent chemical tests of the metal. Pipe that conforms to ASTM A 106 will usually be in compliance with this specification. It is available in Schedules 40 and 80 in sizes from 2 in. (50.8 mm) through 6 in. (152.1 mm), and in Schedules 20, 40, and 80 in larger sizes.

The use of a heavier or better grade of pipe is often advantageous in grouting, although it comes at a higher, and often substantially higher, cost. Flush threaded joints can be machined into standard pipe and are best achieved with Schedule 80 or thicker walls in combination with higher-quality steel, such as seamless pipe conforming to ASTM A 106.

30.2.4 Carbon Steel Tubing

Very high quality carbon steel tubing is manufactured from initially welded seam tubing drawn over a mandrel (WDOM) and is available in many sizes and wall thicknesses. It is usually manufactured in conformance with American Iron and Steel Institute (AISI) Standard MT-1015. The tubing is first formed from a strip with an electric resistance welded joint and then cold drawn over a mandrel to a smaller diameter and wall thickness. The cold drawing works on the welded joint to produce a dense and homogeneous structure that is essentially the same as that of the base metal. For all practical purposes, the welded seam disappears, resulting in a virtually seamless tube.

WDOM tubing is offered in outside diameters of 1/2 in. (12.7 mm) through 10 in. (254 mm) and in a variety of wall thickness for each outside diameter. Between 1 in. (25.4 mm) and 1-5/8 in. (41.28 mm) outside diameters, it is available in 1/16 in. (1.59 mm) increments and in 1/8 in. (3.18 mm) increments in sizes from 1-5/8 (41.28 mm) to 4-3/4 in. (120.65 mm). Although not a complete presentation of all available sizes and wall thicknesses, Table 30.6 provides dimensional data on sizes most commonly used in grouting.

TABLE 30.6 WDOM Carbon Steel Tubing

OUTSIDE DIAMETER		WALL THICKNESS		INSIDE DIAMETER		WEIGHT PER FOOT (0.3 M)	
in.	(mm)	in.	(mm)	in.	(mm)	lb	(kg)
1.0	(25.40)	0.120	(3.048)	0.760	(19.30)	1.128	(0.5121)
		0.219	(5.563)	0.562	(14.27)	1.827	(0.8304)
1.25	(31.75)	0.095	(2.413)	1.060	(26.92)	1.172	(0.5329)
		0.120	(3.048)	1.010	(25.65)	1.448	(0.6574)
		0.188	(4.775)	0.874	(22.20)	2.132	(0.9690)
1.5	(38.1)	0.035	(0.889)	1.430	(36.32)	0.5476	(0.2486)
		0.065	(1.651)	1.370	(34.80)	0.9962	(0.4523)
		0.109	(2.769)	1.282	(32.56)	1.619	(0.7350)
		0.120	(3.048)	1.260	(32.00)	1.796	(0.8164)
		0.156	(3.962)	1.188	(30.18)	2.239	(1.017)
		0.188	(4.775)	1.124	(28.55)	2.634	(1.196)
		0.313	(7.950)	0.874	(22.20)	3.968	(1.801)

TABLE 30.6 WDOM Carbon Steel Tubing (Continued)

OUTSIDE DIAMETER		WALL THICKNESS		INSIDE DIAMETER		WEIGHT PER FOOT (0.3 M)	
in.	(mm)	in.	(mm)	in.	(mm)	lb	(kg)
2.0	(50.80)	0.083	(2.108)	1.834	(4658)	1.699	(0.7723)
		0.095	(2.413)	1.810	(45.97)	1.933	(0.8786)
		0.120	(3.048)	1.760	(44.70)	2.409	(1.094)
		0.125	(3.175)	1.750	(44.45)	2.503	(1.136)
		0.134	(3.404)	1.732	(43.99)	2.650	(12.04)
		0.188	(4.775)	1.624	(41.25)	3.638	(1.652)
		0.219	(5.563)	1.562	(39.67)	4.166	(1.891)
		0.250	(6.350)	1.500	(38.10)	4.673	(2.122)
		0.281	(7.137)	1.438	(36.53)	5.159	(2.342)
		0.375	(9.525)	1.250	(31.75)	6.508	(2.955)
2.5	(63.50)	0.083	(2.108)	2.334	(59.28)	2.143	(0.9741)
		0.095	(2.413)	2.310	(58.67)	2.440	(1.108)
		0.120	(3.048)	2.260	(57.40)	3.050	(1.385)
		0.156	(3.962)	2.188	(55.58)	3.905	(1.773)
		0.188	(4.775)	2.124	(53.95)	4.642	(2.107)
		0.250	(6.350)	2.000	(50.80)	6.008	(2.730)
		0.313	(7.950)	1.874	(47.60)	7.729	(3.513)
		0.375	(9.525)	1.750	(44.45)	8.511	(3.864)
3.0	(161.12)	0.065	(1.651)	2.870	(72.90)	2.037	(0.9258)
		0.095	(2.413)	2.810	(71.37)	2.947	(1.339)
		0.109	(2.769)	2.782	(70.66)	3.365	(1.528)
		0.125	(3.175)	2.750	(69.85)	3.838	(1.742)
		0.188	(4.775)	2.624	(66.65)	5.646	(2.563)
		0.250	(6.350)	2.500	(63.50)	7.343	(3.334)
		0.375	(9.525)	2.250	(57.15)	10.51	(4.772)
3.5	(88.90)	0.065	(1.651)	3.370	(85.60)	2.385	(1.083)
		0.095	(2.413)	3.310	(84.07)	3.455	(1.569)
		0.188	(4.775)	3.124	(79.35)	6.650	(3.022)
		0.219	(5.563)	3.063	(77.77)	7.674	(3.488)
		0.250	(6.350)	3.000	(76.20)	8.678	(3.940)
		0.313	(7.950)	2.874	(73.00)	10.65	(4.835)
4.0	(101.6)	0.120	(3.048)	3.760	(95.50)	4.973	(22.58)
		0.134	(3.404)	3.732	(94.79)	5.533	(2.512)
		0.156	(3.962)	3.688	(93.68)	6.404	(2.907)
		0.188	(4.775)	3.624	(92.05)	7.654	(3.475)
		0.250	(6.350)	3.500	(88.90)	10.01	(4.545)
		0.313	(7.950)	3.374	(85.70)	12.33	(5.598)
		0.375	(9.525)	3.250	(82.55)	14.52	(6.592)

(continued)

TABLE 30.6 WDOM Carbon Steel Tubing (Continued)

OUTSIDE DIAMETER		WALL THICKNESS		INSIDE DIAMETER		WEIGHT PER FOOT (0.3 M)	
in.	(mm)	in.	(mm)	in.	(mm)	lb	(kg)
4.5	(114.3)	0.125	(3.175)	4.250	(107.95)	5.841	(2.652)
		0.134	(3.404)	4.232	(107.49)	6.248	(2.837)
		0.188	(4.775)	4.124	(104.75)	8.658	(3.931)
		0.219	(5.563)	4.062	(103.17)	10.01	(4.545)
		0.250	(6.350)	4.000	(101.60)	11.35	(5.153)
		3.13	(7.950)	3.874	(98.40)	14.00	(6.356)
		0.375	(9.525)	3.750	(95.25)	16.52	(7.400)
5.0	(127.0)	0.125	(3.175)	4.750	(120.65)	6.508	(2.955)
		0.188	(4.775)	4.624	(117.45)	9.662	(4.387)
		0.250	(6.350)	4.500	(114.30)	12.68	(5.757)
		0.313	(7.950)	4.374	(111.10)	15.67	(7.122)
		0.375	(9.525)	4.250	(107.95)	18.52	(8.418)
		0.438	(11.125)	4.124	(104.75)	21.34	(9.688)
5.5	(139.7)	0.125	(3.175)	5.250	(133.35)	7.175	(3.257)
		0.188	(4.775)	5.124	(130.15)	10.67	(4.844)
		0.250	(6.350)	5.000	(127.00)	14.02	(6.365)
		3.75	(9.252)	4.750	(120.65)	20.53	(9.321)
6.0	(152.4)	0.125	(3.175)	5.750	(146.05)	7.843	(3.561)
		0.188	(4.775)	5.624	(142.85)	11.67	(5.298)
		0.250	(6.350)	5.500	(139.70)	15.35	(6.969)
		0.313	(7.950)	5.374	(136.50)	19.01	(8.631)
		0.375	(9.525)	5.250	(133.35)	22.53	(10.229)

Because WDOM tubing is drawn over a mandrel, the steel is of a generally uniform grain structure for machining. It is also readily welded. Whereas flush threads can be machined into the heavier wall tubing, lighter wall can still be used for casing but will require welding of the joints during installation. This is not very practical in most grouting work, but has been done advantageously where very deep holes are required to be drilled to tight verticality tolerances. Where welded joints are used and the casing must be removed either during or after the grout injection, sections can be handily cut with a pipe cutter, usually aided by the rotation of the drill rig, as shown in Figure 30.14.

FIGURE 30.14 Welded casing is easily broken with pipe cutter on removal.

30.3 DRILL ROD

Because different forces are induced into the drilling system by rotary and rotary percussion drills, it follows that different drill rod is required. Generally speaking, drill rod, or *steel,* for percussive drilling is much heavier than that used for rotary drilling. Commonly available rod for rotary drilling was designed for relatively high powered drill rigs. Unfortunately, much grouting is performed in areas that are not accessible to such equipment, requiring the use of handheld drills or mini-rigs. Standard drill rod is quite heavy for such equipment. Special lightweight systems have been developed for such requirements.

30.3.1 Rotary Drill Rod

As mentioned earlier, the DCDMA standards include drill rod and other accessories matched to the casing by size group. There are two main styles of rod, *parallel wall* and *upset wall.* The wall thickness of parallel wall rod is uniform and somewhat thicker than that of upset wall, which adds strength but with greater weight to be handled. Upset wall rod is lighter to handle because of the wall thickness being reduced throughout the body.

30.3.1.1 DCDMA W SERIES ROD

The most common DCMDA drill rod is "W" series drill rod. Both ends of this hollow drill rod are supplied with square female threads, which in driller's jargon are referred to as *box* ends. Individual rods are joined with couplings having male square threads on each end, as illustrated in Figure 30.15 (top). In driller jargon, these male couplings are known as *pins.* Sections, or *joints,* as they are commonly referred to in the field, are added each time the drill head becomes fully lowered on the mast. The rod is broken from the drill head, which is raised to accept another joint. The entire length of the drill rod in a hole at any given time is known as the drill *string.* Table 30.7 provides dimensions and configurations of the standard "W" series drill rod.

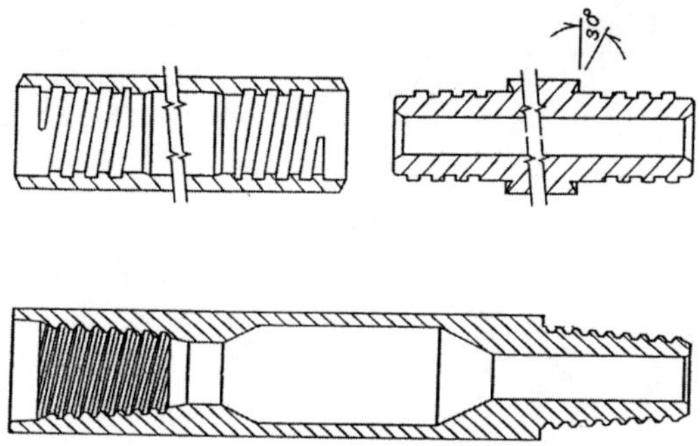

FIGURE 30.15 DCDMA drill rod joints, W series on top, WJ series taper threads on bottom.

The length of the individual rod joints is dependent on the mast height, which is often dependent on the drill rig size. Lengths of 5 or 10 ft (1.5 or 3 m) are most commonly used in grouting; however, they can be either longer or shorter. Many manufacturers also supply standard lengths of 1 and 2 ft (0.3 and 0.6 m). Very short joints may be required in areas of low clearance. Longer lengths require a higher mast, which is usually available only on heavier drill rigs than commonly used.

30.3.1.2 DCDMA WJ SERIES TAPER JOINT ROD

More recently, DCMDA has developed standard rod that is coupled with a taper joint. This joint design is significantly faster and easier to make up and break during drilling. Taper jointed rod, known as the "WJ" series (Figure 30.15, bottom) is readily available in sizes A through H and a new intermediate size, designated K. Unlike the earlier series, this rod is supplied with a box on one end and pin on the other. The particulars of the WJ series rod are shown in Table 30.8.

TABLE 30.7 DCDMA Nomenclature and Dimensions for "W" Series Drill Rod

Size	RW in. (mm)	EW in. (mm)	AW in. (mm)	BW in. (mm)	NW in. (mm)	HW in. (mm)
O.D.	1.09 (27.8)	1.38 (34.9)	1.75 (44.5)	2.13 (54.0)	2.63 (66.7)	3.50 (88.9)
I.D.	0.72 (18.3)	1.00 (25.4)	1.38 (34.9)	1.75 (44.5)	2.25 (57.2)	3.06 (77.8)
I.D. Box	0.41 (10.3)	0.44 (11.1)	0.63 (15.9)	0.75 (19.1)	1.38 (34.9)	2.38 (60.3)
I.D. Pin	0.85 (21.6)	1.06 (27.0)	1.37 (34.9)	1.68 (42.8)	2.22 (56.3)	3.03 (76.9)
Threads/in.	4	3	3	3	3	3

30.3.1.3 WIRELINE DRILL ROD

There are no standards for wireline drill rod or down-the-hole equipment, although the various manufacturers have, in general, conformed to DCDMA standard sizes and nomenclature. Although there are some minor variations, the wireline products of most manufacturers are compatible. Most use the standard size designators of A through H as their prefixes, but the letters that follow vary considerably; WL or Q are most commonly used to designate that the product is for wireline drilling.

Some manufacturers produce this rod in sizes greater than the 3.5 in. (89 mm) outside diameter of standard H rod. For these larger sizes, manufacturers have not always followed standard nomenclature but have developed their own designators. Dimensional data for the most common sizes of A through H wireline rod are provided in Table 30.9. Care must be taken in its use, however, as slight variations may be found between the products of different producers. And as previously mentioned, although WL or Q are the most common designator

TABLE 30.8 DCDMA Nomenclature and Dimensions for "WJ" Series Taper Thread Drill Rod

Size	AWJ in. (mm)	BWJ in. (mm)	NWJ in. (mm)	KWJ in. (mm)	HWK in. (mm)
O.D.	1.75 (44.5)	2.13 (54.0)	2.63 (66.7)	2.88 (73.1)	3.50 (88.9)
I.D.	1.37 (34.8)	1.75 (44.5)	2.35 (57.2)	2.50 (63.5)	3.12 (79.2)
I.D. Box	0.88 (22.4)	1.25 (32.0)	1.13 (28.8)	1.38 (34.9)	1.75 (44.5)
I.D. Pin	0.63 (16.1)	0.75 (19.3)	1.13 (28.8)	1.38 (34.9)	1.75 (44.5)
Threads/in.	5	5	4	4	4

TABLE 30.9 Common Parameters of Wireline Drill Rod

Rod Size	AQ in.	AQ mm	BQ in.	BQ mm	NQ in.	NQ mm	HQ in.	HQ mm
Outside Diameter	1.75	44	2.19	56	2.75	70	3.50	89
Inside Diameter	1.38	35	1.81	46	2.38	61	3.06	78
I.D. Joint	1.30	34	1.81	46	2.38	61	3.06	78
Threads/in. (25.4 mm)	4		3		3		3	

for wireline, other nomenclature is sometimes used.

Wireline drill rod is similar to casing in that it consists of a fairly thin tube with flush threaded joints so as to provide the largest inside diameter possible. Herein lies the advantage of the system, as it allows removal of cores without withdrawal of the entire drill string. Drill rod usually receives greater service wear than casing and is thus often made to higher standards. For this reason, wireline rod is sometimes used for casing by contractors because it will often withstand more abuse and higher withdrawal forces.

30.3.1.4 CUSTOM LIGHTWEIGHT ROD

As discussed earlier, the weight of the rod becomes important when handheld drills are used. Experience has shown that the standard 3/4 in. (19 mm) pipe used in early horizontal boring did not hold up to continued usage, often breaking in the thread root. The breakage was apparently caused by slight bending of the rod as it flexed during drilling. To overcome this weakness, the Melfred Welding and Manufacturing Co. (now Melfred Borzall of Santa Maria, California) collaborated with my former firm in developing a sleeve coupling (Figure 30.16) that prevented flexing.

For the rod, 3/4 in. (19 mm) ASTM A 106 seamless pipe, as described in Section 30.2.3, was used. It was cut into appropriate lengths, one end of which was treated with "Locktite" and threaded tightly into a sleeve coupling so that it was not easily loosened. Thus, each section of rod had a male pipe thread on one end, with the female sleeve coupling on the other. The sleeve couplings were made of very high quality steel that was further heat treated so that their threads were considerably more resistant to wear than those of the pipe. There was thus little wear on the couplings, and any that did occur was to the male thread on the rod, which was easily refurbished by rethreading in a pipe machine.

Two grades of custom rod were made, one with Schedule 40 pipe, dubbed "S," and the other with Schedule 80 pipe, referred to as "SH." The SH rod lasts longer than the lighter S rod and was used for most work, while the lighter S rod was reserved for holes deeper than about 60 ft (18 m) where the weight had to be minimized. Many tens of thousands of grout holes have been drilled in soil with this lightweight rod.

Another lightweight rod manufactured by Melfred is known as No. 301 drill tubing. It consists of a body of 1-1/8 in. (28.58 mm), outside diameter, by 5/32 in. (3.97 mm) wall steel tubing, to which is attached machined couplings

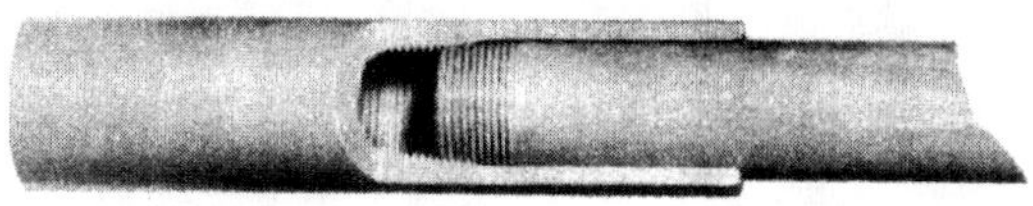

FIGURE 30.16 Sleeved hardened coupling used with a pipe drill rod resists breakage.

TABLE 30.10 Approximate Weight per foot (0.33 m), Including Coupling, of Various Rotary Drill Rods

Rod Type	Pounds	Kilograms
S	1.4	0.65
SH	1.8	0.80
Melfred No. 301	2.1	0.95
EW	3.2	1.45
AW	4.4	2.00
BW	4.8	2.18
NW	6.4	2.90
AWJ	3.2	1.44
BWJ	4.1	1.86
NWJ	5.1	2.32
AQ	3.1	1.41
BQ	4.0	1.81
NQ	5.2	2.36
HQ	7.7	3.49

that are slightly greater in outside diameter. The couplings are machined with a standard 1 in. National Course bolt thread. This rod is supplied in 5, 10, and 20 ft (1.5, 3, and 6 m) standard lengths and can be acquired in any other length on special order. The relative weights of various rotary drilling rods can be observed in Table 30.10.

30.3.2 API Drill Rod

The American Petroleum Institute (API) has established standards for drill rod joints, but not for the shaft of the rod. These joint designs are broadly recognized and widely used by drillers. Because the smallest rod joint has an outside diameter greater than 3 in. (76 mm), they are not used in grouting except in special cases. The API standard calls for three variations of joints, *regular, internal flush,* and *full hole.* The drill rod is described by the dimension of the small end of the threaded pin, together with the joint classification. The dimensions of these drill rod joints are given in Table 30.11.

TABLE 30.11 Dimensions of API Standard Drill Rod Joints

	REGULAR		INTERNAL FLUSH		FULL HOLE	
Size	O.D. in. (mm)	I.D. in. (mm)	O.D. in. (mm)	I.D. in. (mm)	O.D. in. (mm)	I.D. in. (mm)
2-3/8	3.125 (79.37)	1.000 (24.40)	3.375 (85.72)	1.750 (44.45)	—	—
2-7/8	3.750 (95.25)	1.250 (31.75)	4.125 (104.77)	2.125 (53.97)	4.25 (107.95)	2.125 (53.97)
3-1/2	4.250 (107.95)	1.500 (38.10	4.750 (120.65)	2.687 (68.25)	4.625 (117.47)	2.437 (61.89)
4	—	—	5.750 (146.05)	3.250 (82.55)	—	—
4-1/2	5.500 (139.70)	2.250 (57.15)	6.125 (155.57)	3.750 (95.25)	5.750 (146.05)	3.00 (76.20)

Because API specifies only the joint design, there is considerable variation in both the quality of steel and the wall thickness of rod produced by different manufacturers. Most weld the machined joints to otherwise standard tubing, and the joint is commonly machined from a better grade of steel than that of the tube. This provides greater longevity of the joint, which receives more wear and tear in use. This rod is used for both rotary and rotary percussion drilling.

30.3.3 PERCUSSION DRILL ROD

Percussion rod must endure constant pounding and vibration and is thus much heavier than rotary rod. Coupling threads are very coarse, requiring thick, oversized couplings. Breaking the joints is difficult at best, and liberal application of a non-seize lubricant is required as they are made up. Two general types of rod are available, one for lightweight sinker drills, another for heavier operations.

30.3.3.1 HEAVY PERCUSSION ROD

Two different thread configurations are widely available, *rope* and *trapezoidal,* often expressed simply as *R* or *T* prefixing the size (Figure 30.17). Rope design has been around for more than 50 years, whereas trapezoidal is the result of a more recent effort to ease breaking the joints. Other thread designs, usually proprietary to a particular manufacturer, are available but are not widely used.

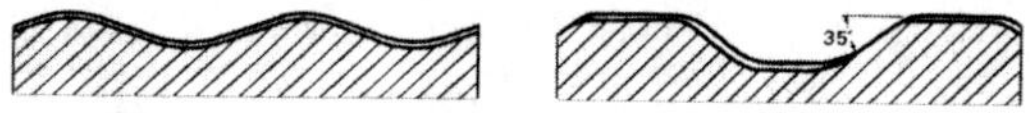

FIGURE 30.17 Heavy percussion rod threads, standard rope (R) on top, trapezoidal (T) on bottom.

Percussion rod is commonly available in diameters of 1-1/4 in. (32 mm) to 2 in. (51 mm) and in standard lengths of 10, 12, and 14 ft (3, 3.6, and 4.2 m). Although not common, 1 in. (25 mm) and 2-3/8 in. (60 mm) diameters can be obtained, as well as lengths varying from about 4 to 18 ft (1.2 to 5.4 m), usually in 2 ft (0.6 m) graduations. Much percussion energy tends to be lost in the rod joints, however, so it is wise to use the longest lengths that can be handled by the drill rig and conform to any site restrictions. Whereas round rod is most common, hex-shaped is available and used, especially in Eastern countries. Both ends of the rod typically have male threads, but sections continuously threaded for their entire length are available. Although these are somewhat harder to handle, they do allow continued use even when broken and can be cut to any desired length. Because the diameter of the couplings is much greater, it controls size selection for a hole of a given diameter. The particulars are provided in Table 30.12. Although precise nomenclature varies with the manufacturer, the basic terms used should be understood by all drillers and equipment suppliers.

TABLE 30.12 Dimensions of Heavy Percussion Drilling Rod and Couplings

DESIGNATION		O.D.		O.D. COUPLING	
Rope	**Trapezoidal**	**in.**	**(mm)**	**in.**	**(mm)**
R-25		1	25	1-5/8	41
R-32		1-1/4	32	1-3/4	45
R-38	T-38	1-1/2	38	2-1/8	54
	T-45	1-3/4	45	2-1/2	64
	T-51	2	51	3	76

TABLE 30.13 Hex Sinker Drill Steel

Thread Designation	STEEL SIZE		HOLE SIZES	
	in.	mm	in.	mm
A or E	7/8	22.2	15/16–1-1/4	2.4–31.8
H	7/8, 1	22.2, 25.4	1-3/8–3	34.9–76.2
D	1	25.4	2-1/4–4	57.2–101.6

As shown in Table 30.12, the smaller sizes, which are the most commonly used in grouting, are rope thread. Although trapezoidal is said to be easier to break, many drillers have indicated that the difference is not noticeable. Of great importance, however, is lubrication, often referred to as *doping*, of the threads prior to makeup.

30.3.3.2 HEX SINKER DRILL STEEL

Sinker drill steel is typically hexagonal-shaped, with both 7/8 and 1 in. (22.2 and 25.4 mm) sizes being common. Although the smaller size is perhaps the most frequently used, it is typically limited to holes smaller than about 2.5 in. (64 mm). Larger-size holes require the 1 in. (25.4 mm) size. The rod size must be the same as that of the drill chuck. Today, drill bits are most commonly secured to the rod with a friction taper. Three different thread sizes were used historically and are still widely employed, as shown in Table 30.13. Adapters from one size to another are also available.

30.4 DRILL BITS

There are many types of drill bits, but all perform similarly in chipping or disintegrating a formation so that it can be expelled in the drill flush. Grout holes have a fundamental requirement for complete expulsion of the cuttings without their entering or blocking any penetrated defects. This is facilitated by producing cuttings of the largest practicable size, which is a function of the particular drill bit used.

30.4.1 Rotary Drilling Bits

Several different types of rotary bits are available. With the exception of core bits, all can be used in soil, although those intended for rock are generally more costly and seldom as efficient. Bit selection must consider the size and depth of the hole as well as the type of drill rig to be used. The reason is that both torque and thrust requirements are dependent on the bit type as well as the formation to be penetrated.

30.4.1.1 DRAG/FISHTAIL BITS

Drag bits, often referred to as fishtail bits, consist of a body that supports individual fingers that chip and loosen soil when simultaneously rotated and thrust at the hole bottom. They can contain two to four cutters and come in many different patterns and configurations. Some are formed as a single piece of hardened cast steel, such as the two-blade bit shown at the left and the three-blade bit in the center of Figure 30.18, and others contain a separate body with replaceable cutters, as depicted at the right in Figure 30.18. Low-cost single-use *knockoff* bits are sometimes used in soil. These are driven by a loose mechanical connection to the drillstring to transfer rotation and thrust, but will drop off when downward pressure ceases.

FIGURE 30.18 Drag bits: single piece, two and three blade on left, replaceable blade on right.

FIGURE 30.19 Tricone rotary bit.

Lower-cost bits have cutting surfaces of common hardened steel, whereas those of higher quality are of set carbide. Obviously, the carbide cutters are more durable and are a virtual requisite for drilling in very hard soils or rock. All can be sharpened, and of course, the removable cutters can be replaced as needed. These bits generally do not reduce the grain size of the soil. When the cutters are sharp, they create relatively large cuttings when penetrating soft rock.

30.4.1.2 ROLLER CONE BITS

For holes greater than about 3 in. (75 mm) in diameter, *roller cone bits* are sometimes used. These are available with either two or three toothed cone-shaped rollers, as shown in Figure 30.19. The *tri-cone roller bit* is the most commonly used. They are actually intended for drilling rock, and efficient use requires a large down pressure on the drill head. As they are rotated, the toothed cones also rotate so as to grind and chip away the rock. This requires a heavier drill rig than often used for small-diameter holes common to grouting, but they are widely used for large and deep holes in both soil and rock.

30.4.1.3 DIAMOND CORE BITS

As discussed in Section 10.2.1.4, there are two very different types of diamond core drills, construction and geotechnical. Bits for construction drilling cut a thin kerf and traditionally were available only in 14 in. (356mm) lengths, for penetration, up to 12 in. (305 mm). Some distributors now stock longer lengths, and on special order any desired length can be obtained. Because of the thin kerf these bits cut, it is not possible to couple extension flush thread core barrel to them.

Construction bits are generally supplied as open tubes with diamond segments on one end, as shown in Figure 30.20 (left). A friction adapter containing a threaded hole for the drill rod is fitted to the opposite end. Other connection adapters include solid built-in and bolt-on ends. They are commonly stocked in sizes ranging from 1/2 in. (13 mm) to 24 in. (60 mm) in

(a) (b) (c)

FIGURE 30.20 Coring bits: (a) thin wall construction type, (b) geotechnical type, and (c) geotechnical reamer.

diameter, but are available in sizes up to 6 ft (1.8 m) or more. Peculiar to construction core drilling is the use of National Course bolt threads for the adapters, extension rods, and so forth. For holes of 1-1/2 in. (38 mm) diameter or less, 5/8 in. -11 thread is used, with 1-1/4 in. - 7 being used for larger holes.

Geotechnical core bits are fitted with flush threaded joints and can be extended to virtually any desired length. To provide sufficient thickness for the threads, a much wider kerf is cut, as shown in Figure 30.19 (center), which results in slower penetration. Because wear causes progressive reduction of the hole diameter, *reaming shells* (Figure 30.19, right) are used to open it as required.

30.4.1.4 CONTINUOUS FLIGHT AUGERS

Continuous flight augers are occasionally used in soil, but in the small sizes of grout holes they are not very stiff and thus tend to wander from the planned trajectory. Consequently, they are not satisfactory for deep holes and usually not very effective even for shallow holes of the small diameters common to grouting.

30.4.2 Percussive Drill Bits

To withstand the constant pounding of percussion drilling, the bits must be much heavier and more robust. These bits can easily penetrate soil, but they are used primarily in rock.

30.4.2.1 EQUIPMENT-MOUNTED DRILL BITS

Several types of bits for larger equipment mounted drills are available, which can be divided into two main groups, fixed-size and expandable. Most holes are drilled with fixed-size bits; however, when a casing is being placed simultaneously with drilling, an expandable bit can undercut the casing. Fixed-size bits can be divided into two main types, button and blade. Button bits, as shown in Figure 30.21 (left), are most frequently used as they last longer and are more economical. They tend to pulverize the rock into very small cuttings, however, which is, of course, a disadvantage in grouting.

A cousin to the most frequently used round button bit is the *ballistic* bit, which contains pointed cones, as illustrated in Figure 30.21 (center). When sharp, these increase drilling speed and produce larger cuttings. Unfortunately, they can wear rapidly, thus eliminating these advantages. Bladed bits generally contain four blades, which are typically of carbide, as shown in Figure 30.21 (right). These tend to not wear as rapidly as ballistic bits, and when sharp they produce relatively large cuttings. With wear, however, the cutting size reduces, as does the penetration rate.

Expandable bits all involve some sort of cam device that swings outward when rotated in the drilling direction and retract when the rotation direction is carefully reversed. Two basic types are widely used. First on the market was ODEX, developed by Atlas Copco, wherein an eccentrically mounted reamer swings out, as illustrated at the left of Figure 30.22. Others are now available, in which either two or three eccentrically mounted surface cutters expand, such as Mitsubishi's SUPER MAXBIT, illustrated at the right in Figure 30.22. Although expanding bits were quite novel only a few decades ago, they are now becoming common and are available from several manufacturers.

FIGURE 30.21 Heavy percussion bits: button on left, ballistic center, and four bladed on right.

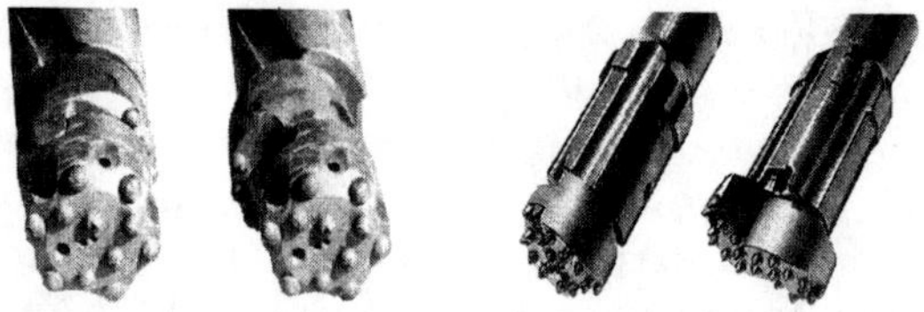

FIGURE 30.22 Expandable bits: ODEX eccentric reamer on left and eccentric surface type on right.

30.4.2.2 BITS FOR SINKER DRILLS

The most common bits for handheld sinker drills are bladed quite similarly to those used with the heavier hammers. Four blades are most common, although six blades are also available, either of hardened tool steel or carbide insert cutters. Much less frequently used are button bits, which are supplied by a few manufacturers. Standard bits are threaded for use with matching drill steel, as outlined in Table 30.13. For holes of less than 1.5 in. (38 mm) diameter, single carbide chisel tooth bits, integral to the appropriate steel, are available. These are commonly stocked in sizes ranging from 5/8 to 1-1/2 in. (16 to 38 mm) and in lengths to 24 in. (51 mm). On special order, they are available in lengths up to 5 ft or more. These have been found especially useful in structural grouting, where they produce clean holes rapidly.

30.5 SLEEVE PORT PIPES (TUBE-À-MANCHETTE)

The use of sleeve port pipes allows almost precise deposition of fluid or pourable suspension grouts. For this reason, they are very often used for fracture grouting, but they have a clear advantage for other types as well, especially where permeability is variable. Sleeve port pipes are simply tubes into which holes have been drilled at regular intervals. Two or more holes, equally spaced around the periphery and about 1/4 in. (6 mm) in diameter, are drilled at each interval, most typically spaced about 1 ft (0.3 m) apart. Standard Schedule 40 or 80 PVC pipe is most commonly used and supplied in 10 or 20 ft (3 or 6 m) lengths, which are joined with either cemented or threaded couplings. Some premium sleeve port pipes are fitted with telescoping milled ends that can be cemented together, without a separate coupling, to provide a flush joint. Diameters are typically within a range of 1 to 2 in. (25 to 51 mm) I.D.

To temporarily seal the holes, each interval is covered with a flexible sleeve, usually made of a short piece of flexible rubber tubing. Premium pipes may have a machined recess into which the sleeve fits. The purpose of the recess is to restrict the sliding of the sleeve along the pipe, although this need not be a problem. Many thousands of sleeve port pipes have been used that simply had a rubber tube stretched over the O.D. of the pipe. A single wrap of plastic electrical tape at each end of the sleeve prevented their displacement. Figure 30.23 shows a pile of 20 ft (6 m) long sleeve port pipes on a job site.

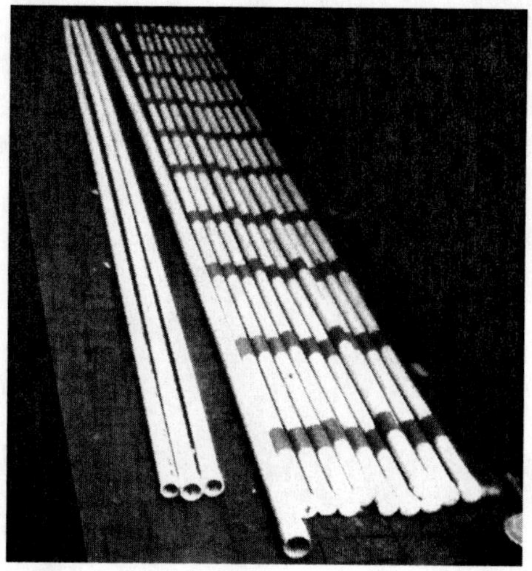

FIGURE 30.23 Sleeve port pipe awaiting installation, solid pieces on left are for nongrouted zone.

30.6 SPECIAL DRIVE NEEDLES

Special needles, which are pushed or driven while grout is simultaneously injected, are sometimes used in permeation and fracture grouting. They must be able to withstand the rigors of repeated driving and withdrawal and are usually custom-made from cold-drawn high carbon steel. There are no standards for this type of probe, and individual designs are usually unique to a particular contractor. Needles are effectively used for injection on the way down, up, or both, but are most advantageously used from the top down.

A design sketch is shown in Figure 30.24 (left), with a photo of the actual needle on the right. Note that the probe is machined, so the initial annulus for the grout is smaller than the rod that follows, to provide a good seal. These needles were made from 1.25 in. (31.75 mm) WDOM tubing with a 0.188 in. (4.78 mm) wall, as discussed in Section 30.2.4. The pointed end was machined from solid stock, with a socket to telescope into the tube where it was welded. The needle was then turned in a lathe while a handheld grinder formed the slight taper shown in Figure 30.24. Note that the recess shown in the photo is noncontiguous, which provides greater strength. These needles have been used in continuous lengths up to 30 ft (9 m), allowing very efficient injection. Needles of the same design, but of 1 in. (25.4 mm), 0.120 in. (3.048 mm) wall tubing, and maximum length of 6 ft (1.8 m) can be used in combination with handheld hammers.

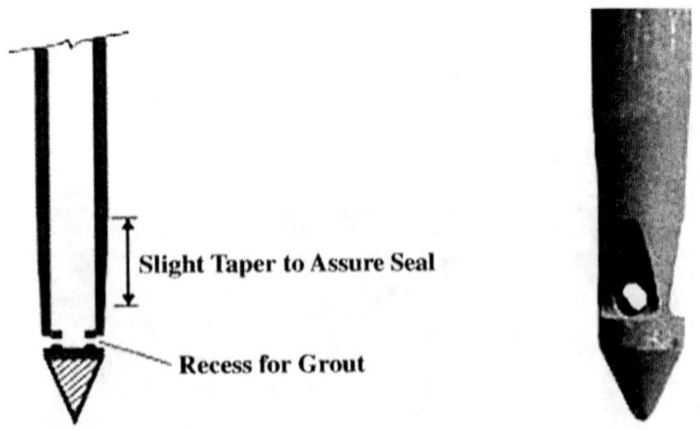

FIGURE 30.24 Taper on custom-made injection needles resists grout leakage around shaft.

30.7 CIRCULATION (MUD) PUMPS

An appropriate tank and pump are often required to pressurize the circulation flush. Traditionally, the tanks have been heavy and the pumps multistage, high-pressure piston pumps, but much lighter and less costly pumps, along with smaller tanks, have been found entirely satisfactory for most grout holes. The air diaphragm pump illustrated in Figure 10.31 is adequate for holes up to about 100 ft (30 m) deep. Small Moyno pumps work well for deeper holes. Piston pumps are, of course, appropriate, but for most work sufficient flow can be maintained by a reasonably light single-stage pump.

CHAPTER 31

Pump Mechanics

31.1 PUMPING FUNDAMENTALS
- **31.1.1** Suction
- **31.1.2** Discharge Pressure Head
  - **31.1.2.1** Output Pressure Capability

31.2 CLASSIFICATION OF PUMPS
- **31.2.1** Pumping Efficiency
- **31.2.2** Pump Throughput

31.3 TYPES OF PUMPS
- **31.3.1** Continuous Flow
  - **31.3.1.1** Progressive Cavity
  - **31.3.1.2** Peristaltic Pumps
  - **31.3.1.3** Gear Pumps
  - **31.3.1.4** Centrifugal Pumps
- **31.3.2** Valved Pumps
  - **31.3.2.1** Piston Pumps
  - **31.3.2.2** Plunger Pumps
  - **31.3.2.3** Diaphragm Pumps

31.1 PUMPING FUNDAMENTALS

Although there are a wide variety of pump types, common principles apply to all. Both an inlet, or *suction*, port and an outlet, or *discharge*, port are provided. During operation, the fluid behind the suction port is under a vacuum whereas that beyond the discharge port is pressurized. The level of each of these conditions dictates the type of pump that can be used, as well as its efficiency.

31.1.1 Suction

The pump must create a vacuum in the inlet port so as to draw the pumped material into the mechanism. Although virtually all pumps are quite efficient in developing pressure at the outlet, few are very effective at creating suction. As discussed in Section 9.7, the maximum suction that can be drawn at sea level is equivalent to a column of water 34 ft (10.2 m) high. As we gain elevation, the atmospheric pressure lowers, and at 10,000 ft (3,000 m), the maximum that can be drawn is only 23 ft (6.9 m). As a practical matter, virtually no pump is capable of drawing a perfect vacuum. Furthermore, even a minute air leak within the suction line will greatly reduce the developed vacuum.

Suction ability varies between different types of pumps, as well as with a given pump, depending on wear. The very best pumps under ideal conditions, will not likely be able to draw a column of water greater than about 25 ft (7.5 m), and the capability of most will be far less. And, of course, performance will decline as the working elevation increases. A pump works best when the inlet is flooded, which is the case with all submersibles and most grout pumps. Even though a pump is said to be capable of a given suction, prudence dictates that the suction head be minimized to the greatest extent practicable.

31.1.2 Discharge Pressure Head

The pressure capability of pumps varies greatly according to the type of pump and its condition. The required pressure output is a function of the pumped fluid density and the vertical distance between the pump and the outlet of the discharge line. This is known as the *static head.* To this must be added any resistance in the discharge line as a result of friction or change in the shape of flow. The combined resistance is known as the *dynamic head.* The pressure existing at the end of the discharge line with the pump at maximum speed plus the total dynamic head (TDH) will be the maximum capability of the particular pump.

31.1.2.1 OUTPUT PRESSURE CAPABILITY

The magnitude of obtainable pressure output varies both among different types of pumps and within a given type. Generally, centrifugal and diaphragm pumps provide the lowest pressures, seldom generating more than about 200 psi (14 bars). Progressive cavity and various rotary pumps commonly produce up to about 350 psi (24 bars), but can be built to develop up to about 5,000 psi (345 bars), whereas very high pressures must be produced by reciprocating piston or plunger pumps. These can achieve in excess of 10,000 psi (690 bars).

31.2 CLASSIFICATION OF PUMPS

Pumps are said to be either *positive displacement* or *non-positive displacement* types. Positive displacement pumps will deliver about the same amount of fluid per pump cycle, regardless of the pump speed or discharge pressure. This results from a very tight fit between the pump element and the case, often in combination with effective seals. Although no pump will deliver 100 percent of its theoretical maximum possible displacement, the loss in positive displacement pumps is negligible as compared with the maximum possible delivery. These pumps will continue to displace the pumped fluid until they stall because of insufficient power or breakage occurs. Piston, plunger, progressive cavitation, diaphragm, and some rotary pumps are considered to be positive displacement pumps.

Non-positive displacement pumps are subject to internal leakage or *slip.* They produce a continuous output; however, volume will vary with changes in pump speed and/or outlet pressure. The discharge volume lowers as the pressure increases, up to the maximum pressure capability of the pump. This pressure level is dependent on the amount of slip in the individual pump. When driven to its maximum pressure, even though the pumping element continues to operate, 100 percent of the fluid is subject to internal slippage and the output is zero, with pressure maintained at the maximum level. The most common non-positive displacement pumps are centrifugal pumps and some other rotary pumps.

31.2.1 Pumping Efficiency

No pump provides 100 percent positive displacement, and all will exhibit at least a little internal slippage, which will vary with operating speed. The *volumetric efficiency* is found by comparing the calculated displacement with the actual displacement. It will generally fall as the pump speed and output pressure rises. In addition to volumetric efficiency, *mechanical efficiency,* which results from friction of the moving parts, must be considered. These two are combined to determine the overall efficiency.

An efficiency curve for a positive displacement pump is essentially a straight line, and varying its speed has little influence on the output pressure. Moreover, because flow is independent of pressure, output change has little effect on efficiency. Because these pumps will continue to build pressure until failure occurs, it is important to provide some type of excess pressure relief. The efficiency of non-positive displacement pumps varies with both speed and head, so the efficiency curve is rounded. Changing the speed will affect both discharge output and pressure. Operating these pumps under varying pressure heads thus greatly affects efficiency.

31.2.2 Pump Throughput

The forces imparted by different types of pumps have varying effects on the pumped product. A centrifugal pump literally throws fluid from the center of the impeller, which imparts high fluid shear and extreme turbulence to the outflow. Most other rotary pumps create pressure by compressing or otherwise changing the shape of the fluid under tortuous deformation. For example, material going through a gear pump is squeezed between the progressively closer spaces of the meshing gears. Any contained solids are thus crushed, and if they are abrasive, will cause rapid wear of the internal parts, increasing slip and negatively affecting efficiency.

Material going through a peristaltic or progressive cavity pump is only gently squeezed, as the cavity geometry does not change from inlet to discharge. Solids are thus relatively unaffected and internal wear is lessened. Because these pumps do not require valves, relatively little fluid shear is imparted and relatively steady flow is produced. Reciprocating pumps have minimal influence on the material when it is in the pumping chamber, but considerable shear is experienced in the valves, especially on the discharge side. And, of course, these pumps produce intense pressure pulsation.

31.3 TYPES OF PUMPS

Pumps can be further classified as continuous flow pumps, for which no valves are required for pressure increase, or valved pumps, in which the pump chamber must be isolated from the intake as the pressure is developed. Continuous flow pumps have the advantage of fewer moving parts to wear and/or malfunction and provide relatively steady output, but are limited in their pressure capability. Conversely, reciprocating pumps can develop high pressure and exhibit little loss through slip, but do cause pulsation of the output. And, of course, the valves are subject to constant wear and maintenance.

31.3.1 Continuous Flow

The actual mechanism of pressurization of continuous flow pumps varies greatly, and other than absence of valves and the resulting constant flow, there are few similarities between different types. Many are suitable only for clear fluids, whereas others can effectively handle both solids and abrasives. Flow and pressurization can be developed either radially or axially, which affects both the forces on the fluid and the requirement for seals and their performance.

31.3.1.1 PROGRESSIVE CAVITY

Progressive cavity pumps were invented by Renée Moineau, an English aircraft designer, in 1936. In the 1950s they became commonly available through the Robbins and Myers Company, which held the manufacturing rights in the United States and produced them under the trade name *Moyno.* The proprietary rights expired long ago and they are now available from several manufacturers, but the common designation, Moyno, continues. The design involves eccentric rotation of a helical metal rotor inside a molded double internal helix, which is twice the pitch length (Figure 31.1). Because the flow is axial and continuous, a smooth flow, free of pulsations, occurs.

As the rotor turns, discrete cavities spiral longitudinally along the stator, which is stationary and typically of an elastomeric material bonded to the interior of a metal tube. The cross section of the pumped material remains constant as it is moves, so that solids are readily handled. The rotor is precisely machined from alloy or stainless steel and heavily plated with chrome. Stators are available in several different compounds to ensure compatibility with the particular material being pumped. For most grouts, a basic stator tube lined with Buna N, nitrile butadiene rubber, is used. The elasticity of the standard stator lining is 70 Durometer, but softer 55 Durometer is available for extreme abrasiveness. The rotor is normally in intimate compression contact with the stator, but the fit can become loosened with wear.

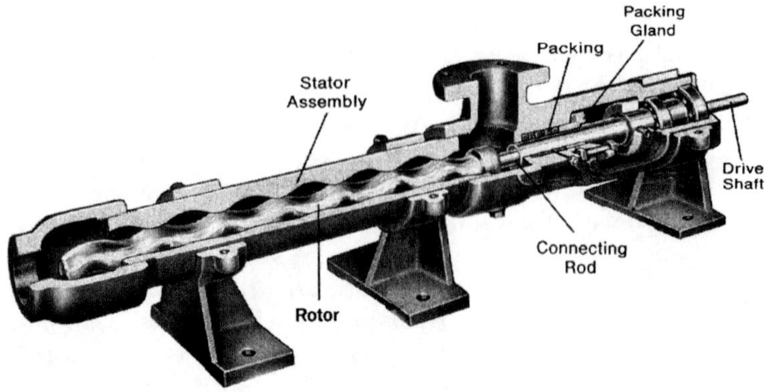

FIGURE 31.1 Cutaway illustration of Moyno pump showing helical rotor.

Although not frequently used, the stator can be of rigid metal, in which case it will be fit with a slight clearance to the rotor. This arrangement has been used successfully for difficult-to-pump resins, but it is not recommended for abrasive materials, as wear would be excessive. This is not a great problem with elastomeric stators, however, as the softer material will tend to deform and envelop the particles until they pass the interface. A word of caution: Compression-fit elastomeric stators cannot be run dry. Even in short periods of dry running, sufficient heat will build up to cause failure of the elastomer. Should this occur, replacement of the stator will be required.

As with all pumps, some slip can be expected. It increases as the interior parts wear, especially at high speed. Although these pumps can be used with abrasives, they will shorten the life of the rotor and the stator, so lower speeds should be used. For example, Robbins and Myers recommends, for its smallest pumps, a maximum speed of 1400 RPM with nonabrasive material but only 350 RPM with heavy abrasives. For its largest models, 600 RPM is the maximum without abrasives, and with heavy abrasive material, 150 RPM should not be exceeded.

Moyno pumps are available in many different sizes, with various numbers of stages, combinations of rotor and stator materials, drive mechanisms, and drive ends. Of particular interest to grouters are two feed types, conventional flange (Figure 31.2, bottom) and open throat (Figure 31.2, top). Thick mixtures, especially resinous mixtures do not run freely and can be difficult to feed in the conventional flange configuration. This is, obviously, less of a problem with open-throat models. These are also

FIGURE 31.2 Open-throat feed on top and conventional flange feed on bottom.

available with a feed auger to handle even the most difficult material.

Pump output likewise varies according to the abrasiveness of the pumped material. These pumps are available in a large assortment of sizes, with pumping rates ranging from less than 1 gpm (3.78 L/min) to about 200 gpm (756 L/min) for typical grouts that are considered moderately abrasive. For nonabrasive fluids, pumping rates approaching 400 gpm (1512 L/min) can be achieved. The pressure developed is a function of the number of stages, which is defined as 1-1/2 flights of the rotor. Although actual levels vary according to pump size and RPM, in typical grouting a pressure capability of about 80 psi (5.5 bars) per stage is reasonable. Pumps are available with as many as nine stages, but these are quite large and heavy to work with. Those with two to three stages are most often used in grouting.

Although Moyno pumps are among the most useful pumps for the grouter, they are subject to wear and do have limitations. By far the most frequent problem is damage to the stator, which results from the pump running dry or at higher speeds and/or pressure with abrasive materials. When this occurs the only solution is replacement. Rotors are subject to wear, but a more serious problem is that once the chrome plating is penetrated, rapid wear of the base metal will occur and the rough interface can destroy the stator. The best way to extend the life of these internal parts is to lower the RPM of the pump, which is easily done by simply selecting a larger-size pumping element. Rotors can be rebuilt and chrome plated if they are not excessively damaged by continuous use once the plating is penetrated. Refurbishing rotors and relining stators are common requirements, and many specialty shops are available for such work.

Because the stator runs eccentrically, a universal joint is needed in the drive. This joint is further required to resist the high thrust of operation. Two types of universal joints are available, gear and sealed-pin types. Sealed-pin joints are far more economical and significantly simpler to maintain. Moreover, replacement parts are widely available, so they are recommended for grouting service. Most manufacturers offer drives that are rated one or more sizes larger than standard for the particular pump size. Grouting service is hard on these pumps, and provision of this extra strength is usually more than cost-effective.

31.3.1.2 PERISTALTIC PUMPS

Although not widely promoted, peristaltic pumps are among the oldest in the world. The ancient Greeks transported fluids, including human waste, by inducing a wave of automatic contractions in channels. Modern peristaltic pumps transmit fluids through a tube that is progressively collapsed and sealed tight by rollers that move along the tube. A continuous stream of material is moved via the use of multiple rollers on a *rotator*, bearing on a hose restrained within a circular chamber, as shown in Figure 31.3. As the compressed tube opens, a vacuum is created that sucks a continuing supply of the material. The constantly moving rollers generate steady suction and outlet pressure. Any discharge pulsation is a function of the number and spacing

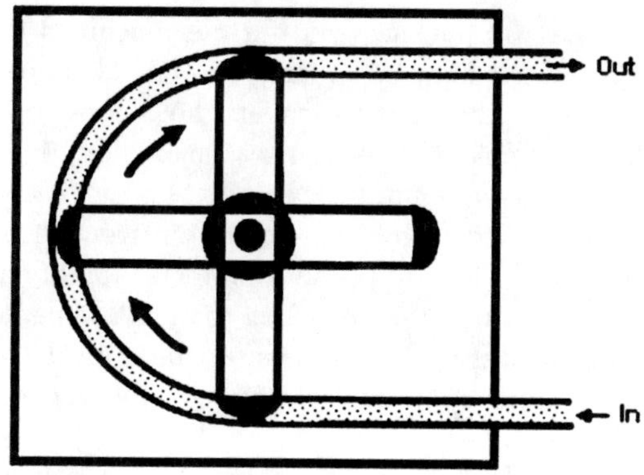

FIGURE 31.3 Travel of rollers over hose pushes material along in peristaltic pump.

of compression rollers, but is generally much less than with reciprocating pumps.

There are many advantages to peristaltic technology, including simplicity and economy. No moving parts contact the pumped material, so messy waste, contamination, and separation are avoided. Positive displacement is provided with extremely little slip, so they are self-priming up to a head of about 30 ft (9 m). Their operating dry causes no damage, and they can actually pump air. Moreover, flow can be reversed so that the lines are emptied back to the source. There was a substantial limitation in the past, however, the short life span of the collapsing hose. This was not much of a problem in the smaller sizes, but was a setback with larger hoses.

One of the first successful concrete pumps was the Challenge Squeezecrete. It did a fine job of pumping even harsh mixes, but the hoses tended to fail suddenly—in as little as several hours of operation—resulting in the early demise of the technology in the United States. Peristaltic concrete pumps are common in Japan, however. The importance of hose quality and dimensional uniformity was not fully understood in earlier times. Rubber is uncompressible, so any overcompression causes damaging deformations. A 1 mm variation in wall thickness is said to reduce hose life by at least 25 percent. Modern peristaltic hose is especially built for precise wall thickness and further machined to a tolerance no greater than 0.25 mm. This precision, combined with modern hose construction technology, results in reasonably long life, and it is now common for even large hose to perform for several months before replacement is required. Modern pump chambers are often enclosed and sealed, enabling the moving parts and the outside of the hose to be constantly flooded in lubricating oil, thus greatly increasing longevity.

Smaller-size peristaltic pumps are widely used in industry where the smaller hoses have performed well. Unfortunately, they have not found such wide use with the larger sizes, probably because of the earlier limitations. Pumps are now available that will move as much as 350 gpm (1,325 L/min) at pressures to about 230 psi (16 bars). Two or more hoses can be combined in the same chamber, with multiple pump heads mounted on a common shaft, to increase the output quantity or proportion multiple component grouts. The more precisely made and durable hose now available provides longevity and low maintenance. There are several different manufacturers of peristaltic pumps that are routinely used to pump concrete and wet-mix shotcrete. Some of the smaller sizes used for feeding and proportioning resin materials and grout admixtures have performed admirably and are to date, underutilized in the grouting industry.

31.3.1.3 GEAR PUMPS

Gear pumps discussed earlier in Section 29.3.3.1 in reference to pressurizing hydraulic oil, are also widely used in industry for transferring and metering fluids of all descriptions. Because the gears must maintain close tolerance with the case, these pumps cannot handle abrasives, so their use is limited to clear liquids only. They perform best with oil and other lubricating products. Very compact in size, they can be mounted in any attitude. These are considered positive displacement pumps, and although there can be considerable slip, depending on the manufacturing tolerances and general quality, a uniform volume of fluid will be moved with each rotation.

Industrial gear pumps are available in many forms with a range of displacement rates and operating pressures. Smaller sizes are widely used for chemical feed or other proportioning use, but high-capacity models are often used for fluid transfer. On the upper end, flow rates of more than 1500 gpm (5670 L/min) and pressures greater than 2000 psi (138 bar) are available. Gear pumps are not widely found in the grouting industry, although they have been used for proportioning chemical solution grouts. They are also used for injection of both solution and thinner resinous materials.

31.3.1.4 CENTRIFUGAL PUMPS

For moving water and other thin fluids, centrifugal pumps are by far the most commonly used type of pump. Rather than impelling the fluid by applied force, they work on the principle of creating a high centrifugal velocity of fluid within the round pump chamber, known as the *volute.* The fluid is fed to the center or *eye* of an *impeller,* which literally throws it outward by centrifugal force. It attains high energy as a result of the elevated speed of rotation in the volute's gradually increasing cross section until forced out by the sudden restriction of the *cutwater,* as illustrated in Figure 31.4 (left). As the fluid exits, its energy is converted from velocity to pressure.

The exact profile of the impeller, as well as number of vanes, vary according to the particular requirements. Economy is gained by minimizing the number of vanes, but efficiency increases with more vanes, as turbulence is reduced. Note that the impeller does not scoop up the fluid, but rather deflects it outward in the circular motion. The discharge flow is dependent on the total dynamic head as well as the design, diameter, and speed of the impeller. All centrifugal pumps are subject to decreasing efficiency with increasing head, which is typically shown by an efficiency curve similar to that shown in Figure 31.4 (right). Although the rotating impeller reduces pressure at the eye, drawing fluid in, centrifugal pumps are not very efficient in suction. The exact capability varies with different designs, but practical suction heads are limited to about 10 ft (3 m).

The output pressure depends on the exact design and manufacturing tolerances, and although some of these pumps will develop pressure in excess of 100 psi (6.9 bars), the capability of most is considerably lower. As illustrated in the efficiency curve (Figure 31.4, right), the volume tends to drop off sharply as the pressure head is increased. Several centrifugal pumps can be assembled in series where high heads must be impelled. These are typically referred to as stages, and are common in submersible well pumps.

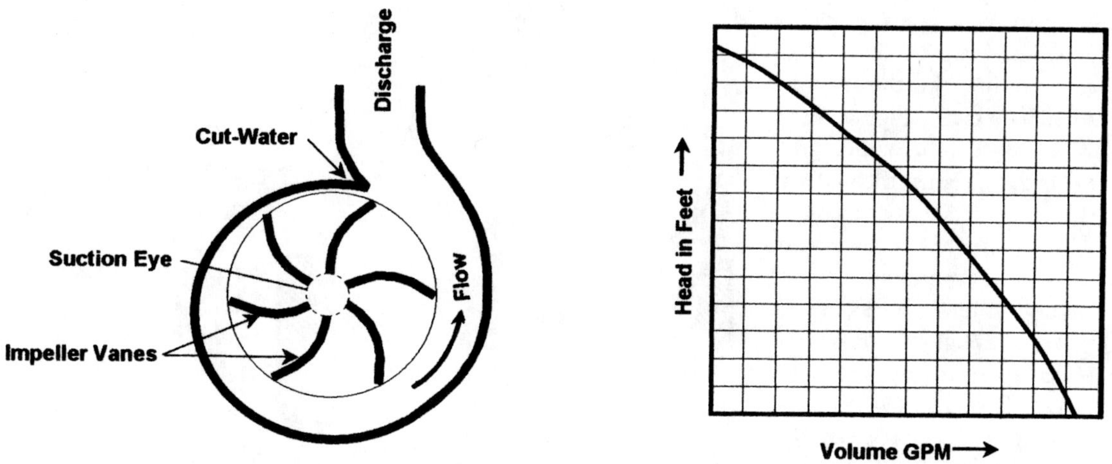

FIGURE 31.4 Fluid is thrown out from the eye of the impeller by vanes of a centrifugal pump.

Centrifugal pumps are not well suited for pumping viscous fluids, as these require very high energy for the needed acceleration. Furthermore, the high speed of the impeller subjects the fluid to very high shear, so they should not be used with shear-sensitive fluids. In general, these pumps are not well suited to passing solids, but some models feature impeller designs that will pass solids up to a few inches in size. Centrifugal pumps can be difficult to prime, especially when drawing fluid some distance. Models are available that are said to be "self-priming." These typically have oversize pump chambers that continue to hold fluid after pumping ceases. The remaining fluid will then be available to provide a prime on restarting.

31.3.2 Valved Pumps

Valved pumps contain chambers that are alternately filled and pressurized with the material to be pumped. In operation, valves on both the suction and discharge sides of the mechanism must operate in perfect sequence with the pumping action. The potential design of the valves is virtually limitless, but can be divided into two major categories, *seated* and *wiper.* Seated valves typically involve a *ball* or *poppet* that reciprocates in line with the seat of a smaller passage (Figure 31.5). When closed, the ball or poppet is tight against the seat and the seal efficiency increases with chamber pressure. Movement of the pumped material usually initiates action of the valve, although it can be impelled or restrained mechanically or by pressure of a spring.

Seated valves are the most commonly used type in industry and perform well with clear liquids and gases. They do, however, have a major limitation when applied to filled materials such as grout, in that solid grains tend to become caught between the valve and the seat. This prevents tight sealing, allowing leakage and preventing positive displacement. Such leakage can also result in pressure filtration, whereby only the liquid phase of a composition is allowed to bypass the seal, with the solids gathering and plugging the passage. This was a serious problem in earlier times with the poppet-valve-equipped air-piston steam pumps often used in grouting. A further shortcoming is that continual passing of solids causes extensive wear of the parts, requiring a high level of maintenance.

The seating area of poppet valves is conical and usually at a 45 degree angle to the material flow so as to encourage the pushing ahead of solids as the valve seats. Moreover, the width of the seat is usually minimized to reduce the available area for solids to gather. The effectiveness of the valve is affected by the velocity of the material being passed through the seat area, which can be controlled through sizing of the passages. The pumped material can also be accelerated mechanically through power driving of the valve. Al-

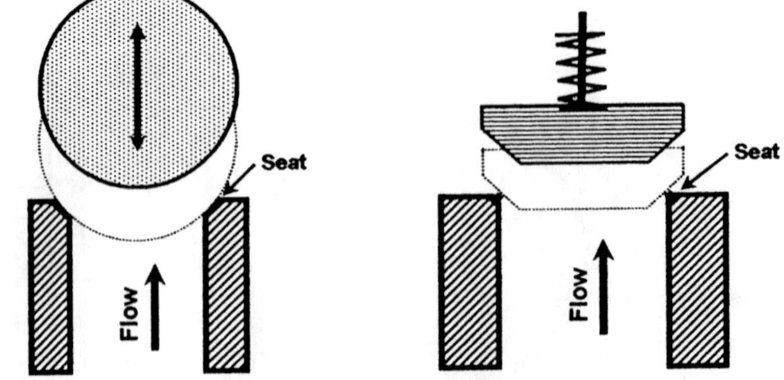

FIGURE 31.5 Typical pump valves: ball on left and poppet on right.

though efficiency can be improved through such design, valves continue to constitute the highest-maintenance area of seated valve pumps.

Wiper valves work by having one body of the valve slide, with a slight clearance, over another. With this action, solids are pushed aside as the parts move. Two major types are *ball wiper valves* and *S-tube valves.* Ball wiper valves involve a semiflexible ball that is forced into a tightly fitting tubular seat. As the ball enters the seat, it pushes any solids ahead as the tubular surfaces are *wiped.* It is restrained by hardened stop pins so that it tends to deform, actually increasing the seal as pressure elevates.

S-tube valves work in combination with a material hopper. The S-tube stretches across the hopper, as illustrated in Figure 31.6. The discharge end swivels from a fixed position on the hopper wall while the valve end reciprocates or *swings* between the pumping cylinders. A replaceable hardened steel *wear plate* is provided on the hopper wall over the cylinders to resist the wear caused by the constantly moving end of the S-tube. Most designs include some sort of floating seal on the end of the S-tube.

Most wear of these pumps is due to the constant S-tube travel. To minimize such wear, generally long cylinder lengths are employed, which reduces the number of cycles required to pump a given amount of material. This presents other difficulties, however, as incomplete filling of the cylinders is frequent. Although this is not a problem with many pumping applications, it does present a significant limitation with grout, inasmuch as maintaining a constant stream of material is often important. Furthermore, stroke counters are often used to determine the quantity or flow rate of the grout, and these determinations are based on complete filling of the cylinders.

The components of pumps with S-tube valves are usually sized so that the tube, reciprocates several times a minute. Valve drive is typically by a hydraulic cylinder and must be timed to properly sequence with the two pump piston cylinders. With the normal rapid cycling, a slight variation of the mechanism movement is not of great consequence. When slowed down, as is common in grouting, however, even slight sequencing variation can have a deleterious effect. Similarly, minor leakage at the valve is usually not a problem at the normal rapid cycling, but can be a major problem when the pump stroking is markedly slowed. This valve design is used almost

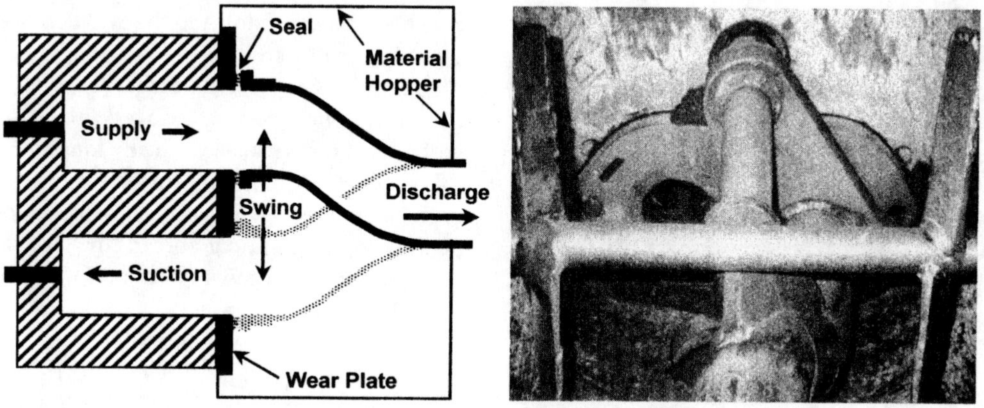

FIGURE 31.6 S-tube valve hugs wear plate with tight tolerance as it reciprocates between pump cylinders.

exclusively for concrete pumps, and these are often employed in grouting, where they are frequently operated at very slow rates. It is important to be aware of the effects of such operation, however, and to mitigate any such deficiencies.

31.3.2.1 PISTON PUMPS

A piston pump consists of a cylinder through which a piston is propelled. The cylinder fills as the piston is retracted, and the material is then pressurized by thrust of the piston. Suitable valves are provided on both the material inlet and discharge. Efficient operation dictates the necessity for an absolute seal between the moving piston and cylinder wall, as illustrated in Figure 31.7. Many types of seals are available, including simple rubber O rings, leather cups, and a variety of special molded polymer seals. Piston pumps are positive displacement pumps and particularly well suited for applications requiring moderate to high pressure and higher flow rates, although they can be used for virtually any rate.

The piston is typically driven by a connected rod. The rod may be reciprocated mechanically

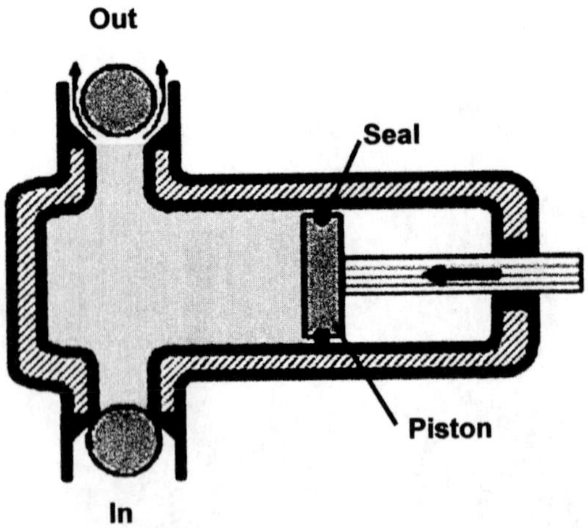

FIGURE 31.7 Piston pump requires tight seal between the moving piston and the cylinder wall.

through a crank or cam, or hydraulically by either oil or compressed air. Pulsation will accompany every stroke, although it can be reduced by use of a pressure accumulator in the discharge line. A disadvantage with all reciprocating pumps is the variation in operating speed of any given stroke as the piston accelerates, decelerates, stops, and reverses direction. Withdrawal must create sufficient suction to fill the cylinder completely. One or more cylinders can be supplied in a pump with a single power source; two-cylinder arrangements are the most common. These can be designed to operate separately or in split phase with each other so as to dampen the pulsation.

Good performance requires that the pistons be continuously lubricated and run cool. A suitable lubricating system should thus be supplied. This can be as simple as drip oilers, a water box where the back side of the piston and drive rod are continuously submerged, or a sophisticated system of spray nozzles fed by a separate lubricating pump. Snug-fitting absorbent swabs are often provided on the back side of the piston to ensure a uniform lubricant film over the entire cylinder wall surface with each stroke. The importance of constant lubrication cannot be overstated, as the principal cause of poor performance is a lack thereof.

A major disadvantage is that once leakage occurs between the piston and the cylinder, effective pumping ceases and very rapid wear can occur. Therefore, these pumps must be shut down promptly upon detecting leakage. With proper lubrication, pump seals are usually long-lived and this is not much of a problem. It becomes a huge problem, however, when lubrication is lacking, especially when pumping abrasive materials. It is the most frequent cause of poor performance and pump downtime in grouting.

Piston pumps come in a wide range of sizes with displacement capability ranging from a few ml to more than 1000 gal (135 ft^3 or 3.8 m^3) per minute and pressure capability to about 5000 psi

(345 bars). Although they can be built to sustain higher pumping rates and pressure, similar reciprocating plunger pumps are probably a better choice for grouting. Very small piston pumps have been used for proportioning resinous grouts, especially for epoxy injection of cracks. On the other end of the scale, most commercially available concrete and mortar pumps use piston pumping elements, and these are widely used for compaction and fill grouting.

31.3.2.2 PLUNGER PUMPS

Like piston pumps, *plunger* pumps, also known as *ram* pumps, involve reciprocation of a plunger within a cylinder, the primary difference being that the plunger does not contact the cylinder wall, but rather maintains at least a measurable distance from it. A seal around the plunger in the end cap of the housing, as illustrated in Figure 31.8 includes a *packing gland.* The pressure of the pumped material pushes on the inside surface of the packing, compressing it ever more tightly with increasing pump pressure. This makes these pumps ideally suited to working under very high levels of pressure.

At pressures below about 800 psi (55 bars), there is insufficient force to optimally assist sealing, and leakage occurs more readily. This should not be allowed to continue for prolonged periods of time, as abrasives in the pumped material can cause rapid wear of the plunger. A huge advantage of plunger pumps over piston pumps is that any leakage developed during pumping can usually be reduced or completely stopped by tightening the packing gland nut, without the need to stop pumping. Thus, it is usually possible to complete the particular pumping operation without delay.

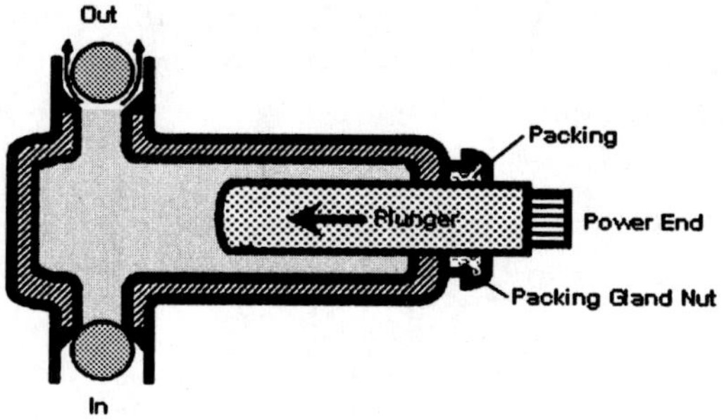

FIGURE 31.8 Only seal of a plunger pump is exterior packing gland around plunger.

Although pumping rates as great as 50 gal (6.6 ft^3) or (190 L) per minute are readily achieved, plunger pumps most often operate at less than 10 gal (3.8 L) per minute. They work best at pressures greater than 500 psi (35 bars), and pressure capability to about 3000 psi (207 bars) is common, with 8000 psi (551 bars) readily available. Two or more cylinders can be assembled and operated in series so as to reach very high pressures. Some manufacturers provide alternate diameter plungers for their machines. Because the output pressure is a function of the area of the plunger and the drive force, higher output pressure is developed by decreasing the plunger diameter. The widest single use of these pumps is in high-pressure water blasters, but they are also used in a wide variety of industrial as well as construction applications.

31.3.2.3 DIAPHRAGM PUMPS

Diaphragm pumps work on the basis of reciprocation of a flexible diaphragm that divides a housing into two chambers. The diaphragm alternately cycles, expanding and reducing the volume of the chambers on each stroke. Either ball or poppet valves are used to control the inflow and discharge of the pumped material. Diaphragm pumps are considered positive displacement pumps, but the volume per stroke can vary, depending on loading conditions. They will pass solids, the sizes of which are limited only by the dimensions of the valve openings and can easily handle abrasive materials.

Although diaphragm pumps can be powered mechanically, hydraulically, or by compressed air, the latter are by far the most efficient and most frequently used. The most efficient arrangement comprises two back to back diaphragm units operating from a common shaft, as illustrated in Figure 31.9. Each unit consists of a housing divided by a diaphragm, which separates the driving air from the pumped liquid. Compressed air is alternately fed from the power side of one diaphragm to the other, expanding and contracting the volume of the material chambers. By connecting two diaphragms with a common shaft, one fills as the other expels the pumped fluid.

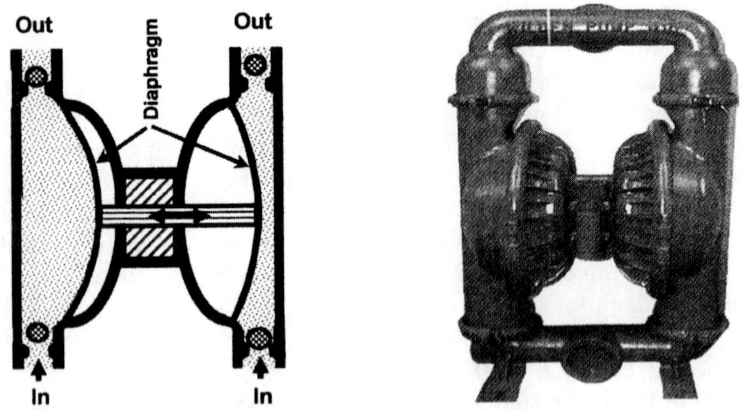

FIGURE 31.9 Duplex pneumatic diaphragm pump is cycled by compressed air against the internal diaphragm.

The absence of close-fitting moving components or shafts requiring mechanical seals negates the requirement for lubrication and allows these pumps to be run dry indefinitely without any damage. The large volume of the chambers ensures good suction capability, although they are most frequently used in a submersible form. Differential pressure across the diaphragm is very low, so the pumping pressure is nearly the same as that of the driving compressed air. Most are rated at 125 psi (8.6 bars) maximum, but some are available to about 300 psi (21 bars). They are also available in many sizes, with inlet piping varying from about 1/4 to 4 in (6.4 to 102 mm), displacing up to about 277 gal. (1048 L) per minute. The flow rate and discharge pressure are easily controlled by variation of the air supply.

To minimize weight and ease handling, the bodies of air diaphragm pumps are most often made of aluminum, but alternate materials, including stainless steel, are available. Diaphragms are made with various types of rubber and thermoplastics, so that resistance to virtually any fluid can be achieved. Surprisingly, the diaphragms are quite durable and long lasting, and their replacement is simple and fast when required. Although better choices for pumping the actual grout are available, these pumps have been found to be reliable and indispensable for handling fluid materials. As illustrated in Figure 10.31, they have been used routinely as mud pumps for the circulation of drill flush.

32

CHAPTER

Grout Pumps

32.1 POWER SOURCE

32.2 TYPES OF GROUT PUMPS
32.2.1 Small Pumps
32.2.2 Medium Pumps
32.2.3 Large Pumps

GROUT PUMPS are a requisite for all pressure grouting. Pump requirements will vary, however, not only for different grout materials but for different applications as well. Although most work can be satisfactorily completed with any one of a number of different types of pumps, some applications are best executed with a given pump type. Valved pumps result in pressure pulsation, whereas continuous pumps will provide an even, constant flow. Which of these flow patterns is best is the subject of much debate among grouters. The fact is that both have been extensively used for at least 50 years with good results.

Pump performance capability must obviously be matched to the injection parameters established for a given application. A constant-output pump is satisfactory for use with circulating delivery systems as long as it provides the maximum output required. Conversely, variable-output pumps are required for direct systems and must provide the full range of pumping rates required.

Reliability and ease of field service and maintenance of grout pumps are essential, as malfunction during injection can cause the loss of a hole and resulting expensive remediation. It is a good idea to maintain a supply of problem-prone spare parts and components on the job site to facilitate rapid repair. In this regard, the ability to quickly change a malfunctioning part is essential, and the use of quick connections and minimal fasteners is strongly recommended. Many holes have been saved by forward-thinking contractors who have fitted their equipment for rapid service or change-out of critical components.

32.1 POWER SOURCE

The pump mechanism can be powered mechanically, with direct connection, hydraulically, or pneumatically. Regardless of the methods of power transmission, the initial energy can be provided by either an internal combustion engine or an electric motor. Electric power is preferred in Europe and much of Asia, whereas either gasoline or diesel engines are more common in the Americas. Even where electric power is used, it is often generated on-site by a diesel-powered generator. Power frequency is important to understand in using electrically powered equipment. Most power in Europe is 50 Hz, whereas in the United States and many other countries, it is 60 Hz. Many European pump manufacturers use motors rated for service at either 50 or 60 Hz, which can operate equally well on either. RPM will increase about 20% at the higher frequency, however. Motors rated for 50 Hz only will likely be damaged if operated at a higher frequency, which should be avoided.

A direct mechanical drive is usually the simplest and least costly option. This will, however, limit the variation of the pump output to the speed range of the drive engine or a variable speed drive, as discussed in Section 29.2.2.1. Speed adjustment can be easily accomplished with hydraulic drives by using either a variable output hydraulic pump or fluid bypass. In addition, adjustable-pressure valves can be included in hydraulic systems that limit the maximum output pressure of the grout pump. The same applies to pneumatic drives, in which speed can be controlled by simply varying the air supply and limiting the maximum output power by the air pressure.

32.2 TYPES OF GROUT PUMPS

Considering the wide range of grouting applications, the variety and capacities of grout pumps is tremendous. To facilitate their description here, grout pumps are divided into three groups: small, medium, and high capacity. Those classified as small include pumps with a maximum output of less than 1 gal (3.78 L) per minute. The medium category includes those pumps most often used in grouting, with an output of about 1 gal (3.78 L) to 40 gal (150 L) per minute. Those with a greater discharge are considered large, which are primarily commercially available concrete and mortar pumps.

32.2.1 Small Pumps

Small pumps are typically used for structural repair, such as epoxy injection in cracks or leakage control in subsurface structures, where relatively small quantities of grout are required. These may be single pumps, but often multipump proportioning units are used. This requires the use of positive displacement pumps, with piston, plunger, peristaltic, and gear pumps being the most common. In structural grouting, unlike much other grouting, the proportion of material costs to the total cost of the work can be high, and this results in many material suppliers providing considerable support. They often market appropriate pumps for their particular materials and, at the very least, will make recommendations on appropriate equipment.

Although common mass-produced pumps are often used, there are many small firms producing (and often selling directly) specialized units for a given requirement. Injection of single-component waterproofing grout is often done with pumps built for airless paint spraying. Likewise, multicomponent compounds are sometimes pumped with equipment developed for spraying plural-component protective coatings. Both of these typically use small plunger-type pumps, which often develop pressures of 1000 to 2000 psi (69 to 138 bars) or more.

Although epoxy injection into structural cracks is sometimes done with batch mixing and a single pump, dual proportioning pumps with in-head mixing are more often employed. Here, accurate proportioning is essential for the injected resin to attain its maximum properties, so the pumps must be very accurate and function precisely. This work typically involves no more than about 1 gal (3.78 L) of material an hour, and the maximum required pumping pressure is only about 100 psi (6.9 bars). Of critical importance, however, is to ensure that there is always proper proportioning of the material. Thus, either transparent material reservoirs or

FIGURE 32.1 Transparent reservoirs or sight tubes promote visual monitoring of material levels.

sight tubes, as illustrated in Figure 32.1, should be required with these pumps. This will facilitate continuous visual confirmation of the relative material levels.

32.2.2 Medium Pumps

Most pumps traditionally used in geotechnical grouting can be classified as medium size pumps, which includes Moyno, plunger, and piston machines. Moyno pumps provide a smooth, uniform output, whereas plunger and piston pumps impart a pulsating pressure to the grout. This is an area of great debate among grouters, some of whom believe that the constant flow of a progressive cavity pump is preferable and others of whom believe that a pulsating pressure provides better injection. I have had extensive experience with both and have not found any substantial difference in the grout behavior.

The only true comparative evaluation may be the extensive test program conducted by Houlsby (1985) described in Section 6.3.1. His many trials were performed with a combination of either high-shear mixing and a progressive cavity pump, or a paddle mixer and a piston pump. As the injection results of the high-shear mixer combined with a progressive cavity pump were better, he concluded that the steady output of the progressive cavity pump was superior. Unfortunately, his tests did not include high-shear mixing with a pulsating pump, so there is a question as to whether the better performance resulted from the pump, or perhaps only from the better-quality grout prepared in the high shear mixer.

Approaching the question from the point of pump mechanics, progressive cavity pumps are simpler, less costly, and can handle large particles in the grout. They have a maximum practical pressure capability on the order of 300 psi (20 bars) and, unfortunately, will self-destruct if allowed to operate dry. Piston and plunger pumps are capable of significantly higher pressures, but both require valves on the intake and output, which limits the size of particle that can be handled. Moreover, pump valves are problematic at best and an inherent maintenance problem. In this regard, some manufacturers provide an extra-fast suction stroke, greatly increasing the grout velocity through the valves. This tends to keep the valves clean and operating more reliably. Plunger pumps are unquestionably superior at high pumping pressures, but seal leakage can be a problem at sustained low-pressure operation.

The pump output should reasonably match the requirements of the individual project as none of these pumps operate very efficiently at either the high or low end of their output range. The drive system must also be considered, especially if the proposed operation is to be performed with a direct delivery system. The ability to vary the output of mechanical drives is limited to the allowable variation of engine speed or requires gear boxes or separate speed controllers. The speed of hydraulic and pneumatic drives, however, can usually be varied by adjustment of the oil or air supply. Tables 32.1, 32.2, and 32.3 provide the operating parameters for commonly used progressive cavity, piston, and plunger grout pumps, respectively. The numbers shown for maximum output and pressure indicate the maximums attainable separately. It is usually not possible to reach the maximum pressure and output simultaneously, however.

TABLE 32.1 Specifications of Some Common Progressive Cavity Grout Pumps

	MAXIMUM OUTPUT		MAXIMUM PRESSURE		
Manufacturer/Model	**gal/min**	**L/min**	**psi**	**bars**	**Drive**
Atlas Copco/Craelius					
ZME 120	32	120	175	12	Electric
ZME 50	13	50	175	12	Electric
Colcrete/High Shear					
Colmono 2000	88	333	145	10	Diesel, electric
Colmono 10FB0	71	267	145	10	Diesel, electric
Colmono 10	35	133	145	10	Diesel, electric
Colmono 4	11	42	145	10	Diesel, electric

32.2.3 Large Pumps

The availability of modern concrete and mortar pumps is of great benefit in low-mobility grouting or where rapid pumping rates are needed. One must keep in mind, however, that the output volume of even the smallest of these large pumps usually greatly exceeds the needs of most common grouting. The hydraulic drive system of most modern concrete pumps incorporates variable-pitch hydraulic pumps, as discussed in Section 29.3.3.3, enabling virtually infinite ad-

TABLE 32.2 Specifications of Some Common Piston Grout Pumps

	MAXIMUM OUTPUT		MAXIMUM PRESSURE		
Manufacturer/Model	**gal/min**	**L/min**	**psi**	**bars**	**Drive**
Atlas Copco					
Pumpac 110B15	26	100	1450	100	Electric
Pumpac 150B22	52	200	796	55	Electric
Pumpac 80 B 22	15	60	2900	200	Electric
Pumpac 110 Basic	26	120	1450	100	Hydraulic
Pumpac 150 Basic	52	230	796	55	Hydraulic
ZB 100 Basic	24	90	1450	100	Hydraulic
ZBA-150-1	40	150	290	20	Air
ZBA-150-2	40	150	580	40	Air
ZBE-100	24	90	1450	100	Electric
ChemGrout					
CG-55	16	60	400	27	
Colcrete/High Shear					
Mocol HL (22 Kw)	53	200	1450	100	Electric
Mocol HL (15 Kw)	40	150	1450	100	Electric
Mocol HS (11 Kw)	34	130	1305	90	Electric
Mocol HS (7.5 Kw)	26	100	1450	100	Electric
Mocol AL	44	167	725	50	Air

TABLE 32.3 Specifications of Some Common Plunger (Ram) Grout Pumps

	PLUNGER DIAMETER		MAXIMUM OUTPUT		MAXIMUM PRESSURE		
Manufacturer/Model	**in.**	**mm**	**gal/min**	**L/min**	**psi**	**bars**	**Drive**
Chemgrout							
HP	3.00		20.0	76	1000	69	Diesel, electric, air
HP	2.25		15.0	57	1600	110	
HP	1.75		10.0	38	2000	138	
Colcrete/High Shear							
Maxicol			42.3	160	1450	100	Electric
Minicol 160DHP			19.8	75	1450	100	Electric
Minicol 160 T			19.8	75	478	33	Electric
Minicol 135T			6.6	25	1450	100	Electric
Minicol 160SHP			9.5	26	1450	100	Electric, air
Minicol 160S			14.5	55	478	33	Electric, air
Colpump			10.8	41	725	50	Electric, air
Hany							
ZMP 610		65	5.3	20	1450	100	Electric
ZMP 625		65	8.8	33	1450	100	Electric
ZMP 710		85	17.6	66	1450	100	Electric
ZMP 710		105	23.3	88	986	68	Electric
ZMP 710		120	35.2	133	725	50	Electric
ZMP 711		85	13.2	50	290	20	Electric
ZMP 711		105	17.6	66	203	14	Electric
ZMP 711		120	26.4	100	145	10	Electric
ZMP 712		85	10.5	40	1305	90	Electric
ZMP 712		105	15.8	60	841	58	Electric
ZMP 712		120	21.1	80	652	45	Electric
ZMP 725		85	25.1	95	1450	100	Electric
ZMP 725		105	37.4	141	986	68	Electric
ZMP 725		120	48.4	183	725	50	Electric
ZMP 726		85	29.0	110	1450	100	Electric
ZMP 726		105	38.7	146	986	68	Electric
ZMP 726		120	57.2	216	725	50	Electric
ZMP 810		60	8.5	32	2900	200	Electric
ZMP 812		60	5.3	20	2610	180	Electric
ZMP 820		60	15.0	57	2900	200	Electric
ZMP 826		60	14.3	54	2900	200	Electric

justment of output speed from 0 to the maximum obtainable. It is thus possible for them to operate very slowly, but efficiency is greatly reduced, output consistency compromised, and the grout mix negatively affected when working at very slow rates.

Many of the smaller concrete pumps have 6 in. (152 mm) diameter material cylinders and use delivery lines larger than 3 in. (76 mm) in diameter. In grouting, delivery systems greater than 2 in. (50 mm) are seldom used, so the extrusion of pressurized grout must be decreased from 6 in. (152 mm) to 2 in. (50 mm), which is torturous for the grout, requires additional pump pressure, and often results in loss of water and fines from the grout through pressure filtration at the pump valve. But of greater significance, in most cases, is that the leaked material is returned to the pump hopper, thinning the grout therein. This results in wet and dry portions of grout going into the delivery system and significant variation in the consistency of the grout entering the header.

Typical operation of these pumps with concrete uses a minimum of about 20 strokes per minute and delivery of about 40 yd^3 (30 m^3) per hour. The maximum pumping rate in grouting is typically much lower, and if such machines are used, they can be operating at less than one stroke per minute. S-tube valves are used on virtually all such pumps and, as discussed in Section 31.3.2.1 although they are the best available, they do not provide a 100 percent positive seal and are subject to excessive leakage and associated abrasive wear. When they are working at one stroke per minute, the total leakage will be at least 20 times as much as the machines were designed to handle, which can result in unacceptable pressure filtration and plugging of the grout.

The force imparted during the pressure strokes work to separate the S-tube from the wear plate, promoting ever-increasing leakage. In a 6 in. (152 mm) pumping system, this force will be 28,274 lb (12851 kg) for each 1,000 psi (69 bars) of pumping pressure. If the size of the system is reduced to, say 3 in. (76 mm), the total force acting to separate the S-tube and wear ring will be reduced to about 7068 lbs (3,213 kg), a four-to-one reduction. In addition, the circumferential length through which leakage can occur will be reduced by half, so both the amount of separation at the valve and the area available for leakage will be greatly reduced.

Other problems develop when the stroking of concrete pumps is excessively slowed. The uniformity of the individual cylinder stroking deteriorates, especially when a closed-loop hydraulic system, as discussed in Section 29.3.1.3, is used. Another problem when large pumps are excessively slowed is filling of the cylinders. Complete filling is not as likely with very slow suction velocity, and this compounds the already difficult filling resulting from the often sticky and/or harsh mixes common in grouting. Although complete filling is usually inconsequential in pumping concrete, it can be a significant liability in grouting, in that air entering the header can damage the formation. Moreover, stroke counters are commonly used for determining both the quantity of the grout pumped and the payment for this material, and this quantity will be incorrect if air is included. As an assist to complete filling, some sort of forced suction feed is needed, but rarely provided. Although concrete pumps often include a "remixer," it is seldom effective in force feeding the pump suction.

Realizing these shortcomings, some pump manufacturers offer machines with smaller cylinders. These are often special versions of otherwise standard 6 in. (152 mm) systems, with the material cylinders reduced to 5 or 4 in. (127 or 102 mm) diameter. They are frequently marketed as high pressure (HP) models. In addition, a few manufacturers now offer 3 in. (76 mm) systems built specifically for specialized work such as grouting. For most grouting, smaller machines that have been designed and are intended for the type of output required, are recommended. Concrete pump salesmen, and the manufacturers as well, seldom have experience in grouting but often promote their machines for the work. And contractors often buy the larger machines believing that they can be effectively used for a wider variety of work, applications that require fast pumping rates, as well as those requiring slower rates. Unfortunately, no one machine is suitable for all work.

In one case a client was advised that the 6 in. (152 mm) pumps with closed-loop hydraulic systems proposed by his contractor were

TABLE 32.4 Specifications for Several Small Concrete Pumps

Model	CYLINDER		S-TUBE		MAXIMUM OUTPUT		OUTPUT PRESSURE		Hydraulic Loop	Horsepower
	in.	mm	in.	mm	yd^3	m^3	psi	bars		
ALLENTOWN POWERCRETER										
Pro	3 × 16	76 × 406	3 × 3	76 × 76	5	3.8	1330	92	Open	23
Magnum	3 × 16	76 × 406	3 × 3	76 × 76	5	3.8	1330/1750	92/120	Open	41
10	3 × 24	76 × 610	3 × 3	76 × 76	10	7.6	1330/1750	92/120	Open	50
20	4.5 × 30	114 × 762	4.5 × 4	114 × 102	20	15.2	1330	92	Open	50
Elite	6 × 30	152 × 762	6 × 5	152 × 127	35	26.6	1150	79	Open	57
PUTZMEISTER										
TK15HP	4 × 39	102 × 990	4 × 4	102 × 102	15	11.4	1550	103	Open	79
TK40	6 × 39	152 × 990	7 × 5	178 × 127	40	30.4	1150	79	Open	75
TK50	7 × 39	178 × 990	7 × 5	178 × 127	54	41.3	1150	79	Open	79
TK50HP	7 × 39	178 × 990	7 × 5	178 × 127	41	31.3	1550	103	Open	79
REED										
B10R	3 × 36	76 × 914	3 × 3	76 × 76	10	7.6	2100	145	Open	84
B20	4 × 36	102 × 914	4 × 4	102 × 102	20	15.2	1664	114	Open	84
B30	6 × 36	152 × 914	6 × 5	152 × 127	30	22.8	1174	81	Open	84
C50HP	6 × 42	152 × 1067	6 × 5	152 × 127	50	38.2	1531	106	Closed	114
C50HPS	6 × 42	152 × 1067	6 × 5	152 × 127	54	41.3	1867	129	Closed	152
TRANSCRETE										
P10	3 × 24	75 × 610	3 × 3	76 × 76	11	8.4	2132	147	Open	30
P20	4.5 × 24	114 × 610	4.5 × 3	114 × 76	25	19.1	928	64	Open	50
P25	5 × 24	127 × 610	4.5 × 3	114 × 76	32	24.4	740	51	Open	49
WHITEMAN CONSPRAY										
SV12	3 × 32	76 × 813	3 × 3	76 × 76	11.5	8.8	2080	143	Open	75
SV20	4 × 38	102 × 965	4 × 4	102 × 102	20	15.2	2200	152	Open	114
SV50	6 × 38	152 × 965	6 × 5	152 × 127	50	38.2	1020	70	Open	114

not satisfactory for an important compaction grouting project where especially low pumping rates were required. The contractor rebelled, even producing a letter from the pump manufacturer stating that those pumps were appropriate for the work, which resulted in their being allowed on the job. After literally months of faulty operation, including tinkering by several "experts" brought to the site to solve the numerous problems, the pumps were replaced by machines with 4 in. (102 mm) cylinders and open-loop hydraulic systems. Larger pumps are fine for fill grouting or in those rare cases in which rapid flow rates can be used. They are simply not satisfactory for most grouting applications.

For compaction grouting, pumps with cylinders no larger than 4 in. (102 mm) and open-loop hydraulic systems should be required, both of which are available from several manufacturers. Even better are 3 in. (76 mm) systems equipped with continuous ribbon blenders oriented to feed the cylinders during their suction stroke. Although the ribbon blenders do not force feed the suction, they do force the grout in the hopper to always be centrally located over the pump suction. Table 32.4 lists the working parameters of common small-line concrete and mortar pumps.

The details of high-capacity concrete pumps are not provided, as they are seldom used in grouting. They are widely available on rental, however, in virtually all metropolitan areas and are very useful for fill grouting, and especially for rapid injection in connection with emergency plugging of water piping failures. These pumps can provide a volume output up to about 250 yd^3 (190 m^3) per hour. They are typically supplied by ready-mix trucks, and the larger ones can pump almost faster than it is possible to position and empty the trucks. Considerable maneuvering room is thus required. Where insufficient space is available at the main pump location, two of these machines can be used in tandem, with one supplying the other. The larger-capacity pumps are typically truck-mounted with a three- or four-section articulated placing boom. Booms that reach as far as 175 ft (52 m) allow rapid setup of the typically 5 in. (127 mm) diameter delivery line to the area of injection.

33

CHAPTER

Batchers, Mixers, and Agitators

33.1 SMALL MIXERS
- **33.1.1** Batch Mixing
- **33.1.2** In-Line Mixing

33.2 MIXING EQUIPMENT FOR POURABLE GROUTS
- **33.2.1** Mixers
  - **33.2.1.1** Paddle Mixers
  - **33.2.1.2** High-Shear Mixers
  - **33.2.1.3** Venturi Jet Mixers

33.3 AGITATORS

33.4 MATERIAL STORAGE AND TRANSFER EQUIPMENT

33.5 EQUIPMENT FOR LOW-MOBILITY GROUTS
- **33.5.1** Paddle Plaster and Mortar Mixers
- **33.5.2** Volumetric Batch Plants

AS FOR ALL aspects of grouting equipment, a variety of machines are available for batching, mixing, and agitation of fresh grout. Because the equipment varies almost as broadly as the purposes for which grouting is preformed, it is divided into three groups here. The first includes all of the small, usually hand-held tools that are used to prepare small quantities of resin and chemical-solution grout for structural repair and water control work. The next includes equipment for working with pourable-consistency grouts, which includes most of the mixers and agitators historically used for geotechnical grouting. The third group includes equipment for batching and mixing stiff mortarlike low-mobility grouts.

33.1 SMALL MIXERS

Depending on the particular grout to be used, it may be mixed in small batches or in continuous in-line mixing heads, usually at the injection header. As discussed earlier, cementitious grouts require considerable viscous shear to completely separate the particles, so some form of power mixing is required, even for small batches. In-line mixing, of course, requires proportional pumping so that the correct ratio of the material is achieved. The amount of shear required for mixing solution and resinous grouts varies with the particular formulation, from virtually none to extreme.

33.1.1 Batch Mixing

Batch preparation is most often accomplished with some sort of blade or paddle mixer, usually powered by a handheld drill motor. With grouts that mix quite readily, *jiffy* mixers, which are typically used to mix paint, are commonly employed. For small quantities, say less than 1 gal (3.8 L), simple fan-bladed mixers turned by a

FIGURE 33.1 Jiffy mixer blending resin in small container (left) and larger paddle mixer preparing 5-gal batches (right).

pistol-grip electric drill, as illustrated in Figure 33.1 (left), are more than adequate. Larger quantities will likely be prepared in 5 gal (190 L) pails with the use of a larger drill motor and mixing head, as shown in Figure 33.1 (right). Double-ribbon blenders, which are commonly used to prepare dough in bakeries, are particularly useful to combine ingredients requiring more thorough integration.

The actual mixing element must be compatible with the container to be used. For work with epoxies and other rapid-setting grouts, one-time disposable containers are often preferable. These are typically of paper or very light plastic composition, the walls of which are easily damaged if hit by a fan-blade-type mixer. Where they are used, sharp edges must be shielded so as not to cut into the container. In all cases, it is best to use a variable-speed drill motor so that the mixing energy can be adjusted to be optimal for the particular material being used.

33.1.2 In-Line Mixing

Most plural-component grouts are pumped separately in the correct proportions and mixed in-line at the header. Solution grouts, such as those based on acrylic, sodium silicate, and some urethanes, are often quite miscible and can simply be stream-mixed with a "T" or "Y" fitting at the header. When this is done, however, it is important to provide a check valve in each of the supply lines to prevent backflow of either component into the other. For many resins and thicker grouts, actual physical mixing is required and is typically supplied through the use of an in-line mixer. Such mixers are available with either static mixing elements or dynamic rotary mixers.

Almost all static mixers consist of a tubular passage filled with some sort of element to cause turbulence and physical combination of the ingredients. In the simplest form, this may be a coarse-textured bottle brush, or it can consist of special elements that have been designed to provide optimal mixing of the particular material. The most common of these is the *Kenics* type, which involves a series of mixing elements. The elements are short helix-shaped pieces with lengths of about 1-1/2 times the tube diameter. Each length is twisted 180 degrees and placed in the tube alternately, so the individual links are in a righthand/lefthand orientation. About six elements are used, but the number is dependent on the difficulty of mixing the particular grout.

Dynamic mixing heads are used for compositions that must receive a relatively high amount of viscous shear. Viscous shear involves continuous rapid physical shearing of the composition so that it becomes thoroughly blended. The amount of shear energy imparted is measured by the temperature increase during mixing. Some hard-to-combine ingredients require dynamic mixing, and most grout will benefit from such mixing, as compositions generally become more penetrable with increasing amounts of dynamic shear as well as increased temperature. An extremely effective dynamic mixing head developed in 1972 uses rapidly turning fan-shaped elements to push the grout back, opposite to the direction of flow. The elements fit snugly into a 1 in. (25 mm) tubular opening with only a small passage between the individual blades, as illustrated in Figure 33.2 (left). Although the mixing elements are directing the

FIGURE 33.2 Dynamic mixing head for difficult-to-combine grouts.

flow backward toward the pump, the injection pressure forces it through the elements, which are spinning at about 2000 rpm, with very high viscous shear. The temperature of the material is increased by as much as 3°F (−16°C) in the 7 to 8 sec travel time through the head during injection. Figure 33.2 (right) shows the assembly in use where a three-component resinous foam system is being injected.

33.2 MIXING EQUIPMENT FOR POURABLE GROUTS

Geotechnical grouting has traditionally been accomplished with pourable grout, be it a cementitious suspension or a chemical solution. This is the only type of grout for which machinery has been specifically manufactured and distributed on a large scale. The grout is initially mixed in a preferably high-shear mixer. If it is a suspension, continual stirring will be required to prevent sedimentation until it is pumped, and this is typically done in an open top *agitator.* The agitator may be mounted on top of the pump suction or it may be separate. Although much grouting in the past has involved bagged cement, the use of bulk cement is ever increasing. This requires storage silos and transfer mechanisms to charge the mixer.

33.2.1 Mixers

Very thorough mixing of pourable grout mixes containing cement, or any other suspended solid, is required to achieve complete wetting and dispersion of all the individual grains. Although a relatively minimal amount of mixing energy will produce a compound in which the grains appear to be suspended, in actuality the cement will be agglomerated, with many particles clumped together as a result of the high polar intensity of common cement, as discussed in Section 6.3.1. To thoroughly disperse the particles, high shear mixing is required, but it is not provided with all available mixers.

33.2.1.1 PADDLE MIXERS

Traditionally in the United States, and unfortunately in some cases even to the present time, grout has been mixed in open-top vertical tanks with a heights equalling about 1-1/2 times the diameter. These *paddle* mixers have a rotating vertical shaft fitted with paddles on the bottom, and sometimes farther up the shaft as well. Most paddle mixers are driven by air motors mounted on the top and very often connected directly to the paddle shaft. A few grouters have adopted similar mixers with one or more horizontal shafts and paddles. There has been no standard for the speed of rotation, but to keep the grout from being thrown out of the tank, the speed is generally low, seldom more than about 60 RPM.

Although this action will disperse cement into a suspension, it will not break up flocks of cement or separate the individual particles sufficiently to form either a quality suspension or a highly penetrable grout. Much of this type of equipment is built by contractors, and it continues to be available from commercial sources.

FIGURE 33.3 Ordinary paddle mixer (left) and high-speed mixer with downward inclined fins (right).

Although it is marginally adequate for bulk placements, it is not very satisfactory for most grouting requirements. The reasons for its being favored by many are its low cost and simplicity. But with these advantages come the substantial disadvantages of producing rather poor quality grout and being quite slow at mixing. Such simple mixers, as illustrated in Figure 33.3 (left), can be enormously improved, however, with rather simple modification.

By increasing the shaft speed to a minimum of about 300 RPM and providing a paddle about 4 in. (102 mm) wide at the base and inclined upward about 30 degree in the direction of rotation, the grout will be forced to the bottom and then up the wall of the tank in a vortex. To contain the grout, downward inclined fins are required on the top of the tank, as illustrated in Figure 33.3 (right). These will cause the grout to turn back, returning toward the center of the tank with continuous circulation. Inclusion of notched fins on the side of the tank will further increase even more the shear energy imparted to the grout. Because such modification imparts much more energy to the grout, a more powerful drive motor than that used for the low-speed mixer will be required. Furthermore, the high forces acting on the blades will require the shaft to be restrained from wandering on the tank bottom.

33.2.1.2 HIGH-SHEAR MIXERS

In 1934, J. P. Morgan invented the so-called *colloidal mill mixer,* the manufacture of which started in 1937 by the Colcrete Company in England. The term "so-called" is used here because the mixers are not mills nor do they produce colloidal grout. Although *colloidal* and *colloidal mill* are frequently used in reference to grout mixers, these terms apply strictly to virtually none of them. Colloidal particles are considered to be finer than 5 microns and they generally float in water. Although it would be nice, reducing grain size during mixing is simply not possible in current practice.

The quality of the grout resulting from such mixers was nonetheless found to be far superior to that produced by the then-popular paddle mixers, and their use became common in the United Kingdom. It was not until 1955, however, that they became available in the United States. Although producing obviously better-quality grout, they were slow in being adopted by U.S. contractors and were not widely used until the 1980s. They are now in use around the world, and several other manufacturers have followed the lead of Colcrete with similar equipment.

In operation, water is injected at the base of a cone-bottom drum, from where it is sucked into a chamber containing a rapidly spinning rotor. The device is actually in the form of a centrifugal pump, the rotor slinging the material against the generally close-fitting case, as illustrated in Figure 33.4. Cement and any other dry materials are deposited over a cone-shaped cap at the top of the drum. The dry ingredients are distributed over the cone so as to flow reasonably evenly into the chamber to prevent clumping. As the ingredients are batched, the mix is circulated, entering the mixer drum tangentially at the top and upper portion to form a vortex in the tank. Circulation continues until the mix is

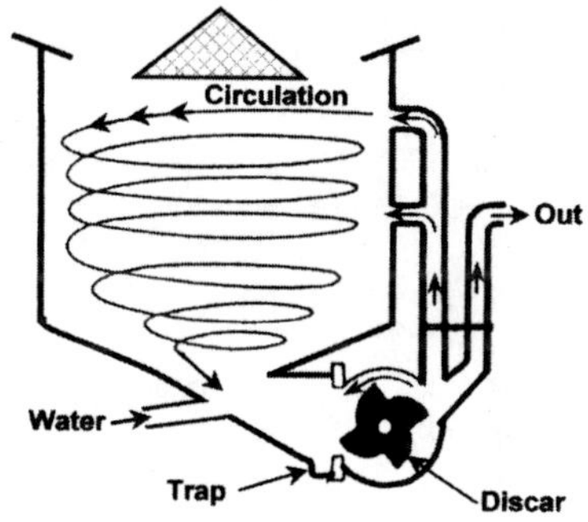

FIGURE 33.4 Colcrete colloidal mill mixer.

thoroughly blended, which occurs rapidly, usually much faster than the mixer can be charged, using bagged cement. Once mixed, the output is diverted from the mixer tank to an agitator by way of a two-way valve. A depressed trap immediately ahead of the rotor is provided to collect any fragments or oversized material in the grout.

The rotor, which typically spins at about 2000 RPM, was originally devised as a smooth-surface disk. Although effective with neat cement, it was unable to deal with sand, so the *Discar* (Figure 33.4, right) was developed and remains quite standard to the present. It can handle mixes containing standard concrete sand, with proportions up to about four times that of the cement. The mixer can effectively combine neat cement grouts with water : cement ratios as low as 0.36. The centrifugal force developed by the rotor provides sufficient pressure to feed the mixed grout into the top of an agitator or to carry it at least 150 ft (45 m) horizontally.

Because high shear imparts considerable heat into the grout and set time is accelerated at higher temperatures, the period of mixing must be controlled. This is usually facilitated by transferring the mixed grout to an agitator within a few minutes of preparation. The Colcrete manufacturers have also developed a combination colloidal mill/paddle mixer for use where the grout might reside in the mixer for an excessive period of time. Here, the initial mixing is by high shear, but once mixed, the paddle agitator takes over to maintain the grout until the drum is emptied.

The Colcrete mixing principle has been imitated to some extent by many through the years and is now available in pumps manufactured by several other firms. Some manufacturers have developed mechanisms that also work on the principle of rapid spinning of other types and configurations of rotors. A few have claimed to produce colloidal mixers by simply circulating the grout through standard centrifugal pumps, which, although beneficial, are not as effective as the better-designed systems of Colcrete and similar mixers. Unquestionably, mixers that impart considerable viscous shear to the grout provide a superior end product. And, as previously discussed, the measure of viscous shear is the increase in temperature of the grout during mixing. Table 33.1 provides the specifics of several high quality high shear mixers.

Discretion must be used in interpreting the various values, and especially the maximum output indicated for the equipment. For example, some manufacturers provide the absolute capacity of the mixer tanks, whereas others provide the maximum amount of grout in a batch, which is typically on the order of 75 percent of the full capacity. In either event, however, the most frequent output limitation is the time required for batching. Simply supplying the necessary water for a typical batch of grout can require a minute or more. In the common case of employing bagged ingredients, several more minutes will be required for picking up, opening, and dumping the numerous bags of cement or other constituents. Then there is the required mixing time, plus that required to transfer the

TABLE 33.1 Operating Parameters of Several Common High-shear Mixers

Manufacturer/Model	BATCH CAPACITY gal	L	MAXIMUM OUTPUT gal/min	m³/hr	Type	Power
ATLAS COPCO CREALIUS						
Cemix 103A	26.4	100	5.3	1.2	Colloidal mill	Air
Cemix 103E	26.4	100	5.3	1.2	Colloidal mill	Electric
Cemix 103H	26.4	100	5.3	1.2	Colloidal mill	Hydraulic
Cemix 203A	52.9	200	13.0	3.0	Colloidal mill	Air
Cemix 203E	52.9	200	13.0	3.0	Colloidal mill	Electric
Cemix 203H	52.9	200	13.0	3.0	Colloidal mill	Electric/Hydraulic
Cemix 403A	106.0	400	21.0	4.8	Colloidal mill	Air
Cemix 403E	106.0	400	21.0	4.8	Colloidal mill	Electric
Cemix 403H	106.0	400	21.0	4.8	Colloidal mill	Hydraulic
Cemix 403WB	106.0	400	22.0	5.0	Colloidal mill	Electric/Hydraulic
COLCRETE/HIGHSHEAR						
SD2000	529.0	2000	176.0	40.0	Colloidal mill	Electric
SD1000	264.0	1000	88.0	20.0	Colloidal mill	Electric
SD700	185.0	700	70.0	16.0	Colloidal mill	Electric
SD400	106.0	400	44.0	10.0	Colloidal mill	Electric, diesel
SD300	79.3	300	35.0	8.0	Colloidal mill	Electric, diesel
SD250	66.2	250	31.0	7.0	Colloidal mill	Electric, diesel
SD200	52.9	200	24.0	5.5	Colloidal mill	Electric, diesel
SD150	39.7	150	20.0	4.5	Colloidal mill	Electric, diesel, air
SD120	31.7	120	15.0	3.5	Colloidal mill	Electric, diesel, air
CP6000	1587.0	6000	110.0	25.0	Colloidal/paddle	Electric
CP4000	1058.0	4000	96.0	22.0	Colloidal/paddle	Electric
CP3000	794.0	3000	88.0	20.0	Colloidal/paddle	Electric
CP2000	529.0	2000	88.0	20.0	Colloidal/paddle	Electric
CP1500	397.0	1500	48.0	11.0	Colloidal/paddle	Electric
CP1000	264.0	1000	44.0	10.0	Colloidal/paddle	Electric
M1000	264.0	1000			Paddle	Electric, air
M500	132.0	500			Paddle	Electric, air
M300	794.0	300			Paddle	Electric, air
M200	52.9	200			Paddle	Electric, air
M90	23.8	90			Paddle	Gas, hand
HANY						
HCM 100	26.4	100	8.8	2.0	High shear	Electric
HCM 300	68.8	260	22.0	5.0	High shear	Electric
HCM 300W	68.8	260	35.0	8.0	High shear	Electric
HCM 600	145.0	550	35.0	8.0	High shear	Electric
HCM 600W	145.0	550	53.0	12.0	High shear	Electric
HCM 800W	212.0	800	88.0	20.0	High shear	Electric
HCM 2500W	661.0	2500	176.0	40.0	High shear	Electric

grout to the agitator. In short, the greatest production limitation is often the charging time required. Of course, this is largely eliminated where bulk materials are used.

33.2.1.3 VENTURI JET MIXERS

Invented and patented by Erle Halliburton in 1924 to rapidly mix a continuous supply of cementitious grout for cementing oil wells, the venturi jet mixer continues to be used this day and remains the standard in the well cementing industry. It involves one or more mixing chamber under bins of cement and/or other finely divided material. In operation, high-pressure water is forced across the mixing chamber, creating a vacuum that sucks in the powder in a controlled manner. Advanced systems spiral the water and cement rapidly, creating a very turbulent flow and thoroughly combining the constituents.

Jet mixers are widely used in industry and are available in many sizes and different configurations from a number of manufacturers. Some utilize multiple jets and, of course, any number can be connected in series to combine multiple materials. When operating continuously, they impart high shear, producing excellent-quality grout. Unfortunately, the quality is not always matched when starting or stopping, so they are best employed where a large continuous flow of material is required, such as for backfilling annular spaces and other voids. They remain the staple of grout mixing for cementing oil wells.

33.3 AGITATORS

Grout agitators are typically open-top tanks with a stirring mechanism, most commonly composed of a rotating vertical shaft to which is attached one or more paddles. Grout often remains in such agitators for an extended time, so to prevent excessive heating, the mixing energy must be restricted to the minimum required to prevent settlement. The typical speed is thus less than 60 RPM, the goal being only to keep the solids in suspension until the grout is pumped. To ensure that a sufficient quantity of grout is always available for the various flow rates commonly experienced, agitators are usually of greater capacity than the typical batch mixers with which they are supplied. Moreover, two or more agitators can be supplied by a single mixer.

Some agitators consist of small propeller-like blades on the end of a shaft, usually turned by a small motor clamped to the side of the tank. Although this can be satisfactory in small tanks, it has not proven advantageous in the more common larger tanks, which often exceed 200 gal (757 L) capacity. Regardless of the exact configuration, all agitator tanks used with circulating injection systems should be fitted with a screen on the return line to trap any debris or partially hydrated grout clumps that the grout has picked up during its travel. Specifications for commonly available agitators are provided in Table 33.2.

33.4 MATERIAL STORAGE AND TRANSFER EQUIPMENT

Although bagged cement is still widely used and perhaps most practical for isolated locations, the use of bulk cement has many advantages and is increasing. It is typically stored in *silos,* which are usually fully enclosed vertical tanks. Because cement is typically transferred from the transport trailers to the silos pneumatically, dust controllers or a *bag house* to prevent solids from escaping to the environment is usually required. Enclosed auger conveyors are often used for transferring the material to the mixer. They can be calibrated to deliver a set amount of cement within a given time to provide accurate volumetric batching. Alternately, a separate weigh hopper and scale is provided with some silos. Portable silos are commonly available, with capacities varying from about 30 to more than 80

TABLE 33.2 Specifications for Commonly Available Agitators

Manufacturer/Model	CAPACITY gal	CAPACITY L	Power
ATLAS COPCO/CREALIUS			
CEMAG 202 A	52.9	200	Air
CEMAG 202 E	52.9	200	Electric
CEMAG 202 H	52.9	200	Hydraulic
CEMAG 402 A	106.0	400	Air
CEMAG 402 E	106.0	400	Electric
CEMAG 402 H	106.0	400	Hydraulic
CEMAG 802 A	212.0	800	Air
CEMAG 802 E	212.0	800	Electric
CEMAG 802 H	212.0	800	Hydraulic
CEMAG 1602 A	424.0	1600	Air
CEMAG 1602 E	424.0	1600	Electric
CEMAG 1602 H	424.0	1600	Hydraulic
COLCRETE/HIGHSHEAR			
A6000	1587.0	6000	Electric
A3000	794.0	3000	Electric
A2000	529.0	2000	Electric
A1500	397.0	1500	Electric
A1000	264.0	1000	Electric, air
A800	212.0	800	Electric, air
A600	158.0	600	Electric, air
A300	79.3	300	Electric, air
A200	52.9	200	Electric, air
HANY			
HRW 160	42.3	160	Electric
HRW 350	92.5	350	Electric
HRW 800	212.0	800	Electric
HRW 1200	277.0	1050	Electric
HRW 2000	528.0	2000	Electric
HRW 3000	794.0	3000	Electric

tons of cement. Two-compartment silos are also available for use with different cement types or other powdery ingredients.

Most transport trailers are set up only for pneumatic transfer, which is the fastest and most economical method. However, because of the dust that must be controlled, it is not very practical for smaller quantities on the job site. Auger conveyors are thus most commonly used, and they can be mounted on transport trailers if desired, allowing transfer directly from transport tanks to the mixer. Separate portable auger units are also available. Typical augers for feeding mixers are 6 in. (152 mm) in diameter and require a 5 hp motor to drive, but much larger-capacity augers are available and are often used.

Figure 33.5 illustrates a contractor-built bulk handling and mixing system for producing cement-flyash grout. The 10,000 gal. (38 m^3) tank has been partitioned into two equal compartments, one for flyash, the other for cement. Each is equipped with a 6 in. (152 mm) feed auger powered by a variable-speed hydraulic motor. The augers are fitted with tachometers so that the relative speed can be set to provide the proper proportion of each ingredient. They both discharge, along with the proper proportion of mix water, into a 9 in. (229 mm) mixing auger (shown extending from the back of the tank), which deposits the mixed grout directly into the hopper of a Moyno pump. The entire operation is simple, fast, and clean. Note the nearly total absence of spillage or other mess in the work area.

FIGURE 33.5 Contractor-built bulk material handling, proportioning, and pumping setup.

33.5 EQUIPMENT FOR LOW-MOBILITY GROUTS

Low-mobility grouts resemble mortar, so mixers manufactured specifically for concrete and mortar are often employed. Grouts, however, are usually stiffer than typical concrete and seldom contain significant large aggregate to assist in dispersing the cement-sand mortar. Drum-type concrete mixers are thus not well suited for grout. Standard horizontal-shaft mortar mixers are, however, quite adequate as are auger-type mixers such as those supplied with volumetric concrete batching plants.

33.5.1 Paddle Plaster and Mortar Mixers

Both plaster and masonry mortar are absent of large aggregate and often contain ingredients such as lime, which make them somewhat sticky. Although they are seldom mixed to the stiff consistency of common grout, they are somewhat like many low-mobility grouts. Horizontal-shaft plaster mixers have been specifically designed to blend these materials. They have horizontal sets of paddles mounted on arms extending from a central shaft. The outermost paddle, which is in close contact with the wall of the drum, is often fitted with rubber wipers so as to pick up and blend the entire contents of the drum. These mixers are widely available in sizes that vary from about 6 to 16 ft^3 (170 to 450 L), the most common having a capacity of 12 ft^3 (340 L).

FIGURE 33.6 Typical auger mixer for low-mobility grouts.

33.5.2 Volumetric Batch Plants

Two types of volumetric batch plant auger mixer combinations are available from several manufacturers. The first is designed for mixing sand-cement mixtures for dry-mix shotcrete, the other for producing concrete. Both contain bins for the various ingredients and a means to deposit them in a controlled, proportionate manner into an attached auger-type mixer (Figure 33.6), which has been found adequate for even the harshest grouts. Water is supplied from either an on-board tank or a hose connected to the machine. Although the dry components are proportioned automatically, the consistency of the mix is dependent on the operator's correct manipulation of a water valve.

The aggregate is typically fed to the mixer by moving belts on the bottom of the bins, which terminate through openings over the base of the mixing auger. The height of the opening, which controls the proportion of the particular aggregate, is controlled by the setting of adjustable *gates*, as shown in Figure 33.7. Some inexpensive models use the same feed mechanism for the cement as illustrated at the right in Figure 33.7, but this is not very satisfactory, as the cement tends to agglomerate and hang up on the gate, thus reducing its proportion in the mix. When

FIGURE 33.7 Adjustable gates control proportions delivered to mixing auger.

FIGURE 33.8 Typical batching/mixer for shotcrete.

this occurs, the mix consistency changes, which is a major shortcoming in grouting where uniformity is requisite. To prevent such occurrences, skilled operators continually prod the cement outlet to keep the cement flowing properly, but this diverts their attention from other tasks and requires exceptionally competent workers if consistent output is to result.

The feed for cement and any other finely divided powdery materials on quality machines is controlled by either an auger or a paddlewheel-type device, both of which provide accurate proportioning. Whereas the machines built for shotcrete usually provide only for the batching of sand and cement, those for concrete have bins for both sand and large aggregate and can have separate bins for fly ash or other powdery material, as well as tanks for liquid admixtures. Concrete batchers are usually truck mounted and self-contained, so they can transport the ingredients and mix any desired amount of concrete independently; they can also be either trailer or skid mounted.

Some units made specifically for shotcrete are usually not intended for transport of the materials, but rather only batching and mixing on-site. They are typically trailer mounted and have their own power source. A bin is provided for the sand, which usually has a capacity on the order of 2 yd^3 (1.5 m^3) and is supplied as needed by a front-end loader. A much smaller cement hopper, which will hold 6 to 12 bags of cement is located in front of the aggregate bin and kept as low as possible to facilitate the breaking of bags and filling by hand labor. The mixer is typically mounted to one side, all as illustrated in Figure 33.8.

There are several manufacturers of common volumetric mixing and batch plants for concrete. They are typically truck mounted (Figure 33.9) and hydraulically powered, with the

FIGURE 33.9 Typical volumetric batcher/mixer for concrete.

truck engine driving the hydraulic pump. In an effort to attain quality performance, major manufacturers of these plants in the United States have formed the Volumetric Mixer Manufacturers Bureau to set standards and disseminate accurate information on use of their products. Their standards for use of these machines are provided in Volumetric (2001). They have further cooperated with the adoption of similar standards, available as ASTM C685/C685M-00.

Made for production rates varying from about 15 to 90 yd^3 (11 to 68 m^3) per hour, volumetric batch plants operate far too rapidly for most grouting work. And here lies a considerable problem in their use. As with concrete pumps, the speed can be varied, but this is almost always achieved through bypassing the pressurized hydraulic fluid. As discussed in Section 29.3.1.2, bypassing causes the fluid temperature to rise, decreasing efficiency. Although the systems usually include heat exchangers and are designed to handle some temperature increase, most are not able to run efficiently at the sustained low rates typical of most grouting.

Most plants have a single hydraulic pump that drives separate motors on the belt feed, cement feed, mixing auger, and other mechanisms. It is the feed motors that must be slowed, whereas the mixing auger speed must be maintained to ensure proper blending. Depending on the particulars of the hydraulic system, this will usually require some alteration. Either very effective fluid cooling or, perhaps, separate hydraulic pumps will be required, one for the mixer and the other for the feed mechanisms. Volumetric batch/mixing plants are widely available for rental, but grouters must realize that problems can develop when high-capacity machines are used, so prudence dictates the use of the lowest-capacity machine available.

CHAPTER 34

Unitized Grout Plants

34.1 BASIC CONSIDERATIONS
34.1.1 Maintenance and Flexibility

34.2 Plants for Pourable Cementitious Grouts
34.2.1 Unitized Plants with Bulk Cement Storage

34.3 GROUT PLANTS FOR LOW-MOBILITY GROUTS

34.4 CUSTOM-BUILT GROUT PLANTS

SELF-CONTAINED UNITIZED GROUT PLANTS are available for various types of grouting. These include the mixer, the pump, and any storage or agitation tanks in a single package. The different elements are typically interconnected with appropriate piping and controls, ready for production. Some have an attached power source, whereas others, such as those powered by electricity or compressed air, usually require connection to an outside supply. Many manufacturers supply unitized grout plants for the pourable cementitious grouts often used in geotechnical grouting. Self-contained mixer-pump combination units are also commercially available for handling low-mobility mortarlike grouts. In addition, many contractors build their own plants using either commercially available elements or those of their own design. Plants specifically made for large-scale handling of chemical-solution grouts are not readily available and so must be custom-built.

34.1 BASIC CONSIDERATIONS

Unitized grout plants are especially practical for work that requires frequent moves and setups for the same general type of grouting. On the downside, however, unitized plants are restricted to the type, size, and particular combination of components included. Although one arrangement may be optimal for a given project, it can be less than efficient for another. For example, a progressive cavity pump may be best for one project, whereas the next will involve very high pressures, requiring a piston or plunger pump. With rapid takes, agitators larger than the mixer are usually desirable, but with slow takes, the agitator should normally be no larger. On some jobs, it may be advantageous to inject two or more holes simultaneously, but without computer monitoring, separate pumps and agitators will be required to facilitate dipstick measurements of the grout take in each hole.

34.1.1 Maintenance and Flexibility

Important criteria in selecting grout plants are ease of maintenance and flexibility to exchange or replace the various elements. For example, the ability to easily change the type of pump by providing compatible drive mechanisms and alternate bolt patterns in the base can make the plant usable for a wider variety of projects. It can also facilitate change-out of the various components. Where a grout plant is intended to be used on many different jobs of short duration, maintenance can be performed in between, in the shop as needed. On large projects that will continue for extended periods of time, however, especially where the grouting continues on multiple shifts and the facilities of a properly equipped shop are not available, the ease of exchanging complete components, especially pumps, which tend to be high-maintenance items, becomes important. Many a grout hole has required expensive replacement because of excessive grout plant downtime during injection. If a defective component can be replaced very quickly, injection into the hole can often be resumed, thus avoiding any rework.

It is preferable for the various components to be mounted with a minimum number of large-size, easily accessible bolts and quick-connect couplings on all power and fluid lines. Large-size bolts have coarser threads, so they are easier and faster to remove and reinstall than small ones. Bolts should be no longer than required for the nut to be fully engaged. Threads that do not extend excessively beyond the nut will not be banged up or filled with grout and thus prevent quick breakdown. Valuable time is also required to unscrew the nut from an excessively long bolt. Even better is to eliminate the nut completely by tapping a hole to receive the bolt in the base, so that only one wrench need be used. This further eliminates the need for nuts and washers, and those that are not used do not tend to get lost. The number of bolt sizes should be minimized to the greatest extent practicable; the use of only one size is optimal. In this way, a minimum number of wrench sizes is required. The seconds or minutes lost looking for the right wrench can be the deciding factor as to whether a hole is lost because of excessive delay.

When larger bolts are used, fewer are required to gain a given value of fixity, although their use will probably require stiffer attachment plates on the base. In addition, base plates must be sufficiently thick to receive a tapped thread. Clamp-type couplings, such as used for concrete-pumping hoses, facilitate very fast breaking and reattachment of fluid lines larger than 1-1/2 in. (38 mm). Brass or steel cam lock fittings are fine for smaller lines, but aluminum cam locks are not recommended, as they are not as strong and the latching ears tend to break in grouting service.

34.2 PLANTS FOR POURABLE CEMENTITIOUS GROUTS

Plants for pourable cementitious grouts typically contain a mixer, agitator, and grout pump, although some have only a single tank used for both mixing and agitation, feeding directly into the pump. The biggest advantage is that all of the elements are piped together and run by a single power source, which is often included. They are thus easily moved and set up at different locations. The capacities of the elements are matched to the particular grouting requirement, and a wide variety are available. Most such plants are constructed on steel pallets for easy handling with a forklift or on steel skids, allowing them to be towed into position as well. Note the lift slots in the skids of the HighShear plant illustrated in Figure 34.1. Many manufacturers also provide custom mounts such as the very portable trailer-mounted Atlas Copco EWB 200-100

FIGURE 34.1 Unitized grout plants should be mounted on skids with forklift slots for easy moving.

FIGURE 34.2 Trailer-mounted Atlas Copco EWB 200-100 grout plant.

plant shown in Figure 34.2. Table 34.1 provides specifications for several available unitized plants.

Because of the increasing use of bulk cement, equipment is becoming available to facilitate its use. The Atlas Copco Crealius unit with a "WB" suffix to the model number is an example. The WB indicates that the unit is furnished with an integral weigh batcher for proportioning. As with mixers, the output capability of a unit must be carefully specified because the rating for such units tends to vary among manufacturers. In the real world, grouting capacity is most often dependent upon the time required to batch and deposit the various constituents into the mixer.

The plants of some manufactures consist of a combination of elements that are sold separately, whereas others possess elements unique to the unitized plant. For example, the agitator of the compact Hany IC 310 is shown in Figure 34.3. Even if the various elements are separately available, they are often welded or firmly fixed to the base frame such that components cannot be easily exchanged. Some manufacturers, such as ChemGrout, prefer to sell all-inclusive plants, such as the CG 600 shown in Figure 34.4. Most suppliers are flexible, however, and will assemble a plant to the buyer's specific needs. An example is the container-enclosed multipump plant by Hany shown in Figure 34.5.

34.2.1 Unitized Plants with Bulk Cement Storage

As already discussed, there are many advantages to using bulk cement. It is generally reserved, however, for larger projects where the necessary silos are set up. HighShear produces plants that are self-contained and include cement storage bins fitted with delivery augers to the mixer. These are large and usually mounted on a semitrailer, as shown in Figure 34.6. The enclosed cement hoppers are available with capacities from 40 to 500 ft^3 (1.1 to 14 m^3) and are

TABLE 34.1 Specifications for Some Commercially Available Unitized Grout Plants

		MAXIMUM OUTPUT		MAXIMUM PRESSURE		MIXER CAPACITY		AGITATOR CAPACITY		
Manufacturer/Model	**Pump Type**	**gal/min**	**L/min**	**psi**	**bars**	**gal**	**L**	**gal**	**L**	**Power**
ATLAS COPCO CRAELIUS										
Grout unit	Piston	24	90	1450	100	53.0	200	105.0	400	Diesel/hydraulic
Unigrout 100-12 EH	Progressive cavity	13	50	174	12	26.0	100	53.0	200	Electric
Unigrout 400-12	Progressive cavity	32	120	174	12	106.0	400	105.0	400	Electric
Unigrout 100-100-E	Piston	24	90	1450	100	26.0	100	105.0	400	Electric
Unigrout 200-100-E02	Piston	24	90	797	55	53.0	200	105.0	400	Electric
Unigrout E 200-100-12	Piston	26/52	100/200	1450/725	100/55	53.0	200	105.0	400	Electric
Unigrout E 200-100-13	Piston	26/52	100/200	1450/725	100/55	53.0	200	105.0	400	Electric
Unigrout E22H	Piston	26/52	100/200	797	55	53.0	200	105.0	400	Electric/hydraulic
Unigrout EWB 200-100	Piston	24	90	1450	100	53.0	200	105.0	400	Electric
Unigrout E 400-55WB	Piston	26/52	100/200	1450/725	100/55	106.0	400	106/212	400/800	Electric/hydraulic
Unigrout E 400-55WBC	Piston	26/52	100/200	1450/725	100/55	2 × 106	2 × 400	106/212	400/800	Electric/hydraulic
CHEMGROUT										
CG-55/030A	Piston	16	60	400	27	45.0	170			Air
GS-600	Progressive cavity	20	75	260	18	97.0	366	97.0	366	Air, electric, hydraulic
GS-680	Progressive cavity	45	170	260	18	127.0	480	127.0	480	Air, electric, hydraulic
CG-500 HP	3.00 in. plunger	20	76	1000	69	70.0	264	97.0	366	Air, electric, hydraulic
CG-500 HP	2.25 in. plunger	15	56	1500	103	70.0	264	97.0	366	Air, electric, hydraulic
CG-500 HP	1.75 in. plunger	10	38	2000	138	70.0	264	97.0	366	Air, electric, hydraulic
CG-600 HP	3.00 in. plunger	20	76	1000	69	97.0	368	97.0	366	Air, electric, hydraulic
CG-600 HP	2.25 in. plunger	15	56	1500	103	97.0	368	97.0	366	Air, electric, hydraulic
CG-600 HP	1.75 in. plunger	10	38	2000	138	97.0	368	97.0	366	Air, electric, hydraulic

(continued)

TABLE 34.1 Specifications for Some Commercially Available Unitized Grout Plants (Continued)

Manufacturer/Model	Pump Type	MAXIMUM OUTPUT gal/min	MAXIMUM OUTPUT L/min	MAXIMUM PRESSURE psi	MAXIMUM PRESSURE bars	MIXER CAPACITY gal	MIXER CAPACITY L	AGITATOR CAPACITY gal	AGITATOR CAPACITY L	Power
COLCRETE/HIGHSHEAR										
Colmono 10/10	Progressive cavity	35	132	145	10	98.0	280	98.0	370	Diesel, electric
Colmono CX 4/10	Progressive cavity	20	75	145	10	31.7	120	50.0	190	Diesel, electric
Colmono 4/6	Progressive cavity	13	49	145	10	31.7	120	43.0	165	Diesel, electric
Colmono 2/4	Progressive cavity	11	41	145	10	15.8	60	18.0	70	Diesel, electric
CMP 1110	Piston	35	132	1450	100	79.2	300	158.0	600	Electric
CMP 1212	Piston	20	75	1305	90	39.6	150	198.0	750	Electric
CMP 1324	Piston	20	75	1450	100	39.6	150	53.0	200	Diesel, electric
CMP 1295	Plunger	20	75	1450	100	31.7	120	53.0	200	Diesel, electric
CMP 1152	Plunger	6.6	25	1450	100	31.7	120	39.0	150	Electric
HANY										Electric
IC 310	Plunger	5.3	20	1450	100	26.4	100	40.0	150	Electric
IC 325	Plunger	8.8	33	1450	100	26.4	100	40.0	150	Electric
IC 650/710	Plunger	22	83	1450	100	68.8	260	92.5	350	Electric
IC 650/711	Plunger	22	83	290	20	68.8	260	92.5	350	Electric
IC 650/712	Plunger	20	75	1276	88	68.8	260	92.5	350	Electric
IC 650/725	Plunger	22	83	1450	100	68.8	260	92.5	350	Electric
IC 650/810	Plunger	8.5	32	2900	200	68.8	260	92.5	350	Electric
IC 650/812	Plunger	5.3	20	2610	180	68.8	260	92.5	350	Electric
IC 650/820	Plunger	15	57	2900	200	68.8	260	92.5	350	Electric
IC 650/826	Plunger	14.3	54	2900	200	68.8	260	92.5	350	Electric
IC 1400/710	Plunger	35	133	1450	100	145.0	550	212.0	800	Electric
IC 1400/712	Plunger	35	133	1276	88	145.0	550	212.0	800	Electric
IC 1400/725	Plunger	35	133	1450	100	145.0	550	212.0	800	Electric
IC 1400/726	Plunger	35	133	1450	100	145.0	550	212.0	800	Electric

FIGURE 34.3 Hany IC 310 plant contains special elements that make it very compact.

FIGURE 34.5 Most manufacturers will build custom units such as this container-enclosed multipump grout plant from Hany.

mounted on load cells that weigh the amount of cement used. The standard plant has a 55-gal- (208-L-) capacity mixer, 198-gal (748-L) agitator, and a progressive cavity pump with a maximum output of 36 gal (136 L) per minute, with a rated top pressure of 250 psi (17 bars). These plants can operate on either electric or hydraulic power with the power source usually furnished on the same trailer. Although not suitable for use in congested areas, these units are very convenient and economical for projects that are sufficiently accessible.

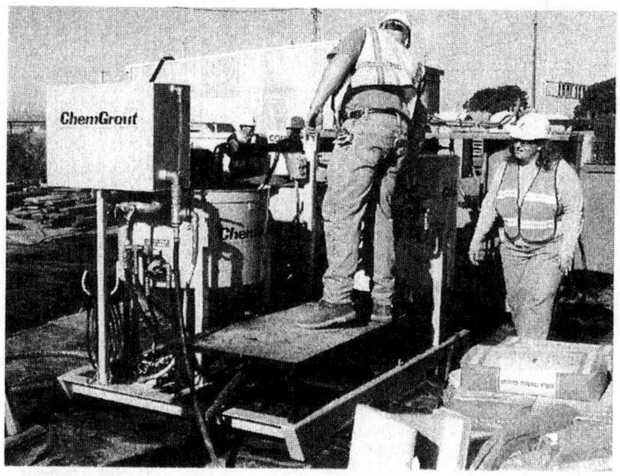

FIGURE 34.4 ChemGrout prefers to supply unitized plants such as this CG 600.

FIGURE 34.6 Trailer-mounted HighShear plant provides for bulk cement storage and handling.

34.3 GROUT PLANTS FOR LOW-MOBILITY GROUTS

Several manufacturers produce self-contained plants with horizontal paddle mixers that feed piston pumps. The predecessor of the modern concrete pump was the plaster pump, which consistently applies sanded plastic mortars. These pumps are now widely used for spraying both plaster and fireproofing materials, which tend to be harsh and hard to pump. Outputs of up to about 250 ft^3 (7 m^3) per hour with pressures up to about 1000 psi (69 bars) are available. The valves of these machines cannot handle particles greater than about 1/4 in. (6 mm), and they do not perform well with very stiff mixes but are quite satisfactory for pumping mortar-like materials. The mixers, which typically have a capacity of about 10 ft^3 (280 L), are designed to mix sticky and harsh materials and are usually more than adequate to mix common grouts.

Several manufacturers of small concrete and mortar pumps offer models with integral horizontal paddle-type mixers. Because these pumps are usually charged by hand height of the mixing drum above the working surface, is always a concern. To lessen this concern, the mixing drums of modern machines, such as the Allentown Magnum, have mixers that are loaded at ground level and then raised hydraulically to dump into the pump hopper, as illustrated in Figure 34.7. This allows laborers to charge the hoppers from ground level. Although the normal maximum pressure rating of the S-tube machine is 1330 psi (92 bars), the unit shown in Figure 34.7, which is used for compaction grouting, has an optional 1750-psi (121 bars) capability as well as a ribbon blending mixer in the pump hopper.

Although ball valve machines are widely available and less costly than those equipped with S-tube valves, they tend to be problematic, especially with thick grouts. In addition they cannot handle the large aggregate preferably used in compaction grouting. Considering the common availability of small-system S-tube pumps, ball valve machines should be avoided except for minor applications.

FIGURE 34.7 Allentown Magnum grout pump features a hydraulically raised mixer.

34.4 CUSTOM-BUILT GROUT PLANTS

Grout plants are often custom-built to provide optimal handling of a given grouting system or to suit the particular needs of a contractor. Figure 34.8 shows a trailer-mounted unit used exclusively for the multicomponent sodium silicate base grouts discussed in Section 7.2.2. All elements of the pump are powered with compressed air. The base solution is prepared in the tanks on the left, which are filled by an on-board air piston pump drawing from a supply tanker. All of the piping is built-in. To be ready for use, only air, water, and a base supply hose need be connected. The smaller tanks on the right are for reactant mixing.

The two main solutions are proportioned by gear pumps that discharge into a flash tank, as discussed in Section 7.2.2.1 and illustrated in Figure 7.4. Another gear pump propels the grout to the header. All pumps are mechanically

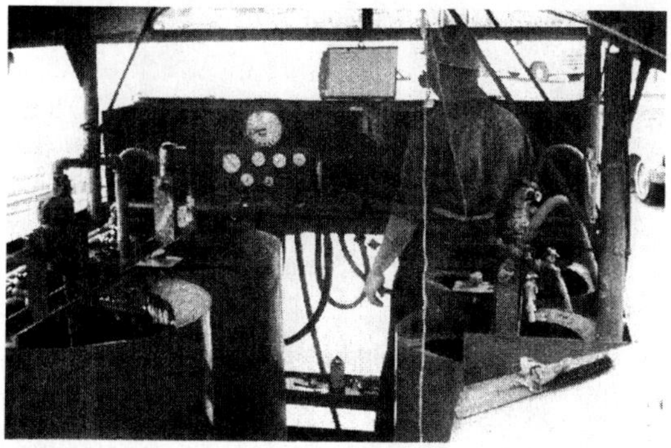

FIGURE 34.8 Custom-built grout plant for multi-component sodium silicate grout.

FIGURE 34.9 Custom-built plant features five mixer-agitators and two progressive cavity pumps.

mounted and secured with only one bolt. Fluid connections are made with flexible hoses fitted with quick-connect couplings. The pumps are driven by belts and variable-speed pulleys. Separate air motors that power the mixers, flash tank, and pumps are all connected with quick-coupling hose. All controls as well as pressure gages and volume meters are located together, as shown in the upper left of Figure 34.8. Barely visible on top, a roof shades the entire plant, reducing ultraviolet sun exposure and providing shade to the grout during mixing.

Figure 34.9 shows another contractor-built plant. It is also trailer-mounted and air-powered, featuring five combination mixer/agitator open-top tanks. Two matching progressive cavity pumps with either separate or keyed power control are mounted under the deck. This double-pump arrangement allows for the preparation and pumping of a wide variety of different grout mixtures, including both single and two-component mixes. For two-component mixes such as the cement-sodium silicate mix being used in Figure 34.9, this contractor always formulates his grouts to a 1:1 ratio, which simplifies handling. In this regard, most chemical-solution and all cementitious grouts contain significant water, allowing flexibility in proportioning, as the water can usually become part of either component.

Finally, any desired components can be grouped together on either a trailer or skid-mounted platform, being jointly powered and piped to allow for optimal operation. This might be in the form of a permanent plant or one for use on a single large project. The individual elements should be sufficiently anchored to prevent displacement during moving. Also, as with all grout plants, it is preferable to provide a roof to shade the mixing tanks and working areas. Such plants can be towed from one location to another, as illustrated in Figure 34.10.

FIGURE 34.10 Separate elements assembled on a skid-mounted platform are easy to move at a job site.

CHAPTER 35

Delivery Lines and Fittings

35.1 DELIVERY LINES
- **35.1.1** Grout Velocity
- **35.1.2** Flow Properties
- **35.1.3** Flexible Hose
  - **35.1.3.1** Hose Construction
  - **35.1.3.2** Hose Selection
- **35.1.4** Rigid Delivery Lines
- **35.1.5** Couplings
  - **35.1.5.1** Cam Lock Couplings
  - **35.1.5.2** Clamp-Type Couplings
- **35.1.6** Reducers and Bends

35.2 VALVES

ALTHOUGH CONVENTIONAL EMPHASIS is typically placed on the pump and perhaps mixers, the delivery line is every bit as important when considering grouting equipment. The line must be able to transfer the grout from the pump to the grout hole without undue frictional resistance, separation of the grout ingredients, or reaction along the way. Obviously, the line must be able to safely withstand the highest pressure that is likely to develop during the injection. Thus, short-term pressure spikes that sometimes occur as pumping is initiated must be considered.

35.1 DELIVERY LINES

The most frequently used delivery lines are flexible hose, but rigid conduit systems are also sometimes employed. Efficiency dictates that either is readily made up, and use of a convenient and easy-to-use coupling is essential. Hoses are always in abundance on grouting jobs, can form literal spaghetti-like masses, and create confusion as to what each is conveying. Fortunately, they can be purchased in many colors, and good contractors use a color-coded identification system—for example, red for air, green for water, black for grout, and so on. Identification can also be by hose size or type of coupling used. Of the five lines attached to the dynamic mixing head illustrated in Figure 33.2, each has a different size or type of coupling and so incorrect attachment is not possible.

35.1.1 Grout Velocity

The delivery line should be sized to provide reasonable velocity of the grout running through it. Generally speaking, the smallest line that will not generate excessive frictional backpressure is desirable. This will hold the least amount of grout, decreasing resident time where separation and/or the start of reaction can occur. Optimal line velocity will vary with type of grout and is controlled by the practical frictional resistance between the conduit wall and the particular grout. As a general rule, this head loss increases with decreasing line size and decreases as the line size increases. Optimal grout velocity creates neither excessive resident time in the hose nor head loss as the grout moves through the line.

35.1.2 Flow Properties

Turbulent flow is desirable for suspension grouts, as it will lessen settlement and separation of the solids from the liquid phase of the grout. Turbulent flow will be created if the cross section of the grout passage is changed, such as by valves or other fittings that have a different cross section than that of the main line. Sharp bends in the travel route, as can result from the use of standard pipe fittings, will also create turbulence. Although this is beneficial to suspension grouts, it is undesirable for low-mobility and some resin grouts that need to travel essentially in a plug flow. Disturbance of plug flow of low-mobility grouts can cause segregation, leading to plugging of the line through pressure filtration. Turbulence will always cause development of heat, and this can be sufficient to accelerate the setting time of some resinous grouts.

35.1.3 Flexible Hose

Literally thousands of different types and sizes of hose are available, as are the types of materials that might be run through them. Although two lengths of hose side by side may look alike, substantial differences can be present. In addition to resisting line pressure, hose for grouting should also be reasonably resistant to *internal expansion.* This becomes especially important for use with low-mobility grout, where the quantity is often based upon the number of piston strokes. For example, a 50-ft (15-m) length of 2-in. (50-mm) hose that expands by only 1/8 in. (3 mm) will consume 0.14 ft^3 (4 L) of grout on each stroke.

Should a 3-in. (76-mm) hose be used, the difference would be about 0.21 ft^3 (6 L). This difference depends on the hose completely deflating on each stroke, which is a function of the stroke rate and does not always happen, but the differential is substantial. Moreover, in the case of very low mobility, such as for compaction grouting, the diameter of the expanded mass will be reduced as the grout encounters couplings, adversely increasing head loss. There are five different common processes by which hoses are constructed, and understanding these is fundamental to the selection for any particular use. Each method imparts different fundamental characteristics, which will affect the hose's use for a given requirement.

35.1.3.1 HOSE CONSTRUCTION

Nearly all hose consists of three basic layers, each of which will affect the overall performance for a given use. The *tube* or *lining* (Figure 35.1) is the innermost element and must provide both chemical and abrasion resistance to the material being conveyed. Normally possessing little strength, the nature of this layer dictates frictional resistance and head loss that will be

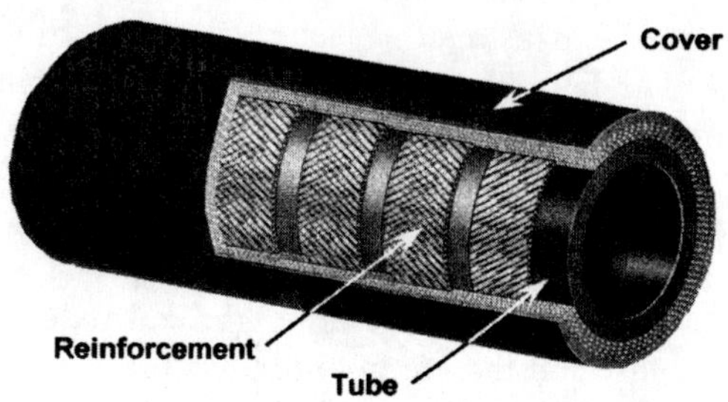

FIGURE 35.1 All hose consists of an inner tube, reinforcement, and cover layers.

experienced. This inner tube is often composed of some type of rubber but can be formed from many polymers—most beneficial is Teflon. While somewhat expensive, Teflon is resistant to nearly every known chemical, and neither viscous nor sticky materials will adhere to it. It thus offers the lowest friction achievable, is almost immune to wear, and is unique in possessing significant strength, which increases resistance to volumetric expansion.

The strength of the hose and the resistance to pressure are provided by *reinforcement*, which can be simple, but for use with grout hose is typically made of a complex combination of different materials. These range from natural and synthetic yarns to wire fabric and even spirally wound wire. When several layers are used, they may be interwoven with rubber to establish a positive bond upon vulcanizing. Herein lies the "guts" of the hose, and the nature of the reinforcement determines not only the pressure capability but also the amount of internal expansion that will occur at high pressures or prior to bursting, as well as resistance to crushing or other abuse. If a hose must withstand a vacuum, the reinforcement will usually also contain a helically wound wire. In general, wire reinforcement can be significantly more resistant to internal expansion than can pure textile reinforcement; however, it will be heavier. And, whereas synthetic textiles are usually stronger than natural ones, they are generally more elastic, allowing greater internal expansion.

The hose *cover* protects the reinforcement from wear abrasion and environmental damage. It is typically some type of rubber, often modified to provide resistance to weather, ozone, and sunlight. Thicker cover provides greater resistance to wear but increases weight and reduces flexibility and the minimum radius that can be bent. Virtually all hose has a cover layer, with the exception of a fire hose, where it is eliminated in order to reduce weight. The cover can be either extruded or built-up and is often wrapped for additional toughness.

Two distinctly different methods are used in hose construction, mandrel and nonmandrel. The simplest and least expensive is nonmandrel wherein an extruded tube is run through a special machine that applies the reinforcement. This may be in the form of spiraled, braided, or knitted layers followed by application of an extruded cover. Further cover protection may be applied in the form of a wrapping. Because the internal tube is unrestrained, the reinforcement cannot be wound very tightly, so the pressure capability is not great. Considerable variation of both internal and external size tolerance can also occur. This is the most economical construction method. However, and because there is no internal mandrel, long continuous lengths can be formed. Lower cost air-and-water hose is often made in this manner.

Two types of mandrels are used for constructing hose, flexible and rigid. Flexible mandrels are made from long lengths of rubber or plastic. The hose is constructed upon the mandrel with the typical layering. Because the mandrel restrains compression of the tube during application of the reinforcing layers, the layers can be wound more tightly, resulting in higher pressure capability and more uniform internal dimensions. The flexible mandrel allows long lengths to be constructed, although the maximum internal diameter is usually limited to about 1 in. (25 mm).

Rigid metal mandrels provide even better support to the core tube, allowing very high winding tension of the reinforcing layers. The internal restraint also produces a very low tolerance internal dimension. Because the individual hose length is limited to the maximum mandrel length, typically about 50 ft (15 m), production of very long lengths is not possible. More costly to produce, rigid mandrel constructed hose is usually reserved for applications requiring very high pressure capability. It is available in diameters up to about 4 in. (102 mm). Variations of mandrel production include spiral ply, wherein all layers are applied in spiral strips on the mandrel, and wrapped ply, which is the oldest method of hose making. In wrapped ply construction, the inner tube is pulled over the mandrel followed by the wrapping of bias-cut fabric layers or plies, usually by a special building machine.

35.1.3.2 HOSE SELECTION

The method of construction and particularly the nature of the reinforcing layers become important in selecting hose for grouting. Although even marginally made hose will often suffice for relatively low pressures, say less than about 100 psi (6.9 bars), as the required working pressure increases, the construction methods become increasingly important. For compaction grouting, 2-in. (50-mm) hose is often used, and pressures of 600 psi (41 bars) or more are common. Because concrete pumps are often used, many grouters simply order concrete pumping hose.

Concrete pumping hose is most commonly textile-reinforced and, while adequate for pumping concrete, may not be appropriate for the slower pumping rates and higher pressures of grouting. Virtually all textile-reinforced hose will undergo significant internal expansion at pressures greater than about 300 psi (20 bars) and some at much lower pressure thresholds. The displacement of smaller cylinder pumps often used in grouting can be completely consumed by the expansion volume of an overly flexible hose. Some manufacturers have data on internal expansion of their hose, but this is nearly impossible to obtain and it is not usually available in any product literature. Even worse, a number of major suppliers do not even provide recommended working or burst pressure values.

Adding to the confusion, there is enormous variability in both performance and safety factor ratings published by different sources. Table 35.1 provides data on several different hoses listed in descending order of reported burst pressure. As can be seen, the safety factor of recommended working pressure varies widely between sources. Also, of interest is that the strongest hose is not necessarily the heaviest. Note that the highest rated burst strength hose is significantly lighter than all but two of those rated lower.

A myth believed by many is that steel-reinforced hose is inherently stiff and heavy. This is not so, and the weight is not very different from that of fabric-reinforced products. As seen in Table 35.1, both steel-reinforced hoses for

TABLE 35.1 Properties of Various 2-in. (50-mm) Hose Marketed for Concrete Pumping

	WORKING PRESSURE		BURST PRESSURE		WEIGHT	
Reinforcement	**psi**	**bars**	**psi**	**bars**	**lb/ft**	**kg/m**
Goodyear Allcrete, fabric	1000	69	4000	276	1.43	0.65
Monster 1200, 6-ply fabric	1200	83	3300	227	1.93	0.88
Con Forms Ultraflex 2-ply steel	1000	69	3000	207	Not given	
Steelcat 900, 2-ply steel	900	62	2850	196	1.61	0.73
Eaglecrete 1200, spiral steel wire	1200	83	2900	200	1.77	0.80
Cobra 800, 4-ply fabric	800	55	2400	165	1.38	0.63
Eaglecrete 800, 4-ply textile	800	55	2200	152	1.39	0.63

which weight is provided are lighter than one of the fabric versions. Because of the inherently poor resistance to internal expansion of virtually all fabric-reinforced hose, the use of steel reinforcement for grouting is recommended. Further, suppliers that fail to provide information as to the composition, burst pressure, and weight of their hose are doing a disservice to potential customers and should be avoided.

35.1.4 Rigid Delivery Lines

Rigid delivery lines provide straight alignment and smooth interiors, so the head loss is significantly less than that of hose. Where long runs are required, rigid delivery lines are often better. Delivery line is available in both steel and aluminum. Although aluminum is lighter and easier to handle, it is best avoided in structural applications. Aluminum reacts with the alkalis of portland cement to produce hydrogen gas, which can result in uncontrollable expansion of the in-place grout. Although standard steel pipe can be used, several suppliers to the concrete pumping industry market *slickline,* usually in 10-ft (3-m) lengths. Slickline can be used with standard clamp-type couplings and is often fitted with heat-treated hardened ends to minimize wear. Standard pipe can be threaded for standard couplings or grooved to receive bolt-on or snap-type mechanical couplings.

Obtaining either working pressure recommendations or bursting pressure capacity of available steel delivery lines is extremely difficult for either standard pipe or the manufactured products. Although the latter are supplied in various wall thicknesses and are often described by words such as *ultra, premium, heavy duty,* and *super,* neither the recommended working pressure nor the bursting pressure is often provided. Unfortunately, the same is true for ordinary pipe.

The bursting pressure of tubes can be calculated by *Barlow's* formula:

$$P = \frac{2St}{D}$$

Where P = bursting pressure in psi
S = tensile strength of tube material
t = wall thickness in inches
D = outside diameter in inches

Table 35.2 provides the bursting pressures of standard steel pipes based on the preceding formula. To confirm the validity of these values, a few pipes were pressure treated to failure, which occurred at somewhat greater pressures than those calculated. Much currently available pipe is imported and may not be produced to established standards that existed when the table was first prepared, so the table must be used with caution.

TABLE 35.2 Calculated Bursting Pressure of Standard Steel Pipe

	SCHEDULE 40		SCHEDULE 80	
Size, in.	**psi**	**bars**	**psi**	**bars**
3/4	2152	148.4	2933	202.2
1	2022	139.4	2722	187.7
1-1/2	1526	105.2	2105	145.1
2	1296	89.3	1835	126.5

The smallest size of commercially available concrete slickline is 2 in. (50 mm), which is about the largest through which grout is pumped. It is lighter than schedule 40 pipe with a common wall thickness of 0.120 in. (3.1 mm). A standard 10-ft (3-m) length weighs about 29 lb (13 kg) and is generally supplied with hardened welded ends to receive clamp-type couplings. Bursting strength levels are not available in either catalogs or component-use guidelines from suppliers. Unless valid pressure capacity data are made available, pressure testing of any line to be used at very high pressures, say more than about 1000 psi (69 bars), is strongly recommended prior to use.

35.1.5 Couplings

Couplings must be able to withstand the rigors and pressures of grouting. Equally as important, they must be easy to make up and break. This may sound elementary, but it can be a serious problem in grouting where the hands of workers are often wet and/or covered with grout. In addition, the grout often penetrates recesses and grooves of the coupling, preventing proper mating. From experience with a number of different types of couplings, I standardized on two—cam lock for hoses smaller than 1-1/2 in. (38 mm) and clamp type as used with concrete pumping for larger sizes, both of which can be used on either rigid pipe or hose.

35.1.5.1 CAM LOCK COUPLINGS

Although cam lock couplings are simple to use, a problem arises when thicker grout gathers on the lower ridge of the female end, preventing complete insertion of the male end. Even with some grout in the joint, it is often possible to engage the cam and force it closed. Unfortunately, excessive force can result in breakage of the locking ears. These couplings are available in several different materials, aluminum being one of the more popular. Its use should be avoided, however, as it is not as strong as either the brass or steel versions. Cam lock couplings are available in sizes from 1/2 in. (13 mm) to more than 4 in. (100 mm).

35.1.5.2 CLAMP-TYPE COUPLINGS

Clamp-type couplings (Figure 35.2) were originally developed by and proprietary to the Victaulic Company. Their main use was in joining pipe that required some flexibility or was of a temporary nature. With the original concept, grooves were placed in the pipe ends to receive the clamp, which often required seating with a hammer rap on the linkage lever. The stated working pressure was 500 psi (34 bars), although in extensive use a failure was not experienced even when the couplings were worked to pressure levels greater than 1000 psi (69 bars). Following the expiration of patents, the developing concrete pumping industry improved somewhat on the original design with longer levers, facilitating hand closure of the couplings.

As required delivery system pressures increased, new generations of couplings were developed, and larger welded-on ends (Figure 35.3, left) appeared. Although these models are said to be stronger, there is no documentation as to the actual pressure capability. Clamps are now available for both grooved (Figure 35.3, right)

FIGURE 35.2 Clamp-type couplings are well established and widely available.

FIGURE 35.3 Heavy-duty weld end on left and conventional groove on right.

and the larger welded ends, often referred to as *heavy duty.* Unfortunately, although they appear similar, the two types are not interchangeable. It is thus important that the ends of all delivery line components be the same as the couplings with which they are secured. Many problems can be created when incompatible components are delivered to job sites.

In addition to the hinged two-part clamps, bolted versions are available and said to be stronger, although here, again, no documentation or established capacities are available. Although the bolted versions require more time to make-up and break, once secured, accidental opening is unlikely. On the other hand, the clamp type can open should the lever be caught on a sharp object. To prevent this, some manufacturers supply *safety pins,* which fit in holes in the couplings to prevent unintentional opening. Where these are not supplied, the levers should be secured with wire, especially when the line will be dragged or moved over objects on which the lever may be caught. The improved clamp-type couplings are available from 2 in. (50 mm) up, whereas the original Victaulic couplings were made for lines as small as 1 in. (25 mm).

35.1.6 Reducers and Bends

Low-mobility mortarlike grouts flow as an extrusion or plug. They are not able to traverse standard pipe fittings with excessive disruption and such will likely cause pressure filtration leading to plugging. Pumped concrete behaves similarly, so wide sweep bends as well as long taper reducers from one size line to another have been developed and are commonly available in sizes of 2 in. (50 mm) and greater. Concrete often contains sharp crushed aggregate and is pumped very much faster than grout, leading to excessive wear of these fittings, so often they are made of hardened steel. In the larger sizes, which are more common in concrete pumping, they are even sometimes lined with carbide or other hard material.

These standard fittings are fine for grouting; however, we sometimes need sizes smaller than 2 in. (50 mm). Fortunately, wear is not as great a problem in grouting, so wide sweep bends made of ordinary pipe generally perform quite satisfactorily. Such bends are readily available from suppliers to the electrical construction industry, where they are known as *rigid conduit* bends. They normally come with threaded ends to which adaptors to the particular coupling must be attached. Functionally this is fine, but it does add considerable length to the bend. More compact versions can be easily made by tooling the appropriate grooves in an appropriate length pipe and bending in a standard pipe bender.

The smallest readily available taper reducers decrease only to 2 in. (50 mm). Commercial *swage nipples* available from large plumbing or industrial suppliers can be used for smaller sizes. These come in many forms, and the *venturi* form, which is basically a uniform taper from one size to the next, is the best to use. Available in standard pipe sizes all the way down to 1/4 in. (6 mm) and in standard, extra heavy, or double extra heavy weights, they come with either standard male pipe threads or tapered ends known as weld fittings. Being a forged product, the outside diameter of the bevel end versions are not always accurate; however, they are easily turned in a lathe and then tooled to the appropriate coupling grooves. Some suppliers offer them with a standard Victaulic groove on special order.

35.2 VALVES

Historically, valves have been a constant problem for the grouter. Grout abrasion wears away metal and damages seals as well as filling grooves and other interstices, making operation difficult and often preventing complete closure. The most used valves historically have been *plug cock* valves intended primarily for natural gas. These typically have a somewhat rectangular opening, changing the cross section of the grout as it passes, are usually hard to turn, and often come without an attached handle. Many field problems have occurred because of delay in operating a valve while looking for a handle or wrench or not knowing which position (open or closed) it is in. Both plug valves and more modern ball valves operate fully with a 90 degree turn of the handle. When properly attached, the handle position will indicate the direction of flow. When perpendicular to the valve axis, it is completely closed, whereas when parallel, it is full open.

Houlsby (1990) strongly favors *Saunders diaphragm* valves, which allow close control of the flow. These have a raised ridge across the flow passage, as illustrated in Figure 35.4, upon which a flexible diaphragm is depressed when closed. Turns of a wheel-type handle slowly raise the diaphragm, opening the passage to flow, with several turns required to open it fully. Intensity of flow is easily controlled by the handle position, but that creates the greatest limitation in that the relative flow amount is not indicated. Moreover, several turns are required to shut off flow when needed. Diaphragm valves hold up well to grouting service except for breakage of the diaphragm, which is easily replaced.

However, I prefer *full-flow* ball valves. Full-flow ball valves are normally supplied with a fixed handle, the position of which indicates the approximate amount of flow. Operating from full open to closed is accomplished quickly with only a 90 degree turn. They can be throttled by manipulation of the exact handle position. Because the flow opening is the same size and cross section as the delivery line, no recesses exist to collect grout, so full-flow ball valves are easy to clean and tend to always work freely. When operating in a partially open position, abrasive wear does occur, but not usually very rapidly. Thus, full-flow ball valves generally provide good service and are very economical when replacement is required.

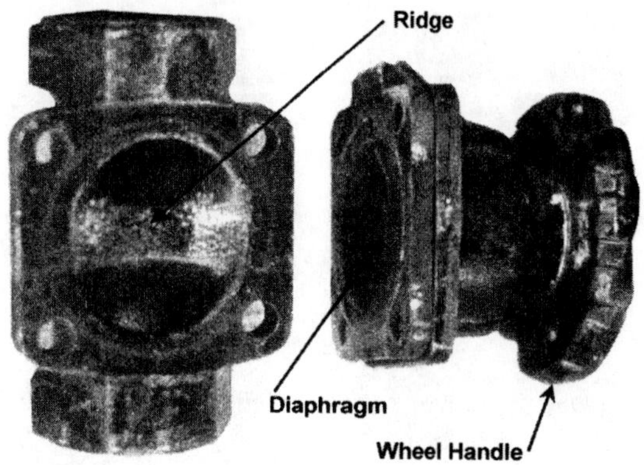

FIGURE 35.4 Disassembled Saunders-type diaphragm valve showing diaphragm and ridge upon which it seats.

There are several different patterns of ball valves, and the flow passage of many is considerably smaller than the line size. Some allow flow in only one direction to avoid damage to the seal. These are not satisfactory for grouting and only those with a *full-flow opening* and *operable with flow in either direction* should be employed. Figure 35.5 illustrates two 2-in. (50-mm) ball valves. The one on the left supports a full 2-in. (50-mm) flow passage while that of the less costly one on the right is substantially smaller and not acceptable. Ball valves operate equally well with air, water, or grout. Thus no other types need be on hand. They should be provided with end nipples set up for whatever coupling will be used, enabling rapid installation and removal.

Other types of valves, such as gate or globe, are not satisfactory for grout and are not very

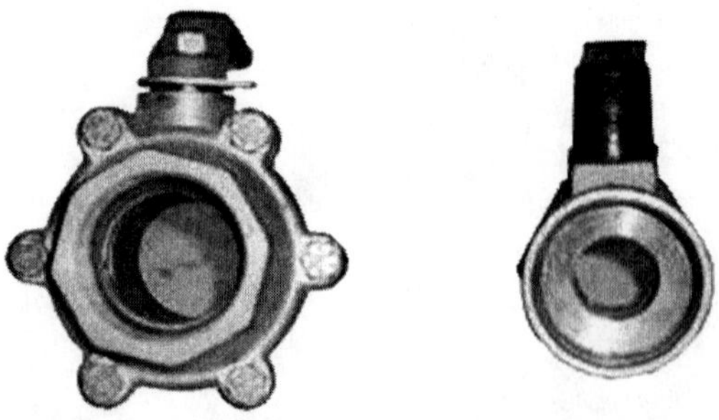

FIGURE 35.5 Two ball valves: full flow on left and reduced flow on right.

efficient for either air or water onsite. The interior passages offer many opportunities for plugging and often cannot be completely shut off because of the buildup of grout on the interior. Further, they are virtually impossible to properly clean out. Either Saunders or ball valves will give satisfactory service. Admittedly repetitious but of great importance, however, is the advice that *all valves should be equipped with attached handles.* For those rare instances where valve operation is to be automated, motorized diaphragm valves will normally be required.

36

CHAPTER

Monitoring Equipment

36.1 PRESSURE GAUGES
- **36.1.1** Gauge Accuracy
- **36.1.2** Dial Size
- **36.1.3** Dial Guards

36.2 GAUGE SAVERS

36.3 FLOW METERS
- **36.3.1** Types of Flow Meters

36.4 AUTOMATED MONITORING EQUIPMENT
- **36.4.1** Data Acquisition
- **36.4.2** Signal Transmission
- **36.4.3** Data Display
- **36.4.4** Batteries
- **36.4.5** System Availability

ALTHOUGH SMALL IN SIZE, gauges and flow meters are among the most important tools on a grouting job, as it is through observation of these that all other equipment is controlled. Therefore, they must be reasonably accurate and of sufficient size to be easily read. To prevent damage, the gauging elements must be protected against direct contact with the grout. In addition to monitoring the grouting parameters, flow meters and pressure gauges are also required for water testing as well as for proportioning the mix water.

36.1 PRESSURE GAUGES

Gauges are delicate instruments but life on a grouting project is anything but delicate. These instruments are easily damaged by shock or other abuse, especially when used by less than conscientious workers. Gauges come in many different sizes, ranges of pressure, degrees of accuracy, and overall quality. Some contractors use very accurate, high-quality gauges and compel their workers to treat them gently, whereas others use the cheapest models available and treat them as expendable items.

36.1.1 Gauge Accuracy

Gauge accuracy is defined as a percentage of the full-scale range of a particular gauge. Typically, the top and bottom 25 percent of a gauge will provide significantly less accuracy than the middle 50 percent. Therefore, gauges should generally be selected such that the working pressures will be within the middle 50 percent range. This means that gauges of varying ranges will be required for different grouting applications. Many degrees of accuracy are standardized in the American Society of Mechanical Engineers (ASME) Standard B40.1, in which accuracy levels are divided into eight grades according to the particular use requirements, as shown in Table 36.1.

Often specifications call for the calibration of gauges used in grouting but do not indicate the standard to be used. Normal pressure control in grouting is far from perfect. Of much greater significance than the pressure level is pressure behavior, that is, the relative changes that occur during injection. This will be evident regardless of gauge accuracy as long as the tolerance is within a range of about 5 percent. Highly precise gauges are thus neither needed nor economically justified for much grouting work. A concern that is real in grouting, however, is pressure pulsation resulting from valved pumps. This will shorten the life of the gauge. Pulsation can be dampened by filling the gauge with glycerin, silicone, or a halocarbon. Some manufacturers also produce *pulsation dampeners* and *pressure snubbers.*

As an example, one grouter used Grade B gauges, and every foreperson carried a supply of gauges with four range scales, as follows:

psi	*bars*
0–100	0–7
0–300	0–20
0–600	0–41
0–1,000	0–69

Gauges were seldom calibrated; instead, their accuracy was checked against a new gauge. This was typically done by connecting at least three or as many as six or eight gauge savers with gauges installed together in series. Any gauges that read more than a few percentage points different than the new one were discarded. Although this could be considered good practice, lower grade (and thus less costly) gauges would have been as satisfactory.

For most grouting work, requiring overly accurate gauges and/or accredited calibration is probably not needed and will result in greater cost. It should be avoided unless fully justified. What is important is that the gauges used have a reasonable dial range for the expected pressures. Grouting is unique in that a high-pressure spike often occurs at the beginning of injection, especially during compaction grouting or groutjacking. The maximum gauge level must take this into consideration. A good rule of thumb for grouting is to use a range scale approximately double the highest expected production pressure.

TABLE 36.1 ASME Accuracy Grades (Percentage of Gauge Scale)

Accuracy Grade	Lower 1/4 of Scale	Middle 1/2 of Scale	Upper 1/4 of Scale
4A	0.1	0.1	0.1
3A	0.25	0.25	0.25
2A	0.5	0.5	0.5
1A	1	1	1
A	2	1	2
B	3	2	3
C	4	3	4
D	5	5	5

36.1.2 Dial Size

Gauges are commercially available in dial sizes of from 1-1/2 in. (38 mm) to 16 in. (406 mm). The smaller sizes are less costly and more readily available. Ability to easily read pressure levels and variations is essential in grouting. Therefore, a 4-in. (102-mm) minimum dial face is preferred but 3 in. (76 mm) is acceptable. The reason is there is a substantial price premium when the face exceeds 3 in. (76 mm). A much larger dial face would allow easier reading and from a greater distance; however, there is a relatively greater risk that gauges with the larger dial face will become struck and damaged.

For most grouting, a dial face between 3 and 6 in. (76 and 152 mm) diameter is suitable. There are exceptions to every rule, however, and smaller faces might be allowed on hand-held injectors such as the dynamic mixing head previously illustrated in Figure 33.2. Conversely, remote gauges in a protected environment can justifiably be much larger.

36.1.3 Dial Guards

Often specifications require that all gauges be guarded to prevent damage during use. Although this may seem reasonable it adds considerable weight to the gauge saver which is a disadvantage where workers must make frequent connections and disconnections, such as in upstage injection. As an example, the 5/16-in (8 mm) thick bar making up the guard illustrated in Figure 36.1 adds about 8 lb (3.6 kg) to the header. Although this might protect the gauge from being physically displaced if stricken or dropped, such is more likely to happen due to the additional weight, and a blow could substantially affect gauge accuracy, which might not be evident. Continued injection with the unknown faulty gauge would then be unsatisfactory.

FIGURE 36.1 Gauge dial protective guards can be heavy and bulky.

Perhaps the most common damage to a gauge is breakage of the glass dial. More robust plastic dials are available but usually only on special order. Occasionally gauges get showered in grout, and this is a particular problem for plastic dials as they are easily scratched and etched, making it hard to read them. Dials are easy to replace in the field, and experience has shown that the standard glass is best. A supply of replacement lenses should be on hand at all times because it is important to replace broken ones promptly. Should operations continue without a lens, the needles will likely be displaced or embedded in grout and permanently ruined.

36.2 GAUGE SAVERS

Virtually all grout will destroy a gauge if allowed to enter the Bourdon tube, which is the internal element that assesses pressure. Some sort of element must thus separate the gauge interior from the grout. This is most often a diaphragm of some sort, which separates a reservoir of glycerin or light oil on the gauge side from the grout.

The diaphragm and filler are enclosed in a *gauge saver*, also referred to as a *gauge protector* or *gauge seal.* (This is not to be confused with dial guards, which protect the gauge from external damage.)

Rather than using gauge savers, some contractors use a short pipe nipple filled with grease between the grout and gauge. As an example, the gauge shown in Figure 36.2 is mounted into one of the continuous legs of an ordinary pipe T fitting. A zerk grease fitting is installed in the branch and the other mainline is attached to the grout line by way of a short nipple. Grease is forced into the zerk fitting before injection until it exits the short nipple on the base of the T. The grout may work its way a short distance into the nipple but will generally not mix with the grease. At the end of the shift, the T is removed and grease is again injected, expelling any grout that has penetrated. While this is somewhat labor intensive during startup and at cleanup, it does effectively protect the gauge.

Gauge savers consist of either a flexible tubular diaphragm the same inside diameter as, and installed in the delivery line, or a flat diaphragm adjacent to the line. In both cases, the diaphragm divides a reservoir of appropriate fluid on which the gauge is installed. The tubular diaphragm, illustrated in Figure 36.3 (left), has the advantage of "feeling" the grout pressure for the entire periphery of the line. It is normally about 2 in. (51 mm) long and contained between two heavy steel flanges much larger than the inside diameter of the line. This is a disadvantage in that the assembly is both relatively heavy and bulky. The gauge illustrated in Figure 36.1 is mounted on such an assembly. Another problem is puncture of the diaphragm by inexperienced workers poking them with prods during cleaning. Although not easy, the diaphragm can be replaced in the field, but at a high cost of about \$100 per replacement diaphragm. These gauge savers are available with rated pressure capability of up to about 720 psi (50 bars).

FIGURE 36.2 Zerk fitting on right is used to keep the gauge and the mounting nipple full of grease.

Two widely used gauge savers with flat diaphragms are available. One offered by many suppliers is known as a *gauge seal* and contains a flat diaphragm about 3 in. (76 mm) in diameter attached to an upper fluid-filled chamber onto which the gauge is installed, as illustrated in Figure 36.3 (center). This type usually contains a female pipe thread in the base that is seldom larger than 1 in. (25 mm) diameter, resulting in the lower chamber being somewhat isolated. Some models contain a connection for flushing to clean the chamber, as illustrated in Figure 36.3. It is usually only about 1/4 in. (6 mm) but can be up to 3/4 in. (19 mm) diameter. The upper and lower body are typically attached with eight or more bolts through mating flanges. These gauge seals are available with capacities up to about 2500 psi (172 bars).

Although commonly available, this type is not the best choice for grouting, especially with thicker grouts, as the grout tends to become impacted in the recesses immediately under the diaphragm. Although it can be flushed out when an appropriate fitting is included as illustrated, this requires time and is messy. Further, the flushing will likely not be complete if the grout

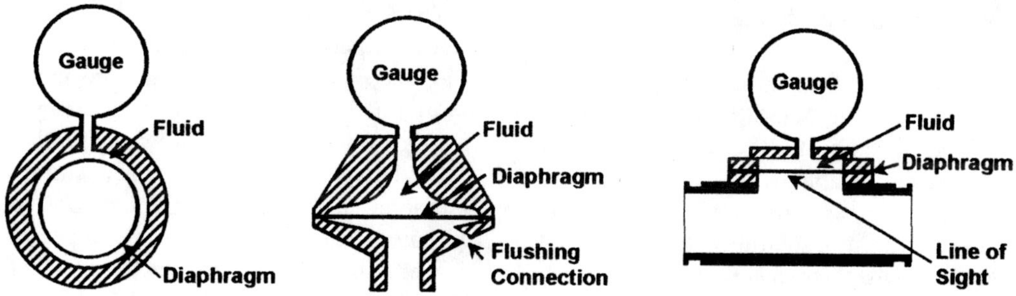

FIGURE 36.3 Three types of gauge savers: tubular diaphragm on left, gauge seal flat diaphragm in center, and preferred full-contact diaphragm on right.

has become tightly impacted, which is common with cementitious grouts. In order to inspect, clean, or replace the diaphragm, many small bolts must be removed. This is time consuming and not easy to do in the field, and the small nuts and washers may be easily lost. Moreover, the diaphragm of some models is bonded to the top body, precluding field replacement. In addition, the distance to the diaphragm is typically more than 1 in. (25 mm) and can itself become impacted with grout.

Far superior but not commonly available is a flat, flexible diaphragm about 4 in. (101 mm) in diameter that divides a chamber with a full-size opening for the grout to contact. The diaphragm is held between two flanges, the bottom of which is welded directly on top of a piece of rigid tube. Ends of the tube are fitted for a standard clamp coupling, and the tube is kept as short as possible, with only enough exposure for connection. This combined with the lowest practicable mounting of the bottom flange allows a line of sight to facilitate cleaning and visual inspection of the diaphragm, as illustrated in Figure 36.3 (right). Because of the short, full-size passage to the diaphragm, plugging virtually never occurs.

After much experience with more complicated and less effective alternatives, this type of gauge saver has been used successfully for many years. The ultimate design involves stiff flanged plates 3/8 in. (10 mm) thick that are sufficiently stiff that only four 1/2-in. (13-mm) bolts are needed for attachment. The bolts do not require nuts as the bottom flange holes are appropriately threaded. This makes field replacement of the diaphragm fast and easy. These custom-made gauge savers have been used in delivery line sizes of 1, 1/12, and 2 in. (25, 38, and 51 mm), although virtually any size will benefit from the design. A picture showing both the assembled unit and the separate pieces is provided as Figure 36.4. The gauge savers were regularly tested to a pressure of 1500 psi (103 bars) without incident.

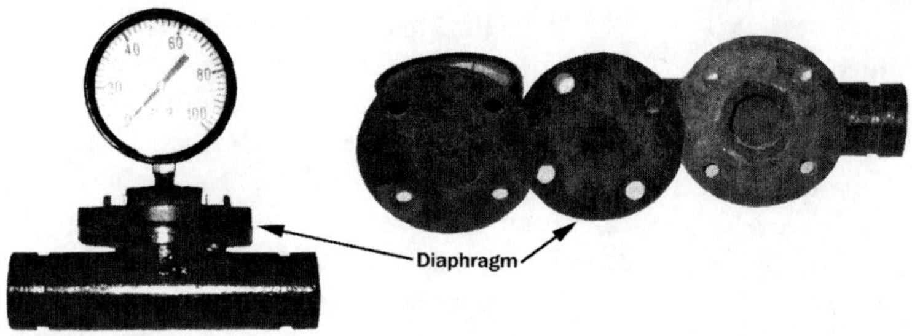

FIGURE 36.4 Preferred full-contact diaphragm gauge saver on left and broken down on right.

36.3 FLOW METERS

Knowing the quantity of grout pumped is fundamental to ensuring satisfactory performance. Unfortunately, compact and direct-reading flow meters that can reliably withstand the rigors of grouting are not easily available. Common meters can be used with true solution grouts, although the accuracy must be frequently checked against the batch volume or with a master meter, as remnants of grout in the chamber will affect accuracy. Stroke counters are often used, especially with low-mobility grouts, but their accuracy is questionable and they nearly always indicate a greater amount than has actually been pumped. This occurs from incomplete cylinder filling as well as internal expansion of many common hoses, as previously discussed in Section 35.1.3.2.

36.3.1 Types of Flow Meters

There are four major types of flow meters: *differential-pressure, positive-displacement, velocity,* and *mass* meters. The differential-pressure meter is simple and perhaps the one most commonly used in industry. It includes a variety of orifices, venturi, and tapered variable-area tubes. It operates on the principle that pressure drop across the meter is proportional to the square of the flow rate. This meter must be matched to the measured fluids properties and flow conditions and its operation can be impaired by the large variations common in grouting.

Positive-displacement meters include a variety of propeller, rotary-vane, and nutating-disk devices. They operate on the principle of separating the passing fluid into accurately measured increments that are counted and displayed by a connected register. Common water meters are of the nutating-disk type. They operate by wobble of a movable disk concentrically mounted on a spherical bearing. Each cycle of rocking movement represents a given unit of flow. A mechanical counter is activated from a shaft mounted perpendicular to the disk. Although these meters are easy to use, they are not appropriate for thick fluids and their accuracy is adversely affected by buildup of grout on the disk.

Velocity meters include turbine, magnetic, and sonic instruments. Turbine meters consist of multibladed rotors that are mounted perpendicular to the flow and spun by the moving fluid. The flow rate is directly linked to the rotational speed, which can be processed either mechanically or electronically. These may be practical for clear-solution grouts but do not perform well with abrasives in the fluid due to rapid wear of the bearings.

Magnetic meters are the most frequently used in grouting and operate well with most fluids and slurries as long as they are electrically conductive. However, when working with low-mobility grouts that are typically near but not at saturation, small changes in water content can result in large changes in electrical potential. The devices work in accordance with Faraday's law of electromagnetic induction, which states that a voltage will be induced when a conductor moves through a magnetic field. The magnetic field is created by energized coils on the flow tube, which has the same cross section as the delivery line. Voltage is measured by spaced electrodes on the tube wall. The moving fluid serves as the conductor, and the differential voltage measured between the electrodes is in direct proportion to the velocity of flow.

Magnetic meters are mounted directly in the delivery line, and the electrodes are set in a tubular chamber of the same diameter. They require a power source that can be supplied by either line or battery power. Quite robust, they can be mounted in any configuration. Virtually any fully saturated fluid can be measured as long as the chamber cross section is completely filled. Output can be directly displayed on an attached readout or wired up to several hundred feet to a remote readout or computer.

Two types of meters operate by ultrasonic pulse: Doppler and travel velocity. Doppler meters read changes in frequency resulting from liquid flow. A transducer directs a pulse of known frequency into the liquid. It is reflected off discontinuities such as solids, bubbles, and such in the fluid and is received by another transducer. A frequency change proportional to the fluid velocity is received and displayed as flow rate. This can be done with separate transducers that are simply strapped to the outside of the delivery line. Milliampere output terminals facilitate connection to a recorder.

Travel time meters use the difference between pulse velocity at different directions from the line of flow. Transducers are placed on opposing sides of the line and at a 45 degree orientation to the direction of flow. Pulses are alternately sent from one transducer to the other, and the travel velocity in each direction is ascertained. The velocity differential is converted to flow rate of the fluid. These meters are very sensitive to solids or entrapped air in the fluid and so are not practical for grouting.

While most meters work on the principle of volumetric flow, Coriolis meters measure the actual *mass* flow. A U-shaped tube is enclosed in the sensor element, which is typically of the same cross section as the line, and the tube is vibrated at its natural frequency by a magnetic device located in the bend. The U tube oscillates at varying frequency depending on the mass of the passing flow. The rate of these oscillations is picked up by magnetic sensors on opposite sides of the flow and it is converted to voltage differentials according to flow rate.

36.4 AUTOMATED MONITORING EQUIPMENT

Automated systems acquire various parameter values and then either transfer and display them on a mechanical instrument or send the data to a computer where software can enhance the output. Automated systems fulfill two functions. First they provide the supervising technician at the header, instant readout of the injection parameters and most significantly, the relationship between grout flow and pressure. Second, they provide permanent records in real time for further analysis and quality assurance. In addition, automated systems allow operators to easily include additional inputs such that the number of different parameters monitored can be easily extended. And of course, pressure can be imported from two or more locations.

36.4.1 Data Acquisition

Pressure is typically acquired by common pressure transducers, which are available in a variety of different grades and accuracies from many manufacturers. Compact transducers with 1/4-in. standard male pipe threads identical to those supplied on most dial gauges can be readily mounted in the same manner as a dial gauge. However, it is recommended that a dial gauge be included at each transducer location. This is easily done either by using multiple ports in the gauge saver or a standard pipe T fitting.

Magnetic flow meters are most often used in automated systems; however, as previously mentioned, they work only with fully saturated grouts. For the stiff mortarlike grouts used in some fill and all compaction grouting, ultrasonic pulse meters are required.

Of particular value when using suspension grouts are the automatic density meters widely available from industrial instrumentation suppliers. Grout density is directly linked to the liquid-to-solid ratio, which will typically be the ratio of water to cement. Continual observation will provide immediate notice of any density changes that have occurred, allowing for rapid remediation.

The acquired data can be provided in real time on either disk or chart recorders or entered into a computer-based data acquisition system for further processing. Computer acquisition is particularly useful as the data can be manipulated and displayed in a nearly endless variety of formats. It can further be stored on disk sent as e-mail, or output as printed hard copies. Regardless of the method of readout, it is essential that both pressure and flow rate be provided in real time, preferably as continuous stacked horizontal curves.

The requirements of grouting present significant challenges for system designers. The primary pressure transducer must be located at or in the header, and the data must be available there for the technician supervising the operation. However, this location is not a particularly friendly environment for the normally sensitive equipment. Additionally, the header is subject to continual disconnection for adding and/or breaking casing as the work progresses, exposing the instruments to potential impact. All equipment at the header is also subjected to frequent movement; exposure to rain, direct sun, dirt, and dust, including occasional grout spray; and less than gentle handling.

Flow meters are generally large, heavy, and quite fragile. They are thus best mounted in an area not subject to frequent movement which is usually near the pump. Their accuracy increases with velocity of the grout, however, so mounting should be in an area of minimal delivery line diameter. Some additional input devices such as pump stroke counters must be mounted at the pump while others can be anywhere in the delivery line.

36.4.2 Signal Transmission

A key element in such systems is the signal transmission from point to point. With input signals from numerous locations, hardwired systems are cumbersome, difficult to protect from damage, and not well suited to service in the real world of grouting. Historically, data telemetry has required Federal Communications Commission (FCC) licensing and has been expensive, power hungry, inflexible, and prone to interference and other sources of gross error. With the recent advent of spread-spectrum unlicensed digital data radio, however, short-range wireless has become convenient, reliable, and economical. Spread-spectrum digital radio operates on unlicensed ultrahigh frequency (UHF) bands, changes operating frequency on-the-fly to avoid interference, and uses error-checked data packets to rapidly and transparently detect and correct any errors.

As a result, even in crowded urban settings with many other users operating in the same band, the data typically get through without a glitch. Spread-spectrum radio is also easily run on batteries, because its throughput is very high compared to the normal data requirements of grouting, permitting power that is automatically controlled to be cycled off much of the time. Spread-spectrum data telemetry is digital in nature, so its use requires "smart" (i.e., microcontroller-based) devices at every node. These are typically data loggers with their own memory, timekeeping, and battery. This approach would have been prohibitively costly in the past, but recent advances have made the approach not only possible but also preferable.

36.4.3 Data Display

Data at the header display must be easily observed; however, the display monitor will be subject to the same abuses as the other header equipment. Additionally, it will often be in direct sunlight. Normal low-power light-emitting diode (LED) screens are not easily read under such conditions, so either they must be shaded or alternate high-power units must be employed. This is not easily provided as either very large batteries or connection to standard line power will be required. Both options further complicate an already difficult application.

The display unit must provide real-time indication of the main grouting parameters, ideally as a multichannel time chart as previously mentioned, to facilitate observation of short-term trends of all measured variables. It should also have the ability for text entry, so that hole-specific data such as number and stage can be input by the technician. This data entry function requires at the very least an abbreviated keypad suitable for site conditions as well as simplicity of operation. With the spread-spectrum digital radio system previously described, the display simply "eavesdrops" on the output function. The data can be similarly acquired in a field office or as many different site locations as desired.

36.4.4 Batteries

The sensors, telemetry, and data loggers described are all power efficient, so relatively small batteries are sufficient for their operation. For most sites, small rechargeable batteries that can be recharged overnight from utility power are the most convenient. Exceptional circumstances such as very low temperatures or inclusion of audible alarms may require larger or heavier primary batteries. The display unit is a special case; if a large display area visible in direct sunlight is required, again either large heavy batteries or connection to line power will be required.

36.4.5 System Availability

Most grouters do not have a good understanding of the equipment and software needed for meaningful computer monitoring. However, numerous monitoring systems as well as many competent consultants and vendors of equipment are readily available. Additionally, several suppliers to the grouting industry have developed packaged systems that are readily available.

- *Atlas Copco:* CFP Meter Units record pressure and flow rate on a strip chart recorder. LOGAC is a computer-based system that presents additional data in real time and monitors multiple delivery lines simultaneously.
- *Hany:* HIR Recording System provides pressure and flow on a strip chart. HFR is a computer-based system that stores the retrieved data on a memory card so it can be downloaded and processed in a computer.
- *Jean Lutz S.A.:* A French firm with U.S. representation, this company is a leader in monitoring systems for both drilling and grouting.
- *RST Instruments Ltd.:* A Canadian manufacturer of a full line of instrumentation for geotechnical applications. With in-house computer application expertise, their custom design and manufacturing of both computer monitoring and special instrumentation are second to none. All of their grout-monitoring systems utilize electronic pulse flow meters, so they can be used with stiff mortarlike grouts.
- *SolData Inc.:* An affiliate of the large French grouting firm Soletanche, SolData provides computer-based systems that include both hardware and the proprietary software to monitor surface improvements as well as grout behavior.

Developments and improvements in automated monitoring equipment are advancing rapidly, and the limited information provided here is likely to soon become obsolete. The commercial sources of equipment mentioned, however, have taken a leadership role in development and will certainly remain good sources of information in the future.

CHAPTER 37

Packers

37.1 TYPES OF PACKERS
37.1.1 Mechanical Packers
37.1.2 Pneumatic Packers
37.1.2.1 Fixed-End Packers
37.1.2.2 Sliding-End Packer
37.1.2.3 Pneumatic Packer Selection
37.1.3 Friction Sealing Packers
37.2 PACKER DEPLOYMENT
37.3 DOUBLE PACKERS

PACKERS ARE expandable plugs formed within an injection pipe that isolate chosen areas of a grout hole or casing. They might also be used to secure a header connection to a hole. They must be sufficiently small to enter and freely move in the hole and yet be capable of expanding to the greatest diameter to provide an effective seal. Most packers are single; that is, they block only one place at a time. Double packers are used when a particular length of hole is to be isolated from the remainder, usually for injection into that portion only. These are special tools of the grouter and were traditionally custom-made, often by the individual operator.

Packer use is prone to problems, and loss of effective seal and/or bypassing by the grout is a constant problem that can cause the packer to become permanently grouted into the hole. Packers have historically been so troublesome to use that many experienced grouters avoid them if possible, except when installed in or very near the hole collar. Although usually larger than those used for grouting, packers are also employed in the well-drilling and service industries. There are now several specialty manufacturers of full lines of packers, and well-engineered and well-built products are commonly available. Although the complications of use have not been completely eliminated, packers are now much more reliable.

37.1 TYPES OF PACKERS

Packers can be expanded mechanically, pneumatically, or seal by friction. Mechanical packers are of limited expansion ability and are thus typically used only at or near the surface in relatively stable, round holes. Alternatively, pneumatically expanded packers are commonly used at virtually any depth, and because they can expand differentially, they are able to seal irregular surfaces. Friction sealing packers are by far the most simple and thus are easy to use and economical. However, they are applicable for use only in smooth, regularly formed tubes such as sleeve port pipes (tube-a-manchette) or pipelines.

37.1.1 Mechanical Packers

Mechanical packers consist of a short elastomeric sleeve over a snug-fitting rigid tube. Mechanically compressing the sleeve longitudinally causes it to increase in diameter, sealing to the wall of the surrounding hole. The most common design involves external threading of the top portion of a rigid central tube such that a nut can be screwed down to compress the elastomeric sleeve. The amount of expansion depends upon the thickness and flexibility of the sleeve, which is typically some sort of hose. Expansion is limited, however, and is typically no more than about 1.2 times the original diameter. For durability, the outer layer should be reinforced; however, the reinforcement needs to be quite flexible or limited so as not to overly restrain good expansion when compressed. In this regard, hose made for dry-mix shotcrete is often used as it contains a thick gum rubber interior tube with an abrasion-resistant, reinforced covering.

The elastomeric sleeve is of course restrained at the base, and an outer steel sleeve is usually placed between the top of the elastomer and the turning nut, which is typically supplied with handles, as illustrated in Figure 37.1 (left). The sleeve becomes longer and increases in weight as the packer lengthens. It is in fact the weight that typically limits the depth of use of mechanical packers. As an example, one with a nominal 3/4-in. (19-mm) central tube will weigh on the order of 3 lb (1.4 kg) per foot. Although possible, providing coupling to extend the independent tubes is complicated, so mechanical packers are typically of a fixed length. Consequently, several different lengths must be available for deeper installations.

FIGURE 37.1 Mechanical packer on left and pneumatic model on right.

The deepest practical depth of use is about 20 ft (6 m). Because the length of the elastomer is usually short, the risk of grout running through formation defects to bypass the seal is substantial. Multiple expansive sections can be provided; however, this will usually increase problems of grout bypassing, and infiltration of the areas between seals becomes a significant risk. Due to these limitations, mechanical packers are commonly reserved for use in shallow holes in competent, continuous formations with hard walls, and uniform roundness.

37.1.2 Pneumatic Packers

Pneumatic packers feature an expandable sleeve that is inflated with sufficient pressure to swell and provide the necessary seal and fixity, as illustrated in Figure 37.1 (right). The inflation medium can be water, air, or gas, nitrogen gas being most common. Because the amount of expansion along the length of the sleeve can be variable, these are far more effective in rough or irregular holes. Further, there is no real restriction to the length of the expanded sleeve, so longer stretches of hole can be sealed off simultaneously, lessening the chance of grout bypass. Although all pneumatic packers are expanded by inflation, there are different types of construction and resulting capability. Also, two or more packer elements can be spaced along a single tube to isolate multiple zones at a time.

37.1.2.1 FIXED-END PACKERS

The simplest pneumatic packer is a *fixed-end* or *balloon packer*. It is formed of a length of flexible hose that is clamped on each end over a rigid central tube. Until recently, this was the only type of pneumatic packer. It was usually made by the user, and the effectiveness depended upon the flexibility and internal expansion potential of the hose used. Traditionally, the primary choice of hose was unreinforced rubber tubing or a fully reinforced hose. Whereas unreinforced tubes were capable of good expansion, they were not very durable, often extruding into the hole or splitting at the end clamp. On the other hand, reinforced hose allowed only restricted expansion.

Most commercial manufacturers of fixed-end packers now use special hose that is especially reinforced to allow maximum expansion as well as strength. Whereas to limit internal expansion the reinforcement of most common hose is pitched at an angle of about 50 degrees to the longitudinal axis, that for packers is much flatter, on the order of 20 degrees, which allows large expansion. Depending upon the particular requirements, the reinforcement might be of steel or a variety of textiles, including nylon, polyester, and Kevlar. Depending upon the pressure potential, from one to several layers of reinforcement might be used. Efficiency and longevity of a packer are dependent upon construction of the flexible element, and this is where commercial manufacturers are able to provide superior products.

Because of their simplicity and lower cost, fixed-end packers are the best choice for some applications. As the diameter of the hose is increased, longitudinal stresses develop, resulting in shortening. In fixed-end packers, this is restrained by the end clamps, restricting the amount of attainable diameter increase. Typical expandability of fixed-end packers is on the order of 1.3 times their initial diameter, with 1.5 being about the greatest that can be achieved. They are therefore best used in relatively uniform holes of generally constant cross section.

37.1.2.2 SLIDING-END PACKER

Only one end of these packers is fixed; the other is allowed to move freely as the rubber sleeve tends to shorten upon inflation, allowing expansion to greater diameters. An O-ring seal is provided on the free end, and special rubber element construction is incorporated, as above discussed, allowing expansion up to about 1.9 times the original diameter. With moving parts, these packers are more expensive to purchase and more prone to operating problems. Satisfactory operation requires the sliding end to continually move freely and the internal seal must remain tight. Higher quality models provide wipers to clean the central tube of grout as the expanded element retracts, lessening abrasive wear of both the O-rings and the central tube.

37.1.2.3 PNEUMATIC PACKER SELECTION

Many things must be considered when selecting a pneumatic packer. Fundamental are hole size and the *differential pressure* to be restrained, which is usually that of the grout. Both the amount of element expansion and its contact length in the hole are important. The ability to restrain pressure is greatest with minimal expansion of the element and becomes less as the inflated diameter increases. It is thus beneficial to use the largest diameter packer that will readily fit in the hole. Frictional resistance is directly related to the contact area, so the restrained pressure capability of the packer increases with increased length of the flexible element. The always present risk of grout escaping through formation defects and bypassing the packer is also reduced with increasing element length. There are thus significant benefits for longer element lengths, but these come at a higher initial cost and longer elements require more time to inflate. Further, longer lengths are more likely to become wedged in irregular holes, and the free end will be required to travel farther, imparting greater wear.

Where the grout holes are of generally uniform size in a reasonably competent formation, minimal annular space is required, so the packer need not be much smaller than the hole diameter. The packer will thus be capable of accepting a higher inflation pressure and provide maximum fixity. If, on the other hand, the holes are likely to contain sloughs or other areas that are significantly larger than the drilled section, the element will be able to expand to its maximum, at which point its ability to withstand grout pressure is the least. Where this occurs, the *maximum unrestrained pressure* capability of the packer element becomes crucial. This is the highest inflation pressure that can be withstood when the element is completely unrestrained.

Packers with unrestrained pressure capability of 500 psi (34 bars) are readily available and, although not required for all work, are a good standard with which to operate. One major manufacturer provides this rating with a safety factor of 2, so the actual burst strength of the packer is 1000 psi (69 bars). Packers up to 4.5 in. (114 mm) uninflated diameter and 40 in. (1 m) uninflated length are rated to withstand differential pressures of up to 1500 psi (103 bars). Of importance to grouters is the internal diameter of the central tube through which the grout will be pumped. In the small-diameter holes used in grouting, this can be only 1/2 in. (13 mm), which is bound to prove troublesome in many situations. The minimum acceptable hole diameter for most grouting is 3/4 in. (19 mm), which is available but might require a special order.

37.1.3 Friction Sealing Packers

Two types of friction sealing packers are available for use in hard, smooth-walled tubing or pipes. These are based on either O-ring or U-cup seals. Both are used extensively for high-pressure hydraulic applications and can resist pressures up to several thousand psi (hundred bars); however, both wear-producing frequent cycling and constant lubrication are common in such usage. Cycling is limited in grouting, but often rather than good lubrication, exposure to the often abrasive grout occurs. Friction sealing packers are thus subject to wear and the seals must be periodically replaced. This is neither costly nor difficult to accomplish in the field, however.

O-rings must be compressed tightly in the tube, so they require considerable force for packer placement and moving. While slightly less costly than U-cups, greater wear will occur due to their constant tightness, and more effort is required for placement and extraction. Thus U-cup seals, which need only be snug during placement, are preferred. They are positioned with the open end to the grout exposure and are hydraulically expanded by the pressure of the grout. When not under pressure, only the outward flared lip is in contact with the casing. As pressure increases, the lip is pushed outward such that it forms a tight seal all the way to the cup base. At this point, sealing is essentially in the central portion of the lip.

37.2 PACKER DEPLOYMENT

Packers are often supported by the grout line, which might be of either rigid pipe or flexible tubing. Except in the case of very deep holes, rigid pipe will usually have sufficient strength to carry the packer safely. This of course depends upon the exact nature of the pipe or tubing used.

TABLE 37.1 Tensile Strength of Schedule 40 and 80 Pipe Joints

	STATIC LOAD			
	SCHEDULE 40		SCHEDULE 80	
Nominal Pipe Size	lb	kg	lb	kg
3/8	540	245	920	418
1/2	740	336	1250	568
3/4	1050	477	1800	818
1	1500	681	2600	1182
1-1/4	2200	1000	3800	1727
1-1/2	2700	1227	4700	2136
2	3900	1773	7000	3182

Couplings are the weakest link of standard pipe when in tension. Approximate tensile capacities of the joints of Schedule 40 and 80 pipe are provided in Table 37.1. Rigid lines are advantageous in that any upward slippage of the packer is easily recognizable during grout injection. Also, down pressure can be exerted on the line to assist in further restraint if needed. Rigid lines are heavier and must be placed in sections that require more time and effort. This is labor intensive and, should long joints be used, will require a drill rig or other lifting equipment.

Flexible lines for inflatable packers might be either standard hose or, more typically, continuous ABS or other plastic tubing, which is much lighter and easier to handle. It can be continuous without joints and be easily handled with take-up reels on the surface. Such tubing and especially its fittings often lack the tensile strength of steel and may not be capable of withstanding the total weight of the packer, especially in deep holes or anywhere it tends to become wedged or stuck. Separate wire rope cables should thus be used where the support capability of the grout delivery tube is in question.

The inflating medium can be water, air, or gas and is typically transmitted to the packer through a separate small tube. Water has the advantage of being incompressible, so it will take on and withstand the grout pressure should it be higher than the inflation pressure. It can be easily pressurized with small hydrowashers, which are widely available, or intensifier pumps, which typically operate on compressed air. Intensifier pumps are available that will develop up to about 10,000 psi (690 bars) water pressure with an input air pressure of only 100 psi (6.9 bars). On the downside, however, the static head of water must be added to the surface gauge pressure to indicate actual downhole packer inflation pressure. This will increase the time required for deflation and may be excessive to allow complete deflation in deep holes.

Compressed air or gas is perhaps the cleanest and easiest to use as it vents rapidly. Air can be conveniently compressed and used for inflation pressures up to about 150 psi (10 bars). Where greater pressures are required, bottled nitrogen gas is the safest and easiest to use. An appropriate pressure-regulating gauge is essential to prevent excessive pressurization. These are readily available from most gas suppliers. Additionally, a valve must be provided between the packer and gauge to vent for deflation, as required.

Regardless of the inflation medium, a means of transporting it to the packer is required. This is typically through 3/16 or 1/4-in. (4.8- or 6-mm) flexible tubing. This is available in several pressure ratings, but nylon is usually the strongest and is readily available to withstand pressures up to about 600 psi (41 bars). Such small tubing is typically quite flexible, so it must be supported and/or withdrawn in perfect unison with the grout line to prevent kinking and possible damage. It is commonly taped tightly to the injection line. A good practice is to place the tape at regular intervals of 1 ft (0.3 m), such that the depth can be easily observed. Movable end inflatable packers should always be placed with their movable end downward.

37.3 DOUBLE PACKERS

Double packers are used where a short segment of conduit or grout hole is to be sealed off from the remainder. The most frequently encountered instances are injection through sleeve port pipes (tube-a-manchette) or into the joints of a sewer or similar conduit. For sleeve port pipes, the sealing elements are typically located about 1 ft (0.3 m) apart on a common central tube. They can be either friction sealing, of which the double cup type (Figure 37.2) is the most frequently used, or hydraulically expanded. A separate fill line that is typically 1/8 or 3/16 in. (3 or 4.8 mm) diameter plastic tubing must be supplied with

FIGURE 37.2 Twin-cup friction-type double packer for sleeve port pipes.

the hydraulic type packer and is usually attached firmly to the injection line. Packers for sleeve port pipe grouting must be accurately set over the desired ports for injection. This can be facilitated by marking the injection tube at the same spacing as the ports. The depth should be indicated on each marked interval.

Double packers used for grouting pipe joints are virtually always of the pneumatic expanding type. They typically operate in concert with a TV camera that can be attached or separately coupled. Pipe grouting packers, especially in the smaller sizes, tend to be sophisticated assemblies, as was previously discussed in Section 15.3 and illustrated in Figure 15.3. They must be either pushed or pulled through the conduit and typically are attached to a support vehicle by a bundle of many hose and other lines. Whereas they are usually supplied with a central tubular opening to allow fluid passage, excessive flows can back up, requiring restraint, which is provided by either the support line bundle or a separate restraint cable. Therefore, they are usually deployed from the upstream access point. These packers are commonly available in diameters from 4 in. (101 mm) up to 36 in. (0.9 m).

CHAPTER 38

Some Closing Thoughts

WHEN I FELL BY CHANCE INTO the grouting industry in 1952, it was virtually undeveloped and almost completely void of any real technology. Grouting of rock was generally limited to dam foundations and tunneling. Drilling, which was typically done by diamond coring, was the main profit center of the fairly small group of active contractors, grout injection being largely ignored. Grouting in soils was not often used in those earlier times, although many attempts had been made to improve soils with the same cementitious suspensions used in rock.

Some chemical-solution grouts had also been developed that would harden after injection into the ground, both increasing the soil strength and reducing its permeability. Technology for these was neither well developed nor understood, and such work was generally done on a hit-or-miss basis. Injection into structures was nearly unheard of with the exception of limited efforts to reintegrate massive masonry sections through injection of fluid cementitious suspensions. So far as grout materials, a rule of thumb was well established: "If you can't pour it, you can't pump it."

Very little equipment built especially for grouting was available anywhere and virtually none in the United States. Grout was most often injected with piston steam pumps, even though they did not perform very well and were subject to frequent downtime. Plugging of the poppet type valves, which were not designed for either thick or abrasive fluids, was a constant problem. Interestingly, the Bureau of Reclamation Technical Memorandum 646 on Pressure Grouting released in 1957 contains a figure illustrating tools for cleaning and maintaining these pumps "when valve trouble occurs". Shown are 14 different pinch bars and custom-made chisellike devices for cleaning. On the other end of the scale, the Mudjack machine, which had been developed some 30 years earlier, was large and awkward and capable of neither much pressure nor control of the mix.

The grout most commonly used might best be described as dirty water, as it was composed of one bag of cement to which from 8 to 80 or more gallons of water was added. Significantly, the 1957 646 document, which was used as a guideline for grouting of most dam foundations in the western United States, says "For tight seams or cracks in the rock, (water cement) ratios of 20:1 (by volume) have been used." Grout with a 20:1 water-to-cement ratio by volume contains 161 gal of water to each bag of cement! Such a mixture results in water that is hardly dirty, and such mixtures would be unheard of with current technology.

Fortunately, I was ignorant of the state of practice and the many misconceptions of the time relative to grouting. But, being familiar with masonry and then available concrete technology, my thinking was, not of pumping fluids, but of pumping mortarlike mixes. Fortunately, I was aware and deeply interested in the then-available admixtures to alter and improve the properties of concrete. And the time was right. Concrete or mortar had not yet been pumped, but some very smart people were working toward that end. I had the good fortune to become acquainted and work with pioneers and leaders in that industry.

I found grouting a fascinating field with great potential that was neither well defined nor developed and for which the limited available equipment was archaic and inappropriate. Certainly this seemed an intriguing area for development, and fortunately, I had a small general contracting company to provide support. As a child, I was fascinated with construction and spent much time around projects in the neighborhood. This led to my becoming acquainted with Max Kemper, an underground and tunneling contractor who built much special equipment in his shop and developed the Flocrete system using pressurized containers to transport and deposit concrete in tunnels. Kemper stated, "the unusual projects were the most rewarding and the ability to build special equipment was required for good performance."

So, with a strong interest in the material aspects and emphasis on developing good equipment, grouting became an important part of my life. My first attraction was the need to inject a suitable material to remediate a void under a small structure. The concrete structure was founded on a sand filter layer within a flood control channel that was under construction. An early rain caused erosion of some filter material, resulting in a void that had to be filled. Unable to find established technology or equipment to fill the void, I constructed a "pump" that consisted of a length of 6-in.- (152-mm-) diameter steel casing about 5 ft (1.5 m) long and positioned vertically in a wooden frame. It had a "piston" consisting of a length of 2 × 4 lumber with a wood disk slightly smaller than the inside diameter of the casing on the end. A few layers of old carpet cut to fit snugly in the casing formed a seal.

The casing was filled with a cement-sand grout similar in consistency to a mason's mortar, the piston was positioned in the top, and pressure was exerted by the weight of two men (boys) hanging onto a short piece of 2 × 4 placed across the top of the 2 × 4 piston rod. It was a memorable lesson in equilibrium, but with persistence, it worked! The void was filled, and I began to develop a suitable powered pump. Fortunately, I was ignorant of the well-established rule of the time, "If you can't pour it, you can't pump it," and with only knowledge of concrete and mortar, my efforts were directed toward pumping those materials. Along the way I came in contact with others who had a similar goal, namely pumping portland cement plaster to elevated scaffolds. Success finally came in 1954 (a working pump is illustrated in Figure 38.1). The major contributor in its development was Marvin Bennett, who along with his brother Richard went on to develop the first plaster pump in 1954 followed by the first small line concrete pump in 1961.

I went my own way, grouting isolated voids but broadening the operation to include jacking of both slabs and structures as well as filling a variety of voids. Along the way, I read anything I could find on grouting. My small general

FIGURE 38.1 The first pump was able to handle mortarlike mixes.

contracting firm grew, but grouting became an ever greater part of our work until 1961, when all operations other than grouting and structural repair were discontinued. By this time we had developed into a full-service specialty contracting firm with experience in all types of grout and grouting. With our origin tied to the use of plastic consistency grouts, much effort was directed to development of structural grouting, and in 1957 we performed perhaps the first application of epoxy injection into cracked concrete.

Examining the state of grouting at the time, I was dismayed by common practice and the complete lack of technology. Pumping *dirty water* in rock defects seemed especially absurd. By simply observing glass jars filled with typical grouts over a period of time, it became clear that most of the volume was being lost as a result of sedimentation and resulting bleed. By simply using a massive dosage of water-reducing admixture, I was able to cut the water required for a similar consistency with dramatic improvement in the grout. Although not understood at the time, we employed viscosity-modifying admixtures not much different than those now becoming well established in the industry.

Similarly, in the area of chemical-solution grouts, there was much speculation but virtually no documented data to support claims of the promoters, especially in relation to the strength of the resulting chemically grouted masses. This led to a major research effort that included more than 2500 laboratory-prepared specimens. Our efforts to understand available material science and improved grout mixes led to ever-expanding opportunities in structural grouting. Because virtually no specialized grouting equipment was commercially available, we went into the manufacturing business as well.

Further development of the mortarlike grouts opened up many new opportunities. These could be proportioned to develop high strength and good stability with little bleed or shrinkage. Where large volumes were to be used, they could be batched and delivered in ready-mix trucks. Because these grouts tended to stay near their point of deposition, groutjacking to very close tolerances became possible. One of the early milestones for use of these grouts and the future of our firm occurred early in 1957. A large concrete box culvert had been undermined by flooding during construction resulting in settlement. When told there was a good chance that the culvert could be jacked back to its original position, engineers from the large public agency that owned it all but scoffed. Certainly, this was not possible, and we were told to just fill any voids!

I requested a survey crew be onsite during grouting just in case jacking might occur but was told bluntly that it could not happen and such monitoring was not justified. While filling the voids, as shown in Figure 38.2, the inspector noticed a separation of form board marks on the culvert with those of an adjacent wall. The culvert was rising! He immediately called his superiors, and in short order a survey party arrived as well as a number of officials, including the manager of inspection and the chief engineer. In spite of the combined weight of all these officials bearing on the invert, the culvert was jacked to its original position. The officials proclaimed both surprise and satisfaction and thereafter directed much work to our firm. This led to many

FIGURE 38.2 A culvert raised to the proper grade.

significant projects, including our first major dam curtain in 1962. And, yes, we included admixtures to reduce bleed, much to the delight of the owner.

In June 1957, the swimming pool in a new apartment complex underwent settlement. The large building completely encompassed the central courtyard in which the pool was founded, precluding access for large equipment. The site had previously been home to a hospital and the area of the pool had been a basement boiler room. Plans called for the bottom slab to be perforated and the entire basement to be filled with *granular* material. However, the slab had apparently not been perforated and the fill was not granular but rather somewhat cohesive clayey silt. The deep end of the pool was situated at about the old floor elevation, and the shallow end was founded on several feet of fill. It was here that the settlement was greatest. Additionally, cracking of the pool shell had allowed pool water to saturate the fill that was contained in the old structure.

For groutjacking, holes were placed on a grid of about 4 ft throughout the bottom of the pool (Figure 38.3). The geotechnical engineer had requested that grout holes in the shallow end be extended to the old floor slab. The engineer recognized that fine-grained soils could not be grouted, but he reasoned a number of small holes filled with grout might provide needed support and prevent further settlement. For us the job was a disaster. As soon as injection started, water ran from the lower grout holes into the pool. The water was removed with buckets but continued to inflow during injection. The job required an amount of grout equal to more than ten times the calculated volume before lift occurred. A horrible mess was made by removal of the muddy water that had been spilled from the pool through the fancy courtyard to an exit, and nearby catch basin for disposal.

FIGURE 38.3 A swimming pool is filled with muddy water that was squeezed out of fill soil during grouting.

As the mess grew uglier, the owners and many of their representatives arrived. Everyone involved was obviously upset and not happy, with one exception, the geotechnical engineer, who exclaimed, "This is wonderful . . . you have squeezed water out of the fill . . . this is not supposed to be possible to do . . . it's wonderful!" And what is now commonly known as *compaction grouting* (we called it *displacement grouting* at the time) was born! The resulting mentoring of that engineer also led to my further study of soil mechanics and foundation engineering (now called geotechnical engineering) and further development of soil grouting, including the large research program on compaction grouting discussed in Section 2.2.1.

These were the beginnings of my grouting career, with contracting continuing for the next 27 years. Throughout its existence, my firm performed continuing research and development, resulting in leadership of many present grouting technologies, including development of compaction grouting (1957) and epoxy crack injection (1958) and extensive evaluation of chemical-solution grouts for soil strengthening (1960s). Other technologies to repair and rehabilitate distressed foundations and structures were developed, including the ability to precisely groutjack settled structures and pavements to proper grade.

Because of the limited availability of documented technology and frequent requests for information from design engineers, a separate group was set up to perform evaluation, testing, and consultation for other engineers within the specialty area of structural and foundation restoration with a strong emphasis on grouting. Through that organization, many large-scale tests were conducted, both to evaluate the existing capacity of structures and to verify adequate improvement.

In 1979, I sold my contracting operations and, following a brief retirement, began consulting services to other design and construction professionals on significant projects around the world. Included has been service as an expert witness in a wide variety of litigation matters, which has provided exposure to the most intimate view of litigants' operations and an unusual understanding of "the way things are done in the real world" of grouting and remedial construction.

Historically, grouters have often hidden behind a veil of mysticism, and this continues to a significant degree to this day. In fact, it is now common to find large "geotechnical contractors" who actively promote their special "proprietary" knowledge in a manner not unlike the past. Only now, the diatride might best be described as "*high-tech* black magic" and often comes from the mouths of engineers who should know better. The fact is, *grouting is a* science, and there is very little special knowledge or ability that is not well established and available. As a registered professional engineer, I believe grouting should always be approached in a scientific manner and in conformance with sound engineering practice.

Grouting has been very good to me. I have succeeded, not through brilliance, but rather through the default of earlier practitioners who failed to recognize the technology and/or put forth the effort to develop it on a sound scientific basis. I can think of nothing as exhilarating as correcting serious situations and saving failing structures. Present grouting technology is used daily to further mankind and it has been fulfilling to be a part of the industry. Hopefully, the experiences provided in this handbook will assist newcomers as well as the experienced to build upon past experience and avoid the errors that brought us to where we are.

Being much more comfortable "onsite solving problems" than in the office writing, compilation of this volume has required an extraordinary effort. Discipline to complete the work has prevailed only because of my disdain of the confusion still created by some special interests, a fervent desire to see grouting recognized for the technology it is, and the fact that no other guide is available.

Unfortunately, not all experiences with grouting have proven positive. There have been many instances of poor performance and in some cases damage to the structures or formations that were to be improved. However, this has virtually always been attributed to violation of well-established basic requirements for proper performance. Some shortcomings are repeatedly experienced, especially injection with little or no advance exploration or engineering knowledge of the site or defects to be corrected. Use of inappropriate grout mixtures, excessively rapid injection rates, and insufficient monitoring and control also contribute greatly. Poor performance is damaging, not only to the affected owner, but also to the reputation of grouting and to all grouting professionals.

It is most unfortunate to see grouters who hide behind the veil of false mysticism, claiming to possess *proprietary* knowledge such that only they are qualified to perform. Worst of all are the large firms that through clever advertising and other marketing efforts lead unknowledgeable professionals to believe they are the very best and represent the state of practice when in reality they are far behind and sometimes incompetent. When clients receive slipshod work from firms they *perceive* are the best, their perception of grouting is bound to suffer. And depending upon how badly they have been burned, hesitancy for further use is bound to occur.

Rather than dwell on the bad actors in our midst, let's rejoice in the many excellent operators who bring us much credit. Although many do little or no advertising and are not well known, they contribute much, by both advancing the technology and performing work that brings us credit. I have had the privilege to know many. They seem to be perpetually busy, always striving for excellence, and enjoy a large amount of repeat business from well-satisfied clients. The strongest form of promotion is excellent work fairly performed for quality clients. As it is said, invent a better mousetrap and the world will beat a path to your door; no advertising is needed: Which begs the question, if a firm really is a leader and the best, why does it need to spend big bucks for clever advertisements in multiple publications to keep reminding us?

Engineers and owners are not as likely to use our services if they don't understand what we are doing as well as the mechanics involved. As we are all aware, there is very little special knowledge in the grouting field that is not commonly available. If grouting is to prosper, we must be honest in our claims, strive for excellence, cooperate with each other, and freely share our technology, especially with designers and engineers.

Of crucial importance, we must provide the necessary continuous real-time monitoring of injection rates, pressures, and other grouting parameters that allow us to better understand the mechanics of our work as well as ensure quality performance. In this regard it is my fervent hope to one day see real-time computer monitoring used on every important grouting project, especially in the United States, which has been very slow in adopting these techniques.

Above all, as grouting professionals, we must apply our knowledge to ensure that all grouting work is performed in a rational, technically correct manner and in conformance with sound engineering principles. Grouting is not magic or a black art but rather is a well-established science, and those of us that are true professionals must abide by the highest standards. We must not hesitate to call the bluff of those among us that choose to hide behind a veil of magic, for only by sharing the fundamentals of the technology with other professionals and following through with the highest quality performance will the role of the grouter shine. Let us all shine and excel in our progress!

Bibliography and References

Acker, W. L., III. 1974. *Basic Procedures for Soil Sampling and Core Drilling.* Scranton PA: Acker Drill Co. Inc.

Al-Alusi, H. R. 1997. "Compaction Grouting: From Practice to Theory." *Grouting: Compaction, Remediation and Testing: Proceedings of GeoLogan Geo-Institute Conference.* ASCE Geotechnical Special Publication No. 66. Reston, VA: ASCE. pp. 43–53.

Al-Manaseer, A., and L. Keil. 1990. "Physical Properties of Cement Grout Containing Silica Fume and Superplasticizer." Proceedings, 1990 CSCE Conference, Hamilton, Ontario, May 15–18, 1990. Montreal: Canadian Society of Civil Engineers.

American Concrete Institute. 1999. ACI 229R-99. *Controlled Low Strength Materials (CLSM),* Farmington Hills, MI: American Concrete Institute.

American Concrete Institute. 1990. ACI 116R-90, Cement and Concrete Terminology. Farmington Hills, MI: American Concrete Institute.

American Concrete Institute. 1989. *ACI 503R, Use of Epoxy Compounds with Concrete.* Farmington Hills, MI: American Concrete Institute.

American Petroleum Institute. 1994. *API RP-43, Standard Procedure for Evaluation of Well Perforators.* 5th ed. Washington, DC: American Petroleum Institute.

American Petroleum Institute. 1974. *Specifications for Oil Well Cements and Cement Additives, Standard 10A.* Washington, D.C.: American Petroleum Institute.

American Petroleum Institute. 1967. *API RP-67, Recommended Practices for Oilfield Explosives Safety.* Washington, DC: American Petroleum Institute.

American Society for Testing and Materials, Committee C-9. 1978. *Significance of Tests and Properties of Concrete and Concrete-Making Materials.* ASTM Special Technical Publication 169B. West Conshohocken, PA: American Society for Testing and Materials.

American Society for Testing and Materials. 1990. *ASTM C 881 Standard Specification for Epoxy-Resin-Base Bonding Systems for Concrete,* West Conshohocken, PA: American Society for Testing and Materials.

American Society for Testing and Materials. 1993. *ASTM D 4219, Standard Test Method for Unconfined Compressive Strength Index of Chemically Grouted Soils.* Designation D 4219–93a. West Conshohocken, PA: American Society for Testing and Materials.

Association Francaise des Travaux en Souterrain (AFTES). 1991. "Recommendations on Grouting for Underground Works." *Tunneling and Underground Space Technology* 6(4):383–461.

Ata, A. and Vipulanandan, C. 1999. "Factors Affecting Mechanical and Creep Properties of Silicate-Grouted Sands." *Journal of Geotechnical and Geoenvironmental Engineering,* ASCE 125(10) (October 1999):868–876.

Baecher, G., and H. Einstein. 1975. "Statistical Description of Rock Properties and Sampling," In *Proceedings, 18th U.S. National Symposium on Rock Mechanics,* pp. 1–7.

Baker, W., E. Cording, and H. MacPherson. 1983. "Compaction Grouting to Control Ground Movements During Tunneling." *Underground Space.* Vol. 7. Oxford, U.K.: Pergamon Press, pp. 205–212.

Bandimere, Samson W. 1997. "Compaction Grouting: State of Practice 1997." *Grouting: Compaction, Remediation and Testing. Proceedings of GeoLogan Geo-Institute Conference,* ASCE Geotechnical Special Publication No. 66. Reston, VA: ASCE. pp. 18–31.

Banfill, P. 1990. *Rheology of Fresh Cement and Concrete.* London: E.&F.N. Spon.

Been, K. and M. G. Jefferies. 1985. A State Parameter for Sands. *Geotechnique* 35:99–112. (Reply to discussion, *Geotechnique* 36:123–132.)

Been, K. and M. Jefferies. 1986. Reply to Discussion of "A State Parameter for Sand," *Geotechnique* 36:127–132.

Been, K., M. Jefferies, J. Crooks, and L. Rothenburg. 1987. "The Cone Penetration Test in Sands, Part 2: General Inference of State," *Geotechnique* 37:285–289.

Bentz, D. P., E. J. Garboczi, C. L. Haecker, and O. M. Jensen. 1999. "Effect of Cement Particle Size Distribution on Performance Properties of Portland Cement-Based Materials." *Cement and Concrete Research* 29. Elsevier, Amsterdam: Pergamon Press, pp. 1663–1671.

Borden, R. H. and D. M. Groome. 1984. "Influence of Bentonite Content on the Pumpability of Low-Slump Grouts." *Innovative Cement Grouting.* SP-83. Detroit: American Concrete Institute, pp. 115–128.

Boynton, R., and J. Blacklock. 1986. *Lime Slurry Pressure Injection Bulletin 331.* Arlington, VA: National Lime Association.

Brown, D., and J. Warner. 1973. "Compaction Grouting." *Journal of the Soil Mechanics and Foundations Division, ASCE,* 99(SM8): 589–601.

Bruce, Donald A. 1994. "Permeation Grouting." Chap. 2 in *Ground Improvement and Control,* edited by Xanthakos, P., Abramson, L., and Bruce, D. New York: John Wiley & Sons, Inc. pp. 536–537.

Burmister, D. M. 1965. *Influence Diagrams for Stresses and Displacements in a Two Layer Pavement System for Airfields.* Contact NBY 13009, Dept. of Navy, Washington, DC.

Bussey, W. H. 1973. "Suspension Grouts and Their Injection" *Bulletin Asso. of Engineering Geologists* 10(4):313–323.

Byle, Michael J. 2003. "Design Considerations for Inclusions by Limited Mobility Displacement Grouting." In *Grouting and Ground Treatment: Proceedings of the Third International Conference,* edited by Johnsen, L., Bruce, D., and Byle, M. Vol. 2. Geotechnical Special Publication No. 120. Reston VA: ASCE Press, pp. 1071–1081.

Byle, Michael J. 2000. "Geotechnical Investigations for Grouting in Soil." *Performance Confirmation of Constructed Geotechnical Facilities: Proceedings of Geo-Institute Specialty Conference on Performance of Confirmation of Constructed Geotechnical Facilities.* Reston, VA: ASCE Press. pp. 427–440.

Byle, Michael J. 1997. "Limited Mobility Displacement Grouting: When 'Compaction Grout' Is Not Compaction Grout." *Proceedings of the Geo-Logan Conference.* Special Geotechnical Publication No. 66. New York: American Society of Civil Engineers. pp. 32–42

Byle, Michael J. 1992. "Limited Compaction Grouting for Retaining Wall Repairs." *Grouting, Soil Improvement and Geosynthetics: Proceedings of the Conference* ASCE Geotechnical Special Publication No. 30. Reston, VA: ASCE. pp. 288–300.

Byle, M. and Borden, R. 1995. *Verification of Geotechnical Grouting.* SP No. 57. New York: American Society of Civil Engineers.

Cambefort, H. 1977. "The Principles and Application of Grouting." *Quarterly Journal of Engineering Geology* 10(2):57–95.

Carter, P., and S. Gurjar. 1987. *Alberta Concrete Patch Evaluation Program.* Edmonton, Canada: Alberta Transportation and Utilities.

Cedergren, H. 1989. *Seepage, Drainage, and Flow Nets.* 3d ed. New York: John Wiley & Sons, Inc.

Clough, G. Wayne, William M. Kuck, and Gyimah Kasali. 1979. "Silicate-Stabilized Sands." *Journal of the Geotechnical Engineering Division, GT1* (ASCE, New York) (January):65–82.

Committee on Grouting. 1980. "Preliminary Glossary of Terms Relating to Grouting." *Journal of the Geotechnical Engineering Division,* ASCE. 106(GT7): 803–815.

Committee on Grouting. 1968. "Guide Specifications for Chemical Grouts." *Journal of the Soil Mechanics and Foundations Division, ASCE* 106(March): 345–352.

Committee on Grouting of the Geotechnical Engineering Division, ASCE. 1977. "Slabjacking—State of the Art." *Journal of the Geotechnical Engineering Division* 103(GT9):803–815.

Committee on Grouting Specifications. 2001. *Guide Specification for Grouting of Post-Tensioned Structures.* Phoenix, AZ: Post-tensioning Institute.

Dershowitz, W. 1995. "Interpretation and Synthesis of Discrete Fracture Orientation, Size, Shape, Spatial Structure and Hydrologic Data by Forward Modeling in Fractured and Jointed Rock Masses." In *Proceedings of the Conference on Fractured and Jointed*

Rock Masses, edited by L. R. Myer and C. F. Tsang. Rotterdam: A.A. Balkema, pp. 579–586.

Dershowitz, W. S., D. A. Shuttle, N. O. Outters, and E. Glynn. 2002. "Differential Settlement of Foundations on Fractured Rock." Proceedings of the Fifth North American Rock Mechanics Symposium, Toronto, July 7–10, 2002. Toronto: University of Toronto Press.

Dershowitz, W. S., and H. H. Einstein. 1988. "Characterizing Rock Joint Geometry with Joint System Models." *Rock Mechanics and Rock Engineering,* 21:21–51.

Diamond Core Drill Manufacturers Association. 1991. *DCDMA Technical Manual.* Dublin, Ohio: National Drilling Association.

"Discussion." 1969. *Journal of the Soil Mechanics and Foundations Division, ASCE* 95(SM2)(March) :625–655.

Donovan, N., A. Becker, and G. Lau, G. 1984. "Liquefaction Mitigation by Site Improvement—A Case Study." *Proceedings, Eighth World Conference on Earthquake Engineering.* Vol. 1. San Francisco, Prentice-Hall.

Essler, R. D., E. R. Drooff, and E. Falk. 2000. "Compensation Grouting, Concept, Theory and Practice." *Proceedings, Advances in Grouting and Ground Modification.* ASCE Geo-Institute Special Publication No. 104, Denver, CO: Reston, VA: ASCE.

Ewert, Friedrich-Karl. 2003. "Discussion of Rock Type Related Criteria for Curtain Grouting." Post Conference Discussion Volume. *Proceedings, Grouting and Ground Improvement, ASCE.* Reston, VA: ASCE.

Ewert, Friedrich-Karl. 1985. *Rock Grouting with Emphasis on Dam Sites.* Berlin: Springer-Verlag.

Fluid Power Society Lightning Reference Handbook. 2001. 8th ed. Tulsa, OK: Berendsen Fluid Power, Inc.

Garner, S., J. Warner, M. Jefferies, and N. Morrison. 2000. "A Controlled Approach to Deep Compaction Grouting at WAC Bennett Dam." Proceedings, 53rd Canadian Geotechnical Conference, Montreal, Quebec, Oct. 15–18, 2000.

Graf, E. 1969. "Compaction Grouting Technique" *Journal of Soil Mechanics and Foundations Division, ASCE,* Vol. 95, No. SM5, Proc. Paper 6766, Sept 1969 pp. 1151–1158.

Graf, E. 1992. "Compaction Grout, 1992." *Grouting Soil Improvement and Geosynthetics.* Geotechnical Special Publication No. 30. Reston, VA: American Society of Civil Engineers.

Graf, T., G. Clough, and J. Warner. 1982. "Long Term Aging Effects on Chemically Stabilized Soil." Proceedings, Conference on Grouting in Geotechnical Engineering, New Orleans, February 1982, American Society of Civil Engineers.

Gustafson, Gunnar, and Åsa Fransson. 2000. "The Use of Transmissivity Data from Probe Holes for Predicting Tunnel Grouting: Analyses of Data from the Access Tunnel to the Äspö Hard Rock Laboratory." *Tunnelling and Underground Space Technology.* 15(4):365–368.

Hakami, E. 1995. "Aperture Distribution of Rock Fractures" (Ph.D. thesis, Division of Engineering Geology, Department of Civil and Environmental Engineering, Royal Institute of Technology, Stockholm, Sweden.

Hakansson, U., Hassler, L., and Stille, H. 1992. "Properties of Microfine Cement Grouts with Additives." *Grouting, Soil Improvement and Geosynthetics.* Geotechnical Special Publication No. 30. New York: American Society of Civil Engineers, pp. 551–563.

Henn, R. W. 2003. *AUA Guidelines for Backfilling and Contact Grouting of Tunnels and Shafts.* Reston, VA: ASCE Press.

Henn, R. W. 1996. *Practical Guide to Grouting of Underground Structures.* Reston, VA: ASCE Press. pp. 24–30 and 133–140.

Heenan, D., and A. Naudts. 2000. "Advanced Grouting Program at Penn Forest Dam." *Geotechnical News,* 18(2): 43–48. (BiTech Publishers Ltd., Richmond, BC, Canada.)

Houlsby, A. C. 1990. *Construction and Design of Cement Grouting.* New York: John Wiley & Sons, Inc.

Houlsby, A. C. 1988. "Improvements in Grouting of Large Ground Anchors." *ASCE Journal of Geotechnical Engineering* 114(4):448–468.

Houlsby, A. C. 1985. "Cement Grouting: Water Minimizing Practices", Proceedings, Issues in Dam Grouting, Denver, April 30, 1985, American Society of Civil Engineers.

Houlsby, A. C. 1982. "Optimum Water: Cement Ratios for Rock Grouting." Proceedings, Conference on Grouting in Geotechnical Engineering, New Orleans, February, 1982, American Society of Civil Engineers.

Ivanova, V. 1998. "Geologic and Stochastic Modeling of Fracture Systems in Rocks." (Ph.D. diss., Massachusetts Institute of Technology, Cambridge, MA.

Jefferies, M. G. 1997. "Plastic Work and Isotropic Softening in Unloading." *Geotechnique* 47:1037–1042.

Jefferies, M.G. 1993. "NorSand: A Simple Critical State Model for Sand." *Geotechnique* 43:91–103.

Jefferies, M. G., and D. A. Shuttle. 2002. "Dilatancy in General Cambridge-Type Models." *Geotechnique* 52(9):625–638.

Karol, Rubin H. 1990. *Chemical Grouting.* 2d ed. New York: Marcel Dekker.

Karol, Rubin H. 1963. *General Specifications for Field Use of AM-9 Chemical Grout.* Soils Engineering Research Center, Princeton, NJ: American Cyanamid Company.

King, J., and E. Bush. 1961. "Symposium on Grouting: Grouting of Granular Materials." *Journal of the Soil Mechanics and Foundations Division, ASCE* 87(Paper 2791):1279–1317.

Kosmatka, S., and W. Panarese. 1990. *Design and Control of Concrete Mixtures, Thirteenth Edition.* Skokie, IL: Portland Cement Association, p. 23.

Kovacevic, N., D. M. Potts, and P. R. Vaughn. 2000. "The Effect of the Development of Undrained Pore Pressure on the Efficiency of Compaction Grouting." *Geotechnique* 50(6):683–688.

Kravetz, G. 1959. Closure to paper "Cement and Clay Grouting of Foundations: The Use of Clay in Pressure Grouting." *Journal of the Soil Mechanics and Foundation Division, ASCE,* Vol. 85 (SM 2), pp. 109–114.

Krizek, R., D. Michel, M. Helal, and R. Borden. 1992. "Engineering Properties of Acrylate Polymer Grout." *Grouting, Soil Improvement and Geosynthetics.* Geotechnical Special Publication No. 30. New York: American Society of Civil Engineers, pp. 712–724.

Lee, H., and K. Neville. 1967. *Handbook of Epoxy Resin.* New York: McGraw Hill.

Lombardi, G. 1985. "The Role of Cohesion in Cement Grouting of Rock." Vol. 3. *Proceedings Fifteenth Congress on Large Dams,* Lausanne, Switzerland: International Commission on Large Dams. pp. 235–261.

Lombardi, G., and D. Deere. 1993. "Grouting Design and Control Using the GIN Principle." *International Water Power and Dam Construction* (June): 15–22.

Long, J. 1984. "Investigation of Equivalent Porous Medium Permeability in Works of Discontinuous Fractures. Ph.D. diss., University of California, Berkeley.

Maag, E. 1938. *Ueber die Vergestigung un Dichtiung das Baugrundes (Injektionen).* Erdbaukers der ETH. Zurich: Eidgerossische Technische Hachshule (Swiss Federal Institute of Technology).

Mehta, P. K. 1991. *Advances in Concrete Technology.* Ottawa, Ont.: CANMET.

Mehta, P. K., and P. J. Monteiro. 1993. *Concrete Structure, Properties, and Materials.* 2d ed. Englewood Cliffs, NJ: Prentice-Hall.

Meyerhof, G. G. 1956. "Pentration Tests and Bearing Capacity of Cohesionless Soils." *Journal of Soil Mechanics and Foundation Engineering, ASCE* 82 (SM 1):1–19.

Meyers, J. G. 1994. "Slabjacking Sunken Concrete." *Concrete Repair Digest* 6:224–288.

Mitchell, J., and F. Wentz. 1991. *Performance of Improved Ground During the Loma Pieta Earthquake.* Report No. UCB/EERC-91/12. Berkeley, CA: Earthquake Engineering Research Center, University of California.

National Directory of Geoscience Data Repositories. Washington DC: American Geological Institute.

Naudts, Alex. 1995. "Grouting to Improve Foundation Soil." Section 5B in *Practical Foundation Engineering Handbook.* 1st ed. New York: McGraw Hill.

Nonveiller, E. 1968. "Grouted Cutoff Curtains in Fissured Rock." *Proceedings, International Symposium on Rock Mechanics.* IV. pp. 388–400. Amsterdam: Elsevier Science.

Partos, A., R. Woods, and J. Welsh. 1982. "Soil Modification for Relocation of Die Forging Operations." *Proceedings, Grouting in Geotechnical Engineering.* New Orleans: American Society of Civil Engineers.

Post-Tensioning Institute Committee on Grouting Specifications (2001).

Powers, T. C. 1968. *Properties of Fresh Concrete.* New York: John Wiley and Sons, Inc.

Randolph, Alan F. 1960. *Plastics Engineering Handbook.* 3d ed. New York: The Society of the Plastics Industry; New York: Reinhold Publishing; London: Chapman and Hall, Ltd.

Recommendations for the Handling of Aromatic Isocyanates. 1976. New York: International Isocyanates Institute, Inc.

Rigny, P. 1974. "European vs. U. S. Grouting Practice." *Proceedings Conference on Foundations for Dams.* Engineering Foundation, ASCE, pp. 37–46. New York: ASCE.

Schmertmann, J. H., and J. F. Henry. 1992. "A Design Theory for Compaction Grouting." In *Grouting Soil Improvement and Geosynthetics,* edited by R. H. Borden. ASCE Geotechnical Special Publication No.30, pp. 215–228. Reston, VA: ASCE.

Shimoda, M., and H. Ohmori. 1982. "Ultra Fine Grouting Materials." *Proceedings. Conference on Grouting in Geotechnical Engineering,* American Society of Civil Engineers, New Orleans, February 1982.

Shuttle, D. A., and M. E. Glynn. 2003. "Grout Curtain Effectiveness in Fractured Rock by the Discrete Feature Network Approach." *Proceedings of Grouting and Ground Treatment 2003—Geotechnical Grouting and Deep Mixing.* February 23–28, 2003. Reston, VA: ASCE.

Shuttle, D., and M. Jefferies. 2000. "Prediction and Validation of Compaction Grout Effectiveness." In *Proceedings Wallace Baker Memorial Symposium,* GeoDenver, ASCE Geo-Institute Geotechnical Special Publication Number 104, Reston, VA: ASCE, pp. 48–64.

Shuttle, D. A., and M. G. Jefferies. 2000. "Prediction and Validation of Compaction Grout Effectiveness." In *Advances in Grouting and Ground Modification. Proceedings of Sessions of Geo-Denver 2000.* edited by Raymond J. Krizek and Kevin Sharp. Geotechnical Special Publication No. 104. Reston, VA: ASCE.

Shuttle, D. A., and M. G. Jefferies. 1998. "Dimensionless and Unbiased CPT Interpretation in Sand." *International Journal for Numerical Methods in Fluids and Numerical and Analytical Methods in Geomechanics.* 22:351–391.

Shuttle, D. A., W. Dershowitz, E. Glynn, S. Burch, and T. Novak. 2000. "Discrete Fracture Network Analysis of Foundation Grouting." In *Proceedings of the Fourth North American Rock Mechanics Symposium,* edited by Girard, J., Liebman, M., Breeds, C., and Doe, T. Rotterdam: Balkema, pp. 1369–1376.

Standard Procedure for Evaluation of Well Perforators. API RP 43. American Petroleum Institute.

Stilley, Alan N. 1982. "Compaction Grouting for Foundation Stabilization." *Grouting in Geotechnical Engineering.* ASCE Geotechnical Special Publication No. 30. pp. 923–937. New York: ASCE.

Swaby, P. A., and K. D. Rawnsley. 1996. "An Interactive 3D Fracture Modeling Environment." SPE 36004. Paper presented at Society of Petroleum Engineers, Petroleum Computer Conference, Dallas, TX, June 2–5, 1996.

Tattersall, G. H., and Banfill, P. F. 1983. The Rheology of Fresh Concrete. London: Pittman.

Taylor, H. W. F. 1992. *Cement Chemistry.* San Diego, CA: Academic Press, Inc.

Technical Memorandum No. 3-408. *Grouting of Foundation Sands and Gravels.* 1956. Vicksburg, MS: U.S. Army Corps of Engineers, Waterways Experiment Station. (June 1955, Report); January 1956, Appendix.).

Volumetric Mixer Manufacturers Bureau. 2001. *Volumetric Mixer Standards.* VMMB 100-01. Silver Spring, MD: Volumetric Mixer Manufacturers Bureau.

Vossoughi, S. 1999. "Flow of Non-Newtonian Fluids in Porous Media." *Advances in the Flow and Rheology of Non-Newtonian Fluids.* Part B. Amsterdam: Elsevier.

Warner, J., and D. Brown. 1974. "Planning and Performing Compaction Grouting." *Journal of the Geotechnical Engineering Division, ASCE,* 100(GT6): 653–666.

Warner, James. 1992. "Compaction Grout: Rheology vs. Effectiveness." *Grouting Soil Improvement and Geosynthetics.* Geotechnical Special Publication No. 30. American Society of Civil Engineers, Reston, VA.

Warner, James. 1972. "Strength Properties of Chemically Solidified Soils" *Journal of the Soil Mechanics and Foundation Engineering Division, ASCE* 98 (SM11):1163–1185.

Warner, James, Norbert Schmidt, John Reed, Don Shepardson, Russ Lamb, and Sam Wong. 1992. "Recent Advances in Compaction Grouting Technology." *Grouting Soil Improvement and Geosynthetics.* Geotechnical Special Publication No. 30. American Society of Civil Engineers, Reston, VA.

Weaver, Kenneth. 1991. *Dam Foundation Grouting.* New York: American Society of Civil Engineers.

Western Construction. 1969. "Grouting Lifts Sagging Pavement," *Western Construction.* King Publications (October 1969). pp. 41–46.

Wilson, D., and T. Dreese. 2003. "Quantitatively Engineered Grout Curtains." *Proceedings of the Third International Conference on Grouting and Ground Treatment.* Geotechnical Special Publication No. 120. Reston, VA: Geo-Institute of ASCE, pp. 881–892.

Index

Academic research, 567
Acceleration, 575, 576
Accelerators, 129
Acetins, 122
Acrylamide grouts, 133–134
Acrylate grout, 134–135
Acrylic resins, 419
Addition reactions, 129
Admixtures (cementitious grouts), 42, 80–87
  air-entraining, 86
  air- or gas-generating, 84–86
  antiwashout/viscosity modifiers, 83–84
  bond strength enhancers, 86
  feed grains, 92
  hydration control, 83
  lost circulation additives, 90
  plastic tabs/punchings, 91–92
  pumping aids, 84
  set-accelerating, 82–83
  set-retarding, 80–81
  special combinations of, 86–87
  synthetic fibers, 90–91
  water-reducing, 80–81
Aeolian soil deposits, 477
Aerial photographs, 488
Agitators, 637, 641, 642
Air-entraining admixtures, 86
Air-generating admixtures, 84–86
Air track drills, 187, 188, 595–597
Alluvial soil deposits, 478
Aluminum powder, 84
Amenability theory, 308–309
Annular spaces, filling, *see* Void areas
Anticlines, 470
Antiwashout admixtures, 83–84, 110
API drill rods, 608–610
API filter press, 147–148
Apparent viscosity, 53
Argus GeoLogistics, Inc., 436–438, 440–450
As-built drawings, 487–488
Ascending stage injection, 319
Asphalt emulsion, 142
Asphalts, 141–142
ASTM standards, *see specific subjects*
Atmospheric pressure, 177–178
Atterberg limits tests, 480, 482
Automated monitoring equipment, 669–671

Backfill grouting, 13, 354, 400–403
Backfill material, densification of, 223–224
Backpack grouting, 8
Bag house, 641
Balanced stable mixes, 309
Ballistic bits, 612
Balloon Injection Concrete System (BICS), 343–344
Balloon packers, 674
Ball valves, 622, 652, 661–662
Ball wiper valves, 623
Baroid Mud Balance, 146
Barometric pressure, 177
Barrel, core, 195
Basic esters, 122–123
Batch mixing, 635–636
Becker hammer drilling, 198–199
Bedding, rock, 468, 474
Bedplates, grouting under, 349–352
Belt drives, 580–581
Bends, delivery line, 660
Benkelman beam, 377
Bentonite, 143, 452
Bingham yield stress, 52, 59
Bitumen, 141
Blaine fineness, 64
Blanket grouting, 298, 302, 514
Bleed, 59–60, 93–95
  and bleed water reduction, 94
  evaluation of, 146–147
  in suspension grouts, 103
  wick bleed, 346
Bleeder valve, 94, 95
Blended hydraulic cements, 75–76
Blocking agents, 89–92
  feed grains as, 92
  lost circulation additives, 90
  plastic tabs/punchings, 91–92
  synthetic fibers, 90–91
Bond strength, 47, 86, 153–155
Bottom-up grouting, 209–211, 252, 265. *See also* Upstage injection
Bucket cleaning, 390
Bucket test, 413

Calcium aluminate cement, 70
Calcium bentonite, 143
Calcium chloride, 82–83
Cam lock couplings, 659
Caps, grout, 312–313, 510
Carbon steel drill hole casing, 602–604
Casing, hole, 311, 448, 597–604
Cation exchange, 287
Cationic asphalt emulsions, 141
Cavities, structural, *see* Void areas
Cellular concrete/grout, *see* Controlled low-strength material grouts
Cements, 40. *See also* Portland cement
  blended hydraulic, 75–76
  hardening rate of, 63
  oil well, 42, 71–72
  specialty, 41–42
Cementitious grouts, 40–42
  admixtures for, 42
  compositions of, 41
  defined, 44
  in extreme environments, 428
  grain size of, 41
  injection of, 289–290
  materials used in, *see* Cementitious grout materials
  mixture formulations for, *see* Cementitious grout mixtures
  oil well cements, 42
  special cements, 41–42
  typical uses of, 46
  unitized grout plants for, 647–651

Cementitious grout materials, 40–42, 62–92
admixtures, 42, 80–87
blended hydraulic cements, 75–76
blocking agents, 89–92
calcium aluminate cement, 70
expansive cements, 70
fillers, 87–89
fly ash, 73
grain size, 41
ground granulated blast furnace slag, 75
masonry cements, 71
natural pozzolans, 73
oil well cements, 71–72
plastic cement, 71
portland cement, 62–70
pozzolanic materials, 72–75
rapid setting and hardening cement, 71
rice husk ash, 75
silica fume, 74–75
special cements, 70–72
special combinations of admixtures, 86–87
supplementary cementing materials, 72–75
ultrafine cements, 76–80
Cementitious grout mixtures, 93–114
and bleed, 93–95
cementitious suspensions, 97–103
expansive and nonshrink grout, 103–107
limited mobility grouts, 107
low-density cellular mortar and concrete, 110–111
low-slump limited-mobility grout, 111–114
mix designs, 96–97
pourable slurries and pastes, 103
ready-mixed mortar and concrete, 107–110
water-to-cement ratio for, 95–96
Cementitious suspension grouts, 5, 97–103
Cement-sodium silicate grouts, 123, 124
Centrifugal pumps, 621–622
Chain-style couplings, 578
Chambers, *see* Underground structures, grouting of
Chemical exposure, extreme, 428–429
Chemical grouts, 45
Chemical solution grouts, 115–117
defined, 44–45
development of, 4
mixing of, 116
for strengthening, *see* Chemical solution grouts (for strengthening)
typical uses of, 46
for water control, *see* Chemical solution grouts (for water control)
Chemical solution grouts (for strengthening), 117–128
durability of, 124–128
long- vs. short-term strength, 124
sodium silicate in, 117–120
sodium silicate reactant systems, 120–124
Chemical solution grouts (for water control), 128–135
polymers, 128–130
polyurethane grouts, 130–133
water-soluble acrylic polymers, 133–135
Circuit grouting, 319, 323
Circulation flush, 187, 199–201
Circulation grout system, 16–17
Circulation pumps, 614
Clamp-type couplings, 659–660
Claquage, *see* Fracture grouting
Clays, 73, 89, 226, 479
Clay grouts, 142–143
Clinker, 63, 78
Closed-circuit TV inspection, 389–391
Closed-loop hydraulic circuits, 585
Coarse-grained soils, 479–481
Coefficient of thermal expansion, 49
Cohesion:
in clays, 479
of grout, 59
as magnitude of force, 52
of soils, increasing, 10–11, 15
Colcrete mixers, 638–639
Cold environments, 68, 430–435
Collar standpipe, 311
Colloidal mill mixers, 638
Columns, structural, 508
Common portland cement, 64
Compaction grout, defined, 54
Compaction grouting, 15, 19–31, 208–260. *See also* Groutjacking
function of, 498
grout holes for, 26–27, 210–213
grout mass shapes, 23–26
grout mixture for, 213–215
guide specifications for, 546–559
history of, 5
improvement of deficient soil, 223–242
improvement of water-retaining embankments, 231–239
injection of grout, 215–223
injection system for, 27–29
limitations of, 29–31
of loose soil/fill/backfill, 223–227
for mass improvement, 500–501
mitigation of liquefaction potential, 227–231
for modulus/strength improvement, 501
monitoring/control of injection, 218–223
numerical analysis for, 538–542
ready-mixed grouts for, 113–114
for settlement control, 502–504
for settlement correction, 242–259
for slabjacking, 380–381
for soft ground tunneling, 259–260
for soil densification, 10
of submerged soil, 239–242
technology development for, 20–26
travel index, 21, 23–24
Compensation grouting, 286
Compressive strength, 47
Computer monitoring, 176
Concrete, ready-mixed, 107–110
Concrete pumps, 632–634, 652
Concrete structures, *see* Preplaced aggregate concrete; Structures, grouting in
Condensation reactions, 129
Conduits, 387. *See also* Pipes and conduits, grouting in
Cone penetrometer test (CPT), 492
Confined spaces, safety in, 388–389
Consistency (of mix), 5, 45, 51. *See also specific types of grouting*
Consolidation grouting, 298, 302, 406, 514–515
Contact grouting, 8, 12, 345, 400–403
Continuous flight augers, 612
Continuous flow pumps, 617–622
Contractors, 559–571
behavior changes after contract award, 568–569
defense of bad work by, 569–570
good, 570–571
marketing by, 565–567
negotiations with, 561
payment basis for, 561–563
prequalification of, 559–561
proposal trickery by, 568
Controlled low-strength material (CLSM), 47, 84–86, 110
Cost plus fixed fee payment, 561–562
Couplings, delivery line, 659–660
Cover grouting, 298
Crack and joint filling, 336–345
cementitious grout for, 337–338
construction joints, 344–345
crack repair, 338–345
epoxy crack injection, 338–344
epoxy resins for, 337

injection materials for, 336–338
joints between precast elements, 345
polyester resins for, 337
urethanes for, 338
Curtain, *see* Grout curtain
Custom lightweight drill rods, 607–608
Cut-and-cover tunnel, 399
Cutting removal, 199–201, 310
Cyclonite, 440

Dams, 299, 304, 514–518. *See also* Water-retaining embankments
DCDMA standards, 597–599, 605, 606
Deep soil mixing, 35, 37. *See also* Mixing/jet grouting
Delivery lines, 654–660. *See also specific types of grouting*
bends, 660
couplings, 659–660
in extreme temperatures, 435
flexible hose, 655–658
flow properties, 655
grout velocity, 654–655
reducers, 660
rigid, 658–659
sizing of, 160–163
Densification, *see* Soil densification
Dental concrete, 312, 515–516
Descending stage injection, 319
Design of grouting (in rock), 301–304, 511–518
and existing geology, 512–514
for seepage control, 516–518
for strengthening, 514–516
Design of grouting (in soil), 497–510
and grout functions in soil, 498–499
for karst mitigation, 508–510
for mass improvement, 499–501
for modulus/strength improvement, 501–505
for permeability reduction, 505
for structural element creation, 505–508
Diamond core bits, 611–612
Diamond core drilling, 192–196
Diaphragm gauge saver, 667
Diaphragm pumps, 625–626
Diaphragm valves, 661
Diatomite, 73
Dibasic esters, 122
Differential-pressure flow meters, 668
Dikes, 468
Dilatant fluids, 53
Diluents, 129
Dimensional stability, 46–47
Dimethyl esters, 122
Dip, 474–475
Diphenylmethane diisocyanate (MDI), 132–133
Direct advertising, 565
Direct grout systems, 16
Direct shear test, 155
Direct tensile test, 154–155
Discar mixer, 639
Discharge pressure head, 616
Discrete fracture network (DFN) models, 533–538
Doppler meters, 669
Double packers, 677
Downstage injection, 27–28, 319, 321
Down-the-hole equipment, 194
Down-the-hole hammers, 597
Drag bits, 610–611
Drawings, construction, 487–488
Drifters, 595
Drill bits, 189, 610–613. *See also specific types of drills*
Drill head, 186–187
Drill hole casing, 597–604
carbon steel tubing, 602–604
DCDMA drill casing, 597–599
special casing systems, 598, 599
standard steel pipe, 598–602
Drilling blind, 200
Drilling equipment/accessories, 589–614. *See also specific types of grouting*
circulation pumps, 614
drill bits, 610–613
drill hole casing, 597–604
drill rigs, 589–597
drill rods, 605–610
sleeve port pipes, 613
special drive needles, 614
Drilling (of grout holes), 186–205, 314–315
Becker hammer drilling, 198–199
and circulation flush, 199–201
and cutting removal, 199–201
diamond core drilling, 192–196
drive drilling, 197–198
drive needles for, 198
dual-head duplex rotary drilling, 191
methods of, 189
quality control for, 520
rotary drilling, 191–192
rotary percussion drilling, 190–191
rotary sonic drilling, 196–197
Drill jumbo, 595
Drill rigs, 187, 188, 589–597
handheld, 590–592
mini, 592–593
pneumatically powered percussive, 594–597
self-contained hydraulic, 594
Drill rods, 186, 187, 605–610
Drill string, 590, 605
Drive drilling, 186, 189, 197–198
Drive needles, 198, 614
Dry mix process, 35
Dry mix shotcrete, 311–312
Dry sand, 89
Dual-head duplex rotary drilling, 191
Dual proportioning pumps, 628
Dyes, water, 299–300, 412–414
Dynamic head, 159, 160, 616

Efficiency, pump, 616–617
Efflux time, 144
Elastic modulus, *see* Modulus of elasticity
Elevation plans, 494–495
Emergency response grouting, 451–462
extinguishing underground fires, 461–462
preparation for injection, 451–453
sinkholes, 453–455
structural failures, 459–461
subsurface soil loss, 458–459
subsurface water flow/piping, 455–458
Emulsion polymerization, 86
Encapsulation, 298
Engineering geologists, 465, 466
Environmental requirements, 489
Epochs, geological, 466
Epoxies, 13–14, 135–139, 337–344, 418
Equipment, 8
drilling, *see* Drilling equipment/accessories
grouting, *see* Grouting equipment
power transmission in, 575–588
and pump mechanics, 615–626
for site surveillance, 523–524
Equivalent porous media (EPM) model, 532–533
Evaluation, *see* Numerical analysis; Tests
Excavation support, grouting for, 507–508
Exfoliation joints, 471
Expandable bits, 612
Expanded head piles, 440–444
Expansive cements, 70
Expansive grouts, 84, 103–107, 150–151
Explosives, use of, 436–450
for accessing subsurface fissures and voids, 445–447
for clearing obstructions to grout flow, 444–445
expanded head pile and ground anchor formation, 440–444

Explosives (*cont.*)
in-place perforation of injection casing, 448–449
principles of, 438–440
Extreme environments, grouting in, 427–435
with cementitious grouts, 428
cold environments, 430–435
and extreme chemical exposures, 428–429
hot environments, 429–430
with noncementitious grouts, 428
provisions for temperature extremes, 429–435
temperature effects on grout, 427–428

Fabric forms, filling, 13, 37–38, 354–355
Faults, 472
Feature grouting, 406
Feed, drill, 187
Feed grains (as blocking agents), 92
Fibrillated fibers, 90–91
Fillers, grout, 87–89, 508–509
Fill grouting, 15, 297
fabric forms for, 13, 37–38
in rock, 297
Fines, 480
Fine-grained soils, 479–482
Fineness modulus (FM), 88–89
Finite element analysis, 539–540
Fires, underground, 461–462
Fishtail bits, 610
Fist-to-the-wall porting, *see* Skip seal porting
Fittings:
delivery line, 659–660
hydraulic system, 585
Fixed-end packers, 674
Flac program, 539
Flange-style couplings, 578
Flexible couplings, 577–579
Flexible hose, 163, 655–658
Flow, 156–164
in delivery lines, 160–163, 655
dynamic head, 159, 160
head pressure, 157–164
hydraulic, 582–584
laminar, 156, 157
nozzle size, 160, 163, 164
properties of, 52–53
and properties of grout, 157–159
static head, 159
turbulent, 156, 157
valves, 160
Flowable fill, 47, 353
Flow breccia, 467
Flow cones, 144–146
Flow meters, 668–669
Fluids, 45, 51–53
Fluidifier, 360
Flush, *see* Circulation flush
Fly ash, 73, 110
Foaming grouts:
low-density air-foamed, 110–111
phenolic resin, 140
resinous, 45, 140, 331–332
for slabjacking, 371–372
urethane, 139
Folds, rock, 470
Fondu cement
Force, 576
Frac, 286
FracMan, 534–535
Fractured rock grouting, 532–538
Fracture grouting, 15, 31–33, 208–209, 286–296
cementitious grout injection, 289–290
history of, 5
lime injection, 287–289
lime injection for, 32–33
limitations/risks of, 33
numerical analysis for, 532–538
for reinforcement of soil, 11
Frictional resistance, 159
Friction sealing packers, 675
Full-flow ball valves, 661
Full-scale trials, 563
Fundamentals of Grouting short course, 7, 22, 78
Fundamental strength, 126, 152
Funnels, 144

Galleries, grout, 313–314, 518
Gas-generating admixtures, 84–86
Gauge savers, 665–667
Gauge seal, 666
Gearboxes, 579–580
Gear pumps, 585, 620–621, 628
Gels, 64, 133–135
Gelman pressure filter, 147
Geologic maps, 486
Geology, 465–483
classification of soil, 479–482
derivation of soil, 476–479
and design of grouting in rock, 512–514
geologic ages, 466–467
geological structure, 469–474
igneous rock, 467–468
metamorphic rock, 468–469
rock configuration/geometry, 474–475
rock types, 467–469
sedimentary rock, 468
stress state, 475–476
Geotechnical core bits, 612
Geotechnical grouting, 399–401
Geotechnical Research Group, UBC, 539–540
Glacial till, 479
Grade N silicate, 119
Grain size:
of cementitious materials, 41
and penetrability, 57–58
in portland cement, 64
Gronowicz, Bob, 436–438
Ground anchors, 347, 440–444
Ground granulated blast furnace slag, 75
Ground-penetrating radar (GPR), 492–493
Grouts, 44–45
materials used in, *see* Materials
temperature effects on, 427–428
terminology used for, 44
types/typical uses of, 46. *See also specific types*
Grout caps, 312–313, 510
Grout curtain, 17, 298, 302–304, 406, 515–519
Grouted masses:
evaluation of, *see* Tests
shapes of, 23–26
strength of, 151–152
Grout holes, 179–206. *See also specific types of grouting*
closure and abandonment of, 205–206
for discrete voids, 183–185
drilling of, 186–205
establishing, 179–186
for grouting in rock, 183, 184, 310–317
identification of, 185–186
for injection in soil, 210–213
size of, 186
for soil strengthening, 18
spacing and layout of, 179–185
split spacing of, 180–181
in structures, 183, 184
Grout hose, 654–658
internal expansion, 655–657
reinforcement, 656–657
burst pressure, 657–658
Grouting, 15. *See also* Pressure grouting; *specific applications, e.g.:* Rock, grouting in; *specific types, e.g.:* Permeation grouting
establishing objectives of, 484–485
as tunneling compensation, 504–505
verifying effectiveness of, 529–530
Grouting equipment. *See also specific types of grouting*
agitators, 641, 642
delivery lines and fittings, 654–662

for low-mobility grouts, 643–645
for material storage and transfer, 641, 642
for mixing pourable grouts, 637–641
monitoring equipment, 663–671
packers, 672–677
pumps, 627–634
small mixers, 635–637
unitized grout plants, 646–653
valves, 661–662
Grouting galleries, 313–314, 518
Grouting intensity number (GIN), 306–308
Groutjacking, 10, 367–386
controlled vs. uncontrolled, 168
monitoring pressure behavior in, 173–174
of pipes and conduits, 394–395
slabjacking, 369–379
of structures, 379–386
Grout plants, 324–325, 646–653
Grout pumps, 627–634
Guide specifications, 543–544, 546–559
G.V.S. chemical solution grout, 126, 127
Gyroscopes, 204

Half-size slump test, 148
Handheld drills, 590–592
Hardness, rock, 469
Headers, grout, 16
Head loss/gain, 157, 159–163
Head pressure, 157–164
Heat-deflection temperature (HDT), 137, 138
Heaters, 432–433
Heavy percussion drill rods, 609, 610
Hexogen, 440
Hex sinker drill steel rods, 610
High alumina cement, 70
High-early-strength cement, 64
High-range water reducers, 318
High-reactivity metakaolin, 73
High-shear mixers, 638–641
High-volume low-pressure (HVLP) systems, 583
Honeycomb, 13, 330, 333–334
Horizontal hydrofracture, 381–382
Hot asphalt, 141–142
Hot environments, 66–68, 429–430
Hydrates, 64
Hydration, 63–65
admixtures for control of, 83
and mix water, 69–70
and temperature, 65–68
Hydraulic cements, 40, 41
Hydraulic fracturing (of soil), 20, 22–25, 30–31, 54, 107, 169, 172, 208, 210, 213, 231
Hydraulic power transmission, 581–587
circuitry, 584–585
flow and pressure, 582–584
fluid, hydraulic, 584
pumps, 585–587
rams and jacks, 587
rotary motors, 587
valves and fittings, 585
Hydraulic pumps, 630, 631
Hydrophilic grouts, 117, 418
Hydrophilic urethanes, 131–132, 418–419
Hydrophobic grouts, 117
Hydrophobic urethanes, 130–131, 419

Igneous rock, 298, 467–468, 471
Illite, 142
Infilling, rock, 474
Inhibitors, 130
Injectability of grouts, 116
Injection, 156–178. *See also specific types of grouting*
flow principles, 156–164
grout penetration, 165
grout rheology effects, 163, 165
pressure, 166–170
pressure behavior during, 171–176
pumping rate, 170–171
quality control, 521–523
records of, 524–529
in rock, *see* Injection (in rock)
and setting time, 60–61
in soil, *see* Injection (in soil)
vacuums, 177–178
Injection (in rock), 319–328
connections between holes, 326–327
conventional technology for, 304–306
grout plant, 324–325
grout supply, 325
hole staging, 319–324
injection process, 325–328
monitoring, 325–326
preparation prior to hookup, 324
pressure, 326
refusal criteria, 328
sequencing, 319
surface leaks, 327–328
Injection (in soil):
compaction grouting, 210–223
permeation grouting, 262–266
In-line mechanical coupling, 577–579
In-line mixing, 636–637
Inorganic salts, 120–122
Intrusion Aid, 360, 449
Intrusive rock, 467
Investigations for grouting, 484–496
data review, 486–488
existing improvements, 489–490
information needed for, 485–486
and objectives of grouting, 484–485
in rock, 495
site access/restrictions, 488–489
in soil, 485–486, 491–495
in structures, 496

Jacks, hydraulic, 587
Jackshafts, 579
Jaw-type couplings, 578–579
Jet grouting, *see* Mixing/jet grouting
Jet mixers, 641
Jiffy mixers, 635
Joints, rock, 470–472
Joint filling, *see* Crack and joint filling
Joosten process, 4, 120
Jumbo Jet system, 34–36

Kaolinite, 142
Karst, 454, 455, 508–510
Kenics mixer, 636
Key trench, 311, 312
Knockoff drill bits, 610, 611

Laminar flow, 156–157
Laser beacons, 222, 524
Leakage control (in sewers), 14
Leakage control (in structures), 14, 410–426. *See also* Water control
application examples, 421–426
grout holes/injection ports for, 414–417
grout injection and monitoring, 419–421
grout material for, 417–419
masonry, 336
path of leakage, 410–414
Lens, 286
Lifter, core, 195
Lignon water reducers, 105
Lignosulfonates, 81
Lime injection, 5, 6, 32–33, 287–289
Limited-mobility displacement (LMD) grout, 112, 508
Limited-mobility grouts, 107, 111–114, 505
Lining systems (underground facilities), 406–408
Liquefaction potential, mitigation of, 227–231
Liquid levels, 220–221
Liquid limit (soils), 480, 482
Logging of drill holes, 315–317
Long chain polymers, 129
Lost circulation additives, 90
Low-density grouts, 84–86, 110–111
Low-mobility grouts, 370–371, 643–645, 652

Low-pressure hydraulic tools (LPHT), 583
Low-slump limited-mobility grout, 111–114
Lugeon unit, 301, 304
Lump sum payment, 562

Machinery bases, grouting under, 349–352
Magma, 467, 468
Magnetic flow meters, 669
Magnetic pressure meters, 668
Manometers, 221–222, 523–524
Marketing, 565–567
Marsh funnels, 50, 145–146
Masonry cements, 71
Masonry sand, 89
Masonry structures, *see* Structures, grouting in
Mast, drill, 187, 590
Materials, 40–49
  admixtures for cementitious grouts, 42
  cementitious, 40–42
  coefficient of thermal expansion of, 49
  dimensional stability of, 46–47
  modulus of elasticity of, 47–49
  noncementitious, 42–43
  properties of, 45–49. *See also* Rheology
  quality control, 520–521
  selection of, 43–45
  strength of, 47
  for water control grouting, 9
Material storage and transfer equipment, 641, 642
Mechanical efficiency (pumps), 616
Mechanical packers, 673
Mechanical power transmission, 577–582
  belt drives, 580–581
  gearboxes, 579–580
  in-line coupled, 577–579
  roller chain and sprockets, 580
  variable speed drives, 581–582
Media advertising, 565–566
Metamorphic rock, 468–469
Micro-Capsule system, 344
Mined tunnel, 399
Mini drill rigs, 592–593
Miscellaneous grouts, 45, 46
Mixers, 635–641
Mixing/jet grouting, 15, 33–37
  applications of, 293–295
  deep soil mixing in, 35, 37
  equipment for, 291–292
  and foreign matter in grout, 292
  function of, 498
  history of, 5–6
  injection parameters, 291
  jet grouting application, 34–36
  for mass improvement, 500
  for permeability reduction, 505
  risks/limitations of, 35
  in soil, 209, 290–295
  strength of soilcrete, 291
  types of soil for, 290–291
  for underpinning, 505
Mixing shear, 59
Mix water, quality of, 69–70
Mobility, 53–55
Modified Earthfirm, 126
Modulus of elasticity, 47–49, 52
Molecular weight (polymers), 129
Monitoring equipment, 663–671. *See also specific types of grouting*
  automated, 669–671
  for drill holes, 204–205
  flow meters, 668–669
  gauge savers, 665–667
  for injection pressure, 176
  pressure gauges, 663–665
Monofilament fibers, 90
Monroe effect, 437–439
Montmorillonite, 142–143
Moraine, 479
Morflex-type couplings, 578
Mortar, ready-mixed, 107–110
Mortar mixers, 643
Mortar pumps, 652
Moyno pumps, 617–619, 629
Mud Balance, 50, 146
Mudjacking, 367
"Mudjack" machine, 5, 367–368
Mud pump, 199, 200, 214. *See also* Circulation pumps
Multiple packer sleeve port pipe (MPSP), 323–324
Multipump proportioning units, 628

Natural pozzolans, 73, 110
Neat cement grout, 93, 509
Negative end/side (leakage), 410
Negotiations, contract, 561
Newtonian fluids, 51–53
N grade silicate, 119, 121
N-methylolacrylamide grouts, 135
Noncementitious grouts, 42–43, 115–143
  asphalts/bitumen in, 141–142
  chemical solution grouts, 115–117
  chemical solution grouts for strengthening, 117–128
  chemical solution grouts for water control, 128–135
  clay grouts, 142–143
  in extreme environments, 428
  resinous grouts, 43, 135–141
  specialty water control grouts, 42–43
Non-Newtonian fluids, 52–53
Non-positive displacement pumps, 616
Nonshrink grouts, 103–107
Normal atmospheric pressure, 177
Normal cure, 125
NorSand, 539, 540
No-slump group mixtures, 54
Nozzles, 160, 163, 164
Numerical analysis, 531–542
  compaction grouting, 538–542
  fractured rock grouting, 532–538

Oil (hydraulic systems), 584
Oil well cementing, 42, 50, 71–72
Open loop hydraulic circuits, 585
Orchard heaters, 432
Ordinary portland cement, 64
Organic/aliphatic esters or amides, 120, 122–123
Outcroppings, 475
Output pressure, pump, 616
Overhangs, 475

Packers, 321–324, 672–677
Paddle mixers, 637–638, 643
Parallel wall drill rods, 605
Pastes, 51, 64, 103
Patents, 567
Payment basis, contract, 561–563
Penetrability, 44, 55–59
Penetration grouting, *see* Permeation grouting
Percussion drill rods, 609
Percussive drills, 594–597
Percussive drill bits, 612–613
Performance specifications, 545
Peristaltic pumps, 619–620, 628
Permeability of soils, 11–12, 116–117, 505
Permeation grouting (generally), 15–19
  function of, 498
  of granular formations, grout holes for, 181–183
  history of, 4–5
  injection systems for, 16–17
  in structures, 18–19
  for water control, 17
Permeation grouting (in soils), 208, 260–286
  for altering soil dynamic properties, 283–286
  for controlling water flow, 279–281
  degree of solidification, 261–262
  grout quantity required, 262–263
  for increasing cohesion, 10

injection holes/probes, 263–266
injection probe layout, 262
for mass improvement, 499–500
for modulus/strength improvement, 501
monitoring and control of injection, 267
for permanent strengthening, 276–279
for permeability reduction, 505
for settlement control, 504
for short-term strengthening, 267–275
for strengthening and water control, 17–18, 281–283
strength test for, 151–152
tests of grouted masses, 151–152
for underpinning, 505
Phenolic resin foams, 140
Pile-supported structures, settlement of, 253–259
Pipes and conduits, grouting in, 387–397
for abandonment, 352–354
access restrictions, 387–388
for jacking to reestablish roundness, 395–396
safety issues with, 388–389
in small inaccessible conduits, 389–392
for stabilization and groutjacking, 394–395
for strengthening surrounding soil, 394
for water leakage control, 393–394
in worker-accessible conduits, 392–397
Piping (soil erosion/relocation as), 456
Piston motors, 587, 588
Piston pumps, 586–587, 624–625, 628–630
Plaster sand, 89
Plastic cement, 71
Plastic consistency, 45, 51
Plasticity index (PI), 482
Plastic limit (soils), 480, 482
Plastic tabs/punchings, 91–92
Plunger pumps, 625, 628–630
Pneumatically powered percussive drills, 594–597
Pneumatic packers, 673–675
Pneumatic power transmission, 587–588
Pneumatic sinker drill, 190
Polar intensity, 55–57
Polyesters resins, 140–141, 337
Polymers, 128–130, 133–135
Polypropylene fibers, 90–91
Polyurethane grouts, 130–133, 418
Poppet valves, 622–623
Porosity, 262
Porphyritic rock, 467
Portals, *see* Underground structures, grouting of
Portland cement, 40, 62–70
bulk density of, 95
expansive, 70
as Group 3 reactant, 123
hydration reaction in, 64–65
mix water for, 69–70
temperature effects on, 65–68
types of, 63, 64, 75–76
and water-to-cement ratio, 96
Positive-displacement flow meters, 668
Positive displacement pumps, 616, 628
Positive end/side (leakage), 411
Postexcavation grouting, 399
Posttensioned concrete work, 12
Posttension tendon grouting, 345–349
Pourable grouts:
mixing equipment for, 637–641
unitized grout plants for, 647–651
Pourable slurries and pastes, 103
Power:
for automated monitoring equipment, 671
defined, 576–577
for grout pumps, 627–628
Power transmission, 575–588
force in, 576
hydraulic, 581–587
mechanical, 577–582
pneumatic, 587–588
power in, 576–577
rate of movement, 575–576
torque in, 576
Pozzolanic materials, 72–75, 109–110
fly ash, 73
ground granulated blast furnace slag, 75
natural pozzolans, 73
rice husk ash, 75
and setting time, 78
silica fume, 74–75
Pre-excavation grouting, 399–401
Prefoam, 111
Preliminary Glossary of Terms Relating to Grouting (ASCE), 44, 54
Preplaced aggregate concrete, 12–13, 355–363
coarse aggregate, 356–357
formwork, 357–358
grout mixture, 359–360
injecting grout, 360–363
injection ports and pipes, 358
placing the aggregate, 358–359
quality control, 363
Prescriptive specifications, 544–545
Pressure:
hydraulic, 582–584
and viscosity, 53
Pressure, injection, 166–178. *See also specific types of grouting*
appropriate levels of, 168–170
grouting control and behavior of, 171–176
pressure–volume relationship, 167–168
and vacuums, 177–178
Pressure filtration, 84, 147–148
Pressure gauges, 216, 663–665
Pressure grouting, 3–14. *See also specific types of grouting*
applications of, 4
history of, 4–8
primary purpose of, 8
in rock, 8–9
in soil, 9–12
structural applications of, 12–14
Pressure snubbers, 664
Pressure–volume relationship, 167–168
Prestressing, 345
Pretensional steel, 345
Primary grout holes, 180–182
Principal stresses, 380
Progressive cavity pumps, 617–619, 629, 630
Promoters, 129
Proposals, contractor, 568
Protection of grouting area, 489–490
Protective shelters, 430–432
Pseudoplastic fluids, 52, 53
Pulsation dampeners, 664
Pumps, 615–626. *See also specific types of drilling and grouting*
centrifugal pumps, 621–622
circulation pumps, 614
classification of pumps, 616–617
continuous flow pumps, 617–622
diaphragm pumps, 625–626
discharge pressure head, 616
efficiency of, 616–617
gear pumps, 620–621
grout pumps, 627–634
hydraulic pumps, 583, 585–587
peristaltic pumps, 619–620
piston pumps, 624–625
plunger pumps, 625
positive vs. non-positive displacement, 616
progressive cavity pumps, 617–619
suction pumps, 615–616
throughput of, 617
valved pumps, 622–626

Pumping aids, 84
Pumping rate, 170–171

Quality control, 519–524. *See also* *specific types of grouting*

Radial grout curtain, 406
Rafts, structural, 508
Rams, hydraulic, 587
Ram pumps, *see* Plunger pumps
Rapid setting and hardening cement, 71
RDX, 440
Reactant systems, sodium silicate, 120–124
Ready-mixed mortar/concrete, 107–110, 112
Records, injection, 524–529
Reducers, 660
Refusal criteria (rock grouting), 328
Regularizing concrete, 312, 515–516
Reinforcement (of soil), 11, 15, 286
Reinforcing steel, 335–336
Repair and rehabilitation applications, 13
Resins, 44, 130, 135
Resinous grouts, 43, 135–141. *See also* *specific types*
  bond strength of, 153
  defined, 45
  development of, 5
  epoxies, 135–139
  phenolic foams, 140
  polyesters, 140–141
  typical uses of, 46
  urethanes, 139–140
Rheology, 50–61, 163, 165
  bleed, 59–60
  cohesion, 59
  consistency, 51
  defined, 50
  flow properties, 52–53
  and injection system flow, 163, 165
  material behavior, 51–59
  mobility, 53–55
  penetrability, 55–59
  setting time, 60–61
  solubility, 61
  temperature, 60
  thixotropy, 52
  viscosity, 53
Rheopectic behavior, 52–53
Rice husk ash, 75
Rigid couplings, 577–578
Rigid delivery lines, 658–659
Robotic tooling, 389
Rock(s), 297–298, 309–310, 512–516. *See also* Geology
Rock, grouting in, 8–9, 296–328
  amenability theory for, 308–309
  blanket grouting, 302
  consolidation grouting, 302
  conventional injection technology for, 304–306
  curtain depth for, 302–304
  design of, 301–304, 511–518
  fill grouting, 297
  grout holes for, 183, 184, 310–317
  grout hole size for, 186
  and grouting intensity number, 306–308
  grout mix for, 317–319
  injection, 319–328
  investigations for, 495
  methodologies for, 304
  for mitigation of sinkhole formation, 9
  numerical analysis for, 532–538
  primary objective of, 296
  and properties of rock, 297–298
  for strengthening, 9, 309–310
  types/uses of grouts for, 46
  for water control, 298–309
  water tests, 299–301
Rock anchors, 347
Rock grouting, 8. *See also* Structures, grouting in
Rock pockets, 330, 333–334
Rodding, 390
Roller chain, 580
Roller cone bits, 611
Rope thread configurations, 609, 610
Rotary drilling, 189, 191–192
  drill bits, 610–612
  drill rods, 605–608
  percussion, 189, 190–191, 594–597
  sonic, 196–197
Rotary motors, 587
Roundness, jacking to reestablish, 395–396

Salamanders, 432
Sampling, soil, 491–492
Sand, 88–89, 102–103, 106
Saunders diaphragm valves, 661
Sealing, grout, 509
Seated valves, 622–623
Secondary grout holes, 180–182
Sedimentary rock, 297–298, 468
Seepage reduction, 9, 11–12. *See also* Water control
Seismic tomography, 493
Self-contained hydraulic drill rigs, 594
Seminar sponsorship, 566
Set accelerators, 82–83
Set-retarding admixtures, 80–81
Setting time, 60–61, 78
  with pozzolanic materials, 78
  rapid setting and hardening cement, 71
  set-accelerating admixtures, 82–83
  set-retarding admixtures, 81–82
  of solution grouts, 117
Settlement control, 502–504
Settlement correction, 243–260
  for common structures, 243–253
  investigations prior to, 494–495
  for pile-supported structures, 253–259
  for soft ground tunneling, 259–260
Sewers, sealing of, 14. *See also* Pipes and conduits, grouting in
Shafts, *see* Underground structures, grouting of
Shales, 73
Shear mixing, 59
Shear rate/shear strain, 52
Shear thinning, 53
Shelters, protective, 430–432
Shotcrete, 311–312
Shrinkage compensated grouts, 103
Shrinkage index (SL), 482
Silica fume, 74–75, 110
Sills, 468
Single-fluid jet grouting system, 34
Single-pass lining systems, 406
Sinker drill bits, 613
Sinker drill steel rods, 610
Sinkholes, 9, 451–455, 473, 508
SIROC, 126, 127
Site surveillance, 523–524
Skip seal porting, 340–341
Slabbing joints, 471
Slabjacking, 369–379
  grout injection, 373–375
  grout mixtures, 370–372
  injection holes, 369–370
  jacking mechanics, 372–373
  monitoring and control, 375–377
  range of use, 377–379
Slag cements, 76, 78
Slant shear test, 153
Sleeve port pipes, 266, 613
Sleeve-type couplings, 577–578
Sliding-end packers, 674
Sliplining, 13
Slip (rock), 472
Slump, 54–55
Slump test, 54–55, 148–149, 521
Slurries, 51, 103, 371
Slush grouting, 311, 515
Smudge pots, 432
Sodium silicate-based grouts, 117–128
  development of, 4–5

durability of, 124–128
group 1 reactants, 120–122
group 2 reactants, 122–123
group 3 reactants, 123–124
reactant groups, 120
Soil(s):
altering dynamic properties of, 283–286
classification of, 479–482
densification of, 9–10
derivation of, 476–479
drained behavior, 237
grouting in, *see* Soil, grouting in
increasing cohesion of, 10–11
injection in, *see* Injection (in soil)
reducing permeability of, 11–12
reinforcement of, 11
subsurface soil loss, 458–459
underground excavations in, 403–404
Soil, grouting in, 9–12, 15, 207–295
compaction grouting, 208–260
design of, 497–510
fracture grouting, 208–209, 286–296
grout holes size for, 186
investigations for, 485–486, 491–495
mixing/jet grouting, 33–37, 209, 290–295
permeation grouting, 208, 260–286
types/uses of grouts for, 46
Soil anchors, 347
Soilcrete, 15, 33, 34, 209, 291
Soil densification, 9–10
in compaction grouting, 19–21
in fracture grouting, 31, 32
submerged soit, 239–242
Soilfrac, 286
Soil strengthening. *See also* Chemical solution grouts (for strengthening)
design of grouting for, 501–505
grouting in pipes and conduits for, 394
permanent, 276–279
permeation grouting for, 17–18, 281–283
pipes and conduits used in, 394
short-term, 267–275
Solidification, degree of, 261–262
Solubility, 61
Solutioning (rock), 472–473
Sounding wells, 362
Special cements, 41–42, 64, 70–72
Specialty grouts, 42–43
Specifications, 543–559
for compaction grouting, 546–559
"guide," 543–544
performance, 545
prescriptive, 544–545
Specific gravity, tests for, 146
Spherical drop, 56
Split ribbed couplings, 578
Split spacing hole layout, 17, 180–181
Splitting tensile test, 153–154
Sprockets, 580
Stability (suspension grouts), 93–94
Stabilizers (reactant systems), 120, 123–124
Staging area, 488–489
Standard Penetration Test (SPT), 482
Standpipes, 311
Static head, 159, 616
Steel pipe (drill casing), 598–602
Stoichiometry, 136
Storm water lines, *see* Pipes and conduits, grouting in
Strata, rock, 468, 474–475
Strength, 47. *See also* Soil strengthening
bond strength, 153–155
bond strength enhancers, 86
of expansive grouts, 150–151
of permeation grouted masses, 151–152
structural, 18
tests for, 149–155
Stress state (rocks), 475–476
Strike, 474–475
Structural columns, 508
Structural damage, evaluating, 495
Structural rafts, 508
Structures, 12–14
concrete, void areas in, 333–334
grouting in, *see* Structures, grouting in
groutjacking of, *see* Structures, groutjacking of
leakage control in, 14, 410–426
repair and rehabilitation of, 13
settlement of common structures, 243–253
settlement of pile-supported structures, 253–259
strength of materials for, 47
Structures, grouting in, 329–366
for abandonment of conduits/tanks, 352–354
for abandonment/removal of radioactive pit, 363–366
under bedplates and machinery bases, 349–352
contact grouting, 345
expansive/nonshrink grouts, 103–107
filling annular space/backfillling, 354
filling cavities and voids, 330–336
filling cracks and joints, 336–345
filling fabric forms, 354–355
grout holes, 183, 184
investigations for, 496
permeation grouting, 18–19
posttension tendon grouting, 345–349
preplaced aggregate concrete, 355–363
tests for, 521
types/uses of grouts for, 46
underground structures, *see* Underground structures, grouting of
for water control, *see* Leakage control (in structures)
Structures, groutjacking of, 379–386
lifting mechanisms for, 380–384
lifting mode for, 380
tolerances for, 384–386
S-tube valves, 623–624, 652
Subsealing, 141, 369
Substructure, protection of, 490
Subsurface fissures and voids, 445–447
Subsurface soil loss, 458–459
Subsurface water flow/piping, 455–458
Suction, pump, 615–616
Super Jet grouting, 34–36
Superplasticizers, 81, 318
Surface grout leaks, 327–328
Surface heave, 217, 219–220
Surface tension, 55
Surfactants, 56, 119
Suspension grouts, *see* Cementitious suspension grouts
Swivel, drill, 187
Synclines, 470
Syneresis, 117
Synthetic fibers, 90–91

Tanks, abandonment of, 352–354
Temperature(s):
effects on grout, 427–428
extreme, 429–435
of grout, 60
and hydration, 65–68
and viscosity, 53
Tensile strength, 47
Terminology (for grouts), 44
Terrace deposits, 478
Tests, 144–145
for bleed evaluation, 146–147
of bond strength, 153–155
of expansive grouts, 150–151
flow cones, 144–146
of grout pumps, 629
for leakage, 412–414
marsh funnel, 50
mud balance, 50
of pressure filtration, 147–148

Tests (*cont.*)
slump test, 148–149
soil, 491–493
for specific gravity, 146
strength, 149–155
of ultrafine cements, 79–80
water testing, 299–301
Test borings, 491
Texture, rock, 469
Thermal expansion, coefficient of, 49
Thermoplastic polymers, 130
Thermosetting polymers, 130
Thixotropy, 52
Three-fluid jet grouting system, 34, 36
Throughput, pump, 617
Time and material payment, 561
Timing belts, 581
Toluene diisocyanate (TDI), 132–133
Tomograms, 493
Top-down grouting, 209–210, 213, 263–265. *See also* Downstage injection
Top drive drill head, 590
Torque, 576, 590, 591
Total dynamic head (TDH), 616
Trademarks, 567
Trapezoidal thread configurations, 609, 610
Travel index (TI), 21. *See also* compaction grouting
Travel meters, 669
Trials, full-scale, 563
Tri-cone roller bits, 611
Tube-à-manchette, *see* Sleeve port pipes
Tunnels, *see* Underground structures, grouting of
Turbulent flow, 156–157, 160
Two-fluid jet grouting system, 34–36
Two-pass lining systems, 406

Ultimate strength, 126
Ultrafine cements, 64, 76–80, 102
Underground fires, extinguishing, 461–462
Underground structures, grouting of, 398–409
for abandoning conduits/tanks, 352–354
excavation methods, 403–406
geotechnical grouting, 399–401
grout plant for, 325
lining systems, 406–409
structural grouting, 400–403
Underpinning, grouting for, 505–507
Unified Soil Classification System, 479–482
Units, geological, 466–467
Unitized grout plants, 646–653
custom-built, 652–653
flexibility of, 647
for low-mobility grouts, 652
maintenance of, 647
for pourable cementitious grouts, 647–651
Unit price payments, 562–563
Unit weight (of grout), 159
Unstable grouts, 93
Unwashed sand, 89
Upset wall drill rods, 605
Upstage injection, 27, 319, 321–322
Urethane grouts, 132, 139–140, 338
Utilities, locations of, 495

Vacuums, (during injection), 177–178
Vacuum grouting, 15, 38–39
Valves, 160
bleeder, 94, 95
delivery line, 661–662
in hydraulic systems, 585
Valved pumps, 622–627
Vane motors, 587–588
Vane pumps, 585–586
Variable speed drives, 581–582
Velocity, 159, 575–576, 654–655
Velocity meters, 668
Venturi jet mixers, 641
Verification of effectiveness, 529–530
Vesicular material, 467
Viscoelastic materials, 52
Viscosity, 51–53, 55, 57, 83–84
Void areas, 13, 19, 330–336
cementitious grouts for, 330–331
in concrete structures, 333–334
discrete, grout holes for, 183–185
fill grouting for, 37
flowable fill materials for, 354
grout holes for, 183–185, 332–333
grout materials for, 330–332
in masonry, 334–336
resinous foams for, 331–332
resinous grouts for, 331
subsurface, accessing, 445–447
Volcanic glasses/tuff/pumice, 73
Volumetric batch plant auger mixers, 643–645
Volumetric efficiency (pumps), 616

Wagon drills, 595
Water control. *See also* Leakage control (in structures)
chemical solution grouts for, 128–135
design of rock grouting for, 516–518
grouting in pipes and conduits for, 393–394
grouting in rock for, 9, 298–309
grouting in soil for, 11–12, 279–283
permeation grouting for, 17, 279–283
for sewers, 14
specialty grouts for, 42–43
in water-retaining embankments, 231–239
Water-deposited soils, 477–478
Water glass, 118
Water quality (in grout mixtures), 69–70
Water-reducing admixtures, 80–81, 110
Water-retaining embankments, 231–239, 299, 456
Water-soluble acrylic polymers, 133–135
Water supply, 434–435
Water tests, 299–301
Water-to-cement ratio, 93–96. *See also specific mixtures*
Weathering (rock), 473–474, 512
Weirs, 457
Wetability, 55–57
Wet head pneumatic drills, 591–592
Wet mix process, 35
Wet mix shotcrete, 311, 312, 515
Wick bleed, 346
Wiper valves, 622, 623
Wireline drill rod, 606, 607
Wire line technique, 195–196
Worker-accessible conduits, grouting in, 392–397
Worm gear drives, 579–580
Wyoming bentonite, 143

Yield, 52